Dennis Cecic
EET034
284-1529

BIPOLAR AND MOS ANALOG INTEGRATED CIRCUIT DESIGN

BIPOLAR AND MOS ANALOG INTEGRATED CIRCUIT DESIGN

ALAN B. GREBENE

MICRO-LINEAR CORPORATION
SUNNYVALE, CALIFORNIA

A Wiley-Interscience Publication

John Wiley & Sons

New York   Chichester   Brisbane   Toronto   Singapore

Library of Congress Cataloging in Publication Data:

Grebene, Alan B., 1939-
    Bipolar and MOS analog integrated circuit design.

    "A Wiley-Interscience publication."
    Includes index.
    1. Integrated circuits. 2. Electronic circuit
design. 3. Metal oxide semiconductors. 4. Bipolar
transistors. I. Title. II. Title: Bipolar and M.O.S.
analog integrated circuit design.

TK7874.G693   1983   621.381'73   83-6563
ISBN 0-471-08529-4

Printed in the United States of America

10 9 8 7 6 5 4

PREFACE

The contents and organization of this book are primarily aimed at the practicing engineer in the field of solid-state electronics. It is intended as a valuable reference for the IC designer and user alike. For the analog IC designer, it provides rigorous design guidelines and examples, while for the user, it offers a detailed analysis of various classes of analog circuits, points out their design philosophy, capabilities, and limitations, and presents application examples and guidelines.

It is intended to be an easy and smooth reading book on a rapidly evolving, high-technology subject. To this end, the lengthy and detailed mathematical treatment of the subject matter is minimized. Long derivations of device or circuit equations are avoided whenever possible; instead, the emphasis is placed on the end result, and the basic design philosophy leading up to it, with a clear understanding of the underlying assumptions and trade-offs. Whenever possible, each new design idea or concept is also demonstrated with a practical example.

The advent of integrated circuit technology has altered many of the established circuit design techniques and principles. This is particularly evident in the field of analog integrated circuits where the designer is faced with a new set of design constraints and ground rules. In writing this book, it is my intention to educate the practicing electronics engineer in the fundamental design principles, capabilities, and applications of monolithic analog circuits. However, the subject matter is treated rigorously and from a fundamental viewpoint, to make this book suitable as a text for graduate study in semiconductor circuits.

This book is an updated sequel to an earlier book by the author, *Analog Integrated Circuit Design* (published by Van Nostrand Reinhold, 1972) which covered the analog IC design technology of the 1960's. Since then, many significant changes have occurred in the world of microelectronics. Perhaps the most important of these has been the "microprocessor revolution," which has resulted in a truly revolutionary growth of digital signal-processing techniques. In turn, this has led to a rapid evolution and advancement of analog circuit methods, particularly in the areas that interface with digital techniques and technologies. As a result, complete LSI systems have evolved which combine complex analog and digital functions on the same chip.

A great deal of this development has been possible by extending the capabilities of MOS devices and process technology to cover analog functions. Consequently, analog IC design using MOS technology has rapidly evolved into a major area of growth. These developments of recent years are profoundly reflected in the contents and the organization of this book.

In the preparation of the text, it is assumed that the reader is familiar with the basic theory and principles of solid-state devices. Therefore, the solid-state device theory, which is already well covered elsewhere in the literature, is reviewed only briefly, and almost all of the space is devoted to circuit approaches unique to monolithic integrated circuits. Hybrid integrated circuits, which represent an area of overlap between discrete and monolithic circuits, are not covered explicitly.

The text of the book is comprised of fifteen chapters which follow a logical sequence in the form of three "sections." The first section of the book, comprised of Chapters 1–3, reviews the basic "tools" of analog IC design and fabrication, namely, process technology, IC components, and techniques for placing these components on the chip, that is, the chip layout. These chapters are intended to familiarize the designer with the physical structures, advantages, and limitations of monolithic components. This knowledge is imperative to an analog IC design engineer since a successful design is one that efficiently utilizes the advantages of monolithic devices while avoiding their shortcomings.

The second section of the text, made up of Chapters 4–6, covers the basic "building blocks," or subcircuits, of analog IC design. One important chapter in this section, Chapter 6, deals with the use of MOS technology in analog or combined analog/digital LSI design. All the subcircuits covered in this section serve as essential building blocks of the complex IC designs that are covered in the remainder of the book.

The third and main section of the book, comprised of Chapters 7–15, covers the entire field of analog integrated circuits by dividing them into functional categories and then examining each category separately. Thus, for example, circuit classes such as operational amplifiers, multipliers, oscillators, phase-locked loops, filters, and data conversion circuits are examined separately. In this section, particular emphasis is given to the recent developments in the field of analog circuits, particularly in the areas of switched-capacitor filters, switching regulators, voltage-controlled oscillators, high-resolution data conversion circuits, and the precision reference circuitry associated with them.

Part of the material in this book is patterned after a sequence of graduate level courses in integrated electronics which I taught at Santa Clara University. Therefore, when preceded by courses on solid-state circuits and semiconductor electronics, this book will be well suited for a senior or graduate level course.

I am grateful to many people who have contributed directly or indirectly to the preparation of this book. In particular, I would like to thank my wife, Karen, who has been a constant source of encouragement for me during the long years of effort that have gone into this book. I would also like to extend my appreciation to many colleagues and associates in the IC industry for their assistance and guidance in the organization and technical accuracy of the text. I am particularly grateful to the management of Exar Integrated Systems, Inc., for providing me the time to work on this book, and to Ms. Sue Wooldridge who has patiently typed and retyped the draft of the manuscript several times over.

ALAN B. GREBENE

Saratoga, California
August 1983

CONTENTS

BIPOLAR AND MOS ANALOG INTEGRATED CIRCUIT DESIGN

CHAPTER ONE

INTEGRATED-CIRCUIT FABRICATION

The aim of this chapter is to familiarize the reader with the fabrication processes for analog integrated circuits. As a designer, one does not need to know the specific details of each and every process step, since these fabrication processes often become, by themselves, areas of specialization and rapid technological development. However, these manufacturing processes determine the capabilities and limitations of monolithic integrated-circuit (IC) products. Therefore, it is imperative at all times that a successful design engineer be familiar with the fundamental properties and the constraints of each of the major process steps in the fabrication of integrated circuits.

1.1. THE PLANAR PROCESS

The fabrication of a monolithic integrated circuit involves a complex sequence of processing steps. Even though the specific nature of these processes is well diversified, the bulk of the manufacturing steps associated with the present-day IC technology can be grouped under the term *planar process*. [1] Prior to the invention of the planar process in 1959, the solid-state electronics field was dominated by germanium devices. Introduction of the planar process has revolutionized the field of microelectronics almost overnight; and silicon, rather than germanium, emerged as the predominant semiconductor material.

When exposed to air, silicon forms an insulating oxide layer, called silicon dioxide (SiO_2). The formation of this oxide layer can be enhanced by exposing the single-crystal silicon wafer to steam or dry oxygen at high temperatures. The electrical and chemical properties of the SiO_2 layer are key to the planar process for the following reasons. First, the SiO_2 layer forms an inert cover over the

1

silicon surface, and protects it from external contamination. Second, it serves as a barrier to the diffusion of impurities into silicon; thus, by etching well-defined patterns or *windows* in the oxide layer, one can diffuse desired impurities into selected areas of the silicon wafer. Finally, the SiO_2 layer provides an insulating surface over which the metal interconnections can be formed.

The planar process technology is comprised of five independent processes: epitaxy, oxidation, photolithography, diffusion, and thin-film deposition. Sometimes a sixth process called *ion implantation* is also used as a supplement or substitute for the diffusion process. A schematic illustration of these basic process steps is given in Figure 1.1 in terms of the cross-sectional view of a silicon wafer. For illustrative purposes, the vertical and horizontal dimensions of the wafer cross section are not drawn to scale.

Epitaxy is a deposition technique during which additional silicon atoms can be deposited on a single-crystal silicon substrate, without changing the crystalline structure of the silicon wafer. In other words, during the epitaxial deposition step, the single-crystal silicon substrate can be extended by the vapor

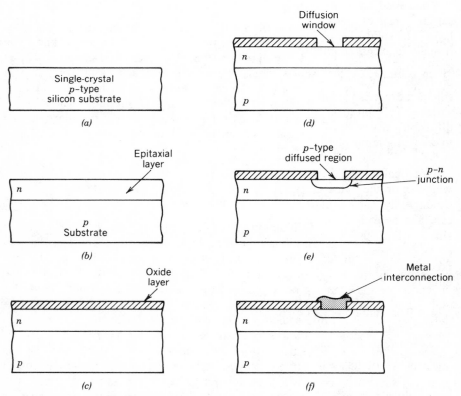

FIGURE 1.1.   Basic steps in planar technology: epitaxy, oxidation, photolithography, diffusion, and metallization.

phase deposition of additional atomic layers of silicon. By controlling the deposition rates, and introducing selected types and amounts of impurities into the carrier gases during the epitaxial deposition process, the thickness and the resistivity type of the epitaxial layer can be accurately controlled. In Figure 1.1b, this is illustrated for the case of an n-type epitaxial layer, deposited on a single-crystal p-type silicon substrate.

Oxidation, or surface passivation, is achieved by exposing the silicon wafer surface to an oxidizing atmosphere at high temperatures. As described earlier, this results in the formation of an oxide layer, SiO_2, which protects the silicon surface from contamination by undesired impurities. This is illustrated in Figure 1.1c.

Photolithography, or a masking technique, is then used to etch selective openings into the oxide layer. These openings serve as diffusion windows from which controlled amounts of impurity doping can be introduced into localized areas of the silicon wafer, as shown in Figure 1.1d. By using photographic techniques, the sizes of the diffusion windows can be greatly reduced without sacrificing the accuracy of alignment and the resolution of the line widths.

Controlled amounts of dopant impurities can be introduced into the pre-selected areas of the silicon surface through the diffusion windows in the oxide layer. The solid-state diffusion of these dopants into silicon at high temperatures (usually in excess of 1000°C) results in the formation of a p-n junction, as shown in Figure 1.1e. Since the diffusion of the impurities from the diffusion window proceeds sideways as well as downwards, the resulting junction edge on the silicon surface is located *under* the oxide layer and is not exposed to air on the surface. This protects the junction from possible contamination. During the diffusion process, the diffusion window can also be closed by exposing the wafer to oxidizing atmosphere, which would result in reoxidizing the exposed silicon surface.

Electrical contact to the semiconductor regions can be formed by opening new windows on the oxide layer and depositing a thin metal film of high electrical conductivity, such as aluminum (Al), over these windows. This conductive film can then be etched into a desired pattern, interconnecting the devices on the silicon wafer, thus completing the monolithic circuit structure (Fig. 1.1f).

As compared with earlier solid-state device fabrication methods, the planar process offers two unique advantages which make it ideally suited for integrated circuits:

1. The semiconductor junctions are protected on the surface by an inert oxide layer and are not exposed to air. This results in very low leakage currents and high reliability.
2. Photolithographic reduction, masking, and etching techniques are used to determine device geometries. This makes it possible to reduce device dimensions greatly and facilitates the simultaneous fabrication, or *batch processing,* of a large number of devices and circuits on the same silicon surface.

The monolithic IC technology owes its rapid growth and acceptance to these two unique features of the planar technology.

1.2. ELECTRICAL RESISTIVITY OF SILICON

The addition of small concentrations of selected impurities, called *dopants,* into otherwise pure single-crystal silicon has a very significant effect on its electrical characteristics. These dopants are normally Group III or Group V elements from the Periodic Table, as shown in Table 1.1. When introduced into the silicon lattice, under proper process conditions, these dopant atoms substitutionally take the place of silicon atoms within the crystal structure. Thus, for the small amount of impurities added, the crystalline structure of silicon remains unchanged, but its electrical characteristics and its resistivity are profoundly affected.

When a silicon wafer is doped with Group III impurities such as boron (B) it is said to be *p* type, where the majority of the carriers available for current conduction are the holes. Similarly, when a Group V dopant such as phosphorus (P) or antimony (Sb) is used, the resulting silicon wafer will be *n* type, and the majority of carriers available for conduction will be the electrons.

For IC fabrication, boron and phosphorus are the most commonly used dopants, with arsenic (As) and antimony (Sb) also finding application in special cases.

In extrinsicly doped semiconductors, it is practical to assume that the majority carrier concentration is approximately equal to the density of the donor or acceptor dopant atom within the crystal. Thus, for example, for *p*- and *n*-type doped material, respectively,

$$p_p \approx N_A \quad \text{and} \quad n_n \approx N_D \tag{1.1}$$

where p_p and n_n are the equilibrium concentrations of the holes and the electrons (expressed in cm^{-3}) and N_A and N_D are the concentrations of the acceptor (Group III) and the donor (Group V) dopants.

The electrical resistivity $\rho(\Omega\text{-cm})$ for a *p*-type doped silicon can be approximated as

$$\rho \approx \frac{1}{q\mu_p N_A} \tag{1.2}$$

TABLE 1.1. *p*- and *n*-Type Dopants for Silicon Device Fabrication

p-Type Dopants		*n*-Type Dopants	
Boron	(B)	Phosphorus	(P)
Aluminum	(Al)	Arsenic	(As)
Gallium	(Ga)	Antimony	(Sb)
Indium	(In)		

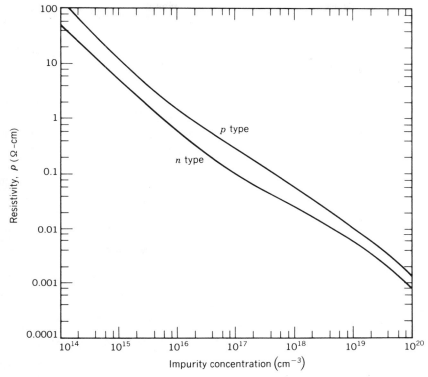

FIGURE 1.2. Resistivity of p- and n-type silicon as a function of impurity concentration.

where q is the electronic charge and μ_p (cm²/V-sec) is the hole mobility. Conversely, the resistivity of a n-type doped silicon can be written as

$$\rho \approx \frac{1}{q\mu_n N_D} \qquad (1.3)$$

where μ_n is the electron mobility.

Figure 1.2 gives a plot of the resistivity of uniformly doped p- and n-type silicon as a funciton of impurity concentration.

In certain calculations, conductivity, rather than resistivity, is used as a parameter. Conductivity σ $(\Omega\text{-cm})^{-1}$ is the reciprocal of resistivity (i.e., $\sigma = 1/\rho$).

1.3. SOLID-STATE DIFFUSION

The diffusion process is by far the most widely used method of introducing controlled amounts of impurities into the silicon substrate. It is a relatively well understood and highly reproducible process step which readily lends itself to the batch processing advantages of the planar technology, since a large number of silicon wafers can be processed simultaneously.

Diffusion, as a general process, is the mechanism by which different sets of particles confined to the same volume tend to spread out and redistribute themselves evenly throughout the confining volume. In the case of solid-state devices and integrated circuits, the relevant diffusion process is the one dealing with the movement and distribution of impurity atoms in a crystalline lattice structure. In crystalline solids, the diffusion process is significant only at elevated temperatures where the thermal energy of the individual lattice atoms has a statistical chance of overcoming the interatomic forces which hold the lattice together.

In the single-crystal silicon lattice, the impurities can move through the lattice by any one or a combination of the two dominant diffusion mechanisms. These physical mechanisms can be briefly outlined as follows.

Substitutional Diffusion.   The impurity atoms propagate through the lattice by replacing a silicon atom at a given lattice site.

Interstitial Diffusion.   The impurity atoms do not replace the silicon atoms at the regular lattice sites, but instead move into the interstitial voids in the three-dimensional lattice structure.

In the fabrication of analog integrated circuits, substitutional diffusion is by far the most important mechanism, since all the impurities intentionally introduced into the silicon substrate to form the junction and the device structures diffuse substitutionally. This is not necessarily the case for digital integrated circuits where controlled amounts of interstitial impurities, such as gold, copper, or nickel, can be introduced into the silicon lattice to reduce the minority carrier lifetime.

Diffusion Theory

Although the diffusion process proceeds in all three dimensions simultaneously, for the analysis of the fundamental properties of that process it is sufficient to consider only one dimension. As will be described in Chapters 2 and 3, the geometry and dimensions of most semiconductor devices fabricated by the planar process justify this one-dimensional assumption.

The fundamental physical property of the diffusion process is that the particles tend to diffuse from a region of high concentration to that of lower concentration at a rate proportional to the concentration gradient between the two regions. This is known as Fick's first law, and can be mathematically expressed as

$$F = -D \frac{\partial N}{\partial x} \qquad (1.4)$$

where F is the net particle flux density, that is, the net number of particles flowing through a unit surface area normal to the direction of flow per unit time, N is the number of particles per unit volume, and x is the distance measured parallel to the direction of flow. D is the diffusion coefficient and has units of (length)^2/time. The magnitude of D gives a measure of the relative ease or difficulty with which the diffusing particle can move about in its environment. The negative sign appears in Eq. (1.4) to indicate that the particle flow is directed from a region of high concentration toward one of lower concentration.

In all diffusion problems, one is interested in the variations of the impurity concentration with time as well as with distance. The fundamental law of diffusion, which relates the time rate of change of concentration to the spatial coordinates of the region in question, can be derived from Fick's law by applying the continuity principle to Eq. (1.4) as follows. Consider a region or a volume in the material enclosed by an area A normal to the flow and a width dx parallel to the flow. The net flow of particles into this volume can be written as

$$FA - \left(F + \frac{\partial F}{\partial x} dx\right)A = -\frac{\partial F}{\partial x} dx\, A \qquad (1.5)$$

This net flow into the volume can be related to the concentration change within the volume as

$$\frac{\partial N}{\partial x} dx\, A = -\frac{\partial F}{\partial x} dx\, A \qquad (1.6)$$

The expression for $\partial F/\partial x$ can be obtained by differentiating Eq. (1.4). Substituting the result into Eq. (1.6), one obtains

$$\frac{\partial N}{\partial t} = D \frac{\partial^2 N}{\partial_x{}^2} \qquad (1.7)$$

This equation is known as Fick's second law, and it defines the impurity distribution at any given point within the semiconductor crystal as a function of time.

The diffusion coefficient D is a property of the particular impurity, and is an exponential function of temperature. Figure 1.3 shows the diffusion coefficients of various dopants in silicon as a function of temperature. Note that the diffusion coefficient increases by approximately one order of magnitude for every 100°C change in temperature. Thus, the diffusion process is effective only at temperatures in excess of 1000°C.

Since the basic equation [Eq. (1.7)], which describes the distribution of the impurities in silicon as a function of both time and distance, is a second-order partial differential equation, its particular solution applicable to a diffusion problem depends on the boundary conditions associated with the diffusion process. Most of the diffusion processes encountered in the fabrication of integrated circuits fall into one or the other of the following two classes of boundary conditions: constant-source (complementary error function) diffusion or limited-source (Gaussian) diffusion. The properties of each of these diffusion profiles are described below.

Constant-Source Diffusion.   In this type of diffusion, the wafer is exposed to the impurity source during the entire duration of the diffusion. Thus, there is

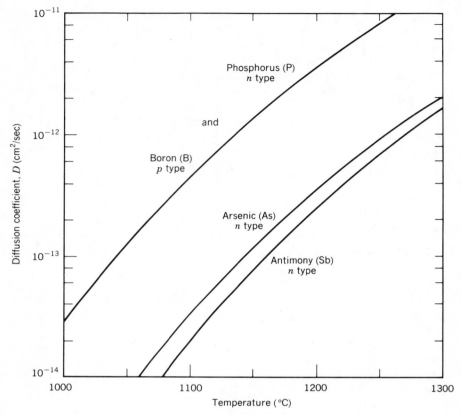

FIGURE 1.3. Diffusion coefficients of various dopants in silicon.[3]

an undiminished supply of impurities at the wafer surface; and the impurity concentration at the wafer surface, N_o is constant, as set by the solid solubility of the particular dopant in silicon. If we also assume that there are no impurities in silicon at time $t = 0$, then the solution describing the distribution of impurities in silicon becomes the *complementary error function,* erfc, where

$$N(x, t) = N_o \, \text{erfc} \left(\frac{x}{2\sqrt{Dt}} \right) \qquad (1.8)$$

In the above expression, N is the density of impurity atoms (atoms/cm^3) at a distance x from the diffusion surface at time t after the start of diffusion. Figure 1.4 shows a logarithmic plot of the impurity concentration as a function of time t, subsequent to the start of the diffusion cycle.

It should be noted that if the diffusion is performed into a wafer of opposite conductive type, then a junction is formed at a depth x_j, where the concentration of the diffused impurity is equal to that of the background concentration N_B.

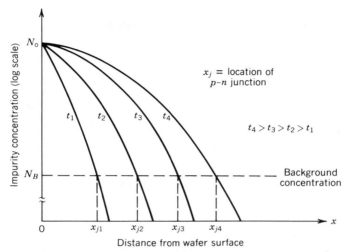

FIGURE 1.4. Constant-source diffusion profiles as a function of time.

Limited-Source (Gaussian) Diffusion. In this type of diffusion, the wafer is exposed briefly to impurities during a so-called predeposition step, where a thin layer of dopant atoms is placed on the silicon surface. After that, the impurity source is turned off, and the total amount of impurities deposited on the surface serves as the impurity source for the rest of the diffusion cycle. In this case, the resulting impurity distribution is approximated by the Gaussian distribution given as

$$N(x, t) = \frac{Q}{\sqrt{\pi D t}} \exp \left(\frac{-x^2}{4Dt} \right) \tag{1.9}$$

where Q (atoms/cm^2), is the initial concentration of impurity atoms deposited on the surface during the predeposition step, which precedes the diffusion cycle. Figure 1.5 shows a logarithmic plot of the impurity concentration into the wafer as a function of increasing time t. Note that since the total amount of impurities, Q, available for diffusion is constant, the surface concentration N_o decreases with increasing time.

Both the complementary error function and the Gaussian distributions are well-defined functions. Figure 1.6 gives a plot of each of these functions for various values of their arguments. Note that, as shown in Figures 1.4 and 1.5, the junction depth x_j is monotonically increasing function of time.

In monolithic IC fabrication, the complementary error-function diffusions are used for either very deep or very shallow diffusions, such as the p-type isolation or the n-type emitter diffusion of npn transistors. The Gaussian diffusion is usually used for medium-depth diffusions, such as the p-type base diffusion of npn bipolar transistors.

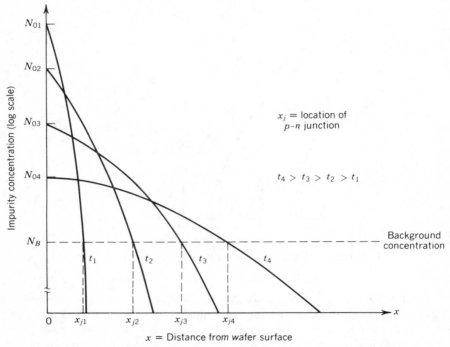

FIGURE 1.5. Limited-source (Gaussian) diffusion profiles as a function of time.

Basic Properties of the Diffusion Process

In the design and fabrication of monolithic integrated circuits, the following three fundamental properties of the diffusion process must be considered:

1. All diffusions proceed simultaneously. The impurities introduced in an earlier diffusion step continue to diffuse during subsequent diffusion cycles. Therefore, when calculating the total effective diffusion time for a given impurity profile, one must often consider the effects of subsequent diffusion cycles. The effects of the subsequent diffusion on a given impurity profile can be estimated by defining an effective Dt product for the particular impurity profile as

$$(Dt)_{\text{eff}} \approx D_1 t_1 + D_2 t_2 + D_3 t_3 + \cdots \qquad (1.10)$$

where $t_1, t_2, t_3, \ldots$ are the different diffusion times, and $D_1, D_2, D_3, \ldots$ are the corresponding diffusion coefficients as determined by the respective temperatures of the diffusion cycles. Thus, for example, in the planar device fabrication, the emitter region of a bipolar transistor is formed by a diffusion process which succeeds the base diffusion step. Therefore, the effective Dt product of the base region contains a finite contribution from the emitter diffusion step.

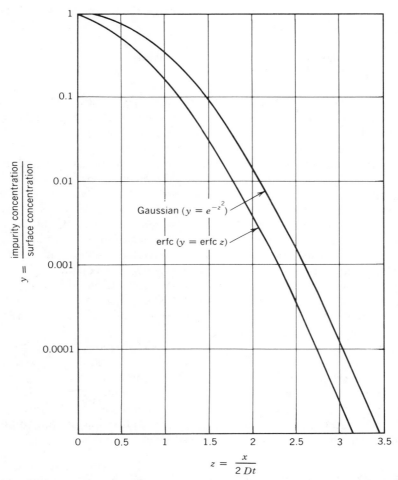

FIGURE 1.6. Values of the complementary error function (erfc) and the Gaussian distribution as functions of their arguments.

2. The diffusion profiles of Eqs. (1.8) and (1.9) are both functions of $(x/\sqrt{Dt})$. Therefore, for a given surface and background concentration, the junction depths x_1 and x_2 associated with the two separate diffusions having different times and temperatures, can be related as

$$\frac{x_1}{x_2} = \sqrt{\frac{D_1 t_1}{D_2 t_2}} \tag{1.11}$$

3. The diffusions proceed sideways from a diffusion window as well as downward. In considering the lateral dimensions of the planar devices, particularly in the case of lateral *pnp* transistors and MOS transistors, these lateral

diffusion effects need to be considered. Typically, the lateral diffusion distance is about 75–80% of the vertical penetration.

1.4. EPITAXIAL DEPOSITION

Epitaxy is a deposition technique where the single-crystal structure of a silicon substrate can be extended by vapor phase deposition of additional atomic layers of silicon. Epitaxial growth, or deposition, is carried out in a special furnace called a *reactor*, where silicon wafers having clean and chemically polished surfaces are heated up to temperatures comparable to those encountered in the diffusion step (i.e., 1000°–1200°C). During the epitaxial growth, vapors containing silicon are passed over the heated substrate. Normally hydrogen is used as the carrier gas, with either silicon tetrachloride ($SiCl_4$) or silane (SiH_4) as the source of silicon. Normally the $SiCl_4$ process requires somewhat higher temperatures than SiH_4 decomposition, and also has a slower growth rate. During the expitaxial growth process, the source compound is chemically reduced, resulting in free silicon atoms, some of which are deposited on the single-crystal substrate. Under proper deposition conditions, the interatomic forces of the single-crystal silicon lattice constrain the deposited silicon atoms to follow the original crystal structure. Thus, structurally, the deposited epitaxial layer forms a continuation of the original crystal structure.

During the process of epitaxial growth, controlled amounts of p- or n-type impurities are also introduced into the carrier gas to control the type and resistivity of the deposited layer. Unlike the diffusion process, epitaxial growth proceeds by uniform addition of atomic layers onto the substrate. Thus, the dopant impurities are uniformly distributed through the epitaxial layer, and do not show a concentration gradient. Furthermore, epitaxial layers can be grown over diffused regions or over other epitaxial layers.

Redistribution of Impurities during Epitaxy

Since epitaxial growth is a high-temperature process, the impurities at the interface of the epitaxial layer (epi) and the substrate tend to redistribute themselves via the diffusion process. For example, in the case of an n-type epitaxial layer grown on a p-type substrate, the epi–substrate interface no longer represents a step junction, but becomes graded due to the diffusion of impurities from the epitaxial layer into the substrate, and vice versa. Consequently, the impurity distribution at the epi–substrate interface may look as illustrated in Figure 1.7. The dashed line shows the ideal p-type impurity distribution at the interface, and the solid line corresponds to the actual distribution.

For relatively rapid rates of epitaxial growth (i.e., > 0.2 μm/min), the impurity distribution $N(x, t)$ across the interface can be written as

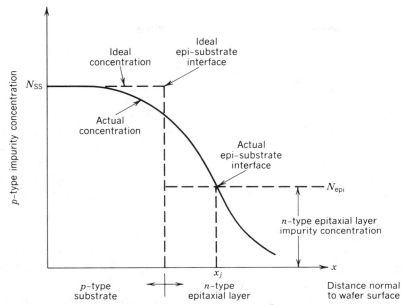

FIGURE 1.7. Impurity redistribution at epi–substrate interface during subsequent epitaxial growth and diffusion cycles.

$$\frac{N(x,\ t)}{N_{SS}} \approx \frac{1}{2}\ \mathrm{erfc}\left(\frac{x}{2\sqrt{Dt}}\right) \qquad (1.12)$$

where N_{SS} is the impurity concentration within the bulk of the substrate. The total amount of interdiffusion across the epi–substrate interface during the entire device fabrication cycle can be estimated by defining an effective Dt product for the out-diffusion process, in accordance with Eq. (1.10).

In practical *npn* bipolar transistor structures (see Figs. 1.11 and 1.13), a heavily doped *n*-type layer, called the *buried layer*, is diffused into selected regions of the *p*-type substrate prior to the growth of the *n*-type epitaxial layer. This buried layer diffusion serves as a low-resistivity conduction path for the collector current of the *npn* transistor. However, during subsequent epitaxial growth or additional diffusion steps, this buried layer tends to out-diffuse into the epitaxial layer and, thus, increase the *n*-type impurity concentration within the region of the epitaxial layer close to the subepitaxial *n*-type buried layer diffusion. In such cases, Eq. (1.12) can also be used to estimate the amount of anticipated out-diffusion from the buried layer into the epitaxial region by replacing N_{SS} by N_o where N_o (atoms/cm^3) is the surface concentration of the *n*-type donor atoms on the buried layer surface, prior to the epitaxial growth step.

In order to minimize the out-diffusion of the buried layer during the epitaxial growth or the subsequent diffusion steps, donor impurities having low diffusion coefficients, such as arsenic or antimony, are used for buried layer dopants.

Crystal Defects in Epitaxial Layers

During the epitaxial deposition process, a number of crystal defects may occur. Depending on their nature and density, these defects and imperfections may affect the overall electrical characteristics of the epitaxial layer and of the device junctions formed in it. When present in large numbers, these crystal defects reduce the minority carrier lifetime within the epitaxial layer, increase the junction leakages, and cause localized voltage breakdowns.

The most commonly encountered imperfections in the epitaxial structures are *dislocations* and *stacking faults*. The dislocation defects are caused by imperfect arrays of atoms within the localized regions of the substrate lattice and by mechanical stress. The dislocations tend to appear in clusters and be oriented along lines known as *slip planes*. They can be reduced by taking additional care during the substrate preparation, prior to epitaxy.

Another prevalent crystal defect in the epitaxial regions is the stacking fault caused by improper stacking of the crystal planes over a localized region. Such a fault may start at the epi–substrate interface and propagate through the deposited layer thickness; or it may originate entirely within the epitaxial region. This latter occurrence is often associated with excessively high growth rates.

Additional types of crystal defects in epitaxial layers are pits or pyramids on the epitaxial layer surface, which are, in general, due to either excessive or parasitic impurities present in the epitaxy system.

Polycrystalline Silicon Growth

If silicon is epitaxially grown on a noncrystalline substrate, such as an SiO_2 layer, the deposited layers of silicon tend to be oriented in random directions, thus leading to a polycrystalline structure. Polycrystalline silicon does not have the electrical properties associated with single-crystal semiconductors, but can be used for a variety of special applications in IC fabrication. One particular application of polycrystalline silicon growth will be described later in this chapter, in connection with dielectrically isolated device structures (see Fig. 1.14). Another commonly encountered application of polycrystalline silicon is in forming the gate electrode of MOS devices (see Fig. 2.50), where it has the effect of reducing the gate threshold voltage.

1.5. OXIDATION OF SILICON

When exposed to an oxidizing atmosphere, silicon forms an insulating silicon dioxide (SiO_2) layer on its surface. This oxide layer is an inert dielectric; in sufficient thickness, it is impervious to impurities or most forms of contamination. Thus, it forms a natural *passivation* layer over the active silicon area within the wafer. In the fabrication of monolithic integrated circuits, this oxide layer performs three fundamental functions:

1. It serves as a diffusion barrier and allows selective diffusions into silicon through the windows etched into the oxide.
2. It protects the junctions from exposure to moisture and other contaminants in the atmosphere.
3. It serves as an insulator on the device surface on which the metal interconnections can be formed.

In addition to these three functions, the SiO_2 layer is often utilized as the dielectric region of monolithic capacitances, or as the gate dielectric of MOS devices. The SiO_2 layers can be formed on the silicon surface by any one of several methods. Some commonly used techniques consist of thermal oxidation, pyrolytic deposition, and anodic or gas plasma oxidation. Among these, thermal oxidation is by far the most commonly used growth technique.

Thermal Oxidation of Silicon

During the thermal oxidation step, an oxide layer is formed on the silicon surface through the basic chemical reaction

$$Si + O_2 \rightarrow SiO_2 \qquad (1.13)$$

In the presence of some water vapor, the oxidation process is significantly accelerated and proceeds in accordance with the chemical reaction

$$Si + 2H_2O \rightarrow SiO_2 + 2H_2 \qquad (1.14)$$

Except for the initial oxidation step (see Fig. 1.1c), the thermal oxide growth is not generally performed as a separate fabrication step. Instead, it is often incorporated into the diffusion process by providing an oxidizing atmosphere during part of the diffusion cycle. This in turn provides sufficient oxide on the previous diffusion windows for the next masking step.

Thermal oxidation normally proceeds at a temperature range of 900–1200°C. During oxidation, a carrier gas containing the oxidizing agent (normally oxygen gas or water vapor) is passed over the heated wafer substrate. The kinetics of thermal oxide growth are well understood and covered in the literature.[4,5] Oxidation proceeds by an inward motion of the oxidizing species toward the silicon–SiO_2 interface. Therefore, as the oxidation process proceeds, it is necessary for the oxygen molecules to diffuse through a thicker layer of SiO_2 to get to the silicon surface where the chemical reactions (1.13) or (1.14) can take place. Consequently, the time rate of oxide growth decreases rapidly with increasing oxide thickness. It can be shown[5] that for very thin layers of SiO_2, the growth rate is linear with time. However, as the oxide thickness x_o increases, the growth rate becomes proportional to $\sqrt{t}$.

The practical thicknesses of thermally grown SiO_2 layers used in monolithic IC fabrication are in the range of 500–20,000 Å (10,000 Å $= 1 \ \mu m$). The lower limit of thickness is often dictated by electrical breakdown or random defect densities (i.e., pin holes) in the oxide layer. The upper limit is set by required oxidation times and the difficulty of etching the oxide layer during the photo-masking step.

Masking Properties of SiO₂

The diffusion coefficients of most dopants in SiO_2 are about two to four orders of magnitude smaller than those in silicon. Therefore, for these impurities, which include all those listed in Table 1.1 with the exception of gallium, an SiO_2 layer of proper thickness can serve as a diffusion barrier. The minimum oxide thickness necessary to mask against a given diffusion step depends to a large extent on the specifics of the diffusion process, such as type of dopant used, surface concentration, predeposition temperature, and time. Figure 1.8 gives some typical curves showing the minimum oxide thickness needed to mask against the two most commonly used dopants, boron and phosphorous.[6]

Several positively charged ionic species, such as sodium (Na^+) or hydrogen (H^+) ions, can diffuse through the SiO_2 layer with relative ease at temperatures as low as 150°C. Therefore, oxide passivation is prone to ionic contamination. These ions tend to generate a positive space charge within the silicon–SiO_2 interface, which in turn leads to an increased free-electron concentration of the silicon side of the interface. As a consequence, the surface layer of silicon directly against the SiO_2 layer tends to appear less p type or more n type than would be expected from the dopant impurity concentration. This effect, when coupled with the depletion of the p-type boron concentration during the oxide growth cycle, may result in the formation of a parasitic n-type inversion layer at the Si–SiO_2 interface. This parasitic inversion layer, known as *channeling*, is a dominant failure mechanism for integrated devices containing lightly doped p-type regions. It can be eliminated by maintaining a relatively high surface concentration of boron within the p-type regions (typically $\geq 10^{17}$ atoms/cm³) and by avoiding ionic contamination of the SiO_2 layer.

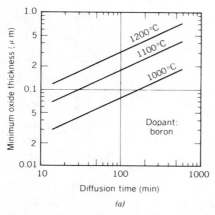
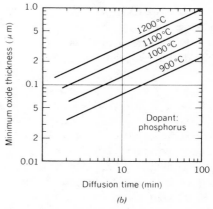

FIGURE 1.8. Minimum oxide thickness required to mask against diffusion: (*a*) Boron and (*b*) phosphorus.[6]

Chemically Deposited Oxide Layers

Sometimes it is advantageous to form an inert dielectric coating over the surface of the integrated circuit by pyrolytically depositing an oxide layer. Such a deposition process is often referred to as chemical vapor deposition (CVD) step. During the CVD step, the silicon wafer is maintained at a relatively low temperature (typically 400°C). Thus, such a step is particularly useful as a final passivation layer over the IC surface, subsequent to the completion of the metal interconnection or the thin-film deposition step, and protects the device surface from mechanical damage or scratches.

Silicon Nitride Passivation

Silicon nitride (Si_3N_4) is far more resistant to ionic contamination than SiO_2. Therefore, it is frequently utilized as a passivating layer for IC structures whose performance can be easily degraded by surface contamination. This is particularly true for analog integrated circuits involving MOS devices or operating at low current levels. An additional advantage of Si_3N_4 over the thermally grown oxide is its superior masking properties against the dopant impurities. Even such dopants as gallium, which readily diffuse through SiO_2, can be effectively masked by Si_3N_4.

The Si_3Ni_4 passivating layer is most conveniently formed by a pyrolytic deposition process at a temperature range of 800–1000°C.

The deposition is obtained by the decomposition of SiH_4 and ammonia (NH_3) in the presence of hydrogen gas, in accordance with the reaction

$$3 \; SiH_4 + 4 \; NH_3 \rightarrow Si_3N_4 + 12 \; H_2 \tag{1.15}$$

Silicon nitride is often used to complement the SiO_2 passivation process. In such an application, a layer of Si_3N_4 (typically 1000 Å thick) is sandwiched between two SiO_2 layers on the wafer surface to provide an added degree of surface passivation. The second layer of SiO_2 over the nitride layer is normally formed by pyrolytic deposition. This second oxide layer also serves as a mask during the photomasking and etching of the contact windows through the nitride layer.

1.6. PHOTOMASKING

The initial layout of an integrated circuit is normally done at a scale several hundred times larger than the final dimensions of the finished monolithic chip. This initial layout is then decomposed into individual mask layers, each corresponding to a masking step during the fabrication process. The individual mask layers are then reduced photographically to the final dimensions of the integrated unit. The reduced form of each of these patterns is then contact-printed on a transparent glass slide to form a photographic *mask* of the patterns to be etched on to the SiO_2 surface. To facilitate batch processing, a large number of such masks are contact-printed on the same glass slide, forming a *masking plate*. The plate is sufficiently large to cover the entire surface of the silicon wafer to be masked. Thus, in a single masking operation, an array of a large number of identical masks can be applied simultaneously over the wafer surface.

During the masking operation, the mask pattern is transferred from the masking plate to the wafer surface by photolithographic techniques. The wafer surface to be masked is initially coated with a photosensitive coating known as *photoresist* or *resist*. The resist-coated wafer surface is then brought into intimate contact with the masking plate and exposed under an ultraviolet light. The portions of the photosensitive resist not covered by opaque portions of the mask polymerize and harden as a result of this exposure. Then the unexposed parts of the resist can be washed away, leaving a photoresist mask on the wafer surface. As a consequence of the masking step, the pattern to be etched through the oxide is transferred to the wafer surface in the form of a hardened etch-resistant photoresist pattern.

The photomasking step is followed by an etching step during which the parts of the SiO_2 layer not protected by the exposed resist mask are etched away, forming the diffusion or the contact windows on the oxide. In this process, a buffered hydroflouric acid (HF) solution is used as the etchant. Following the etching step, the photoresist is washed away by a special cleaning solution, and the silicon wafer is ready for the next diffusion step. A similar photomasking step is also used in forming the metal interconnection patterns. The typical sequence of steps in the photomasking process is illustrated in Figure 1.9.

Dimensional Tolerances

In most monolithic circuit structures, the lateral dimensions of the integrated components are determined by the limitations of the photolithographic reduction, masking, and etching processes. The two fundamental limitations on the photolithography process are the alignment and the resolution of the mask patterns.

Since the monolithic IC fabrication steps require the successive application of a number of masks, it is necessary that each new mask applied to the silicon surface align with the previous set of masks over the entire surface of the wafer. This requires a good degree of dimensional accuracy associated with the initial layout of the circuit. To ensure this dimensional accuracy, the initial layout is carried out at the largest possible magnification within the capabilities of the photoreduction system. Typically, a $500\times$ size is preferred for the initial layout for circuits having final reduced dimensions of up to approximately 70 sq. mils. For larger overall chip dimensions, a smaller initial layout scale, such as $400\times$, may be preferred to avoid optical distortion during the reduction process. Note that the drafting inaccuracies associated with the initial layout are also reduced at the same scale as the original layout. Thus, for example, a 0.01-in. dimen-

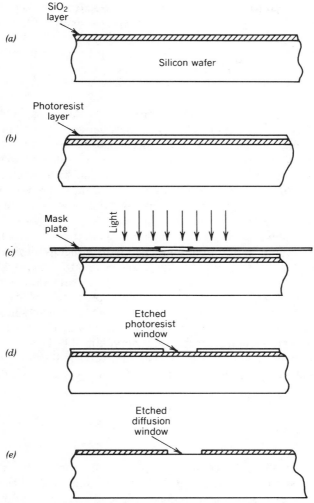

FIGURE 1.9. Typical sequence of steps in the photomasking process which result in a diffusion window pattern in the SiO$_2$ layer that duplicates the pattern on the mask plate. (*a*) Growing SiO$_2$ on wafer surface; (*b*) applying a thin coat of photoresist on oxidized wafer surface; (*e*) exposing photoresist through a mask plate; (*d*) developing and etching photoresist; (*e*) etching the exposed SiO$_2$ layer and stripping the photoresist to end up with a diffusion window in the SiO$_2$ layer.

sional inaccuracy in the initial layout leads to a $\pm$ 0.5-1 μm error in the final dimension at a 500$\times$ reduction.

A possible source of error in the masking step is the tolerance associated with the "step-and-repeat" process in contact-printing the mask array on the masking plate. The source of error in this case is the mechanical advance mechanism involved. An additional factor limiting the alignment tolerances of a mask set is the accuracy of positioning the mask on the wafer surface. This is done with the aid of a mechanical alignment jig under a high-powered microscope. However, it is still subject to some operator error. To minimize the alignment errors at the stage of the masking operation, it is customary to use concentric alignment patterns on successive mask layers. The alignment accuracy for a typical mask set, under production conditions, is approximately ± 1 μm for concentric patterns.

The ability of the mask to define or reproduce fine details on the wafer surface is determined by the resolution of the photomasking step. A good measure of the resolution is the minimum line width needed to resolve, reproducibly, two parallel lines spaced one line width apart. The main limitations to the resolving power of the photomasking techniques are the statistical fluctuations in the molecular structure of photographic emulsions and the diffraction of the light at the mask edges. At present, the minimum line width that can be resolved under production conditions is approximately 2 μm.

The etching step also introduces random irregularities, or errors, which tend to reduce the overall mask resolution. The grainy structure of the exposed and polymerized photoresist does not define a true edge during the etching step, but can cause random irregularities of the order of ± 0.5 μm along straight edges. This effect, along with the nonuniform etching properties of the oxide layers, also tends to round the sharp corners on the masks to a typical radius of 2–3 μm. At present, the minimum dimensions of a diffusion window that can be formed routinely under production environment is approximately 4 μm $\times$ 4 μm. However, these dimensional tolerances can be improved significantly by using more advanced and complex pattern-forming techniques, such as electron-beam photolithography[7].

1.7. ION IMPLANTATION

In the ion-implantation process, the impurities are introduced into silicon by bombarding the wafer surface with high-energy ions of the desired impurity type.[8,9] The implantation operation takes place in a vacuum. Impurity ions are accelerated from an ion source, and a mass spectrometer is used to separate the undesired impurities from the beam. The ion beam is then focused to a small area (typically smaller than $\frac{1}{4}$ in.^{2}) and is scanned across the semiconductor wafer which serves as the target. During the implantation process, the depth of penetration of the impurity ions into the silicon lattice is controlled by their energy, which is set by the accelerating field; and the density of the implanted ions is controlled by the beam current. Typical energy levels used in the ion-implantation process are in the range of 30–200 kilo-electron-volts (keV). When ions penetrate the silicon wafer, they produce lattice defects or dislocations. These are removed by annealing the wafer at temperatures as the order of 500–600°C, subsequent to the implantation step.

The impurity profile resulting from ion implantation has a Gaussian distribu-

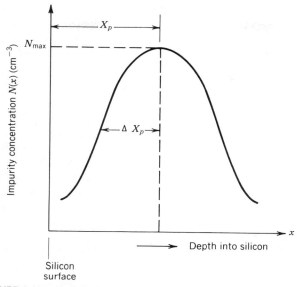

FIGURE 1.10. Typical distribution of implanted impurity atoms in silicon.

tion, with the peak of the distribution appearing *below* the surface of the silcon wafer, as shown in Figure 1.10. The peak of the distribution occurs at a depth X_p, called the *mean range*, which increases with increasing mass and energy of the incident ions. The relative spread of the distribution is measured in terms of its standard deviation, ΔX_p. The relative spread of the distribution $\Delta X_p / X_p$ depends on the ratio between the incident ion mass and the silicon atom mass. Heavier ions produce narrower profiles. Table 1.2 gives some typical values of X_p and ΔX_p associated with various dopants at different energy levels.[8,9]

TABLE 1.2.   Mean Range X_p and Spread ΔX_p of Implanted Impurity Concentration for Various Dopants at Different Implant Energy Levels.[8,9]

Dopant Impurity	Implant Energy Level (keV)				
	30	50	70	100	200
Boron					
X_p (μm)	0.095	1.52	2.10	2.80	4.85
ΔX_p (μm)	0.0375	0.05	0.059	0.07	0.092
Phosphorus					
X_p (μm)	0.036	0.062	0.086	0.122	0.25
ΔX_p (μm)	0.0165	0.022	0.034	0.047	0.078
Arsenic					
X_p (μm)	—	0.031	0.042	0.057	0.11
ΔX_p (μm)	—	0.011	0.015	0.02	0.037

The semiconductor surface can be easily masked against the implanted ions by using a surface oxide (SiO_2) layer as a mask. Thus, the ion-implanted regions can be readily patterned on the silicon surface, in the same manner as the diffused regions, using photomasking techniques. Since ion implantation is a low-temperature step, an unremoved photoresist layer can also be used to mask against implanted ions, instead of an SiO_2 layer.

Ion implantation provides an alternate method to diffusion for the introduction of dopant impurities into silicon. Thus, it gives the device designer an added degree of flexibility in the fabrication of monolithic devices. One key advantage of ion implantation is that it is a low-temperature process. Thus, it can be added to the manufacturing process without affecting the diffusion processes which precede it. It is particularly useful for forming very shallow junctions in silicon, or achieving well controlled impurity concentrations on the silicon surface.

The most frequent application for ion implantation is in controlling MOS device thresholds and fabricating junction-gate field-effect transistors (JFET) or precision resistor-ladder networks. The shallow device junctions formed by ion implantation are also very useful in the design of high-freqency transistors.

1.8. THIN-FILM PROCESSES

After completion of the epitaxy, diffusion, and ion-implantation steps, conductive, resistive, or dielectric thin-film layers can be deposited on the silicon wafer surface. These thin-film layers are then etched and patterned by the conventional photomasking techniques to perform a multiplicity of circuit functions. The conductive thin films, such as aluminum, are used for interconnection of the circuit components. Deposited resistive or dielectric layers can be used in forming thin-film resistors or capacitor structures. The term *thin film* is used to imply deposited film thicknesses of several micrometers or less, as compared with the larger geometry and thicker films associated with hybrid integrated circuits.

The deposition of dielectric films has been described earlier in connection with the surface passivation technology (see Section 1.5). In this section, particular attention will be given to the deposition of conductive and resistive films. Thin-film resistors formed by the deposition and patterning of resistive thin-film layers on the wafer surface have some distinct advantages over the conventional diffused resistors. Thin-film resistors, in general, exhibit lower temperature coefficients and offer a wide range of sheet resistivity values, which can be chosen independently of active device design requirements. Furthermore, in some cases they can be trimmed to a final value by postdeposition heat treatment or anodization techniques. In forming the resistor patterns, resistive thin films, such as tantalum (Ta), nickel–chromium (NiCr) alloys, or tin oxide (SnO_2), are the most commonly used materials. Electrical properties of thin-film resistors are discussed further in Chapter 3.

For interconnection purposes, aluminum is the most commonly used thin-film material because of its high electrical conductivity and good adherence to the SiO_2 surface.

Deposition Techniques

Resistive or conductive thin films can be deposited on the silicon wafer surface by a variety of techniques. Some of these are outlined below.

Vacuum Evaporation. The passivated silicon substrate, together with the source of the material to be evaporated, is placed in a bell jar under high-vacuum conditions (10^{-5}–10^{-6} torr). The material to be evaporated is heated electrically by a tungsten or tantalum filament or by an electron gun until it vaporizes. Under the high-vacuum conditions used, the mean free path of the vaporized molecules is comparable to the dimensions of the bell jar. Therefore, the vaporized material radiates in all directions within the bell jar. Some of the vaporized material then deposits on the substrate, which is placed some distance from the source to ensure uniformity of deposition. The substrate is also maintained at an elevated temperature to provide a good adhesion of the deposited film.

Both conductive and resistive films can be deposited by vacuum evaporation. Aluminum, gold, and silver are among the conductive films formed in this manner. Nickel–chromium resistors can also be deposited by vacuum evaporation techniques, except in this case, due to high power densities required to vaporize the source, electron-beam bombardment, rather than thermal heating of the source material, is used. The films deposited by vacuum evaporation exhibit a fine-grained structure. The grain structure of the film becomes finer as the evaporation rate is increased and the angle of incidence of the radiating vapor on the wafer surface is made steeper. For uniformity and repeatability of film properties, a fine-grained structure is desirable.

Cathode Sputtering. The sputtering process takes place in a low-pressure gas atmosphere. A glow discharge is formed by applying a high voltage (typically 5000 V) between the cathode and the anode sections of the sputtering apparatus. The cathode is coated with the material to be evaporated, and the substrate is attached to the anode or placed within the glow-discharge region. Normally an inert gas such as argon (A) is used as the sputtering medium. The A^+ ions generated by the glow discharge accelerate toward the cathode due to the negative cathode potential. When these high-energy ions impinge on the cathode, they cause the atoms or the molecules of the cathode to break away, or sputter, from the surface. Then some of these cathode particles which float away are intercepted by the substrate and deposit in the form of a thin layer. Under the low-vacuum conditions used in sputtering, the mean free path of the source atoms is much shorter than the source-to-substrate spacing. Therefore, the deposition rates in the sputtering process are much slower than in vacuum evapo-ration.

By adding small amounts of reactive gases, such as oxygen or nitrogen, to the inert argon atmosphere, the chemical composition of the deposited layer can be modified. This is known as *reactive sputtering* and is particularly useful for tantalum deposition.

Vapor Phase Deposition. In vapor or gas phase deposition, halide compounds of the material to be deposited are chemically reduced, and the resulting metal atoms are deposited on the substrate. This basic deposition process very closely resembles the epitaxial growth step of the planar process. Vapor deposition is particularly useful for obtaining thick layers of deposited films (up to 20 μm). It is commonly utilized for forming aluminum oxide ($Al_2O_3 \cdot SiO_2$) dielectric layers or SnO_2 resistive films. The sheet resistance of SnO_2 films can be con-trolled by introducing Group III or Group V ions [such as indium (In) or antimony (Sb)] to increase or reduce the sheet resistance. In this manner, sheet resistances in the range of 100–5000 ohms per square ($\Omega/\square$) can be obtained.

Patterning and Etching of Thin Films

With minor modifications, the basic photomasking and etching techniques de-scribed in Section 1.6 can be utilized in patterning the thin-film components. One significant exception is the case of multiple thin-film layers (such as alumi-num interconnections over thin-film resistors) where additional care should be taken in the choice of the etchant to ensure that the bottom film is not damaged by the patterning of the top layer.

In the case of Al, SnO_2, Ta, or $Al_2O_3 \cdot SiO_2$ layers, patterning and etching can be achieved by direct photoresist techniques. In the case of very thin (300–500-Å) nickel-chromium films, an *inverse metal-masking* technique can be used. In this process, a thin layer of metal film (typically copper) is deposited and etched into an inverse, or negative, of the desired final metal pattern. Then the desired thin-film layer is deposited on this inverse metal pattern. In the final etching step, the inverse metal pattern of the initial metal film is etched away, taking with it the layer of desired metal deposited on it, and only the portions of the desired metal layer that adhere directly to the substrate are left behind.

Interconnections and Ohmic Contacts

The basic prerequisite for the conductive films used for interconnections is that they should make good ohmic contact with the diffused components or other metallic films deposited on the device surface. A good ohmic contact is defined as one that exhibits a linear current–voltage (I–V) relationship which passes through the origin of the I–V characteristic. In a great majority of IC applica-tions, aluminum is used as the interconnection layer. This is because of its ease of deposition and patterning, as well as its high degree of conductivity and its ability to form ohmic contacts with silicon.

The exposure of contact areas to the ambient atmosphere often results in the formation of parasitic oxide layers over the chip areas to be interconnected. Therefore, to provide good ohmic contact, the interconnecting metal must be chemically active so that it can be "alloyed" through these parasitic oxide layers. The most commonly used interconnection metal is aluminum. It can be readily alloyed into the silicon substrate to form ohmic contacts. Since aluminum is a p-type dopant (see Table 1.1), to avoid the formation of a nonohmic rectifying contact, the contact areas on the lightly doped n-type semiconductor regions are n^+ doped prior to metallization (see, for example, the collector contacts of the npn bipolar transistors of Figs. 1.13 and 1.14). The heavy n^+ doping causes a high degree of damage to the silicon lattice at the surface. Therefore, the parasitic pn junction formed by the alloying of a p-type aluminum interconnection into n-type silicon is very leaky and nearly ohmic in its conduction properties.

In conventional monolithic IC fabrication, the alloying of the aluminum interconnections into silicon is the last step of the planar process. It is normally accomplished by a short heat treatment in an inert atmosphere, typically about 10 min at 500°C.

A troublesome metal interconnection problem can occur in devices which employ two active dissimilar metals in their interconnection scheme. At the interface of two dissimilar metals, parasitic intermetallic compounds and oxides can form. A typical example of this is the intermetallic gold–aluminum compounds forming between the aluminum bonding pads and the gold wires that may be used to connect the chip to the package terminals. These compounds, which are brittle and nonconductive, are commonly referred to as the "purple plague" because of their dark color. Under certain circumstances (see Section 1.14) this may be a detriment to the reliability of IC interconnections using gold-wire bonds and aluminum bonding pads.

Although aluminum is a good conductor, it still introduces a finite amount of series resistance into the device interconnections. Typical resistivity of aluminum is of the order of 2.8×10^{-6} ohm-cm. In the case of a typical aluminum interconnection trace of 1-μm thickness, this corresponds to a sheet resistance of approximately 0.03 $\Omega/\square$.

Under very high current densities, conductive thin films such as aluminum exhibit a failure mode due to the so-called *electromigration* effect. It causes the metal atoms to gradually migrate away from the high-current-density points within the conductor. This is a progressive failure mode, under continuous operation, and comes about from the momentum exchange between the conducting electrons and the stationary metal atoms. Electromigration is a slow process which speeds up as the current density or the temperature is increased. It starts as a formation of localized *voids*, or gaps, in the conducting metal strip and eventually leads to a complete open circuit at the point of highest current density within the conductor strip.

To avoid electromigration effects, current densities in the IC interconnection paths are normally kept at less than the 10^5-A/cm^2 level. In the case of a typical aluminum interconnection path of 1-μm thickness, this corresponds to a maximum allowable continuous current of approximately 50 mA per mil width of the interconnection path.

1.9. BIPOLAR INTEGRATED-CIRCUIT FABRICATION STEPS

The fabrication of a bipolar integrated circuit involves a sequence of five to eight masking and diffusion steps. The sequence of some of these basic steps is illustrated in Figure 1.11 for the case of an *npn* bipolar transistor and a *p*-type diffused resistor.

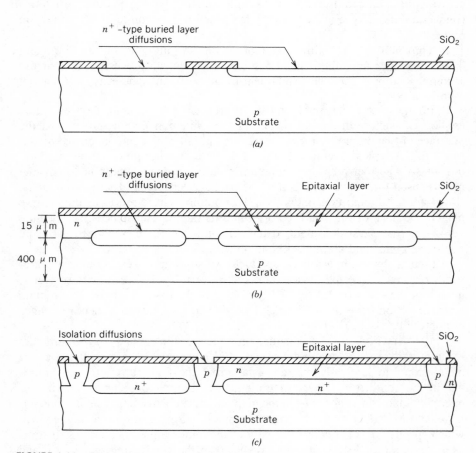

FIGURE 1.11. Basic sequence of steps in the fabrication of a bipolar monolithic circuit: (*a*) Initial oxidation and buried layer diffusion; (*b*) expitaxial layer growth and second oxidation; (*c*) isolation diffusion; (*d*) base diffusion; (*e*) n^+ emitter diffusion; (*f*) contact windows and interconnections.

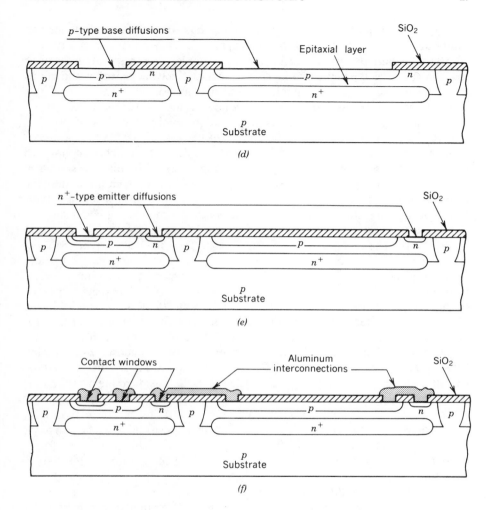

(d)

(e)

(f)

The starting material is a wafer of p-type silicon, typically 4 inches in diameter and approximately 400 μm thick, with a typical impurity concentration of 10^{16} atoms/cm^3. The first masking and diffusion step, illustrated in Figure 1.11a forms a low-resistivity n^+-type layer* which will eventually form a low-resistance current path within the collector region of the resulting npn transistor. Since this heavily doped n-type layer will be covered by the epitaxial layer, it is called a buried layer, and the corresponding diffusion is called the buried layer diffusion. The sheet resistance of the buried layer diffusion is in the range of 10–30 $\Omega/\square$, and the impurity used is either arsenic or antimony. These impurities

*In this context, the + sign is used to imply heavy impurity concentration, not electrical charge.

are chosen because they diffuse slowly and, thus, do not redistribute significantly during the subsequent diffusion steps.

After the buried layer diffusion step, the oxide covering on the wafer is stripped and an n-type epitaxial layer is grown over the entire wafer surface, as shown in Figure 1.11b. The thickness and the impurity concentration of this epitaxial layer determine the breakdown voltage of the resulting transistor structure. Assuming a minimum transistor breakdown voltage LV_{CEO} of 30 V is required, the epitaxial layer will be chosen to be approximately 15 μm thick, with an impurity concentration of 2×10^{15} atoms/cm^3, which corresponds to a resistivity of approximately 2.5 Ω/cm (see Fig. 1.2). Note that, as shown in Figure 1.11b, the n^+-type buried layer also out-diffuses somewhat into the epitaxial layer during this process.

Following the epitaxial growth, an oxide layer is formed on the wafer surface. Then, after a masking step, a p-type (boron) diffusion is made, as shown in Figure 1.11c. The function of this diffusion is to form the deep p-type *isolation walls,* which reach through the n-type epitaxial layer into the p-type substrate. Because of the depth to which this diffusion must penetrate, it requires several hours of diffusion time at temperatures in excess of 1200°C. The sheet resistance of the p-type isolation diffusion is in the range of 20–40 $\Omega/\square$. Note that the n^+-type buried layer diffusion is omitted in the regions directly under the isolation diffusion, and that the n^+-type buried layer does *not* touch the p-type isolation wall. This is done to avoid forming a low-breakdown p-n junction between the n-type *tub* and the p-type isolation; and to ensure that the p-type isolation wall can reach down to the p-type substrate, thus forming a continuous wall surrounding the n-type tub.

The next masking and diffusion step (Fig. 1.11d) forms the p-type base region of the npn transistor. It results in a sheet resistance in the range of 100–200 $\Omega/\square$ and a junction depth of 1–3 μm. Since this diffusion also forms many of the resistors in the circuit, its sheet resistance is closely controlled to be within ±20% of the target value.

Following the base diffusion, the n-type emitter regions of the npn transistor are formed by the emitter mask and the subsequent emitter diffusion step, as shown in Figure 1.11e. Normally, phosphorus is used as a dopant for the emitter diffusion. The resistance of the emitter diffusion is of the order of 2–10 $\Omega/\square$, and the resulting junction depth is in the range of 0.5–2.5 μm. Since the difference in the junction depth of the base and the emitter diffusions determines the base width of the npn transistor, the depth of the emitter diffusion is controlled to be approximately 0.5–1 μm, *less* than that of the base diffusion. The heavily doped (n^+-type) emitter diffusion also serves as a low-resistance contact to the n-type epitaxial layer to form the ohmic collector contact for the transistor. This is necessary because a direct ohmic contact between the aluminum interconnection and the lightly doped n-type epitaxial region is difficult to form (see Section 1.8).

After the emitter diffusion, the wafer undergoes a masking step called the *contact mask*, which opens all the contact windows over all the passive and active devices on the chip. Then the entire wafer is coated with a thin layer of aluminum (0.5–1 μm), which will form a conductive interconnection path between the devices. Then, in the next masking step, called the *metal mask*, the aluminum is etched away, leaving behind the desired interconnection pattern between the components on the chip. The resulting device structure is shown in Figure 1.11*f*.

Figure 1.12 shows a plane view of the completed device structure of Figure 1.11*f*. For illustrative purposes, the collector of the *npn* transistor is assumed to be connected to the resistor to form a simple common-emitter amplifier stage.

Figure 1.13 shows the typical impurity concentration within the vertical cross section of the resulting transistor structure. Assuming a base depth of 3 μm and an emitter depth of 2.5 μm, a base width of $\approx$0.5 μm is achieved. Also, note that the n^+-type buried layer has out-diffused approximately 6 μm from the nominal epi–substrate interface.

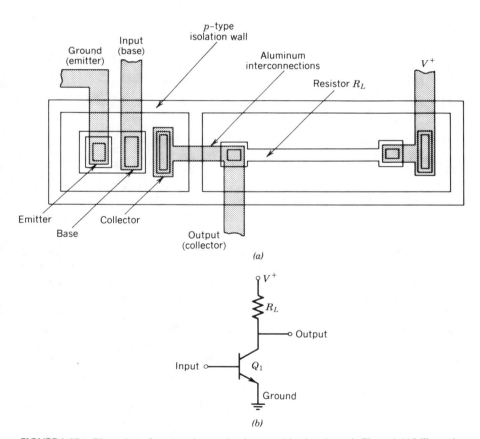

FIGURE 1.12. Plane view of *npn* transistor and resistor combination shown in Figure 1.11*f*, illustrating its connection as a common-emitter gain stage: (*a*) Plane view of circuit layout; (*b*) equivalent circuit.

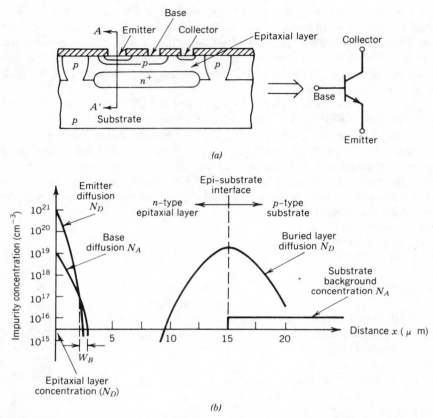

FIGURE 1.13. Device cross section and typical impurity profile for a monolithic *npn* transistor: (*a*) device cross section; (*b*) impurity profile along A–A^+.

In the normal fabrication process, subsequent to the metal mask step, the silicon wafer undergoes an *alloy* or *sinter* step at low temperatures (typically 5–10 min at 450°–500°C) to assure proper ohmic contact between the aluminum interconnections and the contact windows on the wafer surface. Finally, a dielectric oxide layer is deposited on the wafer surface using a low-temperature CVD process (see Section 1.5) to form a passivation layer over the entire chip surface, which also covers the aluminum interconnections. Finally, a passivation mask is applied to the silicon wafer to etch large windows on top of the bonding pads on the IC chip, to which external connections will be made. The purpose of this passivation layer is to protect the chip surface from dust, mechanical damage, and scratches.

After completion of the sequence of fabrication steps described above, the finished silicon wafer is ready for evaluation and the functional electrical testing of the individual circuit chips.

Normally, the wafers proceed through the fabrication process steps in groups called *lots*, with typically 15–25 wafers to a lot. Thus, many wafers, each containing hundreds of complex circuit chips, are manufactured simultaneously. This batch processing capability of the planar process, which enables one to fabricate thousands of circuits in one process sequence, is the key to the economic advantages of integrated circuits.

1.10 MODIFICATIONS OF BASIC PROCESS

The vast majority of the monolithic bipolar analog circuits presently in production are fabricated using the basic sequence of steps illustrated in Figure 1.11. However, when certain performance characteristics are required, additional steps can be incorporated into the basic manufacturing sequence to enhance such characteristics. Some of these modifications will be briefly discussed in this section. Although these modifications improve specific device characteristics, they often require additional process steps or tighter process controls and, thus, add to the manufacturing cost of the monolithic circuit.

Dielectric Isolation

In certain circuit applications, the parasitic capacitances, leakage currents, and breakdown characteristics associated with the conventional junction isolated device structures may not be acceptable. Typical examples of such applications are circuits that must operate at high frequencies or very high voltages (i.e., in excess of 100 V) or circuits that must withstand high amounts of ionizing radiation (such as gamma rays) without suffering permanent damage. In such cases, superior electrical isolation can be obtained by surrounding each n-type pocket or tub not with a reverse-bias p-n junction, but with a dielectric layer. Normally, thermally grown SiO_2 is used as the dielectric material.

In forming the dielectrically isolated pockets on the wafer surface, a number of alternate fabrication techniques can be utilized.[9] Figure 1.14 shows a typical sequence of fabrication steps in forming the dielectrically isolated single-crystal silicon pockets or islands. Starting with an n-type substrate, a nonselective n^+-type layer is diffused into the wafer surface. For reasons which will be explained shortly, a < 100 > oriented crystal is utilized as the starting material, as opposed to the < 111 > oriented crystal normally used for junction isolation.*

Following the initial n^+ diffusion, the wafer surface is oxidized, and a mirror image mask of the desired isolation grid pattern is applied to the wafer to remove the oxide along the isolation grid. The exposed silicon surface is then etched by

---

*During the initial silicon crystal growth, the <111> oriented crystal is easier to grow defect-free than the <100> orientation and is, therefore, somewhat less expensive. In addition, the <111> crystal is less affected by surface contamination than the <100> crystal. These are the main reasons for the choice of the <111> oriented substrate for most bipolar device structures.

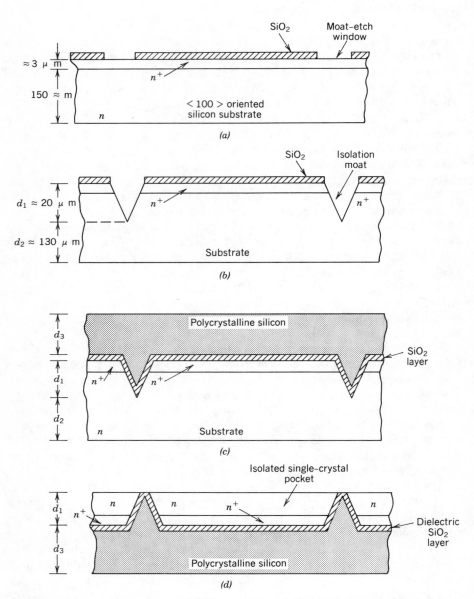

FIGURE 1.14. Sequence of steps in dielectric-isolation process: (*a*) n^+ layer diffusion and opening of moat-etch windows; (*b*) etching of isolation moats; (*c*) forming SiO_2 dielectric layer and growing polycrystalline silicon layer; (*d*) forming isolated n-type pocket by grinding away substrate until peak of V groove appears.

a potassium hydroxide (KOH) based etch. The etchant used in this step etches away the exposed silicon anisotropically, that is, the etch rate is much faster along the [111] planes than along the [100] crystal planes. This preferential etching results in formation of a V-shaped isolation *groove* or *moat* on the wafer surface, as shown in Figure 1.14*b*.

Referring to Figure 1.14*b*, it should be noted that the vertical dimensions of the drawings are not to scale, and the depth d_1 of the isolation moat is a small fraction of the total wafer thickness. After the preferential etching step, the exposed silicon is reoxidized, and a thick layer of polycrystalline silicon is deposited over the oxide layer, as shown in Figure 1.14*c*. The thickness and the electrical characteristics of this polycrystalline silicon layer are of no consequence since its main function is to serve as a mechanical support for the wafer. Then the original wafer is flipped around, and the bottom surface of Figure 1.14*c* is now corresponding to the top of the device structure. Then the single-crystal *n*-type layer of thickness d_2 is backlapped until the isolation grid appears on the wafer surface, resulting in the isolated *n*-type single-crystal pocket shown in Figure 1.14*d*. After the isolated pockets are formed, the fabrication of integrated devices within the pockets is completed by a sequence of conventional masking and diffusion steps, resulting in the isolated device structure of Figure 1.15.

In some cases, it may be desirable to reverse the order of the initial n^+-type layer which covers the sides as well as the bottom of the isolated pocket. Such a structure offers lower ohmic resistance directly below the base. However, in this case a larger clearance is required between the *p*-type base and the dielectric sidewall to prevent the n^+-type layer from touching the base-diffused regions.

The basic sequence of processing steps has been known for quite some time. However, the process has become practical only after the development of anisotropic etch techniques,[10] which allow very accurate control of the etch depths by the width of the initial oxide window.

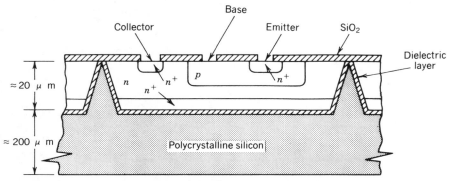

FIGURE 1.15. Cross section of a dielectrically isolated *n*-type pocket containing a *npn* transistor.

Deep n^+ Diffusion

As will be described in Chapter 2, the basic bipolar transistor structure of Figure 1.11 has a relatively large parasitic resistance in series with its active collector area. Most of this resistance is due to the resistivity of the n-type epitaxial region. This parasitic series resistance can be significantly reduced by extending the heavily doped (i.e., low-resistivity) n^+-type collector contact region all the way down through the epitaxial layer into the n^+-type buried layer. In practice, this can be achieved by means of a deep n^+ diffusion which is performed subsequent to the p-type isolation diffusion and prior to the p-type base diffusion (i.e., between steps c and d of Fig. 1.11). The resulting deep diffusion is called an n^+ sinker, plug, or sidewall diffusion.

Figure 1.16 shows the cross section of a npn transistor with such a deep n^+ diffusion. Although this step adds an extra masking and diffusion cycle to the basic fabrication sequence, its effect on the overall manufacturing cost is minimal.

Up-Down-Diffused Isolation

The p-type isolation diffusion is the longest diffusion cycle during the bipolar IC fabrication process. This is because the depth of the isolation diffusion must be sufficient to penetrate the entire thickness of the n-type epitaxial layer, which can be anywhere from 5 to 25 μm, depending on the particular fabrication process used, which in turn depends on the transistor breakdown voltage requirements. Since the diffusion process proceeds sideways as well as downward into the silicon, the side diffusion of the isolation wall in turn determines the base-to-isolation spacing of a npn transistor. As shown in Figure 2.4, the base-to-isolation spacing, in turn, is the major contributor to the chip area taken up by a small-geometry npn transistor.

The minimum base-to-isolation spacing can be significantly reduced by using the so-called *up-down isolation diffusion* technique illustrated in Figure 1.17. In this method, the isolation wall is diffused simultaneously from the epi–substrate interface as well as from the wafer surface. This is achieved by forming a p^+-type subepitaxial layer directly under the isolation wall, prior to the epitaxial

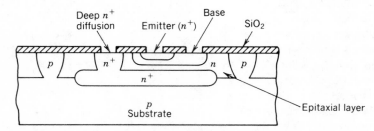

FIGURE 1.16.   Cross section of npn transistor structure using deep n^+ diffusion.

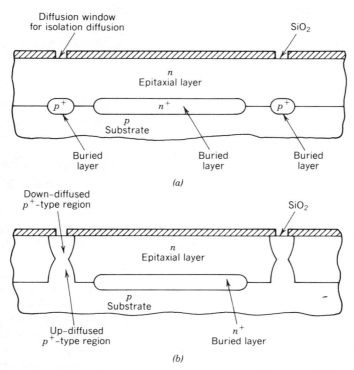

FIGURE 1.17. Forming an isolation wall with simultaneous upward and downward diffusion of p-type impurities. Device cross section (*a*) before isolation diffusion and (*b*) after isolation diffusion.

growth. In other words, both the p^+-type and the n^+-type buried layer regions are formed before the epitaxy process. This results in the device cross section shown in Figure 1.17a, prior to the start of the conventional isolation diffusion step. Then, during the isolation diffusion, the p^+-type subepitaxial layer out-diffuses into the epitaxial layer and meets the down-diffusing isolation wall at approximately halfway in the epitaxial layer to complete the isolation wall, as shown in Figure 1.17b. In this manner, the total depth of the top-diffused isolation wall is only 50–70% of the epitaxial layer thickness, which results in a very significant reduction of isolation side diffusion. As a result, the minimum allowable base-to-isolation spacing can be reduced significantly, allowing a more compact device layout.

An added benefit of the up-down-diffused isolation is that the total isolation diffusion time is reduced, resulting in the reduction of the undesired out-diffusion of the n^+-type buried layer into the epitaxial region. This out-diffusion is undesirable since it reduces the epitaxial layer resistivity and lowers the transistor collector–base breakdown voltage.

The up-down isolation, similar to the case of deep n^+ diffusion, adds one extra masking and diffusion step to the conventional bipolar fabrication process shown in Figure 1.11. However, it is a relatively noncritical additional step and can result in significant chip area reduction (typically of the order of 15–20% of the active chip area) in high-voltage integrated circuits with 20-μm or thicker epitaxial layers.

Two-Step Emitter Diffusion

The common-emitter current gain β_F of bipolar transistors is a strong function of the transistor base width and increases very rapidly as the base width is reduced. Unfortunately, the transistor breakdown voltage BV_{CEO} is also reduced greatly as the base width is reduced (see Section 2.1). In certain applications, such as in designing the input stages of monolithic operational amplifiers, it is necessary to have very high-gain (i.e., *superbeta*) transistors on the same chip as the conventional medium-gain but high-breakdown *npn* transistors. Superbeta transistors exhibit a current gain β_F on the order of 2000–5000, with a breakdown voltage of 2–5 V, whereas the conventional *npn* transistors would have $\beta_F \approx 200–500$, with a breakdown voltage of 30–50 V.

The superbeta transistors can be fabricated simultaneously with conventional *npn* transistors by using a two-step emitter diffusion process: the emitters of superbeta transistors are diffused first, and the emitters of conventional *npn* transistors are then masked and diffused as the next step. Thus, the emitter regions of the supergain transistors undergo a *longer* diffusion cycle, and since both transistors have the same base diffusion, the superbeta transistor ends up with a *deeper* emitter diffusion, resulting in a *narrower* base width. The basic two-step emitter diffusion process is also used for fabricating double-diffused JFETs on the same chip as *npn* bipolar transistors (see Fig. 2.36).

Figure 1.18 shows the structure of a superbeta transistor side by side with a conventional *npn* transistor. The two-step emitter diffusion process requires one additional masking and diffusion step, as well as tighter process controls, than the basic bipolar process.

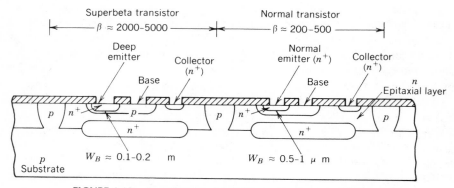

FIGURE 1.18.   Superbeta transistor using two-step emitter diffusion.

In certain applications, it is also possible to replace the two-step emitter process with a two-step base process, where the base can be diffused to a *shallower* depth. Thus, the same emitter diffusion applied to both devices shown in Figure 1.18 would then result in a narrower base width for the superbeta transistor.

The applications and the electrical characteristics of superbeta transistors are described further in Chapter 2.

Ion Implantation

As described in Section 1.7, ion implantation provides another alternate method to diffusion for introducing impurities into silicon. Since it is a low-temperature process, it is normally implemented *after* all the diffusion steps have been completed (i.e., between steps *e* and *f* of Fig. 1.11). It is normally used to form high-value *p*-type resistors, bipolar compatible JFET (BIFET) or MOS structures, or high-frequency integrated injection logic (I^{2}L) gates. Ion implantation normally requires one masking and one implantation step, followed by a low-temperature ($\approx$ 500–600°C) heat treatment. In certain cases, such as bipolar compatible JFETs, two ion-implantation steps are used.

Double-Layer Metallization

In certain complex circuits, it may be necessary to use two separate interconnection layers on the chip. This is done by first forming the initial layer of interconnections (see Fig. 1.11*f*) and then depositing a layer of dielectric. Then interconnecting windows, or *via holes* are etched through this dielectric. Next, a second layer of metal is evaporated and etched to provide a second layer of interconnections, which connect to the first layer through the via holes cut in the deposited dielectric layer. Double-layer metallization is a difficult process requiring careful process control.

Thin-Film Resistors

Thin-film resistors are deposited and etched on the oxidized silicon surface prior to the aluminum evaporation and etching. Formation of thin-film resistors requires a well-controlled deposition step (by either vacuum evaporation or sputtering) followed by a separate masking and etching step.

High-Performance Complementary Transistors

Monolithic bipolar fabrication technology is designed around the *npn* bipolar transistor. This basic process also produces a simple low-frequency *pnp* device, called the *lateral pnp transistor*, which is covered in Section 2.3. However, in certain applications, it is required to have high-performance *pnp* devices with the gain and frequency response characteristics comparable to those of normal *npn* transistors. This can be accomplished by adding a number of additional diffusion and masking steps to the basic fabrication sequence shown in Figure 1.11.[11, 12] However, in practice this is at best a compromise solution which has a significant impact on manufacturing costs and yields because of the following factors:

1. The basic device structure of Figure 1.11 requires only *one* critical diffusion step (i.e., the n^+ emitter diffusion) to control the *npn* transistor characteristics. If additional diffusion steps were provided to form a *pnp* transistor, that transistor would in turn require at least one critical diffusion step to set its base width. Since all diffusions proceed simultaneously, it is very difficult to perform two independent critical diffusion steps under a continuous high-volume production environment.

2. Adding additional masking and diffusion steps increases the amount of handling of the silicon wafer during the fabrication process, which in turn increases the chances of creating circuit defects, parameter variation, wafer surface contamination, wafer breakage, and operator error.

Economic Considerations

Many variations and modifications can be added to the basic bipolar IC fabrication sequence of Figure 1.11 to obtain specialized or improved device characteristics, or to obtain complementary or dissimilar device combinations. However, in practice, particularly in a high-volume production environment, the economic considerations (i.e., the manufacturing costs and the resulting yield of good circuits at the end of the wafer fabrication process) are of paramount importance. Thus, the performance benefit of any process modification has to be carefully judged against the resulting increase in manufacturing cost and complexity.

Table 1.3 gives the approximate relative manufacturing costs of various modifications to the basic process described in this section. For comparison purposes, the basic four-diffusion, seven-mask manufacturing process illustrated in Figure 1.11 is taken as a reference (i.e., as unity). The relative complexity figures indicated give a rough estimate of the relative manufacturing cost of one completely finished wafer of silicon under high-volume manufacturing conditions. It should be noted, however, that the complexity figures given in Table 1.3 are somewhat subjective (i.e., they reflect the view of the author) and are subject to change as the IC fabrication technology continues to evolve.

1.11. ASSEMBLY AND PACKAGING

Fabrication of the monolithic circuit on the surface of a silicon wafer represents only one part of the total manufacturing process. Additional asembly and packaging steps are required to make the circuit electrically functional.

TABLE 1.3.   Estimate of Relative Manufacturing Costs for Various Bipolar IC Fabrication Processes

Process Description	Estimated Relative Cost Factor[a]
Basic bipolar process (Fig. 1.11)	1.0
Dielectric isolation (Fig. 1.14)	2 to 4
Deep n^+ diffusion (Fig. 1.16)	1.1
Up-down isolation diffusion (Fig. 1.17)	1.15
Two-step emitter diffusion (Fig. 1.18)	1.3
Ion implantation (single implant)	1.2
Bipolar compatible JFET	
(using two ion-implant steps)	1.3
Thin-film resistors (untrimmed)	1.8
Thin-film resistors (laser trimmed)	2.5
Double-layer metallization	1.5–2

[a]Relative cost factors do not add linearly. If combinations of these modifications are used in a given process, the resulting relative cost factor is usually higher than the sum of the corresponding cost factors.

The wafer fabrication steps described up to this point represent the most efficient part of the IC fabrication process since they allow thousands of IC chips to be fabricated simultaneously through batch processing methods. However, once the wafer processing is complete, each chip on the wafer must be individually tested. Then the wafer is separated into *chips* or *dice*, and each circuit must be handled and packaged individually. Figure 1.19 gives a flowchart of the manufacturing steps that follow the wafer fabrication operation.

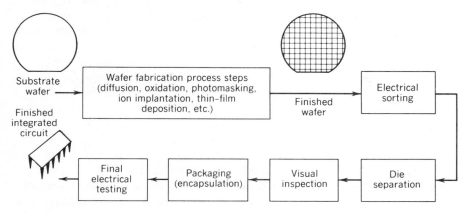

FIGURE 1.19.   Flowchart of IC manufacturing steps.

Electrical Sorting

After the wafer fabrication steps and prior to encapsulation, each of the individual circuit chips on the wafer has to be electrically tested to ensure that it meets the desired electrical performance requirements. Since the assembly process is a rather costly step in the overall IC manufacturing process, it is necessary to screen the monolithic chips for defects and electrical failures while the units are still on the wafer. For this purpose, an automatic probing station is utilized. During the electrical probing operation, the bonding pads on each circuit chip are contacted by needle-like metallic probes, and various current and voltage levels at the terminals of the circuit are measured. Normally, these electrical tests are performed by programmed automatic testers which can perform up to 100 tests per second on a single chip. If the electrical characteristics of the circuit are not acceptable, an automatic marking pen at the probing station is activated to mark the unit as a reject. When the test sequence for a given circuit is completed, the wafer is automatically indexed, and the probes advance to the next chip. In this manner, the entire wafer is automatically sorted, and the rejected devices are marked with ink dots.

Die Separation

Subsequent to electrical sorting, the wafer is cut along the rectangular grids separating the individual chips or dice. A number of methods have been developed for separating the individual chips. The most commonly used ones are scribing the wafer with a diamond-tipped cutting tool, called *scriber*, or using a circular diamond-tipped saw to cut a series of deep grooves into the wafer along the rectangular grid separating the chips. Subsequent to the scribing or sawing operation, the wafer is fractured along the scribed channels, or grooves, into physically separated chips. Recently the use of a laser beam, rather than a diamond saw or scriber, has also gained acceptance in the IC industry. Laser scribing is preferred over conventional scribing techniques since it is a much faster process and eliminates accidental breakage, chipping, or cracking of the individual dice during the die separation step.

Visual Inspection

After die separation, the electrically good chips (i.e., those with no ink marks) are visually separated from the rest of the chips on the wafer. These good chips are then placed in plastic carriers, with proper orientation such that all the chips on the carrier face in the same direction. At this point, each chip is examined under magnification to ensure that it has no visually obvious defects, such as scratches or cracks, or was not damaged by the probe contacts during the electrical sorting operation. The purpose of this visual inspection is to separate out and discard any mechanically damaged or defective chips prior to the assembly step. To avoid potential reliability problems, electrically good chips which may exhibit severe over- or underetching of various mask patterns, and particularly the aluminum interconnection lines, are also discarded. At this point, the remaining chips are ready to be mounted into final circuit packages.

Encapsulation

The packaging or encapsulation of the visually inspected IC chip proceeds in three steps. First, during the *die attach* step, the IC chip is attached or bonded to the gold-plated header, or the Kovar lead frame. Then, in the second step, called the *wire bonding* step, gold- or aluminum-wire leads are used to connect the bonding pads on the circuit chip to the package leads or posts. During this step, any one of several bonding techniques, such as thermocompression bonding (for gold wires) or ultrasonic bonding (for aluminum wires), can be used.

The last step of the encapsulating process is sealing the monolithic circuit package by injection molding (in the case of plastic packages), by soldering or welding a cap (in the case of side-brazed or metal can type packages), or by forming a hermetic glass seal between the package and its cap (in the case of ceramic packages). In most assembly operations, it is also customary to inspect each chip in the package after the wire-bonding step and prior to sealing. In this manner, obvious assembly defects can be detected quickly, and the rejects are discarded.

1.12. INTEGRATED-CIRCUIT PACKAGES

An IC package is expected to satisfy a large number of partly conflicting requirements: low cost, mechanical strength, high packing density, hermeticity, low parasitic reactances, low thermal resistance, and ease of handling and testing. No single circuit package exists which ideally fulfills these characteristics. For a majority of monolithic analog circuits, the package choice is narrowed down to the three most commonly used package types: *dual-in-line packages, metal cans,* and *flat packs.*

The dual-in-line (DIP) package, the dimensional diagrams of which are shown in Figure 1.20, is by far the most commonly used package type because of its relatively low cost and ease of handling. The in-line bent structure of the leads makes it convenient for automatic handling during electrical testing or board-insertion steps. The DIP packages are available in 8-, 14-, 16-, 18-, 20-, and 22-pin versions, with the narrow package dimensions shown in Figure 1.20a. Higher pin count versions of the DIP packages, from 24 to 40 pins, have the wider bodied structure shown in Figure 1.20b.

The DIP packages are available in three types: plastic packages which are nonhermetic and made of injection-molded epoxy compounds, black ceramic (CERDIP) packages, and the combined metal–ceramic or side-brazed packages. The plastic DIP package, which is rated for operation up to 85°C, is the lowest cost package type when hermeticity is not required. Both CERDIP and side-

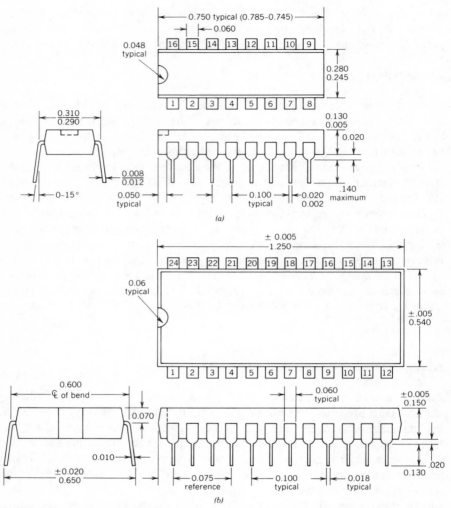

FIGURE 1.20. Dual-in-line (DIP) packages most commonly used in IC packaging (dimensions in inches): (*a*) Narrow-body DIP for 8–22 pins, (*b*) wide-body DIP for 24–40 pins.

brazed DIP packages are hermetic and can operate over a temperature range of −55 to +125°C.

The two most commonly used metal can type packages are the TO-99 and the TO-3, as shown in Figure 1.21. The metal can packages have a welded cap which hermetically seals the chip. Their key advantages are good thermal characteristics, high mechanical strength, and a very high reliability rating. Their disadvantages are (1) the available number of leads is limited (12 pins maximum), (2) leads bend easily and are difficult to insert into sockets, and (3)

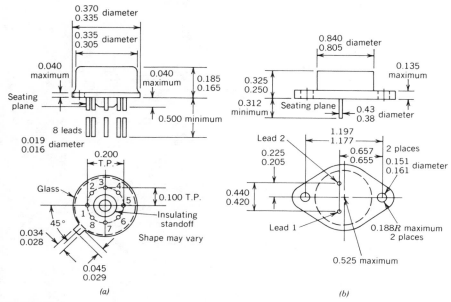

FIGURE 1.21. Metal can type IC packages (dimensions in inches): (*a*) TO-99 type package; (*b*) TO-3 type power package.

higher cost, particularly due to gold plating needed for certain parts of the package.

The flat pack type, small-geometry packages (see Fig. 1.22) were developed to improve on the volume, weight, and pin count limitations of metal can packages. Flat packs with up to 22 leads are commercially available. They have approximately one-fifth the volume and weight of conventional DIP packages and can be produced in both round and rectangular shapes. Their key advantages are lightweight and small volume; their major disadvantages are high cost and difficulty in handling.

Packaging and assembly are the major considerations in the low-cost high-volume manufacture of monolithic circuits. As such, an extensive amount of engineering and development effort has gone into these areas within recent years. The cost element, in particular, has brought about a multiplicity of new, fully automated packaging and assembly techniques, such as *beam-lead*, *flip-chip*, and *spider-bond* methods. A detailed analysis and comparison of these techniques is a highly specialized subject, which is well covered in the literature.[13,14]

Thermal Considerations

One of the basic limitations of integrated circuits is the dissipation of heat produced during operation of the circuit. This heat must be transfered to some

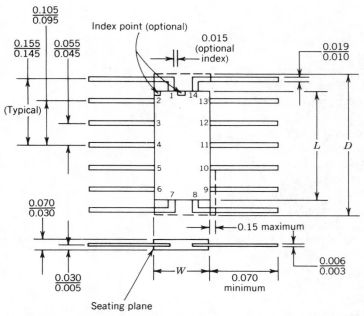

FIGURE 1.22. Physical dimensions of a 14-pin flat pack (dimensions in inches).

sink without causing excessive temperature rise in the circuit elements or inter-connections. For silicon devices, reliability considerations dictate that the junction temperature on the chip T_j be kept below about 150°C. This basic limitation, along with the thermal conduction properties of the IC package, determines the maximum allowable power dissipation or ambient temperature range of operation.

The steady-state thermal behavior of the circuit chip and the IC package can be estimated by using its electrical analog shown in Figure 1.23.[15] In this model, the current is analogous to the flow of heat, and the voltage is equivalent to temperature. The current source P_d represents the power dissipation of the IC chip. The resistors R_{jc} and R_{ca} represent the thermal resistance from junction to

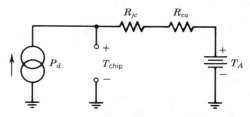

FIGURE 1.23. Electrical analog of package power dissipation.

case and from case to ambient for the package. The sum R_T of the two resistances is the total thermal resistance of the system. From the approximate electrical model of Figure 1.23, the chip temperature can be written as

$$T_{chip} = T_A + (R_{jc} + R_{ca})P_d \qquad (1.16)$$

where T_{chip} and T_A are the temperatures of the chip and the ambient air surrounding the package.

The junction-to-case thermal resistance depends mainly on the particular package type used. However, it is also affected by the chip size and the manner in which the chip is attached to the package. The case-to-ambient thermal resistance is a function of the package surface, external heat sinking, and whether or not forced air cooling is used. Table 1.4 lists the approximate values for R_{jc} and R_{ca} for various package types, assuming no heat sinks and still air ambient.

Many manufacturers also like to state the package power handling capability in terms of the power dissipation capability at $T_A = 25°C$, with a derating factor expressed in mW/°C, for operation at temperatures above 25°C. These typical ratings and derating factors are also listed in Table 1.4 for a maximum allowable chip temperature of 150°C. Since different IC manufacturers may assume different maximum chip temperatures for their products, the 25°C power ratings and the derating factors may vary between different manufacturers.

TABLE 1.4. Approximate Thermal Resistance and Allowable Power Dissipation for Various IC Packages.[a]

Package Type	Thermal Resistance (°C/W)			Allowable Power Dissipation at $T_A = 25°C$ (W)	Derating Factor for $T_A > 25°C$ (mW/°C)
	Junction to Case	Case to Ambient	Total		
Dual in line (DIP), 14/16-pin, plastic (Fig. 1.20a)	70	100	170	0.750	6
Dual in line (DIP), 14/16-pin, ceramic (CERDIP)	40	100	140	0.90	7
Metal can (Fig. 1.21)					
TO-99	50	100	150	0.8	6
TO-3	10	30	40	3.0	25
Flat pack (Fig. 1.22)	100	150	250	0.5	4

[a]$T_{chip} \leq 150°C$, no heat sink, still air ambient.

1.13. TESTING OF INTEGRATED CIRCUITS

Although integrated circuits are batch processed during wafer fabrication, each and every circuit chip must be individually tested at least twice during the manufacturing cycle: once at the electrical sort step prior to scribing and die separation, and a second time after completion of the assembly operation. This second test cycle, called the final test step, includes a detailed testing of all circuit parameters according to the minimum and maximum specifications given in the product data sheet. Unless specified otherwise, both the electrical sort and the final test operations are done at room temperature.

A large number of tests are often necessary to characterize an analog integrated circuit. Consequently, testing is one of the most important, expensive, and time-consuming parts of the overall IC manufacturing process. The types of testing that can be performed on integrated circuits can be divided into three categories:

1. *DC Testing.*   This measures the static parameters of the circuit, such as operating voltage and current levels, input bias currents, offsets, and so on. It is performed by forcing preprogrammed current or voltage levels to various circuit terminals and then sensing the resulting voltage or current levels.

2. *AC Testing.*   This evaluates the circuit performance under operating bias, with sinusoidal ac signals applied to it.

3. *Dynamic Testing.*   This includes testing the circuit in an environment or operating condition which simulates its actual application and includes pulse, amplitude, and time measurements as well as complex waveforms.

In normal production testing of analog integrated circuits, extensive dc testing with some ac testing is utilized. Dynamic testing is usually quite complex and time consuming. Therefore, it is used mainly for very specialized circuits which combine analog and digital functions on the same chip, and whose end applications are very well defined.

Because of the large number and the complexity of the tests required, computer-controlled and fully automated test systems have become a major element in IC testing. Figure 1.24 shows a generalized block diagram of such a computer-controlled test system.[16] The test program can be loaded into the computer by punched cards, paper tape, or magnetic tape. Instructions from the computer are then sent to an interface or control unit which controls the various elements of the system. Stimulus instructions for the integrated circuits are buffered, converted into analog voltages, and delivered to the pins of multiplexed test stations or wafer probes, which are time shared under computer control. Analog-to-digital (A/D) converters convert the output functions of the integrated circuits into digital form.

This information is then buffered and returned to the computer for processing. The computer makes a "go" or "no go" decision based on the test results and

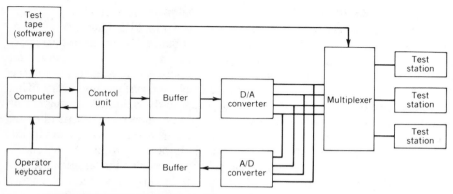

FIGURE 1.24. Functional block diagram of a computer-controlled test system.

automatically sorts the device into the appropriate container. Normally, such a tester would also have a data-logging capability such that preliminary modes of failure can be recorded for failure analysis and yield improvement purposes.

1.14. RELIABILITY CONSIDERATIONS

One of the most significant attributes of monolithic integrated circuits is high reliability. Integrated circuits are far more reliable than their discrete component counterparts, and their reliability is improving rapidly with increased knowledge of the processing techniques and an understanding of the possible failure mechanisms.

Failure Modes and Mechanisms

The most commonly encountered failure mechanisms in integrated circuits can be attributed to one of the following four sources: (1) bulk failures, (2) surface-related failures, (3) failures of metallization or interconnections, and (4) package-related failures.

Bulk Failure. Bulk failure is a relatively unimportant failure mode. Good starting material is essential in the fabrication of reliable integrated circuits. Crystallographic defects, such as dislocations, stacking faults, and growth strains, enhance long-term degradation mechanisms and, therefore, contribute to the unreliability of integrated circuits. Failure modes associated with bulk silicon include die breakage, short-circuits due to secondary breakdown, uncontrolled *pnpn* switching, and degradation of electrical characteristics. Bulk failure mechanisms are accelerated at high operating current densities due to localized heating effects.

Surface-Related Failures.   These failures are statistically second only to failures in the interconnection system. Typically, 35% of all IC failures result from surface effects. Surface effects significantly influence p-n junction characteristics and tend to control transistor gains, junction breakdown voltages, and leakage currents. Charge migration along the silicon surface, especially in the vicinity of a p-n junction, is a major mechanism of surface instability. This instability is often caused by ionic contaminants in the oxide on the surface or near the silicon–oxide interface. The charge buildup due to ionic contamination may be high enough to cause the formation of an inversion layer along the surface, where the resistivity type of the underlying silicon may be reversed. Failures due to surface effects are accelerated by increasing the temperature and reverse biasing the p-n junctions. Both conditions tend to increase the ionic charge mobility and enable ionic contaminants to induce inversion layers near the junctions.

Metallization and Interconnection Failures.   The most common failure modes in integrated circuits are open or short circuits in the circuit metallization and in bonding. These conditions contribute to more failures than all other failure types combined (typically between 50 and 60%).

Under high current densities ($\geq 5 \times 10^5$ A/cm^2), electromigration effects become a dominant failure mode. Electromigration is a mass transport effect which causes the atoms of the interconnection metal to migrate gradually toward the more positive end of the conductor. This mass transport phenomenon takes place along the grain boundaries of the metal interconnections and results in the formation of voids in the interconnection pattern which may eventually lead to an open circuit. Electromigration effects are enhanced at elevated temperatures. An additional failure mechanism associated with the metal interconnections is the formation of micro-cracks in the aluminum interconnections as a metal trace passes over a step on the SiO$_2$ layer.

Package-Related Failures.   One of the serious reliability problems in integrated circuits is associated with the bonding of the wire leads between the package and the chip. A serious failure mechanism associated with gold-wire bonds on aluminum is known as purple plague, and is due to the formation of gold-rich intermetallic compounds such as Au$_2$Al, Au$_4$Al, and Au$_5$Al$_2$. These compounds create porous regions in the bond which are mechanically weak and electrically nonconductive. As these intermetallic compounds are formed, the differences in their structure and thermal expansion can stress the porous interface layer to the point of rupture. However, the "plague" formations are a serious problem only at elevated temperatures (typically $\approx 200°$C) and are more likely to occur in step stressing than in actual use. Therefore, gold-wire–aluminum metallization and bonding systems are still considered to be the most reliable under normal operating conditions.

Testing for Package Reliability

Most packages use a number of different materials, such as metal, glass, ceramics, or plastics, to isolate the IC chip from its environment. Special testing procedures have been developed to test the actual sealing ability and the hermeticity of these packages. One of the tests commonly used for this purpose is the helium-leak test, where the package is immersed into a helium atmosphere under pressure for extended periods of time (usually 1 h). The package is then transferred to a mass spectrometer chamber and tested for helium leaks. Radioactive tracer methods can also be used to detect or trace leaks in the package seals.

Thermal stresses introduced between leak tests can point out losses in the package integrity and the cracking of seals. Thermal shock tests typically consist of cycling the package 10–20 times between the temperature extremes of -55 and $+125°C$. Other structural tests include lead fatigue tests, where the leads are bent back and forth for a given number of times; soldering tests, where the device must withstand typical soldering temperatures applied to the leads; and acceleration and shock tests, where the integrity of the package and the leads is examined under centrifugal or inertial shock conditions. In specialized applications, other parameters of the circuit and package may have to be measured. An example of such a parameter is the radiation resistance test, which tests the package integrity and circuit operation during and after irradiation by neutrons, X rays, and gamma rays.

Reliability Measurements

Providing a quantitative measure of reliability is a difficult task. In general, reliability is measured or compared to standards in terms of a *mean time between failures* (MTBF). The difficulty of demonstrating a given failure rate, or MTBF, becomes apparent from the requirement that the testing time be at least as long as the MTBF. In general, the MTBF for integrated circuits is greater than 10^7 h. To demonstrate this reliability with 90% confidence, approximately 2.3×10^7 h of operational life testing with no failures is required. To reduce this testing time, accelerated life tests may be used. This can be done by aging the integrated circuit at accelerated stress conditions, such as elevated temperatures. As shown in Figure 1.25, the reliability of the circuit decreases rapidly with increasing junction temperature.[17] This effect can be utilized to accelerate the life testing process.

Since a quantitative measure of reliability is very difficult to obtain, a number of relatively simple test or screening procedures have evolved, which provide at least a qualitative measure of reliability. These are briefly described below.

Burn-In. In this test, the units are operated at elevated temperatures (typically 125°C) under static bias or dynamic operating conditions for a relatively short

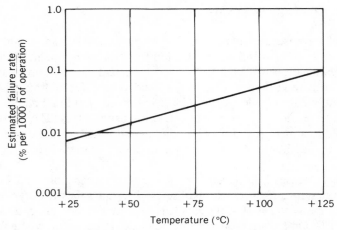

FIGURE 1.25.   Estimated failure rate as a function of junction temperature.[17]

period (typically 168 h). It is primarily intended to detect and weed out the early failures.

Storage Life Tests.   Storage life tests are the basic reliability tests for IC manufacturing. Circuits under test are stored at elevated ambient temperatures for 1000 h and up. Periodically, their characteristics are measured after cooling to room temperature. Usually, storage life test temperatures are between 125°C and 350°C.

Operating Life Tests.   The operating life test is another basic reliability test, which is carried out at 25°C and/or 125°C under power, with the circuit operating under bias conditions similar to those encountered in its actual use. Normally, supply voltage and bias conditions are chosen to provide maximum stress conditions which may be encountered in practice. The key device parameters, such as offset currents and voltages, are then read and recorded at periodic intervals. These tests simulate actual use conditions more closely.

REFERENCES

1.   J. A. Hoerni, U.S. Patent No. 3,025,589, Assigned to Fairchild Camera and Instrument Corp., New York, 1960.

2.   J. C. Irvin, "Resistivity of Bulk Silicon and of Diffused Layers in Silicon," *Bell System Tech. J.* **41,** 387–410 (March 1962).

3.   C. S. Fuller and J. A. Ditzenberger, "Diffusion of Donor and Acceptor Elements in Silicon," *J. Appl. Phys.* **27,** 544–553 (1956).

4.   B. E. Deal, "The Oxidation of Silicon in Dry Oxygen, Wet Oxygen and Steam," *J. Electrochem Soc.,* No. 110, 527 (1963).

5. A. S. Grove, *Physics and Technology of Semiconductor Devices*, Wiley, New York, 1967, Chap. 2.

6. R. M. Burger and R. P. Donavan, *Fundamentals of Silicon Integrated Device Technology*, Vol. 1, Prentice-Hall, Englewood Cliffs, N.J., 1967, p. 254.

7. J. Lyman, "Lithography Chases the Incredible Shrinking Line," *Electronics*, 105–116 (April 1979).

8. J. W. Mayer, L. Eriksson, and J. A. Davis, *Ion Implantation in Semiconductors*, Academic Press, New York, 1970.

9. G. Carter and W. A. Grant, *Ion Implantation of Semiconductors*, Wiley, New York, 1976.

10. H. A. Waggener, R. C. Kragness, and A. L. Tyler, "Anisotropic Etching for Forming Isolation Slots in Silicon Beam-Leaded Integrated Circuits," IEEE Int. Electron Dev. Conf., Washington, DC, 1967.

11. B. Polata, "Compatible High-Performance and Complementary Bipolar Transistors for Integrated Circuits," IEEE Int. Electron Dev. Conf., Washington, DC, 1969.

12. P. C. Davis, S. F. Moyer, and V. R. Saari, "High Slew Rate Monolithic Operational Amplifier Using Compatible Complementary pnp's," *IEEE J. Solid-State Circuits*, **SC-9**, 340–346 (December 1974).

13. A. B. Glaser and G. E. Subak-Sharpe, *Integrated Circuit Engineering*, Addison-Wesley, Reading, MA, 1977, Chap. 10.

14. D. J. Rose, "Packaging and Assembly: The 1980's Semiconductor Technology Forecast," Semiconductor International, 41–50 (January 1980).

15. P. R. Gray and R. G. Meyer, *Analysis and Design of Analog Integrated Circuits*, Wiley, New York, 1977, Chap. 2, pp. 122–126.

16. F. VanVeen, "An Introduction to IC Testing," *IEEE Spectrum,* 28–37 (December 1971).

17. Research Triangle Institute, "Integrated Silicon Device Technology—Reliability," Vol. 15, Rep. ASD-TDR 63-316, May 1967.

CHAPTER TWO

ACTIVE DEVICES IN INTEGRATED CIRCUITS

In designing monolithic circuits, designers do not have the luxury of choosing their devices at will. Instead, they are constrained to use those device types that can be fabricated easily and economically with standard manufacturing processes. The limitations imposed by the fabrication techniques, and the strong interdependence of device parameters and parasitics, often require significant design and layout compromises. Therefore, it is necessary for an analog IC designer to be familiar with the characteristics and limitations of monolithic devices. The purpose of this chapter is to provide a comprehensive overview of the active components available in monolithic circuits. It will be assumed that the reader is already familiar with the fundamentals of semiconductor device theory. In surveying the integrated device and component structures, particular attention will be given to relating the device characteristics and the associated parasitic effects to the physical properties of integrated structures.

An important distinction between IC design and the conventional circuit design with discrete components is that IC designers also have the capability to determine, or specify, the geometry and the layout of the devices they use. This gives IC designers an added degree of freedom in optimizing their circuit performance. Thus, IC design involves a certain amount of device design as well. For example, the need often exists for a transistor with a high current-carrying capability to be used in the output stage of an amplifier. Such a device can be made by using a device geometry other than the standard one, and effectively consists of many standard devices connected in parallel. However, the larger device in turn will exhibit a somewhat different set of parasitics than the small-signal devices due to added stray capacitance effects. Thus it is the responsibility of IC designers to be familiar with the characteristics, the parasitics, and the design trade-offs associated with the active devices at their disposal so that they can make optimum use of the tools available to them.

2.1. *npn* TRANSISTORS

The *npn* bipolar transistor is by far the most significant active component in analog integrated circuits. The basic fabrication steps described in Chapter 1 were initially developed around the *npn* bipolar transistor, and were later extended to other active devices. Consequently, in going from a discrete to an integrated device, the basic *npn* bipolar transistor structure involves the least amount of design compromise. In most analog IC designs, the characteristics of the available *npn* transistors serve as the starting point for the rest of the design; and the remainder of the circuit components are then chosen or designed to be compatible with the fabrication steps required for the *npn* transistor.

Device Structure

The monolithic *npn* transistor differs from its discrete counterpart in one important aspect, as illustrated in Figure 2.1. In the case of the discrete transistor, one can make a direct electrical contact to the collector portion of the device through the backside of the chip. In the case of the integrated *npn* transistor, the device is surrounded by a reverse-biased *p-n* junction (i.e., the isolation wall and

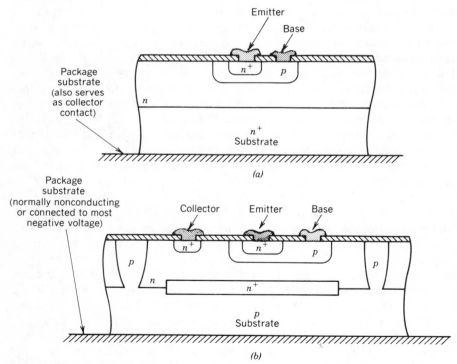

FIGURE 2.1. Comparison of discrete and integrated *npn* bipolar transistor structures: (*a*) Discrete transistor; (*b*) integrated transistor.

the p-type substrate), which electrically isolates it from the other devices on the same chip. As a result, the collector portion of the device is accessible only through an electrical contact at the top surface of the device. This introduces an additional parasitic series resistance r_{cs}, into the collector terminal of the integrated *npn* transistor.

The presence of a reverse-biased junction isolation pocket surrounding the collector region of the integrated device introduces a potentially parasitic *pnp* transistor within the same device structure, as illustrated in Figure 2.2a. This *pnp* transistor is formed by the p-type base of the *npn* transistor along with the n-type collector and the p-type isolation regions. In normal operation, the substrate is *always* biased at a more negative voltage than the n-type collector of the *npn* transistor which in turn causes the emitter–base junction of the parasitic *pnp* transistor to be permanently reversed biased, and thus maintains it in the off or nonconducting state. Under this condition, the isolation pocket can be considered as a reverse-biased diode D_{CS}, with its associated parasitic capacitance C_{CS} as shown in Figure 2.2b.

The subepitaxial n^+-type layer in the integrated *npn* transistor (Fig. 2.1b) serves a dual purpose: it provides a low-resistivity current path from the active collector region to the physical collector contact, and it minimizes the possibility of parasitic *pnp* action by reducing the current gain of the parasitic *pnp* transistor of Figure 2.2a.

Figure 2.3 shows the two possible conditions which may cause the parasitic *pnp* transistor to be active. One such case, shown in Figure 2.3a, is when the collector of the *npn* transistor is pulled to a potential below the substrate voltage. This condition may occur if several IC chips operating with different supply voltages and different substrate bias levels have to be interfaced. The second case illustrated in Figure 2.3b, comes about when the *npn* transistor is driven into saturation (i.e., when its collector–base junction becomes forward biased). This condition may also come about when the *npn* transistor is operated in an inverted

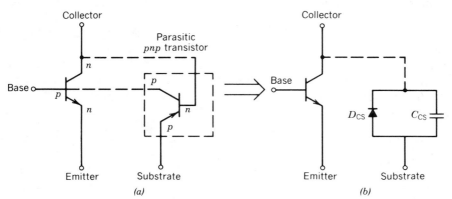

FIGURE 2.2. Parasitic devices associated with a junction-isolated *npn* transistor: (a) Parasitic *pnp* transistor to substrate; (b) its equivalent circuit under proper reverse-bias condition.

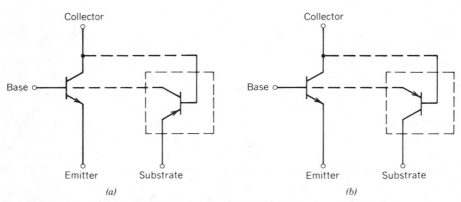

FIGURE 2.3. Two possible conditions where parasitic *pnp* action may result: (*a*) When the collector of the *npn* transistor is pulled to a voltage below substrate potential; (*b*) when the *npn* transistor is saturated so that its collector–base junction is forward biased, or when it is operated in the "inverted" mode with its emitter and collector reversed.

mode, with the roles of its emitter and collector reversed. The parasitic *pnp* action of Figure 2.3*b* is harmful only if the emitter of the *npn* transistor is *not* connected to the same potential as the substrate, since under this condition the parasitic *pnp* transistor would shunt some of the collector current of the *npn* transistor into the substrate. In normal design practices, both of these possible parasitic *pnp* conditions are avoided by proper biasing or design analysis, and the substrate junction remains as a reverse-biased diode, as shown in Figure 2.2*b*.

Figure 2.4 shows the lateral geometry of a typical small-signal *npn* transistor. To give an idea of the lateral dimensions and tolerances, the scale of the drawing, in micrometers is shown alongside the figure. The composite mask layers which form the device are superimposed on the figure, with appropriate coding to identify their functions. The structural cross-section of the same device, sectioned normal to the wafer surface, is also shown. For illustrative purposes, a medium-voltage fabrication process with 15-μm epitaxial layer thickness and 30-V breakdown voltage is assumed (see Fig. 1.13).

The minimum size for the lateral dimensions of the device is limited by two significant factors: (1) masking and mask alignment tolerances and (2) side-diffusion effects. Normally, a clearance of approximately 5 μm is left around an oxide contact cut and the edge of the corresponding diffusion, as is the case for the base and emitter contacts. This tolerance is left to account for any possible misalignment of the mask patterns during the masking operation, or the overetching of the oxide window during the subsequent etching step.

The transistor action takes place directly below the emitter region. Therefore, to be able to supply the collector current with a minimum amount of series voltage drop, it is preferable to locate the collector contact as close to the emitter as possible. The distance between the edge of the base region and the collector contact is chosen to be significantly more than the respective side diffusions of the p-type base and the n^+-type collector contact areas. If this precaution is not

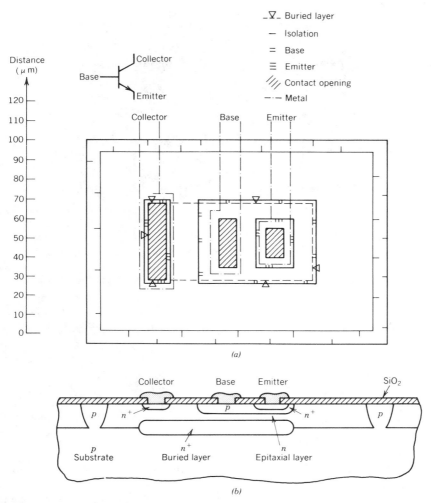

FIGURE 2.4. Lateral dimensions and cross section of a typical small-geometry *npn* transistor. (Mask layers coded for identification as shown.)

taken, the n^+-type collector contact region may touch the p-type base and result in a low collector–base breakdown voltage. In the case of the typical device geometry shown in Figure 2.4, this dimension is approximately 10 μm. The subepitaxial n^+-type layer is located directly below the base region and extends to the area directly below the collector contact.

The distance between the p-type isolation wall and the inner transistor structure is set by the side-diffusion effects, as well as by the thickness of the *depletion layer* associated with the collector–base and the collector–isolation junctions. Since the isolation diffusion is a deep one, it also tends to side-diffuse significantly more than the base. Therefore, it is customary to leave a typical

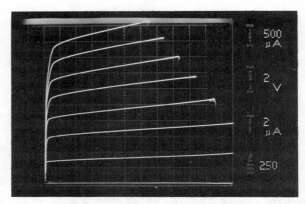

FIGURE 2.5. Typical curve-tracer picture of current–voltage characteristics for the small-geometry *npn* transistor. (*Photo:* Exar Integrated Systems, Inc.)

clearance of 1.5–$1.8X_e$ between the inner edge of the isolation wall and the edge of the base diffusion, where X_e is the thickness of the epitaxial layer. In the case of the 15 μm thickness of the epitaxial layer shown in Figure 2.4, this corresponds to a spacing of approximately 25 μm. The base-to-isolation spacing is the dominant factor in determining the minimum size of a small-signal *npn* transistor. Since this spacing is directly proportional to the epitaxial layer thickness, high-breakdown transistors which require thicker epitaxial layers do not provide as good a component packing density on the chip as those devices fabricated with a low-voltage process and a thinner epitaxial layer.

    Figure 2.5 shows the typical current–voltage characteristics for the small-geometry *npn* transistor of Figure 2.4.

Electrical Characteristics

When biased in its active region, the bipolar transistor functions as a nonideal current-controlled current amplifier. A given amount of base current I_B injected into the base terminal causes a much larger collector current I_C to flow. Figure 2.6 shows a simplified equivalent circuit for an *npn* bipolar transistor which

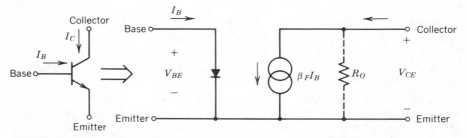

FIGURE 2.6. Simplified equivalent circuit of *npn* transistor for bias calculations. [*Note* $R_O = (V_A + V_{CE})/I_C \approx V_A/I_C$ signifies the finite output resistance due to the Early effect.]

approximates its current–voltage characteristics. When the transistor is in its active region, the base–emitter junction is forward biased and the base–collector junction is under reverse bias.

The collector current I_C is related to the base current I_B as

$$I_C = \beta_F I_B \tag{2.1}$$

where β_F is the forward-current gain. Similarly, the base current I_B is an exponential function of the applied base–emitter voltage V_{BE},

$$I_B = I_{BO} \exp\left(\frac{q\,V_{BE}}{k\,T}\right) \tag{2.2}$$

where I_{BO} = reverse saturation current of the base–emitter diode
q = electronic charge
k = Boltzmann's constant
T = temperature (in °K)

The term kT/q, which has the dimensions of voltage and is proportional to temperature, is often referred to as the thermal voltage V_T,

$$V_T = \frac{kT}{q} \approx 26 \text{ mV at } 25°C \tag{2.3}$$

The collector current I_C is related to the base current through the forward gain factor β_F as

$$I_C = \beta_F I_B = I_{CO} \exp\left(\frac{V_{BE}}{V_T}\right) \tag{2.4}$$

Similarly, the base–emitter voltage to sustain a given collector current can be expressed as

$$V_{BE} = V_T \ln\left(\frac{I_C}{I_{CO}}\right) \tag{2.5}$$

The parameter I_{CO}, which is called the collector reverse saturation current, is given as

$$I_{CO} = \frac{q D_n n_i^2}{Q_B} A = I_S A \tag{2.6}$$

where D_n is the diffusion constant of electrons in the base, n_i is the intrinsic carrier concentration in silicon, Q_B is the total number of dopant atoms in the base region per unit area of the emitter, and A is the area of the base–emitter junction. The main significance of Eq. (2.6) is that the collector reverse saturation current is *directly proportional* to the emitter area. The first term in Eq. (2.6) is a parameter which depends on the particulars of the device impurity profile and the semiconductor material properties. For the IC devices fabricated simultaneously on the same chip, this term will be the same. Thus, as implied by Eq. (2.6), the emitter area A of the *npn* bipolar transistor can be used as a scaling factor such that if the same base–emitter voltage is applied to two transistors on the same chip, their collector currents I_{C1} and I_{C2} will be related to their respective emitter areas, A_1 and A_2 as

$$\frac{I_{C1}}{I_{C2}} = \frac{A_1}{A_2} \tag{2.7}$$

The scaling of transistor currents by scaling their emitter areas is one of the most often used circuit design techniques in analog IC design and will be discussed in more depth in Chapter 4.

The common-base current gain factor α_F is the ratio of the collector and emitter currents, and is given as

$$\alpha_F = \frac{I_C}{I_E} = \frac{I_C}{I_B + I_C} = \frac{\beta_F}{1 + \beta_F} \tag{2.8}$$

In terms of the physical current conduction mechanism within the transistor, α_F indicates the fraction of the carriers injected from the emitter which reach the collector, and is a number very close to unity. It can be expressed as a product of a number of device parameters,

$$\alpha_F = \gamma\beta^*M \tag{2.9}$$

where γ = emitter efficiency
β^* = base transport factor
M = collector avalanche multiplication factor.

The emitter efficiency γ is defined as the ratio of the electron current (for the case of an *npn* transistor), injected into the base from the emitter, to the total hole and electron current crossing the emitter–base junction. It can be closely approximated by an expression of the form[1]

$$\gamma = \left(1 + \frac{\rho_E}{\rho_B}\frac{W_B}{L_E}\right)^{-1} \tag{2.10}$$

where L_E is the diffusion length of the minority carriers into the emitter, ρ_E and ρ_B are the average resistivities of the emitter and base regions within a diffusion length of the junction, and W_B is the width of the base region. As implied by Eq. (2.10), γ approaches unity as the emitter is more heavily doped with respect to the base, and as the base width W_B is narrowed. For an integrated device structure similar to that shown in Figure 2.4, with the impurity profile of Figure 1.13, the emitter efficiency is in the range of 0.992–0.998 for I_E in the low milliampere range.

The base transport factor β^* is the fraction of minority carriers injected from the emitter that reach the collector, and it can be approximated as

$$\beta^* \approx 1 - \frac{W_B^2}{2D\tau} \tag{2.11}$$

where D is the minority carrier diffusion coefficient, and τ is the excess minority carrier lifetime within the base region. For a typical integrated *npn* transistor, β^* is of the order of 0.995.

The collector avalanche multiplication factor M is the ratio of the carriers entering the collector region to the number of minority carriers arriving at the base side of the collector–base junction depletion layer. Because of the secondary ionization effects within the collector–base depletion region, M is usually slightly larger than unity. It can be related to the base–collector reverse bias V_{CB} by the empirical relationship

$$M = \left[1 - \left(\frac{V_{CB}}{BV_{CB}} \right)^m \right]^{-1} \qquad (2.12)$$

where BV_{CB} is the breakdown voltage of the collector–base junction. The exponent m has the approximate values of 4 and 2, respectively, for the *npn* and *pnp* transistors.

Figure 2.5 shows the typical current–voltage characteristics of a monolithic *npn* transistor. The finite slope of the collector current–voltage characteristics is due to the modulation of the effective base width by the widening of the collector–base depletion layer, as the collector–base voltage is increased. This base width modulation, known as the *Early effect*, becomes more significant as the base width W_B is reduced, or as the resistivity of the base region is increased with respect to that of the collector region. For dc modeling of the transistor, the finite slope of the output characteristics can be extrapolated to a common point on the collector–emitter voltage V_{CE} axis as shown in Figure 2.7, and this extrapolated point can be used to define an effective Early voltage V_A. This allows the base width modulation effects to be incorporated into Eq. (2.4) by

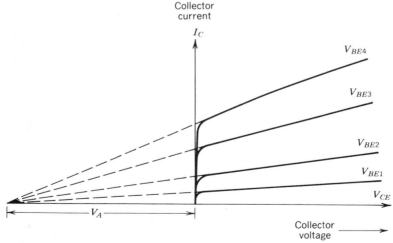

FIGURE 2.7. Bipolar transistor output characteristics showing the Early effect voltage V_A.

rewriting it as

$$I_C = I_{CO}\left(1 + \frac{V_{CE}}{V_A}\right)\exp\left(\frac{V_{BE}}{V_T}\right)$$ (2.13)

For typical *npn* transistors, the Early voltage V_A is in the range of 50–100 V and decreases as the forward-current gain β_F is increased due to the reduced base width.

The simplified equivalent circuit of Figure 2.6 is often sufficient for the first-order analysis of the dc bias conditions for an *npn* transistor stage. Depending on the required accuracy of the calculations, the effect of base width modulation can be incorporated into the model by the addition of the resistor R_O, shown by dashed lines in Figure 2.6.

The forward-current gain β_F varies with both the temperature and the collector current. It exhibits a strong positive temperature coefficient of approximately +5000 to +7000 ppm/°C (where ppm stands for parts per million). This temperature dependence of β_F is primarily due to the increase in emitter efficiency γ [see Eq. (2.10)] with increasing temperature.[2]

Figure 2.8 gives some typical β_F versus I_C curves for the small-signal *npn* transistor of Figure 2.4 at three different temperatures. As shown, the dependence of β_F on I_C can be divided into three regions. In the low-current region, the parasitic surface recombination and the recombination of carriers in the base–emitter depletion region are primarily responsible for the β_F falloff. The low-current β_F can be improved by minimizing surface recombination effects through additional surface passivation steps, such as silicon nitride deposition, or by reducing the emitter periphery and the emitter–base junction area. In the

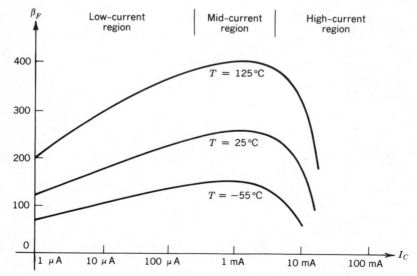

FIGURE 2.8.   Typical β_F versus I_C characteristics for a small-geometry *npn* transistor.

mid-current range, which stretches from approximately 100 μA to 3 mA, β_F is relatively constant.

The decrease of β_F at high currents is due to two dominant factors: decrease of emitter efficiency and emitter-crowding effects. The decrease of emitter efficiency at high currents results from the presence of a large number of excess minority carriers in the base, reducing the effective base resistivity ρ_B near the base–emitter junction. The emitter-crowding effect is caused by the ohmic drop within the active base region due to the flow of base current. As a consequence of this, a voltage gradient is created within the active base region and the edge of the emitter becomes more forward biased than the bottom of the emitter area. Thus, the emitter region injects carriers preferentially along its periphery, and only the edge of the emitter is electrically active. To reduce the β_F falloff at high current levels, it is necessary to maximize the emitter periphery-to-area ratio, and to minimize the base-spreading resistance. This in turn leads to an inter-digitized transistor structure for high-current applications (see Fig. 2.16). Another factor which reduces β_F at high current levels is the onset of the so-called *Kirk effect*[3] which occurs when the minority carrier concentration in the collector becomes comparable to the donor atom density. This causes the effective base with W_B of the transistor to appear to be larger by stretching into the collector region, and this in turn reduces β_F by decreasing the base transport factor β^* of Eq. (2.11).

Voltage Breakdown

As the reverse bias across a *p-n* junction is increased beyond a critical value, the current through the junction increases rapidly. This critical voltage is known as the junction breakdown voltage BV. In silicon, two separate breakdown mechanisms exist. These are the avalanche and the Zener breakdowns.

If the impurity concentration on either side of the junction is less than $\approx 10^{18}$ atoms/cm^3, the breakdown voltage is determined by the onset of avalanche multiplication. It occurs when the electric field within the depletion layer provides sufficient energy for the free carriers to knock off additional valence electrons from the lattice atoms. These secondary electrons, in turn, generate additional free carriers, leading to an avalanche multiplication of the free carriers within the depletion layer. This phenomenon is similar to the ionization breakdown in gases. The avalanche breakdown voltage is normally determined by the impurity concentration on the lighter doped side of the junction. For example, in the case of a monolithic *npn* transistor, the base impurity concentration would have the dominant influence on the emitter–base breakdown voltage; and the collector doping will determine the collector–base breakdown voltage. Figure 2.9 gives the avalanche breakdown voltage versus concentration for a *p-n* junction where the lighter doped side is assumed to have a uniform impurity distribution. This provides a good approximation to the collector–base junction of an *npn* transistor.[1] The avalanche breakdown voltage shows a strong temperature dependence with a typical temperature coefficient on the order of +300 ppm/°C, the dependence being stronger for higher breakdown values.

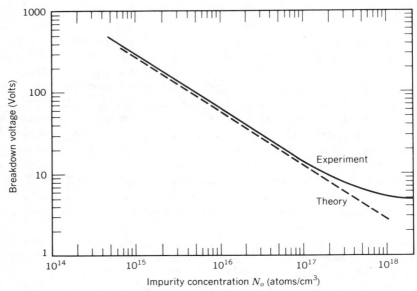

FIGURE 2.9. Avalanche breakdown voltage versus concentration for a step junction.[1]

The basic breakdown voltages associated with the integrated transistor structures are the base–collector BV_{CBO} and the base–emitter BV_{EBO} breakdowns. In designating the transistor breakdown voltage, a triple subscript designation is used where the last subscript O designates that the remaining terminal is open-circuited. Both the BV_{EBO} and the BV_{CBO} values can be closely approximated from the data of Figure 2.9 once the base and the collector impurity levels are known. The typical values of BV_{EBO} are in the range of 6–8 V for integrated transistors, which makes the base–emitter breakdown diode a convenient device for dc voltage reference and level shifting within the circuit.

Due to carrier multiplication effects within the base region of the transistor, the collector–emitter breakdown voltage BV_{CEO}, is smaller than BV_{CBO}. It can be approximated as[4]

$$BV_{CEO} \approx \frac{BV_{CBO}}{(\beta_F)^{\frac{1}{m}}} \tag{2.14}$$

where m is the exponent associated with the avalanche multiplication factor described in Eq. (2.12). Since β_F can vary significantly over the operating current range of a transistor, the lowest value of BV_{CEO} within the current–voltage characteristics of the transistor is called the *lowest sustaining voltage* LV_{CEO}. In normal operation of the transistor, its maximum operating voltage is determined by its lowest sustaining voltage rating. Thus, for example, when one talks about a 25-V or a 40-V transistor process, one is normally referring to the LV_{CEO} value, which can be reliably and repeatedly obtained from that process.

The second type of breakdown mechanism in semiconductors is the so-called

Zener breakdown, which happens if both sides of the junction are very heavily doped. Zener breakdown is a quantum-mechanical phenomenon where the electrons from the valence band on one side of the junction can "tunnel" to an available energy state in the conduction band on the other side. For junctions which breakdown at 5 V or less, the Zener breakdown is the main conduction mechanism. In analog IC structures, impurity concentration levels that can give way to true Zener breakdown are not normally encountered. However, in the IC jargon, the base–emitter reverse breakdown voltage BV_{EBO}, which normally lies in the range of 6–8 V is mistakenly referred to as the Zener diode. As will be described in Chapter 4, this diode is often used in analog integrated circuits to generate reference voltages within the circuit.

If the current through the junction is limited to a safe value, the breakdown is not destructive and the device junctions recover to normal when the voltage level is reduced. However, one exception to this is the case of the base–emitter junction where the repeated avalanche breakdown of the junction will cause β_F to be degraded.[5] If the device is being used as a voltage reference, this is of no consequence; however, if the transistor is being used as a gain stage, particularly at the input of an amplifier, the β_F degradation may seriously affect the performance.

An additional voltage breakdown mechanism in a transistor is the *punch-through* breakdown. In a narrow and lightly doped base structure, the collector depletion layer can extend through the entire base region into the emitter, thus forming a current path between the collector and the emitter. Punch-through is the dominant breakdown mechanism in the superbeta transistors, which will be covered later in this section.

Matching of Device Characteristics

The absolute-value tolerances in integrated devices are, in general, poorer than their discrete counterparts. However, the monolithic components are far superior to discrete devices in the matching and tracking of device parameters. Table 2.1 shows the typical range of values of β_F, V_{BE}, and the base-emitter reverse breakdown V_{EBO} for integrated *npn* transistors. Typical values of the absolute-value and matching tolerances, as well as the temperature coefficients, are also listed in the table. In all cases, the tolerances and the temperature coefficients

TABLE 2.1. Typical Parameter Values and Tolerances for Monolithic *npn* Transistors

Device Parameter	Typical Range of Values	Absolute Value Tolerance	Matching Tolerance	Temperature Coefficient	Thermal Tracking
β_F	50–300	±20%	± 5%	+5000–+7000 ppm/°C	±500 ppm/°C
V_{BE}	0.6–0.7 V	±20 mV	± 1 mV	−2 mV/°C	±10 μV/°C
V_{EBO}	6–9 V	±200 mV	±25 mV	+2–+6 mV/°C	±200 μV/°C

shown refer to a parameter design value falling halfway between the range limits, that is, $\beta_F \approx 150$, $V_{BE} \approx 0.65$ V, and $V_{EBO} \approx 7$ V.

Small-Signal Model

For small-signal applications, the frequency response of the bipolar transistor can be closely approximated by the *hybrid-π* model shown in Figure 2.10.

In the equivalent circuit, the parameters r_b, C_c, r_o and r_{cs} are the parasitics inherent to the basic bipolar transistor structure. The base-spreading resistance r_b is the resistance of the current path which the base current must traverse from the physical base contact to the active base region below the emitter. C_c is the collector–base junction capacitance which depends on the area A_B of the collector–base junction as well as on the reverse bias V_{CB} across it. Assuming a uniformly doped collector region, it can be approximated by an expression of the form

$$C_c \approx A_B \sqrt{\frac{q\epsilon N_C}{V_{CB}}} \qquad (2.15)$$

where ϵ = dielectric constant of silicon
N_C = collector doping concentration
A_B = collector–base junction area

The device transconducture g_m can be expressed as

$$g_m = \frac{\partial I_C}{\partial V_{BE}} \approx \frac{\partial I_E}{\partial V_{BE}} = \frac{qI_E}{kT} = \frac{I_E}{V_T} \qquad (2.16)$$

where V_T is the thermal voltage defined in Eq. (2.3)

The small-signal forward gain β_o is defined as the *incremental* change of the collector current with the change of base current, that is,

$$\beta_o = \frac{\Delta I_C}{\Delta I_B} \qquad (2.17)$$

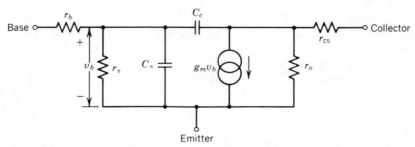

FIGURE 2.10. Small-signal hybrid-π equivalent circuit for a bipolar transistor.

If the dc current gain β_F is relatively constant for small changes of I_C or for I_B, as is the case in virtually all applications, then

$$\beta_o \approx \beta_F \qquad (2.18)$$

If Eq. (2.18) is valid, that is, if the *small-signal* current gain is approximately equal to the *large-signal* current gain, then one often does not differentiate between the two, but uses a single value of β for both ac and dc calculations. In order to simplify the subscript notation, the symbol β will be used to signify β_o and β_F interchangeably.

The dynamic resistance of the base–emitter junction r_π is also related to the direct bias current I_E as

$$r_\pi = \beta_o \frac{kT}{qI_E} = \frac{\beta_o}{g_m} \qquad (2.19)$$

The output resistance r_o is due to the Early effect, and can be related to the Early voltage V_A of Figure 2.7 as

$$r_o = \frac{V_A}{V_T g_m} \qquad (2.20)$$

For typical IC transistors, r_o is in the range of 50–100 kΩ.

The collector series resistance r_{cs}, is the bulk resistance of the semiconductor region through which the collector current must flow from the collector contact to the active collector–base junction. In discrete transistors, r_{cs} is negligibly small (in the 5–30Ω range) due to direct contact to the backside of the chip (see Fig. 2.1a). However, in monolithic IC transistors, r_{cs} is much higher (typically in the range of 100–500 Ω for small-geometry devices), depending on the device layout and the epitaxial layer thickness and resistivity. This will be discussed further in the next section, in connection with other device parasitics which are unique to integrated transistor structure.

C_π represents the total effective capacitance associated with the base–emitter junction and can be expressed as

$$C_\pi = C_D + C_{JE} \qquad (2.21)$$

where C_D is the emitter diffusion capacitance corresponding to the charge in transit through the base region, and C_{JE} is the physical junction capacitance associated with the emitter–base junction. The latter is analogous to C_c of the collector–base junction, and assuming a graded junction, it can be expressed as

$$C_{JE} \approx A_E \left(\frac{q \epsilon N_B}{V_J} \right)^{1/3} \qquad (2.22)$$

where A_E = area of the emitter–base junction
N_B = base impurity concentration at the emitter–base junction
V_J = net voltage across the junction

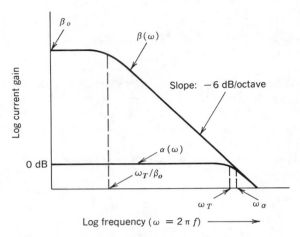

FIGURE 2.11.   Frequency dependence of transistor current gain.

The frequency dependence of the common-base current gain α can be closely approximated as a one-pole gain rolloff function,

$$\alpha(\omega) = \frac{1}{1 + j\omega/\omega_\alpha} \tag{2.23}$$

where $\omega_\alpha = 2\pi f_\alpha$ is known as the *alpha cutoff* frequency. Since the common-emitter current gain is related to α by Eq. (2.8), the frequency dependence of β can be, to a first order, approximated as*

$$\beta(\omega) = \frac{1}{1 - \alpha} = \frac{\beta_o}{1 + j\omega/\beta_o\omega_\alpha} \tag{2.24}$$

However, in the case of graded base transistor structures, it has been experimentally observed that an additional phase delay is introduced into the common-emitter current gain expression above and beyond that predicted by Eq. (2.24). This *excess phase* effect can be incorporated into $\beta(\omega)$ by means of an empirical phase-delay factor by rewriting Eq. (2.24) as

$$\beta(\omega) = \frac{\beta_o \exp(-jm\omega/\omega_\alpha)}{1 + j\omega/\beta_o\omega_\alpha} \tag{2.25}$$

where the excess phase factor m is approximately 0.4 for integrated transistors.

Figure 2.11 shows a plot of the magnitude of $\beta(\omega)$ as a function of frequency in a logarithmic scale. The frequency at which the common-emitter current gain reaches unity is known as the *unity current gain–bandwidth product* ω_T for the

---

*Note that in this analysis, it is assumed that small-signal low-frequency incremental gain is approximately the same as dc gain, that is $\beta_F \approx \beta_o$ similar to the assumption implied by Eq. (2.18).

transistor. It can be shown that ω_T is related to ω_α as

$$\omega_T = \frac{\omega_\alpha}{1 + m\alpha} \approx \frac{\omega_\alpha}{1 + m} \qquad (2.26)$$

Due to -6-dB/octave rolloff of $\beta(\omega)$ at high frequencies, the beat cutoff frequency ω_β can be related to ω_T as

$$\omega_\beta = \frac{\omega_T}{\beta_o} \qquad (2.27)$$

Normally ω_T is measured with a current source input into the base terminal, and with the collector ac grounded. Then, referring to the equivalent circuit of Figure 2.10 and neglecting r_{cs}, ω_T can be expressed as

$$\omega_T \approx \frac{\beta_o}{(C_\pi + C_c)r_\pi} = \frac{g_m}{C_\pi + C_c} \qquad (2.28)$$

Thus, once the ω_T of the transistor is known at a given bias setting, the base–emitter capacitance C_π of Figure 2.10 can be estimated as

$$C_\pi \approx \frac{g_m}{\omega_T} - C_c \qquad (2.29)$$

Figure 2.12 shows the typical f_T versus current characteristics for a small-signal integrated *npn* transistor. The reduction of f_T at low current levels is mostly due to the emitter–base junction capacitance C_{JE}. The drop at high currents is a result of current crowding and the conductivity modulation effects in the base region. For a typical small-signal transistor of the geometry shown in Figure 2.4, the maximum value of f_T is in the range of 600–800 MHz and peaks at a current range of approximately 1–5 mA.

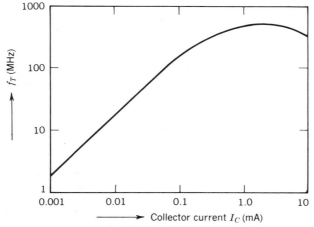

FIGURE 2.12. Typical f_T versus current characteristics of a small-geometry *npn* transistor.

Parasitic Elements in IC Transistors

The simplified hybrid-π model of Figure 2.10 shows the *intrinsic* equivalent circuit associated with *any* bipolar transistor, whether it is discrete or integrated. However, the basic fabrication processes associated with the integrated *npn* transistor also add a number of unique parasitics to the device structure in the form of additional series resistances and shunt capacitances. Figure 2.13 shows the location of these additional parasitics. For illustrative purposes, the figure is not drawn to scale.

The three parasitic resistances associated with the device structure are the base-spreading resistance r_b the collector series resistance r_{cs}, and the emitter series resistance r_{ex}. Their relative values and effects are briefly described below.

Base-Spreading Resistance r_b. This resistance comes about because of the electrical contact to the base region being physically removed from the active base region, directly below the emitter surface, as illustrated in Figure 2.14. It is made up of two sections in series, r_{b1} and r_{b2}, as shown. r_{b1} is the bulk resistance from the base contact to the edge of the active emitter area and can be easily calculated from the knowledge of the sheet resistivity of the base diffusion process. r_{b2} is a nonlinear resistance since it is spread two-dimensionally through the active base region, and the current flow out of it is distributed over the active base area. At low currents, its value is approximately a factor of 5–8 *higher* than r_{b1}; however, it drops to approximately 50% of its low-current value at higher current levels, where the carrier injection from the

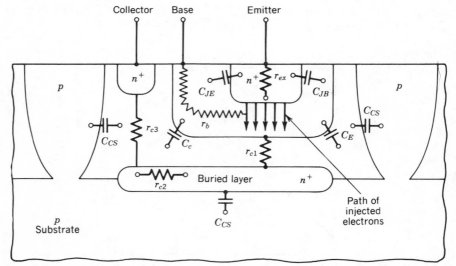

FIGURE 2.13. An illustration of inherent parasitics associated with junction-isolated *npn* transistor. (After Ref. 4).

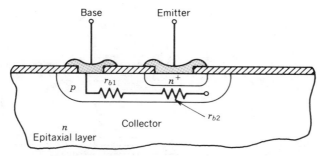

FIGURE 2.14. Components of base resistance r_b.

emitter into the base region takes place primarily along the emitter periphery nearest to the base contact.

In the typical small-geometry *npn* transistor of Figure 2.4, with a typical base sheet resistance of 150 $\Omega/\square$, the value of r_{b1} is approximately 60 Ω, and r_{b2} varies from 400 Ω at low currents to approximately 200 Ω at collector currents in excess of 1 mA.

The base-spreading resistance, r_b is particularly detrimental for low-noise or high-frequency applications. In low-noise transistors, it is often the dominant source of thermal noise. It degrades the transistor frequency response, particularly if the transistor is driven from a low-impedance signal source, since it serves as a voltage divider and reduces the effective signal voltage v_b in Figure 2.10, which reaches the active region of the transistor. Thus, to minimize the above problems, low-noise and high-frequency transistors are designed with multiple base contacts and long, stripe-shaped emitters which maximize the emitter periphery facing the base contact.

Collector Series Resistance r_{cs}. This is the series resistance of the total current path which must be traversed from the physical collector contact to the active collector area, just below the emitter–base junction. As shown in Figure 2.15, it is made up of three components, r_{c1}, r_{c2}, and r_{c3}. In general, r_{c2}, which is set by the sheet resistance of the subepitaxial n^+-type buried layer, is relatively small compared to r_{c1} and r_{c3}, which depend on the epitaxial layer thickness and resistivity. For the small-geometry *npn* transistor of Figure 2.4 with a 20-$\Omega/\square$ buried layer and a 15-μm 2.5-Ω/cm epitaxial layer, the respective values of these resistances are as follows:

$$r_{c1} \approx r_{c2} = 150 \ \Omega \qquad r_{c2} \approx 30 \ \Omega \qquad r_{cs} = r_{c1} + r_{c2} + r_{c3} = 330 \ \Omega$$

The collector series resistance is particularly important for the saturated operation of the transistor, since at high current levels it is the main contributor to the saturation voltage V_{CEsat} across the transistor. In such applications, r_{cs} is often referred to as the *saturation resistance R_{sat}*. In high-current transistors, its effects are minimized by surrounding the active transistor with a large collector-

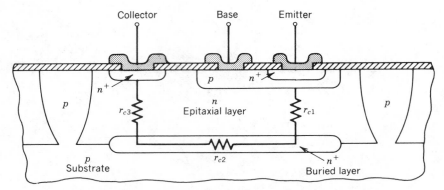

FIGURE 2.15. Components of collector series resistance r_{cs}.

contact area, and by using deep n^+ diffusion (see Fig. 1.16) to eliminate the contribution of r_{c3}.

Emitter Series Resistance r_{ex}.   This is the bulk resistance of the emitter region between the emitter contact and the emitter–base junction. Since the emitter area is very heavily doped, its value is of the order of several ohms. In general, its effect in the ac equivalent circuit is negligible, except in the case of high-current transistors, where it may affect the current distribution within the emitter area. This subject is discussed further, in connection with power control and regulator circuits, in Chapter 10.

Parasitic Capacitances.   As indicated in Figure 2.13, there are several parasitic junction capacitances associated with the integrated *npn* transistor structure. These are the emitter–base junction capacitance C_{JE}, the base–collector junction capacitance C_{CJ}, and the collector–substrate junction capacitance C_{CS}. The values of these capacitors depends on the total area of the junction, its impurity profile, and the *net* applied voltage across the junction that is, the applied voltage, V minus the built-in junction potential ψ_o. As will be described in Chapter 3, the general expression for any arbitrary junction capacitance, C_X can be written as

$$C_X = \frac{C_{XO} A_X}{(1 - V/\psi_o)^n} \qquad (2.30)$$

where V = applied voltage across the junction (positive for forward bias)
  C_{XO} = capacitance per unit area of junction under zero bias (i.e., $V = 0$)
  A_X = total junction area.

The exponent n is equal to $\frac{1}{2}$ for a step junction and equal to $\frac{1}{3}$ for a linearly graded junction. For most semiconductor junctions, either the step or the linear-grading approximation holds very closely.

ψ_o is the built-in junction potential which depends on the impurity concentration on either side of the junction. Assuming a step, or a very steeply graded junction, it can be expressed as

$$\psi_o = V_T \ln \left(\frac{N_A N_D}{n_i^2} \right) \tag{2.31}$$

where N_A and N_D are the acceptor and donor concentrations on the respective sides of the junction, and n_i is the intrinsic carrier concentration in pure silicon ($n_i \approx 1.5 \times 10^{10} \, \text{cm}^{-3}$ at 300°K). For IC transistors, typical values of ψ_o are 0.7 V for base–emitter junctions and 0.5–0.6 V for base–collector and collector–substrate junctions.

Table 2.2 gives typical values of the unit-area zero-bias capacitances C_{JEO}, C_{CO} and C_{CSO}, the corresponding values of the exponent n and the built-in junction potential ψ_o for the three commonly used analog IC processes. Once these values are known and the corresponding junction area is measured from the device layout, the values of each of these three parasitic capacitances can be directly calculated from Eq. (2.30).

Summary of *npn* Device Parameters

Many of the *npn* device parameters, including the device parasitics discussed in the previous sections, depend somewhat on the specifics of the IC fabrication processes and the impurity profiles, even when the lateral dimensions (i.e., the layout) of the device are fixed.

Basic bipolar analog IC fabrication processes available from many IC suppliers differ in their specifics and details. However, the majority of the bipolar IC processes tend to fall into one of the following general categories:

1. *Low-Voltage Process.* This process uses an epitaxial layer thickness in the range of 8–10 μm with an epitaxial layer resistivity of 0.8–1 Ω/cm. It results in an *npn* bipolar transistor with a collector–base breakdown voltage LV_{CEO} of 20 V.

2. *Medium-Voltage Process.* This process uses 2–2.5 Ω/cm epitaxial layer resistivity, with an epitaxial layer thickness of approximately 15 μm, and gives a breakdown voltage of approximately 30 V. The impurity profile of Figure 1.13 and the device geometry of Figure 2.4 correspond to this process.

3. *High-Voltage Process.* This process is used for circuits such as operational amplifiers, which are designed to operate with supply voltages as high as 40 V. It uses 4–5-Ω/cm resistivity, 17–19 μm thick epitaxial layers, and gives a device breakdown voltage in excess of 45 V.

The typical small-geometry device layout is applicable to all three processes. However, in the case of the low-voltage process, a smaller base-to-isolation

TABLE 2.2.   Summary of Typical Values for Parasitic Junction Capacitances for Monolithic *npn* Transistors

Junction Characteristics	Low-Voltage Process: 1 Ω-cm, 9-μm Epitaxial Layer, 20-V Device	Medium-Voltage Process: 2.5 Ω-cm, 14-μm Epitaxial Layer, 30-V Device	High-Voltage Process: 5 Ω-cm, 18-μm Epitaxial Layer, 40-V Device
Emitter–base junction			
Zero-bias capacitance C_{JEO}	10^{-3} pF/μm^2	10^{-3} pF/μm^2	10^{-3} pF/μm^2
Built-in potential ψ_{JEO}	0.7 V	0.7 V	0.7 V
Exponent n	$\frac{1}{3}$	$\frac{1}{3}$	$\frac{1}{3}$
Base–collector junction			
Zero-bias capacitance C_{CO}	2.3×10^{-4} pF/μm^2	1.4×10^{-4} pF/μm^2	10^{-4} pF/μm^2
Built-in potential ψ_{CO}	0.6 V	0.58 V	0.55 V
Exponent n	$\frac{1}{2}$	$\frac{1}{2}$	$\frac{1}{2}$
Collector-isolation junction			
Zero-bias capacitance C_{CSO}	10^{-4} pF/μm^2	10^{-4} pF/μm^2	10^{-4} pF/μm^2
Built-in potential ψ_{CSO}	0.52 V	0.55 V	0.58 V
Exponent n	$\frac{1}{2}$	$\frac{1}{2}$	$\frac{1}{2}$

spacing can be used ($\approx 18 \mu$m versus 25 μm as shown in the figure) to result in a smaller overall isolation pocket size.

Table 2.3 gives a summary of all the relevant device parameters, including the parasitic elements, associated with a small-geometry transistor structure similar to that of Figure 2.4, fabricated with each of these three popular IC fabrication processes.

2.2. npn TRANSISTORS FOR SPECIAL APPLICATIONS

Unlike the discrete circuit designer, monolithic IC designers can exercise control over the geometry and the layout of the devices at their disposal. Thus, by properly modifying and scaling the dimensions of their devices, they can optimize their performance for special applications, such as high-current or high-voltage operation. In this section some of these special *npn* transistor structures will be reviewed.

TABLE 2.3. Summary of Device Parameters for Small-Geometry *npn* Transistor

Device Parameter	Low-Voltage Process: 1Ω-cm, 9-μm Epitaxial Layer, 20-V Device	Medium-Voltage Process: 2.5Ω-cm, 14-μm Epitaxial Layer, 30-V Device	High-Voltage Process: 5Ω-cm, 18-μm Epitaxial Layer, 40-V Device
Forward-current gain β_F	150	200	200
Reverse-current gain β_R	2	2	2
Early Voltage V_A	50 V	80 V	120 V
Collector–base leakage I_{CO}	10^{-7} mA	10^{-7} mA	10^{-7} mA
Emitter–base breakdown BV_{EBO}	6.5 V	6.8 V	7.0 V
Collector–base breakdown BV_{CBO}	40 V	60 V	90 V
Collector–emitter breakdown BV_{CEO}	25 V	40 V	50 V
Lowest sustaining voltage LV_{CEO}	20 V	30 V	40 V
Base resistance r_b			
135-Ω/$\square$ process	200 Ω	200 Ω	200 Ω
200-Ω/$\square$ process	400 Ω	400 Ω	400 Ω
Collector series resistance r_{cs}	70 Ω	120 Ω	200 Ω
Emitter series resistance r_{ex}	2 Ω	2 Ω	2 Ω

High-Current Transistors

There are two parameters which degrade the transistor characteristics at high current levels. These are the β falloff at high currents and the large voltage drop across the collector series resistor r_{cs}. As described earlier in connection with Figure 2.8, β falloff at high currents is due to the decrease of emitter efficiency and the crowding of the emitter current to the periphery of the emitter nearest to the base contact. These effects can be minimized by using an interdigitated transistor structure, as shown in Figure 2.16 which is made up of alternating stripes of emitter and base contact "fingers." Such a structure maximizes the emitter periphery-to-area ratio and reduces the base-spreading resistance r_b.

The collector series resistance r_{cs}, which is also referred to as the saturation resistance R_{sat}, is detrimental to high-current operation because of excessive voltage drop and/or power dissipation generated across it. This resistance can be minimized by using a "wrap around" collector contact structure, as shown in Figure 2.16, as well as using a deep n^+ diffusion from the collector contact to the n^+-type buried layer (See Fig. 1.16).

High-Voltage Transistors

In certain applications, such as interfacing with high-voltage display systems, analog integrated circuits are required to operate with supply voltages in excess

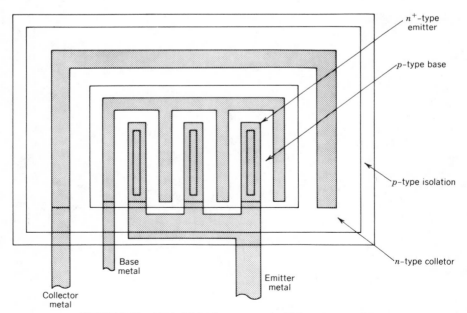

FIGURE 2.16. Typical lateral geometry of a high-current transistor.

of 50 V. By proper device design and layout, it is possible to extend the breakdown voltage characteristics of monolithic *npn* transistors up to 100 V. There are two main limitations to high-voltage operation: (1) the breakdown voltages of semiconductor junctions and (2) the surface inversion problem where a parasitic *p*-channel MOS transistor may be formed on the epitaxial layer surface.

Breakdown Voltage. The junction breakdown voltage imposes a fundamental limit on the high-voltage capability of monolithic devices, as shown in Figure 2.9. However, the data of Figure 2.9 only refer to the avalanche breakdown within the "bulk" silicon and neglect the surface breakdown effects. The discontinuity of the silicon lattice at the device surface, as well as the presence of electrostatic charge trapped at the Si–SiO$_2$ interface, can modify the electrostatic field distribution within the junction at or near the device surface. This can lead to a narrowing down of the junction depletion layer at the surface, or at the sharp corners of the device, and results in a localized surface breakdown at a lower voltage than that predicted in Figure 2.9. This effect is illustrated in Figure 2.17*a* in terms of a simple *p-n* junction. This surface breakdown can be greatly reduced or eliminated by using an electrostatic shield, or *field plate*, over the junction surface, as shown in Figure 2.17*b*. When connected to a more negative potential than the *n* side of the *p-n* junction, this field plate tends to spread out the depletion layer near the surface and, thus, avoids surface breakdown.[6] In the case of a high-voltage transistor, this field plate can be formed by simply overlapping the base and the emitter metal interconnection traces over the base–collector junction.

In fabricating a high-voltage transistor with a breakdown voltage well in excess of 50 V, one would use a high-resistivity epitaxial layer (typically 5–8 Ω/cm) with extended thickness of the order of 25 μm. The base and emitter junctions are made relatively deep, typically on the order of 3.5 and 2.5 μm, respectively, to avoid surface breakdown effects, and sufficient space is pro-

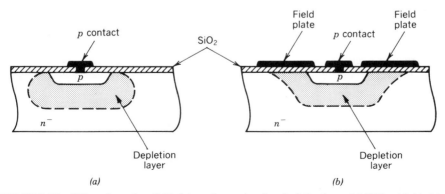

FIGURE 2.17. Effect of a surface field plate on the *p-n* junction depletion layer: (*a*) Without field plate and (*b*) with field plate.

vided within the collector pocket for the extension of the collector–base and the collector–isolation depletion layers, without touching each other or the n^+-type buried layer.

Figure 2.18 shows the typical layout of such a high-voltage transistor. Note that the base and the emitter metal traces are made to overlap the base–collector junction to serve as the field plate to minimize the surface breakdown effects. Also, the base-to-isolation spacing is increased significantly beyond what is shown in Figure 2.4 to account for increased isolation side diffusion and collector–base and collector–isolation depletion layers.

Surface Inversion.   One of the problems associated with high-voltage integrated circuits is the formation of a parasitic p-channel MOS transistor on the surface of the lightly doped n-type epitaxial layer surface. The presence of a conducting metal layer above the oxide layer, biased at a *negative* voltage relative to the epitaxial layer, creates an electric field within the oxide layer

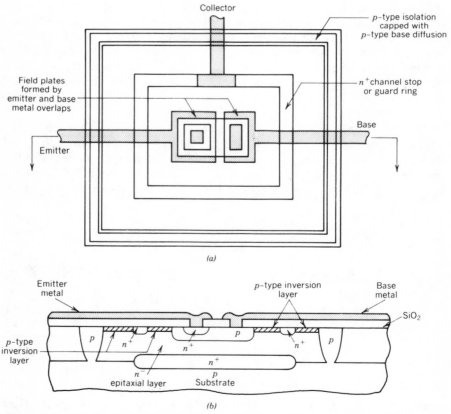

(a)

(b)

FIGURE 2.18.   Device layout and cross section of a high-voltage *npn* transistor using field plate and n^+ channel stops.

which pushes the free electrons *away* from the epi–oxide interface, and creates a p-type inversion layer along the epitaxial layer surface directly under the metal trace. Typically, with lightly doped epitaxial layers, with resistivities ≥ 5 Ω/cm, this inversion layer sets in at voltage differentials of 35–40 V across the SiO_2 layer. If this parasitic p-type "channel" is not blocked by a heavily doped n^+ guard ring, or *channel stop*, completely surrounding the base region, it would result in base–isolation short circuit. Since the heavily doped n^+ ring, which also serves as the collector contact, does not invert, it serves as a channel stop to inhibit this parasitic p-channel MOS transistor action. Since this parasitic channel appears only under the metal traces, in many cases it is not necessary to encircle the entire base area with a n^+ channel stop, but instead, the channel stops can be placed in discrete segments under the metal traces. The layout example of Figure 2.18a shows a device structure with a continuous guard-ring-type n^+ channel stop.

It should also be noted that in high-voltage integrated circuits it is customary to deposit an additional p-type layer on top of the isolation during the subsequent p-type base diffusion. This second p-type layer on the isolation surface, which is often called a *p-cap*, also serves as a channel stop to avoid parasitic n-type channel formation over the isolation region. Such a *parasitic* channel may be formed when a metal trace with a positive potential, such as the V^+ line, passes over the p-type isolation wall.

Superbeta Transistors

In certain analog circuits, such as the input stages of operational amplifiers, it is necessary to have a very high input impedance and low input bias currents. For such an application, the β of a typical integrated *npn* transistor is not high enough, since the actual device design requires a compromise between the current gain and the voltage breakdown requirements [see Eq. (2.14)].

It is possible to increase the value of β by improving the base transport efficiency β^* [see Eq. (2.11)]. This can be done by using a very narrow base structure ($W_B \approx 2500$ Å). However, the collector–emitter voltage breakdown of such a device structure is limited to the 3–4 V range because of the collector–base depletion layer punching through the active base region into the emitter. Figure 2.19 shows the typical current–voltage characteristics of such an ultrahigh-gain, or superbeta, transistor. Such transistors exhibit typical β values of 2000–5000 at collector current levels of 10–20 μA. Due to excessive base width modulation effects, the Early voltage V_A and the output resistance r_o of a superbeta transistor are typically an order of magnitude lower than those of the conventional *npn* transistor. Typical values of V_A are in the 4–10 V range.

The superbeta transistors can be fabricated simultaneously with conventional *npn* bipolar transistors, with the addition of one extra masking and emitter diffusion step. This is done using the *two-step emitter diffusion* process described in Section 1.10 (see Figure 1.17).

In circuit design, superbeta transistors are often used together with a conventional IC transistor which can provide the necessary voltage protection.

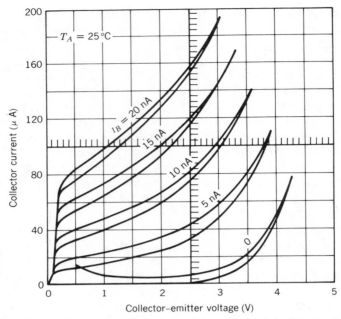

FIGURE 2.19. Curve-tracer display of typical current–voltage characteristics for a superbeta transistor.

Figure 2.20 shows a composite connection of a superbeta transistor with a lateral *pnp* transistor where the base–emitter diode of the *pnp* transistor clamps the voltage swing across the superbeta transistor to a level below its breakdown voltage. Thus, the resulting transistor structure has the effective β of the punch-through *npn* transistor and the breakdown voltage of the lateral *pnp* transistor. The current source I_X shown in the figure is included to ensure proper direction

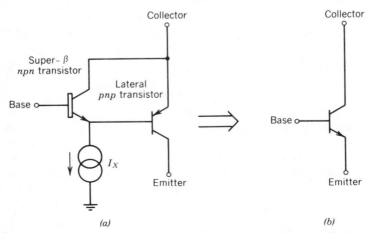

FIGURE 2.20. Composite connection of a superbeta *npn* transistor with a lateral *pnp* transistor to improve breakdown characteristics: (*a*) Actual connection; (*b*) electrical equivalent circuit.

of current flow out of the base of the *pnp* transistor; it is normally a small fraction of the collector current of the composite device.

Multiple-Emitter Transistor

One can easily enlarge the base area of the basic *npn* transistor structure and insert a number of separate emitter regions. This results in a multiple-emitter *npn* structure, as shown in Figure 2.21. Since the emitter region of an integrated *npn* transistor takes up the smallest amount of device area, the resulting device size and the required chip area are not increased significantly. Such a multiple-emitter *npn* transistor, which was traditionally used in TTL-type digital circuits, also finds a range of applications in analog IC design. Some of these applications, such as providing multiple-output emitter-follower circuits, buffered biasing, and temperature-compensated avalanche diodes, are discussed in more detail in Chapter 4.

One word of caution is in order regarding multiple-emitter transistors. If one of the emitters is biased in the off condition relative to the base potential while the other emitter is conducting, a parasitic lateral *npn* action may take place between the two emitters. In that case, a parasitic *npn* transistor will be formed, where the off emitter will serve as the collector for the electrons injected by the on emitter.

Inverted Operation of *npn* Transistor

In certain analog circuit applications, it may be advantageous to reverse the emitter and the collector terminals of an *npn* transistor and operate it in an inverted or reverse mode. This inverted operation is also the key concept behind the integrated injection logic (I²L) circuits.

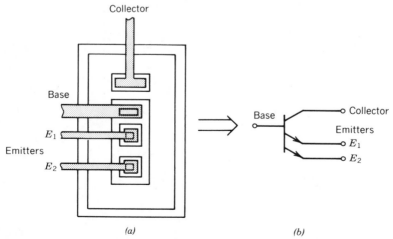

FIGURE 2.21. Multiple-emitter *npn* transistor: (*a*) Device layout; (*b*) equivalent device symbol.

One advantage of the inverted operation of the *npn* transistor is that the normal emitter now becomes the collector of the inverted transistor, and that one can form a multiple-collector transistor in a common isolation tub by using the device structure shown in Figure 2.22.

In inverted operation, the transistor characteristics are very poor. The current gain, which is now given by the inverse beta β_R, is only a small percentage (typically 1–5%) of β_F. This is due to the following reasons. The emitter efficiency γ [see Eq. (2.10)] is degraded due to the emitter being doped lighter than the base, the base transport factor β^* is reduced due to the retarding field gradient in the base region; and a number of carriers injected into the base region have to diffuse a long distance to reach the collector region, thus making the effective base width appear to be much larger than W_B. In addition, a parasitic *pnp* transistor is also formed in the isolation tub between the base of both the *npn* transistor and the *p*-type isolation wall and the substrate, as shown in Figure 2.3*b*. However, the effects of this parasitic *pnp* transistor can be greatly reduced, or nearly eliminated, by surrounding the entire *p*-type base region in an n^+-type pocket by means of a deep n^+ diffusion, as shown in Figure 2.22. The collector–base breakdown of the inverted transistor is also limited to the 6–7 V range by the BV_{EBO} of the normal base–emitter junction.

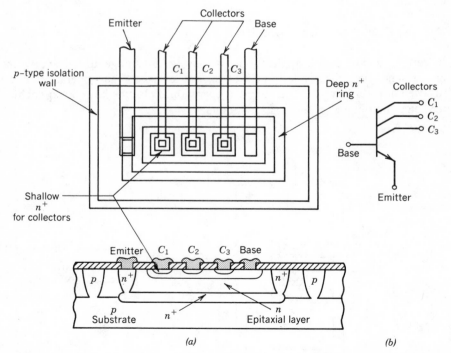

FIGURE 2.22.   Device structure of a multiple-collector inverted *npn* transistor: (*a*) Device layout and cross section; (*b*) equivalent device symbol.

In spite of all the drawbacks outlined above, the inverted transistor still offers some unique advantages in selected analog circuit applications. Some of these are listed below:

1. Under saturated conditions, its offset voltage is lower than that of the normal *npn* transistor. Thus, it makes a good grounding switch for shunting low-level currents (i.e., ≤1 mA) to ground.

2. Its collector–base capacitance is lower, since it is equal to C_{JE} under reverse-bias conditions. Thus, it can be used as a low-capacitance current source for biasing differential amplifier stages.

3. It allows a large number of independent *npn* current sources to be formed within a common isolation tub, using the *current mirror* configuration shown in Figure 4.2.

4. In the multiple-collector device structure of Figure 2.22, the fraction of the total emitter current reaching each of the collectors is proportional to the area of the collector–base junction. Thus, by varying the geometry of the collectors, the current distribution between them can be scaled.

2.3.   *pnp* TRANSISTORS

Some analog circuit functions may require the use of complimentary bipolar transistors. In such cases, it is necessary to fabricate functional *pnp* transistors on the same substrate with the *npn* devices. For this purpose, a number of monolithic *pnp* transistor structures have been developed which are totally compatible with the standard *npn* bipolar process technology, that is, they can be fabricated simultaneously with the *npn* transistors without requiring additional diffusion or masking steps. Two such *pnp* transistors commonly used in analog IC design are the *lateral pnp* and the *substrate pnp* transistors. Both of these devices use the *lightly doped n*-type epitaxial region of the *npn* transistor as the base region. They are, in general, inferior to the basic *npn* transistor in their current-handling or high-frequency characteristics. However, they are extremely useful in biasing, dc level shifting and serving as active load devices in gain stages.

Lateral *pnp* Transistor

Figure 2.23 shows the plan view and the structural diagram of a typical lateral *pnp* transistor. The base region of the device is formed by the *n*-type epitaxial layer which serves as the collector of the *npn* transistors. The *p*-type base diffusion of the *npn* transistor is used to form the emitter and the collector regions of the lateral *pnp* transistor, the n^+-type emitter diffusion of the *npn* transistor is used to form the n^+-type contact region for the *pnp* base. In such a device structure, the transistor action takes place in the *lateral direction*, that is,

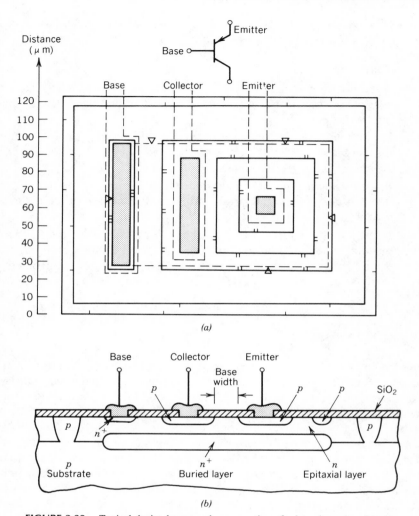

FIGURE 2.23. Typical device layout and cross section of a lateral *pnp* transistor.

parallel to the device surface. The minority carriers injected into the base diffuse laterally toward the collector region. In order to collect the great majority of carriers injected from the emitter, and to minimize parasitic current flow into the substrate, the collector region of the lateral *pnp* transistor is normally made to surround the emitter region along the lateral surface of the device as shown in Figure 2.23*a*. Although the device layout of the figure shows a square emitter surrounded by a square or rectangular collector region, it is also very common to use a circular emitter region surrounded with a circular collector ring.

Figure 2.24 illustrates the direction of current flow in terms of the minority carriers (holes) injected into the base region of the lateral *pnp* transistor. The

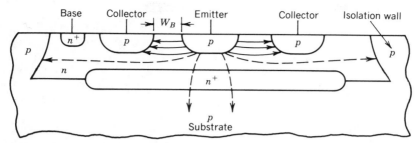

FIGURE 2.24. Minority carrier flow in lateral *pnp* transistor. Solid lines indicate normal carrier flow paths; dashed lines indicate parasitic carrier flow to substrate.

carrier transport across the base region is most efficient at or near the surface of the device, where the separation between the collector and the emitter is minimal. This minimum spacing is the effective base with W_B for the device. Due to masking tolerances, side-diffusion effects, and voltage breakdown requirements, W_B is constrained to be of the order of 6–12 μm. This value for the effective base width is much larger than that of a vertical *npn* transistor ($W_B \approx 0.7 \mu$m). Thus, the current gain and the frequency performance characteristics of the lateral *pnp* transistors are inferior to those of the *npn* devices.

The current gain β of the lateral *pnp* transistor is low (typically in the range of 5–50) due to its poor emitter efficiency and the wide base width. The emitter efficiency of the lateral *pnp* transistor is degraded due to two factors: (1) low impurity concentration in the *p*-type emitter region, and (2) small effective area of the emitter. This latter effect is due to the fact that only the lateral emitter edge facing the collector is active; the holes injected from the rest of the emitter have a much lower probability of reaching the collector. The wide base width W_B of the lateral *pnp* transistor also affects β_F adversely since it lowers the base transport factor β^* [see Eq. (2.11)].

Another effect which adversely influences the lateral *pnp* transistor current gain is the reduction of β^* due to the parasitic minority carrier flow from the emitter directly into the substrate, as shown by the dashed lines in Figure 2.24. Thus, the substrate serves as the collector of a parasitic *pnp* transistor (see Fig. 2.25). The n^+-type buried layer, which is still retained under the emitter and collector regions of the lateral *pnp* transistor, sets up a retarding field to reduce this parasitic *pnp* action. In normal operation, with a wrap around collector geometry and an n^+-type subepitaxial layer, as shown in Figure 2.23, this parasitic substrate current is in the range of 3–5% of the normal collector current.

The current-handling capability of the lateral *pnp* transistor is severely limited by the lightly doped *n*-type epitaxial layer which serves as the base region. Thus, as the collector current is increased, the high-level injection effects set in at the emitter base region, and the emitter efficiency and β_F fall off rapidly. For a small-geometry lateral *pnp* transistor with approximately 100-μm emitter periphery (see Fig. 2.23a), β_F starts falling off at approximately 20–50 μA of collector current. Figure 2.25 shows typical current gain characteristics of a

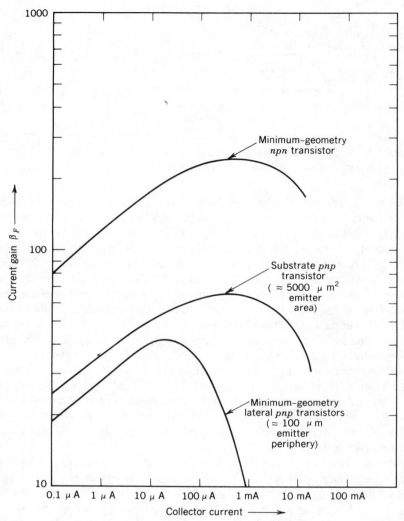

FIGURE 2.25. Comparison of current gain characteristics of the small-geometry *npn*, lateral *pnp*, and substrate *pnp* transistors.

small-area lateral *pnp* transistor as a function of collector current. Note that the current-handling capability of the lateral *pnp* transistor is approximately 50–100 times lower that that of a small-geometry *npn* transistor and its β_F is typically a factor of 10–50 lower.

The frequency response of the lateral *pnp* transistor is limited by the transit time of the carriers through the wide base region. It can be shown that this transit time is proportional to $(W_B)^2$. Similarly, the f_T of the lateral *pnp* transistor is inversely proportional to $(W_B)^2$. Since the base width of the later *pnp* transistor is approximately 10 times larger than that of the normal *npn* transistor (i.e.,

0.8 μm vs. 8 μm), the f_T of the later *pnp* transistor is approximately 100 times lower than that of an *npn* transistor. Typical values of f_T for the lateral *pnp* transistors are in the range of 3–5 MHz.

Parasitics Associated with Lateral *pnp* Transistor. The basic lateral *p-n-p* transistor structure of Figure 2.23 has two inherent parasitic *pnp* transistors associated with it, as shown in Figure 2.26. The first Q_A, is the parasitic *pnp* transistor between the emitter of the *pnp* transistor and the *p*-type substrate and isolation walls. Since the emitter-base junction of the lateral *pnp* transistor Q_1 is normally forward biased, the parasitic transistor Q_A is active at all times. However, its effective current gain is greatly reduced by the use of an n^+-type buried layer and that wraparound collector structure of Figure 2.23. Thus, its net effect is shunting a small amount of the normal emitter current of Q_1 to the substrate. This current, shown as I_X in Figure 2.26, is typically of the order of 3–5% of I_E.

The second parasitic *pnp* transistor, Q_B, formed between the collector of Q_1 and the *p*-type substrate, is normally off since the collector–base junction of Q_1 is reverse biased. However, if the *pnp* transistor Q_1 were operated in its inverse mode (i.e., its emitter and collector functions are reversed), this parasitic transistor would turn on and shunt a substantial amount of current I_Y to the substrate. The parasitic *pnp* transistor Q_B also comes into conduction when Q_1 is driven into saturation. Both of these conditions *must be* avoided by proper circuit design. In certain specialized applications, where the later *pnp* transistor may be driven into saturation, or must operate in an inverted mode, the effects of the parasitic *pnp* transistor Q_B can be reduced by placing a deep n^+ ring around the whole *pnp* transistor and the *p*-type isolation pocket. Such a deep n^+ ring would reduce the value of I_Y to approximately 5–10% of the collector current; however, it would cause a significant increase in the chip area required by the device.

Multiple-Collector *pnp* Transistor. The collector region of the lateral *pnp* transistor can be split into several segments around a center emitter area, as shown in Figure 2.27. This results in a multiple-collector transistor structure

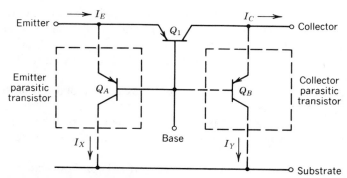

FIGURE 2.26. Parasitic *pnp* transistors associated with the lateral *pnp* transistor.

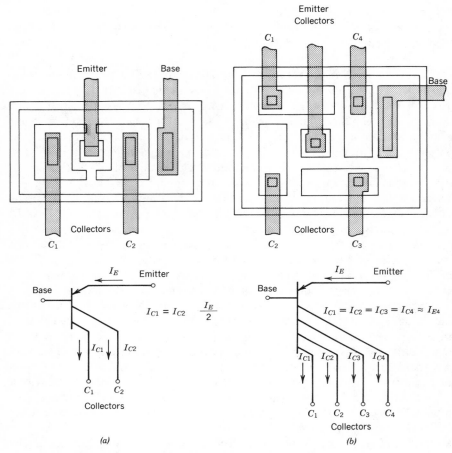

FIGURE 2.27.   Typical multiple-collector lateral *pnp* transistor structures.

which is very useful in a number of design applications. Since the collector current associated with each collector segment is directly proportional to the collector periphery facing the emitter, the partitioning of the emitter current between the different collectors can be controlled and scaled by varying the corresponding collector periphery. As will be described in Chapter 4, this scheme is often used to create accurate and stable ratios of currents which are scaled by the device layout.

When multiple-collector *pnp* transistors are used as ratioed current sources or active loads, care must be taken to ensure that none of the collectors are driven into saturation. Otherwise a parasitic *pnp* action would result between the collector *segments*, as well as between the collectors and the substrate.

Composite Connection of Lateral *pnp* and Vertical *npn* Transistors.   The low β of the lateral *pnp* transistor can be improved by combining it with an integrated *npn* transistor to form a composite transistor, as shown in Figure 2.28. The polarity and the electrical characteristics of such a composite device are equivalent to that of a single *pnp* transistor having a higher current gain β_T, given as

$$\beta_T = \beta_n \beta_p \qquad (2.32)$$

where β_p and β_n are the respective current gains of the *pnp* and *npn* transistors. The transconductance of the composite device is the same as that of the *npn* transistor. Although the emitter of the *npn* transistor acts as the collector of the composite device, the resulting output impedance is still quite high since the base of the *npn* transistor is driven by a current source. Since the lateral *pnp* transistor only carries the base current of the *npn* transistor, the current-handling capability of the composite device is primarily the same as that of the *npn* transistor. However, the frequency response of the composite device is determined by the lateral *pnp* transistor.

For purposes of biasing and network analysis, it is convenient to consider the composite device of Figure 2.28 as a high-gain *pnp* transistor. However, a word of caution is in order. The composite transistor is essentially a cascade of two devices. Therefore, it combines the inherent parasitics and the charge storage effects of both devices. Thus, its frequency response characteristics exhibit a multiple-pole response; and any feedback loop built around it may tend to instability.

A Summary of Device Characteristics. To some degree, the electrical characteristics of the lateral *pnp* transistor also depends on the particular fabrication process used. With reference to the three basic bipolar processes (i.e., the low-, medium-, and high-voltage processes) discussed in Section 2.1, the two key parameters which affect the lateral *pnp* transistor are the epitaxial resistivity and the collector–emitter spacing (i.e., the base width). For low-voltage devices requiring 20-V breakdown and 1Ω-cm epitaxial layer resistivity, the base width W_B can be as low as 6 μm. For devices with 30- or 40-V breakdown, W_B is required to be of the order of 8 μm. Since such devices also require higher epitaxial layer resistivity (i.e., a lighter doped *pnp* base) the β falloff with current comes about at lower current levels.

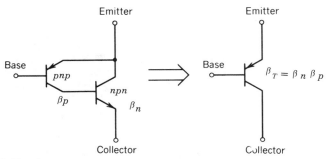

FIGURE 2.28. Composite connection of a lateral *pnp* transistor and a normal *npn* transistor.

Table 2.4 gives a summary of the electrical characteristics and parameters of a small-geometry lateral *pnp* transistor, similar to that shown in Figure 2.23, for each of the three basic fabrication processes discussed earlier. Note that in the table, I_O refers to the current level where β_F starts to fall off, and $I_{1/2}$ refers to the current level where β_F has fallen to 50% of its peak value.

The parasitic capacitance values associated with the lateral *pnp* transistor are not shown separately since they are the same as those associated with the collector–base and the collector–isolation junctions of the *npn* transistor given in Table 2.2.

Substrate *pnp* Transistor

A functional *pnp* transistor can also be obtained by using the base region of the *npn* transistor as the emitter, the *n*-type epitaxial layer as the base, and the *p*-type substrate as the collector. Such a device is known as the *substrate pnp* transistor. The substrate *pnp* transistor is inherently present as a parasitic, associated with the basic *npn* or the lateral *pnp* structures (see Fig. 2.3a or Fig. 2.26).

TABLE 2.4.   Typical Device Parameters for Small-Geometry Lateral *pnp* Transistor of Figure 2.23[a]

Device Parameter	Low-Voltage Process: npn LV_{CEO}, 20 V, 1-Ω-cm Epitaxial Layer, $W_B = 6\ \mu$m	Medium-Voltage Process: npn LV_{CEO}, 30 V, 2.5-Ω-cm Epitaxial Layer, $W_B = 8\ \mu$m	High-Voltage Process: npn LV_{CEO}, 40 V, 5-Ω-cm Epitaxial Layer, $W_B = 10\ \mu$m
Forward-current gain β_F	20	30	40
Peak-gain current I_0	50 μA	30 μA	20 μA
Half-gain current $I_{1/2}$	500 μA	300 μA	200 μA
Early voltage V_A	60 V	50 V	50 V
Collector–base leakage I_{CO}	10^{-7} mA	10^{-7} mA	10^{-7} mA
Breakdown voltages $BV_{EBO} = BV_{CBO}$	50 V	70 V	90 V
Lowest sustaining voltage LV_{CEO}	35 V	50 V	60 V
f_T (at $I_C = 100\ \mu$A)	6 MHz	4 MHz	3 MHz
Base resistance r_b	150 Ω	220 Ω	300 Ω
Collector series resistance r_{cs}	150 Ω	150 Ω	150 Ω
Emitter series resistance r_{ex}	10 Ω	10 Ω	10 Ω

[a]Emitter periphery $\approx 100\ \mu$m.

However now it is put to good use as a functional device. The substrate *pnp* transistor can be fabricated simultaneously with the *npn* bipolar transistors, without requiring additional masking or diffusion steps. However, since the epitaxial layer thickness is now directly related to the effective base width W_B of the *pnp* transistor, a somewhat tighter control of the epitaxial layer thickness is necessary to ensure repeatability of it chararacteristics.

Figure 2.29 shows a typical device layout and the cross section for a substrate *pnp* transistor. There are many variations possible to the basic device layout illustrated. Figure 2.30 shows the basic carrier flow path within the transistor for the minority carriers (holes) injected into the *n*-type epitaxial layer, which serves as the base region of the transistor. Note that the subepitaxial n^+-type layer is omitted between the epi–substrate interface in order to enhance the minority carrier transport through the base region. In the substrate *pnp* transistor, the isolation walls also serve as the collector, provided they are fairly close (i.e., within $20–30\mu$m) to the *p*-type emitter area.

There is one key limitation to the substrate *pnp* transistor: the collector of the substrate *pnp* transistor is formed by the *p*-type substrate, which is common to the rest of the circuit, and is at all times ac grounded. Therefore, the substrate *pnp* transistor is only available in the grounded-collector configuration and cannot be used for level shifting or voltage amplification. However, it still provides current amplification and can be used as a low-impedance output device in class-B complementary stages.

Compared to the lateral *pnp* transistor, the substrate *pnp* transistor can handle a higher amount of current for a given device area. This is because in the case of the substrate *pnp* transistor, the entire emitter–base area, not just the emitter periphery, is active in injecting minority carriers into the base region (see Fig. 2.30). However, since the *n*-type epitaxial layer forming the base region is lightly doped, the current gain drops off rapidly at high currents. For a given emitter area, the current-handling capability of a substrate *pnp* transistor is approximately one-third to one-fifth of that of an equivalent *npn* transistor. Figure 2.25 shows the typical β_F versus current characteristics for the substrate *pnp* transistor of Figure 2.29, which has an effective emitter area of approximately 5000 μm^2.

A few comments on the basic layout of the substrate *pnp* transistor shown in Figure 2.29 are in order. First, note that the base contact is made by means of an n^+ diffusion, made into a "hole" within the *p*-type emitter region. This minimizes the base resistance r_b for the device. Although the presence of this n^+ diffusion, in direct contact with the *p*-type emitter, would cause a low reverse breakdown for the base–emitter junction of the device (typically $BV_{EBO} \approx 7$ V), it would not be a detriment in normal operation of the device since the substrate *pnp* transistor is only used in the grounded-collector (i.e., emitter follower) configuration. Second, an electrical contact to the substrate, or the isolation wall, is placed in the immediate vicinity of the device so that the collector current, which is flowing into the substrate common to the entire circuit, can be extracted with a minimum amount of voltage drop generated within the sub-

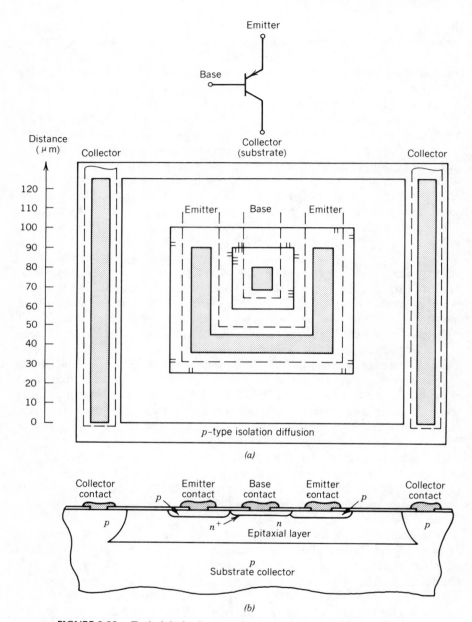

FIGURE 2.29. Typical device layout and cross section of a substrate *pnp* transistor.

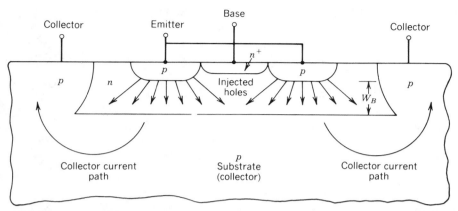

FIGURE 2.30. Direction of minority carrier flow, and the collector current path in a substrate *pnp* transistor.

strate. If this were not done, the voltage gradients generated within the substrate due to collector current of the substrate *pnp* transistor may cause some of the isolation junctions within the chip to be forward biased.

The high-frequency capability of the substrate *pnp* transistor is dominated by the carrier transit time through the wide base region W_B, which is determined by the epitaxial layer thickness. Typical values of the device cutoff frequency are in the range of 8–30 MHz.

Table 2.5 gives a summary of the substrate *pnp* device parameters for a device structure similar to that shown in Figure 2.29 for various basic bipolar manufacturing processes.

High-Performance *pnp* Transistors

In the design of high-performance analog circuits, particularly for operation within a radiation environment, the performance characteristics of the lateral and the substrate *pnp* transistors are not acceptable. A number of high-performance *pnp* transistor structures have been developed for these applications. However, each of these device structures requires additional processing steps above and beyond what is required for the basic *npn* transistor, and they are, therefore, limited to special design applications where the added fabrication cost or complexity can be justified.

The use of dielectric isolation techniques provides an added degree of freedom in fabricating high-performance complementary devices by providing access to the backside of the device structure. Therefore, with dielectric isolation methods, it is possible to fabricate high-performance *pnp* transistors simultaneously with *npn* bipolars, using a device structure as shown in Figure 2.31. The process steps for such a device structure follow the basic sequence of steps shown in Figure 1.14, except that prior to an n^+ deposition on the wafer, a

TABLE 2.5. Typical Device Parameters for the Substrate *pnp* Transistor of Figure 2.29

Device Parameter	Low-Voltage Process: 1 Ω-cm, 9-μm Epitaxial Layer	Medium-Voltage Process: 2.5 Ω-cm, 14-μm Epitaxial Layer	High-Voltage Process: 5 Ω-cm, 18-μm Epitaxial Layer
Forward-current gain β_F	30	40	50
Peak-gain current I_0	5 mA	3 mA	2 mA
Half-gain current $I_{1/2}$	30 mA	20 mA	10 mA
Early voltage V_A	40 V	50 V	50 V
Collector–base leakage I_{CO}	4×10^{-7} mA	4×10^{-7} mA	4×10^{-7} mA
Base–emitter breakdown BV_{EBO}	6.5 V	6.8 V	7.0 V
Collector–base breakdown BV_{CBO}	50 V	70 V	90 V
Lowest sustaining voltage LV_{CEO}	35 V	50 V	70 V
f_T (at $I_C = 1$ mA)	30 MHz	20 MHz	10 MHz
Base resistance r_b	50 Ω	100 Ω	150 Ω
Emitter series resistance r_{ex}	5 Ω	5 Ω	5 Ω
Collector series resistance r_{cs}	50 Ω	50 Ω	50 Ω

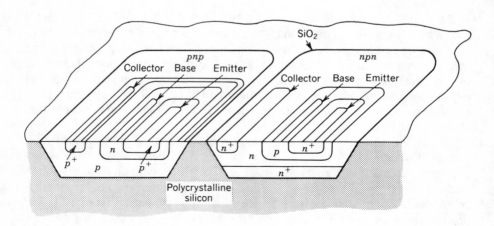

FIGURE 2.31.   Dielectrically isolated vertical *npn* and *pnp* transistors.

selective p diffusion is made from the backside. This is a deep (≈ 20 μm) noncritical diffusion step, forming the p-type collector region for the pnp transistor. This step is then followed by the selective n^+ predeposition for the npn collectors. After moat etching, polycrystalline silicon growth, and backlapping operations, one ends up with dielectrically isolated p- and n-type pockets on the silicon surface. The p-type pockets do not have a buried layer similar to the n^+-type layer below the npn collector. However, due to the backside diffusion step forming the p-type islands, the pnp collector region has a reverse impurity gradient, being more heavily doped near the bottom of the pocket. This reverse impurity profile results in a low collector series resistance for the pnp transistor and eliminates the need for a separate p^+-type buried layer. Once the dielectrically isolated p- and n-type islands are formed, the device structure of Figure 2.31 is completed by the following sequence of diffusion steps: (1) pnp base, (2) npn base, (3) pnp emitter, and (4) npn emitter.

In this process, the pnp diffusion steps are interleaved with the standard npn diffusion cycles without changing the npn impurity profile. Therefore, the electrical characteristics of the resulting npn devices remain unchanged. However, the pnp devices show a significant improvement over the lateral and the substrate pnp transistors, with the following typical performance characteristics:

$$\beta_F \approx 50\text{--}100 \quad (\text{at } I_C = 1 \text{ } mA)$$
$$LV_{CEO} \approx 60 \text{ V}$$
$$BV_{EBO} \approx 9 \text{ V}$$
$$BV_{CBO} \approx 80 \text{ V}$$
$$f_T = 150 \text{ MHz}$$

Typical absolute-value tolerances, matching tolerances, and thermal tracking properties of the high-performance pnp transistors are comparable to those listed in Table 2.1 for the npn devices.

Again, the reader is reminded that the economics and the yield of monolithic IC fabrication are extremely sensitive to additional manufacturing steps. Thus, complex device structures such as those shown in Figure 2.31, which require several additional masking and diffusion steps, are difficult to manufacture, economically, under a high-volume production environment (see Table 1.2 for relative cost factors). They should be used with discretion and limited to those special applications where the added fabrication costs can be justified.

2.4. JUNCTION FIELD-EFFECT TRANSISTORS

The principle of operation of the junction field-effect transistor (JFET) differs significantly from that of the bipolar type. In the case of JFETs, the current conduction mechanism through the device relies solely on the majority carriers, whereas in the bipolar transistor *both* the majority and the minority carriers participate actively in the current transport process. A detailed coverage of JFET device theory is readily available in the literature[7,8] and will not be repeated here. Instead, this section is aimed at covering the salient features of JFET structures which are readily compatible with the basic *npn* bipolar process technology.

The JFET is a voltage-controlled device, where the current conduction through a channel connecting the *source* and the *drain* regions is controlled or modulated by means of a control voltage applied to the *gate* terminal. The JFET can be either a *p*-channel or an *n*-channel device, depending on the conductivity type of the source and drain regions and the channel that connects the two. The gate regions of opposite conductivity type, and a reverse-biased *p-n* junction exists at all times between the channel and the gate regions.

Figure 2.32 shows the basic symbol and the voltage polarities associated with the *p*- and *n*-channel JFET. In monolithic IC design, *p*-channel JFET are more widely used than *n*-channel devices because their structure is more readily compatible with the basic *npn* bipolar fabrication process.

Device Characteristics

Figure 2.33 shows an idealized *p*-channel JFET structure which is helpful in deriving the basic device characteristics and understanding its principle of operation. In the figure, a uniformly doped *p*-type channel is assumed, with symmetrical, heavily doped *n*-type gate regions on either side. With zero applied bias to device terminals, a finite depletion region exists across the *p-n* junction surrounding the channel, due to the built-in junction potential ψ_0 [see Eq. (2.31)]. Assume, for the time being, that the gate region is short-circuited to the source. If a negative voltage V_{DS} were applied to the drain, a drain current I_D would flow from the source to the drain through the channel. This current will produce a voltage gradient within the channel and cause the channel–gate junction to be more heavily reverse biased at the drain side of the channel. This

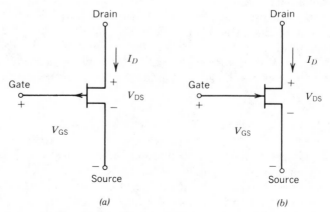

FIGURE 2.32. Circuit symbols and sign conventions for *p*- and *n*-channel JFET: (*a*) *p*-channel; (*b*) *n*-channel.

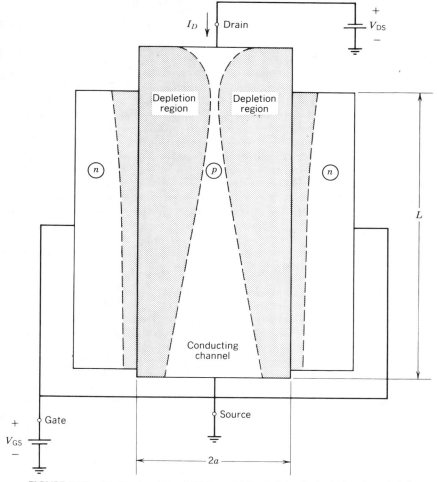

FIGURE 2.33. Idealized p-channel JFET model for deriving basic device characteristics.

reverse bias, in turn, causes the depletion layer to stretch into the channel and constrict the conductive region of the channel further, with the narrowest region appearing at the drain end of the channel.

If the drain–source voltage V_{DS} is increased further, the channel width at the drain end approaches zero. The voltage level at which this happens, that is, when the gate–channel depletion layer reaches completly across the entire width of the channel, is called the pinch-off voltage V_P. If V_{DS} is increased further, beyond V_P, the drain current I_D becomes insensitive to V_{DS} and reaches a saturation value I_{DSS}. Device operation in this region, where the drain current is relatively constant, is called *pinched* operation or operation in the *pinched-off* region.

If an external bias V_{GS} is applied to reverse-bias the gate region with respect to the source, as shown in Figure 2.33, this would create an additional initial depletion layer within the source end of the channel. This additional depletion layer would then cause the drain current I_{DS} to saturate sooner and remain constant at a lower value than the zero gate bias saturation current I_{DSS}. Finally, if the applied gate–source bias V_{GS} is equal to or greater than the pinch-off voltage V_P, the entire channel region becomes depleted of mobile carriers, the drain current becomes zero, and the source becomes electrically disconnected from the drain.

The pinch-off voltage is an important parameter of device design and can be given as

$$V_P = \frac{a^2 q N_A (1 + N_A/N_D)}{2\epsilon} - \psi_o \qquad (2.33)$$

where a is the half-width of the channel region transverse to the direction of current flow, as shown in Figure 2.33, ϵ is the dielectric coefficient of silicon, and N_A and N_D are the doping levels of the p-type channel and the n-type gate regions, respectively. Note that V_P is a very strong function of the channel width and the channel impurity concentration.

Using the idealized device model of Figure 2.33, one can show that I_D can be related to the gate and drain voltages V_{GS} and V_{DS} as[9]

$$I_D = G_o \left[V_{DS} + \frac{3}{2} \frac{(\psi_o + V_{GS} - V_{DS})^{3/2} - (\psi_o + V_{GS})^{3/2}}{(\psi_o + V_P)^{1/2}} \right] \qquad (2.34)$$

as long as the device is not operated in its pinched mode, that is, as long as the total reverse bias across the junction at the drain end of the channel does not exceed V_P.

In Eq. (2.34), the constant G_o is determined by the geometry and the conduc-

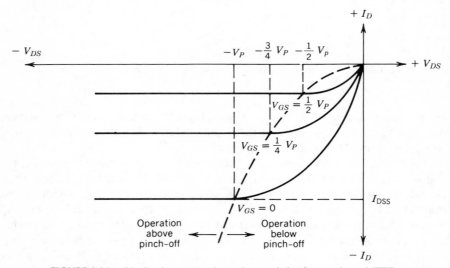

FIGURE 2.34.   Idealized current–voltage characteristics for a p-channel JFET.

tivity of the channel as

$$G_o = \frac{2az}{L} \sigma_c \tag{2.35}$$

where L is the length of the active channel region in the direction of current flow, Z is the depth of the channel (i.e., the dimension going *into* the paper in Fig. 2.33), and σ_c is the conductivity of the channel.

Figure 2.34 shows the idealized p-channel JFET characteristics implied by Eq. (2.34). This equation predicts the current–voltage characteristics to the right of the dashed line in Figure 2.34, corresponding to operation below pinch-off, and assumes that I_D is constant and independent of V_{DS} for higher voltages.

The maximum value of drain current I_{DSS}, which corresponds to $V_{GS} = 0$, is given as

$$I_{DSS} = G_o \left[-V_P + \frac{2}{3} \frac{(\psi_o + V_P)^{3/2} - \psi_o^{3/2}}{(\psi_o + V_P)^{1/2}} \right] \tag{2.36}$$

The saturation value I_{DS} of the drain current in the pinched-off region of the device (i.e., the flat portion of the current–voltage characteristics) is related to I_{DSS} as

$$I_{DS} = I_{DSS} \left(1 - \frac{V_{GS}}{V_P} \right)^2 \tag{2.37}$$

Actual JFET characteristics show a finite slope to the current–voltage characteristics for pinched operation of the device, as illustrated in Figure 2.35. This finite slope comes about from the modulation of the effective channel length L by the drain-to-gate depletion layer, and is analogous to the Early effect in

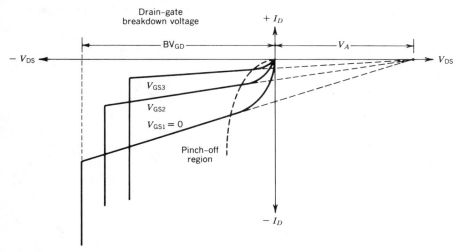

FIGURE 2.35. Actual p-channel JFET current–voltage characteristics showing effects of channel length modulation and gate–channel junction breakdown.

bipolar transistors. Thus, one can similarly extrapolate the device characteristics to an intercept voltage V_A on the V_{DS} axis, as shown in the figure. This allows the idealized current characteristics of Eq. (2.37) to be modeled more accurately as

$$I_{DS} = I_{DSS}\left(1 - \frac{V_{GS}}{V_P}\right)^2\left(1 - \frac{V_{DS}}{V_A}\right) \tag{2.38}$$

For practical JFET structures, V_A is in the range of 50–100 V and increases as the channel length-to-width ratio $L/2a$ is increased.

As shown in Figure 2.35, the maximum value of V_{DS} is limited by the avalanche breakdown of the gate–channel junction. This breakdown occurs at the drain end of the channel, where the total reverse bias across the junction, which is equal to $V_{GS} + V_{DS}$, reaches the junction avalanche breakdown voltage, BV_{GD}.

Integrated JFET Structures

The key requirement for integrated JFET devices is that they should be process compatible with the basic *npn* bipolar technology, such that they can be manufactured with only a minimum number of additional process steps. An additional requirement is that the resulting devices should have a repeatable and predictable pinch-off voltage in the range of 1–5 V so that they can operate with signal and bias levels common to bipolar transistors. These requirements are best fulfilled by *p*-channel devices, made by either the diffusion or the ion-implantation techniques.

Double-Diffused JFET. Figure 2.36 shows the typical cross section of a *p*-channel JFET structure which can be fabricated simultaneously with an *npn*

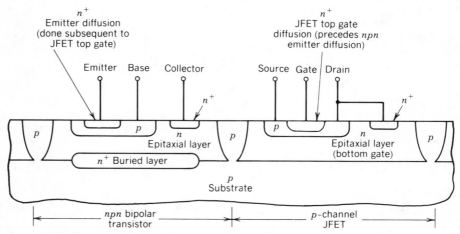

FIGURE 2.36. Bipolar compatible double-diffused *p*-channel JFET structure.

bipolar transistor. This device uses the p-type base region of the npn transistor as the channel, where the n^+ emitter diffusion and the n-type epitaxial region serve as the top and bottom gates. However, if only the normal n^+ emitter diffusion is used to form the top gate, the resulting pinch-off voltage is too high (i.e., in the 8–15-V range). This problem can be avoided by using a separate n^+ gate diffusion step for the JFET which is diffused slightly deeper than the normal n^+ emitter diffusion, thus resulting in a narrower p-type channel. This process is very similar to that described in Section 1.10 in connection with the superbeta transistors (see Fig. 1.17) and results in a V_P in the range of 2–5 V.

Although double-diffused JFET devices have been used in a number of applications such as operational amplifier input stages, they have several drawbacks. The most important limitation is the difficulty of controlling the channel width by means of the critical n^+ diffusion step. This results in poor control of the absolute value of V_P as well as the matching and tracking of V_P between different devices on the same chip. A second limitation is the low breakdown characteristic: the gate–drain breakdown is essentially limited to that of the base–emitter junction of a bipolar transistor, that is, approximately 7 V. Still a third disadvantage is that since the p-type channel is more heavily doped than the n-type epitaxial region, the bottom gate does not contribute significantly to the device transconductance.

Ion-Implanted JFET. The drawbacks and limitations of double-diffused JFETs can be avoided by using an ion-implanted device structure, as shown in Figure 2.37. This device, which is also called a *BIFET*, has gained wide acceptance with the advent of ion-implantation technology, it has replaced the double-diffused JFET in most of the circuit applications.

The ion-implanted JFET structure uses the p-type base diffusion of the npn transistor to form the source and drain contact regions. The channel is then formed by a p-type implant step. This is followed by an n-type implant step to form the gate region. The channel implant is the critical step in the device fabrication. Since the ion-implantation process is capable of very uniform and precise placement of impurities into silicon, it allows the channel resistivity and thickness to be accurately controlled. This in turn allows significant improvement in the control of the absolute values and the matching of the pinch-off voltage and the saturation current across the wafer. The n-type gate implant is a heavy-dose shallow implant. Since it does not extend significantly into the channel, it does not affect the device characteristics significantly.

Figure 2.38 shows the typical impurity profile for the implanted channel and gate regions. Note that the total channel width is ≈ 0.5 μm. The impurity concentration in the channel is kept low (typically in the range of 2×10^{16} atoms/cm^3) in order to maintain a high gate–channel breakdown voltage (typically ≥ 40 V). Typical values of pinch-off voltage in ion-implanted JFETs are in the range of 1–3 V, with a typical matching of $\approx \pm 10$ mV between adjacent devices on the chip.

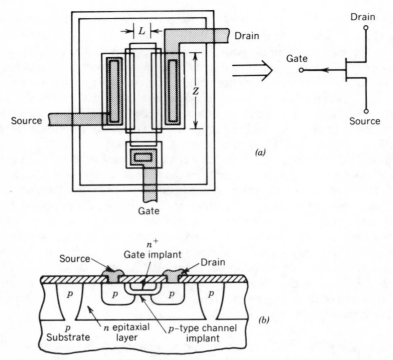

FIGURE 2.37.   Top view and cross section of an ion-implanted *p*-channel JFET.

Small-Signal Model

In most analog circuit applications, the JFET is operated in its pinch-off region, where its current–voltage characteristics approximate that of a voltage-controlled current source. Figure 2.39 shows that the basic small-signal equivalent circuit for a JFET for operation in its pinch-off region. The parasitic capacitances associated with the device structure are also shown.

The small-signal transconductance g_m can be derived from Eq. (2.37) as

$$g_m = \frac{\partial I_{DS}}{\partial V_{GS}} = -\frac{2I_{DSS}}{V_P}\left(1 - \frac{V_{GS}}{V_P}\right) \tag{2.39}$$

which can be rewritten as

$$g_m = g_{mo}\left(1 - \frac{V_{GS}}{V_P}\right) \tag{2.40}$$

where g_{mo} is the zero bias transconductance corresponding to $V_{GS} = 0$, given as

$$g_{mo} = -\frac{2I_{DSS}}{V_P} \tag{2.41}$$

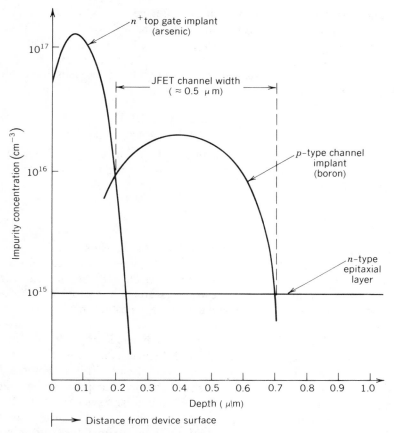

FIGURE 2.38. Typical impurity profile of an ion-implanted p-channel JFET.

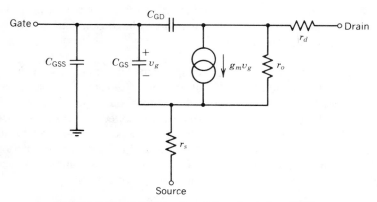

FIGURE 2.39. Small-signal equivalent circuit for a JFET.

It should be noted that for a p-channel device I_{DSS} is negative and V_P is positive, whereas the opposite is true for an n-channel JFET. Thus, g_{mo} is always a positive number. For a pinch-off voltage $V_P \approx 1.0$ V and $I_{DSS} = 250\ \mu A$, g_{mo} is around 0.5 mA/V.

The small-signal output resistance r_o is due to the channel length modulation effect. It can be derived by differentiating Eq. (2.38),

$$\frac{1}{r_o} = \frac{\partial I_D}{\partial V_{DS}} = \frac{I_{DSS}}{V_A}\left(1 - \frac{V_{GS}}{V_P}\right)^2 \qquad (2.42)$$

where V_A is the extrapolated voltage intercept shown in Figure 2.35. Typical values of V_A are in the range of 50–100 V. Assuming that $V_{DS}/V_A \ll 1$, Eq. (2.42) simplifies to

$$r_o \approx \frac{V_A}{I_D} \qquad (2.42a)$$

The parasitic drain and source resistances r_d and r_s, shown in the equivalent circuit of Figure 2.39, are due to the parasitic bulk resistance between the respective drain and source contacts and the active channel region directly below the gate. Typical values of these resistances are of the order of 30–100 Ω, depending on the device geometry and the impurity profile. In most applications, their presence can be neglected.

Compared to the bipolar transistor, one key advantage of the JFET is the high input impedance of the gate terminal. Since the gate is in series with a reverse-biased p-n junction, it is a virtual open circuit at dc. This makes the JFET virtually an ideal device for circuits, such as operational amplifiers which require high input impedance. However, compared to bipolar devices, the JFETs have one serious limitation: in bipolar transistors, the transconductance g_m is determined by the bias current [see Eq. (2.16)] and is approximately 40 millimhos per milliampere of bias current. In JFETs g_m is determined by the channel geometry (i.e., the channel Z/L ratio, as shown in Fig. 2.37) and the channel width and resistivity, and is of the order of 0.5–1 millimho at $I_{DS} \approx 1$ mA. Since the overall voltage gain which can be obtained from the active device is proportional to g_m, a bipolar stage can provide much higher voltage gain than a corresponding JFET stage at similar bias levels.

Parasitic Capacitances.   The parasitic capacitances shown in Figure 2.39 are inherent to the monolithic JFET structures. The gate–substrate capacitance C_{GSS} is the capacitance of the reverse-biased epi–substrate junction of Figures 2.36 and 2.37. This is essentially the same as the collector–substrate capacitance C_{DS} of a bipolar transistor. The capacitances, C_{GS} and C_{GD} are the junction capacitances associated with the gate–source and the gate–drain junctions.

In the bipolar compatible p-channel JFET structures of Figures 2.36 and 2.37, the gate–source capacitance C_{GS} is made up of two sections

$$C_{GS} = C_{GS1} + C_{GS2} \qquad (2.43)$$

where C_{GS1} is the capacitance due to the heavily doped top gate and source junction, and C_{GS2} is the capacitance due to the lightly doped back gate (i.e., the n-type epitaxial layer) and source junction. The capacitance of the gate–channel junction is normally associated with the source terminal.

The drain–gate capacitance C_{GD} is primarily due to the p-type drain and the n-type back gate (i.e., the epitaxial layer) junction. The values of these capacitances depend on the device geometries (i.e., the junction areas) and the applied reverse bias across the junction. Similar to the case of bipolar transistors, they can be evaluated using Eqs. (2.30) and (2.31). The zero-bias junction capacitances, the built-in junction potentials, and the exponent n associated with each junction are given in Table 2.6 for typical double-diffused and ion-implanted JFET structures of Figures 2.36 and 2.37, assuming a 5-Ω/cm n-type epitaxial layer. In both cases, a stripe-type device layout similar to that shown in Figure 2.37a is assumed. Assuming a Z/L of 20, a gate substrate bias of 10 V, and V_{DG} of 5 V, typical values of capacitances calculated from the data of Table 2.6 are

$$C_{GSS} = 2 \text{ pF} \qquad C_{GS1} = 1 \text{ pF}$$
$$C_{GD} = 0.5 \text{ pF} \qquad C_{GS2} = 0.6 \text{ pF}$$

It should be noted that although C_{GD} has the lowest value, it is the most detrimental parasitic to the voltage gain of the device at high frequencies, due to the *Miller effect*, similar to the collector–base capacitance C_c of a bipolar transistor.

TABLE 2.6. Typical Values of Parasitic Capacitances Associated with Double-Diffused and Ion-Implanted JFETS

Parasitic Capacitances	Double-Diffused JFET	Ion-Implanted JFET
Gate–substrate capacitance C_{GSS}		
Zero-bias capacitance	$10^{-4} \text{ pF}/\mu\text{m}^2$	$10^{-4} \text{ pF}/\mu\text{m}^2$
Built-in potential ψ_o	0.52 V	0.52 V
Exponent n	$\frac{1}{2}$	$\frac{1}{2}$
Gate–source capacitance C_{GS}		
1. Top gate capacitance C_{GS1}		
Zero-bias capacitance	$8 \times 10^{-4} \text{ pF}/\mu\text{m}^2$	$5 \times 10^{-4} \text{ pF}/\mu\text{m}^2$
Built-in potential ψ_o	0.7 V	0.5 V
Exponent n	$\frac{1}{3}$	$\frac{1}{3}$
2. Bottom gate capacitance C_{GS2}		
Zero-bias capacitance	$10^{-4} \text{ pF}/\mu\text{m}^2$	$10^{-4} \text{ pF}/\mu\text{m}^2$
Built-in potential ψ_o	0.55 V	0.55 V
Exponent n	$\frac{1}{2}$	$\frac{1}{2}$
Gate–drain capacitance C_{GD}		
Zero-bias capacitance	$10^{-4} \text{ pF}/\mu\text{m}^2$	$10^{-4} \text{ pF}/\mu\text{m}^2$
Built-in potential ψ_o	0.55 V	0.55 V
Exponent n	$\frac{1}{2}$	$\frac{1}{2}$

Summary of Device Characteristics. Table 2.7 gives a comparison of the electrical characteristics of the double-diffused and the ion-implanted p-channel JFET structures. For comparison purposes, both devices are assumed to have the stripe geometry of Figure 2.37a with a Z/L ratio of 20, where the channel length $L = 12\ \mu$m and the width of the device $Z = 240\ \mu$m. The important point to note from the table is that the ion-implanted JFET structure gives a much tighter control of the key device parameters, such as I_{DSS} and V_P and provides a much higher gate–drain breakdown voltage.

2.5. MOS FIELD-EFFECT TRANSISTORS

In the metal–oxide semiconductor field-effect transistor (MOSFET), a thin dielectric barrier is used to isolate the gate and the channel. The control voltage applied to the gate terminal induces an electric field across the dielectric barrier and modulates the free-carrier concentration in the channel region. In the literature, these devices are also often referred to as insulated-gate field-effect transistors (IGFET) or simply as MOS transistors.

MOS transistors are classified as p-channel or n-channel devices, depending on the conductivity type of the channel region. In addition, these devices can also be classified according to their mode of operation as *enhancement* or *depletion* type devices.

In a depletion-mode MOSFET, a conducting channel exists under the gate with no applied gate voltage. The applied gate voltage controls the current flow

TABLE 2.7. Typical Parameter Values and Tolerances for Integrated JFET Devices[a]

Device Characteristic	Double-Diffused JFET	Ion-Implanted JFET
Saturation current I_{DSS}		
Typical value	$-400\ \mu$A	$-250\ \mu$A
Absolute-value tolerance[b]	-200 to $-800\ \mu$A	-150 to $-400\ \mu$A
Matching tolerance[c]	$\pm 10\%$	$\pm 5\%$
Pinch-off voltage V_P		
Typical value	2 V	1 V
Absolute-value tolerance	± 500 mV	± 150 mV
Matching tolerance	$\pm\ 15$ mV	$\pm\ \ 8$ mV
Zero-bias transconductance g_{mo}	800 μmho	400 μmho
Gate–drain breakdown BV_{DGO}	7 V	50 V
Early voltage V_A	100 V	100 V

[a]For comparison purposes, identical device geometries are assumed with $L = 12\ \mu$m, $Z = 240\ \mu$m, $Z/L = 20$.
[b]Sigma limit for absolute-value distribution under production environment.
[c]Sigma limit for matching characteristics of identical device geometries, adjacent to each other.

between the source and the drain by depleting a part of this channel. This is very similar to the operation of the JFET described previously.

In the case of the enhancement-mode MOS transistor, no conductive channel exists between the source and the drain at zero applied gate voltage. As a gate bias of proper polarity is applied and increased beyond a threshold value V_{TH}, a localized inversion layer is formed directly below the gate. This serves as a conducting channel between the source and the drain electrodes. If the gate bias is increased further, the resistivity of the induced channel is reduced, and the current conduction from the source to the drain is enhanced. Figure 2.40 shows a cross-section diagram and the circuit symbol of an n-channel enhancement-mode MOSFET. The particular polarities of the gate and drain bias for the proper operation of the device are also identified. Normally, an SiO_2 layer is used as the gate dielectric. The thickness of this oxide layer is normally much less than that of the oxide layers commonly used for masking or surface passivation.

The enhancement-mode MOSFET is preferred over its depletion-mode counterpart because it is a *self-isolating* device, does not require tight control of

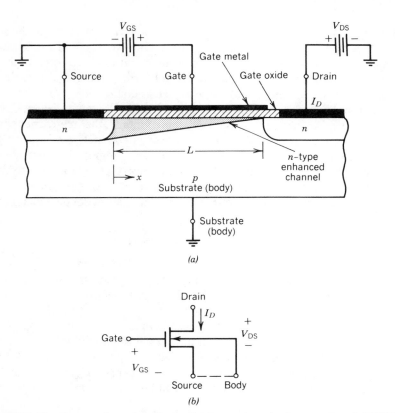

FIGURE 2.40. Cross section and device symbol for a n-channel enhancement-mode MOSFET.

diffusion cycles, and can be fabricated by a single diffusion step forming the source and the drain pockets. Since all the active regions of the MOSFET are reverse biased with respect to the substrate, adjacent devices fabricated on the same substrate are electrically isolated without requiring a separate isolation diffusion. Because of this self-isolation advantage, MOSFET devices offer a much higher packing density per unit area of silicon surface than the bipolar transistors. The depletion-mode MOSFET transistors are often used in conjunction with enhancement-mode MOS transistors to serve as active loads, to improve the gain or the switching speed of the enhancement devices.

The bulk of the semiconductor region, shown as substrate in Figure 2.40, is normally inactive since the current flow is confined to a thin surface channel directly below the gate. This part of the MOS transistor is called the *body* and is normally tied to the same potential as the source. However, in certain circuit applications, particularly in designing analog circuits with MOS transistors, the body may not be connected to the source. This may have a significant effect in the device characteristics, as will be described later.

Device Characteristics

In the device structure of Figure 2.40, with the body short-circuited to the source and with both $V_{GS} = 0$ and $V_{DS} = 0$, there will be no conducting channel between the n-type *beds* which form the source and drain regions. If the gate voltage V_{GS} is raised gradually, the electric field within the gate dielectric layer will increase correspondingly and cause an accumulation of n-type free carriers (electrons) in the region of the device directly below the gate dielectric. If the gate voltage is raised beyond a certain threshold voltage, the free-carrier concentration below the gate will be sufficient to form a conductive n-type channel, which will then connect the n-type source and drain region. In other words, if V_{GS} is raised beyond a threshold voltage V_{TH}, the p-type region directly below the gate would invert its conductivity type and form an n-type channel. It can be shown that this uniform conducting channel has a sheet conductance g_c given as[10]

$$g_c = \mu_s C_{ox}(V_{GS} - V_{TH}) \qquad (2.44)$$

where μ_s = surface mobility of the majority carriers in the induced channel
$ C_{ox}$ = capacitance per unit area of the gate electrode

The value of μ_s is in the range of 150–250 cm²/V-sec for holes and 300–600 cm²/V-sec for electrons, depending on the crystal orientation of slicon and the impurity profile at, or near, the device surface.

The gate capacitance C_{ox} is related to the thickness T_{ox} and the dielectric coefficient ϵ_{ox}, of the gate dielectric, and can be expressed as

$$C_{ox} = \frac{\epsilon_{ox}}{T_{ox}} \qquad (2.45)$$

T_{ox} is normally chosen to be ≥ 800 Å for breakdown purposes. For practical MOS structures, the design value of T_{ox} is in the range of 1000–2000 Å.

If the drain voltage V_{DS} is increased in the polarity shown in Figure 2.40, a finite drain current I_D flows through the induced channel. This current also causes an ohmic drop along the channel which subtracts from the net gate voltage $V_{GS}-V_{TH}$. Since at any point along the channel, the apparent sheet conductance is proportional to this net gate voltage, the voltage gradient due to I_D causes the channel to deplete along its length, and the sheet conductance becomes a function of the distance x along the channel,

$$g_c(x) = \mu_s C_{ox}\left(V_{GS} - V(x) - V_{TH} \right) \qquad (2.46)$$

where $V(x)$ is the channel potential at a distance x from the source. Thus, an increase of V_D causes the drain current to increase, which in turn causes the channel to deplete or pinch off near the drain. For values of the drain voltage in excess of the net gate bias, that is, for

$$V_{DS} \geq V_{GS} - V_{TH} \qquad (2.47)$$

a space-charge layer is formed at the drain end of the channel, since the net gate bias (i.e., the applied gate voltage minus the voltage drop along the channel) at this point is no longer sufficient to maintain an induced channel. This leads to saturation of the drain current, similar to the case of JFETs, and results in a set of idealized drain current versus voltage characteristics shown in Figure 2.41. The voltage values shown on the curve are for illustrative purposes only.

For values of V_{DS} less than $V_{GS} - V_{TH}$, the drain current I_D can be related to V_{DS} as

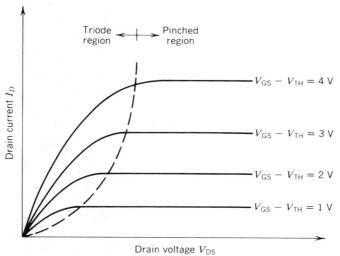

FIGURE 2.41. Idealized current–voltage characteristics of a n-channel enhancement-mode MOSFET.

$$I_D = \frac{\mu_s C_{ox} Z}{L} \left(V_{GS} - V_{TH} - \frac{V_{DS}}{2} \right) V_{DS} \qquad (2.48)$$

where $V_{GS} > V_{TH}$, and Z is the depth of the channel measured normal to the cross section shown in Figure 2.40 (see Fig. 2.47).

The current–voltage characteristics given by Eq. (2.48) correspond to the part of the device characteristics to the left of the dashed line in Figure 2.41. This region of the device characteristics, where I_D increases with increasing V_{DS}, is called the *triode* region of device characteristics. If V_{DS} is increased further, such that $V_{DS} \geq V_{GS} - V_{TH}$, then the channel will be nearly pinched off at the drain end, and the output current level will saturate in a manner similar to the pinched operation of a JFET (see Section 2.4). This region of device characteristics is again called the pinched-off region or the pinched operation.

The saturation value of the drain current I_{DO} can be obtained from Eq. (2.48) by setting $V_{DS} = V_{GS} - V_{TH}$, that is,

$$I_{DO} = \frac{\mu_s C_{ox} Z}{2L} (V_{GS} - V_{TH})^2 \qquad (2.49)$$

Comparing Eq. (2.49) with Eq. (2.37) for the case of the JFET, one sees that the MOSFET also exhibits a *square-law* characteristic, where the controlled current I_{DS} is proportional to the square of the net control voltage $V_{GS} - V_{TH}$. Similar to the case of the JFET, the MOS transistor current–voltage characteristics are linearly dependent on the channel width-to-length ratio Z/L. Thus, for a given channel length L, the current levels within the device can be scaled by making the channel width Z larger or smaller.

In actual MOSFET devices, the drain current I_D does not saturate at its ideal value of I_{DO}, but continues to increase slightly with increasing V_{DS}, as shown in Figure 2.42. This finite conductance of the output characteristics is due to the channel length modulation effects, where the channel drain depletion layer

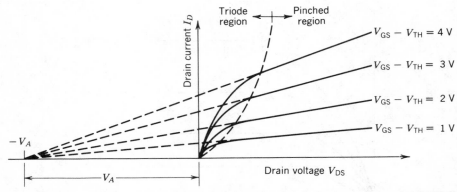

FIGURE 2.42. Actual current–voltage characteristics of a n-channel enhancement-mode MOSFET showing the effect of channel length modulation.

extends back into the channel and causes the effective channel length L to appear lower than its actual value. This effect can be included in the device characteristics by defining a "channel-length modulation voltage" intercept V_A on the voltage axis of the current–voltage characteristics, as shown in Figure 2.42. Then Eq. (2.49) can be written as

$$I_{DO} = \frac{\mu_s C_{ox} Z}{2L} (V_{GS} - V_{TH})^2 \left(1 - \frac{V_A}{V_{DS}}\right) \qquad (2.50)$$

to include the effects of channel length modulation. Note that for an n-channel device, V_A is negative and V_{DS} is positive; for a p-channel device, V_A is positive and V_{DS} is negative. Typical values of V_A are in the range of 30–50 V for MOS transistors with 8-μm channel length.

Depletion-Mode Devices

In the case of depletion-mode MOS transistors, a conducting channel exists under the gate region, with $V_{GS} = 0$. Such devices would exhibit a threshold voltage V_{TH} of opposite polarity. In other words, to bring the channel to the verge of conduction, one has to apply a gate bias of opposite polarity to "deplete" the existing channel from the free carriers.

Figure 2.43 shows the typical current–voltage characteristics of an n-channel depletion-mode MOSFET. For illustrative purposes, a threshold voltage V_{TH} of -2 V is assumed. Note that the device characteristics are virtually identical to those of Figure 2.42, except that the threshold voltage now has an opposite polarity.

All the basic equations derived for the enhancement-mode MOSFET are directly applicable to the depletion-mode devices, with the appropriate change of the polarity of V_{TH}.

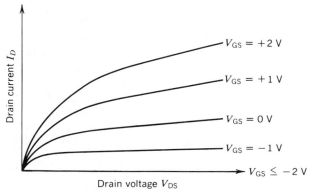

FIGURE 2.43. Current–voltage characteristics of a n-channel depletion-mode MOSFET with $V_{TH} = -2$ V.

Circuit Symbols

Several circuit symbols and sign conventions have been developed for MOS transistors. A set of the most commonly used circuit symbols for various MOS-FET types are shown in Figure 2.44. In the device symbols, the channel conductivity type (i.e., either n or p type) is designated by the direction of the arrow on the body or the substrate terminal. This arrow designates the polarity of the p–n junction formed between the channel and the substrate, and is in the same direction as a forward-biased diode, that is, it points from the p side to the n side of the junction. The depletion-mode MOSFETs are shown with a double line across the source–drain regions to imply the existence of a finite conducting channel under the gate, with zero applied bias.

In many circuit drawings where the body or substrate connection is not explicitly shown, it is easier to describe or identify the MOS devices by a slightly different set of device symbols, as indicated in Figure 2.45. In this case, the source terminal is identified with an arrow, similar to that of the emitter of a bipolar transistor. The polarity of the arrow is in the direction of the p-n junction formed by the source-body junction, pointing from the p region to the n region.

In the rest of this book, unless specified otherwise, the circuit symbols of Figure 2.45 will be used since they are very similar to the basic bipolar *npn* and *pnp* transistor symbols. This makes it easier for an analog circuit designer trained in bipolar IC design to adapt to the use of MOS devices, and to derive many of the basic MOS building blocks, or subcircuits, by direct analogy to their bipolar counterparts. This subject will be covered in depth in Chapter 6.

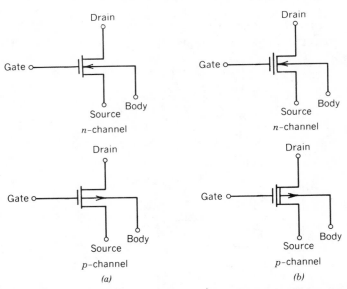

FIGURE 2.44. Set of commonly used circuit symbols for p- and n-channel (a) Enhancement mode; (b) depletion mode. MOS transistors.

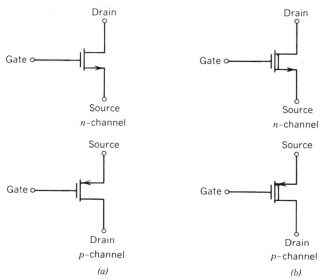

FIGURE 2.45. Alternate set of device symbols for p- and n-channel MOSFETS. These symbols are often preferred in MOS analog design because of their similarity to bipolar transistor notation. (a) Enhancement mode; (b) depletion mode.

Small-Signal Mode

For ac small-signal operation, the frequency and gain characteristics of a MOS-FET can be closely approximated by the equivalent circuit of Figure 2.46 for $V_{DS} > V_{GS} - V_{TH}$, that is, for operation in the pinched region of device characteristics. This ac equivalent circuit is very similar to that for a JFET (see Fig. 2.39).

The device transconductance g_m in the pinched region can be derived by differentiating Eq. (2.49) as

$$g_m = \frac{\partial I_{DO}}{\partial V_{GS}} = \frac{\mu_s C_{ox} Z}{L} (V_{GS} - V_{TH}) \qquad (2.51)$$

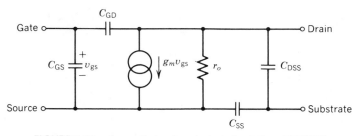

FIGURE 2.46. Ac small-signal equivalent circuit for a MOSFET.

As was the case for JFET, the transconductance is directly proportional to the Z/L ratio of the device layout. Similar to the JFET, g_m also increases linearly with the gate voltage. A device exhibiting this device characteristic is known as a square law device and is particularly useful for multiplication or modulation of ac signals (i.e., mixer applications) or automatic gain control (AGC) functions.

The finite output resistance r_o is due to the channel length modulation effect and depends on the actual length L of the channel. By defining an intercept voltage V_A, as shown in Figure 2.42, the value of r_o can be approximated as

$$r_o \approx \frac{|V_A|}{I_{DO}} \tag{2.52}$$

for the cases where $V_A \gg V_{GS} - V_{TH}$. For $V_A = 30$ V and $I_{DO} = 1$ mA, a typical value for r_o would be approximately 30 kΩ.

The parasitic capacitances C_{SS} and C_{DSS} are the capacitance between the source and drain bed diffusion and the body, or substrate, region. In general, the MOSFET is operated in a grounded-source configuration, with the source and the body regions connected together. In that case, the parasitic capacitance C_{SS} is short-circuited and has no effect on the circuit performance.

C_{GS} is the capacitance from the gate to the source and the unpinched region of the channel. It is an MOS capacitance. Thus, its value is equal to the gate capacitance per unit area C_{ox} of Eq. (2.45) times the area of overlap between the gate oxide and the source region. C_{GD} is the capacitance from the gate electrode to the drain. Its value is equal to C_{ox} times the area of overlap between the gate oxide and the drain region.

The value of C_{ox} is on the order of 4×10^{-4} pF per square micron of gate oxide area. Typical values of the zero-bias capacitances for C_{SS} and C_{DSS} are approximately 10^{-4} pF/μm^2, with a built-in potential ψ_0 of approximately 0.6 V. Thus, for a given device layout and bias condition, their values can be approximated using the generalized junction capacitance expression given in Eq. (2.30) with the exponent n equal to $\frac{1}{2}$.

A basic figure of merit for the high-frequency performance of a MOSFET is its transconductance frequency f_c given as

$$f_c = \frac{g_m}{2\pi C_{in}} \tag{2.53}$$

where C_{in} is the total input capacitance $C_{GS} + C_{GD}$ at the gate terminal, with the drain ac grounded. Since C_{in} is proportional to C_{ox} times the gate area, that is,

$$C_{in} = C_{ox}LZ \tag{2.54}$$

and g_m is given by Eq. (2.51), one can express f_c as

$$f_c = \frac{\mu_s(V_{GS} - V_{TH})}{2\pi L^2} \tag{2.55}$$

Equation (2.55) indicates that for optimizing high-frequency performance, a short channel length is required. As will be described in the next section, such short-channel device structures can be obtained using the so called self-aligned gate MOSFET structures.

Effect of Substrate Bias

In the derivation of the basic device equations and the ac equivalent circuit of Figure 2.46, it was assumed that the substrate, or body, of the MOS transistor is connected to the source. Although this is the case in most digital applications, a number of important exceptions to this case occur in analog IC design using MOSFETs. A typical example of such a case is when two MOS transistors of the same conductivity type are stacked such that one serves as an active load for the other.

The dependence of the MOSFET characteristics on the source-to-body bias V_{SB} is called the *body effect* and may be significant for a number of analog amplifier applications of monolithic MOSFET devices. The primary effect of an added reverse bias, V_{SB} between the body and the source is the increase of the gate threshold voltage V_{TH} above its nominal value. Another undesired result of the body effect is the reduction of the device transconductance and the output impedance when operated in a *cascode* configuration. These effects will be analyzed and discussed in more detail in Chapter 6.

Integrated MOSFET Devices

MOS transistors can be fabricated by the standard IC processing techniques described in Chapter 1. Figure 2.47 shows the cross sections and typical device layouts for p- and n-channel MOSFETs. For a p-channel device, a n-type substrate is used. First the n-type source and drain diffusions are performed. Next, the oxide is removed over the channel region and a thin oxide is grown. Although it is desirable to have a very thin layer of oxide to reduce the threshold voltage and to increase C_{ox}, in production, gate oxide thicknesses less than 1000 Å are generally not used because of reliability and yield problems. Following the thin oxide regrowth, windows are opened in the thick oxide over source and drain regions, and aluminum is deposited for contact and for gate metal. For an n-channel device, a similar procedure is used. The typical resistivity of the substrate region is on the order of 3–5 Ω/cm. The resistivity and the diffusion depth of the source and device drain regions are not critical, except that the side diffusion of these regions under the gate oxide must be taken into account in calculating the effective channel length L. Typically, these regions are diffused to a depth of 2–3 μm, and have a sheet resistivity range of 50–100 Ω/□.

The lateral geometry of the devices is also shown in Figure 2.47 for the so-called stripe geometry. Normally, the channel length L, is fixed by the process and masking tolerances, but the channel width Z is a design parameter which lets the designer optimize the device characteristics by setting the Z/L ratio.

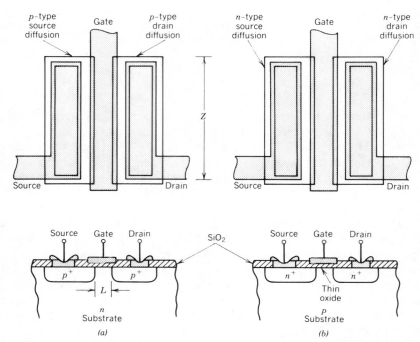

FIGURE 2.47. Typical device layouts and cross sections of p- and n-channel MOS transistors: (a) p-channel MOSFET; (b) n-channel MOSFEt.

Bipolar Compatible MOSFET Structures.   The metal gate p-channel MOSFET transistor structure of Figure 2.47a can be formed with the standard bipolar IC processing steps, with the addition of one extra masking step to form the thin oxide layer over the gate region. The resulting device cross section is shown in Figure 2.48. The source and drain regions are formed by the p-type base diffusion of the *npn* transistor, and the n-type epitaxial layer forms the body of a MOSFET. The p-type inversion layer is then formed under the thin oxide

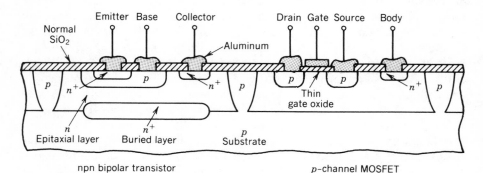

FIGURE 2.48. Bipolar compatible p-channel MOSFET structure.

layer, when the gate voltage exceeds the threshold voltage. For such a device structure, an epitaxial layer resistivity of 3–5 Ω/cm is most readily compatible with either the medium- or the high-voltage process described in Section 2.1.

Although the process complexity of the bipolar compatible p-channel MOS-FET structure of Figure 2.48 is not great, the dependence of the MOS transistor characteristics on the surface properties requires a more careful control of the surface conditions on the wafer than is typically required for standard bipolar IC processing. This is particularly true for controlling the absolute value and the long-term stability of the MOS threshold voltage V_{TH}. Thus, although the inclusion of p-channel MOSFETs along with bipolar transistors is feasible, it is more difficult than it may at first appear.

Self-Aligning Gate Structures. In the basic MOSFET device structures of Figure 2.47, it is necessary that the thin oxide and the metal gate overlap both the source and the drain regions in order to account for registration alignment errors. This means that there will be gate–source and gate–drain overlap capacitances. The latter is the more objectionable in applications where the device is used in the common-source connection, since this may lead to the Miller-effect multiplication of the gate–drain capacitance.

To overcome the capacitance problem associated with the gate-drain overlap, and to minimize the mask alignment problems in defining the gate region, a number of processing techniques have been developed. These techniques make the gate-channel registration automatic and eliminates the mask alignment problems. Figure 2.49 illustrates one such method, using ion implantation. The source and drain regions are diffused with standard processing techniques; however, they are spread somewhat further apart than the final channel length L. Then a thin oxide is regrown between the two p diffusions, and the gate metal is deposited and masked. However, an *undersize* gate metal mask is used such that the gate metal falls in between the two p diffusions but *does not* overlap them. Then the entire wafer surface is implanted with p-type (boron) ions. The gate metal now serves as a mask and blocks the boron ions from penetrating the region below it. Thus, the implantation step causes the p-type source and drain regions to extend to the edge of the gate metal, but not under it, as shown in Figure 2.49b. In this manner, both the gate overlap capacitance and the mask alignment tolerances are greatly reduced.

Silicon Gate Structures. Both the overlap capacitance and the threshold voltage V_{TH} can be reduced by using a gate electrode made up of doped silicon, rather than metal.[12] In this process, after the thin oxide growth, a layer of polycrystalline silicon is deposited over the entire wafer by CVD techniques. This silicon is then etched away from all areas except where the gate is to be located. Next, the windows are opened in the oxide layer to diffuse the p-type source and drain regions. In this process, the polycrystalline silicon gate region becomes p-type doped, and also serves as a self-aligning mask to separate the source and drain diffusions. Next, an oxide layer is deposited on the wafer

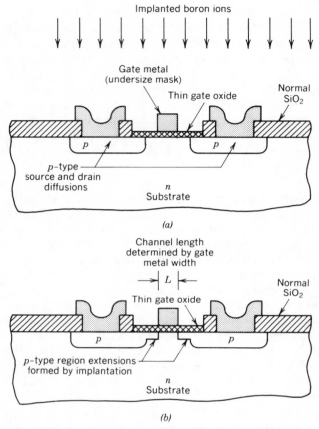

FIGURE 2.49.   Self-aligning gate MOSFET structure using ion implantation and undersize gate metal mask to determine channel length. (Vertical scale exaggerated for clarity) (*a*) Before ion implantation; (*b*) after ion implantation.

surface and the contact windows are opened. This is followed by the conventional metal deposition to form the interconnections. The sequence of the process steps and the resulting device structure are shown in Figure 2.50. Note that the polycrystalline silicon gate electrode is sandwiched within the SiO_2 layer over the gate region.

Often a thin silicon nitride (Si_3N_4) layer is used as a gate dielectric, in silicon gate devices, in conjunction with the thin SiO_2 layer. This additional gate dielectric layer increases the gate capacitance C_{ox} since Si_3N_4 has a higher dielectric constant than SiO_2 and also helps stabilize the threshold voltage against long-term drift.

Complementary MOS (CMOS) Structures.   In a number of circuit applications, it is desirable to have both the *n*-channel and the *p*-channel enhancement-

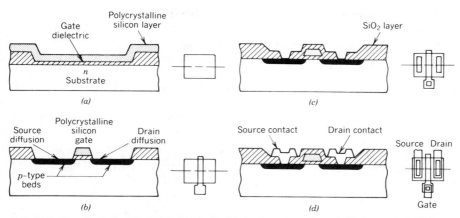

FIGURE 2.50. Typical sequence of steps in the fabrication of a p-channel silicon gate MOSFET: (a) Gate oxide and polycrystalline silicon are grown; (b) source and drain beds are diffused; (c) source and drain contacts are defined; (d) aluminum interconnections are deposited and etched.

mode devices on the same chip. This is particularly true for micropower circuits where low standby power is essential. In the case of analog design, the availability of complementary devices also simplifies the circuit design and biasing significantly.

It is possible to fabricate MOS transistors simultaneously by using the device structure shown in Figure 2.51. The basic process sequence proceeds as follows. First, a deep p-type pocket is diffused into the n-type substrate. The diffusion of this p-type pocket, called a p *well*, is critical since it requires a deep diffusion but low impurity concentration. Its doping concentration affects both the threshold voltage and the breakdown characteristics of the n-channel MOSFET. Next, after a masking step, a second p-type diffusion, called the p^+ *diffusion*, is made to form the source and drain regions of the p-channel MOSFET. At the same time, a ring of this p^+ diffusion is also applied around the periphery of the deep p well to avoid possible parasitic surface inversion along the lightly doped

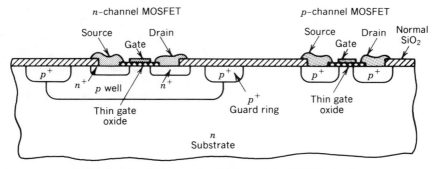

FIGURE 2.51. Complementary MOS transistor (CMOS) structure.

surface of the p well. This p^+ ring around the p well is called a *channel stop*. Next, after a masking step, n^+ diffusion is performed to form the source and drain of the n-channel MOSFET within the p well. Finally, the thin gate oxide is masked and grown over the gate regions, and the metal interconnection pattern is formed.

CMOS has one added advantage over single-polarity MOSFET, it is immune to the substrate bias effect (body effect) in most circuit applications. This makes it particularly suitable for analog design, as will be described in more detail in Chapter 6.

In terms of the number of process steps, the extra masking steps make the CMOS structure somewhat more complex and more expensive to manufacture than the simple p- or n-channel MOSFETs. The side-diffusion effects of the deep p-well diffusion also consume a good fraction of the chip area, similar to the isolation diffusion of the bipolar process. Thus, the packing density of the devices on a CMOS chip is somewhat lower than that of comparable circuits using only p- or n-channel devices.

REFERENCES

1. S. K. Ghandhi, *The Theory and Practice of Microelectronics*, Wiley, New York, 1968, Chaps. 13 and 14.

2. H. DeMan, "The Influence of Heavy Doping on the Emitter Efficiency of a Bipolar Transistor,"*IEEE Trans. Electron Dev.* ED-18, 833–835 (October 1971).

3. R. J. Whittier and D. A. Tremere,"Current Gain and Cutoff Frequency Falloff at High Currents,"*IEEE Trans. on Electron Dev.* ED-16, 39–57 (January 1969).

4. P. R. Gray and R. G. Meyer, *Analysis and Design of Analog Integrated Circuits*, Wiley, New York, 1977, Chap. 1.

5. B. A. McDonald,"Avalanche Degradation of h_{FE}," *IEEE Trans. on Electron Dev.* **ED-17**, 871–878 (October, 1970).

6. A. B. Grebene, *Analog Integrated Circuit Design*, VanNostrand Reinhold, New York, 1972, Chap. 11.

7. A. B. Grebene and S. K. Ghandi,"General Theory for Pinched Operation of the Junction-Gate FET,"*Solid State Electron.* 12, 573–589 (1969).

8. R. Cobbold, *Theory and Applications of Field-Effect Transistor*, Wiley, New York, 1970.

9. P. R. Gray and R. G. Meyer, *Analysis and Design of Analog Integrated Circuits*, Wiley, New York, 1977, pp. 46–53.

10. A. S. Grove, *Physics and Technology of Semiconductor Devices*, Wiley, New York, 1967, Chaps. 8, 9.

11. W. M. Penney, Ed. *MOS Integrated Circuits*, VanNostrand Reinhold, New York, 1972, Chap. 20.

12. F. Faggin and T. Klein, "Silicon-Gate Technology,"*Solid State Electron.* **13**, 1125 (1970).

CHAPTER THREE

PASSIVE COMPONENTS: DIODES, RESISTORS, CAPACITORS

Monolithic IC processes are designed and developed around active devices. In order to fabricate a number of active and passive devices simultaneously, with the same set of processes, a number of design and performance compromises have to be made. This, in turn, results in the limited sizes and kinds of passive components available in monolithic form. Certain passive components, such as inductors, are not compatible with microminiaturization. The resistors and capacitors are available only with poor absolute-value tolerances, and only over a relatively narrow range of values. The diodes are limited to those available from the *p-n* junctions used in basic active device structures.

In spite of the limitations on their sizes or absolute-value tolerances, passive components in integrated circuits still enjoy some of the basic advantages of monolithic structures, such as close matching and close thermal tracking. In many cases, by proper choice of the available components and imaginative circuit design approaches, it is possible to overcome the limitations of monolithic components. The circuit techniques used for this purpose are covered in the following chapters. However, before starting an actual design problem, it is imperative that the IC designer should know the basic choices of passive component types and values available. In going from a discrete circuit design to its integrated counterpart, the economics of design often become "inverted." In the discrete or nonintegrated circuits, cost and complexity can be related to the number of active devices needed. In the monolithic case, it is usually the requirements on the absolute values, or the sizes of the passive components, which can make the design a difficult or uneconomical one.

121

The purpose of this chapter is to familiarize the designer with the important features of passive components, namely, diodes, resistors, and capacitors. Whenever applicable, comparisons between different types and classes of components are presented in either graphic or tabular form for quick reference.

PART I: INTEGRATED DIODES

3.1.   JUNCTION DIODES

Any one of the semiconductor junctions forming the monolithic circuit structure can be used as a junction diode. Figure 3.1 shows the diodes associated with the basic integrated *npn* transistor. D_{BE} and D_{BC} represent the diodes formed by the base–emitter and the base–collector junctions, D_{SS} is the collector–substrate diode in junction-isolated circuits. The resistors r_b and r_{cs} represent the parasitic bulk resistances between the device terminals and the actual diode junctions. Also shown, by dashed lines, is the parasitic *pnp* transistor formed by the base and collector regions of the *npn* transistor and the *p*-type substrate which was described earlier (see Section 2.1).

The collector–substrate diode D_{SS} is not practical for circuit applications since its anode (i.e., the substrate) is common to the entire circuit and must be permanently reverse biased. The collector–base diode D_{BC} must be used with care. If it is forward biased, the parasitic *pnp* transistor is activated. However, the current gain of this transistor is reduced significantly by the presence of the n^+-type buried layer.

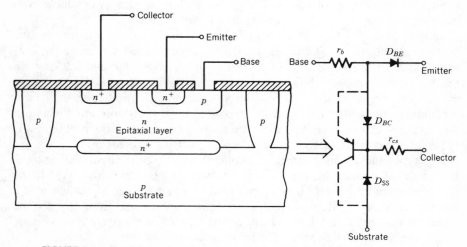

FIGURE 3.1.   Possible junction diodes available in integrated *npn* transistor structures.

The diode current I_D is exponentially related to the voltage V_D applied across the junction diode as

$$I_D = I_S \left[\exp\left(\frac{V_D}{V_T} - 1\right) \right]$$ (3.1)

where $V_T = kT/q$ is the thermal voltage. I_S is the reverse saturation current, and is proportional to the diode junction area.

Note that this equation is the same as that described earlier in connection with the transistor base–emitter characteristics [see Eq. (2.2).] For a typical integrated diode, the forward I_D and V_D characteristics predicted by Eq. (3.1) are valid over six orders of magnitude in current. For a 1-mil² base–emitter junction area, this range covers current values of 10 nA to 10 mA. At high current densities, due to high-level injection effects, the diode forward characteristics can be approximated as

$$I_D \approx \exp\left(\frac{V_D}{2V_T}\right)$$ (3.2)

The dynamic diode forward conductance g_d can be evaluated by differentiating Eq. (3.1) with respect to V_D, that is,

$$g_d = \frac{1}{r_d} = \frac{\partial I_D}{\partial V_D} = \frac{I_D}{V_T}$$ (3.3)

Similarly, the diode voltage drop V_D at any given current level can be written as

$$V_D = V_T \ln\left(\frac{I_D}{I_S}\right)$$ (3.4)

For forward-current levels greater than 10 μA, typical values of V_D are in the range of 0.5–0.7 V, neglecting bulk resistances.

The forward diode voltage V_D shows a strong but highly predictable dependence upon temperature. For the case of the base–emitter diode, which is the most commonly utilized diode connection, the temperature coefficient of V_D falls within the narrow range of -1.8 to -2.2 mV per °C. This temperature coefficient also shows a weak but predictable dependence on the current density through the diode, as illustrated in Figure 3.2. As a rule of thumb, a base–emitter diode with a 1-sq. mil (625 μm²) junction area, biased at $I_D = 60$ μA, corresponds to the -2.0-mV/°C drift point in the figure.

The semiconductor junctions which make up the *npn* bipolar transistor can be interconnected as a diode in any one of the five possible configurations listed below:

1. Base-emitter junction with the collector left open.
2. Base–emitter junction with the collector short-circuited to base.
3. Base–collector junction with the emitter open.

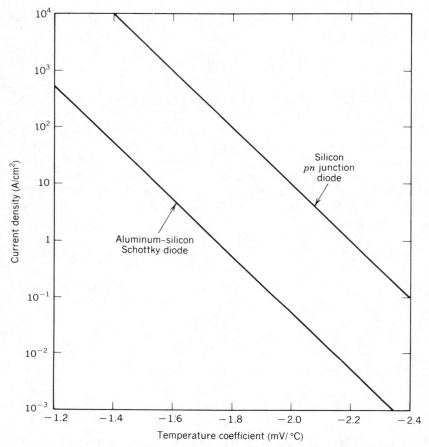

FIGURE 3.2. Temperature coefficient of forward diode voltage V_D for junction and Schottky diodes as a function of current density.[2]

4. Base–collector junction with the emitter short-circuited to base.
5. Base–emitter and base–collector junctions in parallel.

Figure 3.3 gives a relative comparison of each of these diode connections with respect to the parasitics and the breakdown characteristics associated with each configuration. The series bulk resistances associated with a given diode structure are, in general, the most significant parasitics. With respect to the bulk resistances, diode connection 2 has a distinct advantage over the other configurations since, in this connection, any parasitic resistance in the base or terminal of the device appears divided by β of the transistor.

All the diode-connected transistor configurations, which include the base–emitter junction, have low reverse breakdown voltages (typically 6–8 V). The diode configurations which utilize the collector–base junction under forward

Diode Connection	Series Resistance	Reverse Breakdown	Parasitic pnp Action	Storage Time ($I_d = 2$ mA)
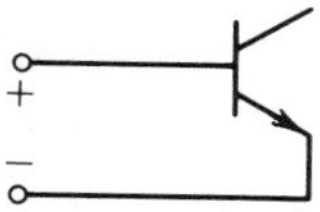	Low ($\approx r_b$)	Low (6–8 V)	No	High (≈ 60 nsec)
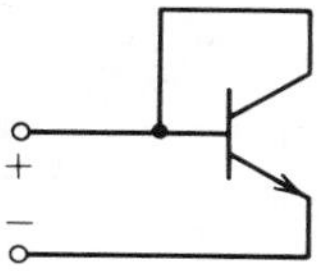	Low ($\approx r_b/\beta$)	Low (6–8 V)	No	Low (≈ 15 nsec)
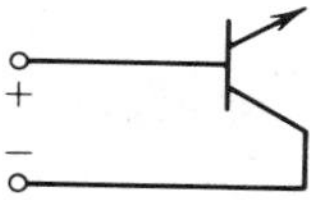	High ($\approx r_b + r_{cs}$)	High (>40 V)	Yes	High (≈ 80 nsec)
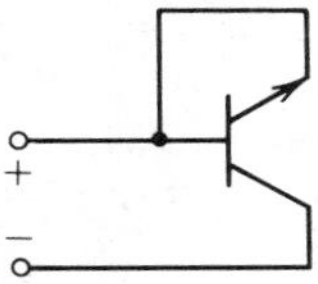	High ($\approx r_b + r_{cs}$)	High (>30 V)	Yes	High (≈ 50 nsec)
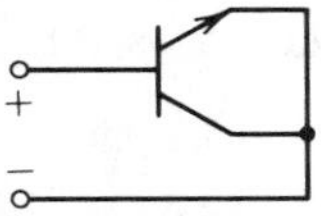	High ($\approx r_b + r_{cs}$)	Low (6–8 V)	Yes	High (≈ 100 nsec)

FIGURE 3.3. A comparison of practical diode connections for an *npn* transistor.

bias can have parasitic *pnp* action between the base, collector, and substrate regions (see Fig. 3.1)

Diode connection 2 in Figure 3.3 combines two desirable electrical properties: low series resistance and no parasitic *pnp* action to the substrate. Therefore, it is the most commonly used configuration for integrated circuits, as long as its low breakdown voltage does not present a problem.

Effect of Bulk Resistance. Figure 3.4 shows the actual equivalent circuit of the preferred base–emitter diode connection. The transistor is assumed to be ideal, and the parasitic resistances r_b and r_{cs} are identified separately. If the transistor is in its active region, the parasitic base resistance r_b appears reduced by the β of the transistor, and the collector series resistance r_{cs} is masked by the high dynamic impedance of the transistor collector. However, if the voltage drop

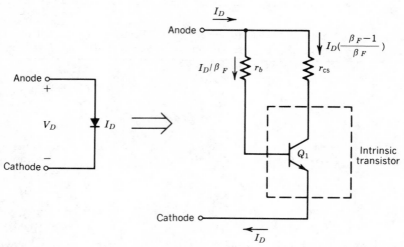

FIGURE 3.4. Effects of parasitic bulk resistances on diode-connected *npn* transistor. (Note that excessive voltage drop across r_{cs} can cause Q_1 to saturate.)

across r_{cs} is sufficiently high, that is, when $I_D r_{cs} \geq 0.5$ V, it may cause the transistor collector–base junction to be forward biased and Q_1 to go into saturation. When this happens, the diode characteristics deteriorate because β of the transistor is reduced, and the parasitic *pnp* transistor of Figure 3.1 is activated and causes a part of I_D to be shunted to the substrate. For example, in a small-area diode, with $r_{cs} \approx 100$ Ω, such a parasitic action may start at current levels as low as 5 mA.

Circuit Layout. Figure 3.5 shows the plane view and the crosssection of a diode-connected transistor. Note that this is basically the same as the small-signal *n-p-n* transistor structure of Figure 2.4, with the n^+-type collector contact region and the collector metal overlapping with the base contact. This is done to reduce the overall device geometry and the collector series resistance, since the collector–base breakdown voltage is of no consideration.

In many circuit applications, most diodes do not share a common anode; thus, they must be placed in separate isolation tubs. In some applications, several junction diodes are used with a common anode, but separate cathodes. In such cases, the multiple-emitter transistor structure of Figure 2.21 can be used, with its collector short-circuited to the base to reduce the chip area.

3.2. SCHOTTKY DIODES

When a metal and a semiconductor material are brought in contact, an electrostatic barrier is formed at the interface, which causes the metal–semiconductor interface to have rectifying properties. In the case of silicon, such a rectifying

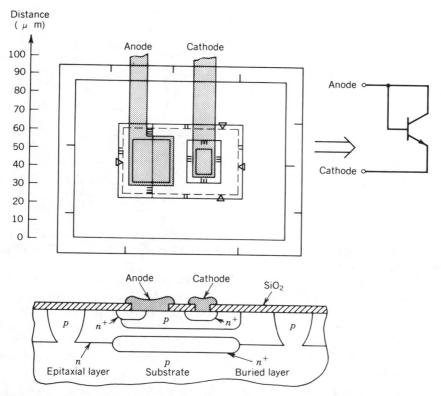

Distance
(μ m)

FIGURE 3.5. Lateral dimensions and device cross-section of a typical diode-connected *npn* transistor. (See Fig. 2.4 for coding of mask layers.)

junction, called a Schottky diode, is formed when a metal such as aluminum is placed in contact with a lightly doped n-type silicon ($N_D \leq 10^{17}$ atoms/cm^3).[1] In such a diode, the metal serves as the anode and the high-resistivity n-type semiconductor region serves as the cathode. Thus, the Schottky diode is forward biased (i.e., conducting) when its metal side is biased positive with respect to the semiconductor side. Schottky diodes can be fabricated with various combinations of metals, such as platinum–silicide (Pt$_5$Si$_2$) or aluminum.

The most commonly used Schottky diodes in analog IC design are the aluminum–silicon type diodes. These diodes can be fabricated simultaneously with ohmic contacts by using conventional aluminum evaporation or sputtering techniques. The nature of the resulting silicon–aluminum contact (i.e., ohmic or rectifying) is determined by the resistivity of the semiconductor region which is in direct contact with the metal. Thus, for example, a Schottky diode is formed by simply bringing aluminum in contact with a lightly doped ($N_D < 10^{17}$ atoms/cm^3) n-type epitaxial region, as shown in Figure 3.6. The polarity of the resulting diode is also indicated. In the simple aluminum–silicon Schottky diode structure of Figure 3.6, the aluminum layer is made to overlap the contact region

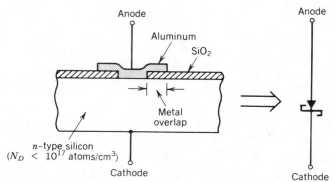

FIGURE 3.6.   Cross section of an aluminum–silicon Schottky diode.

by a small amount (typically $\approx$ 4–6 μm) to serve as a field plate and to reduce the reverse leakage current of the device along the device periphery. This effect is similar to that described in Figure 2.17.

Compared to conventional p-n junction diodes, Schottky diode characteristics have the following significant features:

1. In Schottky diodes, the current flow is by majority carriers, rather than by minority carrier diffusion. Thus, their switching speeds are not limited by the storage time delays and are much faster than those of junction diodes.

2. For a given diode area and the forward-current level, Schottky diodes have a much smaller forward-voltage drop V_F. (For a typical forward current of 10 μA, $V_F \approx 0.25$ V for a Schottky diode and ≈ 0.55 V for a comparable p-n junction diode.)

Because of these advantages, Schottky diodes find a very wide range of application in digital circuits. However, they are also useful in analog circuit applications, particularly in those applications requiring combined linear and digital functions, as a means of "clamping" the collector–base voltage of npn transistors to keep them from saturating.

Electrical Characteristics.   The forward-voltage drop V_F across a Schottky diode depends on the barrier potential ψ_n between the metal and the silicon. In the case of aluminum on n-type silicon, ψ_n is approximately equal to 0.69 V. The forward voltage V_F is related to the forward current I_F as[2]

$$V_F = \psi_n + V_T \ln\left(\frac{I_F}{I_o}\right) \tag{3.5}$$

where $V_T (= kT/q)$ is the thermal voltage. I_o is the reverse saturation current given as

$$I_o = RAT^2 \qquad\qquad (3.6)$$

where R = Richardson constant ($\approx$ 120 Amperes per cm^2 °K^2).

T = temperature (°K)

A = diode junction area in cm^2

The forward current–voltage characteristics for an aluminum–silicon Schottky diode, given by Eq. (3.5), are valid over approximately five orders of magnitude in current densities I_F/A of from 0.1 mA/cm^2 to 10 A/cm^2, neglecting the bulk resistance of the silicon.

The temperature coefficient of an aluminum–silicon Schottky diode decreases with increasing temperature, and can be approximated as

$$\frac{dV_F}{dT} \approx -1.76 + 0.086 \ln \left(\frac{I_F}{A}\right) \qquad\qquad (3.7)$$

where I_F/A is in A/cm^2. This temperature dependence of V_F is compared in Figure 3.2 with that of a comparable p-n diode.

The reverse breakdown of a Schottky diode varies inversely with the square root of the semiconductor impurity concentration. With a 2.5 Ω/cm n-type epitaxial layer, typical measured reverse breakdown values are in the range of 40–50 V.

Schottky Clamped Transistors. The most common application of Schottky diodes in IC design is the clamping of the n-p-n transistor base–collector junction to keep the transistor out of saturation and, thus, improve its switching speed by reducing the minority carrier charge storage in the base region. This is done by connecting a Schottky diode between the base and the collector regions of the transistor, with the anode (i.e., the metal part) connected to the base.

Figure 3.7 shows the layout and the cross section of such a Schottky clamped npn transistor structure. The Schottky diode is formed by simply making direct contact to the n-type epitaxial region with aluminum, without putting a heavily doped n^+ diffusion region at the metal–silicon interface. This is normally done by simply extending the base contact window into the n-type collector region, and covering it with an overlapping aluminum layer. As a result, the part of the aluminum making contact to the p-type base forms an ohmic contact, whereas the aluminum touching the n-type epitaxial layer forms a Schottky diode. The resulting transistor and diode combination is shown in Figure 3.7b, where D_S is the Schottky diode. Note that the normal collector and emitter contacts, which have the heavily doped n^+-type regions in contact with aluminum, are unaffected by this structure.

In the circuit application of the Schottky clamped npn transistor, as the collector voltage drops below that of the base, the Schottky diode D_S of Figure 3.7b starts conducting and cuts off the base current. With this configuration, the

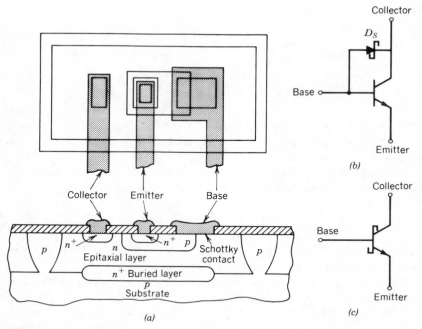

FIGURE 3.7. Schottky clamped npn transistor: (*a*) typical device layout; (*b*) electrical equivalent; (*c*) device symbol.

collector–base voltage cannot drop below $V_{BE} - V_F$, or approximately 0.4–0.45 V at room temperature.

In designing Schottky clamped transistors, a word of caution is in order. The forward voltage V_F is a function of the diode area. Thus, if the transistor to be clamped is operating at high collector currents, then the area of the Schottky clamp diode must be increased in order to maintain a low value of V_F.

Typical applications of Schottky clamped transistors are at the output stages of high-speed analog circuits, such as comparators or oscillators, where large signal swings are required with minimum switching delays.

3.3. ZENER DIODES

The avalanche breakdown of a reverse-biased *p-n* junction can often be used as a voltage reference or for dc level shift purposes. Although the actual breakdown mechanism of semiconductor junctions encountered in IC fabrication is *avalanche* breakdown, rather than the so-called *Zener breakdown*, it is customary to use the term Zener diode to describe this phenomenon.

As discussed earlier, the three basic junctions available in a monolithic integrated circuit are the emitter–base, the base–collector, and the collector–isolation junctions. The reverse breakdown voltage associated with the

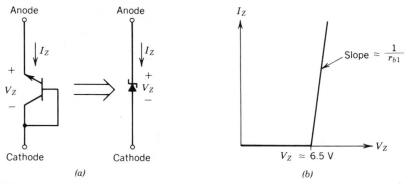

FIGURE 3.8. *npn* transistor base–emitter junction as a zener diode: (*a*) Circuit connection and device symbol; (*b*) current–voltage characteristics.

latter two of these junctions is usually in excess of the power supply voltage, which renders them useless for Zener diode applications. Thus, the base–emitter junction, which has a reverse breakdown voltage BV_{EBO} in the range of 6–8 V, is the most commonly used Zener diode connection. For this purpose, either one of diode configurations 1 and 2 of Figure 3.3 can be used. In general, diode configuration 2 is preferred, since this connection provides a predictable *dc* bias for the *npn* collector region, rather than letting it remain open-circuited, which in turn may lead to parasitic transistor action.

Figure 3.8 shows the basic transistor connection for Zener diode applications, along with its current–voltage characteristics. The breakdown voltage V_Z is equal to the base–emitter breakdown BV_{EBO}. For the case of devices fabricated with the basic bipolar IC processes, this breakdown voltage falls within the narrow range of 6–8 V. The breakdown resistance of the base–emitter Zener is approximately equal to the bulk base resistance r_{b1} of the *npn* transistor (see Fig. 2.14). The typical circuit layout for the Zener-connected *npn* transistor is the same as that of the diode-connected transistor of Figure 3.5, with the emitter biased *positive* with respect to the base region.

The junction breakdown voltage exhibits a temperature coefficient which is positive and increases with increasing breakdown voltage, as shown in Figure 3.9. At very low breakdown voltages, that is, below 5 V, where both sides of the junction are very heavily doped, the true Zener breakdown effect, namely, the carriers tunneling through the junction depletion layer, dominates, and the temperature coefficient of the breakdown changes sign.

The breakdown voltage of the base–emitter junction falls into the rather narrow voltage range of 6–8 V and, thus, exhibits a positive temperature coefficient of +2 to +4 mV/°C.

Temperature Compensation of Zener Diodes. Since the temperature coefficients of the diode forward voltage V_D and the base–emitter breakdown voltage BV_{EB} are of opposite polarity, it is possible to compensate partially the thermal

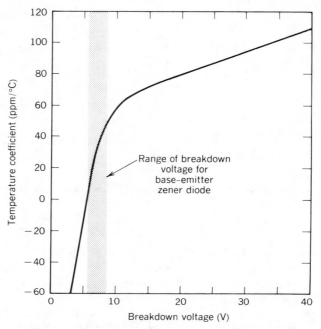

FIGURE 3.9. Temperature coefficient of breakdown voltage.

drift of an avalanche breakdown diode by connecting a forward-biased diode in series with it. The resultant composite diode has a breakdown voltage of $V_D + BV_{EB}$, with a significantly reduced temperature coefficient. In a monolithic circuit, this partial compensation can be obtained with a minimum increase of chip area by connecting two transistors back to back in diode connection 2, as shown in Figure 3.10. Since both transistors now have their collector and base

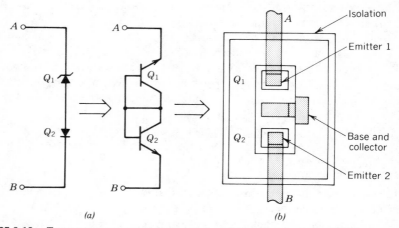

FIGURE 3.10. Temperature-compensated zener diode: (*a*) Circuit connection and (*b*) layout.

regions in common, they can be designed as a single transistor with two separate emitters, as shown in the figure.

Buried-Zener Structures. In the base–emitter junction, the breakdown takes place at the surface of the device, where the doping level of the junction is at a maximum. This makes the actual breakdown voltage difficult to predict and subject to long-term drift, due to surface irregularities and contamination. To avoid this problem, several Zener diode structures have been developed which confine the breakdown mechanism to *below* the surface of the device. Because the actual breakdown takes place in the subsurface region, these Zener diodes are often called *buried Zeners*.

Figure 3.11 shows the simplest buried-Zener structure compatible with bipolar IC fabrication steps. This is the so-called *isolation Zener*, which is formed by placing an n^+ emitter diffusion over a p-type isolation region, such that n^+ completely overlaps the isolation region. Such a Zener diode has its p side permanently grounded, since it is formed by the isolation of the p-type substrate common to the entire circuit. The typical breakdown voltage of the isolation Zener diode is of the order of 5.4 ± 0.4 V, depending on the impurity profile

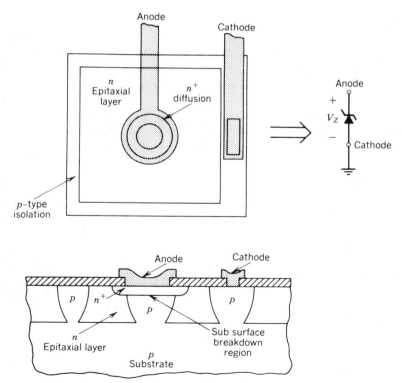

FIGURE 3.11. Layout and cross section of a buried-Zener structure using n^+ emitter and p^+ isolation diffusions. (*Note:* $V_Z \approx 5.4 \pm 0.4$ V.)

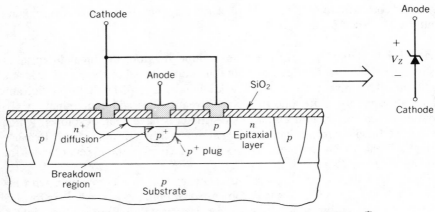

FIGURE 3.12.   Buried-zener structure using p^+ plug diffusion.[3]

of the isolation region. Its temperature coefficient is in the range of $+1$ to $+2$ mV/°C. Note that in the device layout of Figure 3.11, a circular geometry is used to avoid localized breakdown of the device at its corners. This layout technique is often used in designing Zener diodes whose breakdown voltages must be reasonably well controlled.

Figures 3.12 and 3.13 show alternate buried-Zener structures compatible with bipolar IC processing steps. In each case, a circular device layout is assumed. In the device structure of Figure 3.12, an additional deep p^+ diffusion, called p^+ *plug,* is used prior to the p-type base and the n^+ emitter diffusions. The n^+ diffusion completely covers the p^+ plug diffusion. The resulting diode breaks down where the dopant concentration is the greatest, that is, at the intersection of the p^+ and n^+ diffusions. The normal p-type base diffusion is then used to make contact to the p^+-type region (i.e., the cathode) of the Zener diode. Surface

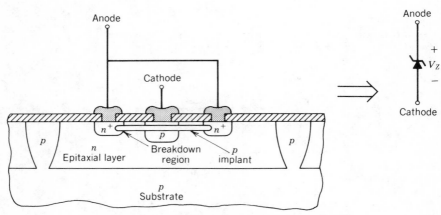

FIGURE 3.13.   Ion-implanted buried-zener structure.[4]

breakdown between the p-type base and the n^+ emitter diffusions does not occur since this breakdown voltage is higher than the p^+–n^+ junction breakdown. This type of buried-Zener structure exhibits a typical breakdown voltage of 6.3 V, with a production tolerance of $\pm$ 200 mV.[3]

An alternate buried-Zener structure is the ion-implanted subsurface Zener diode of Figure 3.13. In this case, first the p-type and the n^+-type regions are diffused during the base and the emitter diffusion cycles. Then a heavily doped p^+-type layer is implanted over the area indicated in the figure. This p^+ implant overlaps the p^+-type base diffusion and touches the n^+ ring surrounding it. Since the implanted p^+-type layer has its maximum impurity concentration *below* the surface, the breakdown occurs along the n^+–p^+ junction below the device surface, with the n^+ and p diffusions serving as the anode and the cathode contacts for the device.[4] With this technique, it is possible to control the absolute value of the breakdown voltage to within $\pm$ 100 mV under production environment.

PART II: INTEGRATED RESISTORS

Two general classes of resistors are available in monolithic circuits: semiconductor and thin film. The semiconductor resistors in turn can be categorized into the following four groups, based on their physical structure:

1. Diffused resistors.
2. Pinched resistors.
3. Epitaxial resistors.
4. Ion-implanted resistors.

Semiconductor resistors are, by far, the most commonly used monolithic resistor structures. With the exception of ion-implanted resistors, they can be fabricated simultaneously with the rest of the circuit elements, without requiring extra processing steps. However, they are, in general, nonideal circuit components, with rather loose tolerances, and by discrete component standards, they have poor temperature and frequency characteristics. The thin-film resistors, on the other hand, offer superior electrical characteristics at the expense of additional processing steps and may be "trimmed," or adjusted, to a specific tolerance after the deposition step.

The electrical parameter that is most often used in characterizing integrated resistors is the *sheet resistance*. In the case of a uniform sample of resistive material having length L, width W, and thickness T, as shown in Figure 3.14, the total resistance R of the resistor can be written as

$$R = \frac{\rho L}{TW} \tag{3.8}$$

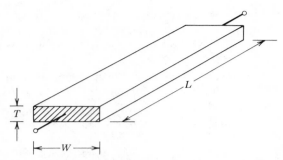

FIGURE 3.14. Calculation of sheet resistance for a uniform sample of resistive material.

where ρ(ohm/cm) is the resistivity of the material. For actual design and layout considerations, it is more convenient to define the parameter R_S (ohms), called *sheet resistance*.

$$R_S = \frac{\rho}{T} \qquad (3.9)$$

In physical terms, the sheet resistance corresponds to the resistance of one "unit square" of the material with $L = W$. For this reason, the value of R_S is often stated in ohms per square ($\Omega/\square$). Once the value of sheet resistance is given, the total value of the resistor is determined only by its length-to-width ratio, namely,

$$R = R_S \frac{L}{W} \qquad (3.10)$$

For example, if one were to make a 1-kΩ resistor using a material with $R_S = 100 \ \Omega/\square$, one would design it to have a length-to-width ratio of 10 between the end contacts.

3.4. DIFFUSED RESISTORS

A diffused resistor structure is formed by the bulk resistance of a diffused semiconductor region. In forming a monolithic resistor, either of the two basic diffusion cycles, that is, the base or the emitter diffusion, can be utilized. The p-type diffused resistor, which is fabricated using the base diffusion cycle of the *npn* transistor, is the most commonly used resistor structure.

Figure 3.15 shows a typical geometry and the cross section of a p-type diffused resistor obtained using the *npn* base-diffusion step. In circuit applications of such a resistor structure, it is necessary that the p-n junction outlining the resistor should, at all times, be reverse biased to confine the current flow to the volume of silicon defined by the resistor geometry. The total resistor value R is given as

$$R = \frac{\bar{\rho}}{x_j} \frac{L}{W} = R_S \frac{L}{W} \qquad (3.11)$$

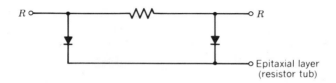

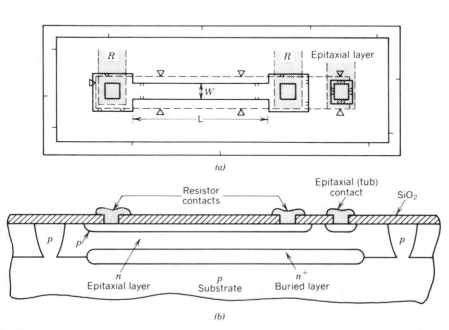

FIGURE 3.15. p-Type base-diffused resistor (see Fig. 2.4 for coding of mask layers): (*a*) Typical device layout; (*b*) Cross section.

where x_j is the junction depth of the diffused resistor, and $\bar{\rho}$ is the *average* resistivity of the diffused layer. In a diffused resistor, due to nonuniform impurity distribution, the average resistivity $\bar{\rho}$ is difficult to calculate analytically. In the literature, detailed tables of $\bar{\rho}$ are available for various impurity profiles and background concentrations.[5]

For the case of base-diffusion profiles commonly used in analog IC fabrication, the available values of R_S are typically in the range of 60–250 $\Omega/\square$, with the most common range of values being between 130 and 200 $\Omega/\square$.

The sheet resistance R_S has a positive temperature coefficient due to the decrease of carrier mobility with temperature. Figure 3.16 shows the typical temperature dependence of R_S for a p-type diffused resistor. As will be described in later chapters, this strong temperature dependence of R_S is one of the dominant thermal drift sources in a monolithic circuit. At high current levels, localized self-heating of the resistor can cause the current–voltage characteristics to be nonlinear. To avoid this effect, it is good design practice to limit the power

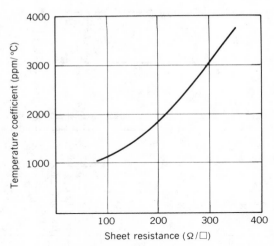

FIGURE 3.16. Typical temperature coefficient of p-type diffused resistor as a function of sheet resistance.

dissipation of a diffused resistor to less than 3 mW per square mil of junction area.

Side-Diffusion Effects

In calculating the value of a diffused resistor from its mask dimensions L and W, it is also necessary to take into account the side diffusion of impurities under the oxide window of width W. Figure 3.17 shows the cross section of a diffused resistor taken transverse to the direction of current flow. Note that the effective cross-sectional area of the resistor is *increased* due to the side diffusion of the impurities under the oxide window, shown as the shaded areas in the figure. The effect of side diffusion can be taken into consideration by defining an *effective width* W_{eff} for the resistor, instead of the actual diffusion window width W where $W_{eff} \geq W$, when calculating the resistor length. Figure 3.18 gives the approxi-

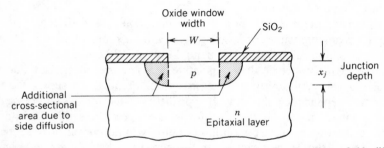

FIGURE 3.17. Transverse cross section of diffused resistor, showing the effects of side diffusion.

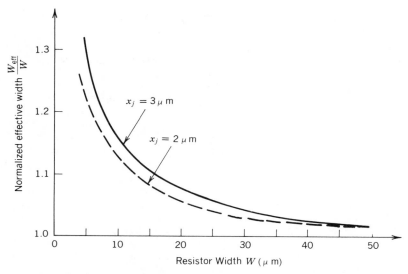

FIGURE 3.18.   Effective width of a resistor, including side-diffusion effects.

mate value of W_{eff} as a function of the window width. Note that the side-diffusion effects are particularly significant for narrow resistor widths (i.e., $W \leq 10\ \mu$m) and deeper junctions.

Resistor Layout and Tolerances

To obtain a given resistor L/W ratio, with efficient use of the chip area, often requires that the resistor follow a folded or zigzag structure. This can be achieved by making square or round bends in the lateral layout of the resistor, as shown in Figure 3.19. The equivalent values of the L/W ratios for such bends (measured between lines A and B) are also shown. The inner separation in the resistor fold, shown as the distance k in the figure, is set by the side-diffusion effects (see Fig. 3.17) and is normally no less than 10 μ. The lower limit of resistor width W is set by the masking tolerances and is approximately 5–7 μm for production circuits. The best matching of resistor values is obtained for the values of W in excess of 20 μm, where the effects of side diffusion or masking irregularities are reduced.

The end points of diffused resistors are normally widened to make sufficient room for ohmic contacts. This often results in an irregular resistor shape near the contact area. Figure 3.20 shows some of the typical diffused resistor endings and their contribution to the total resistor value in terms of *equivalent squares.* [6]

The absolute-value tolerance of diffused resistors is dependent on two factors: (1) control of sheet resistivity and (2) masking tolerances and irregularities. The typical range of sheet resistivity tolerance is approximately ± 20% in the production environment. The masking tolerance is primarily due to the *edge*

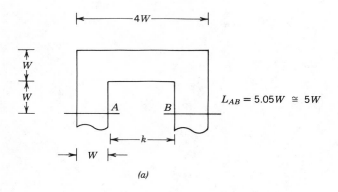

$$L_{AB} = 5.05W \cong 5W$$

(a)

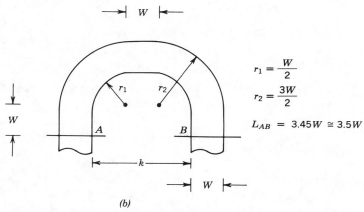

$$r_1 = \frac{W}{2}$$

$$r_2 = \frac{3W}{2}$$

$$L_{AB} = 3.45W \cong 3.5W$$

(b)

FIGURE 3.19. Calculation of effective resistor length around square and round corners.

definition of the resistor pattern, and is typically of the order of $\pm$ 0.5 μm for each edge of the resistor. Table 3.1 gives a summary of some of the absolute-value tolerances associated with various resistor widths. Note that for resistor widths beyond 5 μm, control of the sheet resistance is the dominant factor in determining the absolute-value tolerance.

The matching of diffused resistors, and the tracking of component values over temperature, are excellent for identical devices located side by side on the chip layout. The matching characteristics are primarily limited by random mask irregularities; thus, they improve as the resistor width is increased.

In practice, the typical range of resistor values which can be obtained from a sheet resistance R_S is

$$0.1 \, R_S \leq R \leq 10^3 R_S \tag{3.12}$$

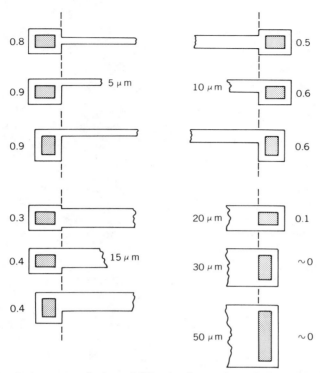

FIGURE 3.20. Resistance contributions of diffused resistor endings. (Units in equivalent squares.)[6]

TABLE 3.1.   Electrical Characteristics of _p_-Type Diffused Resistors

Sheet resistivity range	100–200 $\Omega/\square$
Temperature coefficient	+1500– +2000 ppm/°C
Absolute-value tolerance[a]	
5-μm wide	± 30%
10-μm wide	± 22%
50-μm wide	± 20%
Matching tolerance of identical resistors[a]	
5-μm wide	± 3%
10-μm wide	± 1.2%
25-μm wide	± 0.8%
50-μm wide	± 0.2%

[a]In tolerance distributions, sigma limit is assumed.

In the lower end of this range, the practical limit is set by the contact resistances associated with the resistor contact windows. The practical limit at the upper end is set by the available chip area.

n^+ Diffused Resistors

A low-value resistor can be obtained by using the heavily doped n^+ emitter diffusion of an *npn* bipolar transistor. Figures 3.21 and 3.22 show two such resistor structures commonly used in analog IC design. The sheet resistance R_S of n^+ diffusion is normally in the range of 2–10 $\Omega/\square$, with a typical temperature coefficient of +2000 ppm/°C.

The key advantage of n^+ resistors is the low value of R_S, which makes them feasible for the design of resistors in the range of 1–100 Ω. Such low-value resistors are useful for output short-circuit protection or overload shut-down circuitry in amplifiers and regulators. They also find application as crossunders, or *underpasses,* in circuit interconnections to simplify the layout of the integrated circuit interconnection pattern.

The basic n^+ resistor structure of Figure 3.21 is obtained by putting an n^+ diffusion layer on top of the n-type epitaxial region. Since the epitaxial layer is of much higher resistivity, the resistance between the resistor terminals is almost totally determined by the n^+ diffused layer. The advantage of this device structure is its high breakdown voltage, which is equal to the collector–isolation breakdown. Its disadvantage is that each n^+ resistor would require a separate isolation pocket.

The n^+ resistor structure of Figure 3.22 is imbedded within the p-type base-diffused island. In such a resistor structure, it is necessary that the p-type island be reverse biased with respect to the n^+-type region at all times. It is also

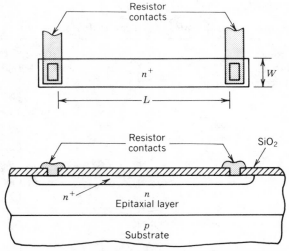

FIGURE 3.21. Top view and cross section of n^+ diffused resistor in epitaxial tub.

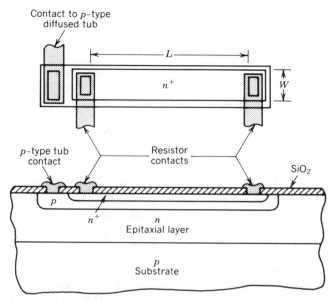

FIGURE 3.22. Top view and cross section of n^+ diffused resistor in p-type diffused tub.

important that the p-type island be either short-circuited or reverse biased relative to the n-type epitaxial region to avoid parasitic *npn* transistor action. The advantage of the n^+ resistor structure of Figure 3.22 is that several such resistors can be placed in the same isolation tub or in the same p-type island and, thus, save on the chip area. Its disadvantage is the low breakdown voltage ($\approx$6.5 V) between the n^+ resistor and the p-type base diffusion.

Frequency Response

The diffused resistor structure has a distributed capacitance associated with it, due to the reverse-biased p-n junction which surrounds it. Figure 3.23 shows the high-frequency equivalent circuit of a diffused resistor of total value R_1, which has a distributed capacitance C_1 associated with it. In the case of p-type diffused resistors formed by base diffusion, this capacitance is equivalent to C_{CO} of Table

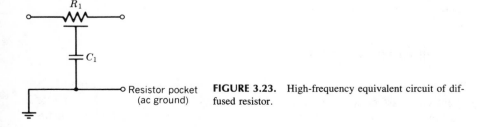

FIGURE 3.23. High-frequency equivalent circuit of diffused resistor.

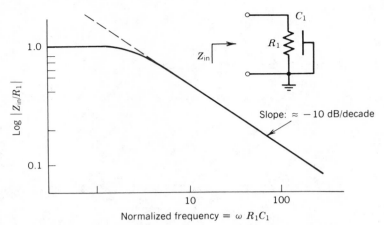

FIGURE 3.24. Normalized frequency response of diffused resistor.

2.2 per unit area of the junction, calculated under appropriate bias conditions. In the case of n^+ diffused resistors fabricated with the emitter diffusion, the junction capacitance per unit area is equal to C_{JEO} for the device structure of Figure 3.21, and C_{CSO} for the device of Figure 3.22, calculated under appropriate reverse-bias conditions.

At high frequencies, the net effect of the distributed capacitance C_1 is to shunt the ac signal to ground and cause excessive phase lag. Figure 3.24 shows the typical frequency response of a diffused resistor in terms of the magnitude of its driving-point impedance. Assuming that the capacitance C_1 is uniformly distributed along R_1, one can show that the magnitude of the driving-point impedance Z_{in} is down by approximately 3 dB at a frequency f_1 given as

$$f_1 \approx \frac{1}{3R_1C_1} \tag{3.13}$$

Replacing the total resistor and capacitor values by the sheet resistance R_S and the capacitance C_o per unit area of the junction, Eq. (3.13) can be written as

$$f_1 \approx \frac{1}{3R_SC_oL^2} \tag{3.14}$$

Note that f_1 decreases as the square of the resistor length. For a typical 10-kΩ diffused resistor structure ($R_S = 200\,\Omega$, $L = 50$ mil, $W = 0.5$ mil), f_1 is approximately 10 MHz.

3.5. PINCHED RESISTORS

The sheet resistivity of a semiconductor region can be increased by reducing its effective cross-sectional area. In a pinched resistor structure, this technique is used to obtain a high-value of sheet resistance from the ordinary base-diffused

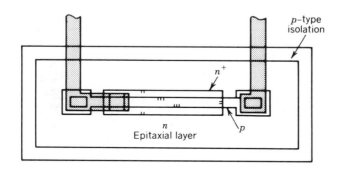

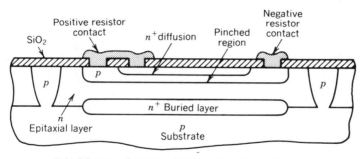

FIGURE 3.25. Layout and cross section of pinched resistor.

resistor. Figure 3.25 shows such a resistor structure formed by placing an n^+ emitter diffusion over the p-type diffused resistor. The emitter diffusion greatly reduces the effective cross-sectional area of the p-type resistor and consequently raises its sheet resistivity. Since the effective area of the resistor is pinched by the n^+ diffusion, such a resistor structure is called a *pinched resistor*.

The sheet resistance R_S of a pinched resistor can be expressed as

$$R_S = \frac{\rho}{x_b - x_e} = \frac{\rho}{W_B} \qquad (3.15)$$

where x_b and x_e are the depths of the base and emitter diffusions. The average resistivity $\bar{\rho}$ for the effective cross section of the resistor can be calculated from the knowledge of the diffusion profiles. It should be noted that the effective depth of such a pinched resistor structure is equal to W_B, the npn transistor base width. Typical values of R_S of a pinched resistor are in the range of 2–10 k$\Omega/\square$.

It is important that the surrounding epitaxial region be biased within the breakdown region of the resistor. Therefore, it is very rarely possible to place a base–emitter pinched resistor in an island common with other resistors.

Figure 3.26 shows the circuit designation and the current–voltage characteristics for a pinched resistor. Normally, the n^- and n^+-type regions surrounding the pinched resistor are short-circuited and connected to the resistor terminal, which is maintained at a more positive potential than the rest of the resistor structure. Such a connection, as shown in Figure 3.26a, ensures that all the junctions forming the resistor are reverse biased. The current–voltage characteristics of the pinched resistor are linear only for small voltage drops across the resistor. In this range of operation, the device behaves as a linear resistor, with the sheet resistivity R_S given by Eq. (3.15). Application of a higher dc voltage results in an increase of the reverse bias between the p-type resistor body and the surrounding n-type island. This reverse bias between causes the junction depletion layer to extend into the resistor and pinch the effective resistor cross section, in a manner similar to the case of a FET. Consequently, for increasing

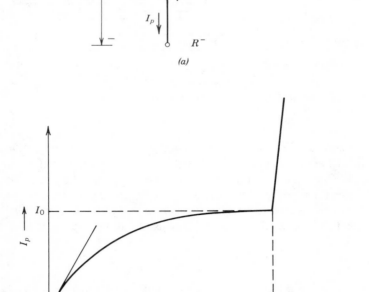

FIGURE 3.26. Pinched resistor. (a) Electrical symbol. (b) Current–voltage characteristics.

voltages, the pinched resistor current I_p tends to saturate to a constant value I_o, as shown in Figure 3.26b. Since the top portion of the pinched resistor is comprised of the heavily doped emitter diffusion, the resistor exhibits a low breakdown voltage, equal to the transistor emitter–base breakdown BV_{EBO} (typically 6–8 V).

The sheet resistance of the pinched resistor structure exhibits a strong temperature coefficient of the order of $+3000$ to$+5000$ ppm/°C. The saturation current I_o is also strongly temperature dependent. Its temperature coefficient is almost identical to that of the sheet resistance, but opposite in polarity. Absolute values of the pinched resistors are difficult to control in fabrication. Typical absolute-value tolerances associated with R_S and I_o are in the range of -50 to $+80\%$. However, the matching and tracking of identical pinched resistor structures on the same chip can be held to within $\pm 6\%$. Since the effective thickness of the pinched resistor is the same as the base width of the npn transistor, the variations in the absolute values of the pinched resistor tend to track the transistor β variations.

When using pinched resistors, a word of caution is in order. The pinched resistor is a nonlinear element. Thus, tapping a pinched resistor will *not* produce a voltage ratio proportional to the geometric aspect ratio, since the two halves of the resistor will not be biased equally.

3.6. EPITAXIAL RESISTORS

The bulk resistance of the n-type epitaxial layer can be used in some applications to form a noncritical high-value resistor. This can be done by using a structure as shown in Figure 3.27. The resulting structure, known as an *epitaxial resistor*. Its sheet resistance can be expressed as:

$$R_S = \frac{\rho_e}{d} \tag{3.16}$$

where ρ_e is the resistivity, and d is the thickness of the epitaxial layer. For a 5 Ω cm epitaxial layer of 10-μm thickness, this results in a sheet resistivity of approximately 5 kΩ/□.

Since the epitaxial resistor is formed by the epi–isolation junction, its breakdown voltage is also substantially higher than that of the diffused resistor structure. The sidewalls of the epitaxial resistor are formed by the deep isolation diffusion which diffuses sideways as well as downward during the diffusion cycles. When calculating the effective cross section of the epitaxial resistor, the side-diffusion effects should be taken into consideration. Figure 3.28 shows the transverse cross section of a 30-μm wide epitaxial resistor, fabricated on a 10-μm thich epitaxial layer, to indicate the significance of the isolation side diffusion on the effective cross section of the resistor. Note that the actual cross section is approximately 35% of the ideal cross section implied by the isolation mask pattern.

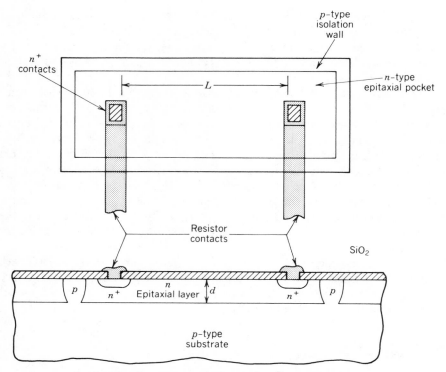

FIGURE 3.27. Layout and cross section of epitaxial resistor.

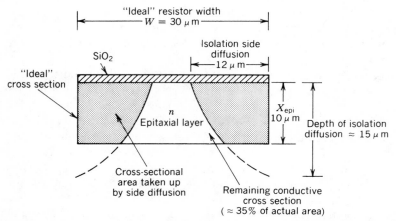

FIGURE 3.28. Transverse cross section of epitaxial resistor showing the effects of isolation side diffusion.

The temperature coefficient of the epitaxial resistor is significantly higher than that of the diffused resistor, due to the strong temperature dependence of mobility at low impurity concentrations. Typical values of the bulk resistor temperature coefficients are in the range of $+3500$ to $+5000$ ppm/°C for epitaxial layers of 1–5Ω cm.

The absolute-value tolerances for an epitaxial resistor are quite loose, due to relatively poor control of the epitaxial resistivity ($\pm 20\%$) and the epitaxial layer thickness ($\pm 10\%$). Thus, the practical absolute-value tolerance for a bulk resistor is typically $\pm$ 30% for resistor widths of 75 μm or more. This tolerance degrades further for narrower resistor structures, due to the side-diffusion effects of the isolation wall illustrated in Figure 3.28.

Epitaxial Pinched Resistors

The conductive cross section of an epitaxial resistor can be reduced further by providing a "blanket" of p-type base diffusion over it. The resulting device structure, which is called an epitaxial pinched resistor, is shown in Figure 3.29.

In the case of the epitaxial pinched resistor, the body of the resistor is formed by the n-type epitaxial region surrounded on all sides by the p-type substrate,

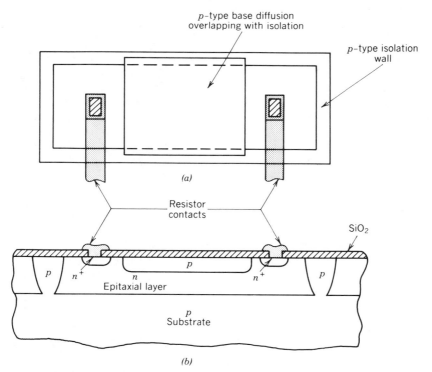

FIGURE 3.29. Epitaxial pinched resistor: (a) Device layout; (b) cross section.

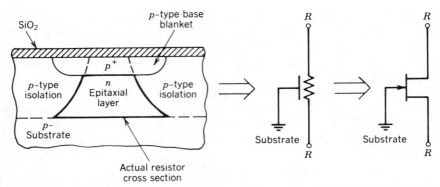

FIGURE 3.30. Effective cross section and electrical equivalent of epitaxial pinched resistor.

isolation, and base diffusion. The sheet resistivity of the resulting device can be expressed as

$$R_S = \frac{\rho_e}{d - x_b} \tag{3.17}$$

where x_b is the depth of the base diffusion, and ρ_e and d are the resistivity and the thickness of the epitaxial layer. For resistivity ranges available in analog circuits, the values of R_S for an epitaxial pinched resistor are in the range of 4–8 kΩ with a temperature coefficient of approximately +4000 ppm/°C.

The epitaxial pinched resistor provides an alternate method of obtaining high-value resistors, without the breakdown limitation of the base-diffused pinched resistor. Its breakdown voltage is basically equal to the base–collector breakdown BV_{CBO} of the n-p-n transistor. The absolute-value and matching tolerances associated with the epitaxial pinched resistor are somewhat poorer than those of the epitaxial unpinched resistor; they are comparable to those of the basic base-diffused pinched resistor.

Figure 3.30 shows the effective cross section of an epitaxial pinched resistor and its electrical equivalent. Note that similar to the case of the normal epitaxial resistor, the epitaxial pinched resistor cross-sectional area is also strongly dependent on the side diffusion of the isolation walls. It is basically an n-channel JFET structure, with the isolation walls, p-type substrate and the p-type base diffusion all together serving as the gate region. Thus, it exhibits nonlinear current–voltage characteristics, similar to an n-channel JFET, with a pinch-off voltage V_p in the range of 25–40 V, depending on the epitaxial layer resistivity and thickness.

3.7. ION-IMPLANTED RESISTORS

Ion-implantation techniques can be utilized to form resistor structures on the semiconductor surface. The fundamental principles of ion-implantation technology were briefly described in Chapter 1 (see Section 1.7) and are well covered in the literature.[7] With this technique the impurities are introduced into the silicon lattice by bombarding the wafer surface with high-energy ions.

The implanted ions lie within a very shallow layer (typically on the order of 0.1–0.8 μm) along the silicon surface. Thus, for similar doping levels, the implanted layers yield a sheet resistivity which is roughly 10–20 times higher than a correspondingly doped diffused layer of 2–4-μm thickness.

In fabricating ion-implanted resistors, normally boron-implanted p-type resistor structures are used. For defining resistor geometries, either a thick oxide or an unremoved photoresist layer can be used as a mask. The sheet resistance of the implanted resistor is *inversely* proportional to the implantation dose, that is, the total number of boron atoms implanted per unit area of the resistor surface.

Figure 3.31 shows the planar layout and the cross section of a p-type ion-implanted resistor. Due to the shallow depth of the ion-implanted resistor, it is difficult to obtain a good ohmic contact to the implanted region. Therefore, p-type diffused beds are used at the contact areas of the resistor, as shown. Unlike diffused resistors, ion-implanted resistors are not influenced by the side-diffusion effects. Thus, the true length and width of an implanted resistor is defined accurately by the window size of the ion-implantation mask.

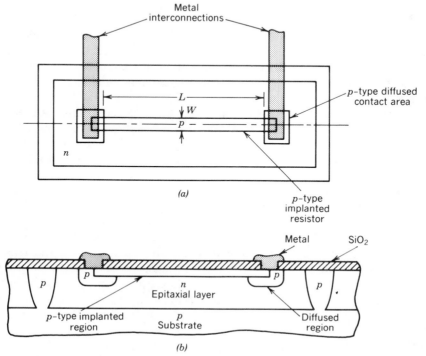

FIGURE 3.31.   Lateral geometry and cross section of p-type ion-implanted resistor.

The final value of the sheet resistance is also affected by the post implant annealing time and temperature, since this heat treatment determines the electrical activity of the implanted dopant atoms. By proper choice of the implantation dose and annealing temperatures, sheet resistivity ranges of $0.1–20$ k$\Omega/\square$ can be obtained. Both the implantation dose and the post implant anneal cycles can be very accurately controlled to approximately $\pm 5\%$, and a uniformity or matching tolerance of $\pm 1\%$ can be obtained across the wafer for resistor widths ≥ 25 μm. The main contributor to the matching error is, in general, the edge definition of the masking step, rather than the uniformity of the ion implant.

After the implantation step, the implanted impurities can be either partially or totally activated by the post implant heat treatment. The choice of this step profoundly affects the resistor characteristics, as described below:

1. *Partial Activation.* In this process, the wafer is annealed at a low temperature (typically 10–20 min at 500–600°C) and only a small fraction (i.e., 50% or less) of the implanted impurities are electrically activated. Thus, for a given implant dose, one can obtain a range of R_S values by simply controlling the annealing time and temperature.

2. *Total Activation.* In this process, the wafer is heat treated at an elevated temperature (typically 10–20 min at 900°C) to activate virtually all of the implanted impurities, and only the implant dose is used to control the value of R_S.

The advantage of partial activation is that it results in a lower temperature coefficient of R_S, typically in the range of $+100–+600$ ppm/°C. However, its disadvantage is poor thermal noise characteristics of the resistor and poor long-term stability. For this reason, in most high-volume production processes, the total activation technique is preferred.

The practical values of R_S obtained by the total activation process is in the $1–10$ k$\Omega/\square$ range, with the most commonly used range of values being $2–4$ k$\Omega/\square$. The temperature coefficient of fully activated ion-implanted resistors is relatively high, typically on the order of 1500–4000 ppm/°C, depending on the value of R_S, as shown in Figure 3.32.

The implanted resistors may also exhibit JFET-like characteristics, that is, become nonlinear, as the voltage across the resistor is increased, due to the pinching off of the resistor cross section by the depletion layer of the *p-n* junction around the resistor. Thus, the resistor may exhibit a voltage dependence and appear to have a higher value than nominal, as the voltage across it increases.

Figure 3.33 shows the typical voltage dependence of a 1.5-k$\Omega/\square$ *p*-type implanted resistor, as a function of applied voltage, for various values of resistor width W. Note that the voltage dependence is much more pronounced for narrow resistors (i.e., $W \leq 15$ μm) than for wider ones, due to the additional pinch-off effect along the resistor edge.

The voltage dependence is also a very strong function of the implant dose, as shown in Figure 3.34. For low-implant doses of 3×10^{13} atoms/cm^2 or less,

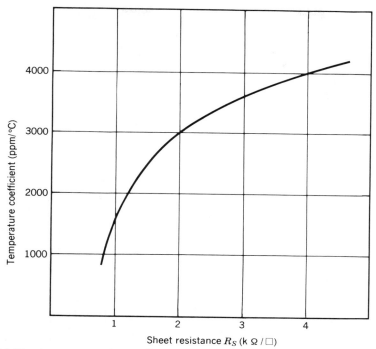

FIGURE 3.32. Average temperature coefficient of p-type ion-implanted resistor as a function of sheet resistance. (All impurities are activated.)

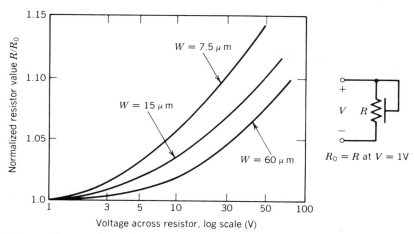

FIGURE 3.33. Voltage dependence of resistor value for boron-implanted 1.5-kΩ/$\square$ resistor as a function of resistor width.

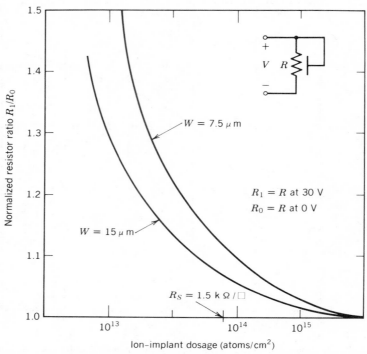

FIGURE 3.34. Dependence of voltage coefficient of resistance on ion-implant dosage for 100-keV boron implant. (All impurities are fully activated.)

which corresponds to $R_S \geq 5 \text{ k}\Omega/\square$, this voltage dependence may become very strong and may lead to complete pinching off of the resistor, causing it to behave as a constant-current source.

3.8. THIN-FILM RESISTORS

Using the film deposition techniques in Chapter 1, resistive thin-film layers can be deposited and patterned on the silicon surface. Compared with the ordinary base-diffused resistors, thin films offer the following advantages:

1. Low temperature coefficient.
2. Tighter absolute-value control by additional trimming steps.
3. Wide choice of sheet resistances.
4. Low stray capacitance.

The main disadvantage of thin-film resistors is the additional process steps required in their fabrication. In some cases, in addition to the basic deposition

TABLE 3.2. Typical Characteristics of Thin-Film Resistors

Resistor Type	R_S ($\Omega/\square$)	Temperature Coefficient (ppm/°C)	Absolute Tolerance without Trim[a]	Matching Tolerance (25-μm wide)[a]
Ta	10–1000	±100	±5%	±1%
Ni-Cr	40–400	±100	±5%	±1%
SnO$_2$	80–4000	0–1500	±8%	±2%
CrSiO	30–2500	±50– ±150	±10%	±2%

[a]Sigma limit is assumed.

and patterning steps, an additional SiO$_2$ deposition is necessary to stabilize the resistor structure by sealing it off from an ambient atmosphere.

The most commonly used thin-film resistors in integrated circuits are tantalum, nickel–chromium (Ni–Cr), cermet (CrSiO), and tin oxide (SnO$_2$). Table 3.2 gives a summary of the basic properties of each of these thin-film resistors. Of the four listed in the table, Ta and Ni–Cr films are by far the most widely used.

In the circuit layout of thin-film resistors, zigzag geometries with sharp corners are avoided since these etch poorly during the resistor masking step. Instead, a circular fold, as shown in Figure 3.19b, is preferred. An alternative way of eliminating sharp corners in thin-film resistors is to layout a number of parallel strips which can then be connected in series by aluminum interconnections at each end of the strips. The thin-film resistors should also be laid out over a smooth region of the surface oxide layer, with no steps or sudden changes of the oxide layer.

3.9. TRIMMING OF RESISTORS

In a number of applications for precision analog circuits, such as converters or precision voltage references, the absolute-value tolerances associated with semiconductor or thin-film resistors are not acceptable. In those cases, additional steps must be taken to "trim" the resistor, after its fabrication, to meet a given absolute-value tolerance. Some of these methods will be outlined in this section.

Adjustment and Trimming of Semiconductor Resistors

The semiconductor resistors, such as the diffused, pinched, epitaxial, or ion-implanted types discussed earlier, cannot be adjusted, once fabricated. One possible exception to this is the case of partially activated ion-implanted resistors which can have their values lowered by additional heat treatment.

A permanent design change, or iteration, in the value of a diffused resistor requires the change and the retooling of one or more mask layers. In such cases,

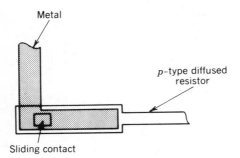

Metal

p-type diffused
resistor

Sliding contact

FIGURE 3.35. Resistor end section layout with sliding contact option.

where a possible change is anticipated, the so-called *sliding contact* method illustrated in Figure 3.35 can be used. This is done by providing a large end-area for the resistor, which is then covered by a metal layer. Then if a design change is needed, the value of the resistor can be changed by sliding, or relocating, the contact window. Such a sliding contact method requires only one mask change, and does not affect the diffusion-related mask layers.

Although the semiconductor resistors cannot be directly adjusted, their values can be indirectly trimmed by open- or short-circuiting various prearranged taps on the resistor, usually at the wafer probing stage. This can be done by any of the following methods.

Zener Zapping.   This method uses the selective short-circuiting of base–emitter Zener diodes across the resistor taps. When a large current, typically on the order of 200–300 mA, is passed through a small-area Zener diode, the localized power dissipation at the junction reaches such a level that the junction is permanently destroyed. In addition, the metal in the contact window melts and gets swept under the oxide layer and across the silicon surface over the junction. This narrow link of metal fuses into silicon and provides a permanent short circuit across the Zener diode, as shown in Figure 3.36. In the industry jargon, this process is called *Zener zapping,* and can be used for providing a short circuit across one or more of the prearranged resistor taps, as shown in Figure 3.37.[8] The zapping of the selected Zener diode is done by applying a current and voltage pulse across the selected trim pads during the wafer probing stage. The unzapped Zener diodes remain as open circuits, providing that the voltage drop across their shunting resistor tabs is below the Zener breakdown voltage.

Fusible Links.   This technique uses narrow fuse links between prearranged resistor taps. These links initially short-circuit all the taps together, but they can be selectively open-circuited by burning them out. When an excessive amount of current is passed through them, they behave as "fuses" and burn out to form a permanent open circuit. The fuse sections can be made either with narrow and thin aluminum strips ($\approx 5 \ \mu$m wide and 1000 Å thick) or with Ni-Cr resistor segments. Figure 3.38 illustrates how various resistor taps can be open-circuited

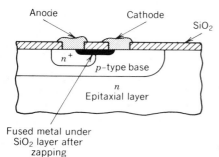

FIGURE 3.36. Cross section of a short-circuited Zener diode after zapping operation.[8]

by selectively blowing such fuse links. Similar to Zener zapping, the blowing of the fuse links is normally done at the wafer probing stage.

Laser Trimming. In this method, a laser beam is used to cut through selected metal links which connect the prearranged resistor taps, as shown in Figure 3.39. Laser trimming is also performed at the wafer probing stage, and is more accurate and reliable than the fusible link technique, since it produces a clean break in the metal link.[9]

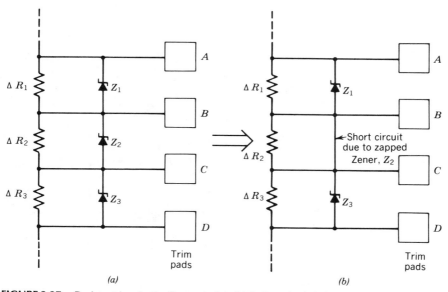

FIGURE 3.37.   Resistor trimming by Zener zapping: (a) Before trim and (b) after trim, with Z_2 zapped.

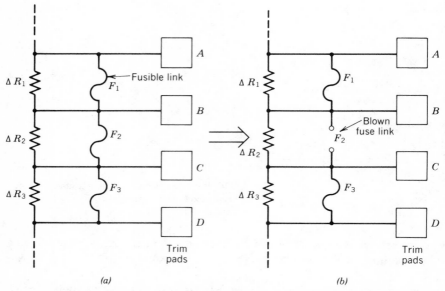

FIGURE 3.38. Resistor trimming by blowing fuse links: (*a*) Before trim and (*b*) after trim.

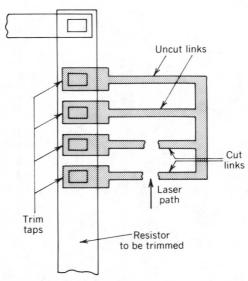

FIGURE 3.39. Laser trimming of resistor by selective cutting of metal links.

Trimming of Thin-Film Resistors. After deposition, the absolute-value tolerances of thin-film resistors can be trimmed to within 1–0.01% of the desired
value by means of one of the following methods:

1. *Oxidation.* By heating certain resistor films in an oxidizing atmosphere, some of the material on the film surface can be converted to a
   nonconductive oxide layer, which increases the total resistor value. This
   method is generally used for tantalum.

2. *Annealing.* Most thin-film materials exhibit a coarse grain structure
   after deposition. A subsequent heat treatment, or annealing cycle, causes
   the grain structure to reorient itself in a more dense fashion and causes
   the sheet resistance to be lowered. Annealing can be done locally by
   applying power to the resistor, or by heating it with a laser beam.
   However, it should be noted that since annealing changes the material
   structure of the thin film, it also affects the temperature coefficient of the
   resistor and the tracking of the resistor values over the temperature.

3. *Laser Trimming.* By selectively evaporating a small portion of the
   thin-film resistor, its effective resistance can be increased. This is normally done by cutting an L-shaped groove into the resistor, as shown in
   Figure 3.40, by means of a finely focused laser beam (spot size $\leq$
   1 μm). This is normally done with an automated test system, while the
   resistance is continuously monitored. The initial cut is made perpendic-
   ular to the direction of current flow. As the resistance approaches the
   desired value, the direction of the beam is moved parallel to the length
   of the resistor for fine adjustment of the resistor value. For accurate
   trimming, relatively wide resistor geometries ($W \geq 50\ \mu$m) are needed.
   Although laser trimming is an accurate technique for fabricating precision thin-film resistors on monolithic chips, it is a relatively slow and
   costly step since each precision resistor must be aligned and trimmed
   individually.

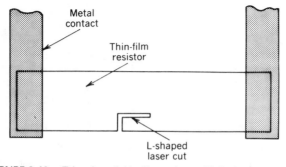

FIGURE 3.40. Trimming of thin-film resistor with L-shaped laser cut.

PART III: INTEGRATED CAPACITORS

The most fundamental limitation on integrated capacitors is size. A general expression for the capacitance of a parallel-plate capacitor can be written as

$$C = C_o A \tag{3.18}$$

where C_o is the capacitance per unit area, and A is the area of one of the plates. The value of C_o is restricted to a narrow range (typically on the order of $0.05-0.5\text{pF/mil}^2$) due to the type of dielectric materials available in monolithic circuits and their voltage breakdown properties. Thus, the chip area requirement increases quite rapidly with the required capacitor value. As a rule of thumb, for every 3–5 pF of capacitance value, an amount of chip area equivalent to an isolated small-signal *npn* transistor is used up. Since the practical chip size of a monolithic circuit is dictated by yield considerations, this puts an upper limit on the value of capacitance that can be economically fabricated in monolithic form.

There are two basic classes of capacitor structures available in monolithic circuits: junction and MOS (or oxide) capacitors. The properties of each of these capacitor types are discussed below.

3.10. JUNCTION CAPACITORS

Application of a reverse bias across a semiconductor junction results in the formation of a depletion layer which is devoid of mobile carriers. Figure 3.41 shows an idealized illustration of such a depletion layer formed across a reverse-biased junction. This structure closely simulates a parallel-plate capacitor with an area equal to the total junction area, and with the two plates of the capacitor being separated by the total depletion layer width x. The depletion layer width can be related to the total voltage V_t across the junction (i.e., the applied voltage *plus* the built-in potential) by the solution of the one-dimensional Poisson equa-

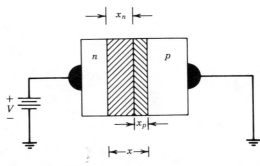

FIGURE 3.41.   Idealized illustration of *p-n* junction depletion layer under reverse bias.

tion.[1] It can be shown that the overall charge neutrality conditions require that an equal number of positive and negative charges be depleted on either side of the junction. This implies that the depletion layer tends to spread more into the more lightly doped side of the junction. For a step-junction structure, with uniform impurity concentrations of N_A and N_D atoms/cm^3 on the p and n sides of the junction, the total depletion layer width can be related to the total voltage V_t as

$$V_t = V + \psi_o = \frac{q}{2\epsilon} \frac{N_A N_D}{N_A + N_D} x^2 \qquad (3.19)$$

where V is the externally applied reverse bias, and ψ_o is the built-in junction potential which depends on the impurity concentration on either side of the junction, and is given by (Eq. 2.31).

The dielectric constant ϵ can be expressed as

$$\epsilon = \epsilon_r \epsilon_o \qquad (3.20)$$

where ϵ_o = permitivity of free space ($= 8.85 \times 10^{-14}$ F/cm)

ϵ_r = relative dielectric constant of the insulating material ($=12$ for silicon)

Thus, in the case of junction capacitors in silicon, $\epsilon = 1.04 \times 10^{-12}$ F/cm. For a step junction, the charge neutrality condition also requires that the total charges on each side of the junction be equal, that is,

$$N_A x_p = N_D x_n \qquad (3.21)$$

Since the capacitance per unit area of a parallel-plate capacitor with a plate separation x is given as

$$C_o = \frac{\epsilon}{x} \qquad (3.22)$$

the capacitance per unit area of the junction can be related to the total reverse bias V_t across the junction, from Eqs. (3.21) and (3.22), as

$$C_o = \sqrt{\frac{q\epsilon}{2V_t} \frac{N_A N_D}{N_A + N_D}} \qquad (3.23)$$

With most junctions, the impurity level on one side of the junction is much higher than that on the other. This is particularly true if one side of the junction, say, the p side, is formed by diffusion into the uniform n background, as is the case with the base–collector junction of the npn transistor. In such a case, one can still invoke the step-junction approximation with $N_A \gg N_D$. Then Eq. (3.23) reduces to

$$C_o = \sqrt{\frac{q\epsilon N_D}{2V_t}} = \frac{C_{XO}}{\sqrt{1 + V/\psi_o}} \qquad (3.24)$$

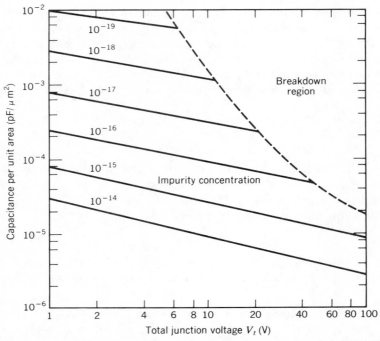

FIGURE 3.42. Capacitance per unit area versus junction voltage for a step junction. (*Note:* Impurity concentration shown corresponds to more lightly doped side of junction.)

where C_{XO} is the capacitance per unit area of the junction with zero applied bias (i.e., $V = 0$ and $V_t = \psi_o$).

Figure 3.42 shows the capacitance per unit area as a function of the total junction voltage for a step junction, for different values of impurity concentration on the lighter doped side. In general, the voltage dependence of the junction capacitances which can be fabricated with the monolithic IC processes can be described by a generalized version of Eq. (3.24) as

$$C_o = \frac{C_{XO}}{(1 + V/\psi_o)^n} \tag{3.25}$$

where the exponent n is equal to $\frac{1}{2}$ for a step junction, and $\frac{1}{3}$ for a linearly graded junction. For most semiconductor junctions, either the step or the linearly graded junction approximation holds very closely. A detailed family of capacitance versus voltage curves is available in the literature.[11]

In a planar epitaxial integrated circuit, there are three separate junctions which can be used as capacitors. As shown in Figure 3.43, these are the base–emitter, the base–collector, and the collector–substrate capacitances associated with the integrated *npn* bipolar structure. Also shown is the resultant interconnection of these three capacitors, along with the ideal diodes in shunt

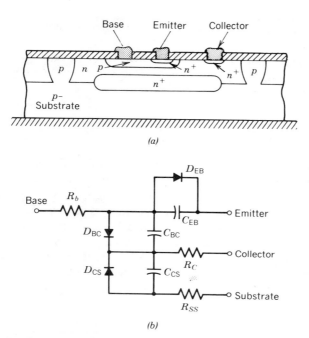

FIGURE 3.43. Junction capacitances in bipolar integrated circuits: (*a*) Physical device structure; (*b*) Equivalent circuit of junction capacitances.

with them, to indicate the bias polarity requirements for each capacitor. Note that each capacitance also has a finite bulk resistance in series with it.

With reference to Figure 3.43, the collector–substrate capacitance C_{CS} has only a very limited application since one of its terminals, the substrate, is common to the rest of the circuit and represents an ac ground point. However, C_{CS} is an inherent part of the device structure, and is always present in any junction isolated structure. The remaining capacitances, C_{EB} and C_{BC}, can be eliminated when not needed by omitting the emitter or the base diffusion.

The emitter–base junction offers the highest capacitance per unit area; however, its low reverse breakdown voltage (≈ 6.5 V) limits its use in some applications. The base–collector capacitance finds a wider range of applications than C_{EB} because of its high breakdown voltage (typically ≈ 50 V). However, its performance is hampered by the collector series resistance R_C and the shunt C_{CS} to ground. An n^+-type buried layer can be used to reduce R_C. However, this causes approximately a 30% increase in the value of C_{CS} by increasing the impurity concentration at the epi–substrate interface.

In order to utilize C_{BC} efficiently for ac coupling purposes, it is necessary that the C_{BC}/C_{CS} ratio be kept as high as possible. This can be done by choosing the reverse bias across the collector–substrate junction. In this manner, C_{BC}/C_{CS} ratios in the range of 3–10 are feasible. In the case of dielectrically isolated integrated circuits (see Fig. 1.14), such bias precautions are not necessary since C_{CS} is negligible.

Table 2.2 gives a summary of the typical junction capacitances associated with each of the three basic junctions available in bipolar monolithic IC structures for various commonly used fabrication processes.

3.11.　MOS CAPACITORS

MOS or oxide capacitors are formed by using a thin oxide layer as a dielectric between two conducting surfaces. Normally, the low-resistivity semiconductor region below the oxide layer serves as one plate of the capacitor, and the aluminum deposited over the oxide forms the second plate. Figure 3.44 shows the cross section and a typical layout for such a capacitor structure.

Normally, n^+ emitter diffusion is used as one plate of the capacitor because of its low sheet resistance. During the fabrication process, once the emitter diffusion is made, a special capacitor mask is applied to grow a well-controlled thin SiO_2 layer over the selected portion of the n^+ diffusion. The thickness of this

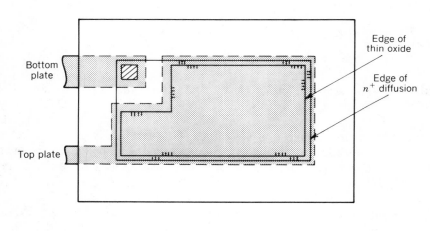

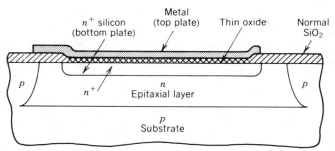

FIGURE 3.44.　Typical layout and cross section of MOS capacitor.

SiO_2 layer, which is in the range of 1000–2000 Å, serves as the dielectric thickness of the parallel-plate capacitor. Finally, an aluminum layer is deposited over the thin oxide region, slightly overlapping the oxide, to form the complete capacitor structure.

The resulting capacitor structure has a capacitance per unit area C_o given as

$$C_o = \frac{\epsilon}{t_{ox}} = \frac{\epsilon_r \epsilon_o}{t_{ox}} \tag{3.26}$$

where t_{ox} is the thickness and ϵ_r is the relative dielectric constant of SiO_2. Typical values of ϵ_r are in the range of 2.7–4.2, depending on the growth rate and the purity of the SiO_2 layer.

In certain applications, Si_3N_4 rather than SiO_2 may be used as a dielectric, because of its higher relative dielectric constant ($\epsilon_r \approx 4$–9), in order to increase the value of C_o. In the case of Si_3N_4, a dielectric thickness of 500–1000 Å is normally used. The lower limit of oxide or nitride thickness is normally set by yield, process control, and breakdown voltage requirements. Normally, Si_3N_4 is not used alone, but in conjunction with an SiO_2 layer to form a two-layer dielectric, thus enhancing breakdown and long-term reliability characteristics of the capacitor.

Figure 3.45 shows the equivalent circuit of the MOS capacitor illustrated in Figure 3.44, in terms of the inherent device parasitics. The resistor R_S represents the access resistance to the bottom plate (i.e., the n-type layer) and is typically of the order of several ohms. C_{CS} is the capacitance of the epitaxial layer–substrate junction, and D_{CS} is the reverse-biased collector–substrate diode. In order to reduce the parasitic effects of C_{CS}, the MOS capacitor is normally fabricated without an n-type buried layer under it.

Unlike their junction counterparts, MOS capacitors can operate with either positive- or negative-polarity voltage applied across the capacitor, and the capacitance value is not voltage dependent. However, MOS capacitors fail by the breakdown of the dielectric layer, when their voltage rating is exceeded. This breakdown is an irreversible (i.e., destructive) failure mechanism, which results

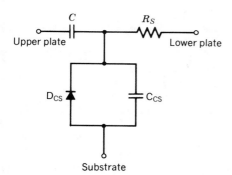

FIGURE 3.45. Electrical equivalent circuit of MOS capacitor.

TABLE 3.3. Typical Characteristics of MOS Capacitors

Device Parameter	Dielectric Material	
	SiO_2	Si_3N_4
Capacitance (pF/mil^2)	0.25–0.4	0.5–1.0
Relative dielectric constant	2.7–4.2	3.5–9
Breakdown voltage (V)	50–100	50–100
Absolute tolerance (%)	±20	±20
Matching tolerance (%)[a]	±1	±1
Temperature coefficient (ppm/°C)	±20	+4 to +10
Q at 10 MHz	25–80	20–100

[a]The matching tolerance can be reduced to as low as 0.1% by careful device layout.

in a permanent short circuit across the capacitor. Therefore, additional care must be taken in the use of MOS capacitors in the circuit, to provide overvoltage protection in a manner similar to the gate protection of conventional MOS transistors.

Table 3.3 gives a summary of the electrical characteristics of MOS capacitors, using either SiO_2 or Si_3N_4 as the dielectric material. The absolute-value tolerance is normally set by the control of the dielectric oxide or nitride layer thickness. MOS capacitors offer excellent matching characteristics, typically on the order of ± 1%. By careful layout of the capacitor geometry, this matching tolerance can be improved to ± 0.1%.[12] This feature makes MOS capacitors extremely useful as precision building blocks in D/A converters and switched-capacitor filters, which are described in Chapters 13 and 14.

REFERENCES

1. A. Y. C. Yuh, "The Metal-Semiconductor Contact: An Old Device with a New Future," *IEEE Spectrum*, 83–89, (March 1970).

2. H. R. Camenzind, *Electronic Integrated System Design*, VanNostrand Reinhold, New York, 1972, Chap. 9.

3. R. Dobkin, "IC Voltage Reference has 1 ppm/°C Drift," National Semiconductor Appl. Note 161, June 1977.

4. J. C. Schmoock, "A Precision Voltage-to-Frequency Converter," *Digest of Tech. Papers*, IEEE Int. Solid-State Circuits Conf., Vol. 21, pp. 136–137, February 1978.

5. J. C. Irvin, "Resistivity of Bulk Silicon and Diffused Layers in Silicon," *Bell System Tech. J.* **41**, 247–410 (1962).

6. H. R. Camenzind, *Electronic Integrated Systems Design*, VanNostrand Reinhold, New York, 1972, Chap. 4.

7. J. W. Mayer, L. Eriksson, and J. A. Davies, *Ion Implantation in Semiconductors*, Academic Press, New York, 1970.

8. G. Erdi, "A Precision Trim Technique for Monolithic Analog Circuits," *IEEE J. Solid-State Circuits*, **SC-10**, 412–416 (December 1975).

9. J. J. Price, "A Passive Laser-Trimming Technique to Improve Linearity of a 10-Bit D/A Converter," *IEEE J. Solid State Circuits* **SC-11,** 789–794 (December 1976).

10. A. B. Phillips, *Transistor Engineering,* McGraw-Hill, New York, 1962, Chap. 5.

11. H. Lawrence and R. M. Werner, "Diffused Junction Depletion Layer Calculation," *Bell System Tech. J.* **39,** 389–404 (1960).

12. J. L. McCreary and P. R. Gray, "All-MOS Charge Redistribution Analog-to-Digital Conversion Techniques—Part I," *IEEE J. Solid-State Circuits,* **SC-10,** 371–379 (December 1975).

CHAPTER FOUR

BIAS CIRCUITS

The starting point of a monolithic IC design is the design of the bias circuitry internal to the chip. This design step is very critical since it determines the internal voltage and current levels over all operating conditions of the integrated circuit as well as over all manufacturing process variations.

For a linear circuit designer trained in the area of discrete circuits, the basic constraints and limitations of monolithic circuit technology often pose a difficult challenge. This is particularly true with regard to the biasing circuitry. Many of the conventional biasing techniques cannot be directly applied to monolithic designs because of the following limitations of IC components:

1. Poor absolute-value tolerances.
2. Poor temperature coefficients.
3. Limitations on component values.
4. Lack of coupling capacitors.
5. Limited choice of compatible active devices.

On the other hand, IC fabrication methods offer a number of unique and powerful advantages to the circuit designer:

1. Availability of a large number of active devices.
2. Good matching and tracking of component values.
3. Close thermal coupling.
4. Control of device layout and geometry.

Fortunately, over the years a number of basic circuit configurations, or subcircuits, have been developed, which make efficient use of the advantages of monolithic technology, while avoiding most of its shortcomings. The purpose of this chapter is to outline some of these biasing techniques, along with the design philosophy and the guidelines associated with them. The basic subcircuits discussed in this chapter deal with the dc design of a monolithic circuit, with regard to the current and voltage bias levels within the circuit, and their drift with temperature or power supply changes. These basic subcircuits form a useful set of building blocks which, along with the basic gain stages described in the next chapter, serve as the starting point of more complex analog functions and subsystems.

4.1. CONSTANT-CURRENT STAGES

In a constant-current stage, the reference current in one branch of the circuit is accurately reproduced or reflected in a second branch, relatively independent of the absolute values of the device parameters. Because of this property, these subcircuits are also known as *current mirror* circuits. Such circuit configurations are particularly useful building blocks for analog circuit design, since they provide a means of establishing the dc bias levels within the circuit, within the accuracy of the matching or tracking properties of the monolithic components.

Basic Current Mirror

The basic current mirror subcircuit is shown in Figure 4.1. It is made up of two matched transistors, where one transistor, Q_1, is connected as a diode and sets the base–emitter voltage V_{BE} of Q_2. The operating principle of the current mirror is derived from the basic properties V_{BE} of the transistor and the collector current I_C as given by Eq. (2.5),

$$V_{BE} = V_T \ln \left[\frac{I_C}{I_{CO}} \right] \tag{4.1}$$

where V_T is the thermal voltage ($=kT/q$), and I_{CO} is the reverse saturation current. The reverse saturation current is related to the area A of the emitter–base junction as

$$I_{CO} = I_S A \tag{4.2}$$

where the constant of proportionality I_S depends on the intrinsic semiconductor parameters, such as the minority carrier diffusion lengths and the concentrations on both sides of the junction. For the IC devices on the same chip, which are fabricated simultaneously, this term will be the same. Therefore, if the two *npn* transistors are operated with the same base–emitter voltage, their collector currents will be related to each other as the ratio of their emitter areas, that is,

$$\frac{I_{C1}}{I_{C2}} = \frac{A_1}{A_2} \tag{4.3}$$

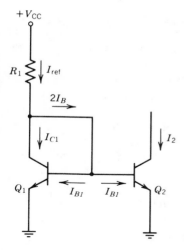

FIGURE 4.1. Basic current mirror.

In the current mirror circuit of Figure 4.1, the base–emitter voltages of the two transistors are forced to be equal. This in turn forces the collector currents of Q_1 and Q_2 to be equal, that is,

$$I_2 = I_{C1} = I_{ref} - I_{B1} - I_{B2} \tag{4.4}$$

where I_{B1} and I_{B2} are the base currents of Q_1 and Q_2. Since the base currents are also proportional to the respective emitter areas, I_2 can be related to I_{ref} from Eq. (4.3) as

$$I_2 = I_{ref}\frac{A_2}{A_1}\left(\frac{\beta}{\beta + 1 + A_1/A_2}\right) \tag{4.5}$$

which, for the case of identical devices (i.e., $A_2 = A_1$) reduces to

$$I_2 = I_{ref}\left(\frac{\beta}{\beta + 2}\right) \approx I_{ref}\left(1 - \frac{2}{\beta}\right) \tag{4.6}$$

If $\beta >> 1$,* both equations converge to

$$I_2 \approx I_{ref} \tag{4.7}$$

Another potential source of error in the current mismatch is the mismatch of the V_{BE} of Q_1 and Q_2. For small values of mismatch ΔV_{BE} this effect can be incorporated into Eq. (4.6) as an additive term,

$$I_2 = I_{ref}\left(1 - \frac{2}{\beta} + \frac{\Delta V_{BE}}{V_T}\right) \tag{4.8}$$

---

*In order to minimize the subscript notation, the symbol β will be used to mean either β_F or β_o, interchangeably.

where $\Delta V_{BE} = V_{BE2} - V_{BE1}$. Note that ΔV_{BE} can be positive or negative, depending on the polarity of the mismatch. The above expression indicates that *each millivolt of mismatch in the V_{BE} of the two transistors results in approximately 4% of current mismatch.*

The basic current–voltage equations (4.1) through (4.8) hold over a broad temperature range (typically -60 to $+150°C$) and over six orders of magnitude of current. Thus, the basic current mirror circuit of Figure 4.1 provides a means of obtaining a current reference level, independent of implicit device parameters, which can be "scaled" by proper choice of the emitter areas of the two transistors. Assuming typical V_{BE} mismatch of $\leq \pm 1$ mV and β in the range of 100–200, the typical output current I_2 would be within $\pm 5\%$ of I_{ref} over a wide range of temperature and current levels.

Since I_{ref} is set by resistor R_1, I_2 can also be expressed in terms of the setting resistor R_1 as

$$I_2 \approx \frac{V_{CC} - V_{BE}}{R_1} \frac{A_2}{A_1} \approx \frac{V_{CC}A_2}{R_1 A_1} \tag{4.9}$$

assuming that $V_{CC} \gg V_{BE}$.

The output impedance of the basic current mirror circuit is primarily due to the Early effect of Q_2 (see Fig. 2.7). This effect also tends to contribute to the current mismatch as the collector voltage V_2 of Q_2 is increased. It can be incorporated into the basic current expression by means of a multiplication term as

$$I_2 = I_{ref}\left(1 - \frac{2}{\beta}\right)\frac{1 + V_2/V_A}{1 + V_{BE}/V_A} \tag{4.10}$$

where V_A is the Early voltage associated with the *npn* transistor (see Table 2.2). It should be noted that for typical values of $V_A \approx 100$ V, this second term can contribute as much as 25% of current mismatch at a collector voltage V_2 of 25 V.

The output resistance R_{out} of the current mirror is equal to the collector

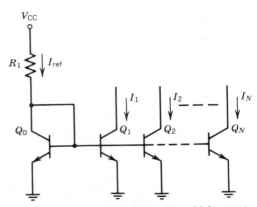

FIGURE 4.2. Basic current mirror with multiple outputs.

resistance r_o of the hybrid-π model, given by Eq. (2.20), which can be written as

$$R_{\text{out}} \approx \frac{V_A}{I_2} \tag{4.11}$$

For $V_A \approx 100$ V and $I_2 = 1$ mA, this results in an output impedance of approximately 100 kΩ. For ac calculations, it should be remembered that R_{out} is also shunted to the substrate by means of the parasitic collector–substrate capacitance C_{CS} of Q_2.

In many applications, it is advantageous to derive not one but several currents from a reference current I_{ref}. This can be done by using a multiple-output current mirror, as shown in Figure 4.2. In this case, assuming N identical transistors and neglecting the Early effect of the output transistors, the resulting currents are related to I_{ref} as

$$I_1 = I_2 = \cdots = I_N = I_{\text{ref}}\left(1 - \frac{N+1}{\beta}\right) \tag{4.12}$$

When using the multiple-output current mirror of Figure 4.2, a word of caution is in order. If any one of the output transistors, Q_1 through Q_N, is driven into saturation, its collector–base junction becomes forward biased and causes a part of I_{ref} to be shunted to ground, through it. This effect, which is known as *current hogging*, then causes saturated transistor to use up more than its share of the available base current. If only a limited amount of base current is available, this causes the currents of the remaining outputs to *decrease* below the nominal value given by Eq. (4.12). As will be discussed later, this phenomenon is sometimes useful as a technique for detecting the onset of saturation.

The basic current mirror configuration is also very useful as a controlled-gain transistor, as illustrated in Figure 4.3. The resulting three-terminal device is

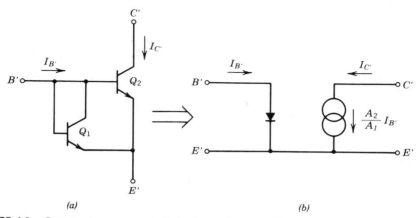

(a)   *(b)*

FIGURE 4.3. Current mirror as a controlled-gain transistor: *(a)* Circuit connection; *(b)* simplified equivalent circuit for large-signal operation. (*Note:* A_1 and A_2 are the respective emitter areas of Q_1 and Q_2.)

effectively equivalent to an *npn* transistor where the effective current gain β_{eff} is equal to the area ratio of Q_1 and Q_2, that is,

$$\beta_{\text{eff}} = \frac{I'_C}{I'_B} = \frac{A_2}{A_1}\left(1 - \frac{1 + A_2/A_1}{\beta}\right) \qquad (4.13)$$

or

$$\beta_{\text{eff}} = \frac{A_2}{A_1} \qquad \text{for} \quad \beta \gg 1 \qquad (4.14)$$

As will be described later, this controlled-gain transistor is particularly useful in designing high-current *pnp* current mirrors (see Fig. 4.19) or output stages with fixed current drive capability.

Base Current Compensated Current Mirror

The main error source in the basic current mirror circuits of Figures 4.1 and 4.2 are the finite values of the base currents which directly subtract from I_{ref}. This source of error can be greatly reduced by adding an extra transistor Q_3 into the circuit, as shown in Figure 4.4. The base currents of Q_1 and Q_3 are then supplied from Q_3, and only the base current of Q_3 is subtracted from I_{ref}. Thus, assuming that Q_1 and Q_2 are well matched, the output current differs from I_{ref} by I_{B3},

$$I_2 = I_{\text{ref}} - I_{B3} = I_{\text{ref}} - \frac{I_{B1} + I_{B2}}{\beta_3} \qquad (4.15)$$

and it can be related to I_{ref} as

$$I_2 = I_{\text{ref}}\left(1 - \frac{1}{2 + \beta\beta_3}\right) \qquad (4.16)$$

where β is the current gain of Q_1 or Q_2, and β_3 is the gain of Q_3. In the above expression, β_3 is identified separately since Q_3 operates at a much lower current

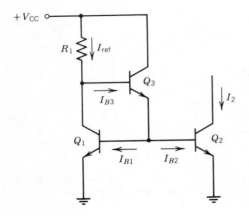

FIGURE 4.4. Base current compensated current mirror.

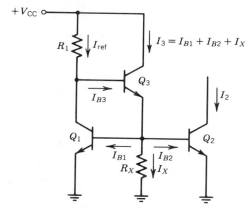

FIGURE 4.5. Base current compensated current mirror with current shunting resistor R_X. (*Note:* Normally I_X is set 5–10 times higher than I_{B1} or I_{B2}.)

level than Q_1 and Q_2 and, thus, may exhibit a lower value of β. In some designs, to avoid this problem, an additional resistance, R_X, is connected from the bases of Q_1 and Q_2 to ground, shown in Figure 4.5, to increase the current in Q_3 by an amount I_X, where

$$I_X = \frac{V_{BE}}{R_X} \qquad (4.17)$$

I_X is the current in R_X. Typically R_X is chosen to set I_X to within 5–10 times I_{B1} or I_{B2}.

Since the accuracy of the base current compensated current mirror is largely insensitive to the base current errors, such a circuit is often used to set up a multiple number of constant-current stages from a given reference current, as shown in Figure 4.6. In a generalized case with N identical outputs, assuming

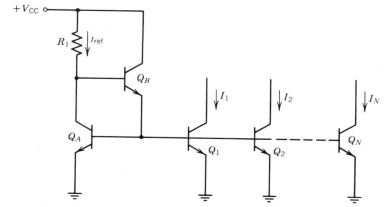

FIGURE 4.6. Base current compensated current mirror with multiple outputs.

all transistors to be matched, the output currents are related to I_{ref} as

$$I_1 = I_2 = \cdots = I_N = I_{ref}\left(1 - \frac{N + 1}{N + 1 + \beta^2}\right) \qquad (4.18)$$

which simplifies to

$$I_1 = I_2 = \cdots I_N \approx I_{ref}\left(1 - \frac{N + 1}{\beta^2}\right) \qquad (4.19)$$

The base current compensated current mirror is also very useful for designing constant current stages that have a much higher output current than I_{ref}. This can be done by making the emitter are of the output transistor Q_2 to be much larger than that of Q_1. Assuming that the two emitter areas are related by the ratio N, such that $N = A_2/A_2$, the output current I_2 can be related to I_{ref}

$$I_2 \approx NI_{ref}\left(1 - \frac{N + 1}{\beta^2}\right) \approx NI_{ref} \qquad (4.20)$$

Note that Eq. (4.20) is very similar to the case of the multiple-output current source [Eq. (4.19)], except that N does not have to be an integer.

To summarize, the key advantage of base current compensation is the greatly reduced dependence of the output current on the transistor β. This makes it particularly suited to the design of multiple-output current mirror circuits, or to the scaling of output currents. However, the output impedance R_{out} is not affected by the base current compensation and is still the same as that given by Eq. (4.11). The methods of increasing this output impedance will be discussed in later sections of this chapter.

Resistor-Ratioed Current Mirror

The basic current mirror subcircuit can be modified to use resistor ratios, rather than the transistor emitter areas, to scale the current. Such a modified version of the circuit, often called the *resistor-ratioed current mirror*, is shown in Figure 4.7.

Neglecting the base currents, the current levels through each of the two transistors are related as

$$I_{ref}R_1 + V_{BE1} = I_2 R_2 + V_{BE2} = V_A \qquad (4.21)$$

The base–emitter voltage drop difference between two identical transistors operating at the respective collector currents I_{ref} and I_2 can be written as

$$\Delta V_{BE} = V_{BE2} - V_{BE1} = V_T \ln\left(\frac{I_2}{I_{ref}}\right) \qquad (4.22)$$

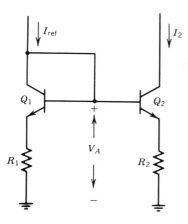

FIGURE 4.7. Resistor-ratioed current mirror.

Thus, from Eqs. (4.21) and (4.22) the ratio of the two currents can be expressed as

$$\frac{I_2}{I_{ref}} = \frac{R_1}{R_2}\left[1 - \frac{V_T \ln (I_2/I_{ref})}{R_1 I_{ref}}\right] \qquad (4.23)$$

If the voltage drop across R_1 is made to be significantly larger than V_T and is comparable to V_{BE}, then the second term within the brackets becomes negligibly small and the two currents are closely related by the resistor ratios, that is,

$$\frac{I_2}{I_{ref}} \approx \frac{R_1}{R_2} \qquad (4.24)$$

For $I_{ref}R_1 \geq V_{BE}$, the two currents follow the resistor ratio of Eq. (4.24) with a maximum error of less than $\pm 10\%$ over two orders of magnitude in current, that is,

$$\frac{1}{10} < \frac{I_2}{I_{ref}} < 10 \qquad (4.25)$$

independent of temperature.

The resistor-biased constant-current stage is preferred over the simple diode-biased current mirror of Figure 4.1 for the cases where the ratio of I_2/I_{ref} is significantly different than unity, since the resistor ratios can be varied over a broader range of values than the emitter areas.

In the resistor-ratioed current mirror circuit, the resistor R_2 in series with the emitter of Q_2, serves as a series feedback resistor and increases the output resistance of the circuit. Using the simple hybrid-π model of Figure 2.10, the output resistance of the resistor-ratioed current mirror can be approximated as

$$R_{out} \approx r_o \frac{R_1 + \beta R_2}{R_1 + R_2} \qquad (4.26)$$

In practical cases, where $\beta R_2 >> R_1$, this expression simplifies to

$$R_{\text{out}} \approx \beta r_o \frac{R_2}{R_1 + R_2} \tag{4.27}$$

where r_o is the collector–emitter resistance of the hybrid-π model given in Eq. (2.20). Comparing Eq. (4.11) with Eq. (4.27), one sees that the output resistance of the resistor-ratioed current mirror is approximately β times higher than that of the ordinary current mirror, provided that R_2 is sufficiently large.

It should be remembered that the basic results given in Eqs. (4.24) and (4.27) are valid only if the voltage drop across R_1 and R_2 is comparable to V_{BE}. In certain design applications requiring very low current levels, in the low microampere range, this requirement may result in excessively large values of R_1 or R_2, and will make the use of resistor ratioing impractical.

Figure 4.8 shows a special case of the resistor-ratioed current mirror, with R_1 set equal to zero. Such a configuration is particularly useful for obtaining very low values of output current with relatively large values of the reference current, and without requiring high-value resistors. Assuming matched device geometries, Q_1 operates at a higher current density than the base–emitter diode of Q_2; and the voltage drop V_2 across R_2 is constrained to be

$$V_2 = I_2 R_2 = V_{\text{BE1}} - V_{\text{BE2}} = V_T \ln \left(\frac{I_{\text{ref}}}{I_2} \right) \tag{4.28}$$

Solving for R_2, we can write

$$R_2 = \frac{V_T}{I_2} \ln \left(\frac{I_{\text{ref}}}{I_2} \right) \tag{4.29}$$

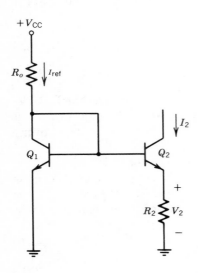

FIGURE 4.8.   Constant-current stage for low current levels.

Since R_2 is proportional to the logarithm of the current ratio, a high degree of current mismatch can be obtained without requiring extremely high resistor values. For example, letting $R_2 = 20$ kΩ, one obtains a $100:1$ ratio between the two currents, with $I_{\text{ref}} = 1$ mA and $I_2 \approx 10$ μA.

An added advantage of the constant-current circuit of Figure 4.8 is that the output current I_2 is relatively independent of the supply voltage, that is,

$$I_2 = \frac{V_T}{R_2} \ln\left(\frac{V_{\text{CC}} - V_{\text{BE}}}{R_o I_2}\right) \tag{4.30}$$

Thus, for $I_{\text{ref}} \gg I_2$, the output current varies as the *logarithm* of the supply voltage. Such a low-value constant-current stage is particularly useful for biasing the input stage of an operational amplifier where it can set quiescent current levels relatively independently of supply voltage changes. However, it should be noted that in Eq. (4.30), V_T is directly proportional to temperature. Thus, I_2 would also exhibit a strong temperature dependence, depending on the temperature coefficient of R_2 and V_T.

The output resistance of the low-current constant-current stage can be easily calculated from the simplified hybrid-π model of Figure 2.10. Assuming that the transconductance g_{m2} of Q_2 is low enough such that $g_{m2}R_2 \ll \beta$, one can show that[1]

$$R_{\text{out}} \approx r_o\left(1 + \frac{I_2 R_2}{V_T}\right) \tag{4.31}$$

Thus, the output resistance R_{out} depends strongly on the value of the voltage drop across R_2. In the current source configuration of Figure 4.8, $I_2 R_2$ is limited to several hundred millivolts for practical current ratios, which limits the maximum value of R_{out} to approximately $10r_o$.

In many applications, a large number of low-value constant-current stages can be driven from a single current reference, as shown in Figure 4.9. Since each

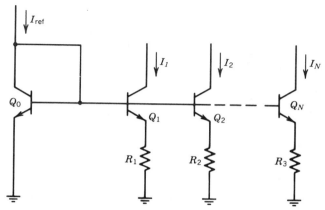

FIGURE 4.9. Driving a multiple number of low-value current sources from a single reference.

of the current outputs operates at much lower current levels than I_{ref}, the base currents of Q_1 throught Q_N have a negligible effect in calculating the output currents. Each of the output currents can be set to its nominal value by calculating the value of the emitter resistor corresponding to it.

Wilson Current Mirror[2]

The β dependence and the low output impedance of the basic current mirror can be greatly improved by using the so-called *Wilson current mirror* configuration shown in Figure 4.10. In this configuration, the base current of Q_2, which is extracted from the reference current I_{ref}, is resupplied back to the base of the reference transistor Q_1, thus keeping the current levels in Q_1 and Q_2 unaffected by the base current changes. The V_{BE} drop across Q_3 sets the bias level for Q_1, which in turn sets the current level in Q_3. It can be shown by straightforward nodal analysis that the output current I_2 is related to the reference current I_{ref} as

$$I_2 = I_{ref} + (I_{B1} + I_{B3} - 2I_{B2}) \qquad (4.32)$$

Note that the additional terms within the parentheses represent the third-order errors due to the base current mismatches, and reduce to zero if the transistor β are matched. If the transistor β are matched to better than $\pm 20\%$, the difference between I_2 and I_{ref} is less than 0.5% for a typical β value of 200.

Further insight can be gained into the operation of the improved constant-current stage of Figure 4.10 by treating it as a special case of the shunt-series feedback, with a unity feedback connection in the shunt branch. This corresponds to the unity current feedback condition from the emitter of Q_2 back to the

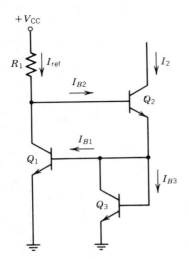

FIGURE 4.10. Wilson current mirror.

base of Q_1. As a consequence of this feedback arrangement, the output impedance of the current mirror stage is greatly increased. By using the hybrid-π
model, the output resistance of the Wilson current mirror can be calculated to
be[3]

$$R_{\text{out}} \approx \frac{\beta r_o}{2} \tag{4.33}$$

One basic drawback of the Wilson current mirror is that it cannot be easily
extended to a multiple number of outputs from a single I_{ref} without upsetting the
base current cancellation feature given in Eq. (4.32). This problem can be
overcome by the modified circuit shown in Figure 4.11, by splitting the output
transistor Q_2 into two matched transistors Q_{2A} and Q_{2B}. Then the base current
cancellation feature is retained, but two matched output currents are provided,
where

$$I_{2A} = I_{2B} = \frac{I_{\text{ref}}}{2} \tag{4.34}$$

Cascode-Connected Current Mirrors

In many applications where a high output impedance is very desirable, it may
be advantageous to cascode or "stack" two current mirrors together, as shown

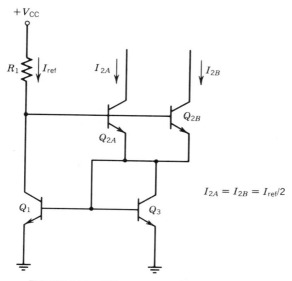

FIGURE 4.11. Wilson current mirror with multiple outputs.

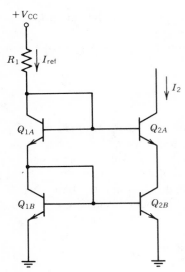

FIGURE 4.12. Cascode-connected current mirror with high output impedance.

in Figure 4.12. In such a configuration, assuming that all the transistors are well matched, the output current I_2 is related to the reference current I_{ref} as

$$I_2 = I_{\text{ref}}\left(1 - \frac{4}{\beta}\right) \tag{4.35}$$

In this configuration, transistor Q_{2B} serves as a constant-current source in series with the emitter of Q_{2A} and forces the output resistance R_{out} to be

$$R_{\text{out}} \approx \beta r_o \tag{4.36}$$

In a typical small-signal *npn* transistor, where $r_o \approx 100 \text{ k}\Omega$ at $I_E = 1$ mA, with $\beta = 200$, this results in $R_{\text{out}} \approx 20 \text{ M}\Omega$.

In certain applications where not one but a multiple number of high-impedance outputs are required, the multiple-output cascode current mirror circuit of Figure 4.13 can be used. In this configuration, Q_0 sets the current levels in the lower sets of current source transistors, with Q_B providing the base current compensation. Q_A provides the base current compensation for the upper set of current source transistors. Since Q_A and Q_B are a set of Darlington, or *common collector*, transistors, they can share the same isolation pocket.

The output currents I_1 through I_N are related to the reference current by a modified version of Eq. (4.19) as

$$I_1 = I_2 = \cdots = I_N = I_{\text{ref}}\left(1 - \frac{2(N + 1)}{\beta^2}\right) \tag{4.37}$$

The output resistance of each of the outputs will still be that of the basic cascode current mirror, that is, $R_{\text{out}} = \beta r_o$.

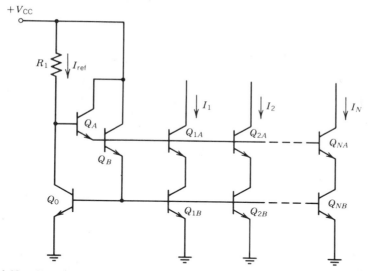

FIGURE 4.13. Cascode-connected current mirror with multiple outputs and base current compensation.

4.2. *pnp* CURRENT SOURCES

In principle, all of the *npn* current mirror or constant-current stage configurations discussed in the previous section apply to the *pnp* transistors, with the appropriate polarity reversal of the currents and the bias voltages. However, in practice, these stages suffer somewhat from the three basic shortcomings of the lateral *pnp* transistor, which are the following:

1. Low value of β. (Typically, β is in the range of 10–50 for lateral *pnp* devices.)
2. Low output resistance. (Since the base region of the lateral *pnp* transistor is lighter doped than its collector, the collector–base depletion region extends mostly into the base region, resulting in excessive base width modulation and a relatively low Early voltage, $V_A \approx 50$ V.)
3. Limited current-handling capability. (For practical lateral *pnp* transistor geometries, β falls off rapidly for I_C in excess of 50–100 μA.)

Thus, in choosing the appropriate *pnp* current mirror configuration, these three basic limitations associated with the lateral *pnp* transistors must be kept in mind. On the other hand, the availability of the lateral *pnp* transistor in a multiple-collector configuration (see Fig. 2.27) provides some unique advantages, particularly in terms of device interconnection and layout. Figure 4.14a shows the basic *pnp* current mirror. Since Q_1 and Q_2 both share a common emitter and a

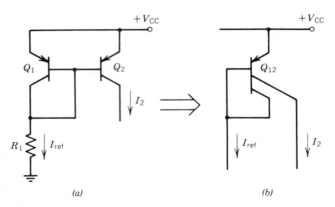

FIGURE 4.14.   Basic *pnp* current mirror; (*a*) Circuit diagram; (*b*) actual implementation using split-collector lateral *pnp* transistor.

common base region, their combination can be replaced by a split-collector *pnp* transistor Q_{12}, as shown in Figure 4.14*b*, which has the device layout of Figure 2.27*a*.

As discussed earlier in Chapter 2, the collector currents in multiple-collector lateral *pnp* transistor split in proportion to the collector *periphery* facing the emitter. This allows the designer to control the ratio of the reference current I_{ref} to the output current I_2 by controlling the periphery ratio of the two collectors,

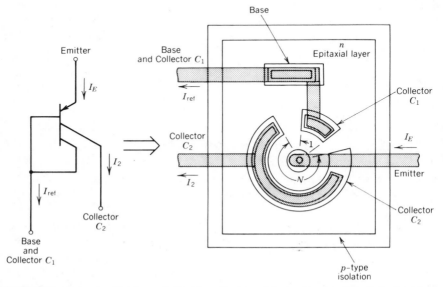

FIGURE 4.15.   Current scaling by controlling the collector periphery ratio of a split-collector lateral *pnp* transistor.

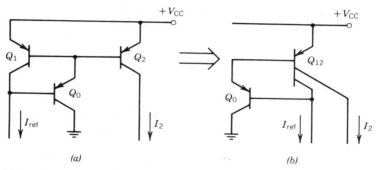

FIGURE 4.16. Base current compensated *pnp* current mirror: (*a*) Circuit diagram; (*b*) actual implementation.

as shown in Figure 4.15. Assuming that the two collectors C_1 and C_2 have a periphery ratio of $1:N$ between them, then by following the basic current mirror equation [Eq. (4.6)], one can relate the output current to the reference current as

$$I_2 = NI_{\text{ref}} \left(1 - \frac{1 + N}{\beta} \right) \qquad (4.38)$$

It should be noted that since the *pnp* transistor β is relatively low, the error term in parentheses in Eq. (4.38) can be significant for large values of N. For example, for $N = 5$ and $\beta \approx 20$, the output current I_2 may be as much as 50% below its ideal value of I_{ref}.

The current mismatches due to low β can be reduced significantly by using the *pnp* equivalent of the base current compensated current mirror shown in Figure 4.16. In this case, the compensating transistor Q_0 is normally designed as a substrate *pnp* transistor (see Fig. 2.29) with Q_1 and Q_2 made up of the split-collector *pnp* transistor Q_{12}, as shown in Figure 4.16*b*.

Figure 4.17 shows the other *pnp* current mirror configurations which are direct derivatives of the *npn* current mirrors discussed in the previous section. Another convenient *pnp* current mirror configuration is the multiple-output Wilson current source configuration shown in Figure 4.18, which can be designed with only two split-collector *pnp* transistors.

One of the major limitations of *pnp* current sources is the limited current-handling capability of the lateral *pnp* transistor. This can be solved by connecting an *npn* transistor to the output port to form a composite *pnp* transistor, similar to that shown in Figure 2.28. Such a transistor basically has the current-handling capability of the *npn* transistor and the polarity of the *pnp* transistor. If a controlled-β *npn* transistor (see Fig. 4.3) is used in place of the normal *npn* transistor, the output current level can be accurately controlled. Figure 4.19 demonstrates a circuit configuration which utilizes these principles. In the circuit, Q_1 and Q_2 from the basic current mirror. The collector current I_2 of

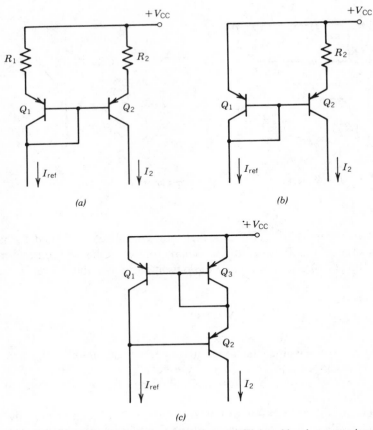

(a)   (b)

(c)

FIGURE 4.17. Other *pnp* current mirror configurations: (*a*) Resistor-biased current mirror; (*b*) low-current current mirror; (*c*) Wilson current mirror.

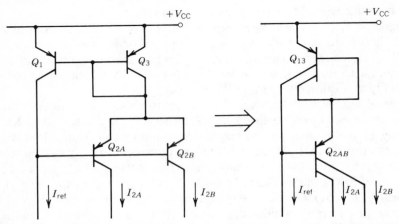

FIGURE 4.18. Multiple-output Wilson current source using split-collector *pnp* transistors. (*Note:* For symmetrical split collectors, $I_{2A} = I_{2B} = I_{ref}/2$.)

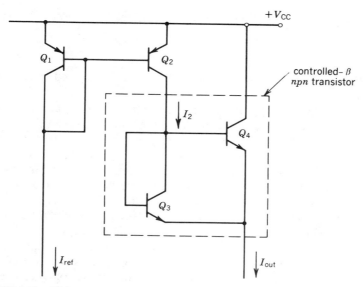

FIGURE 4.19. High-current *pnp* current source with composite controlled-β *npn* transistor.

transistor Q_2 is further amplified by the controlled-β *npn* transistor made up of Q_3 and Q_4.

Assuming that Q_3 and Q_4, have respective emitter areas A_3 and A_4, the output current I_{out} is related to I_2 as

$$I_{out} = I_2 \left(1 + \frac{A_4}{A_3} \right) \tag{4.39}$$

Since I_2 is related to I_{ref} by the basic current mirror equation [Eq. (4.6)] for the matched pair of *pnp* transistors Q_1 and Q_2, the overall current transfer equation becomes

$$\frac{I_{out}}{I_{ref}} = \left(1 - \frac{2}{\beta_P} \right) \left(1 + \frac{A_4}{A_3} \right) \approx 1 + \frac{A_4}{A_3} \tag{4.40}$$

where β_P is the current gain of the *pnp* transistors Q_1 or Q_2. In this manner, accurate current ratios can be maintained while providing a high output drive current.

4.3. VOLTAGE-CONTROLLED CURRENT SOURCES

All of the *npn* or *pnp* current mirror circuits discussed so far can be used in practice as voltage-controlled current sources simply by varying the voltage applied across the resistor R_1 which determines the reference current I_{ref}. For example, in Figure 4.1, if the supply voltage V_{CC} was replaced by the control voltage V_C, then the output current I_2 can be related to V_C as

$$I_2 \approx I_{\text{ref}} = \frac{V_C - V_{\text{BE}}}{R_1} \tag{4.41}$$

where V_{BE} is the base–emitter drop of the diode connected transistor Q_1 of Figure 4.1.

In practice, the control characteristics given by Eq. (4.41) are not very useful because of the dependence of the output current on V_{BE} as well as V_C. This V_{BE} dependence can introduce a very strong temperature drift to the output current, particularly at low values of V_C, approaching V_{BE}. This effect is even more pronounced in the base current compensated current source of Figure 4.5 or in the Wilson current mirror of Figure 4.10, where the collector of the current-setting transistor Q_1 is clamped at not one but two V_{BE} above ground.

Figure 4.20a shows a simple two-transistor voltage-controlled constant-current stage which overcomes this problem. In the circuit, the V_{BE} dependence of the output current is virtually eliminated by effectively canceling the V_{BE} drops of Q_1 and Q_2. Assuming that the V_{BE1} and V_{BE2} associated with Q_1 and Q_2 are approximately equal, the voltage level V_A at the emitter of Q_2 can be written as

$$V_A = V_C - V_{\text{BE1}} + V_{\text{BE2}} \approx V_C \tag{4.42}$$

where V_{BE1} and V_{BE2} are the base–emitter drops of the corresponding transistors. For convenience, in Figure 4.20 V_A and V_C are measured with reference to the positive supply, rather than ground. Assuming that the base current of Q_1 is

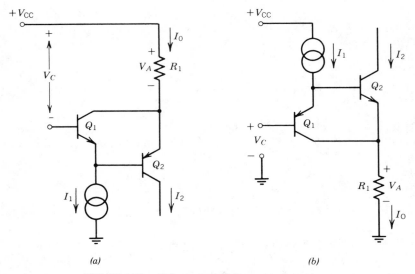

(a)   (b)

FIGURE 4.20. Voltage-controlled current source stages.

negligible, the output current I_2 is related to the other two branch currents as

$$I_2 = I_0 - I_1 \tag{4.43}$$

However, since I_0 is set by the voltage drop across R_1 as

$$I_0 = \frac{V_A}{R_1} \approx \frac{V_C}{R_1} \tag{4.44}$$

the output current then becomes proportional to the control voltage as

$$I_2 = \frac{V_C}{R_1} - I_1 \tag{4.45}$$

where I_1 is a constant bias current.

Figure 4.20b shows the complementary equivalent of the same voltage-controlled constant-current stage. However, in this case, the input stage is a *pnp* transistor, and care must be taken in neglecting its base current. The voltage-controlled current source circuits of Figure 4.20 are useful building blocks in the design of voltage-controlled oscillators or automatic gain control circuits.

4.4. SUPPLY-INDEPENDENT BIASING

The constant-current stages described in Sections 4.1 and 4.2 require the generation of a reference current I_{ref}, which is then reproduced at the output. As shown in the case of the simple current mirror of Figure 4.1, I_{ref} is normally derived by connecting a current-setting resistor R_1 directly to the supply voltage. This results in a supply dependence of the output current I_2 given as

$$I_2 = I_{\text{ref}} = \frac{V_{CC} - V_{BE}}{R_1} \approx \frac{V_{CC}}{R_1} \tag{4.46}$$

In many applications, this supply dependence is not desirable and alternate bias solutions must be found. In this section, some of these alternate approaches shall be examined.

Using V_{BE} as Reference

The supply dependence of output current can be greatly reduced by using the transistor base–emitter voltage V_{BE} as a reference to generate the output current. Figure 4.21 shows two such biasing schemes. In the circuit of Figure 4.21a, the diodes D_1 and D_2, are formed by diode-connected transistors. Assuming that the voltage drops across D_1 and D_2 are each equal to V_{BE}, the base voltage of Q_2 is fixed at $2V_{BE}$, and the voltage drop across R_2 is equal to V_{BE}. Neglecting the base current of Q_2, this results in an output current

$$I_{\text{out}} \approx \frac{V_{BE}}{R_2} \tag{4.47}$$

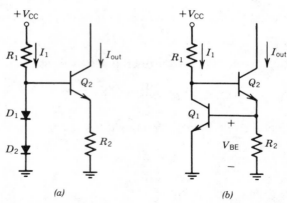

FIGURE 4.21.   Supply-independent current bias sources using V_{BE} as reference.

The circuit of Figure 4.21b also works in a similar manner. The grounded-emitter transistor Q_1 forces a net voltage drop of V_{BE1} to appear across R_2. Neglecting the base current of Q_2, this again results in an output current expression similar to that of Eq. (4.47).

The bias circuits of Figure 4.21 are not completely independent of the supply voltage, since the current I_1, which sets the V_{BE} drops is dependent on the supply voltage. Using Eq. (2.5) of Chapter 2, which relates the V_{BE} drop to the current through the diode, one can write for the case of Figure 4.21, or Eq. (4.47),

$$V_{BE} = V_T \ln\left(\frac{I_1}{I_S}\right) \tag{4.48}$$

where I_S is the reverse saturation current of diodes D_1 and D_2 of Figure 4.21a or the transistor Q_1 of Figure 4.21b. The current I_1 is in turn set by the supply voltage as

$$I_1 = \frac{V_{CC} - 2V_{BE}}{R_1} \approx \frac{V_{CC}}{R_1} \tag{4.49}$$

Thus, taking into account the current dependence of V_{BE}, Eq. (4.47) can be written as

$$I_{out} \approx \frac{V_T}{R_2} \ln\left(\frac{V_{CC}}{R_1 I_S}\right) \tag{4.50}$$

which shows a logarithmic dependence on the supply voltage.

A common problem associated with the V_{BE}-referenced current sources is the strong temperature dependence of V_{BE}, which is of the order of -3000 ppm/°C. If the resistor R_2 of Figure 4.21 is a base-diffused resistor, it would introduce an additional temperature coefficient of $\approx +2000$ ppm/°C into the denominator of Eq. (4.47) which would result in a total temperature coefficient of I_{out} on the order of -5000 ppm/°C. This would result in approximately 50% reduction in the value of I_{out} over a 100°C temperature change.

An alternate approach to supply insensitive biasing is the low-current current source of Figure 4.8, discussed earlier. As indicated by Eq. (4.30), this circuit also exhibits a logarithmic dependence on the supply voltage, where

$$I_2 \approx \frac{V_T}{R_2} \ln\left(\frac{V_{CC}}{R_1 I_2}\right)$$ (4.51)

However, since $V_T (= kT/q)$ exhibits a positive temperature coefficient of approximately $+3000$ ppm/°C at room temperature, its temperature drift can be compensated, in part, by the temperature coefficient of R_2. Assuming that R_2 is a diffused resistor with a temperature coefficient of approximately $+2000$ ppm/°C, the resulting temperature coefficient of I_2 in Eq. (4.51) is on the order of $+1000$ ppm/°C. This is a significant improvement over the simple V_{BE}-referenced bias circuits of Figure 4.21.

Self-Biasing References

The power supply dependence can be greatly improved by using so-called *self-biased* or *boot-strapped* current references. Instead of generating the reference current by connecting a resistor to the supply voltage, the reference current can be made to depend on the output current of the current source. This is done by sensing the output current by means of a current mirror and forcing the reference current to be equal to it. Figure 4.22 shows the application of this technique to the V_{BE}-referenced current source shown in Fig. 4.21b.

Assuming that the effects of finite base currents and the finite output conductances are negligible, the operation of the circuit can be analyzed as follows: Q_1,

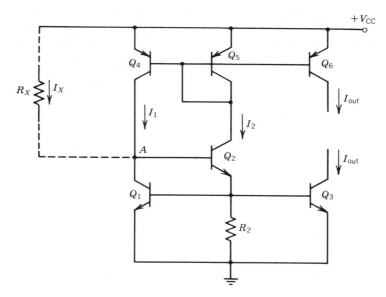

FIGURE 4.22.   Self-biasing V_{BE} reference.[1]

Q_2, and R_2 function as described in Figure 4.21b, and set the collector current I_2 of Q_2. This current is then sensed by the current mirror transistors Q_5 and Q_4, and the reference current I_1 is forced to equal I_2. The operating point of the circuit is determined by two constraints: (1) I_2 must show a logarithmic dependence on I_1 as given by Eqs. (4.47) and (4.48). (2) I_2 must equal I_1 as dictated by the current mirror transistors Q_4 and Q_5. The operating point of the circuit which satisfies these conditions can be graphically determined as the intersection of the two curves shown in Figure 4.23. The intersection of the two curves, point D, determines the desired operation of the circuit. Once the current level I_2 is set, it can be reflected to the outputs of the circuit as the output current I_{out} through the *pnp* or the *npn* current mirrors, as shown in Figure 4.22.

In most of the self-biased circuits of the type described, there are not one but *two* stable operating points. The second operating point, which is not desirable, normally occurs at zero current state. In other words, the circuit will be perfectly stable if all the currents are zero. This can be seen from the graphic solution of Figure 4.23, as well as by examining the circuit of Figure 4.22. For example, if Q_1 was nonconducting, its V_{BE} would be zero, thus the voltage across R_2 would be zero, resulting in $I_2 = 0$, which would cause I_1 to be zero. This problem, which is common to self-biased circuits using internal feedback, is often referred to as the *startup problem*. This problem can be avoided by eliminating the undesired zero-current state. In its simplest form, this can be done by supplying a very small amount of startup current I_X to node A in Figure 4.22 by means of a high-value resistor R_X, connected to the supply voltage, as shown by dashed lines. If the startup current is chosen such that $I_X \ll I_1$, it has a negligible effect on circuit operation.

The circuit of Figure 4.22 is shown in its simplest form to illustrate the principle of self-biasing circuits. If desired, the output of the circuit can be greatly improved by using Wilson current mirrors or cascoded current mirrors. Many varieties of self-biased references, or their startup circuits, have been developed. These are well covered in the literature.[4]

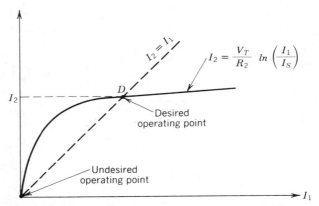

FIGURE 4.23. Graphic solution for operating point of circuit of Figure 4.22.

It must be emphasized once more that when designing self-biased circuits it is *extremely* important to examine the possible operating points of the circuit to avoid startup instability problems.

4.5. VOLTAGE SOURCES

In a variety of circuit applications, it is necessary to establish a low-impedance point within the circuit which can serve as an internal voltage supply. Ideally, such a voltage reference point is required to have both a very low ac impedance and a very stable dc voltage level which is insensitive to power supply and temperature variations. In most applications, however, only one of these two requirements, that is, either the low impedance or the dc voltage stability, is of prime importance. The circuits which primarily fulfill the low-impedance requirements are known as voltage sources, whereas those specifically designed to provide a constant voltage, independent of the supply or the temperature changes, are called voltage references.

The voltage source stages are normally used to provide independent bias levels within the circuit. In such an application, the low ac impedance of the voltage source is necessary to buffer or decouple adjacent gain stages. An example of such an application is the use of voltage source stages to form a common bias point for the inputs of a differential gain stage. Figure 4.24 shows some of the practical voltage source configurations often used in IC design. In the circuit of Figure 4.24a, the low output impedance of an emitter-follower stage is used to simulate a low-impedance voltage source with an output voltage

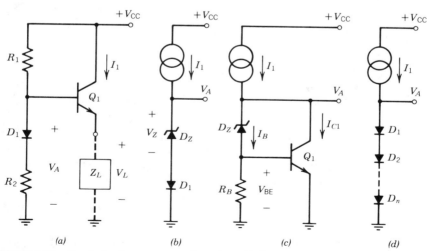

FIGURE 4.24. Practical voltage source configurations: (*a*) Common-collector stage; (*b*) temperature-compensated Zener diode; (*c*) temperature-compensated Zener diode with buffered output; (*d*) diode string.

level V_L given as

$$V_L = (V_{CC} - V_{BE})\frac{R_2}{R_1 + R_2} \approx V_{CC}\frac{R_2}{R_1 + R_2} \qquad (4.52)$$

Note that in this configuration the diode D_1 in the bias string is used to partially offset the dc value and the temperature dependence of the V_{BE} drop across Q_1. The load Z_L represents the rest of the circuitry biased by the current through Q_1. Using the hybrd-π model for the transistor (see Fig. 2.10), the resistance level R_O looking into the emitter of Q_1 can be expressed as

$$R_O \approx \frac{V_T}{I_1} + \frac{R_1 R_2}{\beta(R_1 + R_2)} \qquad (4.53)$$

Due to the resistive bias string in Figure 4.24a, the values of the bias voltage V_A and the output voltage V_L are both dependent on the supply voltage V_{CC}. This dependence can be avoided using the bias circuits of Figure 4.24b, c, or d. In each of these circuits, the impedance level looking into the bias terminal is low enough for most applications to eliminate the need for an additional emitter-follower stage. In many applications, the current source I_1 may also be replaced by a resistor connected between $+V_{CC}$ and V_A.

The circuits of Figure 4.24b and c both provide an output voltage level

$$V_A = V_Z + V_{BE} \qquad (4.54)$$

where V_Z is the breakdown voltage of the Zener diode. The forward-biased diode D_1 of Figure 4.24b provides partial compensation for the positive temperature coefficient of V_Z. In a monolithic structure, the combination of D_Z and D_1 can be conveniently designed as a single transistor structure with two separate emitters, as shown in Figure 3.10. The impedance level measured looking into the output terminal is

$$R_{out} = R_Z + \frac{V_T}{I_1} \qquad (4.55)$$

where R_Z is the dynamic resistance of the base–emitter breakdown diode. For typical integrated device structures, R_Z is in the range of 40–100 Ω.

The circuit of Figure 4.24c also provides an output voltage level equal to $V_Z + V_{BE}$ by impressing a voltage drop equal to V_{BE} across the bias resistor R_B. In this case, Q_1 serves as an active gain stage and automatically adjusts its current I_{C1} to maintain the output voltage level constant. The output impedance of the circuit is approximately the same as that given by Eq. (4.55). However, since the current I_B through the Zener diode is maintained constant, independent of supply voltage changes, the sensitivity of V_A to supply voltage variations is greatly reduced.

Since the base–emitter breakdown voltage is set by the IC fabrication process, the values of V_A available from the circuits of Figure 4.24b and c are restricted to be within the 6.5–8-V range.

Figure 4.24*d* shows how a number of diodes, or diode-connected transistors, can be cascaded in series to simulate a low-impedance output voltage V_A, where

$$V_A = nV_{BE} \tag{4.56}$$

and an output resistance level of

$$R_{out} = \frac{nV_T}{I_1} \tag{4.57}$$

where n is the number of diodes in the string. Such a voltage source has a strong negative temperature coefficient, given as

$$\frac{\partial V_A}{\partial T} = n \frac{\partial V_{BE}}{\partial T} \approx -2n \text{ mV/°C} \tag{4.58}$$

Since each diode in the string requires a separate collector pocket, a string involving large numbers of diodes may occupy a significant portion of the chip area. In many applications, an alternate solution to the diode string is the so-called V_{BE}-*multiplier* circuit shown in Figure 4.25. Such a circuit can produce an output voltage that is an arbitrary multiple of the transistor base–emitter drop. In the circuit of Figure 4.25*a*, the voltage drop across R_2 is constrained to be equal to the transistor V_{BE}. Assuming that the base current of Q_1 is negligible, the current through R_2 is the same as that through R_1. Therefore, the output dc level can readily be related to the transistor V_{BE} as

$$V_A = I_2(R_1 + R_2) = V_{BE}\left(1 + \frac{R_1}{R_2}\right) \tag{4.59}$$

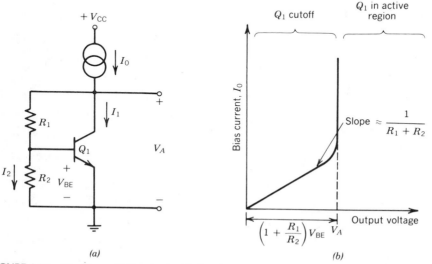

FIGURE 4.25. The V_{BE}-multiplier circuit: (*a*) Circuit connection; (*b*) output level as a function of bias current I_0.

Note that, as long as $I_0 R_2 > V_{BE}$, the transistor is in its active region, and due to the shunt feedback provided by R_1, the transistor current I_1 automatically adjusts itself to maintain I_2 and V_A relatively independent of the supply voltage or current. The corresponding current–voltage characteristics of the V_{BE}-multiplier circuit are shown in Figure 4.25b.

In most applications, the circuit of Figure 4.25a provides a convenient substitute for a diode string, particularly when a large number of diodes or a noninteger multiple of V_{BE} are required. Using the hybrid-π model, the output resistance of the V_{BE} multiplier can be readily calculated as

$$R_O = \frac{R_1}{\beta} + \frac{R_1 + R_2}{g_m R_2} \tag{4.60}$$

where $g_m = (= I_1/V_T)$ is the transconductance of Q_1. For most applications, the value of R_O is in the range of 50–200 Ω.

Figure 4.26 shows a modified version of the basic shunt feedback circuit which is useful for obtaining high-value voltage sources without requiring high-voltage Zener diodes. Assuming that D_Z is the reverse-biased base–emitter diode with a break-down voltage V_Z, the voltage level at the output terminal can be expressed as

$$V_A = (V_Z + V_{BE})\left(1 + \frac{R_1}{R_2}\right) \tag{4.61}$$

provided that $I_0 R_2 > V_Z + V_{BE}$ so that Q_1 is in its active region. The circuit of Figure 4.26 is particularly useful for high-voltage integrated circuits, where it can be used as a high-value voltage source, or it can be substituted for a high-voltage avalanche diode, for overvoltage protection in circuits prone to high-voltage transients.

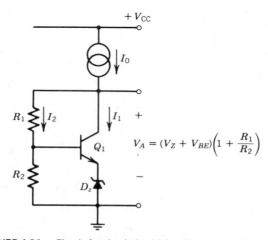

FIGURE 4.26. Circuit for simulating high-voltage breakdown diode.

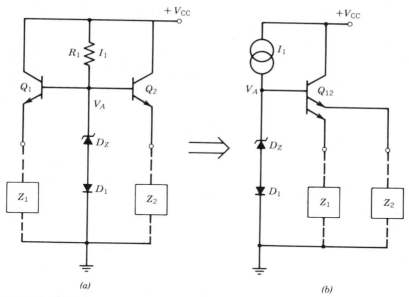

FIGURE 4.27. Multiple voltage sources biased from same reference voltage: (a) Discrete design; (b) its monolithic counterpart using multiple-emitter *npn* transistors.

In providing low-impedance bias points within the circuit, the emitter-follower stage is the most commonly used buffer circuit. Since many emitter-follower stages can share the same isolation pocket, they require a minimum amount of chip area. In some analog circuit applications, it is necessary to provide multiple voltage sources which are biased from the same reference voltage, but are buffered from each other such that the ac signals in one source would be relatively isolated from those in the other. In conventional circuit design using discrete devices, this can be done using two separate emitter-follower stages, as shown in Figure 4.27a. However, using the design and layout advantages of monolithic circuits, the same circuit can be designed using a multiple-emitter transistor (see Fig. 2.21), as shown in Figure 4.27b, with a minimum increase in chip area.

4.6. DC LEVEL-SHIFT STAGES

Since large-value coupling capacitors are not available in monolithic circuits, all broadband gain stages need to be dc coupled. This means that the output dc level of a gain stage should be compatible with the dc level at the input of the next stage. In an *npn* common-emitter gain stage, the output dc level is always higher than the dc level of the input. Therefore, if a number of such gain stages are cascaded, the output dc level rapidly builds up toward the positive supply voltage. This in turn limits the amplitude and the linearity of the available output swing. Ideally, such a dc level buildup can be avoided by using complementary *pnp-npn* gain stages. However, the *pnp* transistors available in monolithic form have relatively poor frequency response and current gain characteristics.

If an analog integrated circuit is comprised of a cascade of *npn* gain stages, this positive dc level buildup can be overcome by using a level-shift stage between each gain stage to shift the output dc level toward the negative supply with minimum attenuation of the ac signal. In general, such a stage also serves as a unilateral buffer between successive gain stages. Therefore, it is required to have a high input impedance and a relatively low output impedance to prevent interstage loading. Figure 4.28 shows some practical dc level-shift stages for monolithic circuit applications. In each case, a common-collector stage is used at the input to avoid loading the output of the gain stage connected to V_1.

The resistive level-shift stage of Figure 4.28*a* provides a simple means of shifting the input level V_1 to a more negative dc level V_2, where

$$V_2 = (V_1 - V_{BE})\frac{R_2}{R_2 + R_1} \qquad (4.62)$$

The main drawback of the resistive level-shift stage is attenuation of the ac signal along with the dc level shift. The ac voltage gain A_V for the resistive level-shift

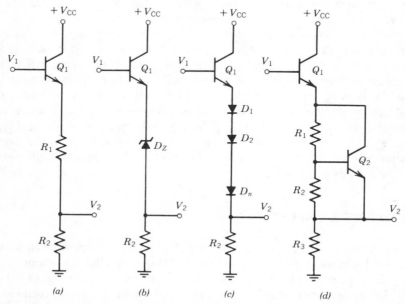

FIGURE 4.28. Some practical dc level-shift stages: (*a*) Resistive; (*b*) Zener diode; (*c*) diode string; (*d*) V_{BE} multiplier.

stage is less than unity, and is given as

$$A_V = \frac{R_2}{R_1 + R_2} < 1 \qquad (4.63)$$

Thus, as the value of R_2 is decreased to improved the net dc level shift from V_1 to V_2, the ac gain of the stage deteriorates rapidly. The output impedance of such a stage is also relatively high, being equal to the shunt combination of R_1 and R_2.

The Zener diode level-shift stage of Figure 4.28b provides an alternate means of shifting the dc level by the amount

$$V_1 - V_2 = V_{BE} + V_Z \qquad (4.64)$$

where V_Z is the breakdown voltage of the Zener diode D_Z. The reverse breakdown characteristic of the base–emitter junction is used to form the Zener diode (i.e., $V_Z \approx 6\text{–}9$ V). If the bulk resistance of D_Z is negligible compared to R_2, the voltage gain for the stage is approximately unity. The two main disadvantages of the Zener diode level-shift stage are the limitations on the values of V_2 available in integrated circuits and the excess noise generated by the breakdown diode D_Z. Therefore, such a level-shift scheme is not suitable for low-level ac signals or for circuits operating with relatively low supply voltages, such as $V_{CC} < 10$ V.

The diode string level-shift stage of Figure 4.28c provides a net dc level shift $V_1 - V_2$ given as

$$V_1 - V_2 = (n + 1)V_{BE} \qquad (4.65)$$

where n is the number of diodes in the string. Normally, the diodes D_1 through D_n would be formed by diode-connected transistors. Assuming that the dynamic impedance of the diodes is negligible compared to R_2, the voltage gain of the stage is approximately unity. The output impedance R_O for the circuit is quite low, given as

$$R_O \approx (n + 1)\frac{V_T}{I_1} \qquad (4.66)$$

where V_T is the thermal voltage. The diode string level-shift stage has two disadvantages: (1) Since each diode requires a separate isolation pocket, such a stage may take up a sizable chip area and have appreciable shunt capacitance to the substrate. (2) The output dc level shows a strong temperature dependence due to the change of the diode voltage V_{BE} with temperature.

The circuit of Figure 4.28d is often used as a substitute for the diode string level-shift stage. In this circuit, the diode chain is replaced by the V_{BE}-multiplier stage described in Figure 4.25. The net dc level change across the stage is

$$V_1 - V_2 = V_{BE}\left(2 + \frac{R_1}{R_2}\right) \qquad (4.67)$$

Assuming that $\beta \gg 1$, the net voltage gain across the stage can be expressed as

$$A_V \approx \frac{R_3 R_2 g_m}{1 + R_3 R_2 g_m} \approx 1.0 \qquad (4.68)$$

where g_m is the transconductance of Q_2. Similarly, the output resistance R_O for the stage can be calculated to be

$$R_O = R_3 \left\| \frac{R_1}{R_2 g_m} \right. \qquad (4.69)$$

The main disadvantages of the level-shift circuits of Figure 4.28c and d are the temperature dependence of V_2 due to the V_{BE} change with temperature.

Another commonly used level-shift stage is the resistor and current source combination shown in Figure 4.29. Since the dynamic output resistance of the current source is much higher than R_1, the voltage gain of the stage is very close to unity. However, the output dc level is shifted toward negative supply (or ground) by the amount

$$V_1 - V_2 = 2V_{\mathrm{BE}} + I_0 R_1 \qquad (4.70)$$

This type of level-shift stage is susceptible to capacitive loading at node A in Figure 4.29, which would limit its frequency response. This capacitive loading is mainly due to the collector–substrate capacitance of the *npn* transistor which forms the current source I_0. Therefore, in the layout of the circuit, care must be taken to minimize the collector area of this transistor.

pnp Level-Shift Stages

In spite of its relatively poor gain and frequency response, lateral *pnp* transistors still find a number of applications as gain or level-shift stages. In general, their

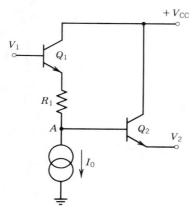

FIGURE 4.29. Level-shift stage using resistor and current source combination.

applications are limited to low-frequency circuits, where the low f_T of the lateral *pnp* transistor (≈ 5 MHz) does not present a problem.

The low-gain characteristics of the lateral *pnp* transistor can be improved by combining with an *npn* transistor to form a so-called *composite pnp* transistor, as described in Chapter 2 (see Fig. 2.28). Figure 4.30 shows the use of such a composite *pnp* transistor as both a level-shift and a gain stage.

Since the composite *pnp* transistor is a feedback circuit made up of two devices, it exhibits a two-pole rolloff in its frequency response and is prone to oscillations. In practical applications, this problem can be avoided by connecting a resistor R_X between the base and the emitter of Q_2 to reduce the overall current gain of the combined Q_1 and Q_2 and to have Q_1 operate at a slightly higher current level than just the base current of Q_2. As a rule of thumb, the current I_X through R_X is chosen to be $\approx 10\%$ of I_1. This is done by choosing R_X such that

$$R_X = \frac{10V_{BE}}{I_1} \tag{4.71}$$

Normally, R_X is designed as a noncritical pinch resistor.

The output dc voltage of the composite *pnp* level-shift stage of Figure 4.30 is given as

$$V_2 = \frac{R_2}{R_1}(V_{CC} - V_{BE} - V_1) \tag{4.72}$$

Similarly, the small-signal ac voltage gain A_V of the circuit can be expressed as

$$A_V \approx -\frac{R_2}{R_1} \tag{4.73}$$

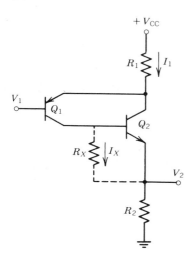

FIGURE 4.30. Level-shift stage using composite *pnp* configuration.

Figure 4.31 shows other commonly used level-shift stages comprised of *pnp-npn* transistors. The circuit of Figure 4.31*a* is derived from the voltage-controlled current source circuit described earlier (see Fig. 4.20). It provides an output voltage level V_2 given as

$$V_2 = I_2R_2 = \frac{R_2}{R_1}(V_{CC} - V_1 - I_1R_1) \qquad (4.74)$$

where I_1 is a constant-current source. The ac gain can be evaluated by differentiating Eq. (4.74)

$$A_V \approx -\frac{R_2}{R_1} \qquad (4.75)$$

In the circuit of Figure 4.31*b*, the lateral *pnp* transistor is operated in its common–base configuration in order to enhance its frequency capability. The output voltage level V_2 is equal to

$$V_2 = \alpha_p I_1 R_2 = \frac{\alpha_p R_2}{R_1}(V_1 - V_{bias} - 2V_{BE}) \qquad (4.76)$$

where α_p is the common-base current gain of Q_2. The ac voltage gain of the circuit is

$$A_V = \alpha_p \frac{R_2}{R_1 + 2/g_m} \approx \alpha_p \frac{R_2}{R_1} \qquad (4 . 77)$$

where $g_m (= I_1/V_T)$ is the transconductance of Q_1 or Q_2.

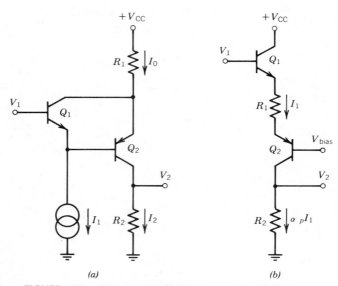

(a)   *(b)*

FIGURE 4.31. Other commonly used *pnp–npn* level-shift stages.

The circuit of Figure 4.31b offers the best frequency performance among the *pnp–npn* level-shift stages discussed so far. However, it has two disadvantages: it requires a low-impedance bias source V_{bias}, and its gain and level-shift characteristics depend on α_p, which may vary between 0.8 and 0.98 in practical lateral *pnp* devices.

pnp current mirrors are also suitable for dc level shifting, provided that the frequency requirements are relatively low (i.e., ≤ 1 MHz). They are particularly suitable for level shifting in differential gain stages. Figure 4.32 shows their application in a different gain stage. The circuit of Figure 4.32a is a basic differential gain stage with a voltage gain of R_L/R_E, and with an output common-mode voltage $(V_{out})_{CM}$, close to the power supply, where

$$(V_{out})_{CM} = V_{CC} - \frac{I_0 R_L}{2} \qquad (4.78)$$

In the level-shifted version of the same circuit (see Fig. 4.32b), the *pnp* current mirrors, made up of D_3, Q_3 and D_4, Q_4, reflect the output current toward ground (or negative supply) so that the voltage gain remains unchanged, but the output common-mode voltage now becomes

$$(V_{out})_{CM} \approx \frac{I_0 R_L}{2} \qquad (4.79)$$

In the actual implementation of the design, D_3, Q_3 and D_4, Q_4 would each be designed as split-collector *pnp* transistors to conserve chip area.

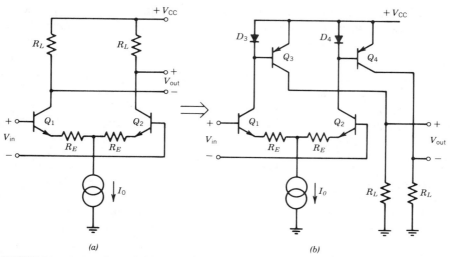

FIGURE 4.32. Differential level shifting using *pnp* current mirrors: (*a*) Basic differential stage; (*b*) its modified version with current mirror level shifting.

4.7.   TEMPERATURE-INDEPENDENT BIASING

In the design of various analog circuits, such as D/A converters, voltage regulators, or low-drift amplifiers, it is necessary to establish a temperature-independent bias reference within the circuit. This stable bias reference can be either a current or a voltage. In most applications, voltage rather than current references are preferred since they are easier to interface with the rest of the circuitry.

In the case of a voltage reference, unlike the case of voltage sources, the main emphasis is not on low output impedance but on the temperature stability of the voltage level. Temperature stability requirements on a voltage reference are typically $\leq$ 100 ppm/°C. In many cases, by careful design and compensation even this figure can be significantly improved to be within the range of 20–40 ppm/°C.

Typical temperature coefficients of monolithic components are in the range of a few thousand ppm/°C or more. However, by making use of the matching and tracking characteristics of monolithic components, and their close thermal coupling on the chip, it is possible to compensate the thermal drifts to a few parts per million. The basic principle of temperature compensation used in monolithic IC design is very simple. One starts with a *predictable* temperature drift, then finds another predictable temperature source of opposite polarity which can be scaled by a temperature-independent scale factor. Then, by proper circuit design, the effects of the two opposite-polarity drifts are made to cancel, resulting in a nominally zero temperature coefficient voltage level.

Among the monolithic components, there are three basic temperature drift sources which are reasonably predictable and repeatable:

1.   The temperature dependence of the base–emitter drop V_{BE} which shows a strong negative temperature coefficient, typically on the order of -2 mV/°C.

2.   The temperature dependence of the V_{BE} difference ΔV_{BE}, which is proportional to the absolute temperature, through the thermal voltage V_T [see Eq (4.28)] and, thus, exhibits a positive temperature coefficient.

3.   The temperature drift of the base–emitter Zener diode V_Z, which is inherently low and positive in polarity (typically on the order of $+200$ to $+500$ ppm/°C.

In designing temperature-compensated bias references, one achieves the required compensation by scaling one or more of these drift sources and subtracting them from each other.

The temperature coefficient of the monolithic resistors (with the exception of thin-film resistors) is too high and too nonlinear to be used for any predictable temperature compensation. However, the resistor ratios show excellent tracking over temperature, with the temperature coefficient of the resistor ratio being on the order of ± 5–± 20 ppm/°C for well-matched resistors. Thus, monolithic resistors are suitable for generating temperature-insensitive scale factors, when used in a ratio rather than in an absolute-value form.

Figure 4.33a shows a simple voltage reference circuit which makes use of the opposite-polarity drift between the Zener voltage V_Z and the forward diode voltage V_{BE}.[5] The base–emitter avalanche diode D_Z is supplied by a constant current I_1 and provides a bias voltage V_Z with a positive temperature coefficient (typically $\approx +3$ mV/°C). The temperature dependence of the V_{BE} drop across Q_1 and D_1 results in a temperature coefficient of about $+7$ mV at the cathode of D_1. Similarly, the thermal variation of the voltage drop across D_2 creates a temperature coefficient of ≈ -2 mV/°C at the anode of D_2. Thus, by tapping the resistor string R_1 and R_2, connecting these two points of opposite temperature drift, a voltage reference V_{ref} can be made to have a nominally zero temperature coefficient. The voltage level of V_{ref} is given as

$$V_{ref} = \frac{R_2 V_Z + V_{BE}(R_1 - 2R_2)}{R_1 + R_2} \tag{4.80}$$

The temperature coefficient of V_{ref} can be nominally set to zero by setting the resistor ratio as

$$\frac{R_1 - 2R_2}{R_2} = -\frac{\partial V_Z/\partial T}{\partial V_{BE}/\partial T} \tag{4.81}$$

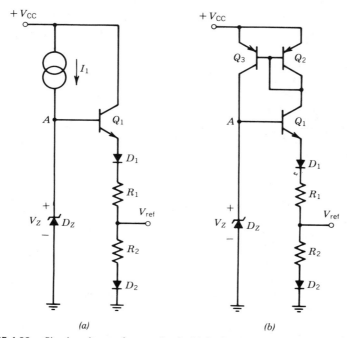

(a) (b)

FIGURE 4.33. Simple voltage reference circuit: (a) Basic circuit; (b) its self-biased version.

For typical values of $\partial V_Z/\partial T$ and $\partial V_{BE}/\partial T$ associated with IC components, the nominal value of V_{ref} for a zero temperature coefficient is in the range of 1.7–2.5 V. Figure 4.33b shows a self-biased version of the same circuit, which can generate the supply-independent constant current I_1 internally by means of the *pnp* current mirrors Q_2 and Q_3. A word of caution is in order, however, regarding the circuit of Figure 4.33b. It is a self-biased circuit; therefore, it may require startup circuitry which will inject an initial startup current to node A in the circuit to avoid the zero-current stable stage (see Figs. 4.22 and 4.23).

Stable voltage references with temperature coefficients on the order of ± 30 to ± 50 ppm/°C have been reported using the basic circuit technique shown in Figure 4.33.[5]

The basic disadvantages of the Zener-referenced bias circuits of the type shown in Figure 4.33 are that they require a relatively high value of power supply (typically ≥ 10 V) and introduce substantial noise into the circuit, due to the avalanche breakdown within the diode, as well as exhibiting some long-term drift of the Zener voltage V_Z. However, in the more recent designs, the long-term drift problems have been largely eliminated by using buried-Zener structures (see Fig. 3.11), which confine the breakdown to the subsurface region of silicon.

Band-Gap Reference Circuits

The high supply voltage and the Zener noise problems associated with the Zener-biased reference circuit can be avoided by using the so-called *band-gap* reference circuit. Such a reference circuit operates on the principle of compensating the negative temperature drift of V_{BE} with the positive temperature coefficient of the thermal voltage V_T. Its principle of operation is illustrated, symbolically, in Figure 4.34. First, we generate a known negative temperature drift due to V_{BE}; next, we produce a positive temperature drift due to the thermal voltage $V_T = kT/q$; and finally, we scale the latter with a constant (i.e., temperature-independent scale factor K) and subtract it from the former to obtain nominally zero temperature dependence. In the simplified model of Figure 4.34, the output voltage V_{out} is given as

$$V_{out} = V_{BE} + KV_T \qquad (4.82)$$

Since each of the two terms in the above equation exhibits opposite-polarity temperature drifts, it should be possible, at least in theory, to make V_{out} nominally independent of temperature. This, in summary, is the principle of a band-gap reference. The temperature-stabilized output dc level, where $\partial V_{out}/\partial T$ is nominally equal to zero, comes about at an output voltage level on the order of $+1.25$ V. One can show mathematically that this voltage level is very nearly equal to the band-gap voltage of silicon.[6] The name *band-gap reference* is derived from this relationship.

Figure 4.35 shows a simple implementation of the band-gap reference concept. In this circuit, Q_1 and Q_2 operate as a low-current bias stage, as shown in

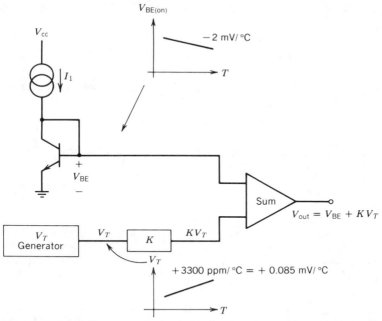

FIGURE 4.34. Symbolic model for illustrating the principle of operation of a band-gap reference.[1]

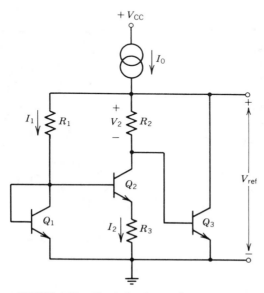

FIGURE 4.35. Simple band-gap reference circuit.

Figure 4.8. The voltage across R_3 is equal to the V_{BE} difference between Q_1 and Q_2, that is,

$$I_2 R_3 = V_{BE1} - V_{BE2} = V_T \ln\left(\frac{I_1}{I_2}\right) = \Delta V_{BE} \tag{4.83}$$

Assuming $\beta \gg 1$, the net voltage drop V_2 across R_2 can be written as

$$V_2 = \frac{R_2}{R_3} \Delta V_{BE} = V_T \frac{R_2}{R_3} \ln\left(\frac{I_1}{I_2}\right) \tag{4.84}$$

The output voltage V_{ref} is then equal to the base–emitter drop of Q_3 plus the voltage drop V_2, that is,

$$V_{ref} = V_{BE} + V_T \frac{R_2}{R_3} \ln\left(\frac{I_1}{I_2}\right) \tag{4.85}$$

Assuming that the ratio I_1/I_2 can be kept relatively insensitive to temperature, Eq. (4.85) is the same form as Eq. (4.82), and the first-order temperature dependence of V_{ref} can be reduced to zero. Practical voltage reference circuits with temperature coefficients in the range of 30–60 ppm/°C can be obtained in this manner.

The main drawback of the basic band-gap reference circuit of Figure 4.35 is the difficulty of maintaining the current ratio I_1/I_2 independent of temperature. Figure 4.36 shows two alternate circuit approaches which largely overcome this problem at the expense of added circuit complexity, that is, by using a high-gain operational amplifier in a feedback loop.

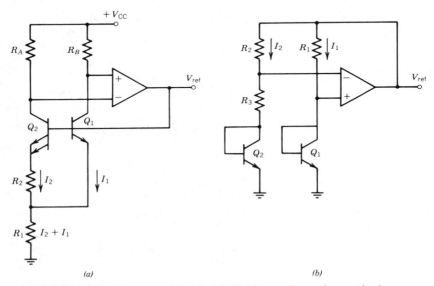

(a)   (b)

FIGURE 4.36. Alternate configurations for band-gap voltage reference circuits.

In the circuit of Figure 4.36a, the emitter area of Q_2 is made to be n times that of Q_1. The output voltage can be expressed as

$$V_{ref} = V_{BE1} + (I_1 + I_2)R_1 \tag{4.86}$$

Assuming that the collector resistors R_A and R_B are identical, the collector currents of Q_1 and Q_2 are *forced* to be equal in order to set the differential voltage at the input of the operational amplifier equal to zero. This in turn forces I_1 to be equal to I_2, and the voltage drop across R_2 is equal to the ΔV_{BE} between Q_1 and Q_2. Thus, Eq. (4.86) can be rewritten as

$$V_{ref} = V_{BE1} + \frac{2R_1}{R_2}V_T \ln(n) \tag{4.87}$$

which is of the form given by Eq. (4.82). In practical applications, the emitter area ratio n is usually taken as 2, that is, Q_2 is made to have *twice* the emitter area of Q_1.

The circuit of Figure 4.36b also operates on a similar principle.[1] Since the differential voltage at the input of the operational amplifier has to be zero, the currents I_1 and I_2 are forced to have the ratio

$$\frac{I_1}{I_2} = \frac{R_2}{R_1} \tag{4.88}$$

Assuming that Q_1 and Q_2 are well matched and neglecting the effect of base currents, the difference in their base-emitter voltages can be expressed as

$$V_{BE1} - V_{BE2} = \Delta V_{BE} = V_T \ln\left(\frac{I_1}{I_2}\right) = V_T \ln\left(\frac{R_2}{R_1}\right) \tag{4.89}$$

Note that this differential voltage ΔV_{BE} appears directly across resistor R_3, that is,

$$V_{BE} = I_2 R_3 = \frac{I_1 R_2 R_3}{R_1} \tag{4.90}$$

The output voltage V_{ref} is equal to V_{BE1} plus the voltage drop across R_1,

$$V_{ref} = V_{BE1} + I_1 R_1 \tag{4.91}$$

Substituting the results of Eqs. (4.89) and (4.90) into Eq. (4.91), one obtains

$$V_{ref} = V_{BE1} + V_T\frac{R_2}{R_3} \ln\left(\frac{R_2}{R_1}\right) \tag{4.92}$$

which is again of the same form as the basic band-gap reference equation [Eq. (4.82)].

The advantage of the circuits of Figure 4.36 over that of Figure 4.35 is the added degree of stability, since the internal current levels and their ratios are well controlled. Their disadvantage is the need for additional resistors, which need to be perfectly matched, and the need for a stable high-gain feedback amplifier.

4.8. STABILIZATION OF CHIP TEMPERATURE

The close thermal coupling between the integrated components on a monolithic chip allows the circuit designer an added degree of freedom in controlling the thermal environment of the IC chip. Because of the small thermal capacity of the chip, the entire chip can be maintained at a constant elevated temperature within the IC package. This can be done by incorporating a heating element, along with a temperature sensor–controller circuit, on the same IC chip, alongside the circuit to be stabilized. This temperature sensor–controller unit is basically a temperature regulator whose function is to maintain the IC chip at a constant elevated temperature with minimum dependence on ambient variations. In other words, the IC package can serve as a miniature *temperature chamber*, where the IC chip itself is both the heating and the temperature-regulating element. Since the thermal capacity of the IC chip is relatively small, such a stabilization scheme can often be achieved without requiring excessive amounts of power dissipation.[8]

Figure 4.37 shows a functional block diagram of a temperature sensor–controller circuit for stabilizing the substrate temperature. All the necessary circuit elements to form such a temperature regulating system are readily available in the form of integrated components. The predictable temperature dependence of the transistor V_{BE} can be utilized as the temperature-sensing element, and a power transistor can be used as the heating element. In order to minimize the thermal gradients through the chip, the circuit to be stabilized is laid out symmetrically with respect to the heating and sensing elements, and the most critical components, such as the input stage of a high-gain amplifier, are placed equidistant from the heating element and located nearest to the sensor. The threshold level of the heater unit is set such that the heater is operative over

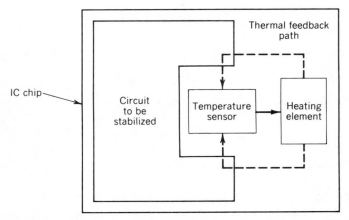

FIGURE 4.37.   Block diagram for substrate temperature stabilization of an IC chip.

the entire temperature range of interest and keeps the chip temperature at a relatively constant level, above the highest ambient temperature to be encountered.

The most important application of temperature-controlled substrate circuits is in the design of precision voltage reference circuits or in very-low-drift differential amplifiers. For example, if the temperature of the chip can be stabilized to $\pm 5°C$ over an ambient temperature change of from $-55°C$ to $+125°C$, an inherently stable voltage reference, which exhibits ± 40-ppm drift per degree celsius change of chip temperature, can appear to be stable to ± 2-ppm/$°C$ change of ambient temperature.

Using this technique, voltage reference circuits with temperature drifts of ≤ 1 ppm/$°C$ have been developed as successful commercial products.[9,10]

Figure 4.38 shows a practical design example to illustrate the principle of operation of a temperature-stabilized integrated circuit. For illustrative purposes, the circuit to be stabilized is assumed to be a voltage reference. In the circuit, the diodes D_1 and D_2, which have a temperature-dependent voltage drop $V_{BE}(T)$, serve as the temperature-sensing elements and generate a temperature-sensitive control voltage V_0, where

$$V_0 = V_1 - V_2 = V_{BE}(T) - V_{ref}\frac{R_2}{R_1 + R_2} \qquad (4.93)$$

This voltage drives a differential amplifier which activates the heater transistor Q_0.

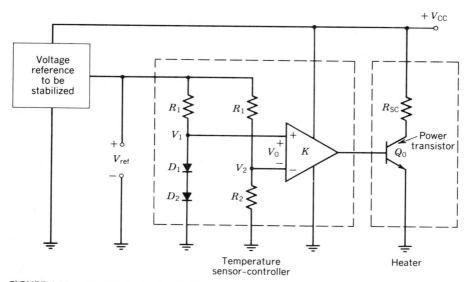

FIGURE 4.38. Simplified example of chip temperature stabilization for precision voltage reference.

The diode voltage drop makes an excellent temperature sensor, since it varies very linearly with temperature T, and is given as

$$V_{BE}(T) = V_{BE}(T_0) + \gamma_D(T - T_0) \tag{4.94}$$

where $V_{BE}(T_0)$ is the nominal diode drop at the reference temperature, which is usually taken as $+25°C$. For base–emitter diodes, the temperature coefficient γ_D is ≈ -2 mV/°C.

In most applications, the chip temperature T_C is normally chosen to be 25–50°C higher than the highest operating ambient temperature. Once T_C is chosen, one would then choose the resistor divider ratio in Eq. (4.93) to have $V_0 = 0$ at $T = T_C$. Thus, the heater will be turned on until the chip temperature reaches T_C, and then would stabilize itself at a power dissipation level sufficient to maintain the chip temperature at a steady-state value very close to T_C. The current-limiting resistor R_{SC}, at the collector of the heater transistor, is used to minimize the surge current transients when the heater transistor is first turned on. The accuracy of the temperature sensor–controller circuit to maintain the chip temperature at or very near T_C over wide changes of ambient temperature is determined by the amplifier gain K. The total power dissipation required, as well as the thermal time constants of the system, are determined by the thermal properties of the IC package. To avoid excessive power dissipation, a thermally insulated package structure is normally used.

At this point, a note of caution is also in order. The temperature sensor–controller circuit of Figure 4.38 is a closed-loop feedback system and, thus, may have stability problems, particularly if the amplifier gain K is set too high.

Although circuits with stabilized substrate temperatures are useful in special circuit applications requiring an unusually high degree of precision, their general acceptance has been somewhat limited due to the following reasons:

1.   Relatively high power dissipation (typically on the order of 500 mW or more).
2.   Need for special packaging with good thermal insulation, in order to minimize power dissipation.
3.   Relatively large surge currents and long time constants (typically on the order of 5–10 secs) needed for the initial warm-up time when the power is first applied.

As mentioned in the beginning of this chapter, the excellent matching of integrated components, as well as the close thermal coupling between them, provides the IC designer with a diverse set of design techniques. With imaginative application of these unique circuit techniques, some of which are outlined in this chapter, it is possible to design monolithic analog circuits which offer superior performance characteristics over their discrete counterparts.

REFERENCES

1. P. R. Gray and R. G. Meyer, *Analysis and Design of Analog Integrated Circuits*, Wiley, New York, 1977, Chap. 4.

2. G. R. Wilson, "A Monolithic Junction FET-NPN Operational Amplifier," *IEEE J. Solid-State Circuits*, **SC-3,** 341–348 (December 1968).

3. J. Davidse, *Integration of Analogue Electronic Circuits*, Academic Press, London, 1979, Chap. 4.

4. A. B. Grebene, Ed., *Analog Integrated Circuits*, IEEE Press, New York, 1978.

5. R. J. Widlar, "A Versatile Monolithic Voltage Regulator," National Semiconductor Appl. Note AN-1, 1967.

6. R. J. Widlar, "New Developments in IC Voltage Regulators," *IEEE J. Solid-State Circuits*, **SC-6,** 2–7 (February 1971).

7. A. P. Brokaw, "A Three-Terminal IC Bandgap Reference," *IEEE J. Solid-State Circuits*, **SC-9,** 388–393 (December 1974).

8. T. F. Prosser, "An Integrated Temperature Sensor-Controller," *IEEE J. Solid-State Circuits*, **SC-1,** 8–13 (1966).

9. R. Dobkin, "IC Reference Has 1 ppm/°C Drift," National Semiconductor Appl. Note 181, 1977.

10. D. P. Laude and J. D. Beason, "5 Volt Temperature Regulated Voltage Reference," *IEEE J. Solid-State Circuits*, **SC-15,** 1070–1076 (December 1980).

CHAPTER FIVE

BASIC GAIN STAGES

The choice of input and output gain stages is among the most critical steps in analog IC design. This chapter will examine some of the basic gain stages very commonly utilized in analog integrated circuits. These gain stages serve as building blocks or subcircuits in almost all of the different classes of analog integrated circuits discussed in the later chapters.

The differential amplifier is one of the most widely used classes of gain stages in analog IC design. The first part of this chapter covers the subject of differential gain stages, with particular emphasis on the use of active devices as "active loads." The current mirror subcircuit, which was covered as a biasing element in Chapter 4, is now utilized as a dynamic load to obtain very high voltage gains from a single-stage amplifier. The second part of the chapter deals with output stages. These are the gain blocks or subcircuits designed to provide large signal swings into an output load, with minimum signal distortion or standby power requirements.

The analyses and design examples presented in this chapter are limited to bipolar technology. However, most of the design criteria and circuit properties are readily applicable to MOS devices, which will be covered in the next chapter.

5.1. DIFFERENTIAL GAIN STAGES

Differential amplifiers represent a broad class of circuits whose basic function is to amplify the *difference* between two input signals. For this reason, they are also referred to as *difference amplifiers*. The bias levels and the gain characteristics of a differential stage, by and large, depend on the symmetry between the two branches of the circuit.[1,2] This balanced nature of the differential amplifier makes it ideal as a gain block for integrated circuits, since close matching is

215

inherent to the monolithic components. In fact, since the matching and temperature tracking properties of monolithic components are far better than those of their discrete counterparts, the performance characteristics of an integrated differential gain stage is, in general, superior to that of a nonintegrated one.

Another important advantage of differential gain stages is that they can be directly cascaded or coupled to one another, without requiring extensive level shifting or interstage coupling capacitors. In the case of monolithic integrated circuits, where coupling capacitors are not available, this is a very important feature.

Figure 5.1a shows the circuit diagram of the basic differential gain stage. This circuit is also referred to as the *emitter-coupled* pair, since the emitters of Q_1 and Q_2 are connected together. The biasing is normally provided by the current source transistor Q_0. In order to keep the analysis general, the current source is assumed to be nonideal, and it is replaced by its Norton equivalent, as shown in Figure 5.1b.

Assuming that the resistors and transistors in both circuits are precisely matched, and both input voltages are equal, the output dc levels V_{o1} and V_{o2} would be equal,

$$V_{o1} = V_{o2} \approx V_{CC} - \frac{I_0 R_C}{2} \tag{5.1}$$

Thus, the differential output voltage, V_{od}, where $V_{od} = V_{o1} - V_{o2}$, would be equal to zero. If the input voltages were made unequal, that is $V_{o1} \neq V_{o2}$, the current division between the two legs of the circuit will no longer be sym-

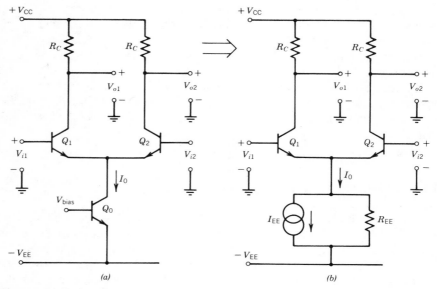

Figure 5.1. Basic differential gain stage: (a) Actual circuit diagram; (b) its equivalent circuit.

metrical, and a differential output voltage, V_{od} will be produced. Therefore, the circuit, in its idealized form, is designed to amplify only the *difference* between the two input signals.

The Half-Circuit Concept

The differential amplifier belongs to a special class of circuits known as *symmetrical networks*. Therefore, for small-signal ac analysis, the performance of a differential gain stage can be analyzed by using a well-known circuit theorem, called the *bisection theorem*. [3] This approach allows one to determine the overall ac performance by examining the behavior of only *one-half* of the circuit for a set of symmetrical (i.e., common-mode) or antisymmetrical (i.e., differential-mode) input signals.

Figure 5.2 illustrates the basic axis for symmetry of a differential gain stage for small-signal ac analysis.* To illustrate the symmetry, the bias current source

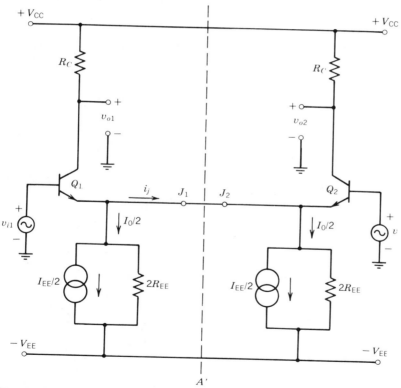

FIGURE 5.2. Axis of symmetry for splitting a differential circuit into its equivalent half-circuits.

*Lowercase letters are used to imply ac small-signal quantities.

is shown as made up of two parallel current sources, each carrying one-half of the bias current, I_0.

If a set of identical common-mode inputs were applied to the circuit of Figure 5.2 (i.e., $v_{i1} = v_{i2} = v_{ic}$), the currents and voltages in each symmetrical half of the circuit would vary identically. The relative voltage levels on either side of the interconnecting link (i.e., the connection between nodes J_1 and J_2) would remain unchanged, and the current i_j, in this branch would be identically equal to zero. Thus, for a common mode input condition, the link $J_1 - J_2$ can be open-circuited, resulting in the equivalent half-circuit of Figure 5.3a.

If a net differential-mode voltage v_{id} were applied to the circuit of Figure 5.2 such that $v_{id} = v_{i1} - v_{i2}$, then the current and voltage levels in each half of the circuit would vary *antisymmetrically*. Under these conditions, the voltage v_j at the interconnecting link $J_1 - J_2$ would remain unchanged; and this point will behave as a *virtual ground node*. Therefore, for differential-mode input signals, the small-signal ac performance can be analyzed using the half-circuit of Figure 5.3b.

The half-circuit concept described so far has been based on a set of purely

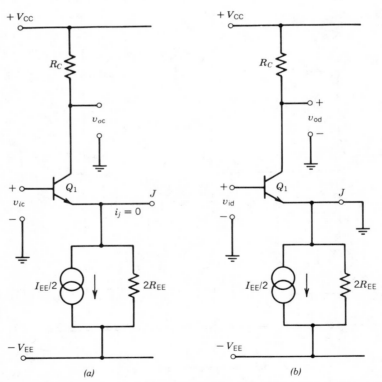

FIGURE 5.3. Equivalent half-circuits for small-signal analysis: (a) Common-mode equivalent; (b) differential-mode equivalent.

differential-mode, or purely common-mode, inputs. However, this analysis technique can be readily extended to any *arbitrary* set of small-signal input voltages v_{i1} and v_{i2} by separating them into their *net* common-mode and differential-mode components as

$$v_{id} = \frac{v_{i1} - v_{i2}}{2} \tag{5.2}$$

$$v_{ic} = \frac{v_{i1} + v_{i2}}{2} \tag{5.3}$$

where the differential-mode component is the *difference* of the two inputs and the common-mode component is the *average* value of the two inputs. Figure 5.4 illustrates the effects of decomposing an arbitrary set of inputs to their common-mode and differential-mode components in terms of the basic differential amplifier circuit. Once this decomposition is done, the analysis can be carried out with the basic half-circuits of Figure 5.3.

The output voltages v_{o1} and v_{o2} of Figure 5.2 can also be expressed in terms of their differential-mode and common-mode components as

$$v_{od} = v_{o1} - v_{o2} \tag{5.4}$$

$$v_{oc} = \frac{v_{o1} + v_{o2}}{2} \tag{5.5}$$

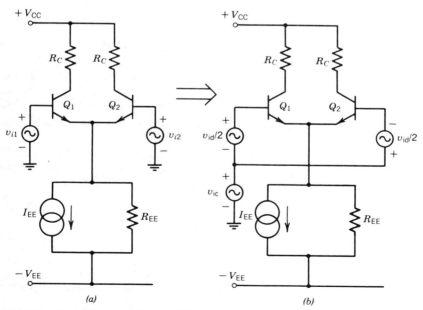

(a)

(b)

FIGURE 5.4. Separating the arbitrary set of inputs v_{i1} and v_{i2} into their common-mode and differential-mode components v_{ic} and v_{id}.

Similar to the definitions of Eqs.(5.2) and (5.3), v_{od} is the *difference* and v_{oc} is the *average* of the two outputs. The differential output v_{od} is often referred to as the "full differential output", to distinguish it from the so-called single-ended outputs v_{o1} and v_{o2}.

The common-mode voltage gain A_{cm} which is the change of the common-mode output voltage per unit change of the common-mode input voltage, can be evaluated from the half-circuit of Figure 5.3a as

$$A_{cm} = \frac{v_{oc}}{v_{ic}} = \frac{-g_m R_C}{1 + 2g_m R_{EE}(1 + 1/\beta)} \approx \frac{-R_C}{2R_{EE}} \tag{5.6}$$

where $g_m(=I_{EE}/2V_T)$ is the transconductance of Q_1.

The differential-mode voltage gain A_{dm} can be evaluated from the equivalent half-circuit of Figure 5.3(b) as

$$A_{dm} = \frac{v_{od}}{v_{id}} = -g_m R_C \tag{5.7}$$

which is the same as that of a grounded-emitter gain stage.

As indicated by Eqs. (5.4) and (5.5), gain characteristics of a differential amplifier are quite different for differential-mode and common-mode signals. The ability of the circuit to amplify differential-mode signals while rejecting common-mode ones is known as the common-mode rejection characteristic. It is quantitatively described in terms of the *common-mode rejection ratio* (CMRR), which is defined as the ratio of the differential gain to the common-mode gain, that is,

$$CMRR = \frac{A_{dm}}{A_{cm}} = 1 + 2g_m R_{EE}\left(1 + \frac{1}{\beta}\right) \approx 2g_m R_{EE} \tag{5.8}$$

and is normally expressed in decibels. Ideally, a differential amplifier is expected to be insensitive to common-mode (i.e., symmetrical) inputs, and amplify only the differential-mode (i.e., antisymmetrical) inputs. Thus, the higher the CMRR, the closer the circuit is to being an ideal differential amplifier which *only* amplifies differential-mode signals.

In the case of the simple differential gain stage of Figure 5.1a, where the Norton equivalent of the bias network corresponds to a grounded-emitter current source, such as the simple current mirror of Figure 4.1, the effective resistance R_{EE} can be expressed as [see Eq. (2.20)]:

$$R_{EE} = \frac{V_A}{I_{EE}} = \frac{V_A}{2V_T g_m} \tag{5.9}$$

where V_A is the Early voltage associated with the current-source transistor, and V_T is the thermal voltage $(=kT/q)$. From Eqs. (5.8) and (5.9), one can express CMRR as

$$CMRR \approx \frac{V_A}{V_T} \tag{5.10}$$

for a simple current-source bias case, such as that shown in Figure 5.1a. For a typical IC *npn* transistor, V_A is of the order of 100V, and $V_T \approx 26$ mV at room temperature, which gives a value of CMRR ≈ 4000, or 72 dB. The common-mode rejection characteristics can be improved significantly by using more complex current-source bias circuits, such as those shown in Figures 4.10 through 4.13, which result in much higher values of R_{EE}.

The different behavior of differential gain stages for common-mode and differential-mode signals makes them ideal for integrated circuits. This is true because the basic absolute-value tolerances and the thermal drifts of component values appear as a common-mode variation, since they simultaneously appear at each branch of the circuit. Therefore, they do not affect the basic operation of the gain stage. On the other hand, component mismatches appear as effective differential-mode signals, since they cause an asymmetry between the two branches of the circuit. Since close matching of the components is an inherent property of monolithic circuits, the differential stage topology is nearly ideal for integration.

Effects of Emitter Degeneration

The basic differential gain stage of Figure 5.1 behaves as a linear amplifier only for a small range of differential input signals. This can be understood by examining the differential half-circuit of Figure 5.3b, which is linear only for signal swings on the order of the termal voltage V_T. In order to increase the range of the differential input voltage over which the differential pair behaves approximately as a linear amplifier, emitter degeneration resistors are frequently used in series with the emitters of Q_1 and Q_2, as shown in Figure 5.5.

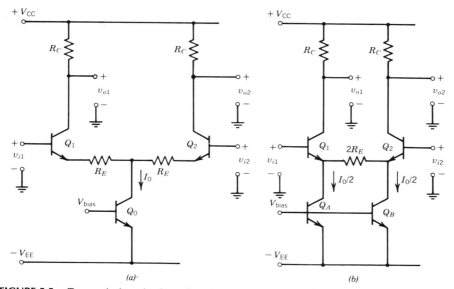

FIGURE 5.5. Two equivalent circuit configurations for differential gain stage with emitter resistors.

Both of circuit configurations shown in Figure 5.5 are equivalent. Using the half-circuit concept, one can readily show that both circuits have the same equivalent half-circuit for differential-mode signals, as shown in Figure 5.6, where R_E serves as a series feedback resistor. Using the hybrid-π model for the transistor, the small-signal differential-mode gain A_{dm} can be calculated from Figure 5.6 as

$$A_{dm} = \frac{v_{od}}{v_{id}} = \frac{-g_m R_C}{1 + g_m R_E} \approx \frac{-R_C}{R_E} \tag{5.11}$$

The common-mode half-circuit is virtually unaffected by the presence of emitter degeneration resistance. Thus, the common-mode gain is still the same as that given in Eq. (5.6), that is,

$$A_{cm} = \frac{v_{oc}}{v_{ic}} \approx - \frac{R_C}{2R_{EE}} \tag{5.12}$$

where R_{EE} is the output resistance of the bias current source Q_0 in Figure 5.5a. From Eqs. (5.11) and (5.12) the common-mode rejection ratio can be expressed as

$$CMRR = \frac{A_{dm}}{A_{cm}} \approx \frac{2R_{EE}}{R_E} \tag{5.13}$$

As indicated by Eq. (5.13), CMRR is reduced by the presence of the emitter degeneration, due to the reduction of A_{dm}. For example, assuming a typical value of $R_{EE} \approx 10$ kΩ for a simple current source ($I_0 \approx 1$ mA) and $R_E = 200$ Ω, one obtains a CMRR of approximately 100, or 40 dB. Thus, if an increased amount

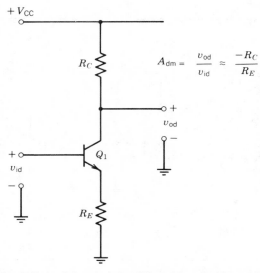

FIGURE 5.6. Equivalent half-circuit for calculating differential-mode voltage gain for circuits of Figure 5.5.

of CMRR is required, one is forced to use an improved current-source bias to increase the value of R_{EE}.

The presence of the emitter degeneration resistance significantly extends the linear range of operation of the circuit. With a sufficiently large emitter degeneration resistor, such that $g_m R_E \gg 1$, the circuit can provide linear amplification with peak-to-peak differential input signals on the order of $I_0 R_E$.

Differential-Mode and Common-Mode Input Resistances

Since the differential amplifier responds differently to differential-mode and common-mode signals, its input resistance for differential-mode and common-mode signals is also different. These resistance levels can be readily calculated from the corresponding half-circuits given in Figures 5.3 and 5.6.

The small-signal differential input resistance R_{id} is the resistance seen directly looking into the input terminals of a differential gain stage, as illustrated in Figure 5.7. Similarly, the common-mode input resistance R_{ic} is the resistance

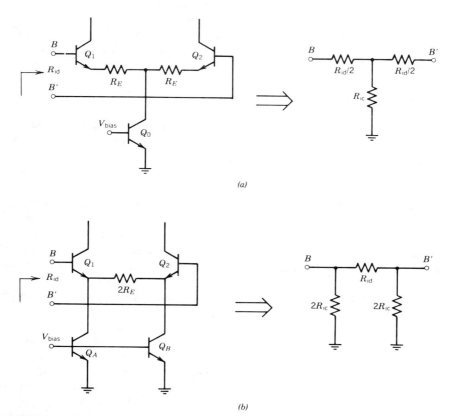

(a)

(b)

FIGURE 5.7. Equivalent circuits for calculating differential-mode and common-mode input resistances.

seen by a common-mode signal source, such as the voltage source v_{ic}, of Figure 5.4b, between *both* input terminals of the amplifier and the ac ground. The small-signal equivalent circuits for calculating the input impedance levels are shown in Figure 5.7. For the case of a generalized differential stage, including emitter degeneration, the values of R_{id} and R_{ic} can be calculated, using the hybrid-π model and the equivalent half-circuits, as

$$R_{id} = 2[r_\pi + (\beta + 1) R_E] \approx 2(r_\pi + \beta R_E) \qquad (5.14)$$

and

$$R_{ic} = (\beta + 1) R_{EE} + \frac{r_\pi}{2} \approx \beta R_{EE} \qquad (5.15)$$

where r_π is the input resistance of the common-emitter transistor stage as given by Eq. (2.19).

In most applications, the differential gain stage normally handles purely differential input signals, thus, R_{id} rather than R_{ic} is the critical parameter. R_{id} can be increased by increasing R_E; however, this may not always be desirable since it lowers both the differential gain A_{dm} and the CMRR. If no R_E is used, an alternate way of increasing the input resistance is to increase r_π by operating Q_1 and Q_2 at lower current levels, or to use a Darlington-connected input stage configuration (see Fig. 5.10).

Common-Mode Range

The common-mode range of a differential gain stage is the maximum range of dc voltage that can be applied, simultaneously, to both inputs without causing the cutoff or saturation of the gain transistors. A wide common-mode range is desirable, since it allows the amplifier inputs to be easily interfaced with input signals of various dc levels.

In the case of the simple differential gain stage of Figure 5.1, the lower limit of the input common-mode range is set by the saturation of the current-source transistor Q_0, or the cutoff of Q_1 and Q_2, which happens when both inputs are lowered, approaching within V_{BE} of $-V_{EE}$. The upper limit of the common-mode range is set by the saturation of Q_1 and Q_2, as both inputs are raised toward $+V_{CC}$.

As will be described in Chapter 7, the wide common-mode range is one of the most important design requirements in operational amplifiers.

Effects of Device Mismatches

In spite of the inherent matching advantages of monolithic devices, small but finite mismatches still exist even between two adjacent, identical transistors or resistors on the same chip. In this section, the effects of these small mismatches will be examined with regard to the current balance in a differential pair. In

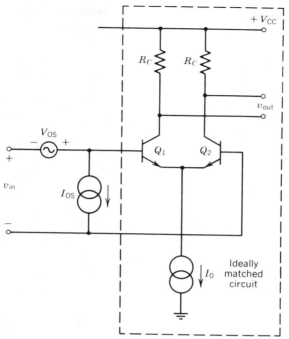

FIGURE 5.8. Analysis of effects of component mismatches by means of offset currents and voltages referred to input.

differential amplifiers, the effects of device mismatches on the dc performance of the circuit are best represented by referring their effects to the input of the amplifier, in the form of an input offset voltage or an input offset current. Thus, as illustrated in Figure 5.8, one can analyze an actual differential amplifier by treating it as an ideal, that is, a perfectly matched, circuit, which has an equivalent offset voltage source V_{OS} and an offset current source I_{OS} across the input terminals. These offset sources, arising from component mismatches, can be defined as follows:

Input Offset Voltage V_{OS}. This is the voltage that has to be applied across the input terminals in order to reduce the differential output voltage V_{out} to zero, with no applied signal.

Input Offset Current I_{OS}. This is the difference between the base currents of the gain transistors Q_1 and Q_2, arising from the device mismatches. In other words, it is the value of the current source that has to be placed in shunt with the two inputs in order to set the output voltage to zero with the inputs open-circuited.

In the normal application of a differential gain stage, the offset sources V_{OS} and I_{OS} are indistinguishable from the signals being amplified. Thus, they present a basic limitation to the resolution of the amplifier for handling low-level signals. Consequently, in analog IC design it is extremely important to take the necessary design precautions to minimize these mismatch-induced offsets. In most applications where the differential gain stage is driven by a low-impedance signal source, the offset voltage V_{OS} is the primary limit to the resolution of the low-level signals. However, in those applications where the input signal is applied from a high-impedance source, the effect of I_{OS} is more detrimental.

The quantitative effects of component mismatches can be calculated from the circuit of Figure 5.9 by summing the voltages around the source loop,

$$V_{OS} - V_{BE1} - \frac{I_0 \Delta R_E}{2} + V_{BE2} = 0 \tag{5.16}$$

and

$$V_{OS} = V_{BE1} - V_{BE2} + \frac{I_0 \Delta R_E}{2} \tag{5.17}$$

where ΔR_E is the mismatch in the emitter degeneration resistors. Depending on the mismatch, it can be either a positive or a negative quantity.

The V_{BE} difference can be expressed in terms of the collector currents and the reverse saturation currents of Q_1 and Q_2 as

$$V_{BE1} - V_{BE2} = V_T \ln\left(\frac{I_{C1}}{I_{C2}} \frac{I_{S2}}{I_{S1}}\right) \tag{5.18}$$

Since the output voltage V_{out} had to be equal to zero, by the definition of V_{OS}, the voltage drop across the collector resistors must be equal, that is,

$$I_{C1} R_C = I_{C2} (R_C + \Delta R_C) \tag{5.19}$$

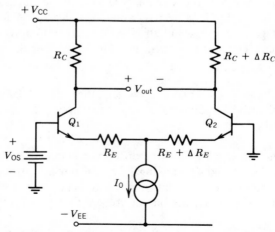

FIGURE 5.9. Circuit for calculating effects of small component mismatches. (*Note:* Delta quantities indicate random component mismatches.)

or

$$\frac{I_{C1}}{I_{C2}} = \frac{R_C + \Delta R_C}{R_C} \tag{5.20}$$

Substituting Eq. (5.20) into Eq. (5.18) and using the approximation that

$$\ln\left(\frac{x + \Delta x}{x}\right) \approx \frac{\Delta x}{x} \tag{5.21}$$

for $\Delta x \ll x$, one can write Eq. (5.17) as

$$V_{OS} = V_T\left(\frac{\Delta R_C}{R_C} - \frac{\Delta I_S}{I_S} + \frac{I_0 R_E}{2V_T}\frac{\Delta R_E}{R_E}\right) \tag{5.22}$$

where $\Delta I_S = I_{S1} - I_{S2}$, and I_S is the nominal value of the reverse saturation current for Q_1 and Q_2. The polarity of the signs in Eq. (5.22) is not important, since each of the delta quantities is independent, and random mismatches can be either positive or negative.

Equation (5.22) points out the three independent contributors of V_{OS}. The first term is due to the mismatches of load resistors, the second term is due to the mismatches of the transistor characteristics and the emitter areas of Q_1 and Q_2, and the third term is due to mismatches in the emitter degeneration resistance.

The mismatch parameters depend on the process control and the device layout. Under a production environment, with careful layout, typically observed values of mismatches are of the order of about 1% for resistors and about 5% for the reverse saturation current, that is,

$$\left|\frac{\Delta R_C}{R_C}\right| \approx \left|\frac{\Delta R_E}{R_E}\right| \approx 0.01 \quad \text{and} \quad \left|\frac{\Delta I_S}{I_S}\right| \approx 0.05 \tag{5.23}$$

It should be pointed out that the mismatch parameters given by Eq. (5.23) are actually random variables which may take on a different value for each circuit fabricated. Thus, the values given by Eq. (5.23) represent a statistical or "typical" case, rather than a worst case parameter mismatch.[4]

If the differential gain stage is used with an excessive amount of emitter degeneration, such that $I_0 R_E \gg V_T$, then the last term in Eq. (5.22) would dominate the offset voltage characteristics, and the offset voltage would be determined primarily by the mismatch in R_E. For example, with $I_0 = 1$ mA and $R_E = 500\ \Omega$, a 1% mismatch in R_E would create a 2.5 mV offset voltage.

For the case of a simple differential gain stage with no emitter degeneration, Eq. (5.22) reduces to

$$V_{OS} = V_T\left(\frac{\Delta R_C}{R_C} - \frac{\Delta I_S}{I_S}\right) \tag{5.24}$$

Using the typical values of component mismatches given by Eq. (5.24) and assuming, for the worst case, that both terms in Eq. (5.24) are additive, one can calculate an estimated value of V_{OS} at room temperature as

$$V_{OS} = (26 \text{ mV})(0.01 + 0.05) \approx 1.5 \text{ mV} \tag{5.25}$$

When differential gain stages are used for amplifying low-level dc signals, the offset voltage may have to adjusted to zero with an external potentiometer. When this is done, the important parameter becomes not the offset voltage itself, but its variation with temperature, often referred to as *offset drift*. For the case of the differential gain stage with no emitter degeneration, offset drift can be easily calculated by differentiating Eq. (5.24) with respect to T, with $V_T = kT/q$. The result is

$$\frac{dV_{OS}}{dT} = \frac{V_{OS}}{T} \tag{5.26}$$

The above expression implies that the offset drift is *proportional* to the offset itself, and that a differential pair exhibits an offset drift of approximately 3.3 μV/°C for every millivolt of offset voltage at room temperature ($T = 300$°K).

The offset current I_{OS} of Figure 5.8 is equal to the difference between the base currents of the two input transistors, that is,

$$I_{OS} = I_{B1} - I_{B2} = \frac{I_{C1}}{\beta_{F1}} - \frac{I_{C2}}{\beta_{F2}} \tag{5.27}$$

where β, with appropriate subscript, is the dc forward-current gain of Q_1 and Q_2. Assuming that I_{C1} and β_1 are equal to the nominal values, but I_{C2} and β_2 differ from them by the small amounts ΔI_C and $\Delta \beta$, one can write Eq. (5.27) as

$$I_{OS} = \frac{I_C}{\beta_F} \left[\frac{\Delta I_C}{I_C} - \frac{\Delta \beta_F}{\beta_F} \right] \tag{5.28}$$

Since the voltage drops across the collector, resistors have to be identical in order to set $V_{out} = 0$,

$$\frac{\Delta I_C}{I_C} = -\frac{\Delta R_C}{R_C} \tag{5.29}$$

Substituting these results back into Eq. (5.27), one obtains

$$I_{OS} = -\frac{I_0}{2\beta_F} \left[\frac{\Delta R_C}{R_C} + \frac{\Delta \beta_F}{\beta_F} \right] \tag{5.30}$$

where I_0 ($= 2I_C$) is the bias current. Typical variations of β between adjacent devices is on the order of $\pm$ 10%, that is,

$$\left| \frac{\Delta \beta_F}{\beta_F} \right| \approx 0.1 \tag{5.31}$$

Since the resistor variations are on the order of $\pm$ 1%, the second term in Eq. (5.30) is dominant, and I_{OS} is primarily determined by the β mismatch between Q_1 and Q_2. For a typical β mismatch of 10%, the value of I_{OS} is typically on the order of 10% of I_B.

Darlington-Connected Differential Pairs

In certain applications where a very high input impedance or low base currents are required, the input transistors of a differential gain stage can be replaced by Darlington-connected transistor pairs, as shown in Figure 5.10. Such a connection increases the value of the differential input resistance R_{id} of Eq. (5.14) by an additional factor equivalent to the current gain β_A of the outside transistors Q_{1A} or Q_{2A}, such that

$$R_{id} = 2[r_{\pi A} + \beta_A(r_{\pi B} + \beta_B R_E)] \tag{5.32}$$

where the subscripts A and B refer to the outer and inner transistor pairs of Figure 5.10. Since the inner and outer transistors of the Darlington pairs operate at drastically different current levels, their current gains may be different, therefore, their parameters are identified separately in Eq. (5.32). This is particularly true in the case of the circuit of Figure 5.10a, where Q_{1A} and Q_{2A} carry only the base currents of Q_{1B} and Q_{2B}.

The differential voltage gain of a Darlington differential pair can be readily calculated from the half-circuit model as

$$A_{dm} = \frac{v_{od}}{v_{id}} = -\frac{g_m R_C}{2} \tag{5.33}$$

where g_m is the transconductance of the inner transistor pair Q_{1B} and Q_{2B}. Note that the use of a Darlington connection results in a factor of 2 reduction in gain over that of a non-Darlington circuit. This reduction comes about because only one-half of the input signal appears across the inner pair of transistors.

In general, the circuit of Figure 5.10b is preferred over that of Figure 5.10a since the outside transistors operate at predictable current levels set by I_1 and,

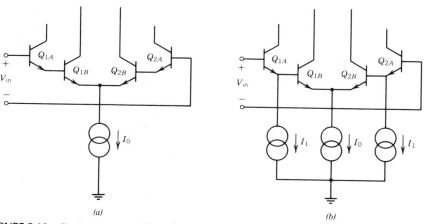

FIGURE 5.10. Darlington-input differential gain stages. Circuit (b) is preferred over circuit (a) since it offers lower input offset voltage.

thus, result in lower offset voltage and device mismatches. Normally, I_1 is chosen to be of the order of 5%–10% of I_0.

The collectors of the outside transistors, Q_{1A} and Q_{2A} can be connected either to $+ V_{CC}$ directly or short-circuited to the respective collectors of the inside pair. In general, the connection of $+ V_{CC}$ is preferred, since this reduces the effective input capacitance of the circuit.

The additional transistors in the Darlington connection also contribute adversely to the offset voltage of the differential gain stage. In the case of the simple Darlington stage of Figure 5.10a, offset voltages on the order of 5–10 mV are common, due to the current mismatches in the outside transistors. In the case of the current-biased Darlington stage of Figure 5.10b, the typical offset voltages are approximately twice as much as those of a simple differential pair.

Internal Biasing of AC-Coupled Differential Gain Stages

In many applications, differential gain stages are required to operate with ac-coupled single-ended inputs. In such cases, the internal biasing of the circuit can be easily achieved by biasing both inputs of the circuit from a low-impedance internal bias source, as shown in Figure 5.11. The input signal can then be ac-coupled to the desired input by means of an external coupling capacitor. The input impedance of the circuit is primarily determined by the bias resistor R_B. Both inputs are normally biased through a matched set of resistors R_B to minimize the dc offsets. The internal bias source V_{bias} is chosen to have sufficiently low impedance such that it serves as an ac ground and uncouples the base of Q_2 from the input signal. Such a bias voltage can be derived from voltage-source circuits, similar to those shown in Figure 4.14.

JFET Differential Gain Stages

JFET's are sometimes used as the input devices for differential gain states in order to increase the differential input impedance and minimize the input bias

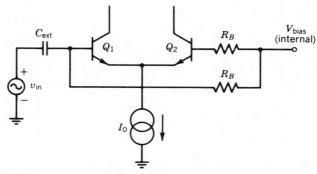

FIGURE 5.11.   Internal biasing of differential stage for ac coupled inputs.

currents. Since the gate input of a JFET is a reverse-biased p–n junction, it offers a very high input resistance (typically on the order of 10^{10} Ω), and its input bias current is the leakage current of the gate–channel junction. As described in Chapter 2 (see Section 2.4), JFET devices can be fabricated simultaneously with bipolar devices, with the addition of several process steps. The inclusion of JFET's in bipolar analog design has become particularly feasible with the advent of new fabrication process steps, such as the ion-implantation techniques. Differential gain stages with JFET inputs are often used as input stages for operational amplifiers, where the high input impedance and low bias current properties of JFET's are particularly useful.

Figure 5.12 shows a simple differential gain stage using n-channel JFET devices. The following analysis is equally applicable to p-channel JFET gain stages, with the appropriate polarity changes. The operation of the circuit is identical to that of the basic bipolar differential stage. If both halves of the circuit are ideally matched, an equal amount of bias current $I_0/2$ flows through Q_1 and Q_2, with equal inputs v_{i1} and v_{i2} applied to the circuit. If v_{i1} and v_{i2} are made unequal, resulting in a differential input voltage v_{id}, then the circuit symmetry is changed; either Q_1 or Q_2 would conduct a bigger share of the current, and a differential output $v_{od} = v_{o1} - v_{o2}$ will be produced.

The maximum amount of drain current which can be conducted by a JFET without causing the gate–channel junction to be forward biased is limited to I_{DSS} [see Eq. (2.36) and (2.37)], which is a device parameter determined by device design and geometry (i.e., channel width-to-length ratio). Thus, in designing a differential gain stage, such as the one shown in Figure 5.12, the bias current I_0 should be chosen such that

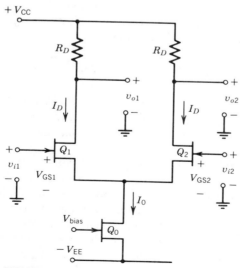

FIGURE 5.12. Differential gain stage using JFETs.

$$|I_0| < |I_{DSS}|$$ (5.34)

in order to ensure that the gate inputs do not become forward biased when either Q_1 or Q_2 is forced to carry the entire bias current I_0.

The differential-mode and common-mode gain characteristics of the JFET differential pair can be calculated by using the small-signal model of Figure 2.39 and the half-circuit concept illustrated in Figure 5.3.

The differential gain A_{dm} can be expressed as

$$A_{dm} = \frac{v_{od}}{v_{id}} = -g_m R_D$$ (5.35)

where the transconductance g_m is given by Eqs. (2.39) and (2.40) as

$$g_m = -\frac{2I_{DSS}}{V_p}\left(1 - \frac{V_{GS}}{V_p}\right) = \frac{2}{|V_p|}\sqrt{|I_D| \, |I_{DSS}|}$$ (5.36)

or, since $I_D = I_0/2$ Eq. (5.36) can also be written as

$$g_m = \frac{\sqrt{2|I_0| \, |I_{DSS}|}}{|V_p|}$$ (5.37)

where V_p is the pinch-off voltage for the JFET and is typically in the range of $1V$–$5\,V$ for bipolar compatible JFET devices.

The common-mode small-signal gain can be calculated from the common-mode half-circuit as

$$A_{cm} = \frac{v_{oc}}{v_{ic}} = -\frac{g_m R_D}{1 + 2g_m R_{SS}}$$ (5.38)

where R_{SS} is the drain-source resistance of the bias JFET Q_0, and is essentially equivalent to R_{EE} of Figure 5.1b. The common-mode rejection ratio can be calculated from Eqs. (5.37) and (5.38) as

$$\text{CMRR} = \frac{A_{dm}}{A_{cm}} = 1 + 2g_m R_{SS}$$ (5.39)

It should be noted that g_m for the JFET, as given by Eq. (5.37), is quite low. For example, assuming some typical values such as $I_0 = 0.5$ mA, $I_{DSS} = -2$ mA, and $V_p = -2\,V$, one obtains a typical value for g_m of $\approx 0.7 \cdot 10^{-3}$ mho. This is approximately a factor of 15 lower than the g_m of a comparable bipolar gain stage. Consequently, both the differential voltage gain and the common-mode rejection range available from a JFET stage are approximately a factor of 10–20 (i.e., 20–26 dB) lower than that obtainable from a similar bipolar design.

The offset voltage of JFET differential pairs is primarily determined by the V_p and I_{DSS} mismatches between the input devices. As described in Section 2.4, both V_p and I_{DSS} are complex device parameters [see Eq. (2.33) and (2.36)]. The pinch-off voltage is a function of channel thickness and doping, and I_{DSS} depends on the channel length, thickness, and doping. It can be shown that the effects of I_{DSS} mismatch can be minimized by operating Q_1 and Q_2 with drain currents much lower than I_{DSS} by choosing I_0 to be approximately 5%–10% of I_{DSS}. This is normally achieved by designing Q_1 and Q_2 to have a much larger channel width-to-length ratio than Q_0.

The effects of V_p mismatches between the input devices can be controlled to a large degree by careful device layout, and by using accurately controlled impurity doping steps, such as ion implantation. Typical V_p mismatches observed under production environment are on the order of 0.5%. With $|V_p| \approx 1$ V, this results in typical V_{OS} values in the range of 5–8 mV.

5.2 GAIN STAGES WITH ACTIVE LOADS

The voltage gain available from a single-stage amplifier circuit with a resistive load R_C such as the one shown in Figure 5.13a is given as

$$A_v = \frac{v_o}{v_{in}} = -g_m R_C = -\frac{I_C R_C}{V_T} \tag{5.40}$$

In order to achieve a high voltage gain, one has to increase either the load resistor R_C or the collector current I_C. In many cases neither of these may be practical since increasing the $I_C R_C$ product requires both very large resistor values and very high supply voltages. Since large resistor values are wasteful of silicon chip area, and high supply voltages may not be practical for device breakdown or power dissipation reasons, an alternate solution has to be found.

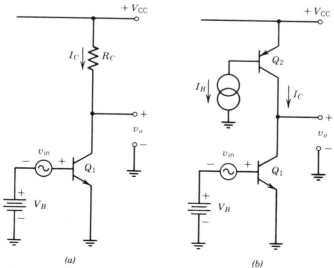

(a)   *(b)*

FIGURE 5.13. Simple common-emitter gain stage: *(a)* Resistive load; *(b)* active load.

Figure 5.13*b* illustrates a practical solution to obtaining high voltage gains by using the dynamic impedance of active devices as active loads. In this simplified illustration, the dynamic output resistance r_{o2}, of the *pnp* transistor Q_2 serves as the load resistance for the gain stage Q_1, providing a small-signal voltage gain

$$A_v = \frac{v_o}{v_{in}} = - g_m (r_{o1} \| r_{o2}) \tag{5.41}$$

which is determined by the g_m of Q_1 and the parallel combination of the output resistances r_{o1} and r_{o2}, associated with Q_1 and Q_2.

The example of Figure 5.13*b* is used for illustration only. Its actual implementation, as shown, is not practical since any mismatch in the operating points of Q_1 and Q_2 will cause one or the other of the two devices to be driven into saturation.

The practical bias instability problem associated with the simple gain stage of Figure 5.13*b* can be avoided by using a differential gain stage with a current mirror subcircuit serving as the active load. This is shown in Figure 5.14. In order to analyze the principle of operation of the circuit, let us assume initially that all transistor pairs are matched and $\beta \gg 1$. With no applied signal, that is, $v_{id} = 0$, the currents in all the branches of the circuit would be balanced, with $I_C = I_{EE}/2$ flowing through Q_1 and Q_2. If an incremental differential signal v_{id} is applied across the inputs, its effect would be to upset the circuit balance such that the collector current of Q_1 will be *increased* by an incremental amount i_1, while the collector current of Q_2 would be *decreased* by the same amount. However, the diode-connected transistor Q_3 would cause the collector current of Q_4 to be equal to that of Q_1. With these conditions, if one sums the currents at

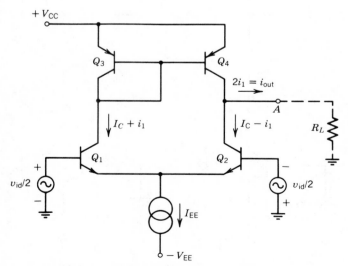

FIGURE 5.14. Differential gain stage with current mirror load.

the output node (node A of Fig. 5.14), one obtains

$$i_{\text{out}} = I_C + i_1 - (I_C - i_1) = 2i_1 \tag{5.42}$$

The incremental current i_1 is directly proportional to the incremental differential input voltage v_{id} as

$$i_1 = g_m \frac{v_{\text{id}}}{2} = \frac{I_{\text{EE}}}{2V_T} v_{\text{id}} \tag{5.43}$$

where g_m is the transconductance of Q_1 or Q_2.

Thus, the circuit functions as a transconductance amplifier, producing an incremental output current from an incremental differential input voltage as

$$i_{\text{out}} = 2g_m \frac{v_{\text{id}}}{2} = g_m v_{\text{id}} \tag{5.44}$$

Figure 5.15 gives a small-signal low-frequency equivalent circuit for the differential transconductance stage of Figure 5.14. Note that the output resistance of R_{out} of the circuit is equal to the parallel combinations of the output resistances r_{o2} and r_{o4} associated with Q_2 and Q_4, respectively,

$$R_{\text{out}} = r_{o2} \,\|\, r_{o4} \tag{5.45}$$

The output resistances of Q_1 and Q_2 can be calculated from Eq. (2.20), which is repeated below:

$$r_o = \frac{V_A}{V_T g_m} \tag{5.46}$$

where V_A is the Early voltage associated with the respective transistors.

Substituting Eq. (5.46) into Eq. (5.45), one can express the dynamic output impedance R_{out}, in terms of the Early voltages V_{AN} and V_{AP} associated with the npn and pnp transistors as

$$R_{\text{out}} = \frac{1}{V_T g_m} \frac{V_{\text{AN}} V_{\text{AP}}}{V_{\text{AN}} + V_{\text{AP}}} \tag{5.47}$$

The maximum amount of small-signal voltage gain available from the circuit

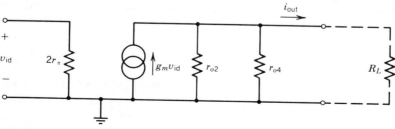

FIGURE 5.15. Low-frequency small-signal equivalent circuit for differential gain stage of Figure 5.14.

(corresponding to $R_L = \infty$, i.e., open-circuit load) is

$$(A_v)\max = \frac{i_{out}R_{out}}{v_{id}} = g_m R_{out} = \frac{1}{V_T}\frac{V_{AN}V_{AP}}{V_{AN} + V_{AP}} \tag{5.48}$$

For a typical value of $V_{AN} \approx V_{AP} = 50\,\text{V}$, the maximum value of voltage gain available is on the order of 1000, or 60 dB. This gain limit can be increased further by using more complex current mirrors and cascode-connected transistors to increase the value of r_o.

Large-Signal Transfer Characteristics

When a large differential signal V_{id} is applied to the differential gain stage of Figure 5.14, such that $V_{id} \gg V_T$, either Q_1 or Q_2 would be turned off depending on the polarity of the input. If V_{id} is positive, Q_2 would be turned off, and Q_1, as well as the current mirror transistors Q_3 and Q_4, would be conducting the full bias current I_{EE}. Neglecting the finite base currents, this would result in an output current $I_{out} \approx I_{EE}$. Conversely, if a large negative differential signal is applied to the input Q_1 along with Q_3 and Q_4 would turn off, and Q_2 would be forced to conduct the full bias current I_{EE}. This would result in an output current of $I_{out} = -I_{EE}$.

Quantitatively, this behavior can be explained by examining the exponential nature of the voltage–current characteristics of the base–emitter junctions. The collector currents, I_{C1} and I_{C2} of Q_1 and Q_2 are related to the differential input V_{id} as

$$I_{C1} = \frac{\alpha_F I_{EE}}{1 + \exp(-V_{id}/V_T)} \tag{5.49}$$

$$I_{C2} = \frac{\alpha_F I_{EE}}{1 + \exp(V_{id}/V_T)} \tag{5.50}$$

where α_F is the common-base current gain of Q_1 or Q_2. Assuming that the current mirror transistors are matched, the output current I_{out} is the *net difference* of the two collector currents, that is,

$$I_{out} = I_{C1} - I_{C2} \tag{5.51}$$

Substituting Eqs. (5.49) and (5.50) into Eq. (5.51) and rearranging the terms, one obtains

$$I_{out} = \alpha_F I_{EE} \ \tanh\!\left(\frac{V_{id}}{2V_T}\right) \tag{5.52}$$

The resultant transfer characteristics are shown in Figure 5.16. Note that the transfer characteristics go through zero with $V_{id} = 0$, so that there is no inherent input offset voltage. The circuit is capable of "sinking" or "sourcing" an amount

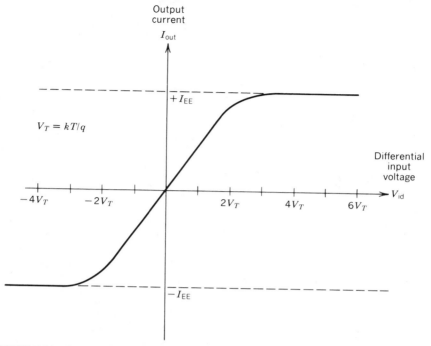

FIGURE 5.16. Large-signal voltage-to-current transfer characteristics for differential gain stage with current mirror load. (*Note:* Low-impedance external load assumed.)

of current, $I_{out} \approx I_{EE}$, into an output load. The slope of the transfer characteristics at the origin, that is, $V_{id} = 0$, is the transconductance g_m as given by the small-signal analysis.

Offset Voltage

The offset voltage of the differential pair with the current mirror load can be calculated from Figure 5.17

$$V_{OS} = V_{BE1} - V_{BE2} = V_T \ln\left(\frac{I_{C1}}{I_{C2}}\right) \tag{5.53}$$

Assuming that I_{C1} and I_{C2} differ by a small amount ΔI_C, the logarithmic term in Eq. (5.53) can be expanded in a power series and the higher order terms neglected, that is,

$$\ln\left(\frac{I_{C1}}{I_{C2}}\right) \approx \frac{\Delta I_C}{I_C} \tag{5.54}$$

where I_C is the nominal value of I_{C1} or I_{C2} (i.e., $I_C \approx I_{EE}/2$).

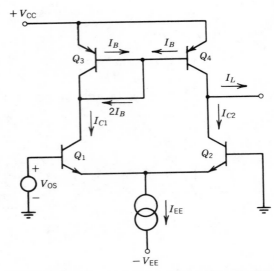

FIGURE 5.17. Circuit for calculating offset voltage of differential stage with simple current mirror load.

Combining Eqs. (5.53) and (5.54), V_{OS} can be expressed as

$$V_{OS} \approx V_T \frac{\Delta I_C}{I_C} = \frac{\Delta I_C}{g_m} \tag{5.55}$$

In the case of the circuit of Figure 5.17, assuming that both Q_1 and Q_2 are ideally matched, the collector current difference ΔI_C is entirely due to the base currents of the current mirror and to the quiescent load current I_L, that is,

$$\Delta I_C = I_{C1} - I_{C2} = 2I_B + I_L = \frac{2I_C}{\beta_p} + I_L \tag{5.56}$$

where β_p is the β_F of *pnp* current mirror transistors. Thus, V_{OS} is equal to

$$V_{OS} = V_T \left(\frac{2}{\beta_p} + \frac{I_L}{I_C} \right) \tag{5.57}$$

Assuming a *pnp* transistor β of 25 with an open-circuit load (i.e., $I_L = 0$), Eq. (5.57) indicates that the circuit of Figure 5.17 would have a predictable offset of approximately $+2$ mV, even when the input transistor pairs Q_1 and Q_2 are perfectly matched. It should be noted that the offset voltage V_{OS} given by Eq. (5.57) is due solely to the *inherent* current imbalance in the circuit. This is *in addition* to the normal offset term [i.e., the second term in Eq. (5.24)] which is due to the *random* mismatches between Q_1 and Q_2.

The inherent offset voltage can be greatly reduced by using an improved current mirror load, such as the one shown in Figure 5.18. In this case, neglecting the load current I_L, the current imbalance is only due to the base current of

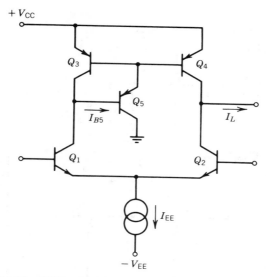

FIGURE 5.18. Differential gain stage with improved current mirror load.

Q_5 [see Eq. (4.16)],

$$I_C = \frac{I_C}{2 + \beta_3\beta_5} \approx \frac{I_C}{\beta_3\beta_5} \tag{5.58}$$

where β_3 and β_5 are the current gains of *pnp* transistors Q_3 and Q_5. Thus, the inherent offset due to current imbalance is

$$V_{OS} \approx V_T \frac{1}{\beta_3\beta_5} \tag{5.59}$$

Assuming typical values of $\beta_3 \approx 25$ and $\beta_5 \approx 5$, the inherent offset predicted by Eq. (5.59) is $\approx 200\ \mu V$, which is negligible in comparison with the random offsets due to device mismatches.

In minimizing the inherent offsets, the effects of the quiescent load current I_L [i.e., the second term in Eq. (5.56)] must also be considered. This can be minimized by using a very high impedance load, such as a Darlington-connected buffer stage. The current mismatch due to I_L can also be cancelled, or neutralized, by subtracting an equal amount of current from the other leg of the differential gain stage by means of an artificial, or "dummy", load connected to that leg.

Other Circuit Configurations

All of the current mirror circuits, both the *pnp* and the *npn* types covered in Chapter 4, can be used as active loads in differential amplifiers. Some of the

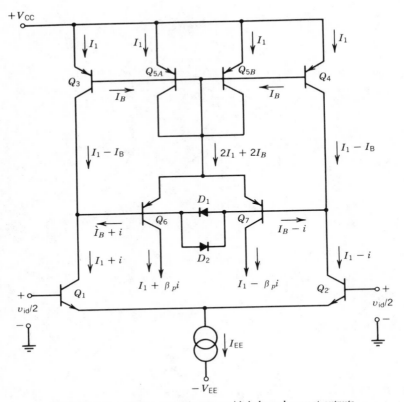

FIGURE 5.19. Differential gain stage with balanced current outputs.

most commonly used circuit configurations are shown in Figures 5.19 through 5.23.[5,6]

In certain cases, a fully differential rather than a single-ended output may be preferred. In such applications, the circuit configuration shown in Figure 5.19 is particularly useful, since it can provide a balanced set of differential output currents i_{o1} and i_{o2}. In the circuit, the input transistors Q_1 and Q_2 convert the differential input voltage v_{id} to a set of differential current mismatches i at their respective collectors, such that

$$i = \frac{g_m v_{id}}{2} \tag{5.60}$$

This current mismatch is, in turn, amplified by the inner differential pair to produce a differential set of output currents,

$$i_{o1} = -i_{o2} = \beta_p i = \frac{g_m \beta_p}{2} v_{id} \tag{5.61}$$

where β_p is the current gain of *pnp* transistors. The dc bias currents and the ac incremental signals are identified in Figure 5.19 with capital and lowercase letters, respectively. The bias for the first-stage active loads, as well as for the second stage, is set by the diode-connected transistor Q_5, which is made up of two identical transistors Q_{5A} and Q_{5B} connected in parallel. Diodes D_1 and D_2 are included in the circuit to keep Q_3 and Q_4 out of saturation under large-signal operation. Since the circuit is fully symmetrical, it has no inherent offset sources other than the random device mismatches. Its main drawback is that the output currents i_{o1} and i_{o2} are a function of the *pnp* transistor β, which cannot be very closely controlled in production.

Figure 5.20 shows an improved version of the basic differential trans-conductance stage, which is modified to increase the maximum voltage gain available from the circuit by increasing the output impedance terms associated with the *npn* and *pnp* transistors driving the output load. This is done by a Wilson current source for the *pnp* active load, and by connecting Q_3 and Q_4 in cascode connection with the input gain transistors Q_1 and Q_2. In this manner, the output resistance terms r_{o2} and r_{o4} of Eq. (5.45) are increased by a factor equal to the

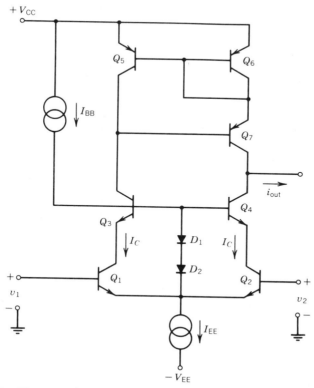

FIGURE 5.20. High-gain differential stage with cascode inputs and Wilson current mirror load.

npn and the *pnp* transistor β, respectively. Note that the biasing of the cascode transistors Q_3 and Q_4 is achieved by means of the bias current source I_{BB} and the two level-shift diodes D_1 and D_2. The actual bias current I_C for the gain transistors is determined by the difference of the two bias sources, that is,

$$I_C \approx \frac{1}{2} \left(I_{EE} - I_{BB} \right) \tag{5.62}$$

Typically, I_{BB} is chosen to be approximately 10% of I_{EE}.

Figure 5.21 shows another commonly used differential gain stage using active loads. This particular circuit configuration has been extensively used as the input stage in a number of popular analog IC products, such as the μA-741 or the LM-101 type operational amplifiers. The operation of the circuit can be explained as follows. Both emitter-coupled transistor pairs Q_1, Q_3 and Q_2, Q_4 are essentially composite transistors, which have the polarity of a *pnp* transistor with a low-frequency gain β equivalent to that of the *npn* transistors Q_1 and Q_2. This combination of *npn* and *pnp* transistors is basically a dc level-shift scheme similar to that illustrated in Figure 4.31*b*. The differential current output of these composite transistors is then converted to a single-ended current output by the

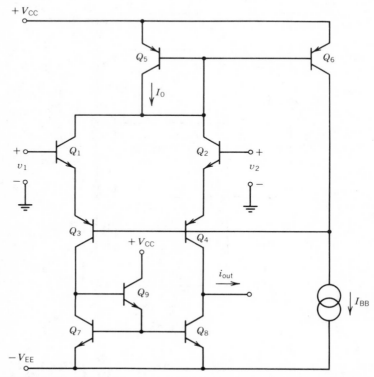

FIGURE 5.21. Differential gain stage using composite *npn–pnp* transistors at inputs.

npn current mirror subcircuit made up of Q_7, Q_8, and Q_9. The incremental output current i_{out} is related to the differential input as

$$i_{out} = -\frac{g_m v_{id}}{2} \tag{5.63}$$

where g_m is the transconductance of Q_1 or Q_2.

Since the absolute values of the lateral pnp transistor characteristics are not well controlled, a feedback arrangement is necessary to stabilize the operating points of Q_3 and Q_4 in Figure 5.21. In the circuit, this feedback is provided by the pnp current mirror comprised of Q_5 and Q_6. The diode-connected transistor Q_5 monitors the current level through Q_1 and Q_2, and correspondingly sets the current level through Q_6, which in turn adjusts the base currents of Q_3 and Q_4 by adding or subtracting from the constant-bias current I_{BB}. Assuming, for the purpose of illustration, that Q_5 and Q_6 are well matched, and neglecting base currents, the current I_0 supplied to Q_1 and Q_2 will be forced to be equal to I_{BB}. For example, if the bias current I_0 of Q_1 and Q_2 were to increase, the current through Q_6 would also increase. Since I_{BB} is constant, the increase of the current through Q_6 would cause the collector voltage of Q_6 to raise toward V_{CC}, which in turn would unbias Q_3 and Q_4 and reduce I_0 until it is very closely equal to I_{BB}. This type of feedback is known as *common-mode feedback*, since the feedback is activated by the common-mode current changes in Q_1 and Q_2. In a differential gain stage, common-mode feedback improves the bias stability as well as the CMRR without affecting the differential gain characteristics.

Figure 5.22 shows another commonly used differential stage configuration

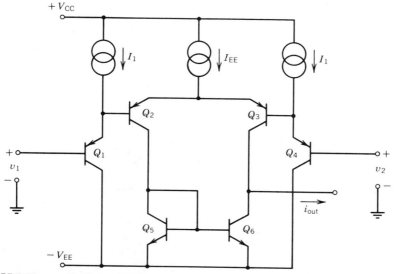

FIGURE 5.22. *pnp* Darlington differential stage with input common-mode range extending to $-V_{EE}$.

with active loads. Although this circuit utilizes the simple current mirror load, its offset voltage due to current mismatches is not significantly degraded because of the high β of the *npn* transistors in the current mirror section [see Eq. (5.56)]. The input stage is the *pnp* version of the Darlington connection illustrated in Figure 5.10*b*. A unique feature of the circuit of Figure 5.22 is that its negative common-mode limit $(V_{CM})^-$ extends to slightly below the negative supply voltage $-V_{EE}$. Because of this property, the circuit is often referred to as *ground-sensing* or single-supply circuit, since with a single positive supply voltage (i.e., $-V_{EE} = 0$), it can sense differential input signals at the ground level.[7]

The value of $(V_{CM})^-$ can be derived from Figure 5.22 as the common-mode input voltage, where Q_2 is at the verge of saturation and Q_5 is at the verge of cutoff,

$$(V_{CM})^- = -V_{EE} + V_{BE5} + V_{sat} - (V_{BE1} + V_{BE2}) \qquad (5.64)$$

where V_{sat} is the saturation voltage of Q_2. For typical device parameters, $(V_{CM})^-$ extends to approximately 200 mV *below* $-V_{EE}$.

Figure 5.23 shows a commonly used *p*-channel JFET differential gain stage with bipolar active loads. This circuit is widely used as the input stage in operational amplifier circuits utilizing combined bipolar and junction FET transistors. The circuit is shown in Figure 5.23 as using *p*-channel JFETs as input devices, since these devices are easier to fabricate simultaneously with the *npn* bipolar transistors, using the so-called ion-implanted BIFET process described in Section 2. The incremental output current i_{out} is related to the differential input voltage v_{id} as

$$i_{out} = - g_m v_{id} \qquad (5.65)$$

where g_m is the transconductance of the input JFET devices. The bias current

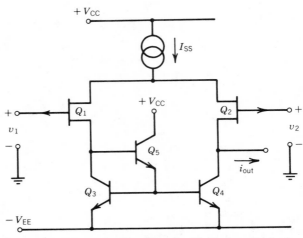

FIGURE 5.23.   JFET differential pair with bipolar active loads.

source I_{SS} of Figure 5.23 is normally designed as a JFET current source so that its saturation current I_{DSS} will track with those of Q_1 and Q_2 to ensure that the inequality of Eq. (5.34) holds through manufacturing tolerances.

Current-Differencing Amplifiers

The differential amplifier circuits discussed so far have one common property: they all amplify *differential input voltages* to produce an output current or voltage. In certain industrial applications, particularly those that require single-supply operation and interface with current-mode transducers, it is advantageous to use a differential current rather than voltage as the input signal. The types of differential gain stages which handle these types of signals are known as *current-differencing amplifiers.* [8]

Figure 5.24 shows the basic principle of operation of such a current-differencing gain stage. The current mirror, made up of matched transistors Q_1 and Q_2, creates a current-differencing bias circuit for the gain stage Q_3, such that both the dc bias currents and the incremental ac signals appearing at the base of Q_3 are given as

$$I_{B3} = I_{in1} - I_{in2} \quad \text{and} \quad i_{b3} = i_{in1} - i_{in2} \tag{5.66}$$

where the uppercase symbols denote bias currents, and the lowercase notation is used for incremental ac signals. The incremental output current i_{out} is then

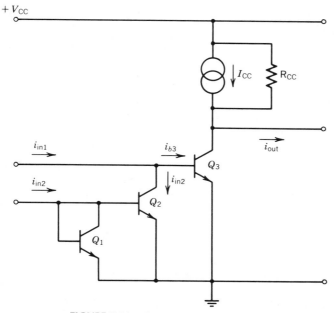

FIGURE 5.24. Current-differencing gain stage.

equal to

$$i_{\text{out}} = -\beta i_{b3} = -\beta(i_{\text{in1}} - i_{\text{in2}}) \tag{5.67}$$

In the simplified circuit diagram of Figure 5.24, the bias circuit made up of I_{CC} and R_{CC} represents the Norton equivalent of a *pnp* transistor current source which would be used as an active load for Q_3. The output resistance of the circuit is essentially the same as that given by Eq. (5.45), that is,

$$R_{\text{out}} = (R_{\text{CC}} \parallel r_{o3}) \tag{5.68}$$

where r_{o3} is the output resistance of Q_3 primarily due to the Early effect. In most applications, the current-differencing gain stage would be used in conjunction with an emitter-follower output buffer stage (similar to that shown in Fig. 5.25), to provide high voltage gain with relatively low values of load resistance.

Features of Differential Gain Stages with Active Loads

The differential gain stages with active loads, particularly those shown in Figures 5.21 through 5.23, are the most widely used building blocks as input stages for monolithic operational amplifiers. This is due to the features summarized below:

1. High differential gains (typically voltage gains ≥ 60 dB are possible with high-impedance loads).
2. High common-mode rejection (typically > 100 dB) and high input impedance.
3. Favorable dc level shift (i.e., output level shifted toward $-V_{\text{EE}}$ to simplify interfacing with the subsequent gain stage).
4. Wide input common-mode range (typically extending to within ± 1V of supply voltages).
5. Differential to single-ended conversion (differential input is converted to a single-ended output, which fits the basic operational amplifier configuration).

These important features of differential stages with active loads will be covered in more detail, in connection with monolithic operational amplifiers, in Chapter 7.

5.3. OUTPUT STAGES

The output stage of an amplifier must be able to deliver a substantial amount of power into a low-impedance load with acceptably low levels of signal distortion. Therefore, in general, it is required to have one or more of the following desirable properties:

1. Large output current swing.
2. Large output voltage swing.
3. Low output impedance.
4. Low standby power.

In addition to these basic properties, an output gain stage is also required to have sufficiently good frequency response such that it will not present a limitation on the rest of the amplifier circuit. Since the output stage is designed to handle large signal swings at high current levels, it should also have a built-in short-circuit protection such that it will not burn out when the output is accidentally short-circuited to ground or to one of the power supply lines.

In this section, several output stage configurations will be examined with regard to their advantages and limitations, starting with the simplest configurations and moving on to more complex designs. Particular attention will also be given to output short-circuit protection and current-limiting techniques.

Emitter Follower as an Output Stage

Figure 5.25 shows a typical output stage using an *npn* emitter-follower configuration. Since the driving stage, which immediately precedes the output stage, has a very strong influence on the performance of the output amplifier, both stages are shown. For illustrative purposes, the internal bias and load circuits are shown as current sources I_B and I_E, since in most IC designs these would be realized as active loads. However, the analysis of the circuit is equally applicable to cases where the current sources were replaced by load resistors.

Neglecting the saturation voltages V_{sat} associated with the current sources I_B and I_E, the output voltage V_L can ideally swing from $-V_{EE}$ to within $1V_{BE}$ of $+V_{CC}$. Its output resistance R_{out} can be readily calculated from the hybrid-π

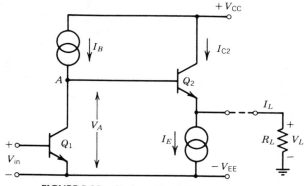

FIGURE 5.25. Basic emitter-follower output stage.

model as

$$R_{\text{out}} = \frac{1}{g_m} + \frac{R_A}{\beta} \tag{5.69}$$

for $\beta \gg 1$. R_A is the effective resistance at the base of Q_2 (i.e., at node A), and g_m is the transconductance of Q_2. If a lower output resistance is required, or if the driving voltage V_A needs additional buffering from the load, the emitter-follower transistor Q_2, can be replaced by a Darlington-connected pair of transistors.

The emitter-follower type output stage has the following advantages:

1. Simple and stable circuit configuration.
2. Low distortion.
3. Low output impedance.
4. Under proper bias condition, wide output swing.

However, the circuit also has the following drawbacks or limitations:

1. High standby power.
2. Unsymmetrical current drive capability.

The high standby power comes about since the circuit has a constant standby current drain of I_E, even when no signal power is delivered to the load. As it will be shown shortly, the value of I_E has to be kept substantially high to ensure proper operation of the circuit, with negative output swings.

Unsymmetrical current drive characteristics of the emitter-follower stage come about because the transistor Q_2 can only *source* current into the load, but it cannot *sink* current. In other words, with reference to the circuit of Figure 5.25, as the drive voltage V_A is increased, the output transistor can supply a maximum load current equal to βI_B, which in general is more than sufficient. However, if V_A is decreased, causing a negative output swing, the maximum output current I_L, which can be *drawn* from the load, is equal to the bias current I_E. This corresponds to the case where Q_2 would be completely cut off, that is, $I_{C2} = 0$.

Figure 5.26 illustrates this asymmetry associated with the current drive characteristics of the emitter-follower stage. For positive values of the drive voltage V_A, neglecting the saturation voltage of the current source I_B, the circuit can deliver a maximum positive output current I^+_{max} given as

$$I^+_{\text{max}} = \frac{V_{\text{CC}} - V_{\text{BE}}}{R_L} \tag{5.70}$$

In the above expression, it is assumed that the output transistor Q_2, is not β limited, that is, I^+_{max} given by Eq. (5.70) is less than βI_B. This is almost always

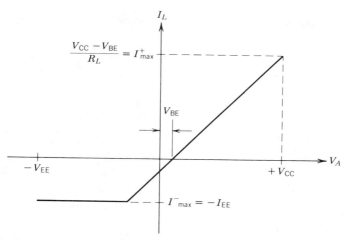

FIGURE 5.26. Output current of emitter-follower stage of Figure 5.25 as a function of drive voltage.

the case for practical values of the load resistance R_L and results in a full positive voltage swing V_L across R_L, equal to $V_{CC} - V_{BE}$.

As the drive voltage V_A is lowered, the load current is reduced. The current–voltage characteristics of Figure 5.26 do not go through the origin, due to the finite V_{BE} of Q_2. The circuit is capable of pulling out, or *sinking,* a maximum amount of load current I^-_{max} equal to I_E.

If the load resistance R_L is sufficiently large such that

$$\frac{V_{EE}}{R_L} < I_E \tag{5.71}$$

then this current drive limitation will not be a problem, and the peak negative load voltage V_L can swing down to $\approx -V_{EE}$. However, if R_L is sufficiently small, such that the inequality of Eq. (5.71) is not satisfied, then the negative output swing will be limited to

$$(V_L{}^-)_{max} = I_E R_L \tag{5.72}$$

which will result in *clipping* of the negative signal swing. This problem can be solved by increasing I_E. However, such a solution is not always practical, since it increases the standby power dissipation.

The unsymmetrical current drive capability of the emitter follower also makes it unsuitable for driving capacitive loads, particularly where large signal swings are required. This problem is illustrated in Figure 5.27. With a capacitive load, the maximum rate of change, or the so-called *slew rate,* of the output voltage is given as

$$\frac{dV_L}{dt} = \frac{I_L}{C_L} \tag{5.73}$$

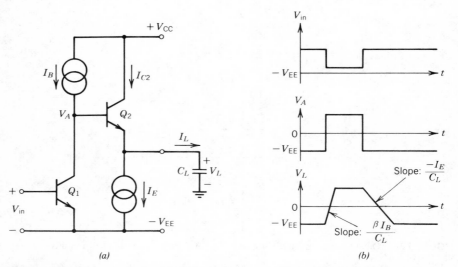

FIGURE 5.27. Step response of emitter-follower output stage with capacitive load: (*a*) Circuit diagram; (*b*) input and output waveforms.

As shown in the waveforms of Figure 5.27*b*, the positive slew rate of an emitter follower can be quite high. However, the negative slew rate is very low, limited by I_E. This asymmetry severely limits the usefulness of the emitter-follower output stage with capacitive loads.

Although the analysis shown in Figures 5.25 through 5.27 was carried out for an *npn* emitter-follower configuration, the results are equally valid for a *pnp* emitter follower with the appropriate polarity changes.

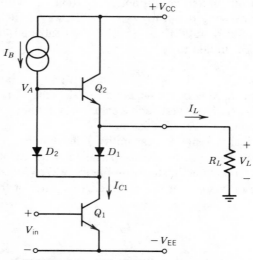

FIGURE 5.28. All-*npn* Class B output stage.

All *npn* Class B Output Stages

Some of the drawbacks of the emitter-follower output stage, such as the excessive standby power dissipation and limited current-sink capability, can be avoided by using Class B type output stages, which draw no quiescent current when the load current I_L is zero. In this section, the basic circuit configuration for an all-*npn* Class B output stage shall be described. The main advantage of this circuit, over the complementary *npn–pnp* type output stages described in the next section, is its high current-handling capability. This advantage comes about since no *pnp* transistors are used in the high current signal path.

Figure 5.28 shows the basic circuit configuration for the all-*npn* Class B output stage. This circuit configuration, often referred to as the *totem pole* connection, was initially developed for digital integrated circuits and has later been adapted to analog IC design. The operation of the circuit can be briefly described as follows. For positive swings of the output voltage (i.e., for negative-going input voltage), the upper transistor Q_2 is active with the diode D_1 reverse biased. Thus, for positive swings of the output voltage, the active part of the circuit is equivalent to that shown in Figure 5.29a. In this manner, the output voltage can swing to within 1 V_{BE} of the positive supply, neglecting the saturation voltage of the current source I_B, and it can supply a positive load current I_L^+ up to a maximum value of βI_B.

For negative swings of the output voltage, the diode D_1 is brought into conduction and causes Q_2 to be turned off. This results in the equivalent circuit

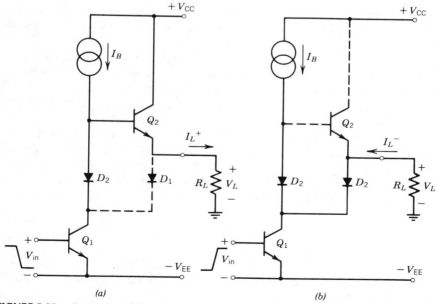

(a)   (b)

FIGURE 5.29. Conducting portions of circuit of Figure 5.28: (*a*) For positive output swings; (*b*) for negative output swings.

of Figure 5.29b, where the negative load current I_L^- is conducted through D_1 and Q_1. In this manner, the output voltage can swing to within 1 V_{BE} (due to D_1) plus 1 $V_{CE\ sat}$ (due to Q_1) of the negative supply, provided that the transistor Q_1 can sink all the necessary current.

The all-*npn* output stage of Figure 5.28 has two basic disadvantages: (1) unsymmetrical gain characteristics, and (2) a 1 V_{BE} *dead band* at the zero crossing of the output voltage. These are discussed below.

Unsymmetrical Gain.   This comes about because the two equivalent circuits for the positive and negative output swings are dissimilar, as shown in Figure 5.29. For positive-going output swings, the circuit has a voltage gain A_V^+,

$$A_V^+ \approx -g_{m1}\beta_2 R_L \qquad (5.74)$$

where the subscripts refer to the respective transistors.

For negative-going output swings, the voltage gain A_V can be calculated from Figure 5.29b as

$$A_V^- \approx -g_{m1} R_L \qquad (5.75)$$

Thus, the two voltage gains differ by a factor of β_2, and the gain during the positive output swing is much higher than the gain during the negative swing.

Output Dead Band.   This comes about because the driving voltage V_A at the base of Q_2 has to swing 1 V_{BE} *above* ground before Q_2 would start conducting, and the circuit of Figure 5.29a becomes operational, whereas the circuit of Figure 5.29b becomes operational for $V_A \leq 0$. The presence of such a dead band is objectionable since it leads to so-called *crossover* distortion as I_L goes through zero. The effects of crossover distortion are particularly undesirable at small output signal swings, centered around zero. In theory, this dead band can be reduced to nearly zero by using a V_{BE}-multiplier circuit (see Fig. 4.25) to replace D_2. In practice, however, this technique must be used with care to avoid having both Q_1 and Q_2 conducting simultaneously. Such a condition would result in excessive current drain between the supply voltages, and the possible burn-out of Q_1 or Q_2.

In certain applications, it may be desirable to have the all-*npn* Class B output stage self-biased with the output at nearly halfway between the power supplies, so that the input signal can be ac coupled through a coupling capacitor. Figure 5.30 shows a practical circuit configuration to achieve this objective.[9] In the circuit, Q_3 and Q_1 form a controlled-β transistor (see Fig. 4.3), with the area ratio

$$\frac{A_1}{A_3} = n \qquad (5.76)$$

Then, ignoring the base currents, the bias currents I_A and I_B through resistors R_A and R_B would be forced to be

$$I_B = n I_A \qquad (5.77)$$

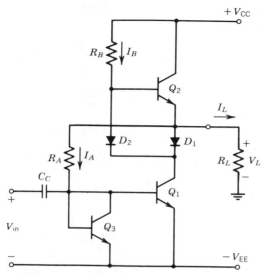

FIGURE 5.30. Self-biasing all-*npn* Class B output stage.

In order to set the output voltage V_L nominally at zero in the quiescent state, the resistor values have to be chosen such that

$$R_B = \frac{V_{CC} - V_{BE}}{I_B} \qquad (5.78)$$

and

$$R_A = \frac{V_{EE} - V_{BE}}{I_A} \qquad (5.79)$$

If symmetrical supplies are used such that $V_{CC} = V_{EE}$, which is often the case, then choosing $I_A R_A$ equal to $I_B R_B$, or

$$R_A = nR_B \qquad (5.80)$$

one can ensure that the quiescent output level is located midway between the two supply voltages for a wide range of supply voltages and resistor values. The scale factor n is chosen to be > 1 since Q_1 is required to handle large output currents. Typical values are normally chosen to be in the range of $5 < n < 20$, as dictated by practical component sizes.

Complementary Class B Output Stages

Figure 5.31*a* shows a simple Class B output stage configuration using complementary transistors. Q_2 is normally formed by a substrate *pnp* transistor (see Fig. 2.29), which has somewhat higher current-handling capability than the lateral *pnp* transistor. As the input voltage is varied to cause the driving-point

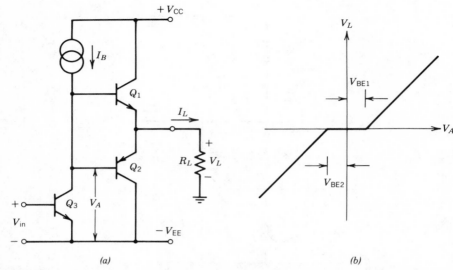

FIGURE 5.31. Simple *npn–pnp* complementary output stage; (*a*) Circuit diagram; (*b*) Voltage transfer characteristics.

voltage V_A to vary, either Q_1 or Q_2 is brought into conduction, and the output voltage V_L follows V_A within 1 V_{BE} drop. Neglecting the saturation voltages of Q_3, or the current source I_B, the output voltage can swing within 1 V_{BE} of either supply voltage. The maximum positive and negative output current is limited by the β of Q_1 and Q_2 and the bias source I_B. The voltage gain of the circuit is reasonably symmetrical for positive and negative swings and can be written as

$$A_V^+ \approx - g_{m3}\beta_1 R_L \tag{5.81}$$

$$A_V^- \approx - g_{m3}\beta_2 R_L \tag{5.82}$$

neglecting the output resistances of the bias current source I_B or Q_3.

Figure 5.31*b* shows the voltage transfer characteristics of the circuit. Note that the circuit exhibits a dead band of 2 V_{BE}, centered around zero, during which neither Q_1 or Q_2 is on.

Figure 5.32 shows a modified version of the complementary output stage, which uses two diodes D_1 and D_2 to offset the V_{BE} drops of Q_1 and Q_2, respectively, and thereby nearly eliminates the crossover distortion. This circuit, or one of its derivatives, is very often used as the output stage for a number of operational amplifier products.

Strictly speaking, the circuit of Figure 5.32 is not Class B, but Class AB in its operating mode, since both Q_1 and Q_2 carry a finite amount of quiescent current at standby state. In the design of the circuit, particular care has to be taken in calculating the voltage drops across D_1 and D_2 and matching them to those across the base–emitter junctions of Q_1 and Q_2, respectively. This is not

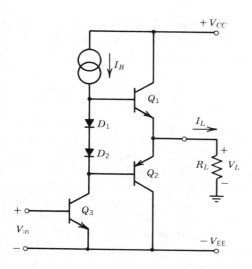

FIGURE 5.32. Complementary output stage with compensation diodes to avoid crossover distortion.

achieved easily, since Q_1 and Q_2 must be capable of carrying large amounts of current. Thus, they need to be large area devices, whereas D_1 and D_2 should be made small to conserve chip area. Overestimating the value of voltage drops across D_1 and D_2 would cause crossover distortion; underestimating the voltage drops would cause excessive quiescent current through Q_1 and Q_2, leading to too much standby power dissipation.

Figure 5.33 shows a practical method of compensating for the crossover distortion, relatively independent of manufacturing tolerances.[10] In the circuit, D_1 and D_2 are replaced by emitter-follower transistors Q_4 and Q_5, respectively,

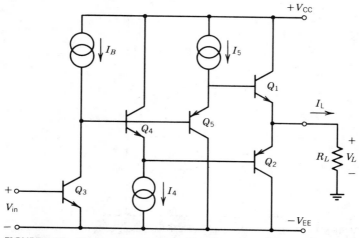

FIGURE 5.33. Improved output stage with crossover distortion compensation.

which operate at fixed bias currents I_4 and I_5. In this manner, the quiescent current through Q_1 and Q_2 can be adjusted by proper scaling of the emitter of Q_4 and Q_5 and/or by the choice of their bias currents I_4 and I_5.

Quasi-Complementary Output Stages

One basic limitation of the complementary output stage shown in Figure 5.32 is the current-handling capability of the substrate *pnp* transistor Q_2. For practical device geometries, the maximum current through Q_2 is usually limited to approximately 20 mA. If higher current levels are required, Q_2 can be replaced by a composite connection of a lateral *pnp* transistor and a high-current *npn* transistor, as described in Section 2.3 (see Fig. 2.28). This results in the quasi-complementary output stage configuration shown in Figure 5.34.[11] In the figure, the interconnection of the lateral *pnp* transistor Q_{2A} with the high current *npn* transistor Q_{2B} results in a composite device which has the polarity of a *pnp* transistor with the current gain and current-handling capability of the *npn* transistor. However, the saturation voltage of the composite transistor is equal to the V_{BE} of Q_{2B} plus the $V_{CE\,sat}$ of Q_{2A}, and this causes a slight reduction in the negative swing of the output voltage.

One major problem with the composite connection of the Q_{2A} and Q_{2B} in Figure 5.34 is the potential instability due to the internal feedback loop formed by the composite connection, particularly with capacitive loads on the amplifier. A practical solution to this problem is to replace Q_{2A} with a controlled gain *npn* transistor as shown in Figure 4.19, which would reduce its β to about 20–30; or connect a bypass resistor R_X from the base of Q_{2B} to its emitter (see Fig. 4.30).

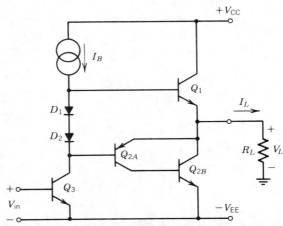

FIGURE 5.34. Quasi-complementary output stage using composite *pnp–npn* transistor.

Short-Circuit Protection

One of the major failure modes of output stages is the possible burn-out of the output devices or interconnections due to accidental or transient overload conditions. A typical example of such an overload condition would be the accidental short circuit of the output to the power supply or the ground terminals. In such cases, the burn-out of the output devices, or other permanent damage to the monolithic circuit, can be avoided by limiting the maximum available output current to a safe value. This safe value is determined by the size and layout of the output devices and by the maximum allowable power dissipation considerations.

The limiting of the maximum available current, which is also called *short-circuit protection,* can be done in two ways: (1) passive current limiting, where passive components such as resistors or diodes are used to limit the output current, and (2) active current limiting, where active devices in a feedback path are used for sensing and controlling the output current.

Passive Current Limiting. Figure 5.35 shows two typical examples of passive current limiting often used in complementary output stages. In the circuit of Figure 5.35a the limiting of the output current under overload conditions is achieved by the resistors R_{E1} and R_{E2}, which are in series with the emitters of the high current output transistors Q_1 and Q_2. The major disadvantage of this approach is, in order to limit I_L to a safe value, that both R_{E1} and R_{E2} have to be relatively large; and this causes the available output swing to be reduced and increases the output impedance. In the circuit of Figure 5.35a, short-circuit protection can also be obtained by using a single resistor in series with the output terminal, rather than two separate resistors R_{E1} and R_{E2}, one at each one of the emitters of the power devices. However, the use of two resistors is often preferred since R_{E1} and R_{E2} also tend to balance the quiescent current through Q_1 and Q_2, and make it less sensitive to absolute values of the V_{BE} drops across D_1 and D_2.

The circuit of Figure 5.35b uses a set of additional diodes D_3 and D_4 to sense the voltage drop across R_{E1} or R_{E2}. If the output current I_L is such that the voltage drop across R_{E1} or R_{E2} exceeds 1 V_{BE}, either D_3 or D_4 isa brought into conduction, and some of the drive current available to the bases of Q_1 or Q_2 is shunted directly to the output. This method works very effectively for limiting the current through Q_1, whose maximum available base drive is limited to I_B. However, it is less effective for limiting the current through Q_2, whose base drive depends primarily on the current which the input transistor Q_3 can sink. Thus, in order to limit the total current which Q_3 can sink, an added resistor, R_{E3}, is included in the circuit in series with the emitter of Q_3. Note that as a result of this current-limiting technique, the peak-to-peak output voltage swing is reduced by 2 V_{BE} drops at high currents, due to the voltage drops across R_{E1} and R_{E2}.

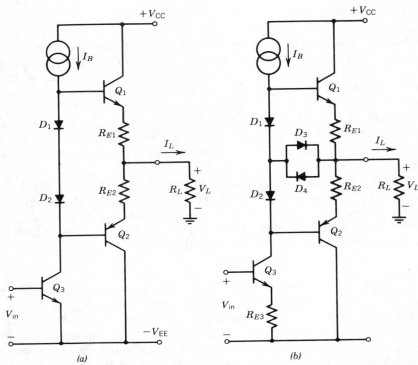

FIGURE 5.35. Methods of passive short-circuit protection: (*a*) Resistive current limiting; (*b*) diode-resistor current limiting.

In many cases, for negative values of the load current, that is, when Q_2 is conducting, the limited current-handling capability of the substrate *pnp* transistor Q_2 can also provide additional short-circuit protection. The current-handling capability of Q_2 is limited both by the *pnp* transistor β fall off at high currents and by the parasitic series resistances associated with its base and collector contacts.

Active Current Limiting. Figure 5.36 shows a typical example of active current limiting. For illustrative purposes, a Class A type output stage using an *npn* emitter follower is shown. In the circuit, if no current limiting is used, the emitter follower can supply a maximum load current of βI_B for positive output swings. In certain cases, this can be more than the transistor Q_2 can handle, and has to be reduced to a safe value. In the circuit of Figure 5.36, this current limiting is achieved by sensing the amount of current through the sensing resistor R_E and using the voltage drop across R_E to turn on a normally off transistor Q_3. The value of R_E is chosen such that the voltage drop across it would equal 1 V_{BE} when I_L reaches its maximum safe value. When this value is reached, Q_3 is

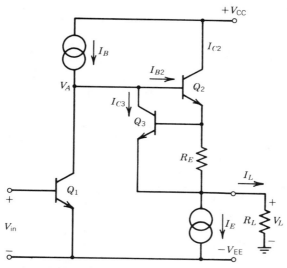

FIGURE 5.36. Active current-limiting technique for short-circuit protection.

brought into conduction, and safely shunts most of the base drive available to Q_2 directly to the output.

One important point to remember is that the active current-limiting technique, such as the one shown in Figure 5.36, works only for limiting the current through devices that have a *limited* and *predictable* amount of base drive. An example of a circuit where such a short-circuit protection will *not* work is shown in Figure 5.37. Note that this is also an emitter-follower output stage, except that a *pnp* emitter follower is used as the output stage, with the same drive circuitry as in the case of Figure 5.36. At the excessive negative swings of the load current I_L, the voltage drop across R_E would activate the current-limiting transistor Q_3. However, if the drive transistor Q_1 has sufficient current-sinking capability, it will still provide Q_2 with sufficient base drive to inhibit the current-limiting action. This is a common design pitfall, often overlooked by an inexperienced designer.

Figure 5.38 shows a more detailed schematic of a complementary output stage which utilizes active current-limiting protection for both positive- and negative-current overloads. The positive-current overload protection is provided by the current limiting transistor Q_5 and the sensing resistor R_{E1}. The negative overload protection is, in part, provided by R_{E2} and Q_6. However, to avoid the problem illustrated in Figure 5.37 (i.e., limit the maximum amount of current which Q_2 can sink), an additional current source resistor R_X is used in conjunction with the third current-limiting transistor Q_8, which becomes activated and shunts the input current I_{in} when the voltage drop across R_X approaches 1 V_{BE} drop.

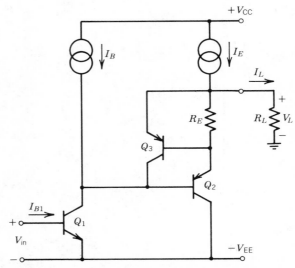

FIGURE 5.37. Active current-limiting circuit that will *not* work.

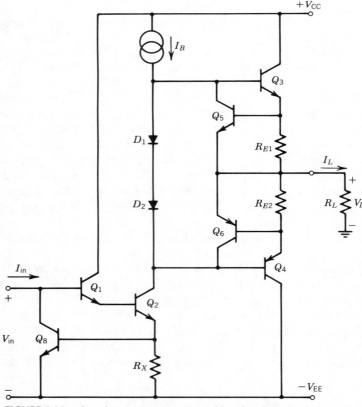

FIGURE 5.38. Complementary output stage with active short-circuit protection.

Thermal Shutdown

An additional overload protection technique commonly used in monolithic integrated circuits is the so-called *thermal-shutdown* technique. In this approach, an internal sensing mechanism such as the temperature dependence of the V_{BE} drop, is used to measure the chip temperature. If the average chip temperature exceeds a predetermined "safe" limit, a shutdown circuitry is activated to limit, reduce, or completely shut off the load current available at the output until the chip temperature drops to a safe level. Basic circuit approaches for designing thermal-shutdown circuitry will be discussed further in connection with voltage regulator designs (see Chapter 10).

Thermal shutdown is primarily intended to protect the circuit against prolonged overload conditions, whereas the short-circuit current-limiting techniques discussed earlier are basically intended to protect the circuit against transient overloads.

REFERENCES

1. R. D. Middlebrook, *Differential Amplifiers*, Wiley, New York, 1963.
2. L. J. Giacoletto, *Differential Amplifiers*, Wiley, New York, 1970.
3. E. J. Angelo, Jr., *Electronic Circuits*, McGraw-Hill, New York, 1964, Chap. 4.
4. P. R. Gray and R. G. Meyer, *Analysis and Design of Analog Integrated Circuits*, Wiley, New York, 1977, Chap. 3.
5. A. B. Grebene, Ed., *Analog Integrated Circuits*, IEEE Press, New York, 1978.
6. J. Davidse, *Integration of Analogue Electronic Circuits*, Academic Press, London, 1979, Chap. 4.
7. R. W. Russell and T. M. Frederiksen, "Automotive and Industrial Electronic Building Blocks," *IEEE J. Solid-State Circuits*, **SC-7**, 446–454, (December 1972).
8. T. M. Frederiksen, W. F. Davis, and D. W. Zobel, "A New Current-Differencing Single-Supply Operational Amplifier, *IEEE J. Solid-State Circuits*, **SC-6**, 340–347, (December 1971).
9. H. R. Camenzind and A. B. Grebene, "An Outline of Design Techniques for Linear Integrated Circuits," *IEEE J. Solid-State Circuits*, **SC-4**, 110–122 (June 1969).
10. K. Fukahori, Y. Nishikawa, and A. R. Hamade, "A High Precision Micropower Operational Amplifier," *IEEE J. Solid-State Circuits*, **SC-14**, (December 1979).
11. E. L. Long and T. M. Frederiksen, "A High Gain 15-Watt Monolithic Power Amplifier with Internal Fault Protection," *IEEE J. Solid-State Circuits*, **SC-6**, 35–44, (February 1971).

CHAPTER SIX

ANALOG DESIGN
WITH MOS TECHNOLOGY

Metal-oxide-semiconductor (MOS) technology was originally developed for digital large-scale integration (LSI) design. The small size and the self-isolating nature of enhancement-mode MOS transistors allow a much higher functional density to be achieved on an IC chip than is possible with bipolar technology. Traditionally, MOS technology has been used extensively in the design of wholly digital system blocks, such as microprocessors and memories. On the other hand, analog circuit functions, such as amplifiers, D/A converters, and active filters, have been designed primarily with bipolar technology.

Technological progress over recent years has rapidly increased the feasible level of integration for complete systems containing both digital and analog functions. Thus, the dividing line between analog and digital LSI technologies has become less distinct. It has, therefore, become necessary to extend the capabilities of MOS technology into analog IC design so that a higher level of system integration can be made economically feasible.

The use of MOS transistors to perform analog functions presents many design challenges and requires a number of design compromises. In most cases, the resulting MOS analog blocks, such as operational amplifiers, comparators, or voltage references, may not be able to meet the performance specifications of their bipolar equivalents, but still perform satisfactorily as a subsystem within the monolithic chip. The high density of MOS transistors makes it possible to integrate analog functions on a much smaller chip area. For example, a MOS operational amplifier requires only 30–50% of the chip area needed for an equivalent bipolar amplifier. This feature makes it possible to increase greatly the density of analog functions on the chip.

This chapter will focus on the application of MOS transistors to analog circuit design. The discussion will be confined to conventional MOS transistors which are fabricated with standard process technologies used for digital LSI. These process technologies use the standard n-channel enhancement-mode process, the complementary-MOS (CMOS) process, and the n-channel process with depletion-mode devices. In these basic processes, monolithic resistor structures of suitable values are not readily available. Therefore, the designer is often forced to perform the entire function using *only* active devices.

In designing analog circuit functions with MOS technology, one very important point must be remembered. Often designers *do not* have the luxury of choosing the best or most versatile MOS technology that suits their analog designs. Instead, they are *forced* to use whatever MOS technology was chosen for the digital part of the circuit. As will be described in later sections of this chapter, this presents a new and unique set of challenges to the analog IC designer.

The purpose of this chapter is to lay the groundwork and develop the basic building blocks for analog MOS LSI design. These building blocks will serve as the subsections of more complex and complete MOS LSI designs, such as operational amplifiers, data converters, and switched-capacitor filters, which will be described in later chapters. The first part of the chapter deals with the basic characteristics of the MOS transistor as a circuit element. The second part covers the analog building blocks for single-channel MOS technology. The third and fourth sections of the chapter focus on analog circuit functions using CMOS or n-channel depletion-mode devices. The fifth section deals with the design of voltage reference circuits with MOS transistors. The last section of the chapter covers the applications of the MOS transistor as an analog switch.

6.1. BASIC CHARACTERISTICS OF MOS TRANSISTORS

Basic device equations of the MOS transistor were derived and discussed in Section 2.5. As a refresher, they will be briefly reviewed in this section. The current–voltage characteristics of an n-channel enhancement-mode MOS transistor are given in Figure 2.42. The transistor has two distinct regions of operation:

1. *Triode Region.* This corresponds to operation of the device with very low source-to-drain voltages, such that $V_{DS} < (V_{GS} - V_{TH})$, where the device behaves in a manner similar to a voltage-controlled resistor.

2. *Pinched Region.* This corresponds to operation of the device with sufficient drain–source voltage, such that $V_{DS} > (V_{GS} - V_{TH})$. In this region of its current–voltage characteristics, the MOSFET behaves as a voltage-controlled current source. This region of operation is also referred to as the *saturated region,* since the drain current I_D is saturated and relatively insensitive to the source–drain voltage V_{DS}.

When MOS transistors are used as active gain elements, they would normally be biased to operate in the pinched region, whereas when they are used as analog switches, they would be operated in their triode region.

Figure 6.1 shows a simple MOS gain stage using n-channel enhancement-mode MOSFET with a resistive load. Assuming that the MOSFET is biased to operate in its pinched region, its drain current I_{DO} can be expressed by Eq. (2.49), which is repeated below:

$$I_{DO} = \frac{\mu_s C_{ox}}{2} \left(\frac{Z}{L}\right) (V_{GS} - V_{TH})^2 \tag{6.1}$$

where μ_s is the carrier mobility in the conducting channel, C_{ox} is the capacitance per unit area of the gate region, and Z/L is the width-to-length, or aspect, ratio of the channel region. The transconductance g_m in the pinched region can be derived from Eq. (6.1) as

$$g_m = \frac{\partial I_{DO}}{\partial V_{GS}} = \sqrt{2\mu_s C_{ox} \left(\frac{Z}{L}\right) I_{DO}} \tag{6.2}$$

or

$$g_m = \mu_s C_{ox} \left(\frac{Z}{L}\right) (V_{GS} - V_{TH}) \tag{6.3}$$

which can also be written as

$$g_m = \frac{2I_{DO}}{V_{GS} - V_{TH}} \tag{6.4}$$

Equation (6.2) is particularly important in illustrating one of the key properties of the MOS transistor: at a given saturated drain current level I_{DO}, the

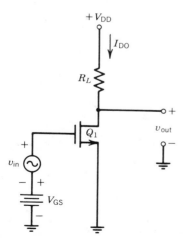

FIGURE 6.1. NMOS gain stage with resistive load.

transcondutance is proportional to the square root of the Z/L ratio, which is determined by the device geometry (i.e., device layout).

The $\mu_s C_{ox}$ product is a process parameter which is fixed for a given MOS fabrication process. For practical device structures with $1000=\text{Å}$ gate oxide thickness, the value of the $\mu_s C_{ox}$ product is about 20 $\mu A/V^2$ for n-channel devices and approximately 10 $\mu A/V^2$ for p-channel devices. It does not change drastically between various MOS fabrication processes.

At this point, it is instructive to compare the g_m of a MOSFET with that of a bipolar transistor, where

$$g_m \mid \text{bipolar} = \frac{qI_C}{kT} = \frac{I_C}{V_T} \tag{6.5}$$

The following differences stand out:

1. The transconductance of a bipolar device is completely defined once the collector current I_C is given. It does *not* depend on the device geometry. This is due to the exponential relationship between I_C and V_{BE} of the bipolar transistor.
2. The transconductance of a MOS device depends on two sets of variables once the process parameters are given. These are: (*a*) the channel width-to-length ratio Z/L and (*b*) the gate–source voltage V_{GS} or the drain current I_{DO}.

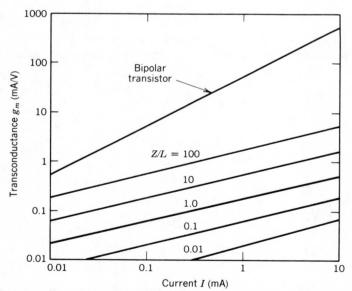

FIGURE 6.2. Comparison of MOS and bipolar transistor transconductance as a function of operating current. (*Note;* For MOS transistor, various Z/L ratios are shown; operation in pinched region is assumed with $\mu_s C_{ox} \approx 10$ $\mu A/V^2$.)

The above differences lead to the following important conclusions:

1. In designing with MOS transistors, the Z/L ratio, that is, the lateral device geometry and size, become a very important design parameter.
2. Quantitatively, the transconductance available from a MOSFET at a given current level is much lower than that of a bipolar transistor. This is illustrated graphically in Figure 6.2 for various Z/L ratios.[1]

The small-signal voltage gain that can be obtained from the simple gain stage of Figure 6.1 can be derived from the ac small-signal equivalent circuit of Figure 2.46 as

$$A_v = \frac{v_{out}}{v_{in}} = -g_m(R_L \| r_o) \qquad (6.6)$$

where r_o is the output resistance of Q_1 due to the channel length modulation effect, as illustrated in Figure 2.46 [see Eq. (2.52)].

In practice, the resistively loaded gain stage of Figure 6.1 is rarely used because the low g_m of the MOS transistor makes it difficult to obtain large voltage gains from a resistive load. This problem is further compounded due to the lack of suitable resistor structures in MOS technology. High-value resistors are either incompatible with the basic MOS processes, or they require far too much chip area to be practical. Thus, in analog MOS design, one is forced to perform the gain functions almost exclusively with active devices.

Figure 6.3 shows a simple gain stage using single-channel technology (i.e., no complementary devices). Neglecting the so-called *body effect,* which will be discussed shortly, and neglecting the finite output resistance, the voltage gain of

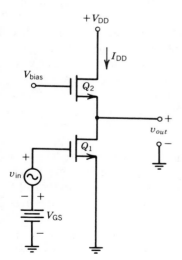

FIGURE 6.3. Simple NMOS inverter stage with active load.

this stage can be written as

$$A_v = \frac{v_{\text{out}}}{v_{\text{in}}} = - g_{m1} r_{s2} \tag{6.7}$$

where r_{s2} is the dynamic resistance looking into the source terminal of Q_2. From the equivalent circuit of Figure 2.46, one can show that this resistance is equal to the reciprocal of the transconductance g_{m2} of Q_2, that is,

$$r_{s2} = \frac{1}{g_{m2}} \tag{6.8}$$

Since both transistors of Figure 6.3 are operating at the same drain current I_{DO}, the approximate voltage gain then becomes

$$A_v \approx - \frac{g_{m1}}{g_{m2}} = - \sqrt{\frac{(Z/L)_1}{(Z/L)_2}} \tag{6.9}$$

where the transconductance ratio is related to the Z/L ratios by Eq. (6.2).

Equation 6.9 also illustrates the limited voltage gain available from the circuit. Even with $Z/L = 100$ for Q_1 and $Z/L = 1$ for Q_2, the maximum available gain is only 10. As will be described in the following section, the actual gain is lower than this amount due to the so-called body effect resulting from the substrate bias.

The Body Effect

The reverse-biased junction formed between the conducting channel and the substrate, or the "body" region, in a MOS transistor (see Fig. 2.40) behaves in a manner similar to the gate of a FET. As the substrate is made more negative with respect to the source of an n-channel MOS (NMOS) transistor, the depletion region between the substrate and the channel experiences a larger potential drop and, hence, becomes wider and contains more charge. This implies that for the same amount of surface charge in the channel of the MOSFET, a higher electric field must exist in the oxide layer between the channel and the gate electrode. This in turn means that the gate voltage must be made more positive to achieve the same number of charge carriers in the channel. The net result of this effect is the apparent increase of the threshold voltage V_{TH} as the reverse bias between the substrate and the source is increased. This effect is often referred to as body effect.

Figure 6.4 shows the inverter circuit of Figure 6.3 which is redrawn to emphasize the presence of the body effect. Both Q_1 and Q_2 share the same p-type substrate, or body, which is biased at the same potential as the source of Q_1. However, the source–body junction of Q_2 experiences an *added* reverse bias which is equal to the voltage drop across Q_1.

The dependence of the threshold voltage on the source–body bias V_{BS} gives way to a body effect transconductance g_{mB} which relates the incremental de-

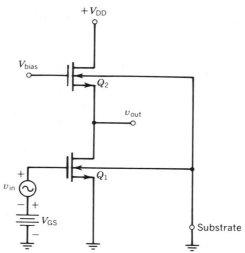

FIGURE 6.4. Simple inverter circuit of Figure 6.3 redrawn to illustrate the body effect due to substrate bias.

pendence of the drain current to the body–source bias. In other words, g_{mB} is the transconductance of the substrate-channel junction. A quantitative analysis of the body effect can be made by defining an incremental body-effect factor λ_B, which is the ratio of g_{mB} to the actual transconductance g_m due to the gate voltage, that is,

$$\lambda_B = \frac{g_{mB}}{g_m} \qquad (6.10)$$

Ideally, with no body effect, $\lambda_B = 0$.

The analytical expression for λ_B can be written as[2]

$$\lambda_B = \frac{1}{2C_{ox}} \sqrt{\frac{2q\epsilon_s N_A}{2\phi_F + V_{SB}}} \qquad (6.11)$$

where ϕ_F is the potential difference between the Fermi level and the intrinsic level in the substrate and is typically on the order of 0.3 V. ϵ_s is the dielectric constant of silicon, and N_A is the substrate impurity concentration. Note that the primary factors in determining λ_B are the substrate doping N_A and the source–body bias V_{SB}.

Figure 6.5 gives a plot of λ_B as a function substrate doping for various values of V_{BS}. The body effect decreases as the substrate bias is increased or as the substrate doping is reduced.

In the case of the inverter circuit of Figures 6.3 and 6.4, a parameter which is useful in evaluating the influence of the substrate bias on the voltage gain is

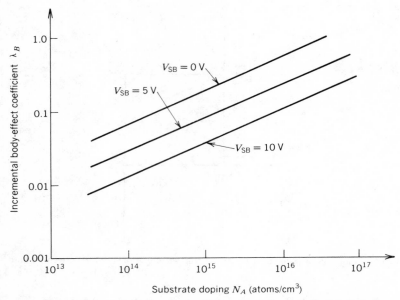

FIGURE 6.5. Value of incremental body-effect coefficient, λ_B as a function of substrate doping and bias level.[2]

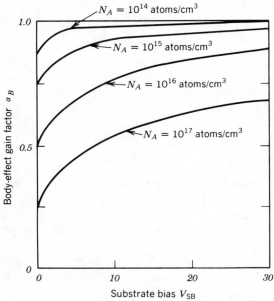

FIGURE 6.6. Body-effect gain factor α_B as a function of substrate bias and doping level.[1]

the body–effect gain factor α_B, which is defined as

$$\alpha_B = \frac{1}{1 + \lambda_B} \tag{6.12}$$

α_B is the ratio of the gain of the basic inverter circuit *with* body effect to the ideal gain *without* body effect.

In terms of the body-effect gain factor, the actual form of Eq. (6.9) becomes

$$A_v = -\alpha_B \frac{g_{m1}}{g_{m2}} = -\alpha_B \sqrt{\frac{(Z/L)_1}{(Z/L)_2}} \tag{6.13}$$

Note that α_B is at all times less than 1. Figure 6.6 gives a plot of α_B as a function of substrate bias for various levels of substrate doping. The body-effect gain factor α_B is reduced significantly at low substrate bias voltages or at high substrate doping levels. The most important consequence of the body effect in MOS circuit design is in the case of circuits utilizing depletion-mode MOS devices as active loads (see Section 6.3).

6.2 BUILDING BLOCKS OF NMOS ANALOG DESIGN

The basic NMOS technology developed around the n-channel enhancement-mode MOSFET has been the workhorse of digital LSI development. This is because, compared to other MOS or bipolar technologies, the basic NMOS process provides the highest functional density on the chip and requires fewer processing steps. Unfortunately, for analog circuit design, the NMOS process is quite restrictive since it provides only one type of device, namely, the n-channel enhancement-mode MOSFET, to perform all the active or passive circuit functions. In spite of these restrictions, a number of design approaches have been developed over recent years for NMOS analog LSI, which have resulted in the successful implementation of many of the analog functions, such as operational amplifiers, voltage references, and active filters, using nothing more than the basic NMOS technology.[3-6]

In this section, some of the basic building blocks or subcircuits for NMOS analog IC design will be reviewed. The resulting conclusions and the circuit configurations are equally applicable to PMOS analog IC design, with the appropriate polarity reversals.

Diode-Connected MOSFETs

One of the simplest building blocks of NMOS analog circuits is the diode-connected MOS transistor, which is shown in Figure 6.7a. Its current–voltage characteristics are shown in Figure 6.7b. The diode connection forces V_{DS} to be equal to V_{GS}. The device does not conduct until $V_{GS} = V_{TH}$. For $V_{GS} \geq V_{TH}$, the

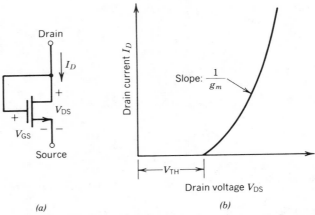

FIGURE 6.7. Diode-connected enhancement-mode MNOS transistor. (*a*) Circuit connection, (*b*) current-voltage characteristic.

current conduction starts, and the current I_D is determined by Eq. (6.1) with $V_{DS} = V_{GS}$.

The dynamic resistance r_{ds} of the current–voltage characteristics is given as

$$r_{ds} = \frac{\partial V_{DS}}{\partial I_D} = \frac{\partial V_{GS}}{\partial I_D} = \frac{1}{g_m} \tag{6.14}$$

In terms of Eq. (6.3), r_{ds} can be written as

$$r_{ds} = \frac{1}{g_m} = \frac{1}{(\mu_s C_{ox} Z/L)(V_{DS} - V_{TH})} \tag{6.15}$$

Thus, the dynamic resistance can be scaled inversely with the Z/L ratio.

Figure 6.8 shows some of the typical circuit applications for the diode-connected NMOS transistor. In the circuit of Figure 6.8*a*, a series of such diode-connected devices are used as a voltage divider to set up various voltage levels V_1 and V_2, which can be scaled by the choice of the Z/L ratios of the devices. One basic drawback of this configuration is that the current increases rapidly with increasing supply voltage, this effect being more severe if many devices are used in series.

In the circuit of Figure 6.8*b*, the diode-connected device is used as an active load to provide voltage gain for the inverter stage formed by Q_1. This circuit configuration is essentially equivalent to that of Figure 6.3, where the need for external bias is eliminated. Its voltage gain is given by Eq. (6.9) in the ideal case and by Eq. (6.13) if the influence of the body effect is considered.

In order to increase the voltage gain available from the inverter circuit of Figure 6.8*b*, the value of Z/L for Q_2 has to be reduced. Increasing the Z/L ratio

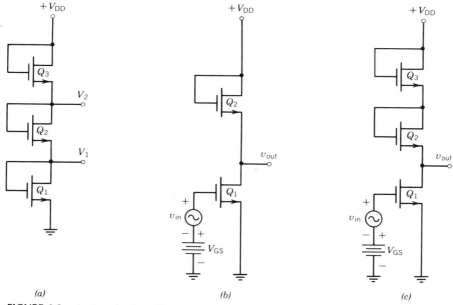

(a)   *(b)*   *(c)*

FIGURE 6.8. Analog circuit applications of diode-connected NMOS transistor: (*a*) Voltage divider; (*b*) active load; (*c*) split-load inverter.

to much below unity, that is, making the channel much longer than its width, is often undesirable since it greatly increases the parasitic gate–source capacitance C_{GS} of Q_2 (see Fig. 2.46), which is in parallel with r_{ds}. This problem can be partly avoided by using a so-called *split-load* connection, as shown in Fig. 6.8*c*, which divides the load into two diode-connected transistors and reduces the effect.

One basic problem associated with the gain stages of Figure 6.8, which use diode-connected NMOS transistors as active loads, is the limited voltage swing available at the output. Since the diode-connected device must maintain a minimum voltage drop $V_{DS} \geq V_{TH}$ across it, the maximum positive output swing is limited to one threshold voltage below the supply voltage, that is, to $V_{DD} - V_{TH}$ for the circuit of Figure 6.8*b* and to $V_{DD} - 2V_{TH}$ for the circuit of Figure 6.8*c*.

Cascode Connection

The active-load inverter circuits of Figure 6.8*b* and *c* present a significant capacitive load to the stages driving them, due to the parasitic gate–source and gate–drain capacitances associated with Q_1. Among these, the gate–drain capacitance is the most troublesome due to the Miller effect. This problem can be reduced by using a cascode connection, as shown in Figure 6.9.

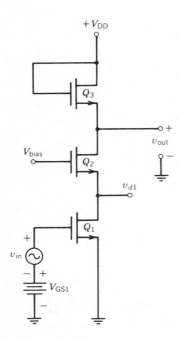

FIGURE 6.9. Cascode connection to reduce capacitive loading at input.

Neglecting body effect,* the voltage gains in the circuit can be expressed as

$$\frac{v_{d1}}{v_{in}} \approx - \frac{g_{m1}}{g_{m2}} = - \sqrt{\frac{(Z/L)_1}{(Z/L)_2}} \tag{6.16}$$

and

$$A_v = \frac{v_{out}}{v_{in}} \approx - \frac{g_{m1}}{g_{m3}} = - \sqrt{\frac{(Z/L)_1}{(Z/L)_3}} \tag{6.17}$$

Since the Miller-effect capacitance reflected to the input depends on the voltage gain at the drain of Q_1, as given by Eq. (6.6), it can be kept low by making $(Z/L)_2 \approx (Z/L)_1$. However, the overall voltage gain, given by Eq. (6.17), can still be maintained by keeping $(Z/L)_1 \gg (Z/L)_3$. In other words, Q_1 and Q_2 are made to have similar geometries, but Q_3 is made much smaller with a long channel.

Source Followers

The source-follower configuration is the MOS equivalent of the well-known emitter-follower configuration of bipolar transistors. It is normally used for level

---

*If desired, the body-effect factor α_B can be directly added into Eqs. (6.16) and (6.17) by calculating it from Figure 6.6.

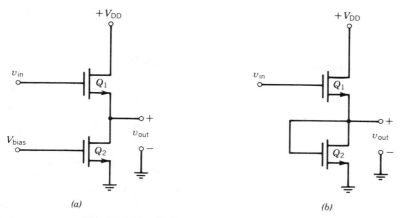

FIGURE 6.10.   Basic source-follower configurations.

shifting and buffering purposes. Figure 6.10 shows two commonly used NMOS source-follower configurations. It can be shown by small-signal analysis, using the equivalent circuit of Figure 2.46, that the low-frequency gain for the circuit of Figure 6.10a is given as[1]

$$A_v = \frac{v_{out}}{v_{in}} = \frac{g_{gm1}}{g_{m1}/\alpha_{B1} + 1/r_{o1} + 1/r_{o2}} \qquad (6.18)$$

where r_{o1} and r_{o2} are the drain resistances associated with Q_1 and Q_2 [see Eq. (2.52)], and α_{B1} is the body-effect gain factor associated with Q_1. In most practical designs, r_{o1} and r_{o2}, which are due to channel length modulation, are such that $r_{o1} \approx r_{o2} \gg 1/g_{m1}$. Then Eq. (6.18) reduces to

$$A_v \approx \alpha_{B1} \qquad (6.19)$$

Note that even in the extreme case, where the bias transistor Q_2 is considered to be an ideal current source (i.e., $r_{o2} \approx \infty$), the voltage gain of the source-follower circuit of Figure 6.10a is still less than unity, due to the body effect on Q_1.

In the case of the diode-loaded source follower of Figure 6.10b, the gain expression of Eq. (6.18) is still valid, with r_{o2} replaced with $1/g_{m2}$. Then, neglecting the channel length modulation effect on Q_1 (i.e., $r_{o1} = \infty$), the low-frequency voltage gain of Figure 6.10b becomes

$$A_v \approx \frac{g_{m1}}{g_{m1}/\alpha_{B1} + g_{m2}} \qquad (6.20)$$

The value of g_{m2} is minimized by making the (Z/L) ratio of Q_2 as small as possible in order to keep the voltage gain close to unity, as given by Eq. (6.20).

The source-follower circuits provide relatively low output resistance. The output resistance r_{out} for the circuit of Figure 6.10a is

$$r_{\text{out}} \approx \frac{1}{g_{m1}} \qquad (6.21)$$

For the circuit of Figure 6.10b, the output resistance is

$$r_{\text{out}} \approx \frac{g_{m1} + g_{m2}}{g_{m1} g_{m2}} \qquad (6.22)$$

MOS Current Mirrors

The MOS current mirrors can be implemented in a manner analogous to the bipolar current mirrors discussed in Chapter 4. In the case of MOS devices, a diode-connected MOS transistor is used to set the gate–source voltage of a constant-current output device. Figure 6.11 shows the basic NMOS current mirror configuration, where the current I_{ref}, forced through the diode-connected MOS transistor Q_1, is used to set up the V_{GS} for Q_1 and Q_2 and to produce a corresponding current I_{out} through Q_2.

The gate–source voltage V_{GS} of both devices is related to I_{ref} by Eq. (6.1) as

$$I_{\text{ref}} = \frac{\mu_s C_{\text{ox}}}{2} \left(\frac{Z}{L}\right)_1 (V_{\text{GS}} - V_{\text{TH}})^2 \qquad (6.23)$$

Similarly, I_{out} is related to V_{GS} as

$$I_{\text{out}} = \frac{\mu_s C_{\text{ox}}}{2} \left(\frac{Z}{L}\right)_2 (V_{\text{GS}} - V_{\text{TH}})^2 \qquad (6.24)$$

Since both devices are fabricated simultaneously, the $\mu_s C_{\text{ox}}$ product and V_{TH} are the same for both devices, and $V_{\text{GS1}} = V_{\text{GS2}}$ by the circuit connection. Then the output current I_{out} will be related to I_{ref} as

$$\frac{I_{\text{out}}}{I_{\text{ref}}} = \frac{(Z/L)_2}{(Z/L)_1} \qquad (6.25)$$

The output resistance r_o of the current mirror circuit is due to the channel length modulation effects of Q_2 and can be approximated from Eq. (2.52) as

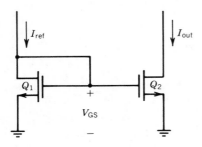

FIGURE 6.11. Simple two-transistor NMOS current mirror.

$$r_o \approx \frac{V_A}{I_D} \tag{6.26}$$

where V_A is the channel length modulation intercept voltage, illustrated in Figure 2.46. Empirical data show that V_A increases approximately linearly with the channel length L for values of $L > 10 \ \mu$. Thus, r_o of the current mirror output can be kept high and the length of the channel of Q_2 larger than that of Q_1, while keeping their ratio constant.

Since the gate of the MOS devices is an open circuit at dc, there are no current errors in MOS current mirrors comparable to the base-current errors in bipolar current mirrors. Therefore, a single diode-connected MOSFET, such as Q_1 of Figure 6.11, can be used to set the output currents in a multiplicity of devices, without introducing additional current errors.

Figure 6.12 shows methods of improving the output resistance r_o of the current mirror by using the NMOS equivalent of the Wilson current mirror. The circuit of Figure 6.12a is the direct equivalent of the Wilson current source. Figure 6.12b shows an improved version of the Wilson current mirror using an additional diode-connected transistor Q_4 to equalize the V_{DS} drop across Q_2 and Q_3.[2]

Using the small-signal equivalent circuit, one can show that the output resistance R_o of the NMOS Wilson current mirror is approximately equal to

$$R_o \approx r_{o1} g_{m3} r_{o3} \tag{6.27}$$

where r_o is the basic output resistance of the NMOS transistor given in Eq. (6.26). Thus, the effect of the Wilson current mirror connection is to increase the output resistance by a factor of $g_{m3} r_{o3}$, which is essentially equivalent to the open-circuit voltage gain of Q_3. This multiplier factor is normally on the order of 50–100.[2]

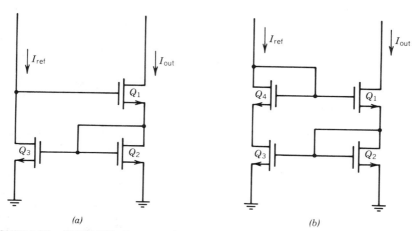

(a)   (b)

FIGURE 6.12. NMOS Wilson current mirror configurations: (a) Basic Wilson current mirror; (b) improved Wilson current mirror with Q_4 added to equalize V_{DS} drops of Q_2 and Q_3.

MOS Differential Pair

Figure 6.13 shows a simple NMOS differential pair biased by a constant-current source I_{SS}. Assuming that both Q_1 and Q_2 are matched devices with equal Z/L ratios, the drain currents I_{D1} and I_{D2} are given as

$$I_{D1} = \frac{\mu_s C_{ox}}{2} \left(\frac{Z}{L}\right)(V_{GS1} - V_{TH})^2 \tag{6.28}$$

$$I_{D2} = \frac{\mu_s C_{ox}}{2} \left(\frac{Z}{L}\right)(V_{GS2} - V_{TH})^2 \tag{6.29}$$

The differential input voltage ΔV_I where

$$\Delta V_I = V_{I1} - V_{I2} = V_{GS1} - V_{GS2} \tag{6.30}$$

would appear as a net voltage difference between V_{GS1} and V_{GS2}, and results in a differential current ΔI_D where

$$\Delta I_D = I_{D1} - I_{D2} \tag{6.31}$$

From Eqs. (6.28) through (6.31), one can derive an expression for ΔI_D neglecting the body effect, as

$$\Delta I_D \approx K\Delta V_I \sqrt{\frac{2I_{SS}}{K} - (\Delta V_I)^2} \tag{6.32}$$

where the constant K is equal to the collection of terms,

$$K = \left(\frac{\mu_s C_{ox}}{2}\right)\left(\frac{Z}{L}\right) \tag{6.33}$$

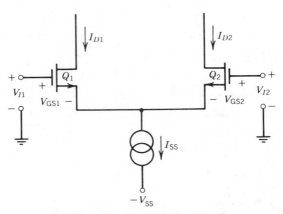

FIGURE 6.13. NMOS differential pair.

Equation (6.33) is valid as long as

$$|\Delta V_I| \leq \sqrt{\frac{I_{SS}}{K}} \tag{6.34}$$

This results in a dc transfer characteristic as shown in Figure 6.14. When $|\Delta V_I| > \sqrt{I_{SS}/K}$, either Q_1 or Q_2 is completely turned off, and ΔI_D is equal to I_{SS}.

The limiting behavior of the MOS differential pair is very similar to that of a bipolar differential stage. However, the range of linear operation is wider, being determined by the value of the biasing current source I_{SS} and the Z/L ratio of the devices. Note that increasing I_{SS} causes the linear region of the transfer characteristics to increase, whereas increasing the Z/L ratio has the opposite effect. The linear operating region is typically in the range of several hundred millivolts to several volts for practical device sizes and bias conditions.

The transconductance G_m of the differential pair can be found by differentiating Eq. (6.32), where

$$G_m = \frac{\partial(\Delta I_D)}{\partial(\Delta V_I)} = I_{SS}(\mu_s C_{ox})\left(\frac{Z}{L}\right) \tag{6.35}$$

or

$$G_m = g_{m1} = g_{m2} \tag{6.36}$$

Thus, the transconductance of the MOS differential stage is equal to the transconductance of Q_1 or Q_2, exactly as in the case of bipolar differential stages.

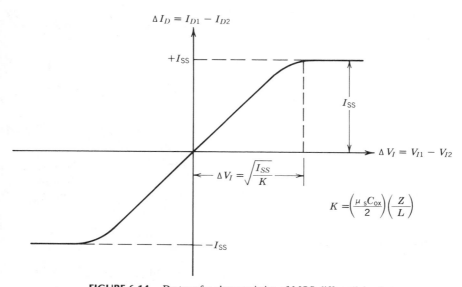

FIGURE 6.14.   Dc transfer characteristics of MOS differential pair.

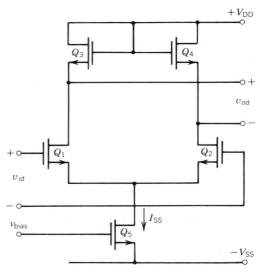

FIGURE 6.15. NMOS differential gain stage with active loads.

Figure 6.15 shows an NMOS differential gain stage using diode-connected transistors as active loads. Assuming $(Z/L)_1 = (Z/L)_2$ and $(Z/L)_3 = (Z/L)_4$, one can calculate the differential gain A_{dm} as

$$A_{dm} = \frac{v_{od}}{v_{id}} = -\alpha_{B3}\frac{g_{m1}}{g_{m3}} = -\alpha_{B3}\sqrt{\frac{(Z/L)_1}{(Z/L)_3}} \qquad (6.37)$$

where α_{B3} is the body-effect gain factor given by Figure 6.6.

Using the half-circuit approach described in Chapter 5, the common-mode gain A_{cm} can be calculated as

$$A_{cm} = \frac{-\alpha_{B1}\alpha_{B3}}{2r_{o5}g_{m3}} \qquad (6.38)$$

and the common-mode rejection ratio, CMRR, is

$$\text{CMRR} = \frac{A_{dm}}{A_{cm}} = \frac{2g_{m1}r_{o5}}{\alpha_{B1}} \qquad (6.39)$$

As expected, for high common-mode rejection, the output resistance r_{o5} of Q_5 must be made as high as possible. This is usually accomplished by making Q_5 to have a long channel length L, or by using a cascode-connected current source or a Wilson current mirror.

Differential to Single-Ended Conversion

Figure 6.16 shows a simple differential-to-single-ended converter circuit which can be used to convert the output of a differential pair to a single-ended output

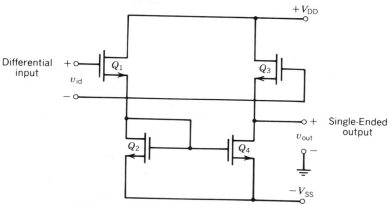

FIGURE 6.16.   NMOS differential-to-single-ended converter.

signal. In this circuit, there are two parallel signal paths: one through the source follower Q_3–Q_4 and the other through the source follower Q_1–Q_2, which is then amplified by Q_4, with Q_3 serving as an active load. The output signal is the *sum* of the signals in these two parallel paths. Then, using the small-signal models, and after some algebraic manipulations, one can arrive at the following gain expression:[1]

$$A_v = \frac{v_o}{v_{id}} = \frac{N+1}{2} \frac{\alpha_{B3}}{1 + (\alpha_{B3}/g_{m3})(1/r_{o3} + 1/r_{o4})} \qquad (6.40)$$

where the quantity N is defined by the collection of terms,

$$N \triangleq \frac{g_{m1}g_{m4}}{g_{m3}(g_{m1}/\alpha_{B1} + g_{m2} + 1/r_{o1} + 1/r_{o2})} \qquad (6.41)$$

Assuming, as a reasonable approximation, that the output resistance r_o is nearly infinite for each of the devices, and neglecting the body effect (i.e., $\alpha_B = 1$), Eq. (6.40) becomes

$$A_v \approx \frac{N+1}{2} \approx \frac{g_{m1}g_{m4}}{g_{m3}(g_{m1} + g_{m2})} \qquad (6.42)$$

In the case where all the devices have the same geometry, Eq. (6.42) gives an approximate voltage gain of $\frac{1}{2}$. Higher values of gain can be obtained by increasing the size of Q_4 to increase g_{m4}.

DC Level-Shift Stages

Similar to the case of all-*npn* gain stages, the output of an NMOS common-source gain stage is at a more positive voltage level than the input signal. Thus, after one or two stages of gain, the output dc level has to be level shifted toward the negative supply. Since there are no practical resistor values available in

NMOS analog design, this level shifting must be accomplished solely with active devices.

One simple and often used level-shift device is the diode-connected NMOS transistor shown in Figure 6.7, which provides a dc level shift of approximately 1 V_{TH} with a dynamic impedance of $1/g_m$.

An alternate method for dc level shifting is the use of source-follower stages similar to those shown in Figure 6.10. If differential signals are present, the differential-to-single-ended converter circuit of Figure 6.16 can also be used to provide a favorable dc level shift.

The shunt-feedback pair of NMOS transistors shown in Figure 6.17a also provide a useful circuit configuration for dc level-shift applications.[7] In this circuit, Q_1 serves as a shunt-feedback path between the gate and the drain of Q_2 and clamps its drain–gate voltage at approximately V_{TH}. Thus, the net voltage drop across Q_2 is set at approximately $2V_{TH}$. Neglecting the body effect, the actual turn-on voltage V_X of the dc level-shift stage can be expressed as

$$V_X = 2V_{TH} + \sqrt{\frac{2I_X}{\mu_s C_{ox}}\left(\frac{Z}{L}\right)_1} \tag{6.43}$$

and the current I_A through Q_2 can be written as

$$I_A = I_{D2} = \frac{\mu_s C_{ox}}{2}\left(\frac{Z}{L}\right)_2 (V_A - V_X)^2 \tag{6.44}$$

where V_A is the net voltage across the level-shift stage. The corresponding current–voltage characteristic for the shunt-feedback level-shift stage is shown

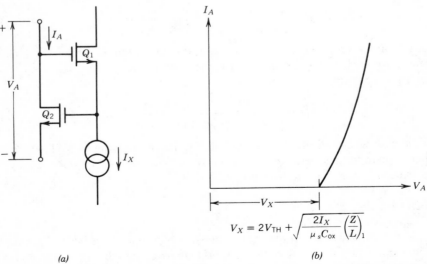

(a)   (b)

$$V_X = 2V_{TH} + \sqrt{\frac{2I_X}{\mu_s C_{ox}}\left(\frac{Z}{L}\right)_1}$$

FIGURE 6.17. Shunt-feedback level-shift circuit; (a) Circuit configuration; (b) current–voltage characteristics.

in Figure 6.17b. Normally, $(Z/L)_2$ is chosen to be $\gg 1$, whereas Q_1 is designed as a long channel device with $(Z/L)_1 < 1$. Under these conditions, the dynamic resistance r_d of the level-shift stage of Figure 6.17a is approximately equal to $1/g_{m2}$.

Output Stages

The choice of output stage configurations in all-NMOS design is severely limited by a lack of complementary devices. The large gate–source voltage drops required for device operation also limit the output voltage swing; and the low value of g_m makes it difficult to obtain low output impedance levels.

The simplest output stage configuration for NMOS design is the basic source–follower stage, as shown in Figure 6.10a. However, this circuit has two drawbacks. The output resistance of the circuit is approximately equal to $1/g_{m1}$, which is low only if $(Z/L)_1$ is quite large, and if the quiescent current is large. Second, similar to the case of a bipolar emitter follower, the circuit has a limited current-sinking capability, limited by the bias current setting of Q_2. Thus, the circuit has a limited drive capability for capacitive loads when the output is swinging toward the negative supply.

Figure 6.18 shows two possible output stage configurations which partially eliminate the drawbacks of a simple source follower. The circuit of Figure 6.18a uses the input device Q_2 as a *phase splitter* to drive the two output devices Q_4 and Q_5 with out-of-phase signals. This improves the output drive capability, however, the output resistance r_{out} is still quite large ($r_{out} \approx 1/g_{m5}$). Another drawback of the circuit is the output positive swing which is limited to approximately $2V_{TH}$ below V_{DD}.

Figure 6.18 b shows a modified output stage configuration which uses internal

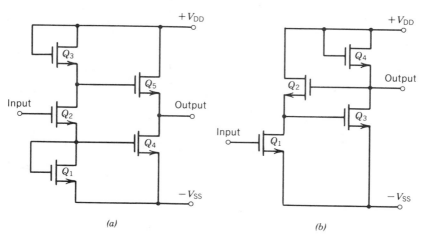

(a)   (b)

FIGURE 6.18. NMOS output stage configurations.

shunt feedback to reduce the output resistance. This internal feedback is provided from the output to the gate of Q_2. Neglecting the body-effect factors, the gain and the output resistance of the circuit can be written as [1]

$$A_v = \frac{v_{out}}{v_{in}} \approx \frac{g_{m1}}{g_{m2}} \frac{g_{m3}/g_{m4}}{1 + g_{m3}/g_{m4}} \tag{6.45}$$

and

$$r_{out} \approx \frac{1}{g_{m4}} \frac{1}{1 + g_{m3}/g_{m4}} \tag{6.46}$$

Another advantage of the circuit of Figure 6.18b is that the output can swing within 1 V_{TH} of V_{DD}.

6.3 ANALOG DESIGN WITH DEPLETION-MODE LOAD DEVICES

Some of the design constraints associated with NMOS analog design can be partially avoided by the inclusion of depletion-mode NMOS transistors on the same chip as the conventional enhancement-mode devices. This can be done by adding an extra masking and ion-implantation step to the basic NMOS process to convert selected MOS transistors to depletion-mode devices. The depletion-mode devices can be used as active loads for the enhancement-mode MOSFETs to provide higher voltage gains from NMOS amplifier stages. In this manner, one can partially make up for the lack of complementary devices, while still retaining the low cost and high functional density of the basic NMOS technology.

Figure 6.19 shows the circuit symbol and the drain current versus gate–source

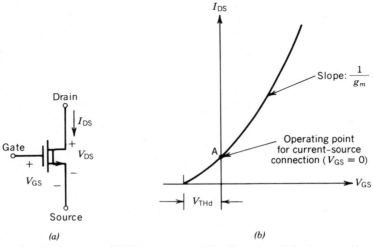

FIGURE 6.19. Depletion-mode NMOS transistor: (a) Circuit symbol; (b) its I_{DS} versus V_{GS} characteristics.

voltage characteristics of an *n*-channel depletion-mode MOSFET. As described in Section 2.5, the depletion-mode devices differ from enhancement-mode devices since a finite channel exists between source and drain at $V_{GS} = 0$, and one has to apply a gate–source bias of *opposite* polarity by an amount equal to V_{THd} of Figure 6.19*b* to *deplete* the existing channel and stop the conduction.

The depletion-mode NMOS transistor can be connected as a high-impedance active load by short-circuiting its source and gate terminals, as shown in Figure 6.20*a*. This automatically biases the device at the $V_{GS} = 0$ condition, corresponding to point *A* in Figure 6.19*b*. Unfortunately, the resulting current source is far from ideal due to the presence of the so-called body effect discussed in Section 6.1.

Figure 6.20*b* gives the Norton equivalent of the current–source-connected depletion-mode NMOS device. The dynamic resistance r_B is due to the body effect and can be expressed as

$$r_B = \left(\frac{\partial I_D}{\partial V_{SB}}\right)^{-1} = \frac{1}{g_{mB}} \tag{6.47}$$

where g_{mB} is the transconductance of the body–channel junction associated with the depletion-mode device. In Figure 6.20*b*, r_o is due to channel length modulation. It is typically much larger than r_B and can be neglected in most voltage gain calculations.

The depletion-mode device is normally used as an active load for the enhancement-mode MOS transistor, as shown in the inverter stage of Figure 6.21*a*. The large-signal voltage transfer characteristics for the circuit are given in Figure 6.21*b*. As the input voltage V_{in} is gradually increased, the behavior of the circuit can be explained as follows. With $V_{in} \leq V_{TH}$ of the enhancement-mode device, Q_1 is off and Q_2 is at its triode region, with $V_O \approx V_{DD}$. With the increase of V_{in}, Q_1 comes into conduction and operates at its pinched region. Initially, Q_1 conducts very little current, so Q_2 stays at its triode region (point

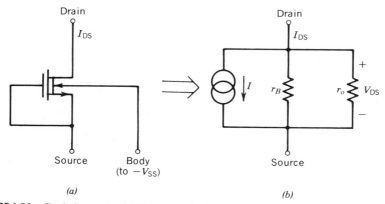

FIGURE 6.20. Depletion-mode NMOS connected as current source: (*a*) Circuit connection; (*b*) approximate equivalent circuit.

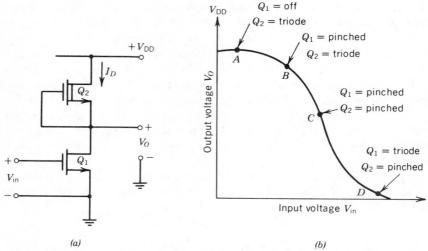

FIGURE 6.21. NMOS inverter stage with depletion load: (*a*) Circuit configuration; (*b*) dc transfer characteristics.

B in the figure). As V_{in} is increased further, both Q_1 and Q_2 operate in their pinched region (point *C*), and as the amount of current required by Q_1 increases beyond that which can be supplied by Q_2, Q_1 enters its triode region and the output approaches ground level.

For small-signal linear operation, both devices would be biased to operate in their pinched regions. Neglecting channel length modulation effects, this results in an available voltage gain A_v, from the inverter as

$$A_v = \frac{\partial V_o}{\partial V_{in}} \approx - g_{m1} r_B = - \frac{g_{m1}}{g_{mB2}} \tag{6.48}$$

where subscripts 1 and 2 refer to Q_1 and Q_2.

Using the incremental body-effect factor λ_B defined by Eq. (6.10), one can rewrite the gain expression as

$$A_v \approx - \frac{g_{m1}}{g_{mB2}} = \frac{g_{m1}}{g_{m2} \lambda_B} = \frac{1}{\lambda_B} \sqrt{\frac{(Z/L)_1}{(Z/L)_2}} \tag{6.49}$$

Comparing Eq. (6.49) with Eq. (6.9), one observes that the gain improvement obtained by the use of a depletion-mode load, over a simple diode-connected enhancement-mode load, is approximately equal to $1/\lambda_B$. For the values of λ_B given in Figure 6.5, with typical substrate doping levels of 10^{15} atoms/cm^3 and $V_{SB} = 5$ V, this results in approximately an order of magnitude improvement in the voltage gain obtainable from the simple inverter stage. For example, choosing a Z/L ratio of 100 : 1 between Q_1 and Q_2, and with $\lambda_B \approx 0.1$, one can now obtain a voltage gain of approximately 100, compared to a voltage gain of 10 with the diode-connected enhancement-mode device shown in Figure 6.8*b*.

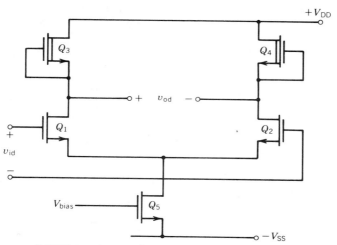

FIGURE 6.22. Differential gain stage with depletion loads.

The depletion-mode devices are also usable as active loads for differential gain stages, as shown in Figure 6.22. Assuming that the circuit is symmetrical, such that the device pairs Q_1, Q_2 and Q_3, Q_4 have matched geometries, the differential gain of the circuit A_{dm} can be expressed as

$$A_{dm} = \frac{v_{od}}{v_{id}} = -\frac{1}{\lambda_B}\sqrt{\frac{(Z/L)_1}{(Z/L)_3}} \tag{6.50}$$

which again offers an improvement in gain by a factor of $1/\lambda_B$ compared to the simple differential stage of Figure 6.15.

Body-Effect Cancellation

One of the limitations in the effectiveness of depletion-mode devices as active loads is the body effect, which leads to a relatively low value of its dynamic resistance r_B [see Eq. (6.47)]. This effect can be greatly reduced by using the circuit configuration shown in Figure 6.23.[8] In this configuration, the entire grouping of the three devices within the dashed area corresponds effectively to a single depletion-mode load Q_2 with virtually no body effect.

The principle of body-effect cancellation illustrated in the figure can be described briefly as follows. The diode-connected enhancement-mode transistor, Q_4 serves as a body-effect-dependent voltage source between the source and the gate of Q_2. The bias current I_X for Q_4 is provided by the depletion-mode current source Q_3. Normally, I_X is chosen to be much smaller than I_D, so that both Q_3 and Q_4 introduce negligible loading at the output node. As the output swings positive, the body effect would tend to reduce the current in Q_2, but the same body effect also causes the voltage drop across Q_4 to increase, which in turn causes the current in Q_2 to increase, thus partly canceling the body effect on Q_2

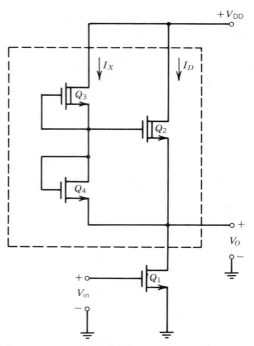

FIGURE 6.23. NMOS inverter stage using body-effect cancellation.[8] (*Note:* Devices within dashed lines are equivalent to a single depletion-mode device with negligible body effect.)

and increasing the value of r_B associated with it. The body-effect cancellation scheme illustrated in Figure 6.23 can increase the value of r_B by a factor of 5–10 by proper choice of the device geometries. Normally, both Q_3 and Q_4 are made as minimum-geometry devices, with small Z/L ratios, to conserve chip area and reduce dynamic loading at the output node. For practical design applications, the entire device combination within the dashed area in Figure 6.23 can be considered as a useful subcircuit to enhance the dynamic resistance of a depletion-mode load device anywhere within the circuit.

One practical limitation associated with the body-effect cancellation circuit of Figure 6.23 is the reduction of the positive output swing. For positive output swings within one depletion threshold plus one enhancement threshold of $+V_{DD}$, Q_3 goes into its triode region and Q_4 cuts off. At this point, the dynamic resistance of Q_2 becomes greatly reduced, since it now appears as a diode-connected MOSFET and the voltage gain is greatly degraded. Thus, the positive swing of the output is limited to approximately one depletion plus one enhancement threshold below $+V_{DD}$.

Transconductance Enhancement

Since the voltage gain of a depletion-mode loaded inverter stage, such as the one shown in Figure 6.21, is proportional to the transconductance of the inverter

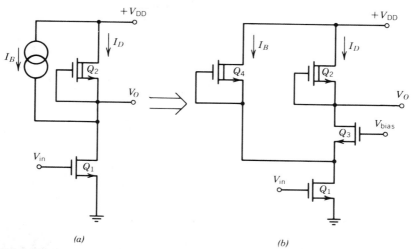

FIGURE 6.24. Transconductance enhancement by current injection: (*a*) Basic technique; (*b*) actual circuit implementation.

transistor, one possible approach to increasing the voltage gain is to increase the transconductance of the inverter device by increasing its operating current without increasing the current in the depletion-mode load device. In this manner, g_{m1} of Eq. (6.49) can be increased while keeping r_B relatively unchanged, which results in higher voltage gain. This transconductance enhancement can be achieved by "injecting" an extra amount of bias current into the inverter device, while bypassing the depletion-mode load.[9] Figure 6.24a shows a method of implementing this technique. The extra bias current I_B is injected into the inverter transistor Q_1 without affecting the load current I_D through the depletion-mode device, Q_2.

The practical circuit implementation is shown in Figure 6.24*b*, where a depletion-mode load device Q_4 is used to simulate the current source I_B and a cascode-connected buffer stage Q_3 is inserted between the inverter transistor and the actual load device Q_2 in order to minimize the capacitive and resistive loading effects of Q_4 at the drain of Q_1. Normally, the Z/L ratios of Q_2 and Q_1 are chosen to set the I_B/I_D ratio in the range of 5–20.

Output Stages

The normally "on" characteristic of depletion-mode devices makes it possible to design Class B or Class AB type output stages, where substantial current can be delivered to a capacitive or resistive load, while maintaining low standby current drain. Figure 6.25 shows a practical circuit implementation for such an output stage.[7] The operation of the circuit can be briefly explained as follows. The differential amplifier and level shifter stage, with unity feedback from the output, has the net effect of causing the voltage difference between the gate and the drain of Q_7 to be equal to the gate–source voltage drop across Q_1. Thus, the

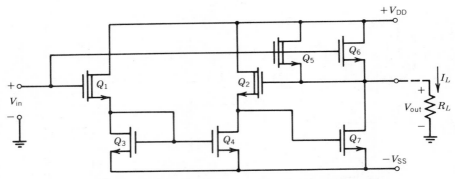

FIGURE 6.25. Class AB output stage using depletion-mode loads.

output will tend to follow the input voltage. If the circuit is designed symmetrically such that the device pairs Q_1, Q_2 and Q_3, Q_4 have matched geometries, one can show that the overall gain is approximately unity, and the circuit behaves as a *voltage follower*.

The depletion-mode devices Q_1 and Q_2 are designed with $Z/L \ll 1$ to minimize the standby current drain. However, Q_5, Q_6, and Q_7 are made with $Z/L \gg 1$ to provide output current drive. In a standby mode, with the input at or near ground, the output is also at nearly zero volts, and the parallel-connected depletion-mode and enhancement-mode pair Q_5, Q_6 are biased at nearly zero gate–source bias. Thus, the enhancement-mode device Q_6 would be off and Q_5 would be conducting a standby current. If the input swings positive, the output will rise until Q_5 cannot satisfy the current drive required, and this will cause the gate–source voltages of Q_5 and Q_6 to increase, causing Q_6 to turn on and supply the additional current. Since Q_6 is normally off in the standby mode, it can be made arbitrarily large to deliver a large amount of current into the load without affecting standby power dissipation.

When the output is driven in the negative direction, the current in the current mirror pair Q_3, Q_4 decreases, causing the gate voltage of Q_7 to increase and, thus, forcing Q_7 to conduct more current. The positive swing is limited to within approximately one enhancement threshold of $+V_{DD}$ for large output currents which require Q_6 to be on. The negative swing is limited by the reduced current-sinking capability of Q_7, as the output swings toward $-V_{SS}$ due to the limited gate–source voltage available to Q_7.

6.4 ANALOG DESIGN WITH CMOS TECHNOLOGY

The CMOS technology offers a much higher degree of design flexibility for analog functions than the single-channel NMOS technology described in the previous section. Although the basic CMOS process requires several additional fabrication steps compared to the NMOS technology, it provides devices with complementary symmetry. Thus, the resulting p-channel devices can be used as active loads, in conjunction with the n-channel devices, in a manner similar to the use of lateral pnp transistors in bipolar design, to obtain both high-voltage gains and dc level shifting.

The basic CMOS device fabrication sequence is described in Chapter 2 and will not be repeated here. The resulting device structure containing electrically isolated n- and p-channel devices is shown in Figure 2.51. It should be noted that the n-channel devices are located in separate p-type islands on the chip surface, which are formed by the p-well diffusion shown in Figure 2.51. Thus, the n-channel devices in the CMOS structure *do not* necessarily have to share a common substrate or body region since each can be placed in a separate p-type island. As a result, the problems associated with the body effect in single-channel designs can be avoided or eliminated in CMOS designs.

Another advantage of the CMOS structure shown in Figure 2.51 is that the n^+ source and drain diffusions for n-channel transistors can be used with the p^+ source and drain diffusions of p-channel transistors to form a functional npn transistor which is available in a common-collector (i.e., emitter-follower) configuration, with the n-type common substrate serving as its collector. This transistor can occasionally be used for a buffer, level-shift, or output driver application (see Fig. 6.33).

CMOS Inverter as a Gain Stage

The most commonly used CMOS gain stage is the basic CMOS inverter, where one device serves as a common-source amplifier and the other device acts as an active load. Figure 6.26 shows both the circuit configuration and the large-signal

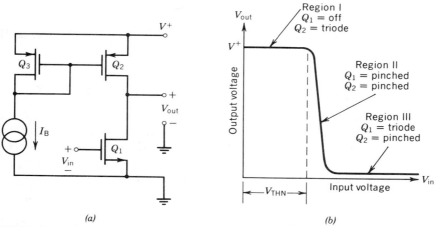

FIGURE 6.26. CMOS inverter with current mirror load: (*a*) Circuit connection; (*b*) transfer characteristics.

voltage transfer characteristics for such a stage. For illustrative purposes, an n-channel device is used as the gain stage, with a p-channel current mirror as its active load. However, with appropriate polarity changes, the analysis is equally valid for p-channel gain stages with n-channel active loads.

The operation of the circuit and its transfer characteristics can be briefly explained as follows. Q_2 and Q_3 form a current mirror whose output current is determined by the bias current source I_B when Q_2 is in its pinched region. If the input voltage V_{in} is less than the threshold voltage V_{THN} of the n-channel device, then Q_1 is nonconducting and Q_2 is in its triode region with virtually a zero voltage drop across it. This corresponds to region 1 of the operating characteristics shown in Figure 6.26b. If V_{in} is increased beyond V_{THN}, Q_1 begins to conduct; however, Q_2 stays in its triode region until the current through Q_1 is approximately equal to I_B. At this point, Q_2 leaves the triode region and enters its pinched region. This corresponds to region II in the transfer characteristics where both Q_1 and Q_2 are in their pinched regions and operate as two current sources in series. If the input voltage is increased further, Q_2 stays in its pinched region and Q_1 is pushed into its triode region with a very low voltage drop across its source–drain terminals. This corresponds to region III in Figure 6.26b, and any further increase in V_{in} will result in a negligible change of output voltage.

For analog applications, the inverter stage is biased in region II where both devices are active and the circuit has a high incremental (i.e., small-signal) gain. This is analogous to the case of a common-emitter npn gain stage driving a pnp current-source load.

The incremental voltage gain for the circuit, for its operation in region II, can be calculated from the small-signal model of the MOS transistor (see Fig. 2.46) as

$$A_v = \frac{\partial V_{out}}{\partial V_{in}} = -g_{m1}\left(r_{o1} \| r_{o2}\right) \tag{6.51}$$

where r_{o1} and r_{o2} are the output resistances of Q_1 and Q_2, and are due to the channel length modulation effects. As indicated by Eq. (2.52), they can be approximated as

$$r_{o1} \approx \frac{|V_{A1}|}{I_{D1}} \quad \text{and} \quad r_{o2} \approx \frac{|V_{A2}|}{I_{D2}} \tag{6.52}$$

where V_{A1} and V_{A2} are the intercept voltages (see Fig. 2.42) for Q_1 and Q_2. The intercept voltages depend on the channel length L of the respective devices and can be increased by making L larger, that is, by using so-called *long-channel* devices.

The transconductance g_{m1} is given from Eq. (6.2) as

$$g_{m1} = \sqrt{2\mu_s C_{ox}\left(\frac{Z}{L}\right)_1 I_{D1}} \tag{6.53}$$

Since both Q_1 and Q_2 are in series in Fig. 6.26a, $I_{D1} = I_{D2} = I_D$, and the incremental gain can be written from Eqs. (6.52) and (6.53) as

$$A_v = \frac{1}{\sqrt{I_D}} \frac{|V_{A1}| \, |V_{A2}|}{|V_{A1}| + |V_{A2}|} \sqrt{2\mu_s C_{ox}\left(\frac{Z}{L}\right)_1} \qquad (6.54)$$

From Eq. (6.54) one can draw the following conclusions:

1. The available incremental gain varies *inversely* with the square root of the drain current. This is because the transconductance decreases with the square root of I_D, whereas the output resistance r_o is inversely proportional to I_D. Therefore, the product shows an inverse square-root dependence on I_D and thus, higher gains can be achieved by operating the stage at lower bias currents. However, if the drain current is reduced to extremely low levels (i.e., $I_0 \leq 0.1 \ \mu A$), then both Q_1 and Q_2 are very nearly cut off. In this very low current range of operation, which is called the *subthreshold operation*, where the conductive channel is barely at the verge of existing, the basic transconductance equation [Eq. (6.53)] is no longer valid. Instead, g_m exhibits a *linear* dependence on I_D in a manner similar to the case of a bipolar transistor. Thus, when the drain current is reduced low enough to bring the device to its subthreshold region, the gain levels off and becomes relatively constant with current. Figure 6.27 shows the typical incremental voltage gain characteristics of a CMOS inverter with a current mirror load as a function of bias current.[2]

2. At any given current level, one can increase the available voltage gain by increasing the channel lengths of Q_1 and Q_2, which causes V_A of each of

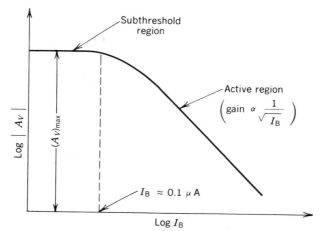

FIGURE 6.27. Open-loop voltage gain of a CMOS inverter with current mirror load, as a function of bias current.

the devices to increase. However, increasing L_1 also has a detrimental effect since it tends to reduce g_{m1}, unless the Z/L ratio of Q_1 is maintained constant. In terms of frequency response characteristics, using long channel devices is not desirable since this increases the gate-channel capacitance of both Q_1 and Q_2.

In practice, typical voltage gains of several hundred to several thousand can be obtained from a simple inverter stage similar to that shown in Figure 6.26a, with the proper choice of device geometries. The voltage gain can also be increased by using more complex circuits, such as a Wilson current mirror or a cascode-connected gain stage, which can result in higher gains by increasing the effective output resistances of the active devices used.

CMOS Differential Stages

The availability of complementary devices in CMOS technology makes it possible to duplicate many of the conventional bipolar differential amplifier configurations with complementary current mirror loads. Figure 6.28 shows the two commonly used CMOS differential gain stage configurations with current mirror loads. In each case, the available voltage gain, assuming an open-circuit load, can be given as

$$A_v = \frac{v_{\text{out}}}{v_{\text{in}}} = -G_m (r_{o2} \| r_{o4}) \qquad (6.55)$$

where G_m is the differential stage transconductance defined by Eq. (6.36), and r_{o2} and r_{o4} are the output resistances of Q_2 and Q_4. With practical bias levels and device geometries, typical voltage gains obtainable from such gain stages are in

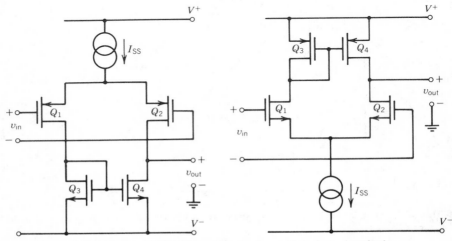

FIGURE 6.28. CMOS differential gain stages with current mirror loads.

the range of 100–1000. The available values of gain can be increased to the range of 1000–10,000 by using a cascode configuration for Q_1 and Q_2, and by using a more complex current source (such as the Wilson current mirror) for Q_3 and Q_4. Yet, it should be noted that the total available gain is still approximately an order of magnitude lower than that of a bipolar stage because of the lower values of G_m.

Push–Pull Gain Stages

Figure 6.29 shows a single-ended high-gain amplifier stage which makes full use of the complementary device symmetry available in CMOS technology.[2] The input stage operates on the push–pull principle and is comprised of cross-coupled devices Q_1 through Q_4. The input pair Q_1, Q_2 operate as source followers and drive the sources of Q_3 and Q_4, which operate as grounded-gate gain stages and drive the opposing current mirror loads made up of the pairs Q_5, Q_6 and Q_7, Q_8.

The grounded-gate transistors are biased symmetrically with internal bias voltage at $+V_B$ and $-V_B$ around the ground reference. This bias voltage is chosen to be approximately equal to one p-channel threshold plus one n-channel threshold voltage, that is,

$$V_B = (V_{TH})_n + (V_{TH})_p \tag{6.56}$$

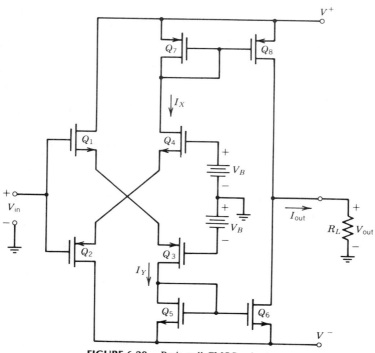

FIGURE 6.29. Push–pull CMOS gain stage.

This bias condition assures that there is no crossover distortion at the input, and that all four input devices are conducting a small standby current with $V_{in} = 0$. Figure 6.30 illustrates a simple circuit configuration for generating the bias voltage V_B.

As the input swings positive, the current I_X conducted by Q_2 and Q_4 decreases, while the current I_Y in Q_1 and Q_3 increases. This produces a net output current I_{out} equal to $I_X - I_Y$.

For small-signal operation, assuming that Q_1, Q_4 and Q_2, Q_3 are matched, the voltage gain can be expressed as

$$A_v = - G_m (R_L \| r_{o6} \| r_{o8}) \tag{6.57}$$

where G_m is the transconductance of the cross-coupled input stage, given as

$$G_m = \left(\frac{g_{m1} g_{m3}}{g_{m1} + g_{m3}} \right) \tag{6.58}$$

Using Wilson current mirrors, the circuit of Figure 6.29 is capable of providing open-circuit voltage gains in excess of 10,000.

The push–pull gain stage has the property that it operates in the Class AB mode, that is, when large input voltages are applied, the current flowing to the

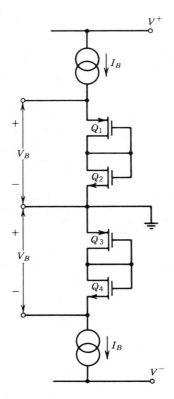

FIGURE 6.30. Generation of internal bias voltages for push–pull CMOS stage of Figure 6.29.

output increases to a value much larger than the quiescent bias current in the circuit. This feature makes it particularly suited for driving capacitive loads.

Output Stages

The availability of complementary devices makes Class B or Class AB push–pull output stages feasible for CMOs designs. Figure 6.31a shows a simple Class B output stage using p- and n-channel devices in a push–pull mode. This configuration is analogous to the simple complementary bipolar stage shown in Figure 5.31a, and also exhibits a high degree of crossover distortion at the output. As shown in Figure 6.31b, the output driver transistors Q_1 or Q_2 are nonconducting until the driving voltage V_A swings an amount equal to the respective threshold voltages V_{THN} and V_{THP} about ground.

Figure 6.32 shows an improved version of the push–pull output stage, which uses diode-connected CMOS transistors Q_A and Q_B to avoid the crossover distortion problem. This circuit configuration is analogous to the complementary bipolar output stage of Figure 5.32. However, compared to its bipolar counterpart, it has two drawbacks: (1) the output resistance is high due to the low g_m values associated with Q_1 and Q_2, and (2) the output swing in either direction is limited to the supply voltage minus the gate–source voltage needed across Q_1 or Q_2 to sustain the necessary load current.

In most CMOS processes which use an n-type substrate (see Fig. 2.50) an additional npn bipolar transistor structure is available as a potential circuit element. This is a parasitic npn transistor and is inherent to the CMOS structure: its collector is formed by the n-type substrate, its base and emitter regions are made up of the p-well diffusion and the n^+-type source–drain diffusions. Since

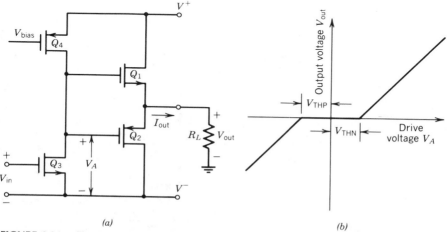

(a)   (b)

FIGURE 6.31. Simple Class B output stage. (a) Circuit diagram; (b) voltage transfer characteristics showing crossover distortion.

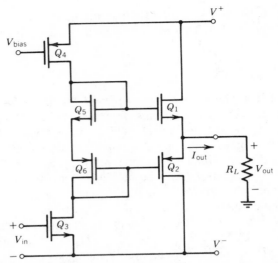

FIGURE 6.32. Class AB output stage with crossover compensation.

the substrate is common to the entire circuit and is connected to V^+, this parasitic *npn* transistor is only usable as an emitter-follower stage and is analogous to the case of the substrate *pnp* transistor in the case of bipolar IC structures. The circuit of Figure 6.33*a* shows the use of this substrate *npn* transistor in a Class A output stage to provide both low output impedance and high current drive capability. However, the output current drive is unsymmetrical and is limited by the bias current in Q_2 for the negative-going output swings. Figure 6.33*b* shows a Class AB output stage using the substrate *npn* transistor. In this circuit, the common-source transistor Q_3 provides both voltage gain as well as an out-of-

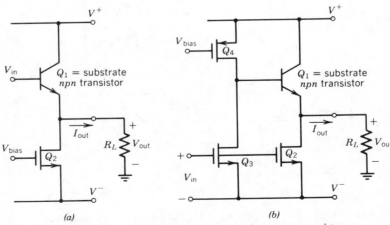

FIGURE 6.33. CMOS output stages using substrate *npn* transistor.

phase signal drive to the base of Q_1, while the gate of Q_2 is driven with the in-phase signal V_{in}. In this manner, the current-sinking capability of Q_2 is enhanced during the negative swing of the output.[10]

6.5   MOS VOLTAGE REFERENCES

Stable voltage references are some of the essential building blocks of analog system design. The basic requirements and general design principles for voltage references were discussed in Chapter 4 (see Section 4.7). In the design of voltage references, the emphasis is on minimizing the temperature drift of the reference voltage. For practical applications, the temperature drift of the reference should be maintained in the range of ≤ 100 ppm/°C over the circuit operating temperature.

As in the case of bipolar circuits, the temperature-compensation method used in MOS references is, at least in principle, a simple one. Starting with a predictable temperature drift source, find another predictable temperature drift source of opposite polarity which can be scaled and added to the first, to result in a nominally zero temperature coefficient of the voltage level. Unfortunately, in the case of MOS circuits, the limited choice of devices, as well as the strong process and temperature dependance of intrinsic device parameters, make temperature compensation a much more difficult problem than in the case of bipolar circuits. In spite of such problems, several circuit techniques have been developed and demonstrated, which make it feasible to design all-MOS voltage references.[11,12] In this section, some of these techniques will be described briefly. Since the design of very accurate temperature references requires extremely lengthy and meticulous calculations of all the temperature drift sources in the circuit, the design approaches described in this section are intended primarily to illustrate the principle of temperature drift cancellation. For this purpose, lengthy calculations will be avoided and the end results, rather than the rigorous derivations leading up to them, will be presented.

CMOS Reference Using Subthreshold Conduction Characteristics[11]

When the MOS transistors are operated at very low currents, the channel which exists under the gate is extremely shallow and contains very few free carriers. This region of operation of the device is called *weak inversion* or *subthreshold* region, and the corresponding device currents are typically on the order of ≤ 10 nA per mil of channel width. In this region of operation, the simplified current and voltage expressions of Eqs. (2.48) and (2.49) are no longer valid. Instead, the drain current exhibits an exponential dependence on the gate and drain voltages. [12] Under subthreshold conditions, one can show that the voltage drop difference between two MOS transistors with current-source bias is nearly proportional to the absolute temperature T. This circuit configuration is illustrated in Figure 6.34 for the case of NMOS transistors.

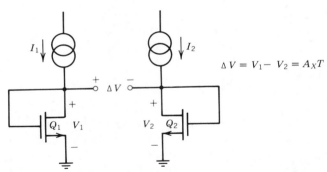

FIGURE 6.34. Generating a voltage difference, ΔV proportional to temperature by operating MOS transistor in subthreshold region.[11]

Following the analysis given in Ref. (11) and assuming that both V_1 and V_2 are much greater than $V_T(= kT/q)$, the voltage differential ΔV can be expressed as

$$\Delta V = V_1 - V_2 = A_X T \qquad (6.59)$$

where A_X corresponds to a collection of terms of the form

$$A_X = \frac{nk}{q} \ln \left[\frac{I_1(Z/L)_2}{I_2(Z/L)_1} \right] \qquad (6.60)$$

In Eq. (6.60), n is a process-dependent parameter with a positive temperature coefficient in the range of $+1500$ ppm/°C.

Figure 6.35 shows a simplified circuit configuration for generating a temperature-compensated voltage reference V_{ref}, using the positive temperature dependence of ΔV, to cancel the negative temperature coefficient of the bipolar transistor V_{BE} voltage drop. In the circuit, the *npn* bipolar transistor Q_0 is formed by the parasitic substrate *npn* transistor inherent to CMOS device structure (see Fig. 6.33). Internal current sources I_{SS} and I_2 force Q_1 and Q_2 to operate at unequal current levels, such that

$$I_1 = I_{\text{SS}} - I_2 \gg I_2 \qquad (6.61)$$

The output voltage V_{ref} is then given as

$$V_{\text{ref}} = V_{\text{BE}} + V_1 - V_2 = V_{BE} + A_X T \qquad (6.62)$$

The voltage expression given by Eq. (6.62) has the same format as that of the basic band-gap reference [see Eq. (4.82)], where the first term, the transistor voltage drop V_{BE}, has a predictable negative temperature coefficient; and the second term, ΔV, has a positive temperature coefficient which can be "scaled" by the choice of currents I_1 and I_2, or by device geometries of Q_1 and Q_2. Thus, by proper choice of the design parameters, the temperature coefficient of V_{ref} can be kept to below 100 ppm/°C over a -55 to $+100$°C temperature range. For

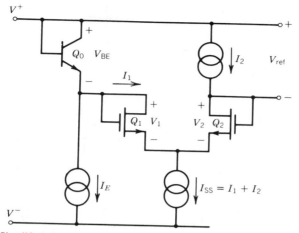

FIGURE 6.35. Simplified circuit schematic for generating temperature-compensated reference voltage.

standard CMOS process technology, the value of V_{ref}, which can be obtained with a nominal zero temperature coefficient at 25°C, is 1.105 V for the identical device geometries of Q_1 and Q_2, and with $I_1/I_2 = 15$.[11]

NMOS Reference Using Threshold Difference between Depletion-Mode and Enhancement-Mode Devices[6]

The threshold voltages associated with depletion-mode and enhancement-mode MOS transistors exhibit very similar negative temperature coefficients. Thus, their difference can be made relatively insensitive to temperature changes. This concept can be utilized, in practice, to fabricate a temperature-stable voltage reference.

Figure 6.36 shows a simplified schematic for a voltage reference circuit, based on the depletion/enhancement threshold difference. The principle of operation for this circuit is similar to that of the band-gap circuits shown in Figure 4.36, except that the threshold difference, rather than the V_{BE} difference, is used to generate the reference voltage level.

In the circuit of Figure 6.36, Q_1 and Q_2 are the enhancement-mode and depletion-mode NMOS transistors with the respective threshold voltages $(V_{TH})_E$ and $(V_{TH})_D$. Q_3 and Q_4 are assumed to be identical in geometry, and form a set of active loads for the differential stage made up of Q_1 and Q_2. The high-gain amplifier A forces the drain voltages of Q_1 and Q_2 to be equal. Assuming that Q_3 and Q_4 are well matched, the drain currents I_1 and I_2 would also be equal.

Summing the voltage drops around the output loop in Figure 6.36, one can show that

$$V_{ref} - (V_{GS})_E + (V_{GS})_D = 0 \qquad (6.63)$$

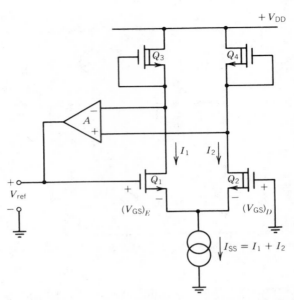

FIGURE 6.36. Generating a temperature-compensated voltage reference equal to the threshold voltage difference between depletion-mode and enhancement-mode NMOS transistors.[6]

or

$$V_{ref} = (V_{GS})_E - (V_{GS})_D \tag{6.64}$$

where $(V_{GS})_E$ and $(V_{GS})_D$ are the gate-source voltage of the enhancement-mode and depletion-mode devices Q_1 and Q_2, respectively.

For saturated operation, the drain current I_{DO} can be related to V_{GS} by Eq. (6.1). Rewriting this equation and solving it for V_{GS}, one obtains

$$V_{GS} = V_{TH} + \sqrt{\frac{2I_{DO}}{\mu_s C_{ox} Z/L}} \tag{6.65}$$

In the circuit of Figure 6.36, both transistors operate at the same current level, with $I_{DO} = I_1 = I_2$.

Assuming that both Q_1 and Q_2 have the same Z/L ratio, the second term in Eq. (6.65) cancels out when substituted into Eq. (6.64), and one gets

$$V_{ref} = (V_{TH})_E - (V_{TH})_D \tag{6.66}$$

The temperature dependence of V_{ref} can be estimated from differentiating $(V_{TH})_E$ and $(V_{TH})_D$ expressions with respect to temperature.

The temperature coefficient of the threshold voltage difference can be related to the ion-implanted donor and acceptor concentrations N_{di} and N_{ai} in the channel region of the depletion-mode and enhancement-mode devices. Normally, N_{di}, the donor implant for the depletion-mode device, is chosen such that $N_{di} \gg N_{ai}$.

With this assumption, the temperature coefficient of the threshold difference can be approximated as

$$\frac{d}{dT}\left[(V_{TH})_E - (V_{TH})_D\right] = -\frac{k}{q}\ln\left(\frac{N_{di}}{N_{ai}}\right) \tag{6.67}$$

Assuming a typical doping ratio of $N_{di}/N_{ai} = 10$, one obtains a typical temperature drift of $-0.2\,\text{mV/}°\text{C}$ from Eq. (6.67). For the case of typical depletion-mode and enhancement-mode threshold values of

$$(V_{TH})_E \approx +1\,\text{V} \quad \text{and} \quad (V_{TH})_D \approx -3\,\text{V} \tag{6.68}$$

this corresponds to typical values of

$$V_{ref} \approx +4\,\text{V} \quad \text{and} \quad \frac{dV_{ref}}{dT} \approx -50\,\text{ppm/}°\text{C} \tag{6.69}$$

It can be shown that the surface mobility term μ_s does not show identical temperature dependence in both enhancement-mode and depletion-mode devices. Its temperature coefficient is negative and somewhat higher in the enhancement-mode device. In practice, this effect results in an additional temperature drift of comparable magnitude but opposite polarity to that given by Eq. (6.69), and improves the reference stability even further.

In conclusion, one additional word of caution is in order. In using the simplified circuit of Figure 6.36 to demonstrate the principle of operation of the reference circuit, the effects of the device mismatches in Q_3 and Q_4 and the offset voltage of the amplifier A were neglected. In a more complete design, all of these factors need to be considered in detail.

6.6 MOS TRANSISTOR AS AN ANALOG SWITCH

MOS transistors make good analog switches for sampling, steering, or multiplexing analog signals. There are two important features of the MOSFET devices which make them nearly ideal switches: (1) when the device is on and conducting, there is no inherent dc offset voltage between the source and drain, and (2) the control terminal (the gate) is electrically isolated from the signal path, thus no dc current flows between the control path and the signal path.

Figure 6.37 shows an idealized equivalent circuit of an MOS transistor as an analog switch. For illustrative purposes, an NMOS device is assumed. When the gate voltage V_G is at or near ground, the device is off, and the source and drain terminals are isolated from each other by a nonconductive channel. When V_G is increased to a level such that $V_G > V_{TH}$, a conductive channel is produced and a low-resistivity path is formed between the source and the drain. Thus, as shown in its idealized form, neglecting the series resistance of the channel, the MOSFET functions as a switch which forms a short circuit between the source and the drain when a positive control voltage $V_G > V_{TH}$ is applied to the gate.

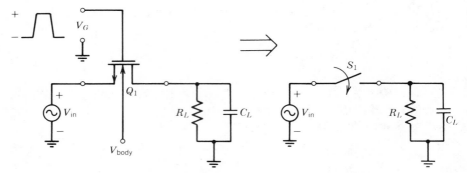

FIGURE 6.37. MOS transistor as ideal analog switch.

In actual operation, the MOSFET is a nonideal switch, having a number of parasitic capacitances associated with it, as shown in Figure 6.38. The capacitances C_{GS}, C_{GD}, C_{SB}, and C_{DB} are the parasitic capacitances between the gate, source, drain, and body (i.e., substrate) regions of the device.

The finite resistance r_{ds} of the channel during the "on" condition can be calculated from Eq. (2.48) as

$$r_{ds} = \left(\frac{\partial I_D}{\partial V_{DS}}\right)^{-1}\Bigg|_{V_{DS} = 0} = \frac{1}{(\mu_s C_{ox} Z/L)(V_{GS} - V_{TH})} \qquad (6.70)$$

Comparing the above expression with Eq. (6.3), one can see that

$$r_{ds} = \frac{1}{g_m} \qquad (6.71)$$

where g_m is the device transconductance in the pinched region at the same applied gate–source bias.

Optimizing the performance of a MOSFET as an analog switch requires a number of trade-offs. If the Z/L ratio is increased to reduce r_{ds}, this also causes

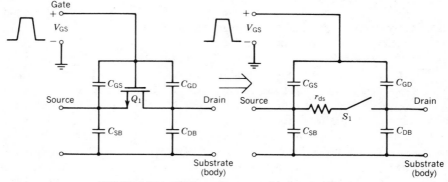

FIGURE 6.38. MOS transistor as nonideal analog switch.

the parasitic capacitances to increase proportionately. Similarly, if the gate signal V_G is increased to make $V_{GS} \gg V_{TH}$, this may result in increased feedthrough of the control signal through the parasitic gate–source and gate–drain capacitances.

In the off state, the total leakage current associated with the source–drain path is on the order of 1 pA for a small-area device ($Z/L \approx 1$). The "on" resistance is typically a few kilohms. Since the "on" resistance is nonlinear and a strong function of voltage and temperature, it should not be used in any critical resistive voltage divider path in the circuit.

In most analog MOS designs, analog switches are used for sampling voltages or transferring charge between capacitors on the IC chip. In such applications, the capacitors encountered are in the range of 1–100 pF, where the time constants due to the finite value of r_{ds} do not present a problem. For example, a 1000-Ω switch charging a 100-pF capacitor can settle to 0.01% of its final value in less than 1 μsec. This is adequately fast for most analog sampling applications.[13]

Effects of Parasitic Capacitances

In designing analog switches or selecting the device geometries and configurations for MOS switches, the relative effects of the inherent device parasitics, primarily the four stray capacitances between gate, source, drain, and body, must be kept in mind. All of these capacitances increase with increasing device geometry and Z/L ratio. Thus, whenever possible, a minimum-geometry device should be used as a switch, as dictated by the maximum allowable r_{ds} requirement.

The relative effects of C_{SB} and C_{DB} can be reduced by increasing the substrate bias. The gate–source and gate–drain capacitances are usually the most troublesome since they allow a finite amount of charge to be transferred from the control path into the signal path. This is particularly bothersome when an analog switch is used to equalize or redistribute the charge across two capacitors connected at both ends of the switch. Although the values of C_{GS} and C_{CD} are of the order of 0.1–0.2 pF for small-area devices, the amount of feedthrough charge for large clock signals can be sufficient to cause several millivolts of voltage level charge across the capacitors connected to the switch, particularly if these capacitor values are small enough to be within an order of magnitude of C_{GS} or C_{GD}.

The feedthrough effect can be virtually eliminated, or canceled, by using a charge-canceling scheme, as illustrated in Figure 6.39, where an additional "dummy" switch is added into the circuit, which is driven by the *complement* of the clock signal.[14] In the figure, Q_1 is the actual analog switch and Q_x, with its source and drain short-circuited, is the dummy switch to cancel the feedthrough effect. The parasitic gate–source and gate–drain capacitances of Q_x, which are driven by opposite-polarity gate signals, virtually cancel out most of the charge

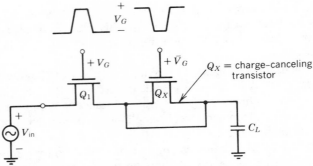

FIGURE 6.39. Minimizing clock feedthrough in analog switch by use of charge canceling device.

feedthrough associated with the gate drive of the actual switch transistor Q_1 by introducing a nearly equal but opposite feedthrough effect.

Double-Throw Analog Switches

One of the most widely used applications of MOS analog switches is the switching of a capacitor between two voltage sources. This is particularly true in the area of switched-capacitor filters (see Chapter 13). Such an application normally requires a single-pole double-throw switch, as shown in Figure 6.40. Using MOS analog switches, this function is normally done by using two switches which are opened and closed sequentially, in a break-before-make manner. This break-before-make condition is essential to ensure that the two voltage levels, between which the capacitor is switched, are electrically isolated from each other during the switching interval. Thus, the complementary clock signals ∅ and ∅̄ applied to both switches must not have overlapping edges. This requires special care in the design of the switch driver circuitry. Figure 6.41

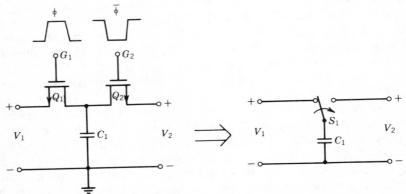

FIGURE 6.40. Use of two analog switches driven by nonoverlapping complementary clock signals to form single-pole double-throw switch.

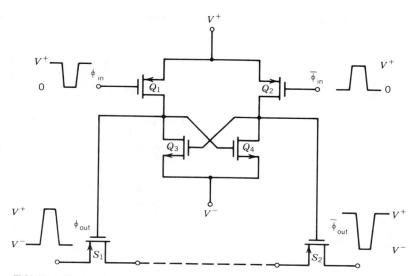

FIGURE 6.41. CMOS analog switch driver circuit for break-before-make switching.

shows a practical CMOS set–reset flip-flop circuit which can be used for such applications.[15] This circuit can take single-polarity clock signals and convert them into nonoverlaping ϕ and $\bar{\phi}$ signals capable of output voltage swing from V^+ to V^-, for driving analog switches S_1 and S_2 in break-before-make sequence.

REFERENCES

1. Y. S. Tsividis, "Design Considerations in Single-Channel MOS Analog Integrated Circuits—A Tutorial," *IEEE J. Solid-State Circuits* **SC-13,** 383–391 (June 1978).

2. P. R. Gray, "Basic MOS Operational Amplifier Design—An Overview," in *Analog MOS Integrated Circuits,* P. R. Gray, D. A. Hodges, and R. W. Brodersen, Eds., IEEE Press, New York, 1980, pp. 28–49.

3. P. R. Gray, D. A. Hodges, and R. W. Brodersen, Eds., *Analog MOS Integrated Circuits,* IEEE Press, New York, 1980.

4. E. Fong, *Design Considerations for MOS Amplifiers,* Prentice-Hall, Englewood Cliffs, NJ, 1981.

5. Y. Tsividis and P. R. Gray, "An Integrated NMOS Operational Amplifier with Internal Compensation," *IEEE J. Solid-State Circuits* **SC-11,** 748–754 (December 1976).

6. R. A. Blauschild, P. A. Tucci, R. S. Muller, and R. G. Meyer, "A New NMOS Temperature Stable Voltage Reference," *IEEE J. Solid-State Circuits* **SC-13,** 767–774 (December 1978).

7. D. Senderowicz, D. A. Hodges, and P. R. Gray, "High-Performance NMOS Operational Amplifier," *IEEE J. Solid-State Circuits* **SC-13,** 760–766 (December 1978).

8. E. Toy, "An NMOS Operational Amplifier," *Digest of Tech. Papers,* IEEE Int. Solid-State Circuits Conf., February 1979.

9. B. J. Hostica, R. W. Brodersen, and P. R. Gray, "MOS Sampled Data Recursive Filters Using Switched Capacitor Integrators," *IEEE J. Solid-State Circuits* **SC-12,** 600–608 (December 1977).

10. Y. A. Hague, R. Gregorian, D. Blasco, R. Mao, and W. Nicholson, "A Two-Chip PCM Codec with Filters," *IEEE J. Solid-State Circuits* **SC-14,** 970–980 (December 1979).

11. Y. P. Tsividis and R. W. Ulmer, "A CMOS Voltage Reference," *IEEE J. Solid-State Circuits* **SC-13,** 774-778 (December 1978).

12. R. M. Swanson and J.D. Meindl, "Ion Implanted Complementary MOS Transistor in Low-Voltage Circuits," *IEEE J. Solid-State Circuits* **SC-7,** 146–153 (April 1972).

13. D. A. Hodges, "Analog Switches and Passive Elements in MOS LSI," in *Analog MOS Integrated Circuits,* P. R. Gray, D. A. Hodges, and R. W. Brodersen, Eds., IEEE Press, New York, 1980, pp. 14–18.

14. R. E. Suarez, P. R. Gray, and D. A. Hodges, "All-MOS Charge Redistribution Analog-to-Digital Conversion Techniques—Part II," *IEEE J. Solid-State Circuits* **SC-10,** 379–385 (December 1975).

15. G. F. Landsburg, "A Charge-Balancing Monolithic A/D Converter," *IEEE J. Solid-State Circuits* **SC-12,** 662–673 (December 1977).

CHAPTER SEVEN

OPERATIONAL AMPLIFIERS

The first six chapters of this book covered the basics of monolithic IC processes, components and the building blocks of IC design. Starting from this chapter, the rest of the book deals with the utilization of these basic building blocks in the design of complete IC products. For convenience, the circuit classes to be discussed are categorized by IC product "families," starting with operational amplifiers and extending to other categories such as wideband amplifiers, analog multipliers, voltage regulators, IC oscillators, phase-locked loops, filters, and A/D and D/A converters.

The monolithic operational amplifier, or op amp, is by far the most widely known and used class of analog integrated circuits. Since its introduction in the mid 1960s, the monolithic op amp has proliferated into many different designs and has found its way into countless applications. In this chapter, we will study some of the fundamental design approaches and techniques for monolithic op amps, and examine some of the commercially successful op amp designs.

The first part of this chapter deals with the very fundamental properties of the basic op amp and the definition of op amp terms. This is followed by an analysis of the op amp architecture and design considerations for input and output stages, frequency stability, and slew-rate requirements. Several commercially successful op amp designs are then reviewed, and their performance characteristics are examined. The MOS op amp designs, which have become the most important building blocks of analog MOS circuits, are also discussed and investigated separately.

The last section of the chapter deals with the other categories of op amp based analog integrated circuits, such as operational transconductance amplifiers, comparators, buffers, and voltage followers.

7.1. FUNDAMENTALS OF OPERATIONAL AMPLIFIERS

The Ideal Op Amp

The "ideal" op amp is a differential input, single-ended output voltage amplifier which offers infinite voltage gain with infinite input impedance, infinite band-width, and zero output impedance. Figure 7.1 shows a conceptual schematic of such an idealized device. Although the actual op amps do not exhibit such idealized characteristics, their performance is often sufficient to approximate the properties of the ideal op amp at low frequencies.

In virtually all applications, the op amp is used in a feedback configuration. Figure 7.2 shows a generalized feedback configuration around an op amp circuit. Neglecting the output impedance of the op amp, the overall voltage gain A_V, of the generalized feedback circuit can be written as:

$$A_V = \frac{V_{out}}{V_1} = -\frac{Z_f}{Z_1} \frac{1}{1 + (1/A)(1 + Z_f/Z_1 + Z_f/Z_{in})} \qquad (7.1)$$

where Z_{in} is the input impedance, and A is the voltage gain of the op amp.

The overall voltage gain A_V given by Eq. (7.1) is called the *closed-loop gain* for the feedback system, to distinguish it from the so-called open-loop gain A, which would correspond to having no feedback at all, that is, $Z_f \rightarrow \infty$.

The most important feature of the closed-loop gain expression given in Eq. (7.1) is that as $A \rightarrow \infty$ for large values of Z_{in}, the closed-loop voltage gain becomes

$$A_V \bigg|_{A \rightarrow \infty} = -\frac{Z_f}{Z_1} \qquad (7.2)$$

Thus, for sufficiently high values of open-loop gain, the closed-loop per-formance becomes determined solely by the feedback elements. In fact, the

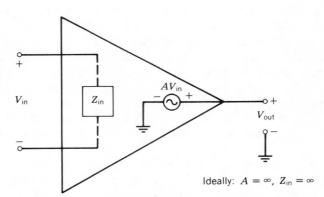

FIGURE 7.1.   Equivalent circuit of ideal op amp.

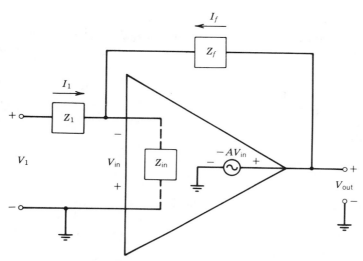

FIGURE 7.2. Basic feedback amplifier.

absolute value of the open-loop voltage gain is not important, provided that it is sufficiently large to reduce the complete closed-loop gain expression of Eq. (7.1) to its simplified form in Eq. (7.2), within a prescribed error tolerance. The versatility of the op amp as a universal building block stems from this feature.

The very high voltage gain and high input impedance of the op amp greatly simplify the analysis of op amp feedback circuits by using the so-called *summing point concept*. [1,2] This concept states that if an op amp is connected in a negative feedback configuration, then for the finite values of the output voltage V_{out}, the voltage V_{in}, appearing directly across the input terminals of the op amp will approach zero, since

$$V_{in} = \frac{V_{out}}{A} \qquad (7.3)$$

and becomes arbitrarily small as $A \rightarrow \infty$. Thus, one can analyze such a circuit by assuming implicitly that the circuit will move its output to *whatever* voltage level is necessary to drive V_{in} to zero. This concept, although workable in almost all practical cases, is not mathematically rigorous since it assumes that a stable and realistic operating point exists for the op amp within the feedback loop. For example, the summing point concept will lead to erroneous results if the output voltage is forced to exceed either the supply voltages or the maximum allowable output swing for the op amp.

Another outcome of the summing point constraint is that no current can flow into the input terminals of the op amp since $V_{in} = 0$ and Z_{in} is large.

The summing point concept greatly simplifies the approximate first-order analysis of an op amp negative feedback circuit, such as that shown in Figure

7.2. Since the inverting input terminal is forced to ground potential, the impedance Z_1 serves to convert the input voltage V_1 to an input current I_1, where

$$I_1 = \frac{V_1}{Z_1} \tag{7.4}$$

Since no input current can go into the op amp terminals, the output voltage V_{out} has to adjust itself such that an equal but opposite amount of current can flow through Z_f such that

$$I_1 = \frac{V_1}{Z_1} = -I_f = -\frac{V_{out}}{Z_f} \tag{7.5}$$

In other words, the inverting input of an ideal op amp has two unique properties:

1. It serves as a "virtual ground" with respect to the noninverting input terminal since $V_{in} = 0$.
2. It serves as an ideal summing node for all the currents coming into it, such that the sum of all currents supplied from the signal source must result in an equal and opposite current in the feedback path (see Fig. 7.3b).

Figure 7.3 shows some of the basic op amp circuit configurations and their gain equations. All of the gain expressions indicated can be derived readily by using the summing point concept. Figure 7.3b illustrates the summing property of the inverting input node of the op amp, where the output voltage is equal to the negative of the weighted sum of the input voltages.

The circuit connection of Figure 7.3e is used to amplify the *difference* of the two input signals V_1 and V_2. Applying the summing point constraint, which requires that $V_{in} = 0$ and $I_1 = I_2$, the operation of the circuit can be briefly explained as follows. The applied voltage V_1 sets up a voltage level V_X at the noninverting terminal of the op amp, where

$$V_X = V_1 \frac{R_2}{R_1 + R_2} \tag{7.6}$$

Since $V_{in} = 0$, the output voltage has to adjust itself to set the voltage level at the inverting input to equal V_X. The currents I_1 and I_2 can be written as

$$I_1 = \frac{V_2 - V_X}{R_1} \tag{7.7}$$

and

$$I_2 = \frac{V_X - V_{out}}{R_2} \tag{7.8}$$

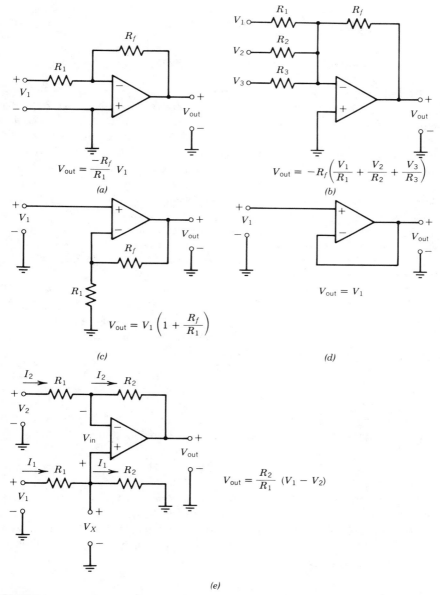

FIGURE 7.3. Basic op amp connections: (*a*) Inverting amplifier; (*b*) summing amplifier; (*c*) non-inverting amplifier; (*d*) voltage follower; (*e*) differential amplifier.

313

Since no current flows into the op amp input terminals, $I_1 = I_2$. Then solving Eqs. (7.6) through (7.8), one obtains

$$V_{\text{out}} = \frac{R_2}{R_1}(V_1 - V_2) \tag{7.9}$$

The op amp can also be used for nonlinear signal processing by applying nonlinear feedback around it. Figure 7.4 shows two commonly used nonlinear signal-processing circuits. The circuit of Figure 7.4a is the so-called *logarithmic amplifier* which produces an output proportional to the natural logarithm of the signal voltage V_1. Such an amplifier, which is also called a *compression circuit*, finds a wide range of applications in instrumentation and recording systems where signals with a very large dynamic range must be recorded or displayed in a display medium having a limited dynamic range. The operation of the circuit can be analyzed by using the summing point concept as follows. Since the voltage across the input terminals of the op amp must be zero, and since no current flows into the op amp input terminals, the collector current I_{C1} of the feedback transistor Q_1 is forced to be equal to the input current I_1, that is,

$$I_1 = \frac{V_1}{R_1} = I_{C1} \tag{7.10}$$

Similarly, the output voltage is forced to equal $-V_{\text{BE}}$. Since the emitter–base voltage of a bipolar transistor, in its active region, is proportional to the *logarithm* of the collector current, one can write

$$V_{\text{out}} = -V_{\text{BE}} = -V_T \ln\left(\frac{I_{C1}}{I_{CO}}\right) \tag{7.11}$$

where V_T is the thermal voltage, and I_{CO} is the collector reverse saturation current of Q_1 [see Eqs. (2.5) and (2.6)]. Combining Eqs. (7.10) and (7.11), one obtains

$$V_{\text{out}} = -V_T \ln\left(\frac{V_1}{R_1 I_{CO}}\right) \tag{7.12}$$

The circuit of Figure 7.4b performs an inverse of the above operation. An input current I_1 is generated by the input voltage V_1, which varies *exponentially* with the input voltage, while the diode D_1 is forward biased. This current is then forced to flow through the feedback resistor R_1 and generate an exponential output voltage, that is,

$$V_{\text{out}} = -I_1 R_1 = -I_S R_1 \exp\left(\frac{V_1}{V_T}\right) \tag{7.13}$$

where I_S is the reverse saturation current of D_1 [see Eq. (3.1)]. The exponential amplifier circuit of Figure 7.4b is often called an *expander* circuit and is normally used to restore the logarithmically compressed input signals back to their original dynamic range.

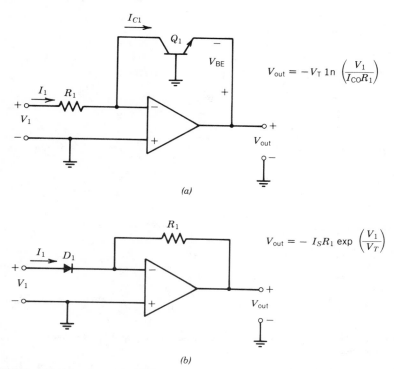

$$V_{out} = -V_T \ln \left(\frac{V_1}{I_{CO}R_1} \right)$$

(a)

$$V_{out} = - I_S R_1 \exp \left(\frac{V_1}{V_T} \right)$$

(b)

FIGURE 7.4. Nonlinear signal processing using op amps. (a) Logarithmic amplifier; (b) exponential amplifier.

In their simplified form, the output voltages of both the logarithmic and the exponential amplifier circuits exhibit very strong temperature dependence due to V_T as well as I_{CO} or I_S. However, this dependence can be significantly reduced by using various compensation techniques.[3]

The op amp can also be used to perform the mathematical operations of integration or differentiation on time-variable input signals. The two basic circuits used for these applications are shown in Figure 7.5. In the case of the integrator, the resistor R_1 is used to develop a current $I_1(t)$ which is proportional to the time-variant input voltage $V_1(t)$. This current is forced to flow through the capacitor C_1. Since the voltage across a capacitor is proportional to the integral of the current through it,

$$V_{out}(t) = -\frac{1}{R_1 C_1} \int_0^t V_1(t)\,dt + V_{out}(0) \qquad (7.14)$$

where $V_{out}(0)$ is the initial value of the output voltage at $t = 0$.

In the case of the differentiator circuit, the capacitor C_1, connected between the input signal and the inverting input of the op amp terminal, generates a

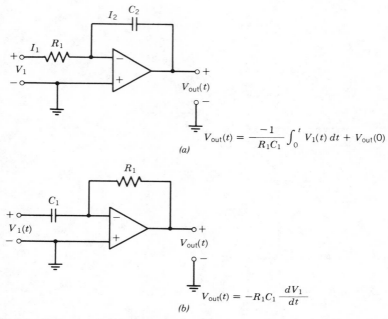

$$V_{out}(t) = \frac{-1}{R_1 C_1} \int_0^t V_1(t)\, dt + V_{out}(0)$$

(a)

$$V_{out}(t) = -R_1 C_1 \frac{dV_1}{dt}$$

(b)

FIGURE 7.5. Performing integration and differentiation with op amps: (a) Integrator circuit; (b) differentiator circuit.

current $I_1(t)$ which is proportional to the time derivative of the voltage across it. This current is then forced to flow through R_1 and produces an output voltage $V_{out}(t)$, where

$$V_{out}(t) = -R_1 C_1 \frac{dV_1(t)}{dt} \qquad (7.15)$$

As will be described in the following sections, the performance limitations of actual op amp circuits restrict the range of input and output swings, and their time rates of change over which the idealized integrator and differentiator equations (7.14) and (7.15) are valid.

In addition to the basic applications listed in this section, the op amp can be used in conjunction with an analog multiplier to perform the mathematical operations of division and the square-root extraction. These applications will be discussed further in Chapter 9.

The Actual Op Amp

The ideal op amp circuit described in the previous section is an excellent tool to illustrate the basic properties and applications of op amps. Even though the actual op amp circuits very closely simulate the important features of the ideal op amp, they may deviate significantly from ideal behavior in many ways. Some of these deviations from ideality are the presence of input voltage and current offsets, finite values of the voltage gain and bandwidth, finite input resistance, nonzero output impedance, limited dynamic range and response speed under large-signal operating conditions. This nonideal behavior places a lower limit on the magnitude of dc signals which can be detected and places an upper limit on the magnitudes of the impedances of the passive elements which can be used in feedback around the amplifier. In addition, these nonidealities also limit the frequency and the time rate of change of the signals which can be accurately amplified.

Figure 7.6 shows the equivalent circuit of an actual op amp, indicating the presence of finite input and output impedances and input voltage and current offsets. The following is a summary of these important deviations from ideality and their effects in actual circuit applications.

Input Offset Voltage V_{OS}. This is the voltage that must be applied across the differential input terminals of an op amp in order to reduce the output voltage V_{out} to zero. As discussed in Chapter 5, the finite value of V_{OS} is due to the device mismatches or the bias unbalances in the op amp input stage. For op amps with bipolar input stages, V_{OS} is in the range of $\pm$ 1–5 mV. For FET input op amps, or for MOS op amps, it can be as high as $\pm$ 20 mV. In dc amplifier applications,

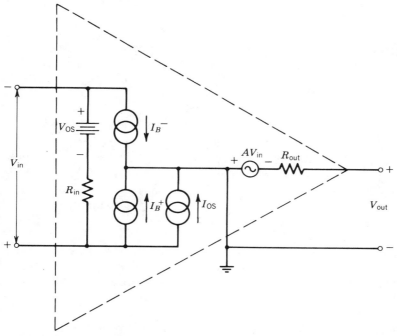

FIGURE 7.6. Equivalent circuit of actual op amp, showing effects of finite input and output impedances and voltage and current offsets.

the offset voltage and its drift with temperature place a lower limit on the magnitude of the dc voltage which can be accurately detected and amplified. The temperature dependence of V_{OS} is called offset drift. The offset drift tends to increase with increasing values of V_{OS} [see Eq. (5.26)] and does not necessarily go to zero when the offset voltage is nulled by an external potentiometer. For the reasons explained in Chapter 5, typical bipolar op amps exhibit approximately $\pm\ 3\ \mu V/°C$ offset drift per millivolt of V_{OS}.

Input Bias Current I_B.   In bipolar op amps, a finite amount of base current I_B must be supplied to the bases of the input transistors. The value of I_B depends on the choice of the input stage quiescent current level and the β_o of the input transistors. Typically, I_B is in the range of 10 nA to 100 nA for most bipolar op amps, and is positive (i.e., into the input terminal) for *npn* input devices and negative for *pnp* input transistors. In the case of JFET input stages, the value of I_B is in the range of 1–10 pA at room temperature, but tends to double for every 10°C change in temperature. The presence of a finite bias current into the op amp terminals violates the summing point concept discussed earlier, since it adds or subtracts from the actual signal current appearing at the *summing node* (i.e., the inverting input) of the op amp.

Input Offset Current I_{OS}.   The input offset current is the mismatch between the bias current at each of the input terminals of the op amp,

$$I_{OS} = I_B^+ - I_B^- \tag{7.16}$$

where the $+$ and $-$ signs refer to the noninverting and inverting inputs of the op amp, respectively. Typical β_o and device mismatches normally lead to a value of I_{OS} on the order of 5–10% of the nominal input bias current I_B, where

$$I_B = \frac{I_B^+ + I_B^-}{2} \tag{7.17}$$

Input Resistance R_{in}.   As described in Chapter 5, the input resistance is a function of the input stage configuration and the bias current levels. In bipolar op amps, typical values of R_{in} are in the range of 0.1–5 MΩ. In JFET and MOSFET op amps the value of R_{in} can be as high as 10^{10} or 10^{12} Ω. As indicated by Eq. (7.1), if the open-loop voltage gain is large, the value of R_{in} has a negligible effect on circuit performance.

Output Resistance R_{out}.   The value of the output resistance depends on the circuit configuration of the output stage and the choice of the current-limiting circuitry used at the output (see Section 5.3). For general-purpose op amps, the values of R_{out} are in the range of 20–200 Ω.

Input Common-Mode Range.   This is the maximum range of the input voltage which can be simultaneously applied to both inputs without causing the cutoff, saturation, or breakdown of any of the gain stages inside the op amp. Typically, the common-mode range approaches to within a few volts of either supply voltage. In the so-called *single-supply,* or *ground-sensing,* op amp stages (see Fig. 5.22), it can extend down to $-V_{EE}$.

Common-Mode Rejection Ratio. The ratio of the differential open-loop gain to the common-mode open-loop gain is called the common-mode rejection ratio (CMRR). This ratio can also be defined as the ratio of the change of V_{OS} for a unit change in the common-mode input voltage. Typical values of CMRR are in the range of 80–120 dB for the most practical op amp circuits. For example, with a CMRR of 80 dB, a 10-V change in the common-mode input voltage can cause a $\pm$ 1-mV change in V_{OS}.

Power Supply Rejection Ratio. The power supply rejection ratio (PSRR,) is the ratio of the change of V_{OS} for a unit change of any one of the supply voltages. Normally, it is specified separately with respect to either one of the supply voltages. Typically, the PSRR is of the same order of magnitude as the CMRR.

Open-Loop Voltage Gain, A. The open-loop voltage gain is the ratio of the incremental change of the output voltage for a unit incremental change of the differential input voltage, with no feedback applied. It is measured at dc or very low frequencies. In practical op amp circuits, the open-loop voltage gain is in excess of 10,000, or 80 dB. Thermal effects and stability requirements limit the maximum practical open-loop gain values to $\leq 10^6$.[4]

Unity-Gain Bandwidth. This is the small-signal 3-dB bandwidth for unity-gain closed-loop operation. For practical op amps, unity-gain bandwidth is in the range of 1–10 MHz.

Slew Rate. This is the maximum time rate of change of the output voltage for a step input. It is normally measured at the zero crossing point of the output waveform, when the op amp is compensated for unconditional stability. Depending on the circuit design, the slew rates for positive- or negative-going output waveforms may be different. The slew rate is normally specified in volts per microsecond. Typical values of slew rate vary from about 1 V/μsec for simple 741-type op amps to in excess of 30 V/μsec for high-speed op amps.

Full-Power Bandwidth. The maximum frequency of input signal over which the full output swing can be obtained for a sinusoidal input is the full-power bandwidth. As will be pointed out in Section 7.4, the full-power bandwidth is directly proportional to the slew rate.

Settling Time. The settling time is the time taken for the op amp output to settle to within $\pm$ 0.1% of its final value for a step change of the input. It is normally measured with the op amp connected as a unity-gain voltage follower (see Fig. 7.3*d*) with a 10-V step input and a 100-pF capacitive load at the output.

Typical values of settling time are in the range of $0.3–5\,\mu$sec for most IC op amps.

In the design of a general-purpose monolithic op amp it is impossible to optimize all of the performance characteristics associated with the device. For example, low-input bias current or high-input impedance requirements may not be compatible with the low offset voltage requirements; or the high slew-rate or unity-gain bandwidth requirements may make the frequency compensation of the circuit difficult. Therefore, a large number of compromises and trade-offs are necessary in the design of a monolithic op amp. In the following sections, some of these design techniques and compromises will be examined in more detail.

7.2. CIRCUIT CONFIGURATIONS FOR MONOLITHIC OPERATIONAL AMPLIFIERS

The basic subcircuits and gain stages described in Chapters 4 and 5 form the building blocks for monolithic op amp design. In analyzing these circuit designs, it is very useful to categorize them according to their "architecture," that is, by the number of gain stages within an op amp. This system of classification is particularly useful and instructive because the number of gain stages tends to determine the dominant open-loop poles which control the ac response and the stability characteristics of the op amp.[5] Using this approach, one can classify the circuit architecture of monolithic op amps as single-, two-, or three-stage designs by referring to the number of high-gain stages in the circuit. As will be discussed below, the great majority of these IC op amps belong to the two-stage category.

Single-Stage Op Amps

The single-stage op amp configuration is used in some of the simple low-performance and low-cost op amp designs,[6] as well as in some very-high-speed op amps. The presence of a single gain stage results in less overall phase shift and a wider unity-gain bandwidth, provided that extremely high voltage gains are not required. Figure 7.7a shows the basic architecture of a single-stage op amp, which is primarily made up of an input transconductance stage that drives a unity-gain broadband buffer stage at the output. In order to obtain sufficiently high dc gain, normally in excess of 10,000, the input impedance of the output buffer stage must be very high. The total low-frequency gain A_v is given as

$$A_v = \frac{v_{\text{out}}}{v_{\text{in}}} = g_m\,(r_{o1}\|r_{i2}) \tag{7.18}$$

where r_{o1} and r_{i2} are the output and input resistances of the first and the buffer stages, respectively, and g_m is the input stage transconductance. The frequency compensation is achieved by narrow-banding the input gain stage with a compensation capacitor C_1 to ground, as shown in Figure 7.7a. Figure 7.7b illus-

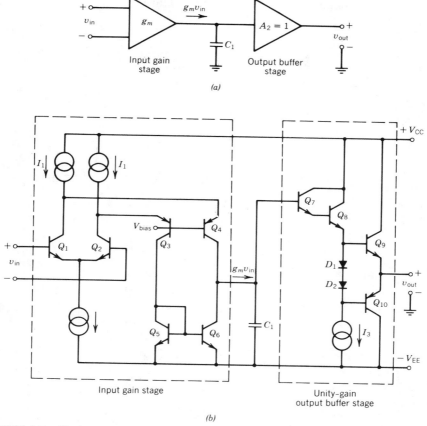

FIGURE 7.7. Single-stage op amp configuration: (*a*) Basic architecture; (*b*) simplified design example.

trates a simplified design example for such a single-stage op amp.[7] For simplification purposes, the current sources, current mirrors, and buffer stages within the circuit are shown in their simplest form, and the nonessential parts, such as the biasing circuits, are not shown. This particular circuit was designed for very-high-speed operation with slew rates in excess of 50 V/μsec. The circuit is frequency compensated with a single capacitor C_1 at the output of the gain stage.

Two-Stage Op Amps

The great majority of the IC op amps belong to this category. As shown in Figure 7.8*a*, the basic architecture of a two-stage op amp is comprised of an input transconductance stage, followed by a high-gain inverting amplifier, and finally an output buffer stage. The frequency compensation can be achieved by a single

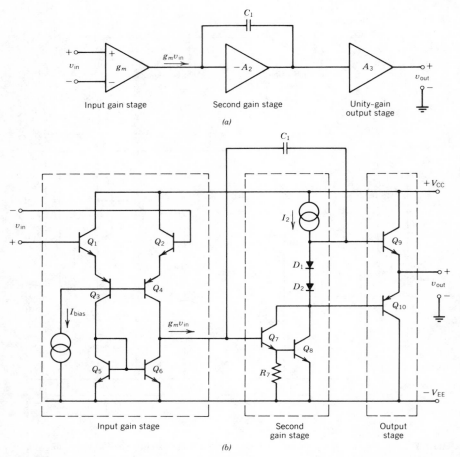

FIGURE 7.8. Basic two-stage op amp configuration: (*a*) Basic architecture; (*b*) simplified design example.

capacitor C_1 connected in feedback around the second stage. As will be described in the next section, C_1 serves as a *pole-splitting* capacitor and causes the entire op amp to exhibit a single-pole frequency rolloff.

A good example of the two-stage op amp is the Fairchild μA-741 op amp whose simplified schematic is shown in Figure 7.8*b*. The input stage is basically a transconductance stage similar to that shown in Figure 5.21. It is then followed by a buffered *npn* inverter stage with an active load, to provide the second stage of voltage amplification. The output is a Class AB complementary stage of the type shown in Figure 5.33. The popularity of the two-stage op amp stems from two factors: (1) it is a relatively simple architecture and can provide large amounts of voltage gain (typically in excess of 100,000), and (2) it can be stabilized with a single low-value capacitor C_1, typically in the range of 10–30 pF, which can easily be incorporated into the chip design. However, it has certain inherent limitations, particularly with regard to large-signal dynamic response parameters, such as slew rate or full-power bandwidth.

Three-Stage Op Amps

Three-stage op amps are primarily used for special design applications where very high voltage gains or low noise characteristics are essential. Since potentially there are three independent dominant poles in such circuits, their stabilization and frequency compensation is quite difficult and requires feedforward as well as feedback circuitry.

Figure 7.9 shows the block diagram representation of some practical three-stage op amp configurations. In the circuit configuration of Figure 7.9a, local feedback and feedforward is used around the second stage, and local feedback is used around a third gain stage to create the dominant system pole. In such a

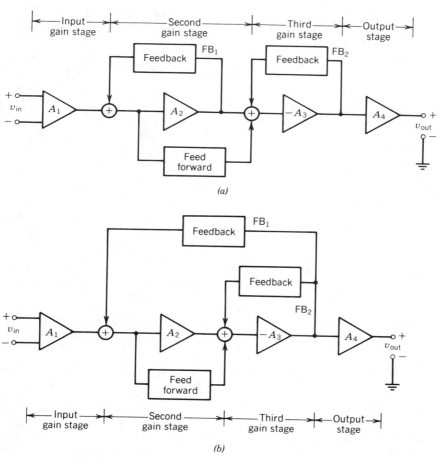

FIGURE 7.9.   Basic circuit architectures for three-stage op amps.

design, normally, the second stage is used with approximately unity closed-loop gain to serve as a broadband level-shift stage.[8]

Figure 7.9b shows an alternate three-stage op amp configuration where each of the three gain stages can have significant voltage gain. In this configuration, an overall feedback loop is used around the second and third gain stages to provide the dominant system pole, and local feedback around the third stage is used to generate a nondominant high-frequency pole. The second gain stage functions as a medium-gain level-shift stage, and its limitation is avoided by bypassing it with a feedforward circuit.

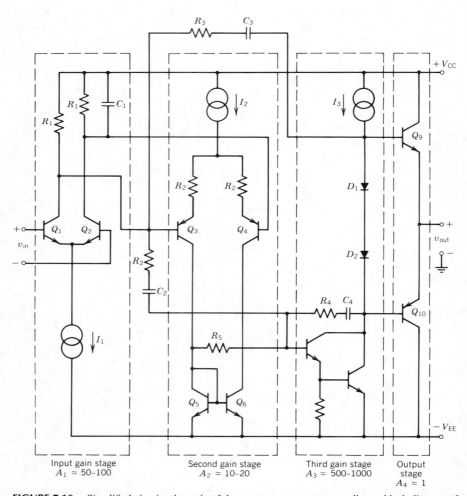

FIGURE 7.10. Simplified circuit schematic of three-stage op amp corresponding to block diagram of Figure 7.9b.[9]

Figure 7.10 shows a simplified design example of a practical three-stage op amp circuit using the block diagram topology of Figure 7.9b. [9] In this circuit, which is designed to have very low noise characteristics, the input stage is operated at high currents and has relatively low gain, the second stage provides some additional gain and dc level shifting, and the third stage provides most of the voltage amplification.

The overall feedback network FB_1 is formed by the series combination of R_3 and C_3, the local feedback around the third stage is formed by R_4 and C_4, and the feedforward around the second stage is provided by R_2 and C_2. The capacitor C_1 at the collector of Q_2 provides an ac ground at the base of Q_4, at high frequencies, so that only a single-ended output is provided from the input stage at the collector of Q_1.

A comparison of the basic two-stage op amp configuration with that of Figure 7.9 indicates clearly that the frequency stability and compensation become a very serious problem as the number of gain stages is increased from two to three. However, as will be discussed in Section 7.6, such added circuit complexity may be justified in special applications which may require very high voltage gains, or very low input noise or offset voltage characteristics. In the following discussions, most of the attention will be focused on the two-stage op amps, since they make up the large majority of the commercially available monolithic op amp products.

7.3. FREQUENCY COMPENSATION

Stability Requirements

In almost all of the circuit applications, the op amp is required to be "unconditionally stable." In other words, the op amp should not break into oscillations for any value of the *resistive* negative feedback applied around it. For unconditional stability, it is necessary for an op amp to approximate a one-pole open-loop transfer function $A(j\omega)$ given as

$$A(j\omega) = \frac{A_o}{1 + j\omega/\omega_o} \qquad (7.19)$$

where A_o is the open-loop gain at dc, and ω_0 is the 3-dB bandwidth. The transfer function of Eq. (7.19) corresponds to a -6-dB/octave rolloff of the open-loop gain for frequencies in excess of ω_o.

Figure 7.11 shows the gain and phase response of a simple idealized op amp circuit, exhibiting the single-pole rolloff characteristics, as given by Eq. (7.19). Note that the phase shift across the amplifier in such an ideal case never exceeds 90°.*

---

*In calculating the phase shift across the amplifier as a function of frequency, the ± 180° phase shift due to the use of inverting input of the amplifier is neglected.

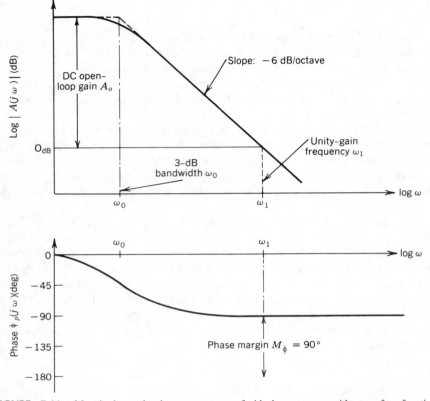

FIGURE 7.11. Magnitude and phase response of ideal op amp with transfer function: $A(j\omega) = A_o/(1 + j\omega/\omega_0)$: (a) Magnitude response; (b) phase response.

In practical op amp circuits, the idealized one-pole rolloff characteristics of Figure 7.11 are not realized since each of the gain stages in the circuit tends to contribute additional poles into the circuit transfer function, and the excess phase shifts associated with the active devices or stray parasitics further complicate the transfer characteristics. Therefore, in the design and application of an op amp chip, it is necessary to compensate the frequency performance of the circuit to avoid oscillations or peaking in the closed-loop response.

In terms of the basic feedback theory, [10,11] it can be shown that for unconditional stability the total phase shift ϕ_p across the amplifier (neglecting the 180° phase reversal associated with the inverting input) must be less than 180° for all frequencies where the magnitude of the gain $|A(j\omega)|$ is greater than 1. The margin of stability in an op amp can be related to its open-loop gain and phase response by the so-called gain and phase margins, defined as follows.

Gain Margin M_G. This corresponds to the amount by which the voltage gain is *below* the unity (0-dB) level, at the frequency when the excess phase shift across the amplifier is exactly 180°. It is measured in decibels and must be positive for unconditional stability.

Phase Margin M_ϕ. This is 180° *minus* the excess phase shift at the frequency where the magnitude of $A(j\omega)$ is equal to 1. It is measured in degrees and must be positive for unconditional stability.

In the case of the ideal single-pole op amp response characteristics shown in Figure 7.11, the phase at the unity-gain crossover frequency ω_1 is $-90°$, which corresponds to a phase margin $M_\phi = +90°$, and since the phase never reaches $-180°$, the gain margin M_G is infinite.

Figure 7.12 shows the typical frequency and phase response characteristics for a multistage IC op amp. The dashed curves correspond to the uncompensated

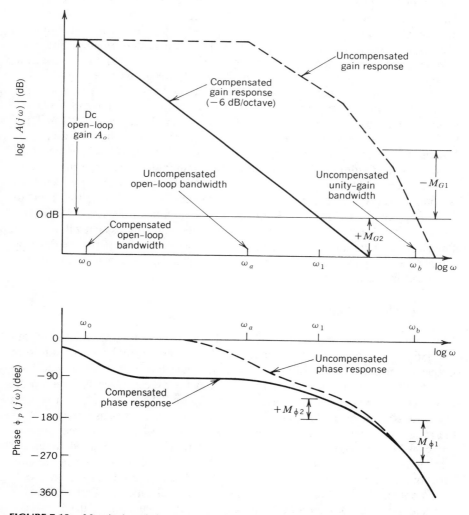

FIGURE 7.12. Magnitude and phase response of typical op amp with and without frequency compensation: (*a*) Magnitude response; (*b*) phase response.

case. Note that the uncompensated op amp exhibits a 3-dB bandwidth of ω_a and a unity-gain crossover frequency of ω_b. However, when the phase shift is 180°, the gain is higher than unity, resulting in a *negative* gain margin $-M_{G1}$, as shown. Similarly, when the gain is equal to unity, the phase shift is in excess of 180°, and the corresponding phase margin $M_{\phi 1}$ is negative. Thus, the op amp is not stable in its uncompensated form since both the gain and the phase margins are negative.

Frequency compensation is done by introducing a dominant pole with the addition of a compensation capacitor. This added dominant pole produces a 6-dB/octave rolloff in gain so that the unity-gain crossover frequency is reached at a lower frequency than the next most important system pole. In the case of the example shown in Figure 7.12, the new dominant pole, produced by the addition of a compensation capacitor, is located at frequency ω_0 and results in a unity-gain crossover frequency of ω_1 for the compensated op amp. Since the gain margin M_{G2} and the phase margin $M_{\phi 2}$ for the compensated op amp are now both positive, the circuit is unconditionally stable for all gain settings.

The frequency compensation technique described above is the simplest and most commonly used technique in monolithic op amps. It is often called *narrow banding* since its main effect is to reduce the amplifier bandwidth and the frequency capability in order to force the phase shift across the amplifier to be *less than* 180° at the unity crossover frequency. Normally, the compensated open-loop bandwidth ω_0 is *several* orders of magnitude lower than the uncompensated bandwidth ω_a, and the compensated unity-gain crossover frequency ω_1 is approximately one order of magnitude lower than ω_b. Thus, this kind of frequency compensation often involves a very significant sacrifice of the frequency capability of the op amp. Some of the problems associated with the simple narrowbanding-type frequency compensation can be partly avoided by using more complex compensation methods such as the feedforward techniques discussed in the next section.

Effects of Phase Margin

Although an op amp can be unconditionally stable, that is, does not break into oscillation under a unity-gain feedback condition, it may still have a nonuniform frequency response and exhibit some peaking near the edge of the closed-loop bandwidth. This peaking is normally caused by the excess phase shift within the amplifier, which reduces the phase margin. Figure 7.13 shows the relative gain characteristics of feedback amplifiers for varying values of phase margin.[2] As the phase margin diminishes, the gain peaks become larger until the gain approaches infinity and oscillation starts at $M_\phi = 0$. The plots are drawn assuming that the response is dominated by the first two poles of the open-loop transfer function, except for the $M_\phi = 90°$ case, which corresponds to the case of a single-pole amplifier. For most practical op amp applications, phase margins of less than 45° are not desirable because of excessive peaking in the frequency response or overshoots in step response.

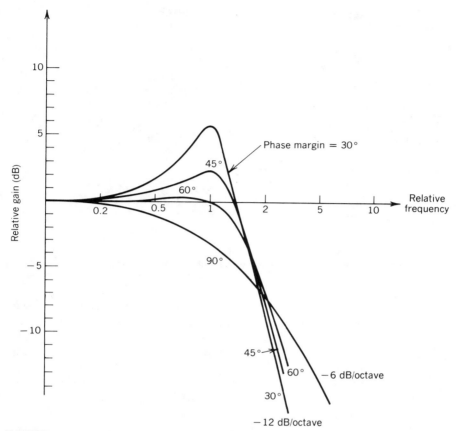

FIGURE 7.13. Normalized gain response of feedback amplifiers for various values of phase margin. Frequency is normalized to unity-gain crossover frequency.[2]

Compensation of Two-Stage Op Amps

The two-stage op amp, which is made up of an input transconductance stage followed by a high-gain inverting stage, is a relatively easy circuit to stabilize. Since the great majority of IC op amps fall into this category, its characteristics justify a more detailed examination than those of other op amp types.

Figure 7.14 shows the basic two-stage op amp in its greatly simplified form.[4] Since the unity-gain output buffer stage is not relevant to this discussion, it is omitted in the figure. The effect of the first stage is to function as a voltage-controlled current source and provide an output current i_{out}, into the input node of the second-stage amplifier,

$$i_{out} = g_{m1} \, v_{in} \qquad (7.20)$$

where $g_{m1} = I_1/V_T$ is the first-stage transconductance.

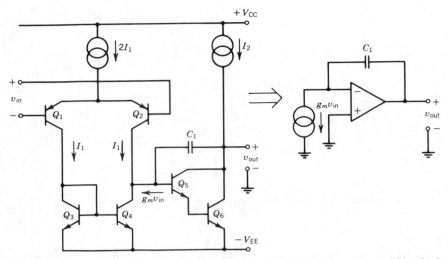

FIGURE 7.14. Simplified model of two-satge op amp for calculating high-frequency small-signal gain.

As a first-order analysis, one can replace the input stage by a voltage-controlled current source, and consider the second stage to be a high-gain op amp with an integrating capacitor C_1 in feedback around it. Neglecting the finite value of gain at dc or very low frequencies, the voltage gain in the mid- or high-frequency range can be expressed as

$$\left| A(j\omega) \right| \approx \frac{g_{m1}}{\omega C_1} \qquad (7.21)$$

which would exhibit a 6-dB/octave rolloff and reach a unity-gain crossover frequency ω_1 at

$$\omega = \omega_1 = 2\pi f_1 = \frac{g_{m1}}{C_1} \qquad (7.22)$$

In practical designs, ω_1 is chosen so that negligible excess phase is built up, in excess of the basic 90° phase shift, due to the integrator pole. This is done by choosing ω_1 to be no higher than the dominant open-loop pole of the uncompensated amplifier. Normally, such an open-loop pole is due to the low f_T of the lateral *pnp* transistors in the input transconductance or level-shift stage, and is of the order of 1–2 MHz. Thus, in general, f_1 is conservatively chosen to be on the order of 1 MHz.

Assuming that the input stage is biased with an operating current of 10 μA per side, one can calculate the required value of C_1 from Eq. (7.22) as

$$C_1 = \frac{g_{m1}}{2\pi f_1} \approx 60 \text{ pF} \qquad (7.23)$$

which is small enough to be incorporated on the chip. As will be shown later, the value of C_1 can be reduced further by reducing the input stage transconductance. For example, in the case of a 741-type composite *npn–pnp* input stage of the kind shown in Figure 5.21 or Figure 7.8, the value of g_m is reduced by one-half [see Eq. (5.63)]; thus only a 30-pF capacitor would suffice for compensation.

Transconductance Reduction [4]

Although the typical values of C_1 predicted by Eq. (7.23) are in the range of 30–60 pF, this capacitor still occupies a significant portion of the chip area. Since the unity-gain frequency ω_1 is fixed by the device characteristics and the circuit configuration, the only way left for reducing the size of C_1 is by reduction of the input stage transconductance. For input stages using lateral *pnp* transistors, such as the one shown in Figure 7.14, transconductance reduction can be achieved by using split-collector *pnp* transistors (see Fig. 2.27). This technique is illustrated in Figure 7.15. Input transistors Q_1 and Q_2 are split-collector transistors with a collector periphery ratio of $1 : m$. Thus, of the total bias current available to each transistor, only a fraction is utilized; the rest is shunted to the

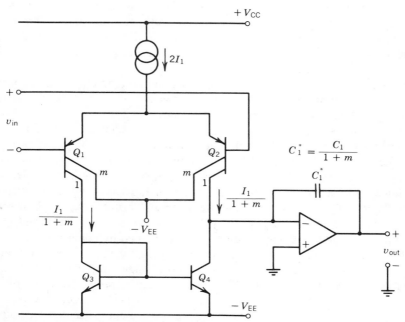

FIGURE 7.15. Input stage transconductance reduction using split-collector *pnp* transistors to reduce compensation capacitor size.

negative supply. As a result of this current splitting, the input stage trans-conductance g_{m1} is reduced to g_{m1}^*, where

$$g_{m1}^* = \frac{g_{m1}}{1 + m} \qquad (7.24)$$

As a result, for the same value of ω_1, a correspondingly smaller compensation capacitor C_1^* can be used, where

$$C_1^* = \frac{C_1}{1 + m} \qquad (7.25)$$

In this manner, the compensation capacitor can be reduced to the 5–10-pF range, resulting in significant savings of chip area.

Feedforward Compensation

The major limitation to the high-frequency capability of IC op amps is the excess phase shift and open-loop system poles due to the poor frequency capabilities of the *pnp* transistors. As described in the previous section, these effects limit the value of the unity-gain frequency f_1 to the 1–2 MHz range. The poor frequency capability of these devices can be overcome, in part, by bypassing them with an ac-coupled signal path at high frequencies. Such an approach is called a *feed-forward* technique and can greatly improve the ac small-signal characteristics at high frequencies.[12] The feedforward technique basically moves the uncompensated open-loop pole of the op amp further out in frequency so that the value of ω_1 ($= 2\pi f_1$) for the unity-gain crossover can be made higher for a given phase margin requirement.

Figure 7.16 shows a simplified diagram of an op amp circuit utilizing feed-forward compensation to enhance the unity-gain bandwidth.[13] This circuit is basically a three-stage op amp of the type described in Figures 7.9 and 7.10, except that the second stage is made up of the *pnp* transistors Q_3 and Q_4, and serves as a level-shift stage for the input differential pair comprised of Q_1 and Q_2. The effect of C_O is to roll off the differential gain of the input stage so that at high frequencies only the signal at the collector of Q_2 is amplified. The feedforward capacitor C_{FF} feeds the input signal around the second stage and goes directly to the third stage at high frequencies. Thus, the excess phase effects due to the poor frequency capability of lateral *pnp* transistors Q_3 and Q_4 are significantly reduced. Compensation is then obtained by connecting C_1 around both the second and the third gain stages which, for the purpose of compensation, function as a single high-gain inverter stage similar to that shown in Figure 7.13. As a result of feedforward, one can improve the value of f_1 by approximately an order of magnitude in the 5–10 MHz range. However, one important problem associated with the feedforward compensation technique is the generation of closely spaced pole–zero pairs in the op amp transfer function. These pole–zero pairs, known as *doublets* do not affect the small-signal fre-

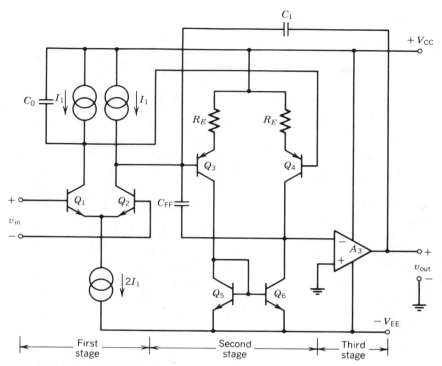

FIGURE 7.16. Simplifed circuit diagram of high-frequency op amp using feedforward to increase unity-gain bandwidth.[13]

quency response, but greatly increase the settling time of the op amp.[14] This effect is very undesirable for such important applications as analog data conversion and data acquisition systems (see Chapters 14 and 15).

7.4. LARGE-SIGNAL OPERATION

In the previous section, the main emphasis was placed on the ac small-signal characteristics of the monolithic op amp. However, in many applications the op amp is used with large input signals (either step inputs or sinusoidal signals), where the circuit performance differs significantly from the ac small-signal characteristics. Two important parameters of an op amp for large-signal ac operation are the slew rate and the full-power bandwidth, both of which are directly related to the frequency compensation techniques used. In this section, the interrelation between these parameters for various circuit configurations will be investigated.

Slew-Rate Considerations

As defined in Section 7.1, the slew rate of an op amp is the *maximum* rate of change of the output voltage for a step change at the input. Figure 7.17 shows the typical step response of an op amp to a large step change of the input voltage and illustrates the effects of a finite slew rate on the resulting output waveform. Due to the finite response speed of the op amp, the output does not follow the input instantaneously. Instead, the input stage becomes overdriven, and the output is forced to ramp, or *slew,* at some limited rate determined by the internal currents and capacitances. Depending on the circuit configuration, the slew rates for positive and negative output swings may be different and may need to be identified separately, as shown in Figure 7.17. Normally, the slew rate is measured under unity-gain operating conditions where the stability requirements are the most stringent.

The simplified model of the two-stage op amp shown in Figure 7.14 can again be used to calculate the slew-rate limitations.[4] When the input is overdriven with a large step input (i.e., $V_{in} \gg V_T$), either Q_1 or Q_2 will cut off, depending on the input signal polarity. In that case, a total current of $\pm 2I_1$ will be available to charge or discharge the compensation capacitor C_1. This condition is illustrated in Figure 7.18. The resulting slew rate is then

$$\text{slew rate} = \left(\frac{dV_{out}}{dt}\right)_{max} = \frac{2I_1}{C_1} \qquad (7.26)$$

Since C_1 is related to the unity-gain crossover frequency ω_1 [see Eq. (7.22)], one can rewrite Eq. (7.26) as

$$\text{slew rate} = \left(\frac{dV_{out}}{dt}\right)_{max} = \frac{2\omega_1 I_1}{g_{m1}} \qquad (7.27)$$

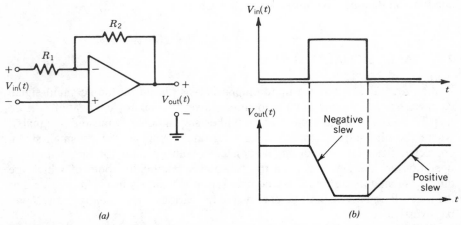

FIGURE 7.17. Slew-limited response of op amp output to input step: (*a*) Circuit connection; (*b*) input and output waveforms.

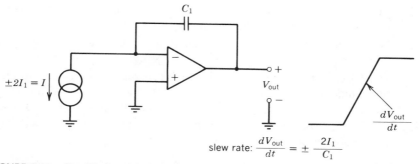

slew rate: $\dfrac{dV_{out}}{dt} = \pm \dfrac{2I_1}{C_1}$

FIGURE 7.18. Simplified model of op amp for calculating slew rate under large-signal conditions.

where g_{m1} is the input stage transconductance. This expression is a very important one since it illustrates one of the fundamental limitations or trade-offs of op amp design. For a given crossover frequency ω_1, it illustrates that the slew rate is determined entirely by the ratio of the first-stage operating current I_1 to the first-stage transconductance g_{m1}. Since ω_1 is determined by small-signal response and stability requirements (i.e., gain and phase margins), the only degree of freedom left to the designer to improve the slew rate is to *increase* the I_1/g_{m1} ratio of the first stage.

In the case of a simple bipolar input stage, such as the one shown in Figure 7.14, this degree of freedom is not available, since g_{m1} is proportional to I_1 as

$$g_{m1} = \frac{I_1}{V_T} \tag{7.28}$$

Then

$$\text{slew rate} = 2\omega_1 V_T = 4\pi f_1 V_T \tag{7.29}$$

Thus, for the simple bipolar input stage circuit of Figure 7.14, or the 741-type op amp of Figure 7.8 which has a typical value of $f_1 = 1$ MHz, the corresponding value of the slew rate is on the order of 0.5 V/μsec. In this simple amplifier scheme, the only way to increase the slew rate is to increase ω_1. Even if ω_1 were increased by an order of magnitude, say, by feedforward techniques, the resulting slew rate would only be of the order of 5 V/μsec, which may still be too low for many applications.

Slew-Rate Enhancement

The basic slew-rate expressions of Eq. (7.26) and (7.27) lead to several design approaches which can be used for increasing the slew rate. These can be summarized as follows:

1. Increase ω_1 by simplified design and by high f_T *pnp* transistors. This can be done by using high-performance complementary *npn* and *pnp* transistors at the expense of increased device fabrication complexity.[7]

2. Increase ω_1 by feedforward techniques. This method has the possible drawback of increasing the op amp settling time.

3. Use emitter degeneration in the input stage to reduce the input stage I_1/g_{m1} ratio.

4. Use Class B input stage design, which can supply additional current to the compensation capacitance under large input swing conditions.

5. Use low-transconductance devices at the input stage, such as JFETs or MOSFETs, whose transconductance is not directly proportional to the input stage current.

The first two techniques, which rely on the increase of the unity-gain rolloff frequency ω_1, have been discussed earlier in this chapter. In this section, we will examine the remaining three design approaches:

Emitter Degeneration. The use of the emitter degeneration resistance R_E, as shown in the differential gain stage of Figure 7.19, allows one to decrease the input stage transconductance without reducing the bias current. The inclusion of R_E in the circuit reduces the input stage transconductance from g_{m1} to an effective transconductance G_m, where

$$G_m = \frac{g_{m1}}{1 + g_{m1}R_E} = \frac{I_1/V_T}{1 + I_1R_E/V_T} \approx \frac{1}{R_E} \tag{7.30}$$

assuming $g_{m1}R_E \gg 1$. As an example, if R_E were chosen such that $I_1R_E = 500$

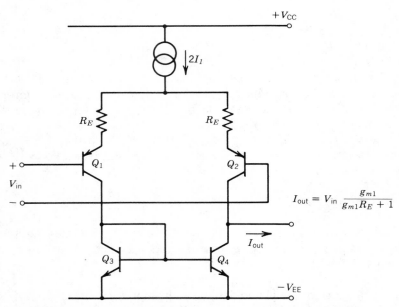

FIGURE 7.19. Input stage transconductance reduction using emitter degeneration.

mV $\approx 20 V_T$, then $G_m \approx 0.05$ and the I_1/G_m ratio would increase by a factor of 20. Under these conditions, one would obtain a slew rate of approximately 10 V/μsec from the simple two-stage op amp configuration. The fundamental reason for this increase can also be explained as follows. The increase of R_E reduces the input stage transconductance and, therefore, the overall voltage gain of the op amp. Thus, a smaller compensation capacitance C_1 can be used. Since the maximum current that can be delivered to C_1 is still unchanged, the slew rate is increased.

The use of emitter degeneration to improve the slew rate may not be desirable in all applications since this technique reduces the open-loop gain and may cause an increase in the input offset voltage due to the resistor mismatches. Furthermore, the presence of the emitter resistors also deteriorates the noise characteristics of the input stage.

Class B Input Stages. In this approach, one leaves the input stage transconductance relatively unchanged, but provides an alternate current path for charging and discharging C_1. This alternate path operates in the Class B mode, that is, it comes into conduction only when large signal swings are involved. Figure 7.20 shows the simplified schematic of such a Class B input stage. With reference to the figure, the operation of the circuit can be explained as follows.

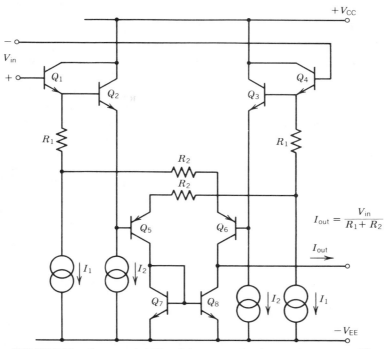

FIGURE 7.20. Cross-coupled Class B input stage for slew-rate improvement.[15]

For small-signal operation, the circuit functions as a normal differential gain stage with a small amount of emitter degeneration due to resistors R_1 and R_2. However, for large signal swings, the Class B action takes over, and the full differential voltage swing appears across resistors R_1 and R_2. The circuit can then supply a charging current I_{out} equal to

$$I_{out} \approx \frac{V_{in}}{R_1 + R_2} \tag{7.31}$$

to the compensating capacitor C_1. Using this circuit approach, slew rates in excess of 30 V/μsec have been reported.[15]

Slew rate improvement is also possible by using other cross-coupled input stage configurations which can reduce the transconductance while maintaining a high value of bias current.[16] However, all of these circuit approaches require relatively complex designs which tend to increase the input offset voltage.

FET Input Stages. Another method of improving the I_1/g_{m1} ratio is by using JFET transistors as the input devices (see Fig. 5.23). Since, at a given current level, the transconductance of JFETs is approximately 3–5% that of bipolar transistors, this results in a factor of 20–30 improvement in the I_1/g_{m1} ratio and slew rate. Thus, slew rates in the range of 10–30 V/μsec are readily achievable with JFET input op amps.

The main drawback of JFET input devices is increased offset voltage. Compared to bipolar devices, the offset voltages associated with JFET devices are worse by a factor of 3–5 and are typically on the order of 2–10 mV.

Full-Power Bandwidth. The full-power bandwidth $\omega_p = 2\pi f_p$ is defined as the maximum frequency at which the peak output swing can be obtained. It is normally measured with a sinusoidal output swing of 10 V peak, with $\pm$ 15-V supplies, and with the op amp connected as a voltage follower.

For a sinusoidal output waveform of the form

$$V_{out}(t) = E_o \sin \omega t \tag{7.32}$$

the peak output voltage swing is obtainable as long as the output rate of change dV_{out}/dt does not exceed the slew rate. From Eq. (7.32) one gets

$$\frac{dV_{out}}{dt} = E_o \, \omega \cos \omega t \tag{7.33}$$

which has its maximum value at $\cos (\omega t) = 1$, that is,

$$\left(\frac{dV_{out}}{dt} \right)_{max} = E_o \, \omega \tag{7.34}$$

The full-power bandwidth can be expressed from Eq. (7.34) as the frequency where the maximum rate of change of the output is equal to the slew rate. Setting $\omega = \omega_p$ in Eq. (7.34), one obtains

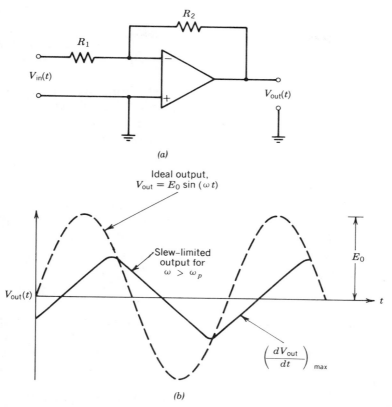

FIGURE 7.21. Slew-limited distortion resulting from large-signal operation beyond full-power bandwidth.

$$\omega_p = 2\pi f_p = \frac{\text{slew rate}}{\text{peak output swing}} \tag{7.35}$$

Once the full-power bandwidth is reached, the output waveform becomes distorted. Since the output rate of change is no longer sufficient to follow the signal swing, the output waveform becomes "triangulated," as shown in Figure 7.21, and rapidly drops in amplitude with increasing frequency.

For 741-type general-purpose op amps, with a slew rate of 0.5–1 V/μsec, the full-power bandwidth is on the order of 8–15 kHz for $\pm$ 10-V output swing. As indicated by Eq. (7.35), the only way to extend the full-power bandwidth is by improving the slew rate.

7.5. INPUT STAGE DESIGN

The input stage is by far the most critical section of a monolithic op amp, since it determines the majority of the key performance parameters for the amplifier. Over the years, a wide range of input stage circuit configurations have been developed for monolithic op amp design which efficiently utilize the inherent advantages of monolithic components in optimizing the circuit performance.[17,18]

The basic op amp input stage is a differential gain block similar to those described in Section 5.1. To be suitable for op amp design, the basic differential gain stage must fulfill some of the important requirements listed below. Note that, when applicable, the "bare minimum" requirements are also listed in parentheses as typical upper or lower bounds.

1. High input resistance (> 100 kΩ).
2. Low input bias current (< 500 nA).
3. Small input voltage and current offset.
4. High common-mode rejection ratio (> 60 dB).
5. High common-mode range (≥ one-half of total supply voltage).
6. High differential input range (≥ one-half of total supply voltage).
7. High voltage gain (> 40 dB).

A well-designed op amp input stage normally offers performance characteristics far in excess of those listed above.

The first four requirements stem from the basic ones associated with op amp performance discussed in the previous sections. The fifth and sixth requirements ensure the large input signal handling capability of the circuit. The last requirement, high voltage gain, is necessary for the input stage for two reasons: (1) it decreases the gain requirements associated with the successive stages, and (2) it reduces the contribution of the second-stage imbalance to the effective input offset. The small current- and voltage-offset requirement also dictates that the input devices should be matched in size and geometry, and be located in close proximity on the chip to ensure close thermal coupling.

In the following sections, as we examine various input stage circuits, their advantages and limitations, one important point will be clear to the reader: there is no "ideal" or "optimum" input stage. Each configuration has its own unique advantages and drawbacks. So the best that designers can do is to make an intelligent choice of many of the trade-offs involved and try to reach an "optimum" design for their particular requirements.

Basic Circuit Configurations

Virtually all of the differential gain stages described in Section 5.1 can be used as an op amp input stage, provided that the component and bias values are chosen to satisfy the important requirements, such as input impedance, offset, gain, and common-mode range. Since the basic characteristics of these circuits have already been discussed in Chapter 5, in this section we shall only review them briefly in terms of their applications in op amp design.

pnp **Input Stages.** Input stages using lateral *pnp* transistors as input devices, with *npn* current mirrors as active loads, are some of the simplest input stage configurations often used in medium-performance low-cost op amp designs. Such a circuit configuration is essentially the *pnp* input equivalent of the circuit configurations shown in Figures 5.17 and 5.18. This type of input stage provides convenient dc level shift of the signal toward the negative supply and makes it easy to interface with the next stage (see Fig. 7.14). It can also be designed with split-collector *pnp* transistors to reduce the input stage transconductance and the size of the compensation capacitance needed (see Fig. 7.15).

When used with a Darlington connection (see Fig. 5.22), the *pnp* input stage can have its common-mode range extended all the way to $-V_{EE}$, which makes it particularly suitable for operation with a single power supply.

Compared to the *npn* input differential stages, the *pnp* input circuits have poorer frequency response and higher input bias currents due to the low f_T and β_o of the lateral *pnp* transistors. The low *pnp* β_o problem can be avoided by using a composite *pnp–npn* input stage similar to that shown in Figure 5.21. However, this requires additional biasing circuitry and had relatively poor offset voltage characteristics due to the use of composite devices at the input.

npn **Input Stages.** Most high-speed and high-performance op amps use *npn* transistors as input devices, with either active or passive loads. Unless special performance requirements, such as very low input noise or very high slew rate, are required, active loads are preferred at the input since they provide a very high amount of voltage gain and excellent common-mode rejection. Figures 5.14, 5.18, 5.19, and 5.20 show typical input stage configurations with *npn* input devices and *pnp* current mirror loads. Except for Figure 5.19, these circuit configurations need dc level shifting before interfacing with an *npn* input second stage.

The device matching and temperature tracking characteristics of *npn* transistors are better than those of lateral *pnp* devices. Therefore, the offset voltages and offset currents of *npn* input circuits are generally superior to their *pnp* counterparts, and since the *npn* β_o is much higher than the *pnp* gain, the input resistance and input bias current levels can be made much higher.

For precision op amps, where input offset voltage and noise must be kept extremely low, *npn* input stages with resistive loads are used. In this manner, the trimming of the offset voltage can be simplified, and the additional noise component due to the current mirror loads can be avoided.

JFET Input Stages. The key advantages of JFET input stages are their very low input bias currents and low transconductance. JFET input devices are normally used in conjunction with bipolar current mirror loads, as shown in Figure 5.23. Although either *p*- or *n*-channel JFET devices can be used at the input, *p*-channel devices are normally preferred since they can operate with *npn* current mirror loads to provide appropriate dc level shift at the input. The low transconductance of JFET devices makes them well suited for high-slew-rate op amp design, as discussed in Section 7.4.

Compared to bipolar input devices, JFETs have several disadvantages. Their offset voltages as well as their offset drift characteristics are approximately 3–5 times worse than those of bipolar transistors, and their noise characteristics are also inferior to those of bipolar transistors. Another limitation on the performance of JFET op amps is the increase of input bias current with temperature: JFET input currents *double* for every 10°C change in temperature, whereas bipolar input currents tend to actually *decrease* with temperature due to the increase of β_0 with temperature. Figure 7.22 gives a comparison of JFET and

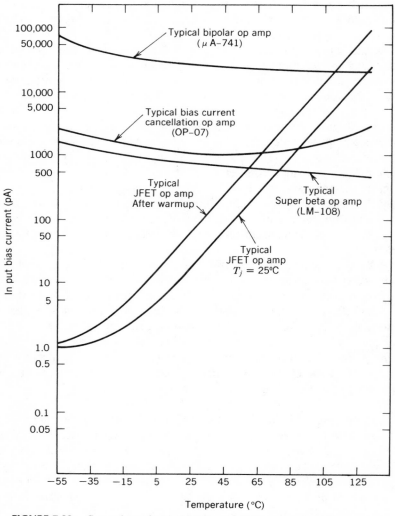

FIGURE 7.22. Comparison of typical bipolar and JFET op amp input bias currents.

bipolar op amp input currents for typical general-purpose op amps as a function of temperature.[19] Compared to a standard 741-type op amp, the JFET op amp offers approximately 1000:1 improvement at room temperature, yet this improvement is virtually gone at 125°C. Compared to superbeta input op amps, JFET input currents are actually higher for operation at temperatures in excess of 60°C. In fact, even the self-heating of a JFET chip during normal operation causes a significant deterioration of its bias current characteristics.

The inclusion of JFET devices on the same chip along with the bipolar transistors requires additional critical processing steps. However, with the advent of modern device fabrication techniques, such as ion implantation, the fabrication of JFET and bipolar devices on the same chip has been greatly simplified.

Low Bias Current Input Stages. In many of the op amp applications interfacing with low-level measuring instruments, such as photodetectors, or high-impedance transducers, the input bias currents associated with the op amp must be kept as low as possible. If the input offset voltage requirements are not a problem, this can often be achieved using JFET input stages. However, the requirement of low offset voltages, along with low input bias currents, can only be met by using special bipolar circuit configurations as input devices, such as superbeta transistors, or using current-cancellation techniques to minimize external bias current (see Fig. 7.24).

Superbeta Input Stages. The superbeta transistors described in Section 2.2 can be used as input devices for op amps. These transistors can be fabricated simultaneously with conventional *npn* transistors, with the addition of one extra diffusion step. These transistors exhibit β_o values of 1000–3000, but have very low breakdown characteristics, typically on the order of 3–5 V (see Fig. 2.19). Thus, the total voltage swing across them must be limited.

Figure 7.23 shows a practical input stage configuration using superbeta transistors along with regular *npn* transistors.[20] To avoid the low breakdown limitation associated with superbeta transistors Q_1 and Q_2, they are connected in a cascode configuration with conventional *npn* devices. The common-mode feedback network formed by the bias string Q_5 and Q_6 "bootstraps" the voltage swing across the superbeta transistors Q_1 and Q_2 to less than 1 V_{BE} under all input conditions. Thus, the cascode-connected transistor pairs Q_1, Q_3 and Q_2, Q_4 correspond to a single *npn* device with the breakdown characteristics of the conventional *npn* transistor Q_3 and the β_o of the superbeta transistor Q_1. Using this type of input stage, input bias currents can be kept in the few nanoampere range over the entire operating temperature, while maintaining offset voltages in the $\leq \pm$ 1-mV range. Since β_o has a positive temperature coefficient, the input bias currents actually tend to decrease with temperature. Thus, as shown in Figure 7.22, the superbeta input stages can provide lower input bias currents than JFET input stages for operation at elevated temperatures.

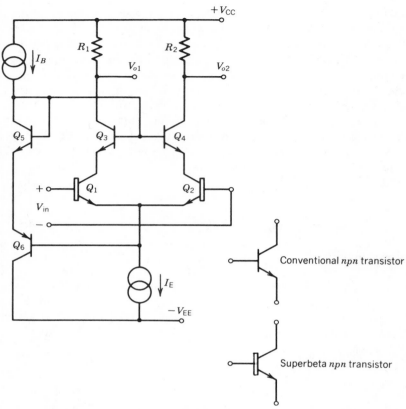

FIGURE 7.23. Input stage configuration using superbeta transistors.[20]

Bias Current Cancellation.   The input bias current of bipolar op amps is equal
to the input stage collector current divided by the β_0 of the input transistor. If
one can "measure" the quiescent collector current, scale it down by a factor of
β_0, and feed it back to the input terminal to subtract from the input bias current,
one can, at least in principle, cancel out the bias current. Figure 7.24 shows a
practical circuit configuration which makes use of this "cancellation" concept.
Assuming that the transistor α is very close to unity, the base currents of the
cascode-connected devices Q_3 and Q_4 duplicate the base currents of Q_1 and Q_2,
respectively. These base currents are then sensed by the *pnp* current mirrors Q_5,
Q_6 and Q_7, Q_8 and are fed back to the input bases as currents I_1 and I_2. Thus,
the total external bias currents I_{in1} and I_{in2} required are

$$I_{in1} = I_{B1} - I_1 \quad \text{and} \quad I_{in2} = I_{B2} - I_2 \qquad (7.36)$$

Thus, if the transistors are well matched, I_1, I_2, I_{B1}, and I_{B2} would all be very
nearly equal to each other, and the input bias currents will be nearly canceled
out.

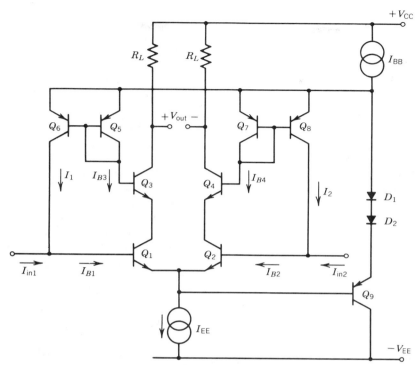

FIGURE 7.24. Typical input bias current cancellation circuit.[21]

In the circuit of Figure 7.24, the bias current source I_{BB}, along with the diodes D_1 and D_2 and the transistor Q_9, form a "bootstrap" bias circuit similar to that described in Figure 7.23, and limit the common-mode voltage swing across the bias cancellation circuit.

Typically, the biggest contributor to cancellation errors in the circuit is the low β_0 of the *pnp* current mirror transistors, which limits the total bias current reduction obtainable by this technique to approximately 10% of the uncanceled current. It should also be pointed out that the bias current cancellation technique described above reduces *only* the input bias currents and not the input offset currents. The input offset current, which is the difference between I_{in1} and I_{in2} in Eq. (7.36), is actually *degraded* by a factor of 2–4 due to the additional mismatches introduced by the bias cancellation circuitry. In such circuits, typical values of the bias and offset currents are approximately equal and are normally in the 1–5 nA range (see Fig. 7.22).

Low-Offset Input Stages. In low-level dc signal detection applications, the major limitation to the op amp performance is the input offset voltage. When the input signal amplitude becomes comparable to the op amp offset voltage, the circuit is no longer able to distinguish the signal from the inherent offset voltages.

As described earlier (see Section 5.1), almost all of the offset voltage is due to the device and biasing mismatches in the op amp input stage. This offset voltage can be minimized by one or both of the following methods:

1. By careful design and layout of the input stage to minimize component mismatches.
2. By use of on-chip or off-chip trimming techniques.

In a precision op amp circuit, both of these steps will be used. In other words, one would minimize the initial offset by proper choice of the circuit configuration and layout of the input stage, and then reduce it further by trimming techniques. Trimming techniques primarily reduce the initial offset, but not the offset drift with temperature and time. Therefore, in order to minimize the offset characteristics with component trimming, it is essential that the proper design precautions be taken to minimize the initial offset voltage prior to the trimming operation.

Layout and Device Matching Considerations

The primary sources of offset voltage are the mismatches between the supposedly identical resistors and transistors due to the random (i.e., statistical) variations during the device fabrication processes. These result from two factors:

1. Variations due to mask resolution. This causes relatively poor edge definition during the photomasking process and would result in emitter-area mismatches in transistors and length-to-width mismatches on resistors.
2. Variations due to diffusion processes. The sheet resistance and junction depth may vary slightly across the wafer surface, due to irregularities or nonuniform process conditions, during the impurity predeposition or diffusion steps.

In addition to the above random sources of mismatches, "systematic" errors may also occur due to thermal gradients or the careless layout of devices. However, these can be eliminated by proper design and layout practices and will not be considered separately.

The effects of mask resolution errors can be reduced by increasing the device geometry and size. In this manner, for example, if the transistor emitter area is made larger, or if a resistor is made wider, then the irregularities in the edge definition due to poor mask resolution would have a lesser effect. In the case of a transistor, this effect is also enhanced when sharp corners are avoided in the device layout, particularly in the emitter area, by using circular or oval emitters.

Figures 7.25 and 7.26 show experimentally observed improvements in device

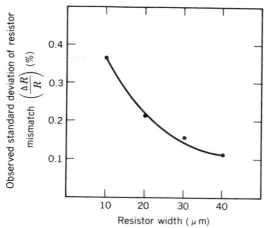

FIGURE 7.25. Experimentally observed standard deviation of mismatch distribution in ion-implanted resistors. (Sheet resistance 500 $\Omega/\square$ resistor length 500 μm.)[2]

matching due to mask-resolution-related mismatches.[2] In both cases, the mismatches are given in terms of the anticipated standard deviation, the sigma limit.

The mismatches due to process-related gradients, or thermal gradients and mask pattern misalignments across the chip, can be partially eliminated by a symmetrical device placement technique using the so-called *common-centroid layout*. Such a layout or placement provides a multifold symmetry for the devices involved such that various gradient or mask misalignment effects would nearly cancel out.

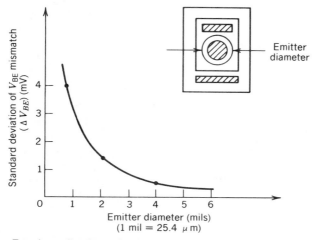

FIGURE 7.26. Experimentally observed standard deviation of V_{BE} mismatch of identical-geometry transistors as a function of circular emitter diameter.

Figure 7.27 illustrates an example of such a common-centroid layout for the case of input transistors. In this case, each transistor is split into two identical devices in parallel, which are then located diametrically from each other. The resulting "quad" of transistors then exhibits a multifold symmetry about the center point $\times$, shown in the layout. In this manner, the gradients of resistivity or temperature as well as the mask alignment effects, would tend to *subtract* from each other between the diametrically opposite halves of the device. Thus, using device geometries with emitter diameters in the 2–3 mil range and a common-centroid layout, input offset voltages can be held to within $\leq \pm 1$ mV.

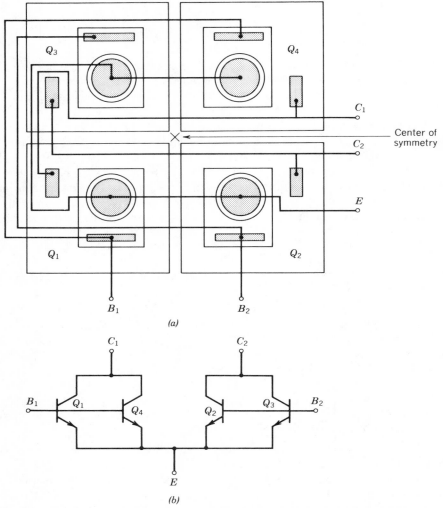

FIGURE 7.27. Typical common-centroid layout for emitter-coupled pair; (*a*) Device layout; (*b*) equivalent connection diagram.

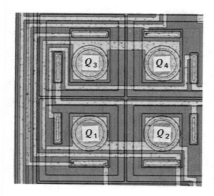

FIGURE 7.28. Photomicrograph of a cross-coupled in-put stage using common-centroid layout topology of Figure 7.27. *(Photo:* Precision Monolithics, Inc.)[21]

Figure 7.28 shows a closeup photomicrograph of a cross-connected input stage using the common-centroid layout.[21]

Offset Trimming Techniques

The initial offset voltage due to the statistical process-related variations can be reduced by approximately another order of magnitude by on-chip trimming techniques. This is normally done during the wafer probing stage, where the offset voltage of each chip is measured and then automatically trimmed by means of a laser trim or by open-circuiting or short-circuiting the *trimming links* built into the circuit. Normally, trimming with links is preferred since this allows one to use diffused or ion-implanted resistors as passive loads.

Figure 7.29 shows a simplified circuit schematic of an input stage designed for on-chip trimming of the initial offset voltage using trimming links. Normally, one would design the resistively loaded gain stage to have a nominal voltage gain in the range of 20–100. This is done by the choice of matched load resistors R_C. Then two identical, binarily weighted trimming resistor ladders, made up of multiples of low-value resistors of unit value R', are connected in series with the load resistors. R' is chosen to be several orders of magnitude smaller than R_C. The segments of the trimming ladder are either selectively short-circuited (by using Zener zapping) or selectively open-circuited (using fusible links) during the wafer-probing stage, in order to trim the value of V_{OS}. Since all the resistors in the collector circuit are of the same type and track with temperature, such a trimming technique is far more stable with temperature than off-chip trimming with an external potentiometer.

The methods of resistor trimming using either fusible or shortable links was covered in Section 3.9. Using this technique, monolithic op amps with worst case offset voltage of ± 25 μV and offset voltage drifts of ± 0.6 μV/°C and ± 1 μV/month have been reported.[22]

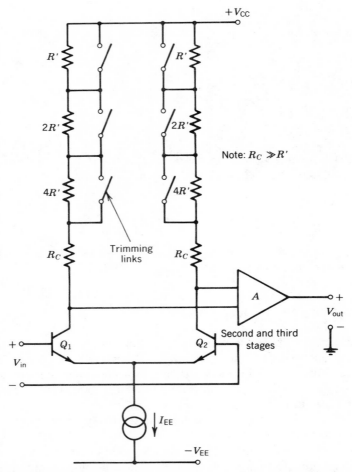

FIGURE 7.29. On-chip trimming of input stage offset using fusible or shortable trimming links.

7.6. PRACTICAL OP AMP CIRCUITS

Over 200 different monolithic op amp circuits are commercially available at the present time. By far the largest group of present-day monolithic op amps are designed as general-purpose devices suitable for a variety of applications. Although a large number of different designs exist for general-purpose op amps, most of these can be related to one or the other basic designs which have become the industry standards. In this section, the circuit design, the layout, and the electrical characteristics of some of these op amps will be examined, and relative merits and shortcomings of each will be compared. All of these circuits are chosen because they represent significant "milestones" in the field of monolithic op amp design, and because they provide excellent examples of some of the basic design techniques and configurations discussed earlier in this chapter.

μA-741 General-Purpose Op Amp

The μA-741 general-purpose op amp introduced by Fairchild Semiconductor in 1966[23] has become one of the most widely accepted op amp products. It was also one of the first op amp configurations to utilize the basic two-stage op amp architecture shown in Figure 7.8, with internal (i.e., on-chip) compensation capacitor. Its popularity stems from the relative simplicity of its design, which makes it fit on a relatively small chip (typically on the order of 40–55 mils square, depending on the particular device geometries or the layout rules used), which can be manufactured with high yields. It is internally compensated and provides a wide common-mode range, and a voltage gain in excess of 100 dB.

The simplified circuit schematic of the μA-741 op amp is shown in Figure 7.8b. Its complete circuit diagram is given in Figure 7.30. The input stage is comprised of the composite npn–pnp transistor pairs Q_1, Q_3 and Q_2, Q_4 and the npn current mirror made up of Q_5, Q_6 and Q_7. The basic biasing arrangement and

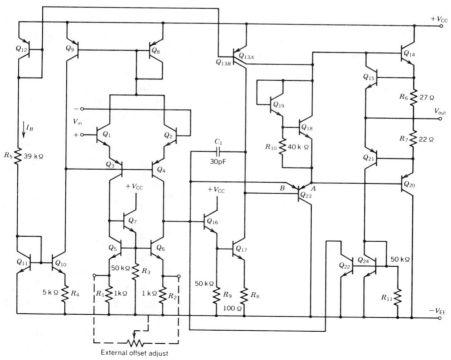

FIGURE 7.30. Circuit schematic of μA-741 monolithic op amp.

the principle of operation for such a gain stage were discussed in detail in Section 5.2 (see Fig. 5.21) and will not be repeated here. In the μA-741 design, the input stage is biased at a nominal quiescent current level of $\approx$ 20 μA through the diode-connected transistor Q_8. This current level is set by resistor R_5, and the low-current current source is made up of Q_{11} and Q_{10} (see Fig. 4.8). The bias current I_B through R_5 is also used for setting the bias current levels in the second stage and the output stage through the multiple-collector *pnp* transistor Q_{13}.

The second stage is formed by the Darlington input transistors Q_{16} and Q_{17} and the current source Q_{13B}, which serves as the active load. For $\pm$ 15-V operation, the split-collector transistors Q_{13B} and Q_{13A} are designed to carry nominal bias currents of 550 and 180 μA, respectively. Frequency compensation is achieved by a single 30-pF compensation capacitance C_1, connected in feedback around the second stage.

The first stage has a voltage gain of approximately 500 (54 dB) at the collector of Q_6, and the second stage provides another 50–60-dB additional gain, giving a total voltage gain in excess of 100 dB at the collector of Q_{17}.

The output is a Class B output stage driven by the *pnp* emitter follower Q_{23}. The quiescent bias is provided by the split collector of Q_{13}, and the Darlington diode-connected transistors Q_{18} and Q_{19} serve as the crossover compensation diodes (see diodes D_1 and D_2 in Fig. 5.32). The output drive is then provided by high-current transistors Q_{14} and Q_{20}, each of which is designed to handle at least 20 mA of load current. Q_{20} is designed as a substrate *pnp* transistor.

The extra emitter shown on Q_{23} is essentially a clamp diode which prevents Q_{23} from going into saturation by stealing base current from the base of Q_{16}. In normal circuit operation, this diode is reverse biased and is out of the circuit. Q_{23} cannot be allowed to saturate, since its saturation will cause Q_{17} to saturate, and this will in turn cause the current in Q_{16} to increase drastically.

Output short-circuit protection is provided by the active current-limiting circuit made up of Q_{15} and R_6 for positive swings of the output voltage (see Fig. 5.36). For negative swings, the output current limiting is provided by R_7 along with Q_{21}, Q_{24}, and Q_{22}. When Q_{21} conducts, Q_{24} and Q_{22} function as a current mirror and steal base current away from the base of Q_{16} to shut off the second stage.

The input stage common-mode range is nearly equal to V^+ for positive common-mode inputs, and is approximately $3V_{BE}$ above $-V_{EE}$ for negative common-mode inputs. The output can swing to within 1 V_{BE} of the supply voltages. Typical values of input offset voltage are in the $\pm$ 2–3 mV range and can be nulled by an external potentiometer, as shown in Figure 7.30.

Figure 7.31 shows a photomicrograph of the μA-741 op amp chip. The locations of various circuit terminals and key components are also identified. Note that the inputs and outputs are placed at the opposite ends of the chip to minimize thermal coupling or undesired feedback effects. Also, the positive and negative supply terminals are located between the outputs and inputs in order to minimize parasitic feedback from the output to the input due to the package pin capacitances.

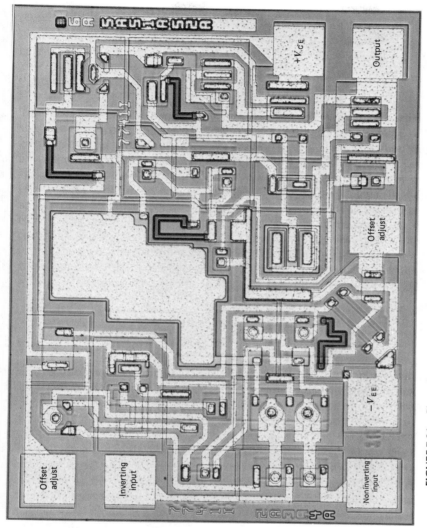

FIGURE 7.31. Photomicrograph of μA-741 op amp. Chip size: 38 mils $\times$ 50 mils. (*Photo:* Fairchild Semiconductor.)

353

LM-108 Superbeta Op Amp

The LM-108 op amp introduced by National Semiconductor Corporation in 1969[20] is an excellent example of using superbeta transistors at the input stage of an op amp to reduce the input bias current requirements. The simplified diagram of the LM-108 circuit is shown in Figure 7.32 in terms of its essential components. The input stage uses the bootstrapped circuit configuration shown in Figure 7.23, which combines the superbeta devices Q_1 and Q_2 in a cascode connection with conventional *npn* transistors, Q_5 and Q_6. The common-mode feedback formed by the combination of current source I_1, and transistors Q_3 and Q_4, limits the total voltage across Q_1 and Q_2 to approximately 1 V_{BE} drop. Q_1 and Q_2 are operated at very low collector currents (≈ 3 μA per transistor) in order

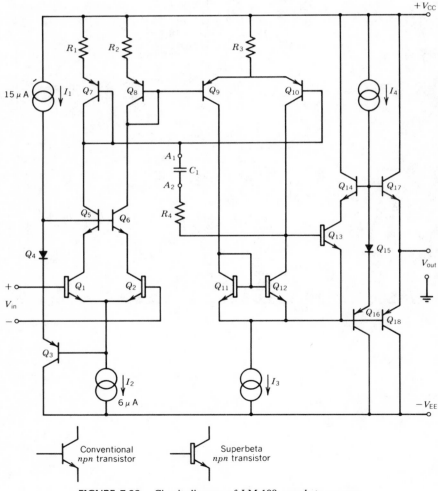

FIGURE 7.32.   Circuit diagram of LM-108 superbeta op amp.

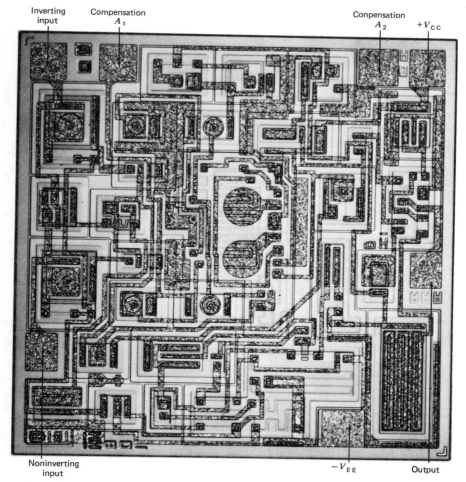

Inverting input

Compensation A_1

Conpensation A_2

$+V_{CC}$

Noninverting input

$-V_{EE}$

Output

FIGURE 7.33. Photomicrograph of LM-108 superbeta op amp. Chip size: 59 mils × 59 mils. (*Photo:* National Semiconductor Corporation.)

to minimize the input bias currents. Since superbeta transistors can maintain current gains in excess of 1000 at such low currents, this results in typical input bias currents of 2–3 nA.

The second stage is formed by the lateral *pnp* differential pair Q_9 and Q_{10}, which uses a superbeta current mirror made up of Q_{11} and Q_{12} to provide both the voltage gain as well as differential to single-ended conversion. The current source I_3, along with the bootstrapped superbeta transistor Q_{13}, forms a unity-gain buffer stage in conjunction with the substrate *pnp* transistor Q_{16} and the bias current source I_4. This buffer stage then provides the drive for the Class AB output stage made up of Q_{17} and Q_{18}.

The LM-108 op amp is not internally compensated. It requires an external 30-pF capacitor C_1, connected as shown in Figure 7.32. Figure 7.33 shows a photomicrograph of the LM-108 op amp chip.

LM-118 High-Speed Op Amp

The LM-118 op amp introduced by National Semiconductor Corporation in 1971 is a good example of an op amp design for high-frequency capability and high slew rate.[13] It is basically a three-stage op amp which uses feedforward compensation around the low-frequency lateral *pnp* gain stages to enhance the high-frequency capability of the circuit.

The basic architecture of the circuit is shown in Figure 7.16. A somewhat more detailed circuit diagram of the LM-118 is given in Figure 7.34. The transistors Q_1 and Q_2 form a conventional differential input stage with the emitter degeneration resistors R_1 and R_2 and the collector loads R_3 and R_4. This stage provides a nominal dc voltage gain of 20. The slew rate is enhanced by the relatively high first-stage current and by the presence of the emitter degeneration resistors (see Section 7.4). Q_3 and Q_4 form the second gain stage, which also

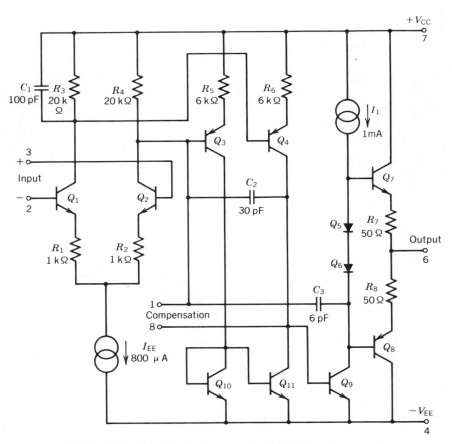

FIGURE 7.34. Simplified circuit schematic of LM-118 high-speed op amp.

provides a dc level shift toward $-V_{EE}$ as well as differential to single-ended conversion. The third gain stage is formed by Q_9 with an active load and the crossover compensation diodes Q_5 and Q_6. The output is a conventional Class AB push–pull stage, with a substrate pnp transistor for Q_8. The frequency compensation is achieved by the feedback capacitor C_3 around the second and third gain stages, and C_2 provides the feedforward function around the low-frequency pnp transistors Q_3 and Q_4. Since the feedforward path is single ended, rather than differential, the capacitor C_1 rolls off one-half of the first stage gain such that at the frequencies where the feedforward is effective, the output of the first gain stage appears to be single ended (i.e., the base of Q_4 is at virtual ground). The collector of Q_9 is also brought out separately, along with the base of Q_3, so that an additional external capacitor can be added in parallel with the feedforward capacitor C_2 to improve the settling time. As discussed in Section 5.4, relatively poor settling time is a problem common to most op amp designs which use feedforward compensation,[8,14] and the LM-118 op amp is no exception to this problem.

The complete circuit schematic of the LM-118 op amp is given in Figure 7.35. Compared to the simplified schematic of Figure 7.34, a few additional points should be noted. In the actual circuit, a Darlington-connected input stage is used to reduce the input bias current requirements. Also, diode-connected transistors are used as clamps across the input terminals and across the collectors of Q_1 and Q_2 at the input stage to limit the differential voltage swings.

The inverted transistor Q_5, which provides the current bias for the outer Darlington pair Q_1 and Q_2 of Figure 7.35, is actually a multiple-emitter transistor operated in the inverted configuration as a multiple-collector device (see Fig. 2.22). This configuration is used to reduce the parasitic capacitances introduced by Q_5. The third-stage gain transistor Q_9 of Figure 7.34 is actually a Darlington device in order to minimize interstage loading between the second and third gain stages. The JFET Q_8 in Figure 7.35, which is used as a high-value resistor, is actually an epitaxial pinched resistor of the type shown in Figures 3.29 and 3.30.

The most important features of the LM-118 op amp are its high slew rate ($\approx$ 50 V/μsec) and its wide unity-gain bandwidth ($\approx$ 10 MHz), which are achieved without requiring special manufacturing processes for high-frequency pnp transistors or other special circuit components.

HI-2510 High-Speed Op Amp

The availability of high-performance pnp transistors with current gain and high-frequency characteristics similar to the npn devices on the chip can greatly simplify circuit design and enhance performance. The high-performance pnp transistors can be fabricated simultaneously with npn bipolar devices by using the dielectric isolation process described in Section 1.10. The resulting device structure is shown in Figure 2.31. Although this process is relatively complicated in terms of the necessary manufacturing steps, it can provide vertical pnp transistors with $f_T \approx 250$ MHz and $\beta_0 > 50$. The availability of such high-

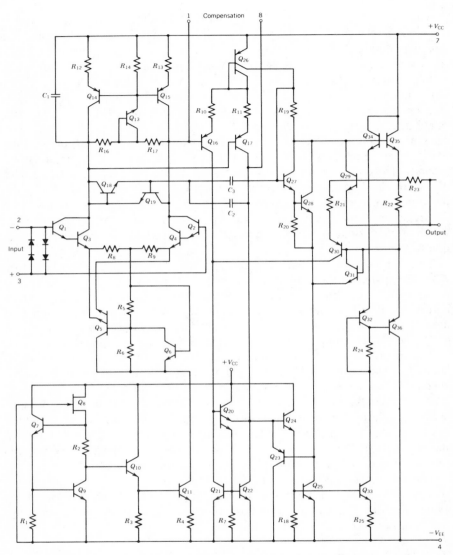

FIGURE 7.35. Complete circuit schematic of LM-118 high-speed op amp.

frequency *pnp* transistors greatly simplifies op amp design and allows one to obtain high slew rates and unity-gain bandwidths without requiring feedforward compensation. The HI-2510 high-speed op amp introduced by Harris Semiconductor Corporation[7] is a good example of such a "high-technology" design, where the circuit performance is achieved by a relatively simple design at the expense of manufacturing process complexity, relying heavily on the availability of high-performance complementary transistors.

Figure 7.36 shows the circuit schematic of the HI-2510 op amp. The circuit has the single-stage op amp architecture shown in Figure 7.7. The input is an *npn* Darlington differential stage with cascode-connected *pnp* level-shift transistors Q_5, Q_7 and Q_6, Q_8 feeding into a cascode-connected active load stage made up of Q_{27}, Q_{28}, Q_{29}, and Q_{30}. The very high output resistances of these cascode level-shift and active load stages result in a total voltage gain of approximately 80 dB at the output of the gain stage (i.e., at the base of Q_{26}). The feedback from the emitter of Q_{26} to the base of Q_{27} greatly enhances the impedance level at the input of the unity-gain buffer output stage, which is then followed by the emitter-follower stages of Q_{33} and Q_{34} that drive the complementary output transistors.

The internal capacitor C_1 rolls off one-half of the differential gain at high frequencies and makes the signal path single ended. C_2 provides the actual frequency compensation. Since only a single gain stage is employed, compensation can be achieved by a relatively small value of C_2 (≈ 10 pF).

The circuit provides a unity-gain bandwidth of 10 MHz with a slew rate of 50 V/μsec and a full-power bandwidth of 1 MHz. Compared to the LM-118 op amp described in the previous section, the HI-2510 op amp has somewhat lower gain (80 dB versus 100 dB). However, it has a more superior settling-time characteristic (typically 0.25 μsec versus several microseconds), since it does not use feedforward compensation.

At this point it is also instructive to compare the circuit approaches employed in the LM-118 and the HI-2510 high-speed op amps. The LM-118 design makes use of standard processing technology and relies on circuit complexity; the HI-2510 relies on high-performance complementary devices obtainable by a fairly complex manufacturing process and uses a relatively simple circuit configuration.

LF-156 FET Input Op Amp

As described in Section 7.5, JFET input stages offer two significant advantages over bipolar inputs: (1) higher slew rate due to the low g_m/I_1 ratio of the JFET devices, and (2) low input bias currents due to the reverse-biased gate–channel junction of a JFET. The ion-implantation technology has made it relatively simple to build bipolar compatible JFET structures which exhibit matching, tracking, and breakdown voltage characteristics compatible with those of conventional bipolar devices. The LF-156 JFET input op amp introduced by National Semiconductor Corporation in 1974 is a good example of a monolithic design which illustrates the use of the bipolar compatible JFET technology in high-speed op amp design.

Figure 7.37 shows the simplified circuit schematic of the input stage and the bipolar second stage for the LF-156 op amp. The circuit uses ion-implanted *p*-channel JFETs Q_1 and Q_2 (see Fig. 2.37) as input devices as well as differential JFET current-source loads Q_3 and Q_4. The use of JFET rather than bipolar current-source loads in the first stage results in lower input noise and offset

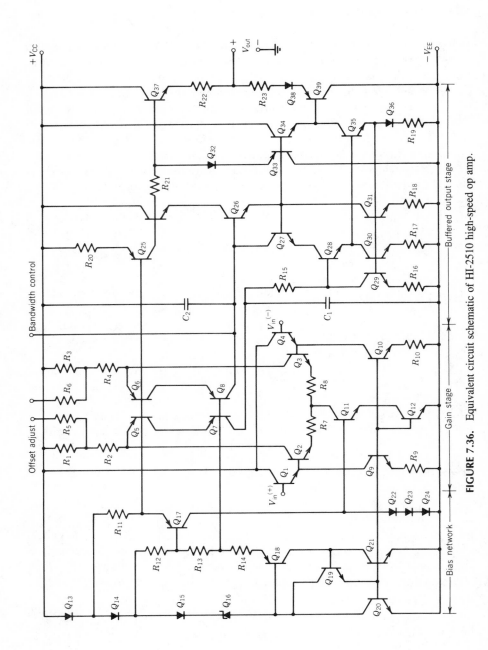

FIGURE 7.36. Equivalent circuit schematic of HI-2510 high-speed op amp.

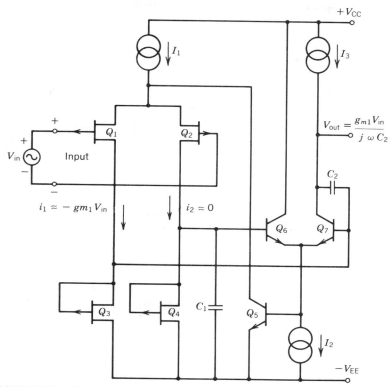

FIGURE 7.37. Simplified circuit diagram of gain stages of LF-156 JFET input op amp.

characteristics.[24] The capacitor C_1 is used to roll off one-half of the differential gain at the input stage to simplify overall compensation by means of a single feedback capacitance C_2 around the second stage.

An important feature of the circuit is the common-mode feedback loop formed by the bipolar transistor Q_5, which regulates the input stage bias current I_1 by sensing the voltage level at the emitters of the second-stage differential pair and clamping this voltage to $1V_{BE}$. In this manner, the circuit provides a very high degree of common-mode rejection (typically > 100 dB).

The complete circuit schematic for the LF-156 op amp is given in Figure 7.38. In the actual circuit, a Darlington input second stage is used to reduce the loading between the first and second gain stages. The substrate *pnp* transistor Q_{13} serves as a clamp to keep Q_{10} out of saturation during large voltage swings. The output is a quasi-complementary Class AB gain stage, where the combination of the *p*-channel JFET Q_{24} and the *npn* Darlington pair Q_{25}, Q_{26} behave as a *pnp* pull-down transistor. The circuit provides voltage gains in excess of 100 dB with 5-MHz unity-gain bandwidth and a slew rate in excess of 10 V/μsec. Typical input offset voltage and drift are on the order of 3 mV and 2 μV/°C, respectively.

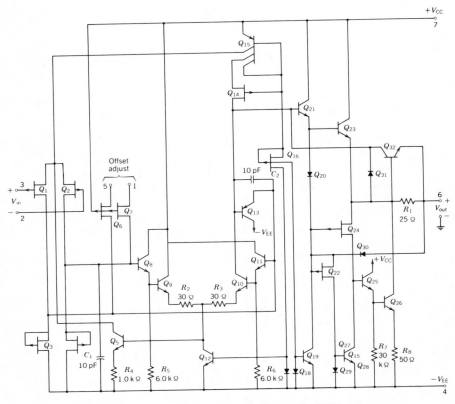

FIGURE 7.38.   Detailed circuit schematic of LF-156 JFET input op amp.

OP-07 Precision Op Amp

The OP-07 precision op amp circuit introduced by Precision Monolithics, Inc., in 1975 is a good example of monolithic op amp design which incorporates on-chip resistor trimming and bias current cancellation techniques to minimize both the input offset voltage and the input bias currents.[21] The simplified circuit schematic of the OP-07 op amp is shown in Figure 7.39.

Input Bias Current Cancellation.   Bias current cancellation at the input terminals is accomplished by using the circuit configuration shown in Figure 7.24. This cancellation reduces the input bias currents to approximately $\pm$ 4 nA. However, it has no effect on the offset current, which is of the same order of magnitude as the residual input current.

Offset Voltage Trimming.   The input offset voltage V_{OS} is measured and nulled by trimming the low-value resistors R_{2A} and R_{2B} during the wafer-probing stage, using the Zener-zapping technique (i.e., shortable links) to short-circuit selected

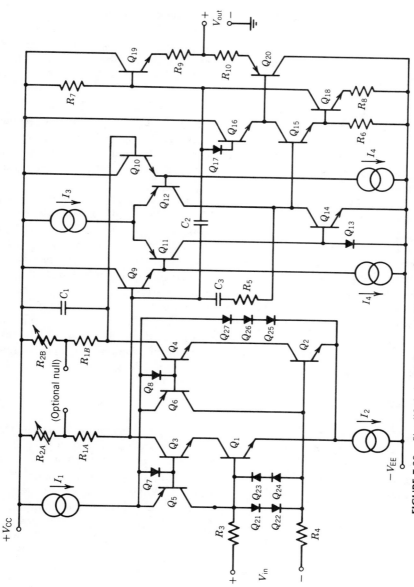

FIGURE 7.39. Simplified circuit diagram of OP-07 precision op amp. (*Note:* Resistors R_{2A} and R_{2B} are trimmed at wafer probe for minimum V_{OS}.)

resistor segments, as shown in Figure 7.29. In the case of the OP-07, the main gain resistors R_{1A} and R_{1B} are chosen to be 50 kΩ, and the trim resistors R_{2A} and R_{2B} are made up of a binarily weighted string of resistors with a unit resistor value of 230 Ω. In this manner, the initial offset can be trimmed to less than $\pm$ 50 μV, with a temperature drift of less than 0.6 μV/°C.

Circuit Operation.   The basic architecture of the OP-07 is that of a three-stage op amp shown in Figure 7.10. Other than the bias current cancellation and the offset trim capability, the input stage is a differential amplifier with resistive loads. The second stage is a *pnp* differential stage with a current mirror load. To avoid the loading between the first and second stages *npn* emitter-follower buffers are used.

The third gain stage is formed by the Darlington inverter made up of Q_{15} and Q_{18} with a load resistor R_7. Output is a Class AB stage with Q_{16} and Q_{17} setting the standby current at the output stage. Frequency compensation is achieved by C_2, with C_3 and R_5 providing the feedforward compensation. The capacitor C_1 rolls off one-half of the differential output of the first stage so that only the ac signal path with feedforward compensation is effective at high frequencies.

Figure 7.40 shows a photomicrograph of the OP-07 chip. The four pads in the upper left-hand corner of the chip are the trimming pads for R_{2A} and R_{2B}. The input transistor pair Q_1 and Q_2 is laid out on the left-hand side of the chip in the form of a common-centroid layout similar to that shown in Figure 7.27. This layout configuration minimizes the effects of thermal gradients due to the output transistors located on the opposite end of the chip.

A Comparison of Op Amp Characteristics

Table 7.1 shows a summary of some of the device and performance parameters of the practical op amp circuits reviewed in this section. The data have been compiled from manufacturers' data sheets or characteristic performance curves. In each case, whether a given parameter is typical or a guaranteed minimum or maximum specification, is indicated by (T) or (M). Since many grades of each amplifier type are available, only the prime grade was considered for comparison purposes. The reader should be cautioned that each manufacturer's measurement method may differ slightly, and, as such, the parameters listed in Table 7.1 must be used mainly for relative comparison purposes.

Each of the op amp design examples discussed in this section was chosen either because it has become a well-recognized industry standard, or because it is demonstrative of particular design or fabrication techniques.

The μA-741 is chosen because it has become the industry standard for low-cost medium-performance op amps in general-purpose circuit applications. The LM-118 represents a circuit design with conventional process technology which is optimized for high-speed operation; this is achieved by feedforward techniques and at the expense of power consumption and settling time. The HI-2510 is an example of what one can do if high-performance *pnp* transistors

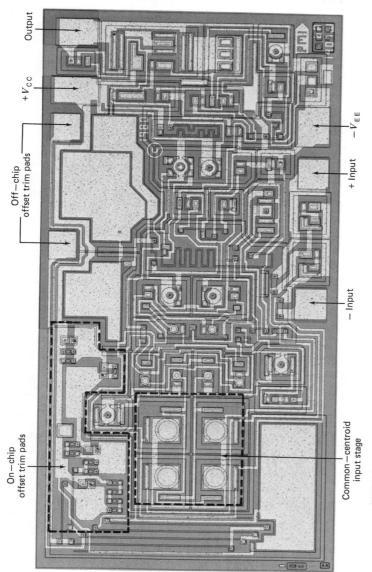

Output

$+V_{CC}$

Off—chip
offset trim pads

On—chip
offset trim pads

$-V_{EE}$

+ Input

— Input

Common—centroid
input stage

FIGURE 7.40. Photomicrograph of OP-07 op amp. Chip size: 100 mils × 53 mils. *(Photo: Precision Monolithics, Inc.)*

365

TABLE 7.1. Electrical Characteristics of Various Monolithic Op Amp Types[a]

Device Parameter	Units	μA-741	LM-118	Hi-2510	LF-156	LM-108	OP-07
Offset voltage V_{OS}	mV	5 (M)	4 (M)	8 (M)	5 (M)	2 (M)	0.07 (M)
Offset voltage drift dV_{OS}/dT	$\mu V/°C$	40 (T)	30 (T)	50 (T)	5 (T)	15 (M)	0.6 (M)
Input bias current, I_B	nA	500 (M)	250 (M)	200 (M)	0.05 (M)	2 (M)	4 (M)
Input offset current I_{OS}	nA	200 (M)	50 (M)	25 (M)	0.01 (M)	0.4 (M)	4 (M)
Open-loop voltage gain A_v	V/mV	50 (M)	50 (M)	10 (M)	50 (M)	25 (M)	300 (M)
Input resistance R_{in}	MΩ	0.3 (M)	1 (M)	50 (M)	10^6 (M)	30 (M)	15 (M)
Common-mode range	±V	12 (T)	11.5 (M)	10 (M)	11 (M)	13.5 (M)	13 (M)
Common-mode rejection	dB	70 (M)	80 (M)	80 (M)	85 (M)	85 (M)	106 (M)
Output swing ($R_L \geq 2$ kΩ)	±V	12 (M)	12 (M)	10 (M)	12 (M)	13 (M)	12.5 (M)
Supply current I_{CC}	mA	2.8 (M)	8 (M)	6 (M)	7 (M)	0.6 (M)	4 (M)
Slew rate	V/μsec	0.5 (T)	50 (M)	50 (M)	10 (M)	0.2 (T)	0.17 (T)
Unity-gain bandwidth	MHz	1 (T)	15 (T)	12 (T)	4 (T)	1 (T)	0.6 (T)
Full-power bandwidth	kHz	10 (T)	1000 (T)	1000 (T)	200 (T)	4 (T)	3.5 (T)
Settling time	μsec	1.5 (T)	4 (T)	0.25 (T)	1.5 (T)	1 (T)	1 (T)

[a]Test conditions: $T_A = 25°C$, $V_{CC} = +15$ V, $-V_{EE} = -15$ V. (T) designates typical value, (M) designates guaranteed minimum or maximum limit.

are available on the same chip, at the expense of process complexity. It is primarily optimized for bandwidth, slew rate, and settling time, at the expense of input offset, gain, and power dissipation characteristics.

The LF-156 represents a good example of the use of ion-implanted JFET transistors, together with conventional bipolars, to improve op amp bias current, slew rate, and bandwidth characteristics. The LM-108 is a precision op amp circuit designed for low input bias currents and low offset voltages with minimum power dissipation, and is an excellent example of the use of superbeta transistors in circuit design. The OP-07 demonstrates the use of input bias cancellation and the on-chip offset trimming techniques for achieving permanently-trimmed offset voltage levels of less than 100 μV.

These design examples illustrate that there is no such thing as the "optimum" op amp. Each design involves extensive trade-offs of two or more parameters against each other to obtain a particular "optimum design" for a specific purpose within the given device or fabrication process constraints.

Other Op Amp Configurations

Programmable Op Amps. In most op amp configurations, one can trade power dissipation for speed, that is, for bandwidth and slew rate. For example, by referring to Figure 7.30, one can see that by increasing the bias resistor R_5 in the circuit, one can reduce the reference bias current I_B and all the other bias currents derived from it. Since the compensation capacitor C_1 is fixed, reducing I_B causes the slew rate and the gain–bandwidth product to be reduced proportionally.

In the so-called programmable op amps, a separate bias setting pin is brought out, as shown in Figure 7.41, which can be used to set the internal bias currents by means of an external bias-setting resistor R_{ext}. In this manner, the

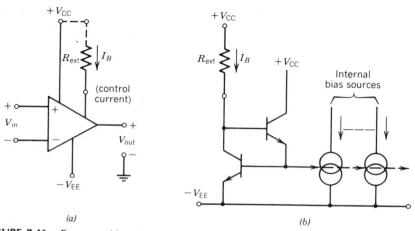

FIGURE 7.41. Programmable op amp configuration which allows the user to set up internal bias levels with external resistor.

gain–bandwidth products and the power dissipation can be "programmed" to a desired level. These programmable op amp circuits are particularly useful for very-low-power applications, where the power dissipation can be trimmed down, provided that the gain–bandwidth and the slew-rate requirements are not stringent.

Multiple Op Amps.   Since the op amp chip size is small (typically ≤ 2000 mil^2), several op amp circuits can be easily combined on a single chip. In this manner, they can also share the same IC package and thus reduce the manufacturing cost. Since for low-cost integrated circuits the package cost is a significant portion of the total cost of a finished device, this may result in significant cost savings. This approach is also practical from an applications point of view, since in many applications one uses not one but several op amps.

If an op amp is internally compensated and has no offset adjustment, it requires only three circuit terminals (i.e., two inputs and an output) in addition to the two power supplies. Thus, a dual op amp chip can be packaged easily in a low-cost 8-pin package; and a quad op amp chip can fit into a 14-pin standard DIP package. In this manner, the cost per op amp to the user can be reduced significantly.

A large number of dual or quad versions of the popular 741 type, or the JFET input type op amps are commercially available. Because of their popularity and widespread usage, the pin configurations of these dual and quad op amps have been standardized in the industry for 8-pin and 14-pin DIP package configurations, as shown in Figure 7.42.

7.7.   MOS OPERATIONAL AMPLIFIERS

As discussed in Chapter 6, MOS technology can also be applied to the design of analog circuits, particularly in the case of complex systems where both analog and digital functions must be combined on the same chip. The basic building blocks of analog MOS design were described in Sections 6.2, 6.3, and 6.4. In this section, the actual utilization of these building blocks in MOS op amp design will be examined.

In general, the performance characteristics of MOS op amps are relatively poor compared to their bipolar counterparts. However, they offer two very important practical advantages:

1.   They can be fabricated with standard, high-density digital LSI technologies.
2.   They require typically *one-third* to *one-fifth* the chip area of a comparable bipolar design.

Because of these two features, most MOS op amps are designed to be used as integral parts of a large monolithic LSI system, rather than being sold as com-

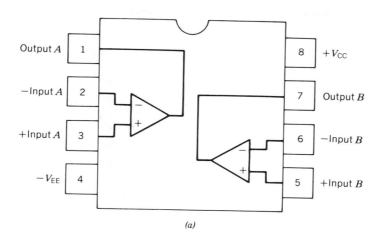

(a)

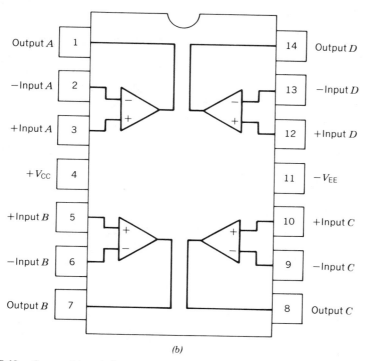

(b)

FIGURE 7.42. Commonly used pin-out configurations for dual and quad op amp integrated circuits.

mercial op amps. As such, the gain, offset, and frequency performance characteristics of many MOS op amps would not need to be as tight as those of
general-purpose bipolar op amps, such as those discussed in Section 7.6 (see
Table 7.1). In most system applications where the op amp is used as an integral
part of a monolithic LSI system, relatively modest performance characteristics
would suffice. Most of the op amp configurations which will be discussed in this
section are designed with this consideration in mind.[25]

CMOS Op Amp Design

Because of the availability of complementary symmetry devices, CMOS technology is by far the best suited MOS process for op amp design. Most CMOS
op amps use the standard two-stage op amp configuration shown in Figure 7.8a.
In many cases where the output loading is negligible, a separate buffer output
stage is not used, but the output is taken directly from the second gain stage.

Figure 7.43 shows a simplified circuit schematic of a CMOS op amp.[26] Q_1
and Q_2 form a differential gain stage, with Q_3 and Q_4 as the current mirror loads
in a circuit configuration similar to that shown in Figure 6.28a. Q_5 and Q_6 form
the second gain stage, which is essentially an inverter circuit with a current-
source load similar to that shown in Figure 6.26. Frequency compensation is
provided by the pole-splitting capacitor C_1. In many applications, it is advantageous to buffer C_1 from the output by means of a source follower. In the circuit
of Figure 7.43, this is done by means of the source follower Q_6 with a current-
source load Q_8. The buffering of the compensation capacitance from the output
avoids parasitic feedthrough signal transmission through C_1 at high frequencies

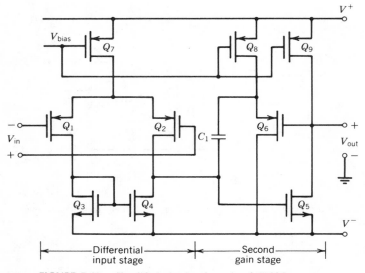

FIGURE 7.43. Simplified circuit schematic of CMOS op amp.

TABLE 7.2. Comparison of Relative Performance Characteristics of CMOS and NMOS Op Amp Circuits Shown in Figures 7.43 and 7.45.

Device Parameter	Units	CMOS Op Amp of Figure 7.43	NMOS Op Amp of Figure 7.45
Open-loop voltage gain A_v	V/V	1500	2500
Input offset voltage V_{OS}	mV	20	25
Unity-gain bandwidth	MHz	3.2	3
Slew rate			
Positive slew	$V/\mu sec$	+8	+5
Negative slew	$V/\mu sec$	−33	−2
Common-mode rejection ratio	dB	60	72
Common-mode range			
($V^+ = V^- = 10$ V)	V	±7.5	+5, −8.5
Power dissipation	mW	22	15
Rms input noise	μV	51	80
Active chip area used	mm^2	0.35	0.32
Fabrication technology		Metal-gate CMOS	Metal-gate NMOS

and thus improves the op amp phase margin.[27] In circuit applications where capacitive loading at the output is minimal, this buffer stage may be eliminated by connecting C_1 directly to the drain of Q_5 and leaving Q_6 and Q_8 out of the circuit. Typical voltage gains available from a simple CMOS op amp circuit, of the type shown in Figure 7.43, are in the range of 1000–5000, with input offset in the range of ± 10 to ± 20 mV. Although modest by conventional bipolar op amp standards, these performance characteristics are adequate for many on-chip signal-processing applications. Table 7.2 gives some of the typical characteristics of the basic CMOS op amp circuit of Figure 7.43 and compares it with other comparable op amps fabricated with single-channel MOS technologies.

It should also be noted that the basic op amp circuit of Figure 7.43 is shown in its simplest form and is designed for driving high-impedance loads, such as the gates of MOS transistors or analog switches. If a low output impedance or high output current drive is required, a Class AB output stage, such as the one shown in Figure 6.32, or an output stage using the substrate *npn* transistor, can be utilized (see Fig. 6.33).

NMOS Op Amp Design

Single-channel technologies such as NMOS or PMOS are not well suited for analog design for two reasons: (1) due to a lack of complementary devices to use as active loads, and (2) due to parasitic feedback caused by the body effect.

These two limitations, combined with the low transconductance value inherent to FET devices, make it difficult to obtain substantial voltage gains from an amplifier stage. These problems were discussed in detail in Chapter 6. However, in spite of these drawbacks, one can still achieve acceptable op amp performance by combining some of the basic building blocks of NMOS design discussed in Section 6.2.

Figure 7.44 shows the basic architecture of a practical NMOS op amp circuit which offers acceptable circuit performance for on-chip signal-processing applications.[28] The detailed circuit schematic is shown in Figure 7.45. As described in Section 6.2, the device geometry ratios (i.e., the Z/L ratios of the MOS devices) are the main design parameters for single-channel analog MOS circuits, since all of the stage gains directly relate to the Z/L ratios [see Eq. (6.9)]. Therefore, in the circuit diagram of Figure 7.45, the Z/L ratios of the respective devices are also shown in parentheses.

With reference to Figures 7.44 and 7.45, the basic NMOS op amp design can be quantatively analyzed as follows. The input stage is a differential input, differential output gain stage with diode-connected NMOS transistors as active loads (see Fig. 6.15). The gain transistors Q_1 and Q_2 are designed to have a high Z/L ratio, whereas Q_3 and Q_4 are the so-called long-channel devices with a very small Z/L ratio. From Eq. (6.37), neglecting body effect, the predicted gain of the stage is 16. The adverse influence of the body effect reduces this gain to approximately 12. The biasing of the input stage is achieved from the voltage divider formed by Q_5, Q_6, and Q_7 (see Fig. 6.8a). In order to improve common-mode rejection, Q_5 is made up of two parallel devices, Q_{5A} and Q_{5B}, which are biased from the first-stage outputs. This configuration cancels our differential signals, but provides a common-mode feedback to the bias string in order to increase the CMRR of the input stage.

The primary output of the first stage (at the drain of Q_1) is level shifted by the source follower made up of Q_9 and Q_{10} and the constant-current transistor Q_{11}. The voltage gain of this stage is slightly less than 1 (typically 0.9) due to the body effect. Its output then drives the grounded-source gain stage Q_{14} with Q_{13} and Q_{12} as active load devices. From Eq. (6.9), the approximate voltage gain of the stage is 30 for the given choice of device geometries. The biasing of Q_{12} from the secondary output of the first stage (i.e., the drain of Q_2) provides a slight amount of signal feedforward around the gain stage (see Fig. 7.44), which

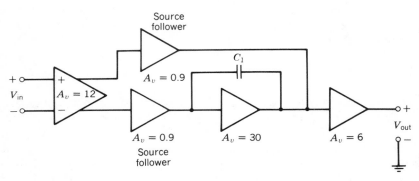

FIGURE 7.44. Typical circuit architecture of NMOS op amp.

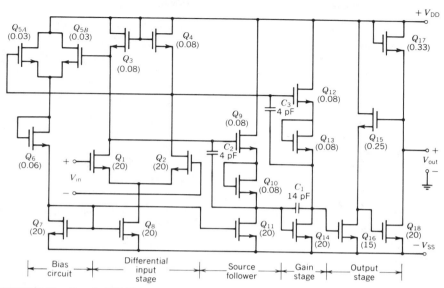

FIGURE 7.45. Practical NMOS op amp circuit. (*Note:* Numbers in parentheses indicate device Z/L ratios.)[28]

makes frequency compensation easier, as will be discussed shortly. The output stage made up of Q_{15}, Q_{16}, Q_{17}, and Q_{20} has the circuit configuration shown in Figure 6.18b. For the device Z/L ratios chosen, this stage provides an additional voltage gain of ≈ 6.

The primary frequency compensation is achieved by the pole-splitting capacitor C_1. However, due to the low transconductance of MOS devices, the second-stage gain is not high enough, and the parasitic signal feedthrough due to the presence of C_1 also causes a right-hand plane zero, which then creates excessive phase shift. The additional feedforward signal provided by Q_{12} tends to cancel this effect and improve the phase margin. The bypass capacitors C_2 and C_3 also provide additional feedforward paths at high frequencies around the long-channel devices Q_{10} and Q_{13}, and thus enhance the phase margin further. The circuit provides approximately 55° of phase margin at its unity-gain crossover frequency of 3 MHz.

Table 7.2 gives a summary of the performance specifications for the NMOS op amp circuit of Figure 7.45 and compares it with the CMOS op amp described in the previous section.

Some of the difficulties associated with NMOS op amp design can be partially avoided by the use of depletion-mode devices as active loads in order to increase the voltage gain of the amplifier stages. This technique was discussed in detail in Section 6.3. However, the use of depletion-mode devices requires additional processing steps, which detracts from the inherent simplicity of NMOS technology. Furthermore, the depletion-mode device threshold voltage is poorly controlled due to the body effect, and it causes the op amp to require a more complex biasing and level-shifting circuitry in order to tolerate large supply voltage changes.

Self-Correction of MOS Op Amp Offset Voltages

Although the offset voltage of matched MOS pairs can be reduced to within the $\pm$ 10– $\pm$ 20 mV range by careful processing and use of the common-centroid layout (see Fig. 7.27), this may still be too high for some of the precision analog applications. However, by using sampled-data techniques, it is possible to measure the op amp offset voltage periodically, store it in the form of a voltage across a holding capacitor, and then subtract it out from the combined signal *plus* the offset. Such sampled-data techniques are known as *auto-zero* or *chopper-stabilization* methods and allow additional reductions of the initial op amp offset of one to two orders of magnitude.

Analog switches with typical "on" resistances of several kilohms and "off" leakage currents of 1 pA or less are readily available in MOS technology (see Section 6.5). The dc open-circuit nature of a MOS gate terminal makes it convenient to sample and hold analog voltages with small capacitors (typically < 100 pF) with MOS analog switches. Thus, MOS technology is ideally suited to the self-correction of offset voltages with auto-zero techniques.

Figure 7.46 shows the simplified diagram of a self-correcting offset-nulling technique which is readily compatible with MOS op amp designs.[29] When the switches S_1 through S_4 are all closed during the initialization cycle, the amplifier offset voltages are stored on the sampling capacitors C_{S1} and C_{S2}, as seen at the output. When the switches are opened during the measurement cycle, only the voltage changes at the input are amplified with zero effective offset.

Figure 7.47 shows a circuit implementation of the offset-nulling technique for eliminating the equivalent input offset of the first stage of a multistage differ-

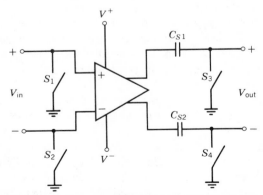

FIGURE 7.46. Block diagram of sampled-data offset-cancellation technique. (*Note:* Switches are closed during the initialization cycle to sample and store the offset voltage on C_{S1} and C_{S2}; switches are then opened during the subsequent measurement cycle.)

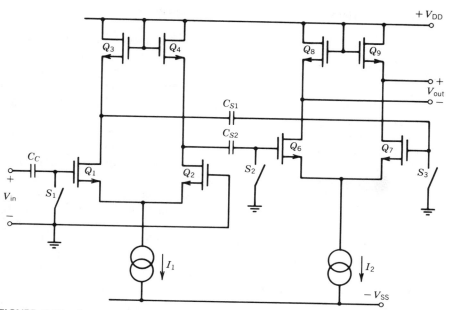

FIGURE 7.47. Auto-zero technique for canceling first-stage offset voltage of MOS differential amplifier.[29]

ential amplifier. During the initialization cycle, S_1, S_2, and S_3 are all closed. The input offset voltage of the first stage, as seen at its output, is then stored in C_{S1} and C_{S2} during this cycle. After the switches are opened, in the subsequent measurement cycle, only the voltage changes of the ac coupled input signal are amplified, with effective cancellation of the dc offset at the op amp input. Since in high-gain amplifiers, almost all of the offset is contributed by the input stage, such a cancellation scheme will eliminate almost all of the offset voltage associated with the entire op amp. Using such a dynamic offset-nulling or auto-zero technique, the effective input offset voltages of MOS op amps or comparators can be kept below 1 mV. The use of sampled-data techniques for offset correction is discussed further in Chapter 15 (see Figs. 15.21 through 15.24).

7.8. SPECIAL-PURPOSE OP AMPS

There are several classes of circuits which perform functions similar to an op amp, and their design concepts are often based on op amp designs. Some of these are instrumentation amplifiers, transconductance amplifiers, and current-differencing amplifiers. In this section, some of the features of these special-purpose circuits will be discussed.

Instrumentation Amplifiers

A common requirement in instrumentation and transducer applications is the need to amplify a very small differential signal (often of the order of microvolts) in the presence of large common-mode signals (often in the range of volts). This requires extremely high common-mode rejection ratios in the amplifier (typically in excess of 100 dB). If a true differential input is required, it will be difficult to achieve with regular op amps.

The instrumentation amplifiers are special classes of op amps developed for such applications. These differ from regular op amps in that true differential operation is achieved with very high differential input impedance and very high common-mode rejection.

An instrumentation amplifier differs from an op amp because it is designed to be used in the open-loop condition. The input signal is applied directly between the high-impedance differential inputs. It is amplified by a well-defined gain, and the output signal is generated from a very low impedance source. No feedback network is required between the input and the output of an instrumentation amplifier, and the gain is generally determined by the ratio of two resistors, normally called the *gain* and *sense* resistors R_G and R_S, connected inside the amplifier. These resistors are isolated from the inputs so that they do not load the input signal source.

Figure 7.48 shows a block diagram representation of an instrumentation amplifier. Such amplifiers are normally required to provide accurate values of gain A_V in the range of 1.0–1000 as a function of the resistor ratios, as

$$A_V = \frac{V_{\text{out}}}{V_{\text{in}}} = \frac{R_S}{R_G} \tag{7.37}$$

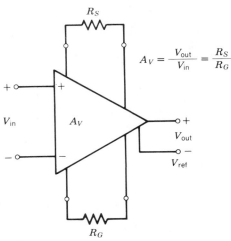

FIGURE 7.48.   Block diagram of instrumentation amplifier.

Other important performance characteristics of an instrumentation amplifier are low noise, low drift, and low offset voltages. Since the gain of an instrumentation amplifier is not determined by external overall feedback, circuit performance cannot be readily predicted by using an amplifier model which refers all errors and noise to the input. In an instrumentation amplifier, there is a component of error and noise which appears at the output, independent of gain. These components are called the *output error* and the *output noise*, and they can be measured by setting gain to zero (i.e., $R_S = \infty$). A second component of error and noise appearing at the output is the one that varies in proportion to the gain. These components are called the *input error* and the *input noise*. They are stated separately and are referenced to the input, in a manner similar to the case of conventional op amps.

The output voltage of an instrumentation amplifier is normally measured against an external reference voltage V_{ref}. In many applications, a ground-referenced output may be needed. This can be obtained by grounding the reference terminal.

A number of different circuit configurations have been developed for monolithic instrumentation op amp design. Figure 7.49 shows one such circuit configuration, which can be used to illustrate the principle of operation of an instrumentation op amp.[30] The circuit uses matched differential current sources

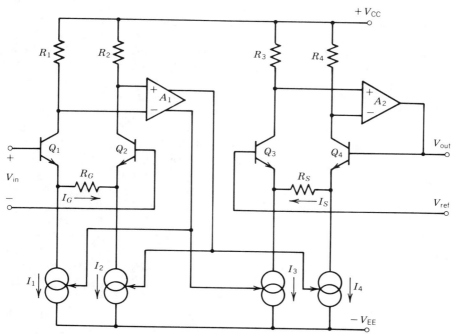

FIGURE 7.49. Possible circuit configuration for monolithic instrumentation amplifier.[30]

I_1, I_2 and I_3, I_4. The input signal V_{in} causes the currents in Q_1 and Q_2 to be unbalanced by an amount inversely proportional to R_G, provided that $R_G \gg 1/g_m$ of the input transistors. The resulting imbalance drives the amplifier A_1, which forces the current sources I_1, I_2 and I_3, I_4 to vary such that the collector currents of Q_1 and Q_2 are equalized. Under this condition, when the currents through Q_1 and Q_2 are forced to be equal, then the difference between I_1 and I_2 must be whatever is required to duplicate the input voltage across R_G, that is,

$$I_G = \frac{I_1 - I_2}{2} = \frac{V_{in}}{R_G} \qquad (7.38)$$

In a similar manner, A_2 forces the collector currents of Q_3 and Q_4 to be equal, such that any current imbalance caused by A_1 must flow through R_S to generate an output voltage, that is,

$$I_S = \frac{I_4 - I_3}{2} = \frac{V_{out}}{R_S} \qquad (7.39)$$

Assuming that the differential current sources match and track, such that $I_G = I_S$, then one can combine Eqs. (7.38) and (7.39) to obtain

$$\text{gain} = \frac{V_{out}}{V_{in}} = \frac{R_S}{R_G} \qquad (7.40)$$

This important result is the basic principle of operation of most instrumentation amplifiers. In other words, the differential input signal is reproduced at the output multiplied with a gain factor determined by the ratio of R_S and R_G, and the circuit is virtually immune to common-mode changes. The output voltage is generated with respect to a reference terminal so that it can be referenced to any desired voltage level. Normally, the reference terminal is one input terminal of the output differential amplifier, while the actual circuit output is connected in feedback to the other input, as shown in Figure 7.49.

Figure 7.50 shows the simplified circuit schematic of a monolithic instrumentation amplifier using the basic configuration shown in Figure 7.49. For convenience in level shifting, a *pnp* differential gain stage is used for the output differential amplifier. In the simplified schematic shown, the details of biasing and buffering circuitry are omitted for simplicity. Transistors Q_1 and Q_2 form the input differential pair, with the lateral *pnp* transistors Q_3 and Q_4 forming the output differential pair.

The two *pnp* differential pairs Q_{13}, Q_{14} and Q_{15}, Q_{16} form the input feedback amplifier A_1 of Figure 7.49. The current output of these gain stages corresponds to currents I_1, I_2, I_3, and I_4. The output amplifier A_2 is a high-gain inverter stage made up of Q_{23} with a current-source load Q_{20}. Its output is buffered by a Class AB stage made up of Q_{24} and Q_{25} and is connected back to the input of the second differential pair (i.e., to the base of Q_4). The frequency compensation capacitor C_1 assures unconditional stability of this feedback loop.

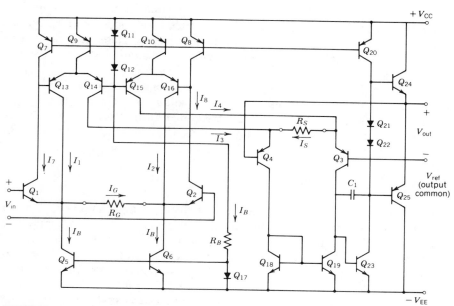

FIGURE 7.50. Simplified circuit schematic of instrumentation amplifier using circuit architecture of Figure 7.49.

With no differential input voltage (i.e., $V_{in} = 0$), the circuit is fully balanced, assuming zero offset errors. In such a case, $I_1 = I_2$ and $I_3 = I_4$ and no current flows across R_G or R_S. If a finite differential input voltage is applied, a differential current I_G and I_S is forced to flow through R_G and R_S such that

$$I_G = \frac{I_1 - I_2}{2} = \frac{V_{in}}{R_G} \tag{7.41}$$

and

$$I_S = \frac{I_4 - I_3}{2} = \frac{V_{out}}{R_S} \tag{7.42}$$

Assuming that the currents I_1, I_3 and I_2, I_4 are well matched and track each other, then $I_G = I_S$, and

$$A_V = \frac{V_{out}}{V_{in}} = \frac{R_S}{R_G} \tag{7.43}$$

which is of the same form as given in Eq. (7.37).

A more detailed version of the simplified circuit schematic shown in Figure 7.50 has been realized as a commercially successful monolithic IC product.* In

*Monolithic amplifier type AD-521, manufactured by Analog Devices, Inc., Norwood, Mass.

the actual complete design, composite *pnp–npn* transistors are used for Q_{13} through Q_{16} to avoid the base current errors, a base-current-compensated current mirror is used for Q_{18} and Q_{19}, and Q_{23} is replaced with a Darlington stage. The key transistors whose offset voltages are critical to the circuit, such as Q_1, Q_2, the bias transistors Q_5, Q_6, and the current mirror pair Q_{18}, Q_{19}, are laid out as cross-coupled transistor pairs (i.e., the common-centroid layout) to minimize offsets due to thermal gradients and mask misalignment. The circuit provides adjustable gains of 0–60 dB, with a CMRR > 120 dB. Its input and output offsets are 0.5 and 30 mV, respectively, with a small-signal bandwidth of 2 MHz and a slew rate of 15 V/μ sec.

Transconductance Amplifiers

Transconductance amplifiers are essentially voltage-controlled current amplifiers which produce a current output proportional to a voltage input. The output current is the product of the input voltage times the transconductance g_m of the amplifier. If such an amplifier exhibits significant transconductance and high input impedance, it is often called an *operational transconductance amplifier* (OTA).

Figure 7.51 shows the basic circuit diagram of a transconductance amplifier which produces an output current I_{out},

$$I_{\text{out}} = g_m V_{\text{in}} \qquad (7.44)$$

and

$$g_m = \frac{I_B}{2V_T} \qquad (7.45)$$

where I_B is the externally supplied bias current and $V_T = kT/q$. In actual designs, Wilson or cascode current sources are used to avoid the current mirror errors and to increase the output resistance.

If an output load resistor R_L is used such that R_L is much lower than the impedance of the output current sources, then the circuit can provide a voltage gain A_V where

$$A_V = \frac{v_{\text{out}}}{v_{\text{in}}} = g_m R_L = \frac{I_B R_L}{2V_T} \qquad (7.46)$$

which is proportional to the bias current I_B.

The main applications of transconductance amplifiers are in voltage- and current-controlled amplifier or attenuator design. These so-called voltage-controlled amplifiers (VCAs) find a wide range of applications in audio sound-effect generation and electronic music synthesis.

Another major application of operational transconductance amplifiers is in sample-and-hold circuits, where the control terminal can be biased on and off to sample and hold an input signal across a holding capacitor, as shown in Figure 7.52. In this circuit, in the sample mode, the control voltage V_C is high, that is,

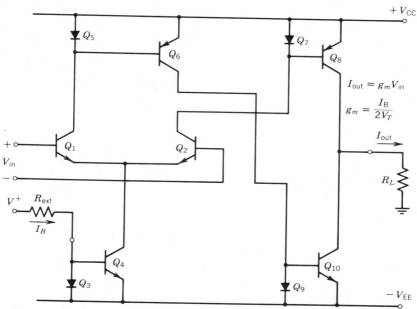

FIGURE 7.51. Simplified circuit diagram of operational transconductance amplifier (OTA). (*Note:* In actual design, Wilson current mirrors would be used for better current matching and higher output impedance.)

at or near $+V_{CC}$, and the output of the transconductance amplifier would charge the holding capacitor, C_H to a voltage equal to V_{in}. In the hold mode, I_B would be reduced to zero, thus the output of the operational transconductance amplifier would be a virtual open circuit and the sampled input voltage will be held on C_H. The decay of the voltage across C_H during the hold mode would depend on the

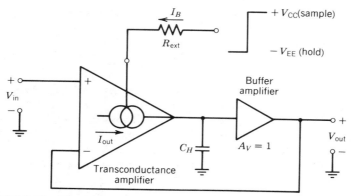

FIGURE 7.52. Sample-and-hold circuit using operational transconductance amplifier.

output impedance of the operational transconductance amplifier and the input resistance of the buffer stage.

Still another area of application for operational transconductance amplifiers is in nonlinear signal processing, such as compressing and expanding analog signals for transmission through a limited dynamic range signal path. Such a circuit function, called a *compandor*, is often used in telephone applications.[31]

Current-Differencing Amplifiers

Current-differencing amplifiers are essentially current-controlled voltage amplifiers which amplify the *difference* of two input currents into a voltage signal. Since the inputs operate in a current rather than a voltage mode, this type of amplifier is often called a *Norton amplifier*, after the Norton equivalent circuit.

The basic principle of operation of a current-differencing differential gain stage was discussed in Section 5.2 (see Fig. 5.24). Figure 7.53 shows the simplified circuit diagram of an op amp type circuit based on the same principle.[6] The circuit converts the input current difference ΔI_{in} into a voltage V_{out} at the base of Q_4, given as

$$V_{out} = -\beta_3 \Delta I_{in} r_{out} \qquad (7.47)$$

where β_3 is the current gain of Q_3, and ΔI_{in} is the input current *difference*, that is,

$$\Delta I_{in} = I_{in1} - I_{in2} \qquad (7.48)$$

r_{out} is the total output resistance at the collector of Q_3, that is,

$$r_{out} = r_{o3} \| r_{in4} \qquad (7.49)$$

where r_{o3} is the output resistance at the collector of Q_3 (due to the Early effect), and r_{in4} is the resistance looking into the base of the *pnp* emitter follower Q_4. The *pnp* and *npn* emitter followers Q_4 and Q_5 provide two stages of buffering for the load R_L connected to the output.

The circuit can be stabilized for closed-loop operation by means of a small capacitor C_1 in the range of 3–10 pF.

The main advantages of current-differencing amplifiers is their circuit simplicity and their ability to operate with a single power supply. The single-supply operation is particularly advantageous in certain industrial or automotive applications where only a single 12- or 15-V power supply is available. By connecting series resistors to the inputs, a differential input voltage can be converted into a differential current and amplified by the circuit. Since the input current terminals are biased 1 V_{BE} above ground, and input voltages are converted to currents by means of series resistors, there is virtually no limit to the positive input common-mode range. This is particularly useful for applications as a high-voltage comparator.

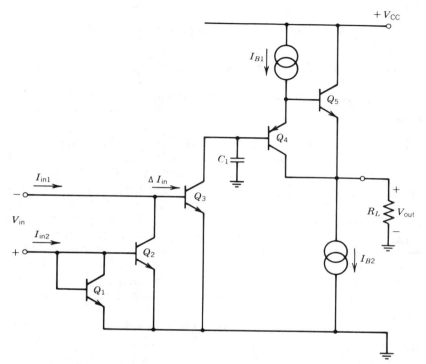

FIGURE 7.53. Simplified schematic of a current-differencing amplifier.[6]

7.9. OTHER OPERATIONAL AMPLIFIER-BASED CIRCUITS: BUFFERS AND COMPARATORS

Buffer Amplifiers (Voltage Followers)

Voltage buffers, or *buffer amplifiers,* are used in interfacing high-impedance signal processing or sample-and-hold circuitry with low-impedance loads. Although buffer amplifiers can be either inverting or noninverting and can have various values of gain, by far the most commonly used buffer configuration is the unity-gain noninverting amplifier. Such amplifiers are also called *voltage followers,* since their function is to reproduce an input voltage signal accurately in both amplitude and phase.

Normally, a voltage follower or buffer amplifier is designed as a single-ended input and output amplifier, with the following characteristics:

1. High input impedance with very low input bias current.
2. Low input/output offset voltage.

3. Accurate gain (typically $1.00 \pm 0.1\%$).
4. Wide bandwidth for small- and large- signal operation.
5. High slew rate and low settling time.
6. Very low impedance output with high current drive capability.

There are two design approaches to buffer amplifier design: (1) closed-loop or feedback amplifiers, and (2) open-loop or nonfeedback amplifiers.

Closed-Loop Voltage Followers.   These circuits normally use a high-gain op amp type circuit in a unity-gain voltage-follower configuration, as shown in Figure 7.3d. The unity feedback can be applied either externally or internally within the IC chip. Figure 7.54 shows a commonly used circuit configuration for implementing closed-loop-type buffer circuits. Normally, such circuits require a compensating capacitor C_1 to assure unconditional stability and rapid settling time. The advantage of closed-loop type buffer amplifiers is the low offset and very precise values that can be obtained. However, their slew rate, large signal bandwidth, and settling time are relatively poor due to the frequency compensation requirements. Typically, slew rates in excess of 40 V/μsec and settling times of less than 300 nsec are difficult to obtain with this approach.

Figure 7.55 shows the simplified circuit schematic of a closed-loop-type buffer amplifier, which uses the basic circuit configuration of Figure 7.54 in conjunction with superbeta transistors.[32] The use of superbeta transistors at the input results in very low bias current and extremely high input impedance. A

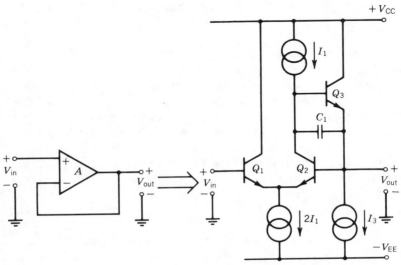

FIGURE 7.54.   Commonly used circuit configuration for closed-loop-type voltage followers.

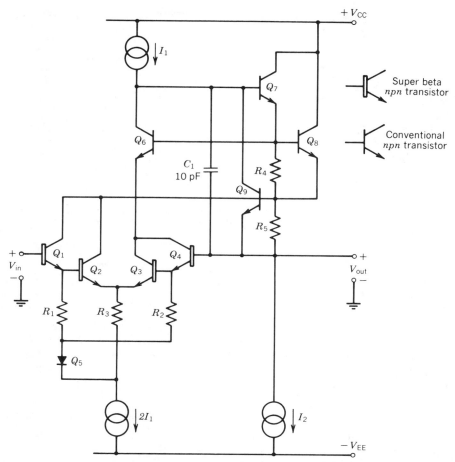

FIGURE 7.55. Simplified circuit diagram of closed-loop-type voltage follower using superbeta transistors. (National Semiconductor LM-110.)

bootstrap bias arrangement is used such that the total voltage across the super-beta transistors Q_1 through Q_4 never exceeds $3V_{BE}$. The circuit has an internal open-loop gain well in excess of 60 dB, so the overall gain accuracy of $1.00 \pm 0.1\%$ is easily achieved. The electrical characteristics of a complete monolithic IC buffer amplifier (National Semiconductor LM-110), based on the simplified circuit schematic of Figure 7.54, are given in Table 7.3.

Open-Loop Voltage Followers. These types of circuits utilize a cascade of emitter followers and source followers to accomplish the buffer function without any feedback loop around the internal gain stages. This scheme is inherently faster because the signal propagates directly from the input to the output, without circulating inside a feedback loop. The main difficulty of this approach is in

TABLE 7.3.   Comparison of Performance Characteristics of the LM-110 and BUF-03 Voltage-Follower Circuits[a]

Performance Parameter	Units	Closed-Loop Buffer (LM-110)	Open-Loop Buffer (BUF-03)
Input offset voltage, V_{OS}	mV	4 (M)	6 (M)
Offset voltage drift dV_{OS}/dT	$\mu V/°C$	12 (T)	100 (M)
Voltage gain A_v	V/V	0.999 (M)	0.994 (M)
Input resistance R_{in}	$M\Omega$	10^4 (M)	10^5 (T)
Output resistance R_{out}	Ω	2.5 (M)	2 (T)
Input bias current T_B	nA	3 (M)	0.4 (T)
Supply current I_{CC}	mA	4 (M)	22 (M)
Slew rate	$V/\mu sec$	30 (T)	220 (M)
Small-signal bandwidth	MHz	20 (T)	60 (T)
Full-power bandwidth	MHz	0.25 (T)	9 (T)
Settling time (to 0.1%)	nsec	400 (T)	90 (T)

[a] Test conditions: $T_A = 25°C$, $V_{CC} = V_{EE} = 15$ V. (T) designates typical value, (M) designates guaranteed minimum or maximum limits.

achieving the required gain accuracy and low offset voltage without the error-reducing benefits of feedback. However, a combination of ion-implanted JFET devices on the same chip as a bipolar device and the availability of on-chip offset-trimming techniques makes such a design approach feasible.

Figure 7.56 shows the simplified schematic of such an open-loop type voltage follower circuit.[33] With reference to the simplified diagram, operation of the circuit can be explained as follows. The p-channel JFET Q_1 is biased by an identical JFET Q_2, which is operated with zero gate–source voltage. Therefore, the nominal value of the gate–source voltage of Q_1 is also zero, and the JFET input stage functions as a source follower. The signal then propagates through the emitter follower Q_3, the level-shift diodes Q_4 and Q_5, and another emitter follower, Q_6. If the current sources are well matched, such that $I_2 = I_3 = I_1/2$, then the 4 V_{BE} drops from the base of Q_3 to the emitter of Q_6 will cancel out and track each other with temperature. In the output, the drain current of the p-channel JFET Q_8 is mirrored by Q_9 and Q_{10}, which bias Q_7. If the two JFETs Q_8 and Q_7 are well matched, then Q_7 would be forced to operate with zero gate–source voltage, and the output voltage will follow the voltage at the emitter of Q_6. The emitter areas of Q_{11} and Q_6 are scaled to provide a higher idling current through Q_{11} with the same V_{BE} drop as Q_6. Another important feature of the output stage shown is that the changes in the output current I_L are absorbed by Q_{12} as long as the load current I_L is not higher than the quiescent current through Q_{12}. This is assured by the feedback provided with the matched JFET devices Q_7 and Q_8. Thus, the V_{BE} of the output emitter follower Q_{11} does not change with the load current I_L, provided that I_L is *less* than the idling current.

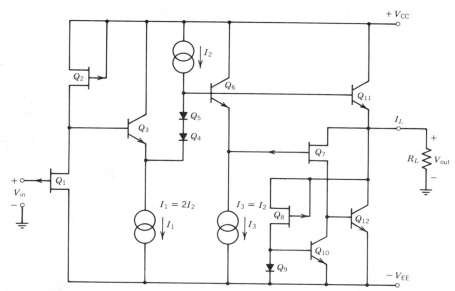

FIGURE 7.56. Simplified schematic of open-loop high-speed voltage follower.[33]

This is important in an open-loop voltage follower since any dc level changes in the input/output signal path will appear as offset voltages.

The finite output resistance of the JFET Q_2 causes the gate–source voltage of Q_1 to be nonzero and can result in as much as 80 mV of offset between the source and the gate of Q_1 for large negative swings of the input. To avoid this problem, a bootstrap bias technique is used in the actual design. The initial offset is trimmed by an on-chip zener-zapping method to reduce the initial offset from its untrimmed mean value of $\pm$ 45 mV to $\leq \pm$ 6 mV.[34]

Typical performance characteristics of an open-loop buffer circuit (BUF-03) based on the circuit configuration of Figure 7.56, are listed in Table 7.3 and are compared with those of a closed-loop-type buffer amplifier. As indicated earlier, the open-loop buffer circuits can provide much higher bandwidth and slew rate than the closed-loop buffers. However, this improvement is gained at the expense of gain accuracy and offset characteristics.

Voltage Comparators

The function of a voltage comparator is to compare the instantaneous value of a signal voltage at one input with a reference voltage on the other input and produce a digital 1 or 0 level at the output when one input is higher than the other. It is normally used to perform any one of the following classes of circuit functions:

1. Variable threshold detector.

   2. Pulse-height discriminator.

   3. Zero-crossing detector.

   4. High noise immunity digital line receiver.

   5. Voltage level comparator (for A/D conversion).

A voltage comparator is designed as a high-gain differential input and single-ended output amplifier with the output swing and dc levels adjusted to be compatible with conventional logic circuits.

   Unlike an op amp, the voltage comparator is designed to be used in an open-loop configuration, and therefore, it does not require frequency compensation. The input impedance, voltage gain, and output voltage swing requirements are somewhat lower for a comparator than for an op amp.

   Since a comparator output is switched between two output states, the voltage gain is necessary only to reduce the differential input level change needed to make the output swing from one extreme level to another. To interface with digital circuits, the required peak-to-peak output swing levels are in the range of 3–5 V. Therefore, a voltage gain of ≥ 1000 is usually sufficient to bring the input signal amplitude needed for full output swing to a level comparable with the input stage offset. Therefore, very high values of voltage gain are not required. The input offset and bias current requirements of a comparator are comparable to those for an op amp.

   The speed of response is a critical parameter for most comparator applications. It is required to switch between two output states in a minimum amount of time, subsequent to an appropriate input level change, and it should exhibit fast rise and fall times at its output. It is also required to have a rapid recovery from cutoff or saturation conditions of the input or output.

   An important parameter, which gives a direct measure of the speed capabilities, is the so-called *response time,* which is defined as the time interval between the application of a predetermined step input and the time when the output crosses the corresponding logic state threshold level. The input step drives the comparator from some initial saturated input condition to an input level *barely in excess* of that required to cause the output to switch state. This excess voltage is called the *overdrive.* The comparator response speeds up as the overdrive is increased. For comparison purposes, response times are normally stated with a 100-mV total input step change and a 5-mV overdrive. If the response time is unsymmetrical for positive and negative transitions, the longer time is normally specified.

   Circuit configurations used for voltage comparators depend strongly on speed requirements. In medium- or low-speed applications, where response times of the order of 1 μsec are tolerated, the design is very simple and straightforward. Virtually any differential gain stage with active loads can be used. Figure 7.57 shows two typical medium-speed voltage comparator circuits. The circuit of Figure 7.57a is basically the so-called *ground-sensing* input stage configuration (see Fig. 5.22), with the addition of clamp diodes Q_5 and Q_6. These diodes

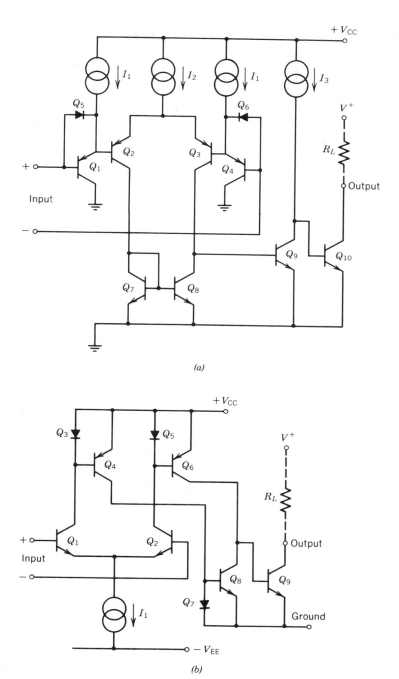

FIGURE 7.57. Simple voltage comparator circuits for low-frequency applications: (a) Single-supply ground-sensing comparator; (b) split-supply circuit with floating ground.

increase the turn-on time of the input transistors by supplying current to charge the stray capacitances associated with the base nodes of Q_2 or Q_3, when these transistors are driven heavily into cutoff. The circuit of Figure 7.57b offers a *floating* ground terminal to which the output logic swing is referenced. This configuration makes it very convenient to interface split supply analog circuits with digital systems which operate with a single supply voltage.

For high-speed comparator designs, one would normally use only *npn* devices and resistive loads in the signal path, and also limit the voltage swings within the circuit by means of clamp diodes to speed up the circuit response. Figure 7.58 shows the circuit configuration for a high-speed voltage comparator (Fairchild μA-710) which operates with a response time of less than 60 nsec. The

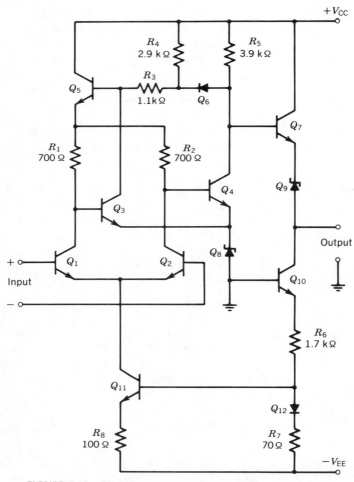

FIGURE 7.58.   High-speed comparator circuit. (Fairchild μA-710.)

operation of the comparator circuit can be briefly described as follows. Q_1 and Q_2 form a differential input stage with balanced collector loads R_1 and R_2, and Q_5 serves as a virtual positive supply for the input stage. Q_3 and Q_4 form the second gain stage with their emitters biased at approximately $+6.5$ V by the Zener diode Q_8. The emitter follower Q_7 provides a low-impedance output which is then level shifted by the Zener diode Q_9 to make the output swings compatible with conventional logic levels. The diode Q_6 clamps the output swing of the second stage, both to increase speed and to provide a logic compatible output swing. Neglecting the interstage loading effects, the overall voltage gain of the comparator is

$$A_V = A_{V1} \, A_{V2} = (g_{m1}R_1)(g_{m4}R_5) \qquad (7.50)$$

where A_{V1} and A_{V2} are the first- and second-stage gains, and g_{m1} and g_{m4} are transconductances of Q_1 and Q_4. For operation of the circuit with $+12$ and -6 V supplies, $A_V \approx 1700$. The response time is approximately 50 nsec with 5-mV overdrive, and the output swing levels are $+3.5$ and -0.5 V, which make it compatible with most IC logic circuits.

Very-High-Speed Comparators. High-speed voltage comparators are some of the essential building blocks of data acquisition systems. In fact, the response time of many of the data conversion techniques is limited by the "decision-making speed" of the voltage comparators within the system. For this purpose, a number of extremely fast comparator circuits have been developed, which have response times on the order of 5–10 nsec. Such high speeds are normally achieved by the use of special device design, and processing methods which result in transistors with f_T of 1–2 GHz, compared to the 500-MHz f_T devices used in conventional analog IC design. The fabrication process and circuit configurations used in these very fast comparators are similar to those used in emitter-coupled-logic (ECL) technology.

Figure 7.59 shows the circuit schematic of such a very-high-speed comparator circuit (Advanced Micro-Devices Am-685). The circuit uses high-frequency transistors with $f_T \approx 2$ GHz which are mainly operated in the current mode (i.e., the voltage swings within the circuit are kept to a minimum). The input is applied to the base of Q_1 or Q_2, which form a differential gain stage with the cascode transistors Q_3 and Q_4 and load resistors R_1 and R_2. The voltage swing at the first-stage output is clamped by the Schottky diodes D_1 and D_2. The differential output of the first stage is then buffered and level shifted by the emitter followers Q_{13} and Q_{14} and the emitter–base Zener diodes D_5 and D_6. The second-stage gain is provided by the differential stage formed by Q_{15}, Q_{16}, Q_{17}, and Q_{18} and the load resistors R_5 and R_6. The output of the second stage is once more level shifted differentially through emitter followers Q_{19} and Q_{20} and Zener diodes D_7 and D_8. The third-stage gain, formed by the differential pair Q_{21} and Q_{22}, provides additional gain and differential outputs through the emitter followers Q_{23} and Q_{24}. These are "open-emitter" outputs which need to be biased by external resistors to $-V_{EE}$. The circuit provides a nominal voltage gain of 1700,

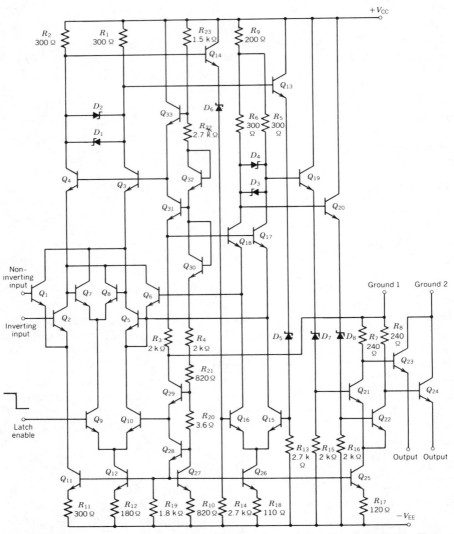

FIGURE 7.59. Very-high-speed comparator circuit with latching outputs. (Advanced Micro-Devices Am-685.)

with a bandwidth of approximately 100 MHz. Its response time is < 6 nsec with 5-mV overdrive.

Table 7.4 gives a comparative listing of various performance characteristics for the comparator circuits discussed in this section.

Latching Comparators. Many high-speed comparator applications, particularly those used in data encoding or A/D conversion systems,[35] also require the

TABLE 7.4. Comparison of Several Classes of Voltage Comparator Performance Characteristics.[a]

Performance Characteristics	Units	Medium-Speed Comparator (LM-139)	High-Speed Comparator (μA-710)	Very-High-Speed Comparator (Am-685)
Input offset voltage V_{OS}	mV	5 (M)	2 (M)	2 (M)
Input bias current I_B	μA	0.1 (M)	3 (M)	10 (M)
Voltage gain A_V	V/mV	50 (M)	1 (M)	1.7 (M)
Output logic swing level		TTL	TTL	ECL
Power dissipation	mW	30 (T)	150 (T)	300 (T)
Response time				
5-mV overdrive	nsec	1500 (T)	50 (T)	6.5 (T)
20-mV overdrive	nsec	600 (T)	40 (T)	4.0 (M)
Latch set up time	nsec	—	—	4 (M)
Latch hold time	nsec	—	—	1 (M)

[a]$T_A = 25$ °C. (T) designates typical value; (M) designates guaranteed minimum or maximum limits.

circuit to have *latching* capability. In other words, subsequent to a latch command, the input stage of the comparator would be disabled and the logic state at the output will be stored indefinitely, until an enable or unlatched command is given. In the very-high-speed converters, the latching function is normally done by using emitter-coupled *R–S* flip-flops in conjunction with differential gain stages, as shown in Figure 7.60. Normally, the latch-control terminal is low; Q_5, Q_8, and Q_9 are all off, and the total bias current I_2 is steered to the input stage. In this condition, the input operates as a differential cascode gain stage. If the "latch-enable" control is raised above the bias voltage V_{B1}, all of the bias current I_2 will be steered away from the input stage, turning Q_2, Q_3, and Q_4 off and turning on the cross-coupled differential pair of Q_8 and Q_9. As Q_8 and Q_9 are brought into conduction, they function as an *R–S* flip-flop and are set to the same state as the output just prior to the application of the latch signal. In this manner, the input stage is disabled and the output is permanently latched in its previous state until the latch signal is disabled.

A very similar latching technique is also used in the very-high-speed comparator of Figure 7.59. In this case, the latch-control signal turns off the differential pair Q_7 and Q_8 and turns on the pair Q_5 and Q_6, which are directly connected to the second-stage output. This positive feedback assures that the circuit will remain latched at its initial state, independent of the input signal, for the duration of the latch-control signal. In the case of the high-speed circuit described, the latch–enable and the latch–setup times are also on the order of several nanoseconds.

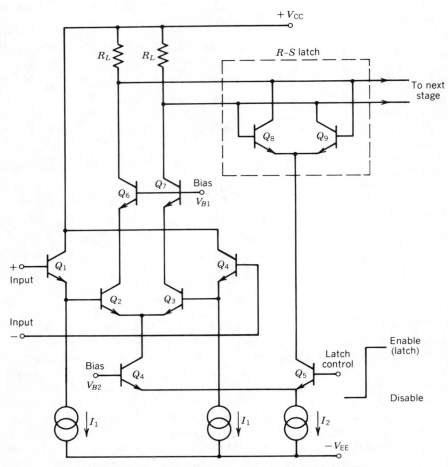

FIGURE 7.60. High-speed gain stage with latching capability.

REFERENCES

1. G. E. Tobey, J. G. Graeme, and L. P. Huelsman, *Operational Amplifiers,* McGraw-Hill, New York, 1971.

2. P. R. Gray and R. G. Meyer, *Analysis and Design of Analog Integrated Circuits,* Wiley, New York, 1977, Chaps. 6, 9.

3. D. H. Sheingold, Ed., *Nonlinear Circuits Handbook,* Analog Devices, Inc., Norwood, MA, 1976.

4. J. E. Solomon, "The Monolithic Op Amp: A Tutorial Study," *IEEE J. Solid-State Circuits* **SC-9,** 314–332 (December 1974).

5. P. R. Gray and R. G. Meyer, "Recent Advances in Monolithic Operational Amplifier Design," *IEEE Trans. Circuits and Systems* **CAS-21,** 317–327 (May 1974).

6. T. M. Frederiksen, W. F. Davis, and D. W. Zobel, "A New Current Differencing Single Supply Operational Amplifier," *IEEE J. Solid-State Circuits* **SC-6**, 340–347 (December 1971).

7. "Harris Semiconductor HA-2510/2512/2515 Operational Amplifier Data Sheet," Harris Semiconductor Corp., March 1973.

8. R. J. Apfel and P. R. Gray, "A Fast Settling Monolithic Operational Amplifier Using Doublet Compression Technique," *IEEE J. Solid-State Circuits* **SC-9**, 332–340 (December 1974).

9. G. Erdi, "Amplifier Techniques for Combining Low Noise, Precision and High Speed Performance," *IEEE J. Solid-State Circuits* **SC-16**, 653–661 (December 1981).

10. K. Ogata, *Modern Control Engineering*, Prentice-Hall, Englewood Cliffs, NJ, 1970, Chap. 8.

11. P. E. Grey and C. L. Searle, *Electronic Principles*, Wiley, New York, 1969, Chap. 20.

12. R. J. VanDePlassche, "A Wideband Operational Amplifier with a New Output Stage and a Simple Frequency Compensation," *IEEE J. Solid-State Circuits* **SC-6**, 347–352 (December 1971).

13. R. C. Dobkin, "LM-118 Op Amp Slews 70 V/μsec," National Semiconductor Linear Appl. Brief 17, September 1971.

14. B. Y. Kamath, R. G. Meyer, and P. R. Gray, "Relationship Between Frequency Response and Settling Time of Operational Amplifiers," *IEEE J. Solid-State Circuits* **SC-9**, 347–352 (December 1974).

15. W. E. Hearn, "Fast Slewing Monolithic Operational Amplifier," *IEEE J. Solid-State Circuits* **SC-6**, 20–24 (February 1971).

16. J. C. Schmoock, "Input Stage Transconductance Reduction," *IEEE J. Solid-State Circuits* **SC-10**, 407–411 (December 1975).

17. R. G. Meyer, Ed., *Integrated Circuit Operation Amplifiers*, IEEE Press, New York, 1978.

18. A. B. Grebene, Ed., *Analog Integrated Circuits*, IEEE Press, New York, 1978.

19. J. Metzger, "Bipolar and FET Op Amps: A Special Report," *Electronic Products Mag.*, 51–58 (June 1977).

20. R. J. Widlar, "Design Techniques for Monolithic Operational Amplifiers," *IEEE J. Solid-State Circuits* **SC-4**, 184–191 (August 1969).

21. G. Erdi, "A Precision Trim Technique for Monolithic Analog Circuits," *IEEE J. Solid-State Circuits* **SC-10**, 412–416 (December 1975)

22. "Precision Monolithics OP-27/OP-37 Data Sheet," Precision Monolithics, Inc., 1981.

23. D. Fullagar, "A New High Performance Monolithic Operational Amplifier," Fairchild Semiconductor Appl. Brief, May 1968.

24. R. W. Russell and D. D. Culmer, "Ion Implanted JFET Bipolar Monolithic Analog Circuits," *Digest of Tech. Papers*, IEEE Int. Solid-State Circuits Conf., pp. 140–141, February 1974.

25. P. R. Gray, D. A. Hodges, and R. W. Brodersen, Eds., *Analog MOS Integrated Circuits*, IEEE Press, New York, 1980.

26. D. A. Hodges, P. R. Gray, and R. W. Brodersen, "Potential of MOS Technologies for Analog Integrated Circuits," *IEEE J. Solid-State Circuits* **SC-13**, 285–294 (June 1978).

27. P. R. Gray, "Basic MOS Operational Amplifier Design—An Overview," in *Analog MOS Integrated Circuits*, P. R. Gray, D. A. Hodges, and R. W. Brodersen, Eds., IEEE Press, New York, 1980, pp. 28–49.

28. I. A. Young, "A High Performance All-Enhancement NMOS Operational Amplifier," *IEEE J. Solid-State Circuits* **SC-14**, 1070–1077 (December 1979).

29. R. Poujois, B. Baylac, D. Barbier, and J. M. Ittel, "Low Level MOS Transistor Amplifier Using Storage Techniques," *Digest of Tech. Papers* IEEE Int. Solid-State Circuits Conf., pp. 152–153, February 1973.

30. A. P. Brokaw and M. P. Timko, "An Improved Monolithic Instrumentation Amplifier," *IEEE J. Solid-State Circuits* **SC-10,** 417–423 (December 1975)

31. C. C. Todd, "A Monolithic Analog Compandor," *IEEE J. Solid-State Circuits* **SC-11,** 754–762 (December 1976).

32. "National Semiconductor LM-110 Voltage Follower Data Sheet," National Semiconductor Corp.

33. G. Erdi, "A 300 V/μsec Voltage Follower," *IEEE J. Solid-State Circuits* **SC-14,** (December 1979).

34. "Precision Monolithics BUF-03 Very High Speed Voltage Follower Data Sheet," Precision Monolithic, Inc., 1981.

35. J. G. Peterson, "A Monolithic Video A/D Converter," *IEEE J. Solid-State Circuits* **SC-14,** 932–937 (December 1979).

CHAPTER EIGHT

WIDEBAND AMPLIFIERS

In the design of broadband high-frequency amplifiers, the inherent limitations of monolithic devices impose severe restrictions on possible design approaches. The most significant design restriction is the lack of integrated inductors which make conventional broadbanding techniques, such as shuntpeaking, not suitable for monolithic wideband amplifiers. Furthermore, large coupling capacitors or impedance-matching transformers are not compatible with integrated circuits. In spite of these basic drawbacks, a large variety of small-signal wideband amplifier configurations are available which are well suited to monolithic integration.

In monolithic amplifiers ac coupling between the stages is, in general, not possible. Therefore, the ac design and the performance of the circuit cannot be considered as a separate problem from dc bias considerations. The availability of a large number of well-matched active devices again provides a distinct advantage for the designer. For example, one can now use a compound connection of two or more devices to replace a single transistor for improved high-frequency performance. Among the monolithic devices, the *npn* bipolar transistor has the best high-frequency capability. For this reason, almost all wideband amplifiers use *npn* devices in the signal path; the lateral or substrate *pnp* devices, when used, are confined to biasing applications. Active loads, which are the key building blocks of low-frequency designs, are almost never used in high-frequency designs because of their poor frequency and noise characteristics. Instead, resistive loads are used almost exclusively.

Perhaps more than any other class of circuit functions, the high-frequency capability of wideband amplifiers will depend on the frequency characteristics of the active devices used. All the circuit techniques and design approaches described in the following sections provide general guidelines for optimizing the high-frequency performance of monolithic amplifiers, starting from a given set of device parameters. However, these circuit techniques are no substitute for high-frequency devices. As a general rule, to design a wideband amplifier with moderate gains (in the 20–30-dB range), one would need transistors with f_T at least 5–10 times higher than the required 3-dB bandwidth of the amplifier.

8.1. GENERAL DESIGN CONSIDERATIONS

In the design of wideband amplifiers, one has to consider most or all of the following requirements:

1. Wide bandwith with prescribed passband response. The amplifier frequency response is required to be flat within the specified tolerance over the entire bandwidth, and particularly near the upper bandedge. Often additional requirements must also be met in terms of attenuation characteristics outside the passband.

2. Accurate gain characteristics. Normally, moderate gain values would be required (typically in the range of 20–30 dB), which must be maintained within a specific tolerance level, over manufacturing process and temperature variations. This is normally achieved by the use of feedback.

3. Linearity of signal transfer. The signal must be amplified with minimum distortion over its specified dynamic range. This is again achieved by using feedback. However, effective suppression of distortion requires high loop gain, which is not readily available in broadband amplifiers due to stability problems. Distortion may come about as harmonic distortion due to device nonlinearities, or it may come about as intermodulation distortion when more than one information signal is applied to the amplifier.

4. Automatic gain control (AGC) and modulation capability. In many wideband amplifier applications, such as intermediate frequency (IF) gain stages in radio and TV receivers, it is necessary to maintain the amplitude of the *output* signal within specific bounds while the input amplitude may vary over a wider dynamic range. This is done by automatic gain control (AGC) circuitry, which can control the gain of the amplifier by controlling the internal bias or impedance levels. This type of electronic gain control capability is often a design requirement.

5. Accurate control of input and output impedance levels. In many cases, wideband amplifiers have to interface with transmission lines or cables which have specific impedance requirements. This may necessitate the use of multiple feedback loops to set the input and output impedance levels.

6. Low noise characteristics. Low broadband noise and noise figure are often among the key requirements of wideband amplifier design. This requires that the devices with poor noise characteristics, such as Zener diodes, active loads, and high value resistors, be eliminated from the signal path, and the parasitic noise sources in active devices, such as the base resistance r_b, should be minimized.

The above list of requirements is primarily intended as a guideline in monolithic wideband amplifier design. An actual design often requires a compromise between various conflicting requirements. Such a compromise, based on the specific performance requirements, then becomes the starting point of IC design and determines the particular circuit configuration to be used.

8.2. HIGH-FREQUENCY TRANSISTORS

In order to optimize the high-frequency characteristics and increase the f_T of the *npn* bipolar transistor, the designer must consider the following:

1. Minimize the emitter–base and collector–base junction capacitances, C_{JE} and C_{CB}, and reduce the transistor base width W_B to decrease the transit time of minority carriers through the base region. This calls for minimizing the emitter–base and collector–base junction areas.
2. Operate the transistor at high emitter currents, where f_T is nearly maximum.
3. Minimize the effects of parasitic series resistances, primarily the base-spreading resistance r_b and the collector series resistance r_{cs}.

These requirements call for various design trade-offs. Junction areas have to be minimized without increasing series resistances, and high emitter efficiency must be maintained at relatively high currents. These requirements are best met by using a narrow "stripe" geometry for transistor emitter regions as well as for the ohmic contacts, as illustrated in the typical device layout of Figure 8.1. The emitter area is normally made up of long and narrow n^+ stripes. The minimum

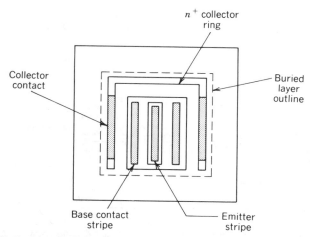

FIGURE 8.1. Typical layout for high-frequency *npn* transistor. (Shaded regions show contact windows whose width is determined by minimum mask dimensions.)

width of the stripe is limited by the minimum mask dimensions for the emitter contact window and the minimum overlap required between the contact and diffusion windows. The length of the emitter stripe, or the number of emitter stripes used, is normally determined by the designer, depending on the current levels in the circuit. The stripe geometry maximizes the emitter periphery-to-area ratio to avoid current-crowding effects while minimizing junction capacitance.

The base contacts are placed beside emitter contacts, again in a stripe or finger geometry, to minimize r_b without increasing the collector–base junction capacitance. Normally, there will be one base contact stripe on either side of the emitter stripe. Thus, a device with n emitter stripes will have $n + 1$ base contact stripes.

The collector contact will be "wrapped around" the base region to minimize collector series resistance. In many cases, a deep n^+ diffusion may be used to extend the collector contact region down to the buried layer (see Fig. 8.2).

The junction depths are kept to a minimum to reduce parasitic capacitances. Typically, an epitaxial layer thickness of 3–5 μm is used, with a 1-μm base diffusion and a 0.5-μm emitter diffusion.

Figure 8.2 shows the device cross section of an ion-implanted high-frequency transistor. Ion implantation is used to form the active base region, directly below the emitter. In this manner, the impurity concentration in the active base region can be kept low to improve emitter efficiency and to reduce C_{JE}. Typically, the depth of this ion-implanted p^--type base region is about 1 μm. The inactive base region below the base contacts is formed by a conventional p^+ diffusion. The n^+-type emitter can be either diffused or implanted.

The minimum mask dimensions are a critical factor in the high-frequency performance of stripe-geometry transistors, since these geometries limit the minimum width of the emitter stripe as well as the spacing between adjacent stripes.

In particular device geometries optimized for very-high-frequency operations or rapid switching, the overlap between the emitter diffusion window and the

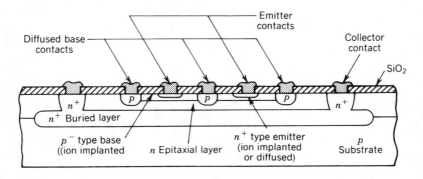

FIGURE 8.2. Device cross section of typical ion-implanted high-frequency transistor. (*Note*: p^--type base is ion-implanted; n^+-type emitter can be either implanted or diffused.)

emitter contact window is altogether eliminated. In such a device structure, the same oxide window is used to perform the emitter diffusion and make ohmic contact to the emitter region, and only the side diffusion of the n^+-type emitter under the diffusion window is used to separate the emitter ohmic contact region from the edge of the n^+-type emitter diffusion. In the fabrication of the device, no separate emitter contact window is opened. Instead, during the contact mask stage, the thin oxide layer over the emitter diffusion window is etched or "washed away" to reopen the entire n^+ diffusion window as the emitter contact area. Because of this processing step, such transistors which have no overlap between emitter diffusion and the emitter contact window are called *washed-emitter* transistors. Using the washed-emitter process, the emitter stripe width can be reduced to be no wider than the minimum allowable mask dimension. Although the washed-emitter process optimizes the high-frequency performance of the transistor, it also requires very tight control of photomasking and etching steps, since the side diffusion of the n^+-type emitter stripe is on the order of a few *tenths* of a micrometer.

Compared to the standard low-voltage process discussed in Chapter 2 (see Section 2.1), which gives an f_T of about 400 MHz, the shallow-junction (i.e., 3-μm epitaxial layer) high-frequency transistors with 2.5-μm minimum mask dimensions provide a factor of 3–5 improvement in the f_T of the transistors, and extend f_T to nearly 2 GHz. However, it should again be pointed out that such a fabrication process requires very tight process controls.

8.3. HIGH-FREQUENCY DEVICE MODELS

An efficient analysis of a wideband amplifier requires a good knowledge of the monolithic device parameters and circuit parasitics. The starting point of the design is the small-signal high-frequency equivalent circuit for the active elements. In the case of the *npn* bipolar transistor, the most convenient model is the hybrid-π circuit. This model was described in Chapter 2 (see Section 2.1) and will be briefly reviewed in this section. Figure 8.3 shows the small-signal hybrid-π equivalent circuit for a bipolar transistor. The components r_b, r_{cs}, and

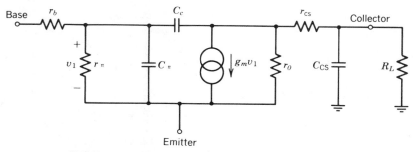

FIGURE 8.3. Basic hybrid-π model for bipolar transistor.

C_c are the parasitics inherent to the bipolar transistor structure; r_o is the output resistance due to the Early effect, and the C_{CS} is the collector–substrate junction capacitance.

The intrinsic transistor parameters r_π and C_π are related to the readily measurable parameters of the transistor, such as the low-frequency small-signal current gain β_o, the gain–bandwidth product f_T, and the transconductance g_m, as given in Eq. (2.15) through (2.29). These are briefly repeated below:

$$r_\pi = \frac{\beta_o}{g_m} \tag{8.1}$$

$$C_\pi = \frac{g_m}{\omega_T} - C_c \approx \frac{g_m}{\omega_T} \tag{8.2}$$

and

$$\omega_T = 2\pi f_T \tag{8.3}$$

The simplified equivalent circuit of Figure 8.1 ignores the lateral current flow in the transistor base as well as the excess phase shifts due to the distributed nature of the actual transistor parameters. Therefore, it can be quite inaccurate at frequencies approaching the f_T of the transistor. For this purpose, a more accurate model, which takes into account the distributed nature of C_c across r_b and C_{CS} across r_{cs} must be utilized.

In the hybrid-π model of Figure 8.3, the presence of base–collector capacitance C_c causes the model to be bilateral, and makes manual calculations quite cumbersome. For an order-of-magnitude estimation of the device performance, the hybrid-π model can be further simplified into its approximate equivalent shown in Figure 8.4, where

$$C_t = C_\pi + C_c(1 + g_m R_L) \tag{8.4}$$

or, for $g_m R_L \gg 1$,

$$C_t \approx C_\pi + C_c g_m R_L \tag{8.5}$$

This is the so-called Miller-effect approximation, [1] which assumes that the load is purely resistive, and the capacitive loading effects at the output are negligible. Although the unilateral hybrid-π model of Figure 8.4 is very convenient to use in manual calculations, it can introduce significant errors even at frequencies as low as 5–10% of f_T, especially when capacitive loads are present. Since it does not indicate the excess phase shifts and nondominant poles associated with the transistor, it should not be used when calculating the response of feedback circuits.

The characterization and accurate modeling of active circuit components at high frequencies is a difficult problem. A number of modeling and measurement techniques have been developed which characterize the performance of bipolar devices to a higher degree of accuracy than the simple models of Figures 8.3 and

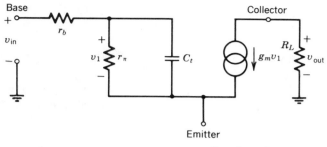

$$C_t = C_\pi + C_c(1 + g_m R_L)$$

FIGURE 8.4. Simplified hybrid-π model using Miller approximation.

8.4. However, these models of the devices require a large number of additional parameters to be defined and measured. Therefore, they are better suited to numerical rather than analytical computations. With the advances in the state of the art and the availability of computer-aided analysis and optimization techniques, the additional complexity of these models does not often present a serious problem.[2,3]

8.4. FREQUENCY RESPONSE OF SINGLE-TRANSISTOR GAIN STAGES

Common-Emitter Stage

The common-emitter (CE) gain stage shown in Figure 8.5 is by far the most commonly used gain block in analog design since it provides both voltage and current amplification. The basic circuit of Figure 8.5 can be used to describe either a single-ended inverting amplifier stage or the differential-mode half-circuit of a differential gain stage (see Fig. 5.3).

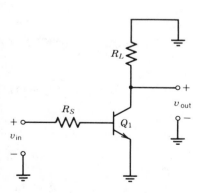

FIGURE 8.5. Basic common-emitter gain stage.

Its frequency response can be approximated using the simplified hybrid-π model of Figure 8.4 with Miller approximation,

$$A_v(s) = \frac{v_{\text{out}}}{v_{\text{in}}} = \frac{A_{v0}}{1 - s/p_1} \tag{8.6}$$

where $s = j\omega$ is the complex frequency variable, A_{v0} is the low-frequency small-signal gain, and p_1 is the location of the dominant pole. From Figure 8.4, the expressions for A_{v0} and p_1 can be derived as

$$A_{v0} = -g_m R_L \frac{r_\pi}{R_S + r_b + r_\pi} \tag{8.7}$$

For the case where $R_S + r_b \ll r_\pi$, this expression reduces to its familiar form

$$A_{v0} \approx -g_m R_L \tag{8.8}$$

The dominant pole p_1 can be expressed as

$$p_1 = -\frac{R_S + r_b + r_\pi}{(R_S + r_b)r_\pi} \frac{1}{C_t} \tag{8.9}$$

For a single-pole response, the -3-dB bandwidth $\omega_{3\text{dB}}$ is equal to the magnitude of p_1, that is,

$$\omega_{3\text{dB}} = |p_1| \approx \frac{R_S + r_b + r_\pi}{(R_S + r_b)r_\pi} \frac{1}{C_\pi + C_c g_m R_L} \tag{8.10}$$

where C_t is replaced by its approximate expression given by Eq. (8.5) for $g_m R_L \gg 1$.

It is instructive to investigate the implications of Eq. (8.10) for large and small values of load resistance R_L. In the case of the large values of R_L, the low-frequency voltage gain A_{v0} is high and the Miller-effect component of C_t dominates the frequency response. Conversely, if the $g_m R_L$ product is low, then the Miller effect is negligible and the 3-dB bandwidth is determined primarily by C_π. Each of these cases is briefly investigated below:

CASE 1.   Miller-Effect Dominant

The Miller effect is dominant when $g_m R_L C_c \gg C_\pi$, or from Eq. (8.2) this corresponds to the condition

$$R_L C_c \gg \frac{1}{\omega_T} \tag{8.11}$$

In this case, Eq. (8.10) simplifies to

$$\omega_{3\text{dB}} \approx \frac{R_S + r_b + r_\pi}{(R_S + r_b)r_\pi} \frac{1}{C_c g_m R_L} \tag{8.12}$$

and the amplifier bandwidth becomes *inversely* proportional to R_L or to the low-frequency gain A_{v0}.

CASE 2. Miller-Effect Not Dominant

If R_L is sufficiently low such that

$$\frac{1}{\omega_T} \gg R_L C_c \tag{8.13}$$

then $C_t \approx C_\pi$, and the Miller effect can be neglected. In that case, substituting Eq. (8.2) into Eq. (8.10) and rearranging terms, one obtains

$$\omega_{3dB} \approx \left(1 + \frac{r_\pi}{R_S + r_b}\right) \frac{\omega_T}{\beta_0} \tag{8.14}$$

which indicates that the bandwidth is determined primarily by the ω_T of the transistor, particularly if the source resistance is high such that $R_S \gg r_\pi$.

Emitter-Follower Stage

The emitter-follower, or common-collector (CC), stage is widely used in analog IC design as a buffer, level-shift, or output stage. The ac circuit schematic and the small-signal equivalent circuit of the emitter-follower stage are shown in Figure 8.6 For simplicity, the effects of the source resistance R_S and the r_b are combined into a single resistor R_B, where

$$R_B = R_S + r_b \tag{8.15}$$

Since the collector of the emitter follower is ac grounded, the Miller effect does not present a problem. The effect of C_c is to form a low-pass circuit with effective source resistance R_B, which would cause the voltage gain to decrease at very high frequencies.

Neglecting the effect of C_c, the voltage transfer function of the emitter

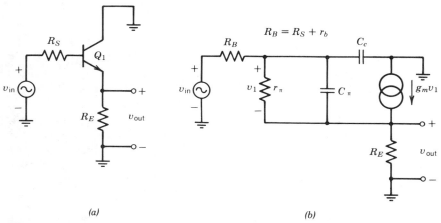

(a)   (b)

FIGURE 8.6. Emitter-follower stage: (a) Ac schematic; (b) small-signal equivalent circuit.

follower can be calculated from Figure 8.6*b* as an expression of the form [4]

$$A_v(s) = \frac{v_{\text{out}}}{v_{\text{in}}} = A_{v0} \frac{1 - s/z_1}{1 - s/p_1} \tag{8.16}$$

where the low-frequency gain A_{v0} can be written as

$$A_{v0} = \frac{g_m R_E + R_E/r_\pi}{1 + g_m R_E + (R_B + R_E)/r_\pi} \tag{8.17}$$

For practical values of $g_m R_E$, the product reduces to its familiar form

$$A_{v0} \approx \frac{g_m R_E}{1 + g_m R_E} \approx 1.0 \tag{8.18}$$

The transfer function of Eq. (8.16) has a zero at z_1,

$$z_1 \approx -\frac{g_m}{C_\pi} = -\omega_T \tag{8.19}$$

and a pole at p_1,

$$p_1 = -\frac{1}{C_\pi R_A} \tag{8.20}$$

where R_A is an effective resistance

$$R_A = r_\pi \left\| \frac{R_B + R_E}{1 + g_m R_E} \right. \tag{8.21}$$

For $g_m R_E \gg 1$, Eq. (8.21) reduces to

$$R_A \approx \frac{R_B + R_E}{g_m R_E} \tag{8.22}$$

and

$$p_1 \approx -\frac{R_E}{R_E + R_B} \omega_T \tag{8.23}$$

From Eqs. (8.19) and (8.23) one sees that the transfer function of an emitter follower contains a closely spaced pole–zero pair in the vicinity of $-\omega_T$, with the pole being closer to the origin. This pole moves closer toward the origin as R_S is increased.

Since the emitter follower circuit is often used as a buffer stage between two gain stages, its input and output impedance levels are also of interest. Starting with the equivalent circuit of Figure 8.6, neglecting the effects of C_c, and assuming that $g_m R_E \gg 1$, one can show that the small-signal input impedance can be described by the circuit model shown in Figure 8.7. The input impedance is almost purely resistive, and at low frequencies it reduces to

$$Z_{\text{in}} \bigg|_{\omega=0} = r_{\text{in}} \approx \beta_0 R_E \tag{8.24}$$

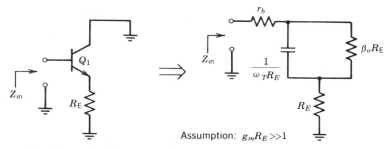

FIGURE 8.7. Equivalent circuit for calculating input impedance of emitter follower.

The output impedance of the emitter follower can also be calculated from the equivalent circuit of Figure 8.6 as

$$Z_{out} = \frac{Z_\pi + R_B}{1 + g_m Z_\pi} \tag{8.25}$$

where

$$Z_\pi = \frac{r_\pi}{1 + sC_\pi r_\pi} \tag{8.26}$$

is the impedance of the parallel combination of C_π and r_π.

Before simplifying Eq. (8.25) further, it is instructive to examine its values at very high and very low frequencies:

$$Z_{out}\bigg|_{\omega=0} = \frac{1}{g_m} + \frac{R_S + r_b}{\beta_o} \tag{8.27}$$

and

$$Z_{out}\bigg|_{\omega=\infty} = R_S + r_b \tag{8.28}$$

Equations (8.27) and (8.28) indicate that at low collector current levels, where g_m is low such that

$$\frac{1}{g_m} > R_S + r_b \tag{8.29}$$

$|Z_{out}|$ decreases monotonically with increasing frequency, and the output impedance appears capacitive. However, at moderate or high current levels, where g_m is high and the inequality (8.29) is *not* satisfied, $|Z_{out}|$ would be higher at high frequencies than at low frequencies. In other words, the output impedance would *increase* with increasing frequency and appear to be inductive. Following this reasoning, one can show that for the condition

$$\frac{1}{g_m} < R_S + r_b \tag{8.30}$$

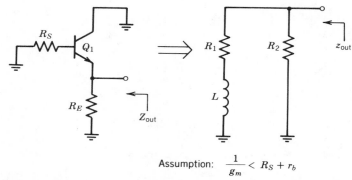

$$\text{Assumption:} \quad \frac{1}{g_m} < R_S + r_b$$

FIGURE 8.8. Output impedance of emitter follower under bias condition of $1/g_m < R_S + r_b$.

the output impedance can be approximated by the equivalent circuit of Figure 8.8, where

$$R_1 = \frac{1}{g_m} + \frac{R_S + r_b}{\beta_o} \tag{8.31}$$

$$R_2 = R_S + r_b \tag{8.32}$$

$$L = \frac{1}{\omega_T} (R_S + r_b) \tag{8.33}$$

One important conclusion of the above derivation is that when an emitter-follower output is forced to drive capacitive loads, the inductive component of its output impedance may cause peaking or instability in the overall circuit response.

Common-Base Stage

The common-base gain stage offers nearly unity current gain and very wide bandwidth along with low input impedance and high output impedance. The ac circuit schematic and its small-signal equivalent circuit are shown in Figure 8.9.

Comparing Figure 8.9 with Figure 8.8, one observes that the input impedance of the common-base stage is essentially the same as the output impedance of the emitter follower with $R_S = 0$ [see Eqs. (8.31) through (8.33)]. At low frequencies, the input impedance is approximately $1/g_m$. However, at high frequencies it can become inductive due to the presence of r_b. The output impedance of a common-base stage is approximately $\beta_o r_o$ at low frequencies, where r_o is the common-emitter stage output resistance due to the Early effect. At high frequencies, this resistance is shunted by C_c and the collector substrate capacitance C_{CS}.

The current transfer characteristics of the common-base stage can be closely approximated as

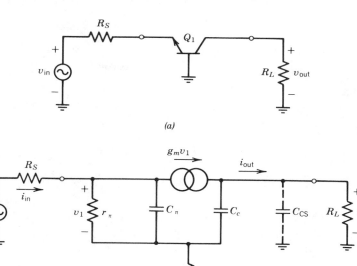

(a)

(b)

FIGURE 8.9. Common-base gain stage: (*a*) Ac circuit schematic; (*b*) Small-signal equivalent circuit.

$$A_i = \frac{i_{\text{out}}}{i_{\text{in}}} \approx \frac{\alpha_o}{1 + s/\omega_T} \qquad (8.34)$$

where

$$\alpha_0 = \frac{\beta_o}{1 + \beta_o} \approx 1.0 \qquad (8.35)$$

is the current gain. This corresponds to a current gain of nearly unity, with a pole p_1 at $-\omega_T$.

The voltage transfer function can be readily derived from Eq. (8.33) as

$$A_v = \frac{v_{\text{out}}}{v_{\text{in}}} = \frac{i_{\text{out}} Z_L}{i_{\text{in}} R_S} \qquad (8.36)$$

where Z_L is the effective impedance at the output, made up of R_L shunted by C_c and C_{CS}, that is,

$$Z_L = R_L \parallel (C_c + C_{\text{CS}}) = \frac{R_L}{1 + sR_L(C_c + C_{\text{CS}})} \qquad (8.37)$$

This results in a two-pole frequency response of the form

$$A_v = A_{v0} \frac{1}{1 - s/p_1} \frac{1}{1 - s/p_2} \qquad (8.38)$$

with

$$A_{v0} = \frac{\alpha_o R_L}{R_S} \tag{8.39}$$

where $R_S \gg 1/g_m$ is assumed.

The first pole at p_1 is due to the current transfer function, where

$$p_1 \approx -\omega_T \tag{8.40}$$

The second pole at p_2 is due to capacitive loading at the output,

$$p_2 \approx -\frac{1}{R_L(C_c + C_{\text{CS}})} \tag{8.41}$$

To summarize, the common-base stage is basically a wideband unity-gain current amplifier with low-input and high-output impedance. It can be converted to a voltage amplifier by appropriate choice of R_S and R_L. It provides no phase inversion between the input and output signals. Since the Miller-multiplication effect is not present for C_c, the circuit provides good signal isolation between input and output terminals.

8.5. COMPOUND DEVICES

The frequency performance of an amplifier stage can be significantly improved by using multiple active devices as direct replacements for single transistor stages. In monolithic IC design, where additional active devices are readily available, the use of such compound devices is much more feasible than in discrete designs. The most commonly used compound device configurations are the common-collector–common-emitter (CC–CE) and the common-emitter–common-base (CE–CB) connections shown in Figures 8.10 and 8.11. The transistor pairs within the dashed lines essentially correspond to a common-emitter stage with improved high-frequency performance. The two compound device configurations of Figures 8.10 and 8.11 are often referred to as the Darlington and the cascode connections. Some of their features and characteristics are reviewed below.

Common-Collector–Common-Emitter Stage

The common-collector–common-emitter (CC–CE) stage shown in Figure 8.10 is essentially a direct extension of the Darlington configuration with the collector of Q_1 connected to an ac ground point. Normally, Q_1 is biased to operate with a relatively high collector current, comparable to that of Q_2, in order to optimize its f_T. In this compound device connection, Q_1 effectively buffers the gain stage Q_2 from the input. Therefore, the capacitive loading of the input due to the Miller capacitance of Q_2 is greatly reduced. This results in an improved

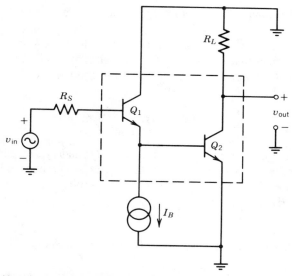

FIGURE 8.10. An ac circuit diagram for common-collector–common-emitter stage.

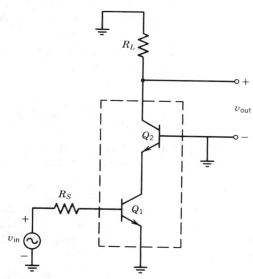

FIGURE 8.11. Cascode or common-emitter–common-base stage.

gain–bandwidth product for the compound device over a single common-emitter stage.

The low-frequency gain of the common-collector–common-emitter pair is somewhat lower than that of a single common-emitter stage, due to the scaling down of the input voltage across the base–emitter junctions of two transistors, and can be expressed as

$$A_{v0} = \left.\frac{v_{out}}{v_{in}}\right|_{\omega=0} \approx \frac{-g_{m1}g_{m2}}{g_{m1} + g_{m2}} R_L \qquad (8.42)$$

where g_{m1} and g_{m2} are the transconductances of Q_1 and Q_2, respectively.

The 3-dB bandwidth of the common-collector–common-emitter stage is significantly better than that of a single common–emitter stage, particularly for large values of R_S and R_L.

Common-Emitter–Common-Base Stage (Cascode Connection)

The cascode connection formed by the combination of common-emitter and common-base stages, shown in Figure 8.11, is very often used in wideband amplifier design. The improvement in high-frequency performance is obtained by the impedance mismatch between the output of the common-emitter stage (i.e., the collector of Q_1) and the input of the common–base stage. The collector load of Q_1 is formed by the input impedance of Q_2. Since Q_2 is operated in a common-base configuration, its input impedance is approximately equal to $1/g_m$. Thus, the influence of the Miller effect on Q_1 is minimal even for large values of R_L. As a result, the cascode connection has significantly better frequency response than the single common-emitter stage, especially for large values of R_L.

The low-frequency voltage gain of a cascode circuit is approximately the same as that of a common-emitter stage, as given in Eq. (8.7), with the addition of the factor α_0, that is,

$$A_{v0} = \frac{v_{out}}{v_{in}} = -\frac{\alpha_o g_{m1} r_{\pi 1} R_L}{R_S + r_{b1} + r_{\pi 1}} \qquad (8.43)$$

where α_o is the low-frequency common-base current gain [see Eq. (8.35)].

For the cases where $R_S + r_b \ll r_\pi$, this expression reduces to

$$A_{v0} \approx -\alpha_o g_{m1} R_L \approx -g_{m1} R_L \qquad (8.44)$$

The frequency response of the cascode stage can be approximated from the transfer characteristics of the common-emitter and common-base stages discussed in the previous section. The overall transfer function can be approximated with a three-pole response of the form

$$A_v = A_{v0} \frac{1}{1 - s/p_1} \frac{1}{1 - s/p_2} \frac{1}{1 - s/p_3} \qquad (8.45)$$

where p_1 is the pole contributed by the common-emitter stage, and p_2 and p_3 are the poles contributed by the common-base stage.

The value of p_1 is closely approximated by Eq. (8.14) as

$$p_1 \approx -\left(1 + \frac{r_1}{R_S + r_{b1}}\right)\frac{\omega_T}{\beta_o} \qquad (8.46)$$

The values of p_2 and p_3 are those given by Eqs. (8.40) and (8.41), respectively,

$$p_2 \approx -\frac{1}{R_L C_c + C_{CS}} \qquad (8.47)$$

and

$$p_3 \approx -\omega_T \qquad (8.48)$$

Note that the common-emitter stage dominant pole is influenced primarily by R_S, whereas the common-base stage pole is influenced by R_L. Normally, for operation with large values of R_L and R_S, only p_1 and p_2 are dominant, and p_3 can be neglected.

The cascode configuration exhibits a higher degree of isolation (i.e., a minimum amount of reverse signal transmission) between its input and output terminals than a simple common-emitter transistor. This isolation is obtained because the reverse transmission across the compound device stage is greatly reduced by the impedance mismatches between Q_1 and Q_2. In other words, Q_2 performs the function of an impedance transformer. This effect makes the cascode configurations particularly attractive for the design of high-frequency tuned amplifier stages where the parasitic cross-coupling between the input and output tuned circuits can make the alignment of the stage difficult.

Since both of these compound-device configurations result in an inverting gain stage, they are often used as a direct replacement for a single common-emitter stage at critical points within the circuit. However, when using the compound device connections, one word of caution is in order. The use of a second transistor adds an additional time constant or a pole to the overall circuit response. In the case of the common-collector–common-emitter pair, the magnitude of this pole becomes comparable to the dominant pole of the stage as the source resistance R_S is increased. The same is also true for the cascode circuit for increasing values of R_L [see Eq. (8.46)]. The common-collector–common-emitter pair and the cascode stage offer the most improvement in their 3-dB bandwidth for large values of R_S *and* R_L, respectively. This implies that when these compound devices are used for gain–bandwidth optimization, they can cause the circuit to exhibit rapid high-frequency rolloff and excess phase lag. Therefore, additional care must be taken in applying feedback around the amplifier stages containing compound devices.

Figure 8.12 gives a practical example of a wideband amplifier (RCA CA3040) which uses compound devices for improved high-frequency performance. This circuit also illustrates the possibility of interconnecting both the

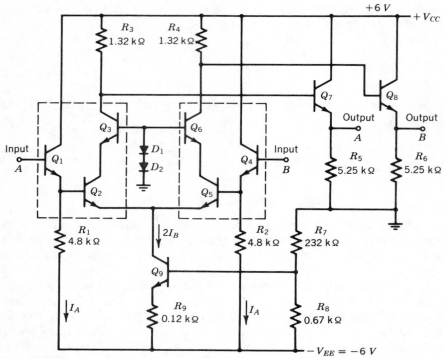

FIGURE 8.12. Wideband amplifier circuit using compound devices. (RCA CA-3040.)

Darlington and the cascode pairs into a three-transistor compound device shown within the dashed lines. These compound devices are then formed into a fully differential gain stage. The common-collector stages Q_7 and Q_8 are then used to provide low-impedance differential outputs for the circuit.

The low-frequency voltage gain of the circuit for single-ended operation can be expressed as

$$A_{v0} \approx - g_{mc}R_3 \qquad (8.49)$$

where g_{mc} is the transconductance of the compound transistor made up of Q_1 and Q_2. From Eq. (8.42),

$$g_{mc} = \frac{g_{m1}g_{m2}}{g_{m1} + g_{m2}} \qquad (8.50)$$

where the subscripts refer to Q_1 and Q_2, respectively.

For the component and bias values shown in Figure 8.12, the circuit exhibits a single-ended voltage gain of 30 dB with a -3-dB bandwidth of 55 MHz, with 500-MHz f_T transistors. Due to the use of compound devices, the frequency response exhibits a rapid falloff (typically -18 dB/octave) at high frequencies.

Since the circuit of Figure 8.12 does *not* use any feedback for gain or bias stabilization, the gain characteristics given by Eq. (8.12) exhibit poor temperature stability due to variations of g_m and R_3 with temperature. In the design, an attempt is made to partially compensate for this drift by causing the bias current $2I_B$ in Q_9 to exhibit a *positive* temperature coefficient.

8.6. NEUTRALIZATION OF COLLECTOR–BASE CAPACITANCE

The parasitic collector–base feedback capacitance C_c is the most significant limitation in the frequency response of a common-emitter gain stage. In the case of a differential amplifier stage, this parasitic capacitance can be reduced significantly by utilizing a bridge-neutralization scheme, as shown in Figure 8.13.[5] In a balanced differential stage, the ac voltages at the collectors of Q_1 and Q_2 have the same amplitude, but differ in phase by 180°. Therefore, the feedback to the bases of Q_1 and Q_2 through C_2 and C_1 is out of phase with the internal feedback due to the C_c of each transistor. By choosing C_1 and C_2 to have the same value as the C_c, it is possible to neutralize the parasitic feedback through C_c by feeding in an equal but opposite signal.

In discrete (nonintegrated) circuit design, this neutralization technique is not practical because an efficient neutralization of C_c requires that feedback capacitors C_1 and C_2 be matched to C_c to better than $\pm$ 10%. In practical devices, C_c is less than 1 pF, and its absolute value is predictable, at best, to $\pm$ 20%. Furthermore; C_c is a function of the device operating point, device geometry, and temperature. Discrete capacitors cannot match or track these variations of the base–collector capacitance and cannot be effectively used in the neutralization scheme of Figure 8.13*a*.

Since the C_c neutralization scheme requires matching and tracking of the components, rather than accurate absolute values, it is ideally suited to monolithic circuit design. Feedback capacitances C_1 and C_2 can be realized in integrated form as the collector–base capacitances of additional "dummy" transistors Q_3 and Q_4, as shown in Figure 8.13*b*. All four transistors in the figure can be designed to have identical geometries and are located in close proximity on the same chip. Therefore, they exhibit nearly ideal matching and thermal tracking of their characteristics. The collector–base capacitances of Q_3 and Q_4 very nearly match the C_c of Q_1 and Q_2, and when cross-coupled as shown in Figure 8.13*b*, they provide an efficient neutralization of C_c associated with Q_1 and Q_2. The reverse transmission through C_c is most troublesome in the design of bandpass intermediate-frequency amplifiers where the reverse interaction between the input and the output of the stage makes the alignment or tuning of multiple gain stages quite difficult. Using the neutralization scheme of Figure 8.13, the reverse transmission through C_c can be reduced by as much as 95% over a wide range of operating conditions.

It should be noted that in the case of a monolithic circuit, the added cost or complexity due to the neutralizing devices is negligible. Q_3 and Q_4 have their

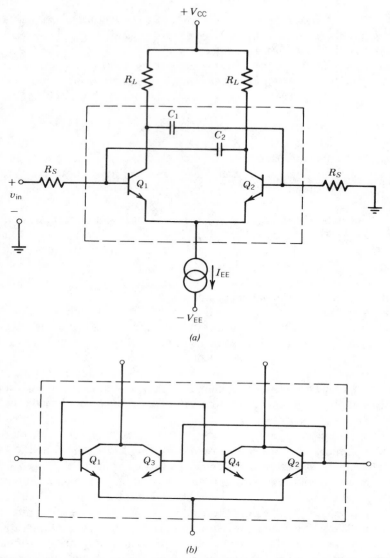

FIGURE 8.13. Neutralization of collector–base capacitance in differential designs: (*a*) Circuit schematic; (*b*) actual implementation of C_1 and C_2 with additional transistors.[5]

collectors common with Q_1 and Q_2, respectively. Therefore, they do not require separate isolation islands. Thus, the increase of the circuit chip size due to the addition of Q_3 and Q_4 is negligible.

The neutralization scheme of Figure 8.13 has one serious drawback. It relies on the presence of symmetrical dc and asymmetrical ac signals in the circuit. This condition is realized only in differential gain stages. Therefore, the application of the C_c neutralization technique is limited to differential designs where asymmetrical input or output loading effects are not present.

8.7. AMPLIFIER CIRCUITS USING LOCAL FEEDBACK

In designing wideband amplifiers, one relies heavily on the use of feedback to exchange gain for bandwidth and to optimize the high-frequency performance. In an amplifier circuit involving multiple gain stages, one has the choice of applying the feedback locally (i.e., around each individual gain stage) or around the overall circuit. Local feedback is somewhat easier to apply from an overall frequency stability point of view, since it allows each gain stage to be designed relatively independent of the others.

Local feedback is a direct carryover from discrete design techniques where each individual feedback stage can be independently biased and ac coupled to each other. However, in monolithic circuits where no ac coupling is available, the number of separate feedback stages which can be directly cascaded without requiring extensive dc level shifting is limited to two or at most three gain stages. Most of the dc level-shift circuits discussed in Section 4.6 are not suitable for wideband amplifier design because of their frequency limitations or their noise characteristics.

Single-Transistor Feedback Stages

The single-transistor amplifier stage can be used with either shunt or series local feedback around it. Figure 8.14 shows the basic shunt-feedback gain stage and its low-frequency small-signal equivalent circuit. The shunt-feedback stage is basically a *transimpedance* amplifier which provides a voltage output from a

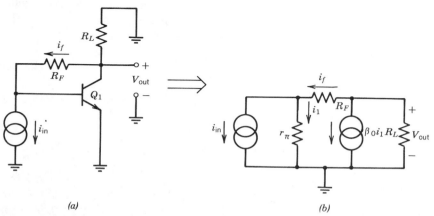

(a)   *(b)*

FIGURE 8.14. Single-stage shunt-feedback circuit: (*a*) ac circuit connection; (*b*) low-frequency small-signal equivalent circuit.

current input, and has low input and output impedance levels. Using the low-frequency equivalent circuit of Figure 8.14b, one can calculate its transimpedance gain as

$$A = \frac{v_{\text{out}}}{i_{\text{in}}} = \frac{-R_F}{1 + \dfrac{(R_L + R_F + R_L r_\pi / R_F)}{\beta_o R_L}} \tag{8.51}$$

For the case where $\beta_0 \gg 1$, this expression reduces to

$$A = \frac{v_{\text{out}}}{i_{\text{in}}} \approx - R_L \tag{8.52}$$

Its output resistance can be calculated from the small-signal equivalent circuit as

$$R_{\text{out}} = \frac{1}{g_m} + \frac{R_F}{\beta_o} \tag{8.53}$$

where $\beta_o \gg 1$ is assumed.

The input resistance R_{in} can also be calculated in a similar manner,

$$R_{\text{in}} \approx r_\pi \left\| \frac{R_F + R_L}{g_m R_L} \right. \tag{8.54}$$

where $g_m R_L \gg 1$ is assumed. In the case where $\beta_o \gg 1$, the above expression further simplifies to

$$R_{\text{in}} \approx \frac{R_F + R_L}{g_m R_L} \tag{8.55}$$

As indicated by Eqs. (8.53) and (8.55), both the input and the output resistances of the circuit are quite low.

Due to its low input resistance, the shunt-feedback stage is most advantageous to use with a high-impedance signal source. Assuming that the circuit of Figure 8.14 is driven from a signal source of internal resistance R_S such that $R_S \gg R_{\text{in}}$, then the input current i_{in} is primarily determined by R_S as

$$i_{\text{in}} \approx \frac{v_{\text{in}}}{R_S} \tag{8.56}$$

and the total low-frequency voltage gain A_v of the stage becomes, from Eq. (8.52),

$$A_v = \frac{v_{\text{out}}}{v_{\text{in}}} = \frac{v_{\text{out}}}{i_{\text{in}}} \frac{i_{\text{in}}}{v_{\text{in}}} \approx - \frac{R_F}{R_S} \tag{8.57}$$

which is essentially the same as that of an op amp with feedback around it [see Eq. (7.2)]. However, it should be remembered that since the overall open-loop gain levels obtainable from a single transistor are much lower than those from complete op amp circuits, the approximations made in deriving Eqs. (8.52) and (8.57) are much *less* accurate than those for an op amp.

The local series-feedback gain stage shown in Figure 8.15 is virtually a complement of the shunt-feedback stage since it offers an almost totally opposite set of performance characteristics. Such a series-feedback stage is basically a *transconductance* amplifier which provides a current output from a voltage input, and has high input and output impedance levels. Using the low-frequency equivalent circuit of Figure 8.15b, one can calculate the transconductance gain as

$$A = \frac{i_{out}}{v_{in}} = \frac{1}{R_E} \frac{1}{1 + (1/R_E)[1/g_m + (r_b + R_E)/\beta_o]} \qquad (8.58)$$

In all practical cases where $\beta_o \gg 1$ and $g_m R_E \gg 1$, Eq. (8.58) reduces to its simplified form

$$A = \frac{i_{out}}{v_{in}} \approx -\frac{1}{R_E} \qquad (8.59)$$

The input resistance of the circuit R_{in} looking into the base of Q_1 can be calculated from Figure 8.15b as

$$R_{in} = r_b + r_\pi + (\beta_o + 1)R_E \qquad (8.60)$$

which reduces to

$$R_{in} \approx \beta_o R_E \qquad (8.61)$$

for the cases where $g_m R_E \gg 1$ and $\beta_o \gg 1$.

The output resistance R_{out} looking into the collector of Q_1 can also be calculated from Figure 8.15b as

$$R_{out} \approx r_o \left(1 + \frac{\beta_o g_m R_E}{\beta_o + g_m R_E} \right) \qquad (8.62)$$

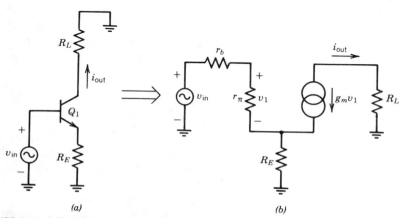

FIGURE 8.15. Single-stage series-feedback circuit: (*a*) ac circuit connection; (*b*) low-frequency small-signal equivalent circuit.

where r_o is the normal output resistance of a common-emitter stage due to the Early effect. The presence of a series-feedback resistor R_E causes the output resistance of the stage to increase, as indicated by Eq. (8.62).

Assuming that the load resistance $R_L \ll R_{out}$, the output voltage is

$$v_{out} \approx i_{out} R_L \tag{8.63}$$

and the total voltage gain of the stage becomes, from Eq. (8.59),

$$A_v = \frac{v_{out}}{v_{in}} = \frac{i_{out}}{v_{in}} \frac{v_{out}}{i_{out}} \approx -\frac{R_L}{R_E} \tag{8.64}$$

Feedback Cascades

In cascading individual feedback stages, it is advantageous to alternate between series-feedback and shunt-feedback stages such that the impedance level requirements for each type of stage will be satisfied. In this manner, the shunt-feedback stage can provide the low-impedance drive for the series-feedback stage, and conversely the series-feedback stage output can serve as the high-impedance drive for the shunt-feedback stage succeeding it.

Figure 8.16 shows the two basic feedback cascade pairs used in wideband amplifier design. The series–shunt cascade of Figure 8.16a operates best as a voltage amplifier since it offers relatively high input impedance and low output impedance.

From Eq. (8.59) and (8.52), the overall voltage gain A_v for the series–shunt cascade can be written as

$$A_v = \frac{v_{out}}{v_{in}} = \left(\frac{i_{o1}}{v_{in}}\right)\left(\frac{v_{out}}{i_{o1}}\right) \approx \frac{R_F}{R_E} \tag{8.65}$$

where i_{o1} is the output current of the series–feedback stage. The circuit offers a high input resistance and a low output resistance, as given by Eqs. (8.61) and (8.53), respectively.

The shunt–series cascade of Figure 8.16b is useful as a wideband current amplifier since it offers low input impedance and high output impedance. Its overall current gain can be approximated from Eqs. (8.52) and (8.59) as

$$A_i = \frac{i_{out}}{i_{in}} = \left(\frac{v_{o1}}{i_{in}}\right)\left(\frac{i_{out}}{v_{o1}}\right) \approx \frac{R_F}{R_E} \tag{8.66}$$

where v_{o1} is the output voltage at the collector of Q_1.

Feedback cascades show somewhat poorer dc stability than other types of wideband amplifier configurations which use overall feedback. In monolithic integrated circuits, where a large number of active devices are readily available, this dc bias stability problem can be partly eliminated by using a differential circuit configuration. Since a differential stage inherently employs a high degree of common-mode feedback (see Chapter 5), it offers a higher degree of dc bias stability and also simplifies dc level-shift problems. Both of the feedback cascade configurations of Figure 8.16 are readily adaptable to differential design.

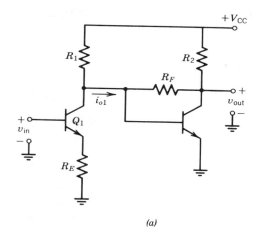

(a)

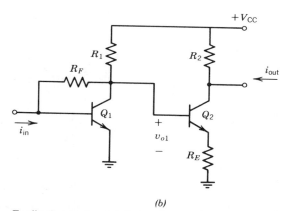

(b)

FIGURE 8.16. Feedback cascades: (*a*) Series–shunt cascade; (*b*) shunt–series cascade.

Figure 8.17 shows a differential series–shunt cascade circuit especially designed for wideband applications. The overall single–ended gain of the circuit can be approximated as

$$A_v \approx \frac{R_{11}}{R_3 + R_4 + 1/g_m} \qquad (8.67)$$

where g_{m1} is the transconductance of Q_1.

The gain of the circuit can be adjusted externally by changing the amount of series feedback in the first stage. This can be done by externally short-circuiting the resistor taps A, A' and B, B', which cause R_4 and $R_3 + R_4$ to drop out of the series feedback and, thus, increase the gain.

Figure 8.18 shows the single-ended frequency response for the wideband amplifier stage of Figure 8.17 with various values of gain for $\pm$ 6-V operation of the circuit. The photomicrograph of the monolithic circuit chip is shown in

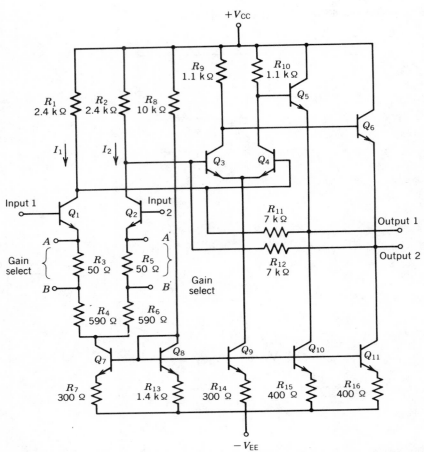

FIGURE 8.17. Differential wideband amplifier using series–shunt feedback cascade configuration (Fairchild μA-733.)

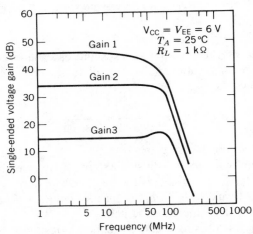

FIGURE 8.18. Typical frequency response for series–shunt cascade amplifier of Figure 8.17, using transistors with $f_T \approx 800$ MHz. (*Note*: Gain 1, A–A' short-circuited; gain 2, B–B' short-circuited; gain 3, all taps open.)

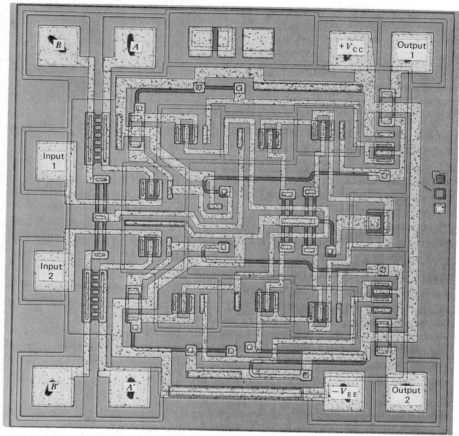

FIGURE 8.19. Photomicrograph of differential–shunt cascade amplifier. Chip size: 45 mils × 45 mils. (*Photo*: Fairchild Semiconductor.)

Figure 8.19. Note that all of the transistors in the high-frequency signal path employ stripe geometry in the device layout with double-base contacts in order to optimize high-frequency performance and reduce the noise contribution due to base resistance r_b.

8.8. AMPLIFIER CIRCUITS USING OVERALL FEEDBACK

An alternate approach to the design of wideband amplifiers is the use of overall feedback around a number of gain stages, rather than applying local feedback around each individual stage. As compared with the local feedback cascades described in the previous section, the overall feedback offers a higher degree of bias stability and desensitivity to individual gain tolerances. However, since the overall feedback approach involves a larger number of active devices within the feedback loop, it requires more careful consideration of the nondominant poles and excess phase shifts associated with the active devices to ensure stability. In general, stability considerations limit to three the total number of gain stages which can be enclosed in a single overall feedback loop. Figures 8.20, 8.21, and 8.23 show some of the basic feedback configurations suitable for monolithic design. Depending on the number of active gain stages within the loop, these circuits can be classified as *feedback pairs* or *feedback triples*.

Feedback Pairs

Figure 8.20 shows the two basic feedback pair configurations often used in wideband amplifier design. These feedback pairs are normally described or classified by the type of feedback associated with their input and output ports as the series–shunt pair and the shunt–series pair respectively.

The series–shunt pair of Figure 8.20*a* is a convenient circuit configuration for broadband voltage amplification. Because of the series feedback associated with the input stage, the circuit offers a high input impedance. Similarly, the shunt feedback associated with the output results in a low output impedance. Assuming that transistor collector resistors R_1 and R_2 are chosen to provide a sufficiently large open-loop gain, the overall closed-loop voltage gain, at low frequencies, can be approximated as

$$A_v = \frac{v_{\text{out}}}{v_{\text{in}}} \approx \frac{R_E + R_F}{R_E} \tag{8.68}$$

The shunt-series feedback pair of Figure 8.20*b* has a low input and a high output impedance. Therefore, it is particularly useful as a current amplifier. Assuming that the overall open-loop gain of the stage is sufficiently high, and

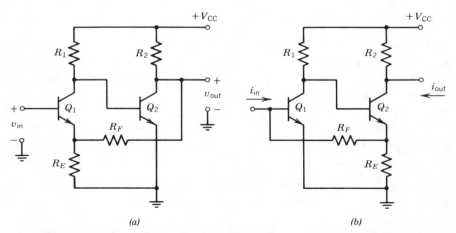

FIGURE 8.20. Basic feedback pairs: (*a*) Series–shunt pair; (*b*) shunt–series pair.

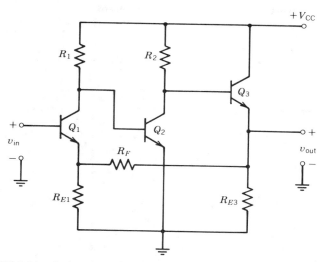

FIGURE 8.21. Series–shunt feedback circuit with emitter-follower buffer stage.

that the circuit is driven from a high-impedance source, its overall current gain is determined by the feedback elements

$$A_i = \frac{i_{\text{out}}}{i_{\text{in}}} \approx \frac{R_E + R_F}{R_E} \tag{8.69}$$

Because of its high input impedance and low output impedance characteristics, the series–shunt feedback stage is more commonly used than its shunt–series counterpart. In most designs, it is also convenient to add an emitter-follower buffer stage into the circuit, as shown in Figure 8.21, both to reduce the output impedance further and to buffer the forward and reverse signal flow paths from each other. In this case, the low-frequency gain of the circuit is still given by the approximate expression of Eq. (8.68), with R_E replaced by R_{E1}.

Figure 8.22 illustrates a practical circuit example (Plessey SL-611C) based on the series–shunt feedback principle. Although the actual circuit diagram of Figure 8.22 contains a number of additional components and features, the basic configuration is based on the circuit topology of Figure 8.21, with Q_1, Q_2, and Q_3 forming the forward signal path and R_F and R_E forming the shunt and series portions of the feedback network. The nominal values of various circuit components are shown on the schematic. Normally, the input and bias terminals are connected together so that the circuit can adjust the internal dc bias levels in accordance with the dc level of the input. This is necessary since the circuit is designed as a single-ended (i.e., nondifferential) amplifier. Therefore, it does not have the inherent common-mode rejection properties associated with differential designs. The current source Q_4 tracks the dc level of the input signal by shunting additional dc current through R_2 to keep Q_2 out of saturation.

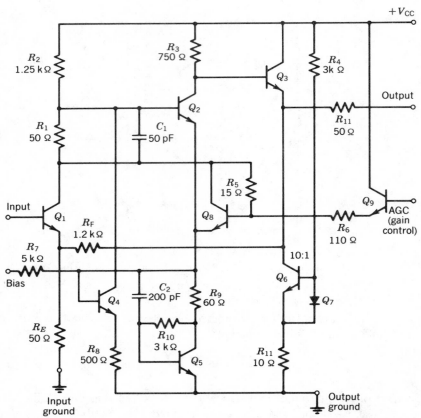

FIGURE 8.22. Wideband amplifier circuit using series–shunt feedback. (Plessey Semiconductor SL-611C.)

The circuit is designed to operate with a single supply voltage in the range of 6–9 V. With gain control off (i.e., the automatic gain control (AGC) terminal grounded), the amplifier provides a nominal voltage gain of 20 with a 3-dB bandwidth of approximately 100 MHz. With the gain control deactivated, the closed-loop voltage gain is set primarily by the choice of R_F and R_E, in accordance with the approximate gain expression of Eq. (8.68).

Feedback Triples

The feedback triples employ three gain stages within the feedback loop. Therefore, they offer a higher open-loop gain and bias desensitivity than the feedback pairs. However, the presence of additional gain stages also complicates the stability problems and makes the optimization of the circuit frequency response more difficult. The most commonly used feedback triple is the series–series triple configuration shown in Figure 8.23. Because of the series-feedback

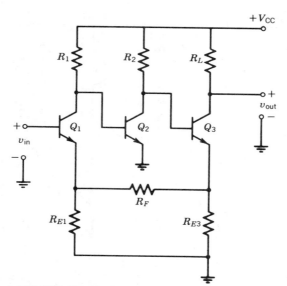

FIGURE 8.23. Basic series–series triple configuration.

configuration used, the circuit has a high input impedance and operates best as a voltage amplifier. When driven from a low source impedance, its low-frequency voltage gain can be approximated as [6]

$$A_v = \frac{v_{out}}{v_{in}} \approx - \frac{R_L(R_{E1} + R_{E3} + R_F)}{R_{E1}R_{E3}} \tag{8.70}$$

The large dc loop gain obtainable in the series–series triple results in a high degree of bias stability. Because of this advantage, the series–series triple can be integrated without using a differential configuration and common-mode feedback for stability.

Figure 8.24 shows a complete circuit schematic for an integrated wideband amplifier circuit using the series–series triple configuration.[7] Note that the circuit contains a large number of additional components and active devices in addition to the basic series–series triple. The triple is formed by transistors Q_1, Q_2, and Q_3 and the feedback resistors R_{E1}, R_{E3}, and R_F. The emitter follower Q_4 is used to provide a low-impedance output for the circuit. C_p is a small ($\approx$ 2-pF) internal capacitance providing additional broadbanding for the circuit. This particular method of broadbanding, called *pole-splitting*, was briefly mentioned in Section 7 and will be described further in Section 8.10. The capacitor C_F across R_F also aids in broadbanding the circuit frequency response by generating a so-called *feedback zero*, as will be discussed in Section 8.10.

In addition to the basic feedback loop, the circuit of Figure 8.24 contains a shunt dc feedback path formed by R_3, Q_6, and R_5, which both provides a stabilized dc bias to input transistor Q_1 and also keeps the output dc level at

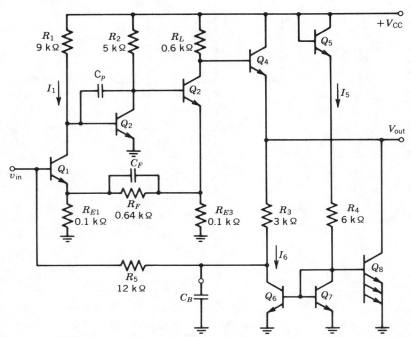

FIGURE 8.24. Wideband amplifier circuit using the series–series triple configuration. (Motorola MC-1553.)

approximately $V_{CC}/2$, relatively independent of component absolute-value tolerances and power supply voltages. The external bypass capacitor C_B is used to eliminate ac coupling between the input and the output through this bias feedback path. If the circuit is driven from a low-impedance source, C_B may not be necessary.

The bias feedback scheme used in the series–series triple of Figure 8.24 gives a good illustration of the efficient use of the two key advantages of monolithic integrated circuits: availability of a large number of active devices and the close matching and tracking of component values. If one makes the reasonable assumption that all transistor V_{BE} drops are well matched, and that the transistor current gain $\beta_o \gg 1$, then with reference to Figure 8.24, the output dc level V_{ODC} can be written as

$$V_{ODC} = I_1 R_{E1} + V_{BE} + I_6 R_3 \qquad (8.71)$$

Since Q_6 and Q_7 are well matched and $\beta_o \gg 1$, then currents Q_5 and Q_6 are approximately equal and given as

$$I_5 \approx I_6 = \frac{V_{CC} - 2V_{BE}}{R_4} \qquad (8.72)$$

Similarly, the collector of Q_1 is only 1 V_{BE} above ground. Therefore, I_1 can be related to V_{CC} as

$$I_1 = \frac{V_{CC} - V_{BE}}{R_1} \tag{8.73}$$

Combining Eqs. (8.72) and (8.73), one can then rewrite V_{ODC} as

$$V_{ODC} \approx V_{CC}\left(\frac{R_{E1}}{R_1} + \frac{R_3}{R_4}\right) + V_{BE}\left(1 - \frac{2R_3}{R_4} - \frac{R_{E1}}{R_1}\right) \tag{8.74}$$

By proper choice of resistor values, the second term of Eq. (8.74) can be made negligibly small, thus making V_{ODC} proportional to V_{CC} and set by the resistor ratios alone. This can be implemented by using the nominal resistor values shown in Figure 8.24.

By substituting these values into Eq. (8.74), one can readily show that the output dc level is almost exactly equal to $V_{CC}/2$, thus ensuring a maximum output voltage swing for the given choice of supply voltage. Once the bias levels within the circuit are established, the closed-loop gain can be determined by the proper choice of the ac feedback resistor R_F, as given by Eq. (8.70). Note that since both ends of R_F are at approximately the same dc potential, negligible bias current flows through R_F. Therefore, dc bias conditions are not affected by the choice of R_F to obtain a predetermined closed-loop voltage gain.

The series–series triple circuit of Figure 8.24 is designed to operate with a nominal current drain of 11 mA at $V_{CC} = 6$ V. When driven from a 50–Ω source, the small-signal 3-dB bandwidth of the amplifier is approximately 45 MHz and 50 MHz for voltage gains of 100 and 50, respectively. As a consequence of overall feedback, the effect of device nonlinearities, temperature dependence of the output dc level, and voltage gain are greatly reduced. Total harmonic distortion at the output is less than 0.2% and the gain variation with temperature is less than ± 0.25 dB over a $-55 + 125°C$ operating temperature range.

8.9. DUAL-LOOP FEEDBACK AMPLIFIERS

Often, as described in Section 8.1, one of the important design requirements for wideband amplifiers is the accurate control of input and output impedance levels. This is especially important if the amplifier is to interface with matched source and load impedances. In order to control both the input and the output impedance levels, one has to use a *dual-loop* feedback configuration comprised of both shunt- and series-feedback paths.

Figure 8.25 shows the basic single-stage dual-loop amplifier and its cascode version. The simultaneous use of both series- and shunt-feedback gives rise to broadband resistive input and output impedances. If the amplifier is required to operate with balanced source and load impedances, such that $R_S = R_L = R$, it

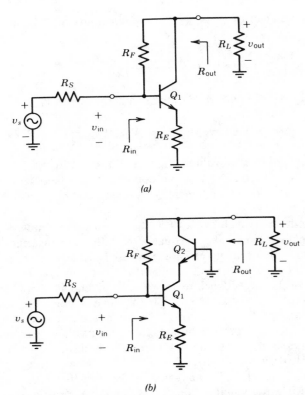

(a)

(b)

FIGURE 8.25. Single-stage dual-loop amplifiers: (a) Basic single-stage amplifier; (b) its cascode version.

can be shown that the approximate expressions for the input and output resistances R_{in} and R_{out} can be written as [8]

$$R_{in} \approx \frac{R_F R_E (R_F + R)}{R_F R_E + R_F R + R_E R} \qquad (8.75)$$

and

$$R_{out} \approx \frac{R_F + R}{1 + R/R_E} \qquad (8.76)$$

In deriving Eq. (8.75) and (8.76), it is assumed that the transistor stage has sufficient gain such that $g_m R_E \gg 1$. Thus, by proper choice of R_F and R_E, both R_{out} and R_{in} can be set equal to R to achieve matched input and output impedances. Under matched conditions, such that

$$R = R_S = R_{in} = R_{out} = R_L \qquad (8.77)$$

one can show that the loaded voltage gain A_v can be given as

$$A_v = \frac{v_{out}}{v_{in}} \approx -\frac{1}{R_E}\frac{RR_F}{R + R_F} \tag{8.78}$$

where $g_m R_E \gg 1$ is assumed.

In many applications, the two-stage dual-loop feedback circuit of Figure 8.26 is preferred over single-stage feedback stages, since the higher open-loop gain available from the two-stage amplifier results in lower distortion and a higher overall gain–bandwidth product. Using the analysis given in Ref. (8), one can show that the approximate expressions for the input and output resistances can be given as

$$R_{in} \approx \frac{(R_{F2} + R_{E2})R_{E1}R}{R_{E1}R + R_{E2}(R_{F1} + R_{E1} + R)} \tag{8.79}$$

and

$$R_{out} \approx \frac{(R_{F1} + R_{E1})R_{E2}R}{R_{E1}(R_{F2} + R_{E2} + R) + R_{E2}R} \tag{8.80}$$

where matched source and load impedances are assumed, such that $R_S = R_L = R$.

The loaded voltage gain A_v is given as

$$A_v = \frac{v_{out}}{v_{in}} \approx \frac{R_{F1} + R_{E1}}{R_{E1}} \tag{8.81}$$

which is essentially the same as that of a series–shunt pair [see Eq. (8.68)].

Figure 8.27 shows an actual circuit implementation of the two-stage dual-loop feedback amplifier.[9] This particular circuit was designed to fit into a four-pin

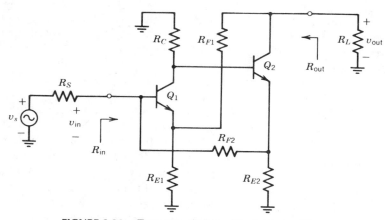

FIGURE 8.26. Two-stage dual-loop feedback amplifier.

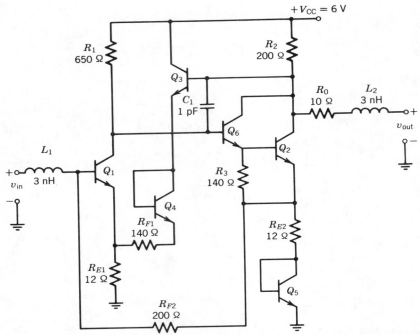

FIGURE 8.27. Circuit example for dual-loop amplifier. (*Note*: Inductances L_1 and L_2 are the lead inductances of bonding wires.)

TO-46 type metal can package and work with 50-Ω source and load imped-ances. The series inductors L_1 and L_2 are due to the package leads and bonding wires. Note that a Darlington-connected output transistor is used to provide additional buffering in the signal path. The output is also buffered from the shunt-feedback path of R_{F1} by means of the emitter follower Q_3. The diodes Q_4 and Q_5 are used for additional dc level shifting. The circuit is designed to produce a nominal loaded voltage gain of 8(18 dB) into a 50-Ω load with a 3-dB bandwidth of 700 MHz, using high-frequency transistors with $f_T \approx 2$GHz.

Figure 8.28 shows a photomicrograph of the monolithic amplifier circuit of Figure 8.27 in terms of the four-pin circuit package. Note that the ground terminals for the input and output stages are brought to separate pads and are bonded with multiple wires to the package surface which serves as the ground plane. The multiple bonding wires are used to minimize parasitic lead in-ductances between ground pads on the chip and the ground plane formed by the package surface. Such precautions are necessary in multistage wideband amplifiers in order to avoid parasitic coupling between gain stages.

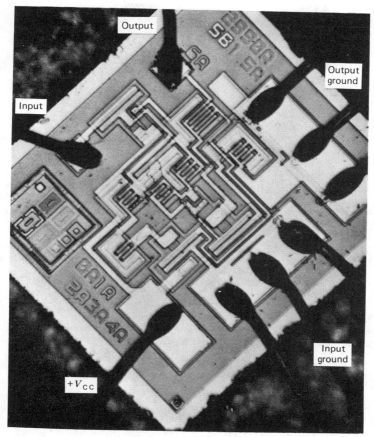

FIGURE 8.28. Photomicrograph and bonding diagram of wideband amplifier of Figure 8.27. Note the separate bonding pads and multiple bonding wires for input and output contacts. (*Photo*: Signetics Corporation.)

8.10. ROOT-LOCUS TECHNIQUES

In optimizing the frequency response of feedback amplifiers, root-locus methods provide one of the most versatile design techniques. Although root-locus methods have been initially developed for the analysis of feedback and servo systems, they are readily adaptable to feedback amplifier design.[10]

In a linear feedback system, root locus is the plot of the location of *closed-loop poles* on the complex frequency plane, as a function of the feedback loop gain. In other words, it provides a graphic illustration of the path along which the closed-loop system poles would move as the feedback loop gain is increased from zero to any arbitrary amount. Since zero-feedback loop gain corresponds to the no-feedback or open-loop condition, at that point the closed-loop poles are the same as the open-loop poles. Then, as the feedback is increased, by increasing the gain around the feedback path, the closed-loop system poles are consequently moved along the root locus as a function of the loop gain starting from the open-loop poles.

In the design of wideband feedback amplifiers using root-locus methods, the first step is to break the overall circuit into two segments $A(s)$ and $F(s)$, which designate the transfer functions of the forward and reverse signal flow paths, respectively, where $s = \sigma + j\omega$ is the complex frequency variable. This is shown schematically in Figure 8.29. Normally, the forward signal path contains active devices, and therefore, its signal transmission properties are nearly unilateral. The feedback path $F(s)$ is generally a passive, bilateral network. In order to analyze the network with basic root-locus techniques, one normally makes the basic assumptions that the forward signal transmission is almost totally through $A(s)$, and the reverse transmission is almost totally through the feedback network $F(s)$. Then one can readily express the transfer function $H(s)$ of the feedback network in terms of the transfer characteristics of the subnetworks $A(s)$ and $F(s)$ as

$$H(s) = \frac{A(s)}{1 + A(s)F(s)} \tag{8.82}$$

where s is the complex frequency variable. The roots of the denominator expression $1 + A(s)F(s)$ can be obtained graphically using root-locus techniques to determine the poles of the closed-loop transfer function. Note that the open-loop transfer function around the entire forward and reverse signal path is the product term $A(s)F(s)$, whose poles are the open-loop poles of the system and whose amplitude is the open-loop gain.

Figure 8.30a shows the typical root-locus diagram for a three-stage feedback amplifier similar to the series–series triple discussed in Section 8.8 (see Fig. 8.24). The open-loop poles are shown with crosses, and the closed-loop poles

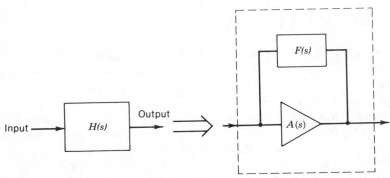

FIGURE 8.29. Decomposition of feedback system into its forward and reverse transmission paths. (*Note: $s = \sigma + j\omega$.*)

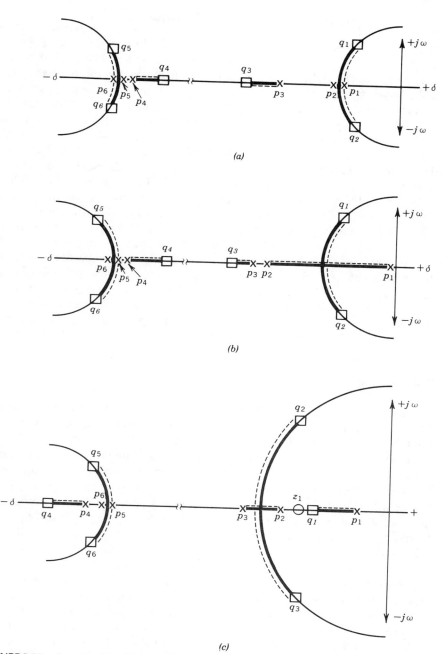

FIGURE 8.30. Investigating effects of broadbanding for series–series feedback amplifier of Figure 8.24, using root-locus techniques:[7] (a) Root locus for basic three-stage amplifier; (b) root locus using pole-splitting capacitor C_P; (c) root locus using feedback zero with capacitor C_F. (*Note:* Crosses indicate open-loop poles; squares indicate closed-loop poles.)

435

are designated as squares. Note that the negative real axis is shown to be discontinuous so that both the dominant and the nondominant time constants, or poles, of the circuit can be included in the same figure. The three dominant open-loop poles p_1, p_2, and p_3 are normally due to the three respective gain stages in the forward path. The nondominant poles p_4, p_5, and p_6 are usually associated with the circuit parasitics and the distributed nature of the circuit elements. The closed-loop poles move along the root loci defined by the solid line, as a function of the overall loop gain. To avoid peaking in the closed-loop response, the closed-loop poles are, in general, required to be located within 45° radially from the origin. Thus, the maximum 3-dB bandwidth is obtained when the closed-loop poles q_1, q_2, and q_3 are located as shown in Figure 8.30a. The resulting 3-dB bandwidth is approximately equal to the distance of q_1 from the origin. If the overall open-loop gain is increased, poles q_1 and q_2 would move further along the locus toward the $j\omega$ axis, which would result in undesirable peaking in the frequency response and possibly instability. It is possible to modify the location of the open-loop poles by introducing a small amount of capacitance into the proper nodes of the circuit, in either the signal or the feedback path. This in turn can result in a modified set of locations for the open-loop poles and zeros and lead to a more favorable set of closed-loop poles. Two such techniques which are readily compatible with monolithic circuits are pole-splitting and feedback-zero methods. These are briefly described below.

Pole Splitting

In general, it is not desirable to have open-loop poles p_1 and p_2 very close to each other. It can be shown that the root loci of q_1 and q_2 can be displaced further away from the origin and the $j\omega$ axis, if these two poles can be split apart.[7] In a practical design, this can be achieved by narrowbanding one of the gain stages in the signal path by introducing some shunt capacitance at a given circuit node. In the case of the series–series triple of Figure 8.24, this corresponds to connecting a capacitor C_p, across the collector–base junction of Q_2. This capacitor splits poles p_1 and p_2 by causing p_1 to move closer toward the origin and p_2 away from the origin. Then, when the overall feedback loop is closed, closed-loop poles q_1 and q_2, move on a new root locus, as shown in Figure 8.30b. Since the new locus is now further away from the origin, the 3-dB bandwidth of the closed-loop amplifier is significantly higher. In most cases, the value of the pole-splitting capacitor necessary for broadbanding is quite small, typically on the order of 1–10 pF. Therefore, it can be readily incorporated in the monolithic design without requiring an excessive amount of chip area. In the case of the circuit of Figure 8.24, the value of C_p is approximately 2 pF, which takes up no more chip area than an ordinary small-signal transistor. At this point, it should be noted that this technique is the same as that used in conjunction with the compensation of two-stage op amp circuits (see Section 7.2).

Feedback Zero

It can be shown that by introducing a transmission zero into the feedback path of a broadband amplifier, the root locus can be significantly altered. This is achieved by placing a small shunt capacitor across the feedback network. In the case of the feedback triple of Figure 8.24, this feedback zero can be obtained by putting a small shunt capacitor C_F across the feedback resistor R_F, which results in a real-axis transmission zero z_1, given as

$$z_1 = -\frac{1}{R_F C_F} \qquad (8.83)$$

If by proper choice of C_F, z_1 is located just to the right of p_2, between p_1 and p_2 on the $-\sigma$ axis, then the root-locus pattern can be modified, as shown in Figure 8.30c. Note that the poles of the complex pole pair now break away from the real axis between p_2 and p_3, instead of p_1 and p_2. Therefore, closed-loop poles q_2 and q_3 on this portion of the axis are now located further away from the origin. In most design examples, the value of C_F necessary to generate a feedback zero is on the order of several picofarads. Therefore, it can also be readily incorporated into the monolithic design without a significant increase in chip area.

In summary, the root-locus technique is a powerful tool for analyzing and optimizing the response of wideband amplifiers using feedback. It provides the following important advantages:

1. It lets the designer spot potential frequency instability problems associated with the circuit.
2. It lets the feedback gain be adjusted to maximize the bandwidth.
3. It indicates where (or when) additional circuit components, such as the pole-splitting or feedback-zero capacitances C_p and C_F, can be added into the circuit to optimize frequency response.

8.11. CURRENT AMPLIFIERS: THE GILBERT GAIN CELL

In the design of very-wide-bandwidth monolithic circuits, current amplifiers have an intrinsic advantage over voltage amplifiers. Since most of the parasitics associated with the monolithic devices are capacitive, the amplifier bandwidths can be improved, if most or all of the signal processing on the chip can be done in terms of current rather than voltage amplification, thus eliminating voltage swings across parasitic capacitances at circuit nodes. For example, as a current amplifier, the transistor is useful for frequencies up to its cutoff frequency f_T. However, the useful range of most voltage amplifiers is significantly below that because of the excessive phase shifts associated with the voltage transfer across a transistor at high frequencies. Therefore, to utilize the maximum frequency capability of a bipolar transistor, it is necessary to utilize it as a current, rather than voltage, amplifier whenever possible. Even in the case where voltage amplification is required, the input voltage signal can be converted to a current, amplified through several current gain stages, and finally converted back to a voltage swing at the output.

The so-called *Gilbert gain cell,* [11] whose basic circuit is shown in Figure 8.31, is an excellent example of such a current amplifier configuration which has flat gain characteristics over a frequency range comparable to the f_T of the individual transistors. The linear operation of the circuit relies very strongly on the close matching and tracking of monolithic device characteristics, particularly the base–emitter voltage drops.

The basic gain-cell structure shown in Figure 8.31 is very closely related to the four-quadrant transconductance multiplier circuit discussed in Chapter 9 (see Section 9.4).

Assuming that the devices are well matched and the transistor current gain $\beta \gg 1$, the operation of the gain cell can be explained as follows. The input drive is provided by differential current sources I_{B1} and I_{B2}, which partition a total bias current I_B as

$$I_{B1} = xI_B \quad \text{and} \quad I_{B2} = (1 - x)I_B \tag{8.84}$$

The unequal partitioning of currents I_{B1} and I_{B2} creates a differential voltage V_1 at the emitters of Q_1 and Q_4, which in turn forces the currents I_2 and I_3 in Q_2 and Q_3 to be unequal. This effect is illustrated in Figure 8.32. The differential

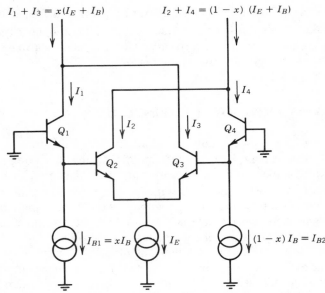

FIGURE 8.31. Basic configuration of Gilbert gain cell.

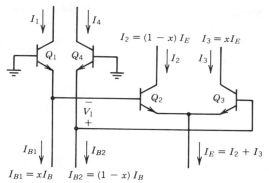

FIGURE 8.32. Redrawing of inner section of Gilbert gain cell to illustrate current partitioning between Q_2 and Q_3.

voltage V_1 is related to the logarithm of the current ratio in Q_1 and Q_4 as

$$V_1 = V_T \ln\left(\frac{I_{B2}}{I_{B1}}\right) = V_T \ln\left(\frac{I_4}{I_1}\right) \tag{8.85}$$

This differential voltage appearing across the bases of Q_2 and Q_3 in turn causes I_2 and I_3 to be partitioned as

$$\frac{I_3}{I_2 + I_3} = \frac{1}{1 + \exp^{(V_1/V_T)}} \tag{8.86}$$

and

$$\frac{I_2}{I_2 + I_3} = \frac{\exp^{(V_1/V_T)}}{1 + \exp^{(V_1/V_T)}} \tag{8.87}$$

Then the ratio I_2/I_3 would be

$$\frac{I_2}{I_3} = \exp\left(\frac{V_1}{V_T}\right) \tag{8.88}$$

From Eq. (8.85), taking antilog of both sides, one can show that

$$\frac{I_4}{I_1} = \frac{I_{B2}}{I_{B1}} = \exp\left(\frac{V_1}{V_T}\right) \tag{8.89}$$

From Eqs. (8.88) and (8.89), one reaches the fundamental result

$$\frac{I_2}{I_3} = \frac{I_{B2}}{I_{B1}} = \frac{I_4}{I_1} \tag{8.90}$$

or

$$I_2 = (1 - x)I_E \quad \text{and} \quad I_3 = xI_E \tag{8.91}$$

In other words, the current partitioning in the inner differential pair Q_2 and Q_3 is the mirror image of the current partitioning of the outer pair Q_1 and Q_4.

Since the collectors of Q_2 and Q_3 are cross-coupled to the collectors of Q_1 and Q_4, currents I_1, I_3 and I_2, I_4 are summed together to produce a set of output currents $I_1 + I_3$ and $I_2 + I_4$, such that

$$I_1 + I_3 = x(I_B + I_E) \quad \text{and} \quad I_2 + I_4 = (1 - x)(I_B + I_E) \qquad (8.92)$$

Thus, a differential input current I_{in}, where

$$I_{in} = I_{B1} - I_{B2} = 2xI_B \qquad (8.93)$$

produces a differential output current I_{out}, where

$$I_{out} = (I_1 - I_3) - (I_2 + I_4) = 2x(I_B + I_E) \qquad (8.94)$$

which results in a net current gain of

$$A_I = \frac{I_{out}}{I_{in}} = \frac{I_B + I_E}{I_B} = 1 + \frac{I_E}{I_B} \qquad (8.95)$$

As shown by Eq. (8.92), the overall current gain of the Gilbert gain cell is determined solely by the ratio of the bias currents associated with the inner and outer transistor pairs. A unique property of the current gain expression of Eq. (8.92) is that it was derived by relying solely on exponential current–voltage characteristics of the transistor base–emitter junctions. Thus, it does not depend on the small-signal hybrid-π model and is equally valid for large as well as small signal analysis, as long as the transistor characteristics match and track over temperature.

Figure 8.33 shows a circuit schematic for a voltage amplifier stage which uses the current gain cell as the intermediate amplifier stage. The emitter degeneration resistors provide a linear conversion of the input voltage to a current signal. Similarly, the matched load resistors R_L convert the output current signal into a differential voltage. The net voltage gain of the stage can be expressed as

$$A_v = \frac{v_{out}}{v_{in}} = \frac{R_L}{R_E}\left(1 + \frac{I_E}{I_B}\right) \qquad (8.96)$$

When higher values of current amplification are desired, a number of identical gain cells can be directly cascaded, as shown in Figure 8.34. Note that in this case, the total current gain A_I for a cascade of n stages becomes

$$A_I = 1 + \frac{I_{E1} + I_{E2} + \cdots + I_{En}}{I_B} \qquad (8.97)$$

Since the basic gain cell is a current amplifier, the cascade of several gain–cells, as shown in Figure 8.34, also provides an efficient use of the supply current, as most branches of the circuit are connected in series with the power supply and use the same supply current. Since the voltage level shifts necessary to bias the successive gain stages are quite low, a diode string similar to that

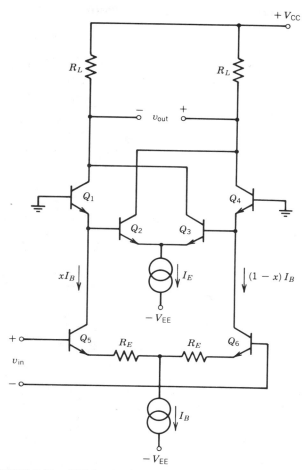

FIGURE 8.33. Wideband voltage amplifier using Gilbert gain cell.

shown in Figure 8.34 can be used to provide sufficient bias to all stages of the cascade, without the need for ac decoupling.

It can be shown that the high-frequency capability of the gain-cell amplifier is directly proportional to the current gain–bandwidth product f_T of the transistors used.[11] For a given value of f_T, the frequency response of the gain cell can be approximated by a single dominant pole p_1,

$$p_1 \approx -\frac{2\pi f_T}{A_I} \tag{8.98}$$

where A_I is the current gain of the cell, given by Eq. (8.95). For a given choice of transistor geometry, f_T is maximum at a given emitter-current level. Therefore, in designing a gain-cell amplifier, care should be taken to scale the device

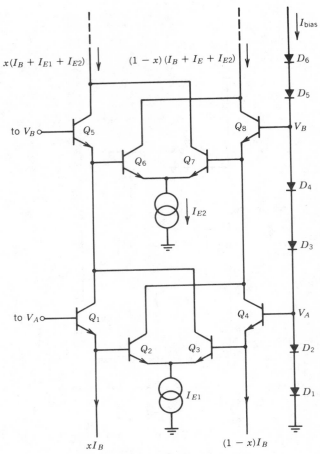

FIGURE 8.34. Cascaded gain cells.

geometries such that the f_T of each of the devices is nearly optimum at their particular bias level of operation. This implies that when cascading a number of gain cells, as shown in Figure 8.34, the emitter areas for the upper transistor pair Q_5, Q_8 should be made larger than those for the lower pairs Q_1, Q_4 by an amount depending on the current I_{E1}

If a number of identical gain-cell stages are cascaded to obtain a desired value of overall current gain, the total 3-dB bandwidth of the cascaded stages is reduced by the bandwidth shrinkage factor γ,

$$\gamma = (\sqrt[n]{2} - 1)^{1/2} \tag{8.99}$$

where n is the number of identical gain stages forming the cascade. Thus, if n identical gain–cells, each with a current gain A_I, were cascaded, the 3-dB bandwidth of the combination would be approximately equal to

$$f_{3dB} = \frac{f_T}{A_I}\gamma \tag{8.100}$$

The gain-cell concept is a good example of one of the many circuit techniques ideally suited to the design of integrated circuits. It again demonstrates that, with some imagination, the circuit designer can overcome the inherent limitations of integrated circuits such as the lack of inductors and the limited choice of active devices, and end up with a design that relies on the advantages of monolithic structures, such as the control of device geometries and close matching and tracking of components.

8.12. ELECTRONIC GAIN CONTROL

In a number of applications utilizing wideband amplifiers, it is often desirable to control the gain of the amplifier electronically, without affecting any other performance parameter. This type of electronic gain control is particularly useful in communication circuits, such as radio-frequency and intermediate-frequency amplifiers, to improve the signal-handling capability or the dynamic range of the amplifier. The electronic gain control capability allows the amplifier gain to be controlled by an automatic gain control (AGC) loop.

In conventional (nonintegrated) solid-state amplifiers, electronic gain control is usually achieved by shifting the dc operating point of one or more transistor stages or by employing diode attenuators between stages. In monolithic integrated circuits, where ac coupling between successive stages is not available, these conventional gain control methods lead to undesirable dc level shifts within the circuit and are generally not suitable. However, by using the matching properties of the integrated devices, some alternate gain control mechanisms can be devised for monolithic wideband amplifiers.

The basic AGC circuits used in monolithic wideband amplifier design are based on the control of the emitter current in a differential pair, or the use of current steering between a number of differential pairs which are stacked on top of each other to share the same operating current.

Two important parameters of an electronic gain control circuit are its AGC range and its dynamic range. These are briefly described below.

AGC Range. This describes the range over which the voltage or the current gain of a stage can be electronically varied without affecting other desired system requirements, such as distortion, noise, and amplifier bandwidth. The AGC range is normally specified in decibels, referenced to maximum gain level. Typical intermediate- and radio-frequency amplifier applications require AGC ranges of 40–60 dB. The maximum value of gain is often determined by bandwidth requirements and maximum allowable signal swings. The lower limit of the AGC range is set by noise, distortion, and signal feedthrough effects.

Dynamic Range. This specifies the maximum range of input signal amplitudes that the circuit can handle for a given set of limits on noise, bandwidth, and distortion. Normally, the upper limit is determined by distortion and the lower limit is set by noise specifications.

In general, the linearity of electronic gain control characteristics is somewhat secondary, since the gain control terminal is normally included in a low-frequency feedback loop. Most AGC circuits which use the current dependence of g_m as the control mechanism exhibit exponential gain characteristics, where the output signal amplitude varies exponentially with the control voltage. This comes about because of the exponential dependence between V_{BE} and the emitter current I_E. Although the control characteristics can be linearized by using current mirror type biasing (see Fig. 8.35), in most applications this is not necessary.

Differential Pair as Variable-Gain Amplifier

The basic differential gain stage shown in Figure 8.35 can be used as an AGC amplifier by controlling the emitter current I_E. The differential small-signal gain is given as

$$A_v = \frac{v_{\text{out}}}{v_{\text{in}}} = g_m R_L = \frac{I_E}{2V_T} R_L \qquad (8.101)$$

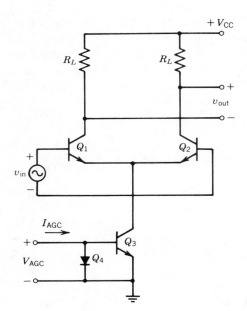

FIGURE 8.35. Differential pair as gain-controlled amplifier.

The emitter bias current I_E is related to the control current I_{AGC} linearly, and to the control voltage exponentially as

$$I_E = I_{AGC} \approx I_{CS} \exp\left(\frac{v_{AGC}}{V_T}\right) \qquad (8.102)$$

where I_{CS} is the collector reverse saturation current.

Inserting Eq. (8.102) into Eq. (8.101), one gets the basic exponential control characteristics,

$$A_v \approx \frac{I_{CS}}{2V_T} R_L \exp\left(\frac{v_{AGC}}{V_T}\right) \qquad (8.103)$$

Although the basic gain expression of Eq. (8.103) offers a wide AGC range, its dynamic range is severely limited since any input signal peak amplitude approaching V_T ($\approx$ 26 mV) would create severe harmonic or intermodulation distortion, as will be described in the next section. This can be avoided by using emitter degeneration resistances R_E in series with the emitters of Q_1 and Q_2. However, this severely limits the AGC range, since g_m is no longer directly proportional to current.

Fully Differential Current Divider as AGC Amplifier

Limited dynamic range problems associated with the simple differential stage can be overcome using a doubly differential current-divider circuit, as shown in Figure 8.36, as a wideband AGC circuit.[12] In this case, the lower differential pair Q_5 and Q_6 is driven by the differential input signal, and the AGC signal is applied to the bases of the upper differential pairs. This gain control voltage V_{AGC} controls the partitioning of the collector currents I_1, I_2 and I_3, I_4 of the upper differential pairs. From Eq. (8.86), the fraction of signal current shunted into the load resistance R_L can be written as

$$\frac{I_2}{I_1 + I_2} = \frac{I_3}{I_1 + I_3} = \frac{1}{1 + \exp(V_{AGC}/V_T)} \qquad (8.104)$$

which results in a voltage gain expression,

$$A_v \approx \frac{R_L}{R_E} \frac{1}{1 + \exp(V_{AGC}/V_T)} \qquad (8.105)$$

The presence of R_E at the emitters of Q_5 and Q_6 linearizes the transconductance characteristics of the input transistor pair, and avoids the dynamic range limitation. Since the gain shows an exponential dependence on the control voltage, the wide AGC range requirements are still met with a relatively small control voltage. If linear rather than exponential AGC characteristics are required, this can be readily achieved in the circuit of Figure 8.36 by driving the bases of the upper differential pair through *predistortion diodes,* as shown in Figure 9.5.

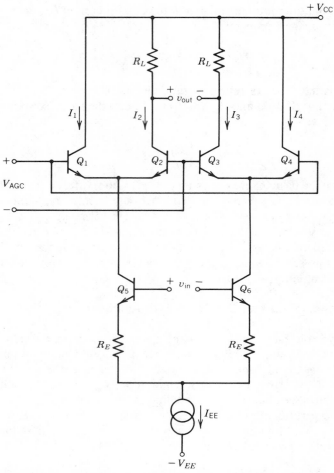

FIGURE 8.36. Wideband AGC amplifier using differential current partitioning.[12]

Other Variable-Gain Amplifier Configurations

The Gilbert gain-cell amplifier shown in Figure 8.34 can be used as a variable-gain stage since its voltage gain is given by Eq. (8.96) as

$$A_v = \frac{v_{out}}{v_{in}} = \frac{R_L}{R_E}\left(1 + \frac{I_E}{I_B}\right) \tag{8.106}$$

Thus, the voltage gain can be varied electronically by controlling either I_B or I_E. However, the AGC range of the circuit is limited due to the *additive* nature of the terms in the gain expression, in other words, the minimum gain value is limited to R_L/R_E, which corresponds to $I_E = 0$. This limits the useful AGC range of the circuit to approximately 20 dB for practical values of bias currents.

Another circuit configuration suitable for electronic gain control is the *balanced-modulator* circuit discussed in Chapter 9 (see Section 9.7). Figure 8.37 shows a variable-gain amplifier based on the balanced-modulator configuration. The principle of operation of the circuit is similar to that shown in Figure 8.36, except that the collectors of the upper differential pairs are cross-coupled to each other so that the currents in opposite branches of each of the differential pairs *subtract* from each other. Thus, the differential small-signal output voltage is

$$v_{\text{out}} = R_L \left[(i_1 + i_3) - (i_2 + i_4) \right] \tag{8.107}$$

Following the current-partitioning analysis given in Section 9.4 [see Eqs. (9.13) and (9.42)], one can write the ac gain expression for the circuit as

$$A_v = \frac{v_{\text{out}}}{v_{\text{in}}} = \frac{R_L}{R_E} \tanh \left(\frac{V_{\text{AGC}}}{2V_T} \right) \tag{8.108}$$

Figure 8.38 shows the resulting gain control characteristics of the circuit. As indicated by the figure, the circuit has two unusual properties which come about

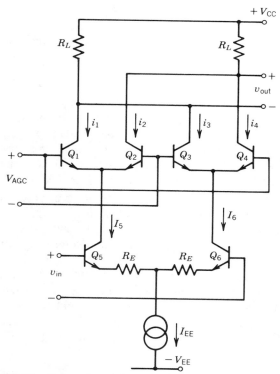

FIGURE 8.37. Balanced modulator circuits as variable-gain amplifier.

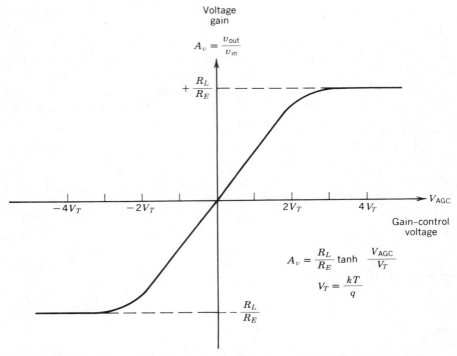

FIGURE 8.38.   Gain characteristics for variable-gain amplifier circuit of Figure 3.37.

because of the subtraction of output currents, namely:

1. The gain is nominally zero when no AGC voltage is applied. This comes about because, assuming all devices are ideally matched, the collector currents will be equal and their difference, which corresponds to the term in brackets in Eq. (8.107), will be zero when $V_{AGC} = 0$.

2. The gain expression would reverse polarity (i.e., 180° phase change) if the polarity of V_{AGC} is reversed. Thus, the circuit would change from inverting to noninverting or vice versa. This again comes about because of the change of polarity in the current difference term in Eq. (8.107).

The main disadvantage of the balanced-modulator circuit as a variable-gain amplifier is the problem of signal feedthrough at high frequencies, due to the imperfect matching of the transistor pairs Q_1, Q_2 and Q_3, Q_4 so that the ideal current cancellation at $V_{AGC} = 0$ implied by Eq. (8.107) does not happen. As a result, the AGC range is limited to a 20–40-dB range at frequencies in excess of 1 MHz.

Noise and Distortion in Variable-Gain Amplifiers

In designing wideband amplifiers, effects of noise and distortion are of particular concern. In the case of variable-gain amplifiers, which require the operating points and dc bias currents within the devices to be varied over a wide range, these problems become more severe.

The effects and the limitations of noise on the performance of broadband amplifiers are well covered in the literature[14,15] and will not be repeated here. However, one significant conclusion, which can be drawn from these analyses, is that the noise performance of monolithic bipolar transistors can be improved by reducing the base series resistance r_b and by increasing the device transconductance g_m. Following the analysis given in Ref. (15), one can approximate the wideband noise density of a bipolar transistor with an equivalent input noise resistance $(R_n)_{eq}$ as

$$(R_n)_{eq} = r_b + \frac{1}{2g_m} \tag{8.109}$$

where r_b is the base series resistance and g_m is the device transconductance.

Normally, in order to minimize $(R_n)_{eq}$, one would design the transistors with low r_b (i.e., use stripe geometry similar to that in Fig. 8.1), avoid using emitter degeneration, and run the transistor at high bias currents to increase g_m. However, in the case of variable-gain amplifiers, the dynamic range requirement necessitates the use of R_E in order to minimize the distortion of the input signal, and the wide AGC range requires the operation of the transistors with very low bias currents (i.e., low g_m) at the extreme end of its AGC range. Thus, in the design of variable-gain wideband amplifiers, there is always a significant trade-off or compromise between the noise characteristics, the dynamic range, and the AGC range requirements, and as either or both of these ranges are increased, noise characteristics become degraded.

REFERENCES

1. J. Millman and C. C. Halkias, *Electronic Devices and Circuits,* McGraw-Hill, New York, 1967, pp. 348–350.

2. B. A. Wooley, "Automated Design of DC-Coupled Monolithic Broad-Band Amplifiers," *IEEE J. Solid-State Circuits,* **SC-6,** 24–34 (February 1971).

3. A. B. Grebene, Ed., *Analog Integrated Circuits,* Part IV, IEEE Press, New York, 1978.

4. P. R. Gray and R. G. Meyer, *Analysis and Design of Analog Integrated Circuits,* Wiley, New York, 1977, Chap. 7.

5. J.A. Mataya, G.W. Haines, and S.B. Marshall, "I-F Amplifier Using C_c-compensated Transistors," *IEEE J. Solid-State Circuits,* **SC-3,** 401–407 (December 1968).

6. P. R. Gray and R. G. Meyer, *Analysis and Design of Analog Integrated Circuits,* Wiley, New York, 1977, Chap. 8.

7. J. E. Solomon and G.R. Wilson, "A Highly Desensitized Wideband Monolithic Amplifier," *IEEE J. Solid-State Circuits*, **SC-1**, 19–28 (September 1966).

8. R. G. Meyer, R. Eschenbach, and R. Chin, "A Wideband Ultralinear Amplifier from DC to 300 MHz," *IEEE J. Solid-State Circuits*, **SC-9**, 167–175 (August 1974).

9. R. G. Meyer and R.A. Blauschild, "A Four-Terminal Wideband Monolithic Amplifier," *IEEE J. Solid-State Circuits*, **SC-17**, 634-638 (December 1981).

10. K. Ogata, *Modern Control Engineering*, Prentice-Hall, Englewood Cliffs, NJ, 1970, Chap. 8.

11. B. Gilbert, "A New Wideband Amplifier Technique," *IEEE J. Solid-State Circuits*, **SC-3**, 353–365 (December 1968).

12. W. R. Davis and J. E. Solomon, "A High-Performance Monolithic I-F Amplifier Incorporating Electronic Gain Control," *IEEE J. Solid-State Circuits*, **SC-3**, 408–416 (December 1968).

13. W. M. Sansen and R. G. Meyer, "Distortion in Bipolar Transistor Variable-Gain Amplifiers," *IEEE J. Solid-State Circuits*, **SC-8**, 275–282 (August 1973).

14. C. D. Motchenbacher and F. C. Fitchen, *Low-Noise Electronic Design*, Wiley, New York, 1973.

15. P. R. Gray and R. G. Meyer, *Analysis and Design of Analog Integrated Circuits*, Wiley, New York, 1977, Chap. 11.

CHAPTER NINE

ANALOG MULTIPLIERS AND MODULATORS

9.1 A CLASSIFICATION OF MODULATORS AND MULTIPLIERS

Modulators belong to a general class of circuits with multiple input terminals, or *ports*, where a control signal applied to one input port can modify, or *modulate*, the signal flow from the second input to the output. Figure 9.1 shows a conceptual description of such a system in a black-box form. The electronic gain control circuits discussed in the previous chapter also belong to this general category.

In a generalized modulator, the output signal $V_{out}(t)$ is related to the input signals by an arbitrary transfer function $F(t)$ as

$$V_{out}(t) = F[V_1(t), V_2(t)] \tag{9.1}$$

A less general but more useful class of modulators are the so-called *product modulators*, where the output is a *product* of two arbitrary functions F_A and F_B, related to each of the two inputs,

$$V_{out} = F_A[V_1(t)] \cdot F_B[V_2(t)] \tag{9.2}$$

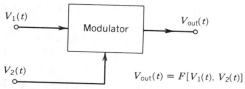

FIGURE 9.1. Generalized black-box description of modulator.

As examples, the automatic gain control (AGC) circuits of Figures 8.36 through 8.38 belong to this category, whereas the Gilbert gain-cell amplifier of Figure 8.34 *does not*, due to the *additive* nature of the terms in its gain expression [see Eq. (8.93)]. The product-type transfer characteristics given in Eq. (9.2) are obtained when a circuit contains a single well-defined signal path from one input to output, where the signal transmission through this *entire* path can be controlled.

A special case of the product modulator arises when the transfer functions $F_A(V_1)$ and $F_B(V_1)$ are *linearly proportional* to the inputs, that is,

$$F_A(V_1) = K_A V_1 \quad \text{and} \quad F_B(V_2) = K_B V_2 \tag{9.3}$$

where K_A and K_B are gain constants. This results in a transfer characteristic of the form

$$V_{\text{out}} = K_A K_B V_1 V_2 = K_1 V_1 V_2 \tag{9.4}$$

where the output is proportional to the *linear product* of the two input voltages. A circuit having this property is called an *analog multiplier*.

In most applications, it is also required that the output voltage V_{out} conserve the polarity relationship between the two inputs such that each of the inputs can be either positive or negative, and the output would be of the proper polarity implied by Eq. (9.4). A multiplier which has this property is called a *four-quadrant* multiplier.

9.2. PROPERTIES OF AN ANALOG MULTIPLIER

The analog multiplier is a circuit block where the output voltage V_z is proportional to the product of the two inputs V_X and V_Y as

$$V_Z = K_1 V_X V_Y \tag{9.5}$$

where K_1 is the gain constant of the multiplier and has the dimensions of V^{-1}. In most analog computation applications, it is customary to set $K_1 = 0.1 \ V^{-1}$, so that

$$V_Z = \frac{V_X V_Y}{10} \tag{9.6}$$

Since an analog multiplier deals with two separate input variables V_X and V_Y for a given output V_Z, its operating characteristics cannot be as readily defined as a single-input system such as an operational amplifier. Instead, a number of separate gain and offset parameters need to be defined to describe the performance characteristics of a nonideal multiplier. In a practical multiplier circuit, the output V_Z is related to any one of the inputs V_X and V_Y by a generalized expression of the form

$$V_Z = \underbrace{K_1 V_X V_Y}_{\substack{\text{ideal} \\ \text{output}}} + \underbrace{[K_X V_Y + K_Y V_X + K_O]}_{\text{offset terms}} + \underbrace{f(V_X, V_Y)}_{\text{nonlinearity}} \qquad (9.7)$$

where only the first term in Eq. (9.7) is the ideal product output, the rest are offset and nonlinearity terms.

The constants K_X, K_Y, and K_O are known as the X, Y, and output offset constants of the multiplier.

Ideally, with any one of the inputs equal to zero, the output *must be* zero for values of the other input. The offset constants K_O, K_X, and K_Y define the amount of deviation from this ideal condition. K_O is the measure of the offset voltage due to the output with *both* V_X and $V_Y = 0$; K_X and K_Y are the offset voltages associated with the X and Y inputs. Assuming that nonlinearities are small, K_X and K_Y correspond to the *change* of the multiplier offset voltage per unit change of V_X or V_Y, with the other input held at zero. Thus, in a high-accuracy multiplier system, at least four adjustments are needed to set the multiplier gain K_1 and to null out the three offset constants.

The last term in Eq. (9.7) represents the amount of deviation from linearity at the output which is irreducible under any combination of input values or offset adjustments. This is called the *nonlinearity* or *feedthrough* error. When the nonlinearity errors are relatively small, the nonlinearity term in Eq. (9.7) can be approximated as

$$f(V_X, V_Y) \approx |V_X|\epsilon_X + |V_Y|\epsilon_Y \qquad (9.8)$$

where ϵ_X and ϵ_Y are the nonlinearity errors associated with the X and Y inputs, and are normally expressed as a *percentage* of the full-scale output.

Definition of Multiplier Terms

Some of the most common terms used in describing the performance of a four-quadrant multiplier circuit are the accuracy, the linearity, and the bandwidth of the multiplier. Each of these terms can be briefly described as follows.

Accuracy is defined as the maximum deviation of the actual output, from the ideal one given by Eq. (9.5), for any choice of X or Y input values within the dynamic range of the multiplier. In other words, it is the measure of the *total error* of the multiplier.

Nonlinearity, which is also called the linearity error, is the maximum difference between the actual and "the best straight-line" theoretical output for all pairs of X and Y input values. Both accuracy and nonlinearity are specified as a percentage of the full scale. Thus, for example, a 1% accuracy specification would mean that for a four-quadrant multiplier with a $\pm$ 10-V output swing, the actual output voltage would be within $\pm$ 100 mV of the ideal level. Similarly, a 1% linearity error means that the output will not deviate more than $\pm$ 100 mV from a best-fit straight line for the maximum deviation of any one input with the other input held constant.

The scale-factor accuracy, or *gain error,* of the multiplier is the difference between the actual and the ideal values of K_1. It is normally specified over temperature.

The X and Y feedthroughs are the measure of the signal at the multiplier output for any value of the X or Y input in the rated range, with the other input set to zero. It has a linear component due to the input offset coefficients K_X and K_Y, which can be nulled out, and a nonlinear component due to nonlinearity factors ϵ_X and ϵ_Y, of Eq. (9.8), which cannot be nulled out. Normally, feedthrough is specified at 50 Hz with a 20-V peak-to-peak sinusoidal signal applied to one input with the other input grounded. Feedthrough increases rapidly with increasing frequency due to dynamic mismatches between devices in the circuit.

The high-frequency capability of an analog multiplier is described in terms of its bandwidth. Depending on the particular application of a multiplier, different definitions of bandwidth may be used. The small-signal 3-dB bandwidth is defined as the frequency at which the output is 3 dB down from its low-frequency value for a constant input level. The 1% absolute-error bandwidth is defined as the frequency where the magnitude of the output $V_Z(j\omega)$ is down by 1% from its low-frequency value. An alternate bandwidth criterion is the 1% vector error bandwidth, corresponding to the frequency where the output phase is shifted by 0.57° from its low-frequency value, resulting in a 1% vector difference between $V_Z(0)$ and $V_Z(j\omega)$. The vector-error bandwidth is always much smaller than the corresponding absolute-error bandwidth.

Temperature stability of a multiplier is also essential for most circuit applications. The stability is normally measured in terms of the temperature drift of the output offset term K_O (mV/°C) and the scale factor K_1 (ppm/°C).

9.3. APPLICATIONS OF AN ANALOG MULTIPLIER

Similar to an operational amplifier, the analog multiplier forms a versatile building block for performing a number of mathematical operations, such as multiplying, dividing, squaring, and square-root extraction. In most of these applications, it is used in conjunction with an operational amplifier to complement its functional capability. Figure 9.2 outlines some of the applications of an analog multiplier in performing basic mathematical operations. The multiplication and squaring functions shown in the figure are self-explanatory. The division operation is performed by placing the multiplier in feedback around an operational amplifier, such that the multiplier output is summed at the operational amplifier input and forced to equal the input signal V_Z. Then the operational amplifier output V_Y necessary to satisfy this condition is

$$V_Y = \frac{V_Z}{K_1 V_X} \tag{9.9}$$

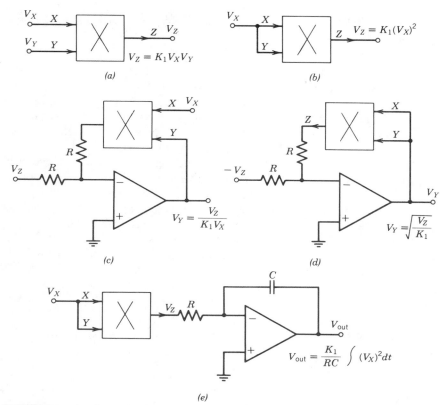

FIGURE 9.2. Some applications of analog multiplier in analog computation: (*a*) Multiplication; (*b*) squaring; (*c*) division; (*d*) square-root generation; (*e*) mean-square calculation.

Similarly, by short-circuiting the X and Y inputs, the same configuration can be used to obtain a square root of the input, that is,

$$V_Y = \sqrt{\frac{V_Z}{K_1}} \qquad (9.10)$$

When using the multiplier in a feedback configuration around an operational amplifier, such as in Figure 9.2*c* and *d*, care should be taken to ensure that $V_Y > 0$ to avoid an unstable positive feedback condition across the operational amplifier. Also, it should be noted that the accuracy of the output in the circuit of Figure 9.2*c* is inversely proportional to V_X. Thus, as V_X decreases, the divider error increases rapidly, which severely limits the dynamic range of the system.

In addition to performing multiplication or division, an analog multiplier can also be used as a modulator, frequency converter, phase detector, or voltage-controlled attenuator. In many of these applications, linearity or the four-quadrant capability are not required. Therefore, the necessary circuit configuration can be greatly simplified. Some of these applications are discussed later in this chapter.

9.4. VARIABLE-TRANSCONDUCTANCE MULTIPLIER

There are a number of diverse circuit techniques which can be used to obtain an output proportional to the product of the two input signals.[1] The multiplication technique, which is most readily suited to monolithic circuits is the so-called *variable-transconductance* method. This method makes use of the dependence of the transistor transconductance on the emitter-current bias. The basic principle of the transconductance multiplier can be readily demonstrated by the differential gain stage shown in Figure 9.3. In order to demonstrate the principle of operation, we shall make some grossly simplifying assumptions.

First, assume that the differential input signal is small, such that $V_1 \ll V_T$, where V_T is the thermal voltage. Then the differential output current ΔI will be related to the input voltage V_1 as

$$\Delta I = I_1 - I_2 \approx g_m V_1 \approx \frac{I_{EE} V_1}{2V_T} \tag{9.11}$$

The transconductance g_m is proportional to I_{EE}, and I_{EE} can be varied by the second input voltage V_2 as

$$I_{EE} = \frac{V_2 - V_{BE}}{R_B} \approx \frac{V_2}{R_B} \tag{9.12}$$

where $V_2 \gg V_{BE}$ is assumed for simplicity. Then, from Eqs. (9.11) and (9.12), one can write

$$\Delta I \approx V_1 V_2 \frac{1}{2V_T R_B} \tag{9.13}$$

which is of the form shown in Eq. (9.5).

Although the simple differential stage of Figure 9.3 illustrates the principle of variable-transconductance multipliers, it has two very serious drawbacks:

1. It can only operate in two quadrants since the polarity of V_2 cannot be reversed.
2. The input dynamic range of V_1 is extremely small. Due to the $V_1 \ll V_T$ assumption in Eq. (9.11), the "linear" range of V_1 is limited to several millivolts only. For higher values of V_1, Eqs. (9.11) and (9.13) have to be represented by the large-signal transfer characteristics of the differential pair as

$$\Delta I = I_{EE} \tanh\left(\frac{V_1}{2V_T}\right) \approx \frac{V_2}{R_B} \tanh\left(\frac{V_1}{2V_T}\right) \tag{9.14}$$

which is no longer a linear product.

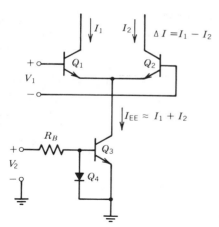

FIGURE 9.3. Differential pair as simple variable-transconductance multiplier.

Both of the shortcomings outlined above can be solved by imaginative design techniques. The two-quadrant operation of the differential pair can be extended to four-quadrant operation by using two differential stages in parallel, and cross-coupling their outputs as shown in Figure 9.4.[2,3] The two differential inputs V_1 and V_2 determine the partitioning of the total current I_T among the different branches of the circuit. Since both V_1 and V_2 are fully differential and can reverse polarity, the circuit enhibits four-quadrant operation capability.

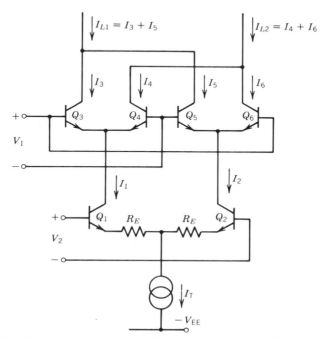

FIGURE 9.4. Variable-transconductance multiplier using cross-coupled differential stages.

Since the circuit configuration of Figure 9.4 is somewhat more complex than that of Figure 9.3, the operation of the circuit as a variable-transconductance multiplier is not as readily apparent. However, in terms of the current partitioning within the various branches of the circuit, the multiplication effect can be explained as follows. Assuming that all devices in the circuit are well matched and neglecting base currents, the current levels within the circuit are related as

$$I_1 + I_2 = I_T$$
$$I_3 + I_4 = I_1 \qquad\qquad (9.15)$$
$$I_5 + I_6 = I_2$$

and the net output current ΔI_L is

$$\Delta I_L = I_{L1} - I_{L2} = (I_3 + I_5) - (I_4 + I_6) \qquad (9.16)$$

Rearranging terms, one can write

$$\Delta I_L = (I_3 - I_4) - (I_5 - I_6) \qquad (9.17)$$

where $I_3 - I_4$ and $I_5 - I_6$ represent the current partitioning through the upper differential pairs Q_3, Q_4 and Q_5, Q_6, respectively. This current partitioning is due to V_1 and can be expressed as

$$I_3 - I_4 = I_1 \tanh\left(\frac{V_1}{2V_T}\right) \qquad (9.18)$$

and

$$I_5 - I_6 = I_2 \tanh\left(\frac{V_1}{2V_T}\right) \qquad (9.19)$$

Thus, from Eqs. (9.17) through (9.19), one can write

$$\Delta I_L = (I_1 - I_2) \tanh\left(\frac{V_1}{2V_T}\right) \qquad (9.20)$$

If the emitter degeneration resistor R_E is sufficiently large such that $I_2 R_E \gg V_T$ and $I_1 R_E \gg V_T$, then the lower differential pair Q_1, Q_2 functions as a voltage-to-current converter, and

$$I_1 - I_2 \approx \frac{V_2}{2R_E} \qquad (9.21)$$

If, as an added restriction, $|V_1|$ is kept very small, such that it is much less than V_T, then

$$\tanh\left(\frac{V_1}{2V_T}\right) \approx \frac{V_1}{2V_T} \quad \text{for} \quad |V_1| \ll V_T \qquad (9.22)$$

and Eq. (9.20) becomes

$$\Delta I_L = V_1 V_2 \frac{1}{4R_E V_T} \qquad (9.23)$$

which is of the basic form shown in Eq. (9.5).

The conclusion of the above analysis is as follows. The cross-coupled trans-conductance multiplier circuit of Figure 9.4 can provide four-quadrant operating capability. However, it still has an extremely small dynamic range for the V_1 input, due to the assumption made in Eq. (9.22). Thus, for linear operation of the circuit, the values of V_1 are restricted to be no more than several millivolts. If V_1 is comparable to or larger than the thermal voltage V_T, transistor pairs Q_3, Q_4 and Q_5, Q_6 function as synchronous switches which turn on and off, depending on the polarity of V_1. Then the circuit functions as a *balanced modulator* rather than a linear multiplier. This application of the circuit will be discussed further in Section 9.7.

9.5. FOUR-QUADRANT MULTIPLIERS WITH WIDE DYNAMIC RANGE

In trying to increase the dynamic range of the multiplier circuits described in the previous section, one faces a fundamental problem: the linear multiplication effect is obtained because the transconductance of transistors Q_1 and Q_2 in Figure 9.3 (or of the Q_3, Q_4 and Q_5, Q_6 pairs in Fig. 9.4) is proportional to their emitter current. This is only true if the transistors are operated with no emitter degeneration and under small-signal conditions, that is, $|V_1| \ll V_T$. With reference to Figure 9.3, one can see that under large-signal conditions the voltage-to-current transfer characteristics of the differential pair Q_1 and Q_2 are no longer linear. Instead, collector currents I_1 and I_2 are related to the applied voltage V_1 as

$$\frac{I_1}{I_2} = \exp\left(\frac{V_1}{V_T}\right) \tag{9.24}$$

and their difference ΔI is related to V_1 from Eq. (9.14) as

$$\Delta I = I_1 - I_2 = I_{\text{EE}} \tanh\left(\frac{V_1}{2V_T}\right) \tag{9.25}$$

In order to obtain linear multiplication over a wide dynamic range, it is necessary to reduce, somehow, the exponential current–voltage transfer characteristics of Eqs. (9.24) and (9.25) to linear ones. This would be possible if one were to process the actual input V_1 in a nonlinear manner with a predetermined and well-controlled nonlinearity before applying it to the bases of Q_1 and Q_2 in Figure 9.3. In other words, suppose one were to take the input signal V_1, obtain from it a new signal ΔV, which is *logarithmically* related to V_1, and apply this new voltage into the circuit in place of V_1. Then the actual differential voltage ΔV appearing at the bases of Q_1 and Q_2 would be logarithmically related to V_1, and the exponential relationship of Eq. (9.24) would be linearized. This is known as the *predistortion* technique and is the key to the basic design philosophy for monolithic four-quadrant multipliers.[2]

Figure 9.5 shows a simplified circuit scheme for obtaining a logarithmic input voltage ΔV in order to linearize the transfer characteristics of a differential gain

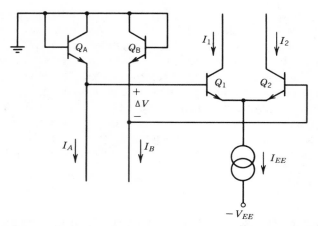

FIGURE 9.5.   Generation of logarithmic input voltage for differential stage.

stage. The diode-connected transistors Q_A and Q_B are driven by differential current sources I_A and I_B. The net voltage difference ΔV appearing across them is then derived from the difference between currents I_A and I_B. Assuming that all transistors are well matched, ΔV can be expressed as

$$\Delta V = V_T \ln\left(\frac{I_B}{I_A}\right) \tag{9.26}$$

Substituting Eq. (9.26) and Eq. (9.24), one can show that currents I_1 and I_2 of the differential stage are directly related to currents I_A and I_B, which drive the diode-connected transistors Q_A and Q_B, as

$$\frac{I_1}{I_2} = \frac{I_B}{I_A} \tag{9.27}$$

Similarly, the *current differences* in respective branches of the circuit are related as

$$I_1 - I_2 = I_{\text{EE}} \tanh\left(\frac{\Delta V}{2V_T}\right) \tag{9.28}$$

and

$$I_A - I_B = (I_A + I_B) \tanh\left(\frac{\Delta V}{2V_T}\right) \tag{9.29}$$

It should be noted that in deriving Eqs. (9.24) through (9.29), no assumptions were made to restrict the amplitudes of the device currents. Therefore, these sets of equations are valid over a broad range of current levels as long as the device characteristics are well matched and V_{BE} drops continue to obey the simple diode equation. Since the close matching is an inherent property of monolithic devices, this former requirement is readily met in integrated circuits.

The fundamental current relationships given by Eqs. (9.26) through (9.29) can be used to increase the dynamic range of the basic transconductance multiplier given in Figure 9.4. This can be done by adding a predistortion stage into the circuit to drive the upper cross-coupled differential pair with a logarithmic voltage signal by using the circuit configuration given in Figure 9.6.

The upper group of six transistors, made up of the cross-coupled differential pairs Q_3, Q_4 and Q_5, Q_6 and the logarithmic voltage-generating diode-connected transistors Q_A, Q_B are assumed to be ideally matched devices. This grouping of six devices is referred to as the *multiplier core*.

From Eqs. (9.20), (9.28), and (9.29), the *current differences,* or the differential currents going in and out of the multiplier core, can be written as

$$\Delta I_L = I_{L1} - I_{L2} = \frac{(I_1 - I_2)(I_A - I_B)}{I_A + I_B} \qquad (9.30)$$

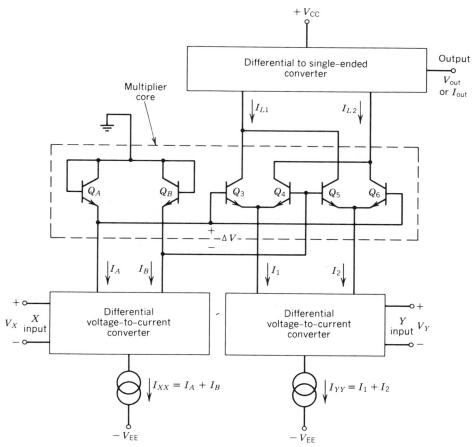

FIGURE 9.6. Basic architecture of linear four-quadrant transconductance multiplier.

In other words, the multiplier core shown within the dashed lines produces an output current difference which is a linear product of the two input current differences.

If the differential voltage-to-current converters in Figure 9.6 were assumed to have conversion gains (or transconductances) K_X and K_Y, respectively, such that

$$I_A - I_B = K_X V_X \quad \text{and} \quad I_1 - I_2 = K_Y V_Y \tag{9.31}$$

then the overall transfer function of the multiplier would be

$$I_L = \frac{K_X K_Y}{I_{XX}} V_X V_Y \tag{9.32}$$

which is of the same form as Eq. (9.5) and does not have the dynamic range limitations for either X or Y inputs within the accuracy limitations of the input voltage-to-current converter stages, and within the matching tolerances of the multiplier-core transistors.

9.6. PRACTICAL ANALOG MULTIPLIER CIRCUITS

At present, a wide range of monolithic analog multiplier products are commercially available, based on the transconductance-multiplier architecture shown in Figure 9.6. The range of performance characteristics also covers a broad spectrum from low-cost general-purpose devices to expensive precision circuits which utilize on-chip laser trimming for gain and offset adjustments.

Figure 9.7 shows the circuit diagram of a relatively simple transconductance-multiplier integrated circuit (Motorola MC-1595) based on the circuit configuration of Figure 9.6. In the circuit, the voltage-to-current conversion is achieved by differential pairs Q_1, Q_2 and Q_3, Q_4 along with emitter degeneration resistors R_X and R_Y, such that

$$I_1 - I_2 = \frac{2V_X}{R_X} \quad \text{and} \quad I_3 - I_4 = \frac{2V_Y}{R_Y} \tag{9.33}$$

where the subscript designations refer to parameters shown in Figure 9.7. Then, from Eq. (9.32), with

$$K_X = \frac{2}{R_X} \quad K_Y = \frac{2}{R_Y} \quad I_{XX} = 2I_X \tag{9.34}$$

the differential output current ΔI_L can be written as

$$\Delta I_L = I_{L1} - I_{L2} = \frac{2}{I_X R_X R_Y} V_X V_Y \tag{9.35}$$

and the differential output voltage V_Z is

$$V_Z = \Delta I_L R_L = \frac{2R_L}{I_X R_X R_Y} V_X V_Y \tag{9.36}$$

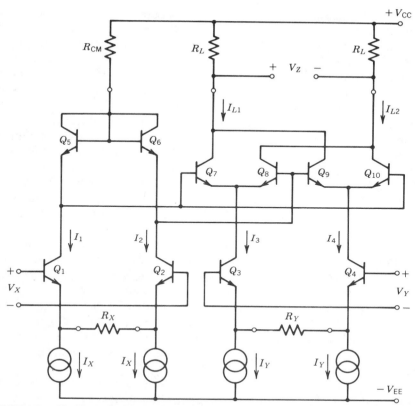

FIGURE 9.7. Simple analog multiplier integrated circuit based on transconductance-multiplier principle (Motorola MC-1595.) (*Note:* Resistors R_X, R_Y, R_L, and R_{CM} are external.)

The multiplier scale factor K_1 [see Eq. (9.5)], is given as

$$K_1 = \frac{2R_L}{I_X R_X R_Y} \tag{9.37}$$

and is set by the choice of the bias current I_X or the resistor values in Eq. (9.37). As described in Section 9.2, for most multiplier applications K_1 is chosen to be equal to 0.1. This choice of scale factor is preferred because it makes the input and output dynamic range of the multiplier compatible with other analog function modules that operate with $\pm$ 10-V signal swings.

The circuit of Figure 9.7 can handle $\pm$ 10-V input signals with a typical nonlinearity of $< 1\%$, with the following nominal bias setting and component values:

$$V_{CC} = 30 \text{ V} \qquad V_{EE} = 15 \text{ V} \qquad I_X = I_Y = 1 \text{ mA}$$
$$R_X = R_Y = 15 \text{ k}\Omega \qquad R_L = 11 \text{ k}\Omega$$

The gain of the circuit is set by adjusting I_X.

The simple multiplier circuit of Figure 9.7 is designed as a low-cost building block rather than a self-contained function module. As a result, the resistors R_X, R_Y, R_L, and R_{CM} are left external to the integrated circuit so that the gain, dynamic range, and common-mode bias levels can be adjusted externally. Although this enhances the versatility of the circuit, it also detracts from its practicality for any one specific application due to the extensive amount of external components and trimming adjustments required.

In more advanced monolithic multiplier designs, all or nearly all of the components are placed on the chip using thin-film resistors for gain stability and utilizing on-chip laser trimming for offset-null and gain accuracy setting. In many cases, it is also advantageous to include an operational amplifier on the chip along with the analog multiplier circuit. Such an operational amplifier can be used for differential-to-single-ended conversion of the multiplier output (see Fig. 9.6) as well as for performing a number of the functions illustrated in Figure 9.2.

Figure 9.8 shows the simplified circuit schematic of a fully self-contained monolithic multiplier circuit (Analog Devices AD-534). The circuit utilizes an on-chip bias reference for stable and accurate setting of bias currents. Thin-film resistors are used for scale-factor accuracy as well as for offset nulling. The basic circuit configuration of the analog multiplier is similar to that shown in Figure 9.6, except that careful design considerations are given to identifying, analyzing, and, to a large degree, compensating the error sources which affect the overall accuracy of the circuit.[4] Some of these special design and layout considerations will be examined later in this section.

The differential current output of the core ΔI_{out} is converted to a single-ended voltage output by means of the summing operational amplifier A_1. One of the main contributors to the full-scale error is the nonlinearity associated with the differential voltage-to-current converter stages at the X and Y inputs. In the design example of Figure 9.8, this nonlinearity is virtually canceled by "generating" an equal and opposite nonlinearity. This is done by using a third voltage-to-current converter stage (made up of Q_{11} and Q_{12}) which has identical distortion characteristics to X and Y input stages in an active feedback loop around the operational amplifier, A_1.

The gain setting resistors R_X, R_Y, and R_W, are all made to be equal and are made up of segments R_1 and R_2,

$$R_X = R_Y = R_W = 2R_1 + R_2 \tag{9.38}$$

This has no effect on the gain-setting characteristics, but improves the slew rate of the input stages by buffering the emitters of input transistors from the parasitic capacitances associated with the collectors of current-source transistors.

The differential output current ΔI_{out} coming out of the multiplier core is given by Eq. (9.35) as

$$\Delta I_{out} = \frac{2}{I_X R_X R_Y} V_X V_Y \tag{9.39}$$

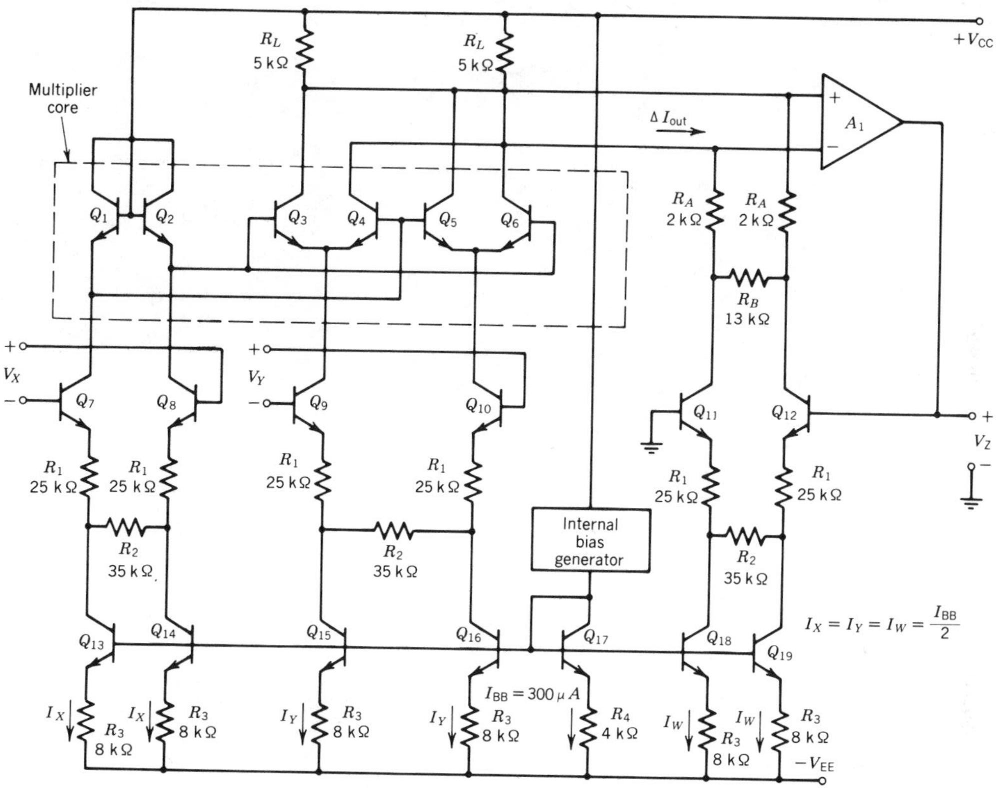

+VCC
Multiplier core
RL 5 kΩ
RL 5 kΩ
Δ Iout
A1
RA 2 kΩ
RA 2 kΩ
RB 13 kΩ
Q1
Q2
Q3
Q4
Q5
Q6
VX
VY
Q7
Q8
Q9
Q10
Q11
Q12
VZ
R1 25 kΩ
R2 35 kΩ
Internal bias generator
Q13
Q14
Q15
Q16
Q17
Q18
Q19
IBB = 300 μA
IX
IY
IW
R3 8 kΩ
R4 4 kΩ
−VEE

$$I_X = I_Y = I_W = \frac{I_{BB}}{2}$$

$$V_Z = \frac{1}{I_X R_X R_Y} \frac{R_W(2R_A + R_B)}{R_B} V_X V_Y$$

$$R_X = R_Y = R_W = 2R_1 + R_2$$

FIGURE 9.8. Simplified circuit schematic of precision analog multiplier integrated circuit using active feedback. (Analog Devices AD-534.)

This differential current is then converted to a single-ended voltage V_Z by means of the summing operational amplifier A_1 and the voltage-to-current converter stage of Q_{11} and Q_{12} in feedback around it, as

$$V_Z = \Delta I_{\text{out}} \frac{R_W}{2} \frac{2R_A + R_B}{R_B} \tag{9.40}$$

which results in the overall transfer equation

$$V_Z = \frac{R_W}{I_X R_X R_Y} \frac{2R_A + R_B}{R_B} V_X V_Y \tag{9.41}$$

The resistive network made up of R_A and R_B at the collectors of Q_{11} and Q_{12} provides a convenient method of adjusting the scale factor without changing the bias currents or the matching gain resistors associated with the input stages.

Figure 9.9 shows the chip photograph of the precision analog multiplier circuit of Figure 9.8. The locations of various sections of the circuit on the chip are also identified. Note that the X and Y inputs and the multiplier core are laid

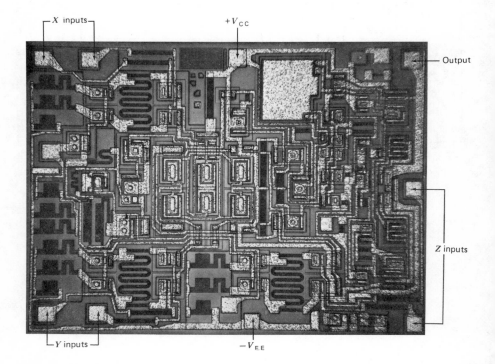

FIGURE 9.9. Photomicrograph of precision multiplier circuit of Figure 9.8. Chip size: 76 mils × 100 mils. (*Photo*: Analog Devices, Inc.)

out symmetrically, relative to each other and to the output, in order to minimize the effects of thermal gradients.

The complete monolithic chip exhibits a total error or less than $\pm$ 0.5% of full scale over the entire operating temperature range, with a scale-factor temperature coefficient of $+$ 75 ppm/°C. The nonlinearity errors are less than 0.15% of full scale. The circuit is designed to operate with $\pm$ 15-V supplies and $\pm$ 10-V input dynamic range.

Error Sources in Multiplier Design

Although the basic four-quadrant multiplier expression given by Eq. (9.32) is valid for large-signal operation, it is derived with the following simplifying assumptions: (1) ideal matching and tracking between the transistors, and (2) negligible base currents, that is, $\beta_F = \infty$. Even though both of these assumptions are reasonable, they each contribute a finite amount of error and nonlinearity into the overall multiplier transfer function. Therefore, in precision IC designs, where total errors must be kept to less than 1% of full scale, over all the operating conditions, the effects of these error sources must be identified and minimized.

Device Mismatch Errors. The main source of mismatch between monolithic transistors of similar geometry are the unequal emitter areas, mask misalignment, process tolerance gradients, or the thermal gradients across the chip surface. To avoid or minimize these effects, the multiplier core transistors are normally designed as large-emitter-area devices and are located within close proximity to each other on the chip layout, as shown in Figure 9.9. To minimize thermal mismatches, the core transistors are also located centrally on the chip, on a thermal *plateau*, equidistant from potential heat sources on the chip, such as output transistors. It should be remembered that a V_{BE} mismatch of 250 μV is enough to cause an output current error of 1%. Similarly, a temperature gradient of 0.25°C between any one of the matched pairs in the multiplier core would result in approximately 1% full-scale error.

Figure 9.10 shows a closeup of the layout of the matched pairs of transistors forming the multiplier core for the design example of Figure 9.8. The emitter areas are designed as wide oval regions with no sharp corners, with a length of approximately 5 mil and a width of 2 mil. Such a large geometry with no sharp corners minimizes the errors due to misalignment or etching during the photomasking step. The common-centroid layout used for operational amp lifter inputs (see Fig. 7.27) is not practical for multipliers, since a larger number of transistors are involved and the interconnection schemes become too complex. However, using the careful layout practices described above, the mean value of offset distribution between each adjacent pair can be kept as low as 50 μV, which is sufficient for the overall accuracy requirements.[4]

Effects of Series Resistances. Any mismatches in the values of parasitic resistances at the emitters of core transistors directly translate into offset errors,

Q_4 Q_2 Q_5

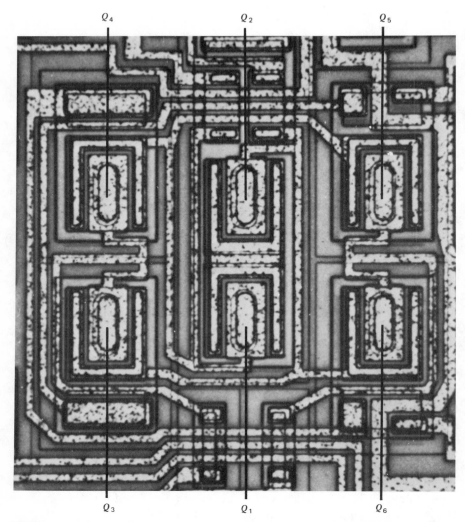

Q_3 Q_1 Q_6

FIGURE 9.10.   Closeup of layout of matched transistor pairs for multiplier-core section of Figure 9.8. (*Photo*: Analog Devices, Inc.)

depending on the value of the bias currents. For example, at an emitter current level of 250 μA, a 1-Ω resistor mismatch directly in series with the emitters of the core transistors would result in apparoximately 1% output error. To avoid this problem, the aluminum interconnection paths (which have a sheet resistance of 0.03–0.05 $\Omega/\square$) at the emitters of core transistors are maintained at equal lengths, and wraparound base contacts are used to minimize the base resistance r_b. (See Figure 9.10).

Effects of Finite β. The basic current transfer function of the multiplier core, given by Eq. (9.30), assumes that the transistor emitter and collector currents are equal (i.e., $\alpha_F = 1$ and $\beta_F = \infty$). One can show that the finite value of β_F introduces an approximate scale-factor error to Eq. (9.30) such that the transfer function now becomes[4]

$$\Delta I_L = \frac{(I_1 - I_2)(I_A - I_B)}{I_A + I_B} \left(1 - \frac{3}{\beta_0}\right) \qquad (9.42)$$

In a precision design, such as the one shown in Figure 9.8, this error is easily corrected by adjusting the reference current I_{BB} to compensate for it.

Effects of Nonideal Voltage-to-Current Converters. Normally, differential gain stages with emitter degeneration are used as input voltage-to-current converter stages. However, in such circuits, the variations of emitter impedances and transistor V_{BE} drops at the extremities of signal swings cause the transfer function to be nonlinear, particularly as one or the other transistor nears cutoff. This effect can be reduced by limiting the remaining current in one or the other of the input transistors to no less than 25% of the quiescent bias setting for a ± 10-V full-scale input signal. If additional refinements are warranted, more complex voltage-to-current converter circuits can be used, [5] or an active feedback scheme similar to that shown in Figure 9.8 can be employed.

9.7. BALANCED MODULATORS

In a number of applications, the four-quadrant capability of a linear multiplier is not necessary. Typical examples of this class of circuit are the *modulator*, or *mixer*, circuits. In a modulator circuit, a linear response is required with respect to only one of the inputs. This linear input is known as the *modulating input*. The second input of the circuit, often referred to as the *carrier input*, is driven by a constant-amplitude ac signal. In certain applications, it is also desirable that the output of the modulator should be zero when the voltage applied to the modulating input is zero. This property of the circuit is called *carrier suppression*, and the modulators which exhibit such a characteristic are called *balanced modulators*.

A circuit configuration which is ideally suited for balanced-modulator applications is the cross-coupled differential stage shown earlier in Figure 9.4. It is repeated in Figure 9.11a. Following the results of Eq. (9.14), the transfer characteristics of the circuit can be written as

$$V_o(t) = \frac{R_L}{R_E} V_2(t) \tanh\left(\frac{V_1(t)}{2V_T}\right) \qquad (9.43)$$

Assuming that R_E is sufficiently large, the response of the circuit to the modulating input $V_2(t)$ is linear. However, if a high-level ac input signal $V_1(t)$ is

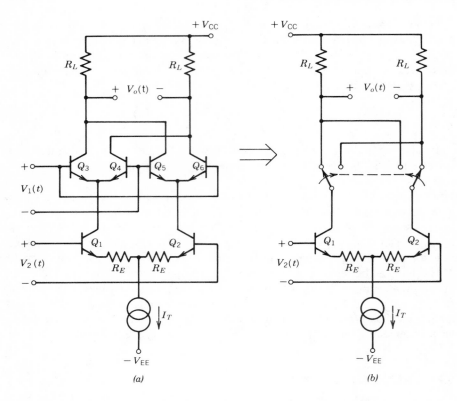

FIGURE 9.11.   Approximate equivalent circuit for balanced modulator with $\left|V_1(t)\right| \gg V_T$.

applied to the carrier input, such that $|V_1(t)| \gg V_T$, the hyperbolic tangent function reduces to a switching waveform $S_c(t)$, as shown in Figure 9.12a. Physically, this means that for a high-level ac input, Q_3, Q_4 and Q_5, Q_6 of Figure 9.11a function as two synchronous single-pole double-throw switches which operate out of phase from each other, as shown in Figure 9.11b. As a result, the modulating input $V_2(t)$ is "chopped" by the carrier input signal to produce an output waveform of the form

$$V_o(t) = \frac{R_L}{R_E} V_2(t)S_c(t) \tag{9.44}$$

Figure 9.12 shows the typical output waveform corresponding to a relatively slowly varying modulating waveform. Note that the output corresponds to a chopped version of the modulating waveform, which is alternately multiplied by $+1$ and -1 at every *half-cycle* of the carrier input signal.

A fundamental property of the balanced-modulator circuit is its *carrier–suppression* capability, which causes the output to be equal to zero when one or the other of the two input signals is equal to zero [see Eq. (9.43)]. In an actual

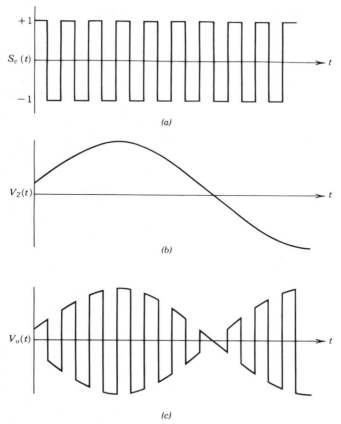

FIGURE 9.12. Modulated output waveform generated by high-level switching signal applied to balanced-modulator circuit of Figure 9.11: (*a*) Carrier signal; (*b*) modulating waveforms; (*c*) output waveform.

circuit, inherent component mismatches and offsets of monolithic components unbalance the circuit and cause the carrier–suppression capability to be somewhat degraded. For example, if the circuit of Figure 9.11*a* had inherent device mismatches, which can be represented as effective dc offset voltages E_{o1} and E_{o2}, at the corresponding terminals, then assuming that $E_{o1} \ll V_T$, the output can be expressed as

$$V_o(t) = \frac{R_L}{R_E}\left[V_2(t)S_c(t) + S_c(t)E_{o2} + \frac{V_2(t)E_{o1}}{2V_T} \right] \qquad (9.45)$$

where the last two terms represent the so-called carrier and modulation feedthrough due to circuit unbalance.

A large portion of the effective offset voltages E_{o1} and E_{o2} is due to device mismatches and can be nulled out at the low frequencies by means of external adjustments, similar to the case of the multiplier offsets. However, a portion of each of these "effective" offsets is due to inherent device nonlinearities and nonlinearity mismatches which cannot be nulled out [see Eq. (9.8)]. These effects cause the carrier-suppression characteristics to deteriorate at high frequencies, in spite of dc offset adjustments.

Carrier suppression is normally defined as the ratio of the rms output voltage with maximum modulation to the ratio of the rms output voltage with zero modulation [i.e., $V_2(t) = 0$], and is expressed in decibels. A balanced-modulator circuit of the type shown in Figure 9.11 can provide a carrier suppression of 50 dB or better at low frequencies (i.e., $f_{carrier} < 10$ kHz) with proper dc offset adjustments. However, at higher frequencies (in the megahertz range), carrier-suppression characteristics deteriorate to a 20–30-dB range due to frequency-dependent device mismatches and nonlinearities.

9.8. APPLICATIONS OF BALANCED MODULATORS

Balanced modulators find a wide range of applications in analog communications or frequency translation circuits. The most direct application of modulator-type circuits is in the generation of amplitude-modulated (AM) signals. In addition to AM signal generation, they can be used for automatic gain control, frequency multiplication, phase detection, synchronous AM or FM demodulation, and frequency discrimination. As described in the previous section, in most of these applications, the carrier input is normally driven with a high-level signal, such that the modulator circuit functions as a set of synchronous switches, and effectively chops the modulating signal. This results in an output of the form given by Eq. (9.44).

A more commonly used terminology in modulator applications is to refer to inputs $V_1(t)$ and $V_2(t)$ as the carrier and modulation inputs, respectively, and use subscripts c and m to describe them. Using this designation, and letting K_m be the gain constant of the modulator, the output voltage under high-level carrier input can be expressed from Eq. (9.44) as

$$V_o(t) = K_m V_m(t) S_c(t) \qquad (9.46)$$

The spectrum of the output waveform can be developed from the Fourier spectrum of the two inputs. The Fourier spectrum of the square-wave input $S_c(t)$ with an amplitude of ± 1 can be expressed as an infinite sum of discrete frequencies at the integral multiples of the carrier frequency ω_c. Furthermore, if $S_c(t)$ is a *symmetrical* square wave, as is the case in Figure 9.12, it contains only odd harmonics and has a spectrum made up of odd multiples of ω_c. Thus, $S_c(t)$ can be expressed as

$$S_c(t) = \sum_{n=1,3,5,\ldots}^{\infty} A_n \cos(n\omega_c t) \qquad (9.47)$$

where the amplitude A_n of the nth Fourier components is given as

$$A_n = \frac{\sin(n\pi/2)}{n(\pi/4)} \qquad (9.48)$$

The frequency spectrum of the carrier waveform is shown in Figure 9.13b. Note that the amplitudes of the higher harmonics decrease rapidly since A_n is inversely proportional to n, as given by Eq. (9.48).

Assuming that the input is a low-frequency modulating sine wave of the form

$$V_m(t) = E_m \cos(\omega_m t) \tag{9.49}$$

Then the output signal can be written as

$$V_o(t) = K_m E_m \sum_{n=1,3,5,\ldots}^{\infty} A_n \cos(n\omega_c t) \cos(\omega_m t) \tag{9.50}$$

which can be simplified to an expression of the form

$$V_o(t) = \frac{K_m E_m}{2} \sum_{n=1,3,5,\ldots}^{\infty} A_n [\cos(n\omega_c + \omega_m)t + \cos(n\omega_c - \omega_m)t] \tag{9.51}$$

which corresponds to the frequency spectrum shown in Figure 9.13c. It has

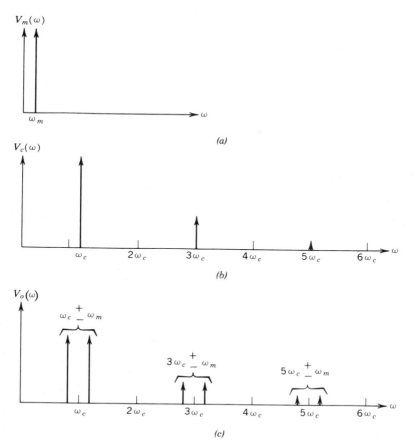

FIGURE 9.13. Input and output signal spectra for balanced modulator: (a) Modulating signal spectrum; (b) carrier signal spectrum; (c) output signal spectrum.

components located at frequencies ω_m *above* and *below* each of the harmonics of ω_c, but no component at ω_c or its harmonics. The lack of a frequency component at carrier frequency is the outcome of the carrier-suppression property of balanced modulators. Normally, the signal is low-pass filtered following the modulation process so that only the frequency components in the vicinity of ω_c are retained, as illustrated in Figure 9.14. Thus, the resulting filtered output of the system corresponds to the first term of the spectral expression given by Eq. (9.51), that is,

$$V_o(t) = \frac{K_m E_m}{\pi} \left[\cos(\omega_c + \omega_m)t + \cos(\omega_c - \omega_m)t \right] \qquad (9.52)$$

where, for simplicity, it is assumed that the low-pass filter has negligible attenuation within its passband.

Amplitude-Modulated (AM) Signal Generation

If a dc component is added to the modulating signal, this appears as an external offset to the balanced modulator and essentially corresponds to the coefficient E_{o2} in Eq. (9.45). As a result, the circuit operates in an unbalanced manner and does not provide carrier suppression. Consequently, the output spectrum also contains a signal component at the carrier frequency and its harmonics.

In the generation of conventional AM signals, the presence of the carrier is often required. This can be done by providing a modulating signal of the form

$$V_m(t) = E_o[1 + M \cos(\omega_m t)] \qquad (9.53)$$

which is essentially a sinusoidal signal superimposed on a dc offset. The parameter M, which is a measure of the amplitude of the modulating sinusoid *relative* to the dc offset, is called *modulation index,* and is normally restricted to be $0 < M < 1$.

The resulting output voltage, after filtering, is

$$V_o(t) = \frac{4K_m E_o}{\pi} \left[\cos(\omega_c t) + \left(\frac{M}{2}\right)\cos(\omega_c + \omega_m)t + \left(\frac{M}{2}\right)\cos(\omega_c - \omega_m)t \right]$$
$$(9.54)$$

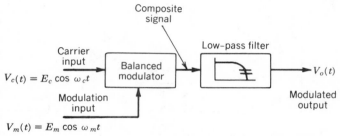

$$V_o(t) = K_G E_1 A_1 (\cos \omega_m t)(\cos \omega_c t)$$

FIGURE 9.14. Use of low-pass filter to eliminate higher order harmonics in modulator output.

which corresponds to a conventional AM signal with upper and lower sidebands centered symmetrically about the carrier frequency. This output spectrum is shown in Figure 9.15 in the vicinity of ω_c. Note that the relative amplitudes of the upper and lower sidebands are proportional to the modulation index.

Frequency Translation

The balanced modulator is essentially a frequency-translation circuit, where the information contained in a modulation signal at frequency ω_m at the input is translated to a frequency in the vicinity of the carrier frequency ω_c at the output. This property can be used for translating an input signal from frequency ω_m to any one of the two upper or lower sideband frequencies ω_u and ω_l where

$$\omega_u = \omega_c + \omega_m$$
$$\omega_l = \omega_c - \omega_m \tag{9.55}$$

In frequency translation applications, signals at frequencies ω_c and ω_m are applied to the inputs of the system, and the desired frequencies ω_u and ω_l are obtained at the output with proper filtering. This property of the circuit makes the balanced modulator one of the basic building blocks of frequency-synthesizer systems.

If both ω_c and ω_m are made identical, the circuit functions as a *frequency doubler,* as shown in Figure 9.16a. If both the carrier and the modulation inputs are driven with the same ac signal, and the output is filtered in the vicinity of $2\omega_c$, then one obtains from Eq. (9.52), for $\omega_c = \omega_m$,

$$V_o(t) = \frac{K_m E_m}{\pi}\left[1 + \cos(2\omega_c t)\right] \tag{9.56}$$

which has the ac component at twice the input frequency.

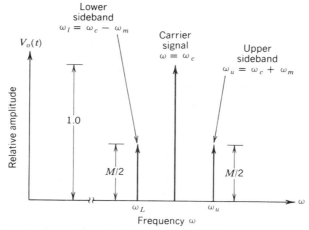

FIGURE 9.15. Frequency spectrum of conventional AM signal with modulation index M.

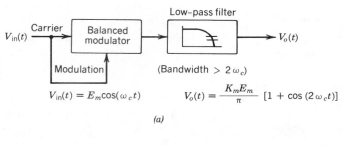

$$V_{in}(t) = E_m \cos(\omega_c t) \qquad V_o(t) = \frac{K_m E_m}{\pi} [1 + \cos(2\omega_c t)]$$

(a)

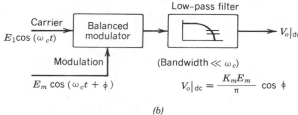

$$E_m \cos(\omega_c t + \phi) \qquad V_o\Big|_{dc} = \frac{K_m E_m}{\pi} \cos \phi$$

(b)

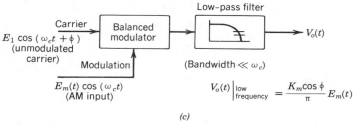

$$E_m(t) \cos(\omega_c t) \qquad V_o(t)\Big|_{\substack{\text{low} \\ \text{frequency}}} = \frac{K_m \cos \phi}{\pi} E_m(t)$$
(AM input)

(c)

FIGURE 9.16. Applications of balanced modulator: (a) Frequency doubling; (b) phase detection; (c) synchronous AM detection.

Phase Detection

If the two input signals applied to the balanced modulator were both at the same frequency ω_c, but had a relative phase difference ϕ, as shown in Figure 9.16b, then the output of the modulator would be of the form

$$V_o(t) = \frac{K_m E_m}{\pi} \left[\cos \phi + \cos(2\omega_c t + \phi) \right] \qquad (9.57)$$

If only the dc component of the output is taken, then

$$V_o\Big|_{dc} = \frac{K_m E_m}{\pi} \cos \phi \qquad (9.58)$$

which is proportional to the cosine of the phase difference between the input signals, as well as to the amplitude of the modulation signal.

In certain phase-detection applications, such as those used in phase-locked-loop (PLL) design, the phase-detection characteristic given by Eq. (9.58) is not very desirable due to the dependence of the output on the input amplitude, and on the cosine ϕ rather than ϕ. As will be described in Chapter 12 (see Section 12.5), these disadvantages can be circumvented by operating both the carrier and the modulation inputs in a *switching mode* (i.e., by setting $R_E = 0$ in Figure 9.11, with $|V_2| \gg V_T$).

AM Demodulation

As shown in Figure 9.16c, the balanced modulator can be used as a so-called *synchronous AM* detector. In this application, the amplitude-modulated input signal with the carrier frequency ω_c which is of the form

$$V_2(t) = E_m(t) \cos (\omega_c t) \qquad (9.59)$$

is applied to the modulation terminal. The carrier input is driven by an unmodulated carrier signal of the form

$$V_1(t) = E_c \cos (\omega_c t + \phi) \qquad (9.60)$$

which is at the same exact frequency as the modulated signal, except for an arbitrary phase difference ϕ.

After filtering the high-frequency components due to harmonics, one obtains

$$V_o(t) = \frac{K_m E_m(t)}{\pi} \left[\cos \phi + \cos (2\omega_c t + \phi) \right] \qquad (9.61)$$

which is the same as Eq. (9.57), except that $E_m(t)$ is now a slowly varying function of time, due to the modulation on the AM input. Assuming that $E_m(t)$ is at a much lower frequency that $2\omega_c$, the second term in Eq. (9.61) can be filtered out with a low-pass filter whose bandwidth is much elss than $2\omega_c$, and the resulting signal is

$$V_o(t) \bigg|_{\text{low frequency}} = \frac{K_m \cos \phi}{\pi} E_m(t) \qquad (9.62)$$

which is the demodulated information signal since its magnitude is proportional to the instantaneous amplitude $E_m(t)$ of the input AM signal.

It should be noted that the demodulated output amplitude is also proportional to the cosine of the phase difference between the modulated signal and the unmodulated signal. Therefore, maximum output is obtained when both signals are in phase or out of phase (i.e., $\cos \phi = \pm 1$), and no output is obtained when the carrier and the AM signal are $\pm 90°$ out of phase (i.e., $\cos \phi = 0$).

Figure 9.17 shows a simple synchronous AM detection system which uses a high-gain limiter amplifier to remove the modulation and generate an unmodulated carrier signal. Then the unmodulated carrier signal is applied to the carrier terminals of the balanced modulator as the switching waveform $S_c(t)$, which the actual AM signal is applied to the modulation terminal. After low-pass filtering, the resulting output is the same as Eq. (9.62) with $\cos \phi = 1$.

The synchronous AM detection technique is particularly useful in phase-locked-loop (PLL) AM detectors or tone decoding systems. The applications of

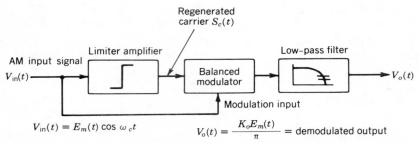

$$V_{in}(t) = E_m(t) \cos \omega_c t \qquad V_o(t) = \frac{K_o E_m(t)}{\pi} = \text{demodulated output}$$

FIGURE 9.17. Synchronous AM detector system using limiter amplifier to regenerate carrier signal.

synchronous AM detection in conjunction with phase-locked-loop circuits is discussed further in Chapter 12.

Frequency-Modulated (FM) Signal Detection

The phase-detection property of a balanced modulator given in Eq. (9.58) can also be used for frequency discriminator applications to detect FM signals. For such an application, the balanced modulator is used in conjunction with a limiter amplifier and a narrowband bandpass filter, as shown in Figure 9.18.[3]

The limiter circuit removes all the amplitude modulation from the input FM signal and generates a high-level switching signal $S_1(t)$ at the same frequency as the input signal. The bandpass phase-shift network is tuned to the input center frequency ω_c. It produced a second high-level signal $S_2(t)$, to drive the second input of the balanced modulator. In this application, the second input is also driven in a switching mode.

If a high-Q bandpass network is used, then the phase shift ϕ introduced by the bandpass network with selectivity Q for small frequency deviations $\Delta\omega$ in the vicinity of ω_c can be expressed as

$$\phi \approx -\frac{\pi}{2} \pm \frac{2Q\,\Delta\omega}{\omega_c} \qquad (9.63)$$

with the assumption that $2Q\,\Delta\omega \ll \omega_c$. Then the low-pass filtered output of the modulator is related to the frequency deviations of the input signal as

$$V_o(t)\bigg|_{\text{low-pass filtered}} \approx K\frac{\Delta\omega}{\omega_c} = K\frac{\Delta f}{f_c} \qquad (9.64)$$

where K is the combined gain of the filter and detector sections, and $f = \omega/2\pi$. Since in an FM signal the frequency deviation of the carrier $\Delta\omega$ represents the information, the output given by Eq. (9.64) represents the demodulated signal.

The balanced modulator circuit of Figure 9.11 is used as a basic building block in monolithic systems designed for commercial radio or TV receiver applications. For example, the FM detection scheme of Figure 9.18 is widely

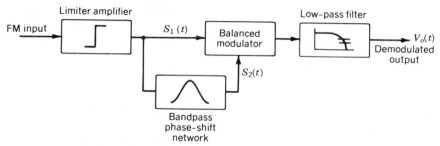

FIGURE 9.18. Application of balanced modulator as FM discriminator.

used today as a monolithic replacement for the detector section of commercial FM receivers or TV sound systems. In addition, many of the monolithic circuits designed for FM stereo decoding, color TV chroma-demodulation, and video detection all use the balanced-modulator circuit as one of the essential subblocks in the complete monolithic system design.

REFERENCES

1. A. B. Grebene, Ed. *Analog Integrated Circuits,* Part V, IEEE Press, New York, 1978.

2. B. Gilbert, "A Precise Four-Quadrant Multiplier with Subnanosecond Response," *IEEE J. Solid-State Circuits* **SC-3,** 365–373 (December 1968).

3. A. Bilotti, "Applications of a Monolithic Analog Multiplier," *IEEE J. Solid-State Circuits* **SC-3,** 373–380 (December 1968).

4. B. Gilbert, "A High-Performance Monolithic Multiplier Using Active Feedback," *IEEE J. Solid-State Circuits* **SC-9,** 364–373 (December 1974).

5. J. Davidse, *Integration of Analogue Electronic Circuits,* Academic Press, London, 1979, Chap. 6.

CHAPTER TEN

VOLTAGE REGULATORS

The function of a voltage regulator is to provide a well-specified and constant output voltage level from a poorly specified and sometimes fluctuating input voltage. The output of the voltage regulator would then be used as the supply voltage for the other circuits in the system. In this manner, the fluctuations and random variations of a supply voltage under changing load conditions are essentially eliminated.

Since the regulation and control of supply voltage is one of the most fundamental and critical requirements of any electronic system design, the monolithic voltage regulator or power control circuits have become some of the essential building blocks of any analog or digital system. As a result, the monolithic voltage regulators, similar to the case of monolithic operational amplifiers have gained wide acceptance and have greatly simplified the tedious task of designing power supply circuits.

Today, there are two very distinctly different types of IC voltage regulators which have gained wide acceptance and popularity. These are the *series regulators* and the *switching regulators*. The series regulators control the output voltage by controlling the voltage drop across a power transistor which is connected *in series* with the load. The power transistor is operated in its linear region and conducts current continuously. The switching regulators, on the other hand, control the flow of power to the load by turning on and off one or more of the power switches connected in parallel or series with the load, and make use primarily of inductive energy storage elements to convert the switched current pulses into a continuous and regulated load current.

Since the principles of operation and the design considerations associated with each of these classes of regulator circuits are quite different, this chapter is divided into two parts, covering each of these categories of regulator integrated circuits.

The first part of the chapter deals with the series-type voltage regulators. The design of series regulators often requires the inclusion of high-current power transistors on the chip, which are capable of dissipating significant amounts of power. The presence of low-level signal-processing circuitry in very close proximity to a power device on the same chip poses a number of special design problems due to thermal drifts and nonuniform temperature distribution on the chip, and therefore requires special layout and packaging considerations. In addition, special fault-protection circuitry is also needed on the chip to prevent the circuit burn-out due to accidental overloads or excessive heating. These problems and other related design considerations will be covered in the first part of this chapter. Several design examples associated with the commercially available series regulator integrated circuits will also be examined in order to illustrate the design principles and trade-offs involved.

The second part of the chapter will deal with the pulse-width-modulated power control systems, or switching regulators. This part will cover both their principle of operation as well as the special design considerations, trade-offs, and protection circuitry associated with them.

PART I: SERIES REGULATORS

10.1. FUNDAMENTALS OF SERIES REGULATORS

Principle of Operation

The series, or series-pass, type voltage regulator is connected in series between the load and unregulated supply line. It is a feedback circuit comprised of three main sections, as shown in Figure 10.1. These are the reference voltage element, the error amplifier, and the series-pass element. In most cases, a fourth section,

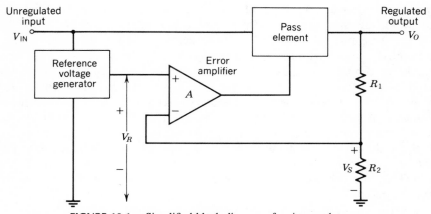

FIGURE 10.1. Simplified block diagram of series regulator.

the overload-protection circuitry, is also included in the system to prevent against burn-out under accidental overload conditions.

With reference to the simplified block diagram of Figure 10.1, the principle of operation of a series regulator can be briefly described as follows.[1] The internal voltage reference generator generates a reference voltage level V_R, which is independent of the unregulated supply voltage or the temperature changes. The error amplifier compares V_R with the sampled and scaled output voltage V_S and generates a corrective error signal to regulate the voltage drop across the pass element such that the $V_R = V_S$ condition is fulfilled. The scaled voltage V_S is derived from the actual output voltage by means of the sampling resistors R_1 and R_2.

Assuming that the error amplifier voltage gain A is sufficiently high, the voltage drop across the pass element will vary with the changes or fluctuations of the unregulated input voltage to maintain the output voltage V_O constant and equal to

$$V_O = \frac{V_R A}{1 + \alpha A} \tag{10.1}$$

where α is the feedback scale factor, determined by the sampling resistors, that is,

$$\alpha = \frac{V_S}{V_O} = \frac{R_2}{R_1 + R_2} \leq 1.0 \tag{10.2}$$

Assuming that the error amplifier gain is sufficiently high, such that $\alpha A \gg 1$, then Eq. (10.1) simplifies to

$$V_O = \frac{V_R A}{1 + \alpha A} \approx \frac{V_R}{\alpha} \tag{10.3}$$

Thus, the circuit produces an output voltage which is, to first order, independent of the input voltage and proportional to V_R.

Figure 10.2 shows a simplified circuit diagram for a practical series regulator. For most applications, the open-loop voltage gain A of the error amplifier is on the order of 60–70 dB, which can be obtained from a single high-gain differential stage using active (current mirror) loads.

The voltage reference is normally either a temperature-compensated Zener reference (see Fig. 4.33) or a band-gap reference (see Figs. 4.35 and 4.36). The temperature stability characteristics of these reference sources were described in Chapter 4, and will not be repeated here. Typical values of V_R obtainable from these circuits are in the range of 1.2–2.5 V, with temperature coefficients on the order of 30–100 ppm/°C.

The pass element is normally a high-current transistor. This corresponds to Q_5 of Figure 10.2. Often a Darlington connection of two transistors is used as the pass element in order to provide adequate buffering of the error amplifier from the load.

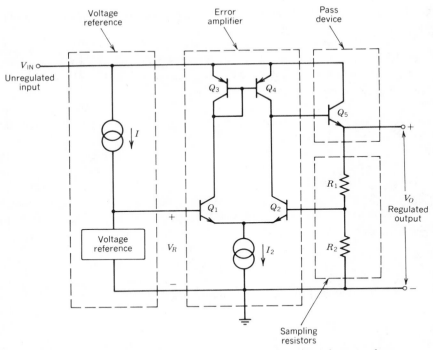

FIGURE 10.2. Simplified circuit diagram of practical series voltage regulator.

In most cases, the high-gain error amplifier may have to be frequency compensated to ensure stability under all operating conditions. This is normally done by providing a compensation capacitance on the IC chip, similar to the case of internally compensated operational amplifiers.

Performance Parameters

The most important performance parameters for series regulators are *line* and *load* regulation and *ripple rejection* characteristics. These basic performance parameters are defined as follows.

Line Regulation.   The amount of change in output voltage for a specific change in input voltage. It is a measure of the circuit's ability to maintain a constant output level under changing values of input voltage.

Load Regulation.   The amount of change in the output voltage for a specified change in the load current. It is a measure of the circuit's ability to maintain a constant output voltage level under changing load conditions.

Ripple Rejection. The ratio of the input peak-to-peak ripple to the output ripple voltage. It is a measure of the circuit's ability to reject periodic fluctuations of the rectified ac voltage signals at the input.

The line regulation is directly related to the sensitivity of the voltage reference to the changes in the line voltage. For example, if the reference voltage changes by an amount ΔV_R for a specified change ΔV_{IN} of input line voltage, then the line regulation is, from Eq. (10.3),

$$\text{line regulation} = \frac{\Delta V_O}{\Delta V_{IN}} = \frac{\Delta V_R}{\Delta V_{IN}} \frac{A}{1 + \alpha A} \approx \frac{\Delta V_R}{\Delta V_{IN}} \frac{1}{\alpha} \qquad (10.4)$$

In practical voltage regulator circuits, typical values of line regulation are on the order of 1–5 mV/V, measured under rated load conditions.

Load regulation is directly related to the output impedance of the regulator. Since the series-pass regulator circuit of Figure 10.1 or 10.2 is basically a shunt feedback circuit, its output resistance R_{out} is given as

$$R_{out} = \frac{R_o}{1 + \alpha A} \approx \frac{R_o}{\alpha A} \qquad (10.5)$$

where R_o is the output resistance of the pass element, with the feedback loop open. For example, in the case of the circuit of Figure 10.2, R_o is the resistance looking into the emitter of Q_5, assuming that no feedback is present. The presence of shunt feedback greatly reduces this output resistance, as indicated by Eq. (10.5), by an amount equal to the error amplifier gain times the feedback factor.

Load regulation, which is a measure of the relative change of output voltage for a given change ΔI_L of load current, can be written as

$$\text{load regulation} = \frac{\Delta V_O}{V_O} = \frac{\Delta I_L R_{out}}{V_O} \qquad (10.6)$$

or from Eqs. (10.3) and (10.5)

$$\text{load regulation} = \frac{\Delta V_O}{V_O} \approx \frac{\Delta I_L R_o}{A V_R} \qquad (10.7)$$

In practical IC regulator circuits, load regulation is typically $\leq 0.5\%$ for changes of the load current from its minimum to its maximum allowable value.

Ripple rejection characteristics depend strongly on the frequency response characteristics of the closed-loop feedback system in the regulator. If, for example, excessive amounts of frequency compensation are needed to ensure the stability of the error amplifier, then the rapid rolloff of gain $A(j\omega)$ at high frequencies can cause the ripple rejection characteristics to deteriorate. Normally, ripple rejection is specified at 120 Hz, and is on the order of 60–70 dB.

As indicated by the block diagram of Figure 10.1, the range of the regulated output voltage is $V_R \leq V_O \leq V_{IN}$. In practice, the upper bound of V_O is always less than V_{IN} due to the finite voltage drop necessary for current conduction across the pass element. This *minimum* voltage drop between V_{IN} and V_O necessary for circuit operation is called the *drop-out voltage* and is typically in the range of 1.5–3 V. The drop-out voltage sets the upper limit of regulated output voltage, which can be obtained from a given line voltage V_{IN}.

Negative-Voltage Regulators

The basic voltage regulation principle shown in Figure 10.1 can be readily applied to negative supply voltages by simply reversing the appropriate voltage polarities, and by using a high-current *pnp* transistor as the pass element. However, in monolithic design, where high-current *pnp* transistors are not available, this direct approach is not feasible. In limited applications, it is possible to use a composite connection of a lateral *pnp* and a high-current *npn* transistor (see Fig. 2.28) to simulate a high-current *pnp* transistor. However, since the series-pass transistor is inside the feedback loop of the regulator, the use of such a composite *pnp—npn* transistor creates stability problems under fluctuating load conditions. A more commonly used approach is a Darlington *npn* power transistor, taking the output from the collector rather than the emitter, as shown in Figure 10.3. Note that since this connection involves a phase inversion of the error signal in the pass transistor, the sampled output voltage V_S is now applied to the noninverting input of the error amplifier.

In a negative regulator, the open-loop output resistance R_o of Eq. (10.5) is higher since the output is taken from the collector of the pass device. However, the closed-loop output resistance R_{out} can still be kept low by increasing the overall loop gain. The fact that the pass transistor also functions as an inverting gain stage within the feedback loop of the regulator makes the frequency compensation of negative-voltage regulators more difficult than that of positive regulators, particularly under heavy load current.

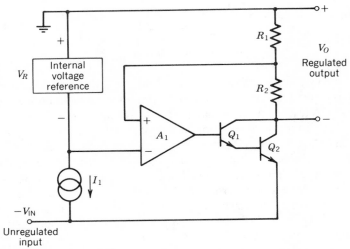

FIGURE 10.3. Simplified block diagram of negative-voltage regulator.

Dual-Tracking Regulators

In certain circuit applications, it is necessary to derive a symmetrical set of positive and negative supply voltages of equal value but opposite polarity, which would track each other and stay symmetrical under line voltage or load current changes. Such supply voltages are required for a wide range of analog circuits, including operational amplifiers, multipliers, and many of the data conversion circuits.

A regulator circuit which can produce a set of symmetrical output voltages $\pm V_O$ from unregulated positive and negative input voltages is called a dual-tracking regulator. Figure 10.4 shows the basic concept and the simplified circuit implementation of a dual-tracking regulator using one fixed reference voltage.[2] The circuit is basically a positive regulator whose output voltage is inverted and duplicated as the negative output. An alternate approach, using a negative regulator and inverting its output polarity, is equally feasible.

With reference to Figure 10.4, the output voltages are

$$V_O{}^+ = V_O{}^- = \frac{V_R}{\alpha} = V_R \frac{R_1 + R_2}{R_2} \tag{10.8}$$

Thus, the output voltages $\pm V_O$ track each other and are both set with a single adjustment (i.e., choice of R_1 and R_2), as long as the voltage drops across the pass transistors are in excess of the drop-out voltages.

Efficiency Considerations

The efficiency η of a regulator is defined as the ratio of the output power to the input power, that is,

$$\eta = \frac{P_o}{P_{\text{in}}} \tag{10.9}$$

where P_o is the power supplied to the load and P_{in} is the power delivered to the regulator from the power supply lines. In the case of a series-pass type voltage regulator, the simplified bias model of the regulator shown in Figure 10.5 can be used to calculate the efficiency.[3] In this model, the current source I_B represents the total bias and operating current consumed in the regulator circuitry, and V_B represents the voltage drop across the pass transistor. For proper operation of the circuit, V_B is restricted to be greater than the drop-out voltage.

From the simple model of Figure 10.5, the input and output power levels can be written as

$$P_{\text{in}} = V_{\text{IN}}(I_B + I_L) \tag{10.10}$$

and

$$P_o = V_O I_L \tag{10.11}$$

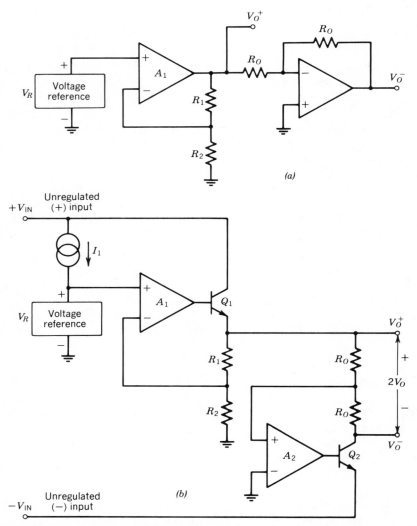

FIGURE 10.4. Dual-tracking regulator: (a) Basic concept; (b) simplified schematic for practical design.

Then the efficiency η can be expressed as

$$\eta = \frac{P_o}{P_{\text{in}}} = \frac{1}{(1 + V_B/V_O)(1 + I_B/I_L)} \tag{10.12}$$

As given by Eq. (10.12), the regulator efficiency depends directly on the ratio of the load voltage and load current to the bias voltage and bias current. Since the bias current I_B is more or less fixed by the regulator design, the maximum

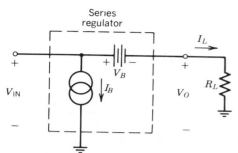

FIGURE 10.5. Simplified circuit model for calculating efficiency of a series regulator.

efficiency η_{max} is obtained when the regulator is delivering its maximum rated current with the minimum input–output voltage differential. If the input contains large ac components, or large minimum–maximum fluctuations, the maximum efficiency is reduced since the average level of the input voltage would have to be increased to ensure that the instantaneous value of V_B is greater than the drop-out voltage.

10.2. PROTECTION CIRCUITS

In order to prevent voltage regulator circuits from burning out or suffering permanent damage under accidental overload conditions, a number of fault-protection precautions are necessary.

In monolithic series regulators, normally three different types of protection circuitry are employed: (1) short-circuit or current-limit protection against accidental current overloads, (2) protection against excessive input–output voltage differentials (i.e., so-called safe operating area protection), and (3) protection against excessive junction temperatures (thermal overload protection). In mono-lithic designs where additional active or passive components can be added to the chip with negligible increase in cost or complexity, it is customary to include all or most of these protection circuits on the chip, alongside the regulator control circuitry.

Figure 10.6 shows a simplified circuit diagram of a series regulator integrated circuit, including the three basic protection circuits listed above. Normally, the output pass transistor, which can be either a single device or a Darlington-connected transistor, is driven from an internal current source I_B, and the error amplifier A regulates the output voltage by shunting away controlled amounts of this bias current I_B. The protection circuitry is designed to be inactive under normal operating conditions. However, it becomes activated when a particular fault threshold is reached and turns off or current-limits the output pass transistor Q_O by shunting away additional drive current I_B. In this manner, the output power device is protected from permanent damage. The particular design considerations for each of these three classes of protection circuits are discussed below.

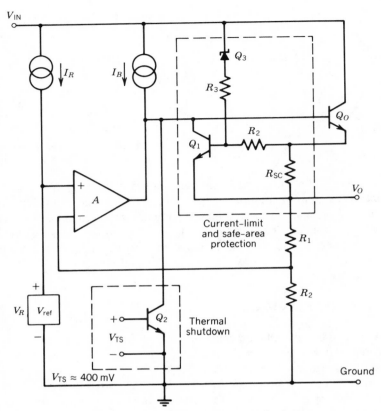

FIGURE 10.6.   Typical fault-protection circuitry.

Current-Limit or Short-Circuit Protection

The short-circuit protection requirements of series regulators can be met by either passive or active current limiting at the output, in a manner similar to that described in Section 5.3 (see Fig. 5.36) for the case of output gain stages. In the case of regulator circuits, passive current limiting using a resistor in series with the output is rarely used since it greatly deteriorates the load regulation characteristics. Thus, almost all of the series regulator integrated circuits rely on some form of active short-circuit protection.

Figure 10.7 shows some of the basic active current-limiting circuits used in series regulator integrated circuits. In all cases, the principle of operation is the same: the total current through the output transistor Q_O is sensed through a sensing resistor R_{SC}, and the resulting voltage drop is used to turn on the normally-off protection transistor Q_1 when the voltage drop across R_{SC} becomes sufficiently large. When Q_1 is turned on, it shunts away the base drive I_B from the output transistor and causes the output current to limit itself at a safe maximum value of

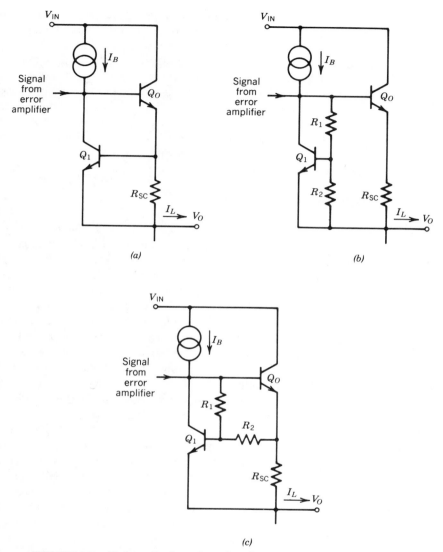

FIGURE 10.7. Various circuit configurations for output short circuit protection.

$$(I_L)_{\max} \approx \frac{V_{BE}}{R_{SC}} \qquad (10.13)$$

The basic drawback of this scheme is that it requires a full V_{BE} drop across R_{SC} before the current-limit mechanism is activated. In many cases, this is not desirable since it deteriorates the load regulation and also adds to the drop-out voltage. The need for a V_{BE} drop across R_{SC} can be avoided by prebiasing the protection transistor near conduction with a resistor string made up of R_1 and R_2, as shown in Figure 10.7b and c. In this manner, the necessary voltage drop across R_{SC} to initiate current limiting can be reduced to a fraction of V_{BE}.

In designing active current-limit circuits of the type shown in Figure 10.7, a word of caution is in order. An active current-limit circuit is basically a closed-loop feedback system by itself. Under heavy current loads, particularly if a Darlington-connected output power transistor is used, the circuit may be prone to oscillations when it reaches current-limit conditions. This problem can be avoided by either adding a compensation capacitance from the collector of Q_1 to ground, or reducing the current gain of the Darlington pass transistor. The gain reduction in the Darlington pass transistor is normally achieved by connecting a bypass resistor from the base to the emitter of the transistor.

Current-Foldback Limiting

In high-power regulator circuits, where the exact nature of the load is not defined, an accidental short-circuit or current-overload condition may exist for a prolonged duration before it is detected. This may cause excessive heating and unnecessary power dissipation in the system. To overcome this problem, most monolithic regulators also incorporate a so-called *current-foldback* mechanism, which operates as a latching switch and reduces the output current to a fraction of its rated value if the current-overload condition persists. In most designs, the active devices necessary for current-limit and current-foldback protection are incorporated into the monolithic design, with sensing resistors left external to the circuit to allow the user to determine the limiting threshold and the amount of foldback according to his application.

Figure 10.8 shows a typical example of the current-limit and current-foldback protection circuitry suitable for general-purpose regulator applications. In the circuit, the lateral *pnp* and the *npn* transistor combination Q_1 and Q_2 form a current-sensing switch with latching capability. If, for the moment, one assumes that the foldback resistor R_{FB} is set equal to zero, the circuit functions as a basic active current limit and limits the output current I_L to a value that will cause 1 V_{BE} drop across the sense resistor R_{SC}. However, if a resistor R_{FB} is added in series with the base of Q_2, as shown in Figure 10.8, this causes an additional voltage drop $I_F R_{FB}$ once Q_1 and Q_2 are conducting, which is in series with the voltage drop across R_{SC}. In this manner, once triggered, the current-limit transistors Q_1 and Q_2 can remain in conduction with a lower amount of output load current. In other words, the output current can increase to a maximum level of I_{max}, where:

$$I_{max} = \frac{V_{BE}}{R_{SC}} \tag{10.14}$$

and then decrease or be "folded back" to a lower value of sustaining current I_{SC}, where

$$I_{SC} = I_{max} - I_F \frac{R_{FB}}{R_{SC}} \tag{10.15}$$

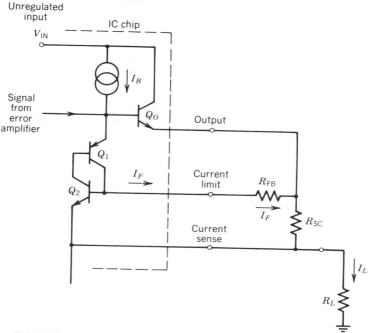

FIGURE 10.8.   Example of short-circuit protection with current foldback.

where I_F is primarily the collector current of Q_1. Thus, increasing the foldback resistor R_{FB} causes the current-foldback effect to increase. Once the circuit goes into the current-foldback mode, it can be brought back to normal operation either by an external reset or by reducing the load current to below I_{SC}. In actual designs, Q_1 would be made up of a split-collector *pnp* transistor, with one of its collectors short-circuited to its base, to operate as a controlled-gain transistor.

Safe-Area Protection

Power transistors, whether in discrete or monolithic form, can exhibit a potentially destructive failure mechanism called *secondary breakdown*. As will be described in Section 10.5, secondary breakdown is caused by the thermal instabilities within the power transistor which cause the formation of so-called *hot spots* within the localized areas of the device surface. This phenomenon causes the current conduction through the transistor to be nonuniform and concentrated at these hot spots, which will then cause further heating and eventual burn-out of the power transistor. In order to avoid the secondary breakdown problem, it is necessary to ensure that the device is operated within its so-called *safe area* under all operating conditions. Figure 10.9 gives an illustrative example of the safe operating area for continuous operation of a power transistor. The figure indicates that, as an example, a power transistor capable of safely handling 3 A

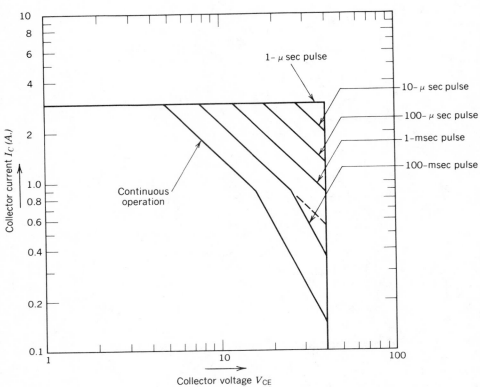

FIGURE 10.9.   Example of safe operating area for 3A, 40V IC power transistor.

of current at $V_{CE} = 5$ V can only handle 1 A of current at $V_{CE} = 15$ V without risking secondary breakdown failure.

In the design of high-current regulators, with output current capabilities in excess of several amperes, it is fairly common to include a safe-area protection circuit into the regulator to ensure that the power transistor current and voltage levels stay within the safe area of the device, particularly in the presence of unpredictable high-voltage transients on the unregulated input line. Figure 10.10 shows a simple safe-area protection circuit often used in high-current regulator design. As indicated, safe-area protection circuitry is usually combined with the active short-circuit protection circuitry. In the circuit, when the voltage differential between V_{IN} and V_O is sufficiently high to turn on the Zener diode Q_2, the safe-area protection becomes activated and the additional current I_X through R_3 and R_2 would increase the prebias at the base of the short-circuit protection transistor Q_1, thus causing the output current limiting to take place at a lower current level. In this manner, as the input–output voltage differential across the pass transistor becomes larger, the maximum available output current is reduced to keep the operating point of the pass transistor within its safe operating area.

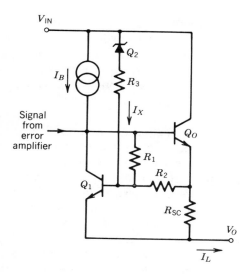

FIGURE 10.10. Combined safe-area and short-circuit protection circuit.

It should be noted that combining the safe-area and current-limit protection into one circuit, as shown in Figure 10.10, has an additional advantage since both have the same temperature dependence characteristics. Both exhibit a negative temperature coefficient of current-limit characteristic, due to the negative temperature coefficient of the V_{BE} of Q_1. This results in a lower current-limit setting at high ambient temperatures, which is often desirable.

Thermal Shutdown

Thermal overload protection is necessary to prevent a high-power regulator circuit from sustaining permanent damage due to prolonged operation at elevated temperatures. For reliable and continuous operation of bipolar integrated circuits, the junction temperatures on the chip should not exceed $+200°C$. In most bipolar and IC designs, this maximum junction temperature is chosen conservatively to be in the range of $+150 - +175°C$. This can be resolved by incorporating into the design a *thermal-shutdown* mechanism which can sense the chip temperature and automatically shut the power off until the chip temperature drops back to a safe level, or until an external reset signal is applied.

In designing a thermal-shutdown circuit, one normally uses the predictable temperature dependence of the diode voltage or the transistor V_{BE} drop as the temperature-sensing element. A simplified schematic for a thermal-shutdown circuit is shown in Figure 10.6. The thermal-shutdown transistor Q_2 in the figure has its base biased from a temperature-stable bias point at a voltage level of V_{TS}. Typically, $V_{TS} \approx 400$ mV at room temperature, and is *not* sufficient to keep Q_2 in conduction. However, since V_{BE} has a negative temperature coefficient of ≈ -2 mV/°C, a temperature increase of 120–150°C above the room temperature would cause Q_2 to come into conduction and shut off the power transistor by shunting away its drive current I_B.

Figure 10.11 shows a typical circuit implementation for deriving a temperature-stable reference bias voltage V_{TS} for a thermal-shutdown circuit. By carefully characterizing the fabrication processes, the thermal-shutdown temperature T_{TS} can be maintained to within $\pm 10°C$ over successive production runs. Normally, in the chip layout, the thermal-shutdown transistor is placed with close proximity to the output pass transistor where the highest chip temperatures are likely to occur.

In high-current series regulators capable of handling load currents of 5 A or more, the simple thermal-shutdown circuit may not respond fast enough or may not be as accurate due to large thermal gradients, which may be produced across the chip in the short duration immediately after an overload condition. In order to overcome such a problem, sometimes a so-called *temperature-gradient-sensing* thermal-shutdown circuit is used, which responds to the *thermal gradients* on the chip, along with the absolute temperature value. As will be described in conjunction with the design examples of Section 10.3 (see Fig. 10.18), such a gradient-sensing shutdown circuit utilizes two sensing devices, one located very close to the power device and one located at the opposite extremity of the chip, away from the power device. In this manner, the thermal shutdown is activated faster by responding to the resulting thermal gradient transient, rather than waiting for the entire chip to heat up. Typical time constants associated with conventional thermal shutdown are on the order of several minutes, whereas gradient-sensing shutdown systems can respond within several tens of milliseconds.

In designing thermal-shutdown circuits, a word of caution is in order. Most thermal shutdown circuits require either a built-in hysteresis of several degrees

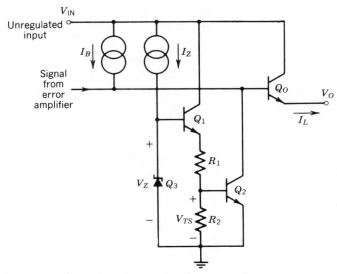

FIGURE 10.11. Thermal-shutdown circuit.

or use an external reset to reactivate the circuit after a shutdown, in order to avoid thermal oscillations in the vicinity of the shutdown temperature, where the circuit may oscillate indefinitely by turning itself on and off by means of the thermal-shutdown control. Such thermal oscillations are undesirable since they disrupt the circuit operation and induce mechanical stresses at the chip–package interface, due to differences of thermal-expansion coefficients between various parts of the package and the IC chip.

10.3. PRACTICAL SERIES REGULATOR CIRCUITS

With the exception of the monolithic operational amplifiers, the series regulator circuits are among the most widely used IC types. Their availability, versatility, and low cost greatly simplify power supply design and power distribution considerations in both analog and digital systems. The low cost of monolithic IC regulators often makes it advantageous to use a separate regulator on each circuit card, instead of designing a single precision voltage supply for the entire system.

Today, well over 100 different series IC regulator designs are commercially available. These circuits vary from the low-power regulators designed for handling currents in the few hundred milliampere range to high-current circuits capable of delivering continuous load currents of 5 A or more.[4-6]

In this section, some of these practical monolithic regulator circuits will be examined. These examples are chosen both because they represent commercially successful and widely accepted products, and because they also provide an excellent demonstration of some of the design considerations discussed in the previous sections.

A summary of the electrical characteristics of these circuit examples is given in Table 10.1.

μA-723 General-Purpose Regulator

The μA-723 regulator circuit, introduced by Fairchild Semiconductor in the mid-1960s, has become one of the most widely used general-purpose regulator integrated circuits for medium-accuracy low-cost power supply design. Figure 10.12 shows a functional block diagram of the μA-723 regulator chip. Since the circuit is primarily intended as a general-purpose building block, many of its subcircuits are not internally connected but are brought out to various package terminals. For example, the reference terminal, the inputs of the error amplifier, as well as the base, collector, and emitter of the pass transistor Q_O and the current-limit transistor Q_1 are brought out to separate package pins in order to enhance the versatility of the circuit. This design philosophy is different from the more recent designs, where almost all of the circuit connections are performed on the chip to minimize the package pin count or the number of external components used.

TABLE 10.1.   A Comparative Listing of Performance Characteristics of Various Series-Regulator Integrated Circuits. Data were Extrapolated from Refs. (4) and (5).

Performance Parameter	Units	μA-723	μA-7800 (μA-7805)	LM-138
Normal operating current range	mA	1–50	5–750	10–5000
Line regulation ($T_A = 25°C$)	%	0.2 (M)[a]	1 (M)	0.05 (M)
Load regulation ΔI_L from I_{min} to I_{max} ($T_A = 25°C$)	%	0.5 (M)	2 (M)	0.3 (M)
Temperature stability (total drift from −55 to +125°C)	%	1.5 (T)[a]	1.5 (T)	1 (T)
Peak output current (pulsed)	A	0.15 (T)	2 (T)	10 (T)
Short-circuit current limit ($T_A = 25°C$)	A	0.065 (T)	0.75 (T)	5 (M)
Drop-out voltage	V	3 (M)	2 (T)	2.5 (T)
Ripple rejection (50 Hz–10 kHz)	dB	74 (T)	68 (M)	60 (T)
Output noise voltage (BW 100 Hz–10 kHz)	μV rms	20 (T)	40 (T)	100 (T)
Output impedance ($f \leq 1$ kHz)	Ω	0.05 (T)	0.05 (T)	0.01 (T)
Input voltage range	V	9.5–40	7–24	3–36
Output voltage range	V	2–37	Fixed	1.2–32
Package pin count		10 or 11	3	3
Number of external components needed for typical application (excluding filter capacitances)		5 or 6	None	2
Thermal overload protection		No	Yes	Yes

[a](T) designates typical value, (M) designates minimum or maximum limit.

In the block diagram of Figure 10.12, the external circuit connections and the external components necessary for circuit operation are shown with dashed lines. The Zener diode D_1 provides a temperature-compensated voltage reference in feedback around the internal reference amplifier. The output of this amplifier is connected to the noninverting input of the error amplifier. The inverting input of the error amplifier is biased from the sampling resistors R_A and R_B to provide an output voltage level,

$$V_O = V_{ref}\left(1 + \frac{R_A}{R_B}\right)$$ (10.16a)

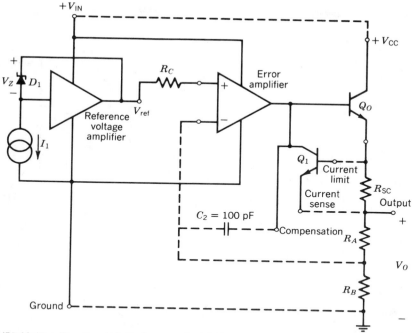

FIGURE 10.12. Functional block diagram of μA-723 general-purpose regulator integrated circuit. (*Note:* Dashed-line connections, resistors, and capacitors shown are external.)

The external resistor R_C between V_{ref} and the noninverting input of the amplifier is set equal to the parallel combination of sampling resistors,

$$R_C = \frac{R_A R_B}{R_A + R_B} \tag{10.16b}$$

in order to equalize the source resistances in series with each of the two inputs of the error amplifier, and avoid offset voltages due to input bias currents.

Figure 10.13 shows the complete circuit diagram of the μA-723 regulator integrated circuit. External components are identified in boxes. The n-channel JFET Q_3 is basically an epitaxial FET (see Fig. 3.29) which keeps the Zener diode D_2 in conduction. The voltage drop across D_2 is then used to set the currents in the *pnp* current sources Q_5, Q_9, and Q_{10}. The reference amplifier is formed by Q_8 and Q_5, with the Darlington pair Q_6 and Q_7, providing a low-impedance output. The value of V_{ref} is approximately equal to the voltage drop across D_1 plus the V_{BE} of Q_8. This results in a reference voltage level of approximately 7.0–7.2 V. The V_{BE} drop of Q_8 compensates for the positive temperature coefficient of D_1, resulting in a typical temperature stability of 50–100 ppm/°C for V_{ref}.

The error amplifier is formed by the different pair Q_{11} and Q_{12}, and its single-ended output drives the Darlington-connected pass transistor. The sam-

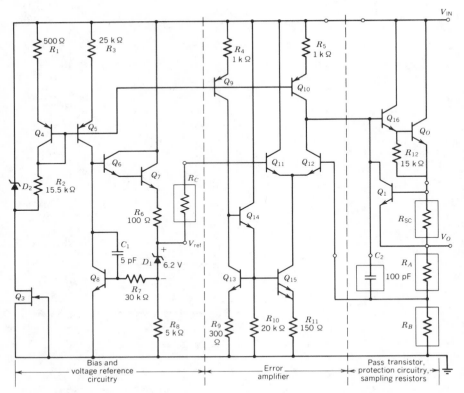

FIGURE 10.13. Circuit schematic of μA-723 voltage regulator. (*Note*: Components shown in boxes are external to IC chip.)

pling resistors, the short-circuit protection resistor R_{SC}, and the error amplifier compensation capacitor C_2 are external to the monolithic circuit.

The circuit configuration of Figure 10.13 is primarily intended for output voltage levels in excess of 7 V. If a lower output voltage is needed, this can be done by scaling V_{ref} to a lower value by means of an external resistor string from the V_{ref} terminal to ground.

The circuit is designed for low-power applications and can supply a maximum load current of 150 mA. If higher amounts of load current are required, an external pass transistor must be used.

Other electrical characteristics of the μA-723 are summarized in Table 10.1.

Three-Terminal Fixed-Voltage Regulators

In a wide variety of linear or digital circuit applications, one normally works with certain "standardized" supply voltages such as +5 V for TTL and NMOS logic, and +6, +9, +12, and +15 V for a number of linear functions. The three-terminal fixed-voltage regulators are primarily intended for such standardized supply voltage applications as the so-called "on-card" regulation. In other words, they are intended to be directly placed on a circuit card to provide local regulation of the supply voltage with no additional adjustments.[7]

There are two unique advantages to three-terminal fixed-voltage regulators:

1. They require no external adjustments or external components with the possible exception of line or load filter capacitances.
2. They can be packaged in low-cost power transistor packages such as the T0-3 type metal can or the T0-220 type molded plastic packages. Since the package cost is a very significant fraction of the cost of a power integrated circuit, this is a significant cost advantage.

At present a number of positive and negative three-terminal fixed-voltage regulator integrated circuits are available at varying power levels from a multiplicity of manufacturers. The most widely used families of three-terminal fixed-voltage regulators are the μA-7800 series for positive regulators and the μA-7900 series for negative-voltage regulators. Both of these families were originally introduced by Fairchild Semiconductor in the mid-1970s.

Figure 10.14 shows both the block diagram and the circuit schematic of the μA-7800 series three-terminal voltage regulator which is designed to handle load currents in excess of 1 A.[5] In the circuit, transistors Q_1 through Q_7 and their associated resistors form a band-gap-type voltage reference with a nominal value of 5.0 V at the base of Q_6. The Darlington-connected transistors Q_3 and Q_4 function as the error amplifier. The error amplifier gain is increased by the *pnp* transistor Q_{11}, which serves as a buffer and drives the active current mirror load made up of Q_8 and Q_9. The collector of Q_9 then drives the Darlington pass transistor, made up of Q_{16} and Q_{17}. The feedback loop is closed by feeding back a portion of the output voltage through the sampling resistors R_{19} and R_{20} to the base of Q_6. The regulated output voltage is

$$V_O = V_{ref}\left(1 + \frac{R_{20}}{R_{19}}\right) \qquad (10.17)$$

which is adjusted by the choice of the proper "taps" in R_{20}. These taps on R_{20} are selected during fabrication, at the metal interconnection stage. Thus, for example, by setting $R_{20} = 0$, one has the lowest fixed output voltage, where $V_O = V_{ref}$. Similarly, choosing $R_{20} = 10$ kΩ, would given an output voltage of $V_O = 3V_{ref} = 15$ V.

In the circuit of Figure 10.14, Q_{15}, along with R_{11}, serves as the short-circuit current-limit circuit; R_{12}, R_{21}, R_{13}, and D_2 serve as the safe-area protection circuit in a manner similar to that shown in Figure 10.10. Thermal-shutdown protection is provided by Q_{14}, whose base is prebiased at approximately +400 mV above ground. Thus, Q_{14} turns on when the nominal junction temperature reaches approximately 175°C.

The Zener diode D_1, along with the emitter follower Q_{12} and its associated resistor string, serves as the thermal shutdown bias source to Q_{14}; and they also

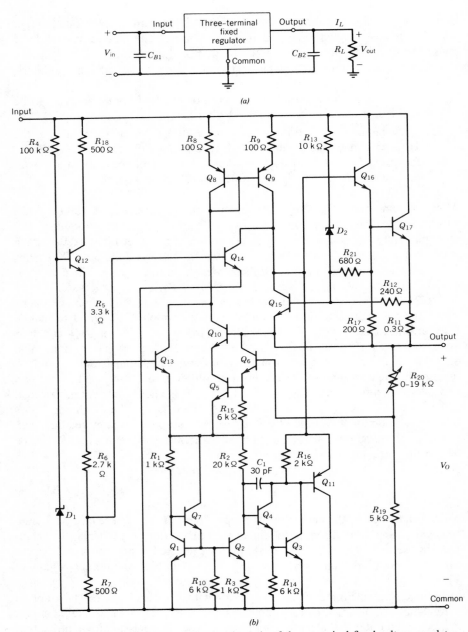

FIGURE 10.14. Block diagram and circuit schematic of three-terminal fixed-voltage regulator. (Fairchild μA-7800 series.)

function as the start-up circuitry for the voltage reference section. When the circuit is initially powered on, Q_{13} comes into conduction and activates the current mirror transistor pair Q_8 and Q_9. Once the circuit starts regulating, Q_{13} is turned off and remains inactive.

The 7800 series regulators are available in T0-3 or T0-220 type power packages, with a variety of fixed output voltages from $+5$ V to $+24$ V. A summary of their electrical characteristics is given in Table 10.1, for $V_O = +5$ V case.

Adjustable Three-Terminal Regulators

By proper design of the regulator circuit, it is possible to adjust the output voltage level of a three-terminal regulator by "floating" its common terminal. This method is illustrated in the block diagram of Figure 10.15 for the case of a positive regulator.[8] The actual regulator integrated circuit is shown within the dashed area. Assuming that the error amplifier gain is sufficiently high, the output voltage can be written as

$$V_O = V_{ref}\left(1 + \frac{R_2}{R_1}\right) + I_C R_2 \tag{10.18}$$

where I_C is the bias current drawn from the common terminal of the regulator. By proper design of the regulator control circuitry, I_C can be kept very low,

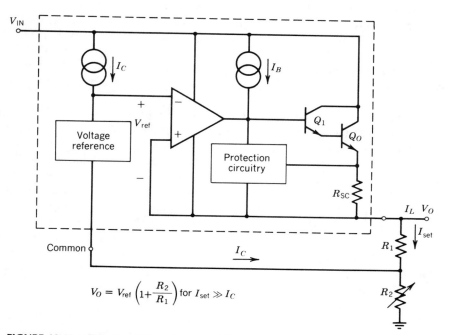

$$V_O = V_{ref}\left(1 + \frac{R_2}{R_1}\right) \text{ for } I_{set} \gg I_C$$

FIGURE 10.15.   Functional block diagram of three-terminal adjustable voltage regulator.

typically in the $50\text{--}100\,\mu\text{A}$ range. If the level-setting current I_{set} is chosen to be such that

$$I_{set} = \frac{V_{ref}}{R_1} \gg I_C \tag{10.19}$$

then the last term in Eq. (10.18) is negligible, and the output voltage can be written as

$$V_O \approx V_{ref}\left(1 + \frac{R_2}{R_1}\right) \tag{10.20}$$

which can be adjusted by the proper choice of R_2.

Figure 10.16 shows the complete circuit diagram of a high-current three-terminal adjustable regulator integrated circuit (National Semiconductor LM-138). The circuit is designed to handle rated load currents up to 5 A, and utilizes the basic block diagram configuration of Figure 10.15. With reference to the complete schematic, and neglecting the basic bias circuitry, the operation of the circuit can be analyzed by separating it into two functional sections: (1) control circuitry and (2) protection circuitry.

Control Circuitry. Figure 10.17 shows the simplified circuit diagram of the feedback-control loop forming the voltage regulator. The voltage reference is a basic 1.2-V band-gap reference.[9] It is made of transistors Q_{17} and Q_{19}, which are forced to operate at equal currents but with unequal V_{BE} drops by means of the current mirror transistors Q_{16} and Q_{18}. The feedback loop formed by the buffer amplifier (made up of Q_{12} through Q_{15}) and the Darlington pass transistors Q_{25}, Q_{26} maintains equal currents through Q_{17} and Q_{19} to result in a stable 1.2-V reference level between the base of Q_{19} and the common terminal. The principle of operation of this control loop is similar to that illustrated in Figure 4.36a.

Protection Circuitry. Because of the high-current (5 A) handling capability of the LM-138 regulator, additional care is taken in the design and layout of the protection circuitry on the chip. Figure 10.18 shows a simplified schematic of the overload protection circuits for the LM-138 regulator. Both the current-limit and the thermal-shutdown functions are performed by the *pnp* transistor Q_{20}, which is driven by the gain stage Q_{21}. Both of these transistors are normally off, and are brought into conduction when an overload condition is present. The transistor Q_{24} is normally on and produces a temperature-compensated prebias of about 400 mV at the base of Q_{21}, which is approximately 100–150 mV less than the amount needed to activate Q_{21} at room temperature. Short-circuit current limiting is obtained by sensing the voltage drop across R_{26} and turning on Q_{21} when this drop exceeds approximately 100 mV. Safe-area protection is provided by the 6.2-V Zener diodes D_2 and D_3 and their associated resistors. When these diodes conduct, the voltage drop across R_{24} and R_{25} will decrease the current in Q_{24} and increase the prebias on Q_{21}. This results in Q_{21} turning on

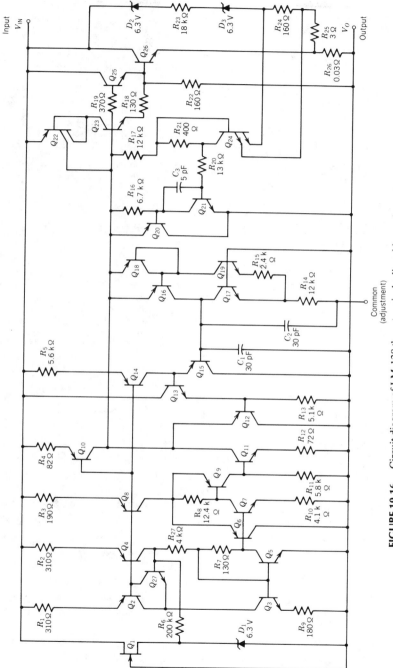

FIGURE 10.16. Circuit diagram of LM-138 three-terminal adjustable regulator. (*Note:* All capacitors shown are on chip.)[4]

505

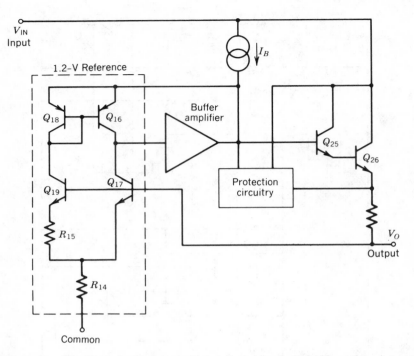

FIGURE 10.17. Simplified circuit diagram of control circuitry in LM-138 regulator integrated circuit.

sooner and limiting the current at a lower voltage drop across R_{26}. Thus, the total current capability is reduced as the input–output voltage differential is increased in order to keep the power transistor within its safe operating area.

Thermal shutdown is obtained when the V_{BE} of Q_{21} decreases with temperature so as to cause it to turn on. This is designed to happen at a junction temperature of approximately 170°C. Another interesting feature of the shutdown circuit is its response to temperature gradients. This is done by placing the shutdown transistor Q_{21} adjacent to the power device while the prebias transistor Q_{24} is set further away, as shown in the chip photograph of Figure 10.19. This causes the shutdown or the current-limit circuit to be sensitive to the thermal gradients on the chip, as well as to the average chip temperature. Under instantaneous overload condition, the local heating of the power transistor can cause sufficient thermal gradient between Q_{21} and Q_{24} to activate the current-limit function without waiting for the average chip temperature to rise. In this manner, the response time of the thermal current-limit function can be greatly speeded up.[10]

Figure 10.19 shows the photomicrograph of the LM-138 regulator integrated circuit, which is designed to handle a rated current load of 5 A and operate over a supply voltage range of 3–40 V. Some of the important portions of the circuitry

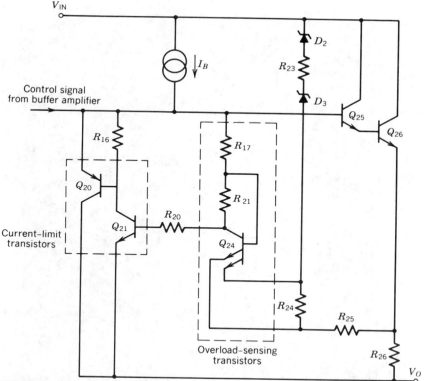

FIGURE 10.18. Simplified schematic of overload protection circuitry for LM-138 regulator integrated circuit.

are identified on the chip layout. It is worth noting that the power transistor takes up approximately 70% of the chip area, while the control and protection circuits occupy a relatively insignificant amount of chip surface. Also note that the gradient-sensing transistors Q_{21} and Q_{24} are located at dissimilar distances from the power transistor to enhance the thermal-gradient effects.

A Comparison of Performance Characteristics

Table 10.1 shows a comparative listing of various performance characteristics for the many series regulator integrated circuits reviewed in this section. The data given in the table have been compiled from the manufacturers' technical literature.[4,5] In important cases, the parameters are either typical or fully guaranteed maximum or minimum specifications and are thus identified with (T) or (M). The reader is cautioned that each manufacturer's measurement method or test condition may differ slightly. As such, although a reasonable effort has been made in checking the accuracy of the data presented in Table 10.1, the reader

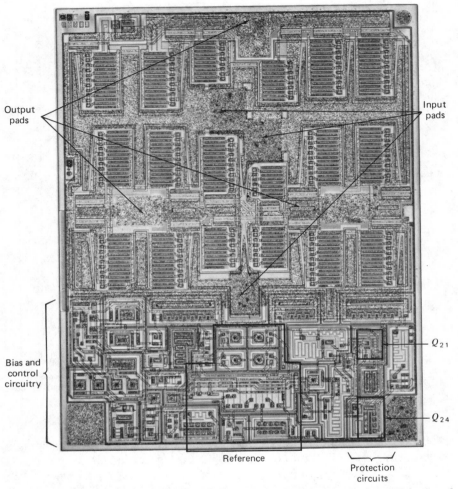

Output pads

Input pads

Bias and control circuitry

Q_{21}

Q_{24}

Reference

Protection circuits

FIGURE 10.19. Photomicrograph of LM-138 adjustable three-terminal regulator. Note size of output the transistor designed to handle 5 A of load current. Chip size: 96 mils × 121 mils. (*Photo:* National Semiconductor Corporation.)

must use the information mainly for relative comparison purposes. Since many grades of each regulator type are commercially available, only the prime grade was considered.

Each of the three basic regulator designs reviewed in this section were chosen because they have become the recognized industry standards and are demonstrative of particular design objectives or approaches. The μA-723 was chosen because of its versatility as a low-cost power-supply building block. The μA-7800 family of fixed regulators were chosen because they represent a very-low-cost solution to on-card regulation in various distributed logic subsystems.

The LM-138 was chosen because it extends versatility of the adjustable three-terminal regulators to high-current applications and demonstrates sophisticated design techniques for efficient overload-protection circuitry.

10.4. LAYOUT CONSIDERATIONS FOR POWER CIRCUITS

In a monolithic power circuit, the output power transistors are the key to the circuit performance. Therefore, the design and layout of the power transistors require additional care and attention. In most cases, more than half of the chip area is devoted to the power transistors (see photomicrograph of Figure 10.19), with the low-power control or signal-processing circuitry taking up a minor portion of the silicon area.

In a power transistor operating at high-current levels, almost all of the current injection occurs along the edge or periphery of the emitter. To increase the emitter periphery, and to minimize the debiasing effects due to base-spreading resistance, one normally uses an interdigitated structure for the emitter and base regions. To reduce the collector series resistance, a deep n^+ *sinker* diffusion is used along the collector contact of the *npn* power transistor (see Fig. 2.16). In the layout of the emitter stripes, voltage drops along the metal interconnection paths also need to be considered. Due to the exponential current–voltage dependence across the base–emitter junction, the injected current density in the emitter will decrease by approximately a factor of 2 for every 18 mV drop along the metal conductor interconnecting the emitter stripes. With a typical aluminum interconnect sheet resistivity of about 20–30 mΩ/$\square$, care has to be taken to keep the entire emitter surface injecting uniformly by using large metal strips over the emitters. To keep the current density reasonably uniform, the forward voltage across the base–emitter junction should not vary by more than 10 mV over the entire transistor.

To prevent the power transistor from being triggered into a secondary breakdown, it is necessary to use small *current-equalizing* or *ballast* resistors in series with each one of the individual emitter areas such that no localized heating or hot spots are formed due to unequal current concentration along the active emitter–base junction. Because of the distributed nature of the base–emitter junction, such a ballast resistor can be added into the circuit as a distributed resistor, made up of the bulk resistance of the n^+ diffused emitter region. This can be done by leaving a substantial distance (typically 15–30 μm) between the metal emitter contact and the active emitter periphery. This is shown schematically in Figure 10.20. If any one point along the emitter periphery tends to conduct heavily and cause localized heating, the voltage drop across the ballast resistor will tend to reduce the bias of that particular emitter region and equalize the current distribution.

The size of the power transistor is determined by the safe operating area requirements. Normally, a *modular* transistor structure is used, made up of a number of identical sections connected in parallel. Each section is made up of

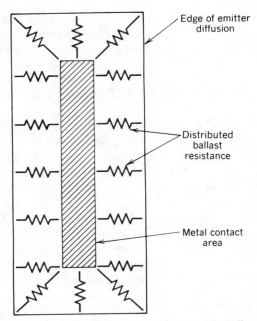

Edge of emitter
diffusion

Distributed
ballast
resistance

Metal contact
area

FIGURE 10.20.   Distributed emitter ballast using emitter bulk resistance.

an individual base region with a number of individually ballasted emitter stripes.
All the base sections then share a common collector tub, with deep n^+ diffused
collector contacts wrapping around each of the base sections. The advantage of
such a modular power transistor structure is that once one of the sections is
characterized for its safe operating area and current-handling capability, the
entire transistor design becomes relatively easy to scale up or down in size by
the choice of such sections connected in parallel.

As a general rule of thumb, a high-current transistor with a saturation voltage
of approximately 1.5 V at the rated current requires about 2000–3000 mil^2 of
chip area per ampere of rated current load.

In the case of a high-current series regulator integrated circuit, similar to that
shown in Figure 10.19, the collector of the power transistor is directly connected
to the unregulated input. Normally, the collector is connected to the input
terminal by means of not one but several bonding wires in order to reduce the
ohmic drops within the collector contacts and the aluminum interconnection
paths. In circuits designed to handle currents in excess of 1 A, 3- or
4-mil-diameter gold bond wires are used for chip-to-package interconnection,
rather than the conventional 1-mil bonding wires.

The emitter of the power transistor is brought out through a low-value sensing
resistor as the output terminal for the regulator. The short-circuit sensing resistor, which is normally less than 1 Ω, is formed either by means of an
emitter-diffused n^+-type resistor or by a metal interconnection strip of a prede-

termined length. The output is normally brought out to several bonding pads, in parallel, which are then bonded to the output terminal by individual bonding wires in order to reduce the parasitic series resistances associated with package bonding wires.

In a monolithic power circuit, the presence of a power device in close proximity to the low-level signal-processing and control circuitry can require additional layout precautions. The power device serves as a concentrated heat source and can cause severe thermal gradients within the chip. This problem is particularly serious in the case of monolithic regulator circuits which also must contain a temperature-independent voltage reference on the chip. Therefore, during the layout of the circuit, additional care must be taken in selecting the location of the control circuitry, and particularly the internal voltage reference, so as to minimize the effects of the temperature gradients within the chip.

Figure 10.21 shows a practical block diagram for the physical layout of a high-power monolithic regulator circuit. Normally, the distributed power source, the pass transistor, is located along one edge of the chip. The localized heating of the pass transistor generates a set of isothermal contours over the remainder of the chip, and these are more or less parallel to the longitudinal edge of the power device, as shown. The thermal drifts of monolithic components can be minimized by laying out the control circuitry so that the components to be matched are along the same isothermal contour as—or equally distant from— the power dissipation source on the chip. Since the voltage reference is most affected by thermal gradients within the chip, it should be located as far away

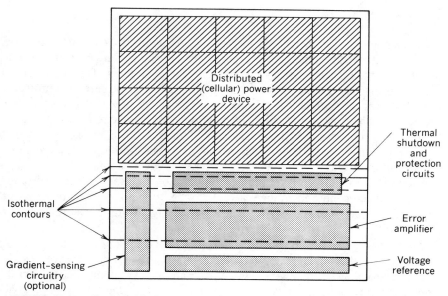

FIGURE 10.21. Typical chip layout topology for high-current regulator integrated circuit.

as possible from the pass transistor. Thermal-shutdown circuitry, which is required to sense the junction temperature of the power device, is located near the output transistor. The rest of the protection circuitry, such as the safe-area and current-limit protection, is also located in close proximity to the power device. One exception to this general rule is the case of "gradient-controlled current limiting" described in connection with the circuit of Figure 10.16, where the presence of the thermal gradients across the chip is used to activate the current-limit circuitry. In such a case, the gradient-sensing devices should be laid out along the direction of heat flow, that is, in *transverse* direction to the isothermal lines, as is the case with transistors Q_{24} and Q_{20} in the chip photograph of Figure 10.19.

10.5. FAILURE MECHANISMS IN POWER DEVICES

There are two separate failure modes that are unique to monolithic devices and circuits, designed to operate at high power levels. These are the so-called *secondary breakdown* and *electromigration* effects. Secondary breakdown is a thermal instability effect encountered in bipolar power transistors; it causes the transistor to switch to a low-voltage high-current mode, which can result in permanent deterioration of the device characteristics or total burn-out. Electromigration is a mechanism which results in the physical movement of metal atoms in the metallic interconnection layers, under high-current-density operation, and can eventually cause an open-circuit in the interconnecting layer.

Secondary Breakdown

This phenomenon comes about because of nonuniform current distribution within the transistor junctions at high-power levels, which cause nonuniform heating and the formation of localized hot spots in junctions. This failure mechanism was discussed in Section 10.2 in conjunction with the safe-area protection circuitry. Under secondary breakdown, the localized power density and temperatures can be very large despite any external stabilizing circuit. If the junction temperature, under a given current and voltage level, reaches a *triggering temperature*, the power device becomes a virtual short-circuit, which results in permanent burn-out.

Secondary breakdown is a complex function of both the collector current I_C and the collector-emitter voltage V_{CE}.[11,12] The collector current necessary to trigger secondary breakdown is much lower at high voltages, as illustrated in Figure 10.9.

Thus, for a given power transistor design, one has to define a safe-operating-area limit on the device current–voltage characteristics and ensure that the current and voltage levels within the device are always maintained within this limit. Since such a safe area is a complex function of the fabrication process and device layout, it is normally defined, somewhat empirically, by careful characterization of the device performance prior to circuit implementation. The safe operating area of the transistor is greatly enhanced by using a modular, or segmented, device structure, made up of a number of individual transistors connected in parallel, and by the use of ballasting resistors in each of the emitters to prevent current crowding, as was described in the previous section.

Electromigration

Electromigration is a mass-transport effect observed in metals under high current densities. It causes the metal atoms to migrate away from a high-current-density point in the metal and can cause a conductor, such as an aluminum interconnection on the chip, to eventually become an open circuit at the point of highest current density. Electromigration is a slow process which speeds up as the current density or the temperature is increased, or as the lateral dimensions of the conductor normal to the current path are diminished. It starts as a gradual formation of voids on a conducting metal strip and eventually leads to a complete open circuit.

An approximate expression for the median time to failure (MTF) of a conducting metal film can be given as[13]

$$\text{MTF} = \frac{KWd}{J^2} \exp\left(\frac{\phi}{kT}\right) \tag{10.21}$$

where K = is a constant of proportionality
J = current density
W = conductor width
d = conductor thickness
ϕ = activation energy, depending on type and grain size of the metal
k = Boltzmann's constant
T = temperature (Kelvin)

In the case of aluminum, the activation energy for electromigration effects has been observed to be in the range of 0.48 eV (for small-grain structures) to 1.2 eV (for large well-ordered grain structures). The electromigration effect is a slow wear-out type of failure effect, similar in many ways to the creep failure in metals due to sustained mechanical stress. Because of the exponential nature of Eq. (10.21), it becomes significant particularly at high temperatures. It should also be noted that since J is inversely proportional to the cross-sectional area, at any given current level, electromigration effects increase inversely as the third power of the cross-sectional area.

As a general rule, the current densities in the interconnecting metal layers are maintained below the 10^5-A/cm$_2$ level. In practical design terms, this implies a maximum continuous current density of approximately 25 mA per mil width of a 10,000 Å-thick (i.e., 1 μm) aluminum interconnection path. In terms of Eq. (10.21), this leads to a predicted median time to failure of well in excess of 100 years.

The thickness of a metal interconnection layer may be reduced at the points where the metal trace is forced to run over oxide steps on the chip surface, due to different mask layers. This results in a reduction of the conductor cross section, and correspondingly, is an increase of the current density within the metal trace at these localized spots. In order to avoid potential metal-migration problems at such points within the circuit, the high-current interconnection paths should not be routed over such oxide steps.

PART II: SWITCHING REGULATORS

10.6. FUNDAMENTALS OF SWITCHING REGULATORS

The switching regulators, which are also called *switch-mode regulators*, find a wide range of applications in power supply design where high power and high efficiency are important. The principle of operation of a switching regulator differs significantly from that of a conventional series regulator circuit discussed in the first part of this chapter. In the case of series regulators, the pass transistor is operated in its linear region to provide a controlled voltage drop across it with a steady dc current flow. In the case of switching regulators, the pass transistor is used as a "controlled switch" and is operated at either the cutoff or the saturated state. In this manner, the power is transmitted across the pass device in discrete current pulses, rather than as a steady current flow.

The most important advantage of switching regulators over the conventional series regulators is their greater efficiency, since the pass device is operated as a low-impedance switch. When the pass device is at cutoff, there is no current through it, thus it dissipates no power. When the pass device is in saturation, it is nearly a short circuit with negligible voltage drop across it, thus it dissipates only a small amount of average power, provided that it can handle the peak current loads. In either case, very little power is wasted in the regulator and pass devices, and almost all the power is transferred to the load. In this manner, a very high degree of regulator efficiency is achieved, typically in the range of 70–90%, relatively independent of the input–output voltage differentials. The efficiency of switching regulators is particularly apparent when there is a large input–output voltage difference across the regulator. For example, if one considers the case of a regulator operating with a 28-V input and delivering a 5-V output at 1-A current, a conventional series regulator would require a drop of 23 V across the series pass transistor. Thus, a total of 23 W of power is wasted in the regulator, resulting in an overall regulator efficiency of approximately 18% [see Eq. (10.12)]. As will be described in later sections, a switching regulator can be readily designed to perform the same function with greater than 75% efficiency under similar operating conditions.

Another important advantage of the switching regulator circuit is its versatility. It can provide output voltages which can be less than, greater than, or of opposite polarity to the input voltage, as determined by the mode of operation of the circuit. In this manner, one can step up, step down, or invert the polarity of an input voltage to generate any arbitary set of dc voltages within a system.

Switching regulators also have some drawbacks. They are more complex and require external components such as inductors or transformers. They generate more noise and output ripple than conventional series regulators, and are slower responding to transient load changes. One area of caution, when using switching regulators, is the generation of electromagnetic and radio-frequency interference (RFI). This interference problem is usually solved by the use of feedthrough low-pass filters isolating the power lines into the regulator, and by using ground shields around the regulator to suppress the interference. However, even with these precautions, switching regulator circuits are not recommended for powering very-low-level signal-processing circuitry, where noise characteristics are very critical.

A switching-regulator power supply system is made up of three basic blocks: (1) the switching element which is normally a power transistor, (2) control circuitry which sets the duty cycle (i.e., on–off time) of the switching element, and (3) output circuitry which converts the pulsed input power to a steady output power flow. The switching-regulator power supplies are classified into three categories, depending on the type of output circuitry used. These classes are

1. Single-ended inductor circuits.
2. Diode–capacitor circuits.
3. Transformer-coupled circuits.

Figure 10.22 shows the three basic configurations for the output circuitry of single-ended inductor-type switching regulators. These are among the most frequently used output circuit configurations, since they are by far the easiest to design and control for medium- and high-current applications.

Figure 10.23 shows the switching regulator configurations using diode–capacitor circuits. These circuits are typically used for very-low-current applications, and primarily as voltage multipliers, to increase the voltage level available from a low-voltage battery. Examples of such applications are the generation of voltage drive for LCD displays in watch circuits from a 1.5- or 3-V battery and in low-voltage hearing aid amplifiers.

Figure 10.24 shows the two basic configurations of transformer-coupled output circuits. The circuit of Figure 10.24a is the so-called *push–pull circuit* used in conventional dc-to-dc converters, with each switch controlled for 0–45% duty cycle modulation. The configuration of Figure 10.24b is the so-called single-ended *flyback converter*, which is useful at low- to medium-current loads.

The design of power supply systems using discrete circuits, and the various types of output circuitry shown in Figures 10.22 through 10.24, are well covered

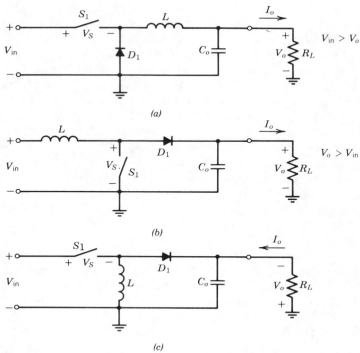

(a)

(b)

(c)

FIGURE 10.22. Inductive switching regulator configurations: (*a*) Step-down; (*b*) step-up; (*c*) polarity-inverting.

in the literature.[14] In the following discussions, we will primarily focus on switching regulator circuits using the single-ended inductor-type output circuit shown in Figure 10.22, since they represent by far the most common and general categories of application, and are especially suited to high-current applications.

The control circuitry section of a switching regulator system, which controls the on–off duty cycle of the switch transistors, can be readily integrated in a monolithic IC form. In many cases, the switching transistors up to 1-A current rating can also be incorporated into the monolithic chip. If higher power levels are required, the switch transistor on the chip is used as a drive for an external high-current switch.

Figure 10.25 shows a simplified block diagram of a typical switching regulator integrated circuit used in conjunction with the single-ended inductor configuration of Figure 10.22*a*. The circuit generates a stream of pulses which turn switch S_1 on and off. The output dc level is sensed through the sampling resistors R_1 and R_2 and compared against an internal voltage reference V_{ref}, and the on–off time on the duty cycle of the switch S_1 is varied accordingly to keep the output voltage constant under changing load conditions.

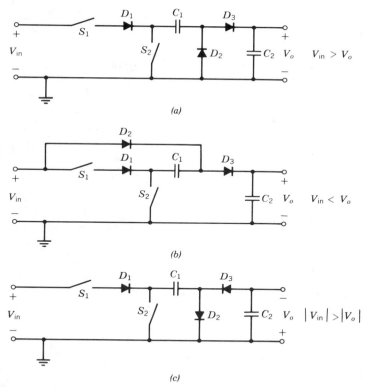

FIGURE 10.23   Diode–capacitor-type switching regulator configurations: (*a*) Step-down; (*b*) voltage multiplier; (*c*) polarity inverting.

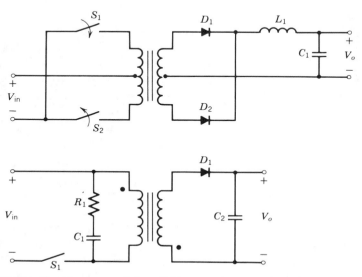

FIGURE 10.24.   Transformer–coupled switching regulator configurations: (*a*) Push–pull; (*b*) flyback.

517

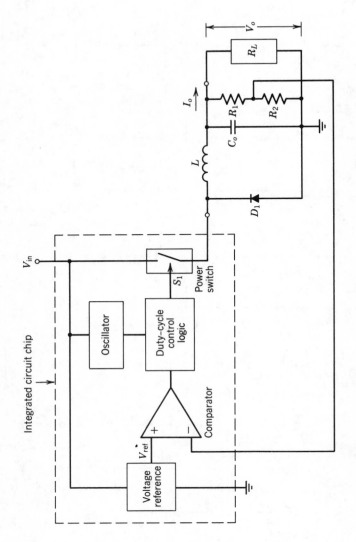

FIGURE 10.25. Simplified block diagram of switching regulator integrated circuit in step-down configuration.

Neglecting the current in the sampling resistors, the average or dc value of the output current I_o delivered to the load is proportional to the duty cycle of the power switch S_1, as shown in Figure 10.26. If the sampled output voltage is lower than V_{ref}, the polarity of the comparator output signal causes the control logic to increase the duty cycle of S_1 and, thus, causes the output voltage level to increase until the equilibrium is reached such that the output voltage, scaled down by the sampling resistors, is equal to the internal reference voltage. Similarly, if the output load current I_o is decreased, this would cause the output voltage to increase, which in turn would be sensed by the control circuitry and would reduce the duty cycle of the switch accordingly.

Control of the Duty Cycle

The switch duty cycle t_{on}/T can be controlled by pulse-width modulation methods at a fixed frequency, or by fixing the on or off time and controlling the frequency. The relative merits and disadvantages of these techniques are briefly examined below.

Fixed-Frequency, Variable-Duty Cycle Operation.   In this type of a switching regulator, the operating frequency is fixed, and the duty cycle of the pulse train is varied to change the average power. This method is often referred to as

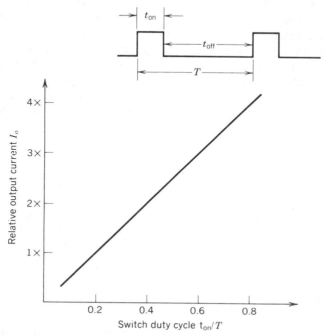

FIGURE 10.26.   Output load current I_o as a function of switch duty cycle.

pulse-width modulation (PWM). The fixed-frequency concept is particularly advantageous for systems employing transformer-coupled output stages. The fixed-frequency aspect enables the efficient design of the associated magnetics. In addition, filtering or shielding the surroundings from the radio-frequency or electromagnetic interference generated by the regulator is somewhat simplified because of the fixed frequency of switching. Because of these features, the majority of the monolithic switching regulators utilize the fixed-frequency, variable-duty-cycle control method.

Fixed-On-Time, Variable-Frequency Operation.   In this method, the switch has a fixed or predetermined on time; and the duty cycle is varied by varying the frequency or repetition rate of the control pulses. This method provides ease of design in voltage conversion applications using the single-ended inductive output circuit configurations of Figure 10.22, and simplifies design calculations for the inductor value. The fixed-on-time method is also advantageous for inductive output circuitry since a consistent amount of charge is developed in the inductor during the fixed on time. This eases the design or the selection of the inductor by defining the operating area to which the inductor is subjected under transient load conditions. Figure 10.27a shows the typical frequency versus load current characteristics of a fixed-on-time, variable-frequency regulator, where the frequency increases *linearly* with increasing load.

Fixed-Off-Time, Variable-Frequency Operation.   In this type of a voltage regulator, the dc voltage at the output is varied by changing the on time t_{on} of the switch while maintaining a fixed off time T_{off}. As shown in Figure 10.27b, the fixed-off-time switching regulator behaves in an opposite manner to the fixed-on-time system: as the load current increases, the on time becomes longer, thus decreasing the frequency. This approach is advantageous for the design of switching regulators that will operate at a well-defined minimum frequency and low-ripple current under full-load conditions. One basic drawback of the

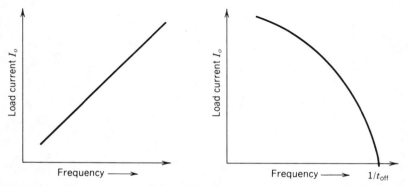

FIGURE 10.27.   Typical load current versus frequency characteristics of variable-frequency switching regulators: (a) Fixed on time; (b) fixed off time.

fixed-off-time system is that the maximum current in the inductor, under transient load conditions, is not well-defined. Thus, additional care is required to ensure that the saturation characteristics of the inductor are not exceeded.

10.7.   MODES OF OPERATION WITH INDUCTIVE OUTPUT CIRCUITS

Two of the most important advantages of switched regulators are their high efficiency and their ability to step up, step down, or change polarity of an input voltage. These basic features can be best understood by examing the voltage and current waveforms at the output of the regulator. In this section, some of the waveforms and key design equations associated with the inductive output circuits of Figure 10.22 will be examined for various modes of operation under steady-state load conditions. The rigorous derivations of the circuit equations associated with the basic current and voltage waveforms is straightforward and is available in the literature.[14,15] Therefore, for the sake of brevity, rigorous derivations will be omitted and only their conclusions will be presented.

Step-Down Operation

In the step-down mode of operation, the switching regulator produces an output dc voltage V_o which is *lower* than the input voltage V_{in}. Figure 10.28 shows the basic voltage and current waveforms associated with the circuit under steady-state operation. The switch S_1 is assumed to have a voltage drop of V_{sat} in its on condition, and the diode D_1 has a forward drop of V_D when it is conducting.

When S_1 is closed or on, D_1 is off and the current I_L in the inductor rises linearly from zero to its peak value I_{pk}, with the slope

$$\frac{dI_L}{dt} = \frac{V_L}{L} = \frac{V_{in} - V_{sat} - V_o}{L} \tag{10.22}$$

At the end of the switch on-time t_{on}, this current reaches its maximum value I_{pk}. When S_1 is off, the inductor generates the necessary voltage to forward bias D_1 and keep I_L from changing instantaneously. During this portion of the cycle, I_L linearly decays to zero at the end of the off time t_{off}.

At steady state, average, or dc current through C_o is zero; therefore, the output dc current I_o is equal to the average value of I_L or

$$I_o = (I_L)_{av} = \frac{I_{pk}}{2} \tag{10.23}$$

From the waveforms shown in Figure 10.28, the switch on and off times can be related to the voltage levels at the input and output of the circuit as

$$\frac{t_{on}}{t_{off}} = \frac{V_o + V_D}{V_{in} - V_{sat} - V_o} \tag{10.24}$$

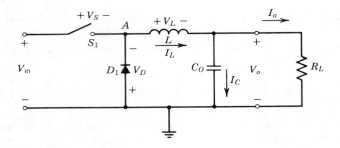

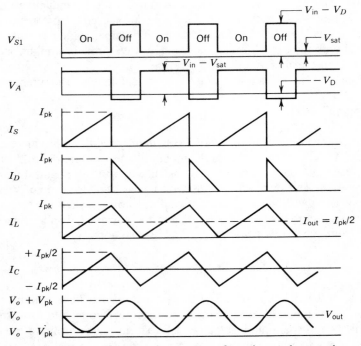

FIGURE 10.28. Voltage and current waveforms in step-down mode.

or, V_o can be related to the rest of the voltages as

$$V_o = \frac{t_{on}}{T}(V_{in} - V_{sat}) - \frac{t_{off}}{T}V_D \qquad (10.25)$$

Assuming the ideal case where both the saturation and the diode voltages are zero or negligible, this reduces to

$$(V_o)_{ideal} = \frac{t_{on}}{T}V_{in} \qquad (10.26)$$

Equation (10.26) implies that, ideally, the switching regulator in its step-down mode provides a down scaling of the input voltage by a scale factor equal to the duty cycle of the switch transistor.

Another important parameter of the step-down regulator is the peak-to-peak output ripple voltage $(\Delta V_o)_{pp}$. Assuming that C_O is sufficiently large so that the ripple voltage is much lower than the average or dc value of the output, $(\Delta V_o)_{pp}$ can be expressed as

$$(\Delta V_o)_{pp} \approx \frac{I_{pk}}{8C_O}(t_{on} + t_{off}) = \frac{I_{pk}}{8C_O f} \tag{10.27}$$

where $f = 1/T$ is the frequency or the repetition rate at which the switch opens and closes.

Step-Up Operation

In the step-up mode, the switching regulator produces an output dc voltage V_o which is *higher* than V_{in}. The circuit configuration for this mode of operation, along with the associated voltage and current waveforms, is shown in Figure 10.29 under steady-state operation.

With reference to Figure 10.29, the operation of the circuit can be summarized as follows. Assuming that S_1 is open and closes at the moment $I_L = 0$, the current in the inductor rises linearly from zero to a peak value I_{pk} during t_{on}. At the end of the on time, S_1 is opened. Since I_L cannot change instantaneously, the inductor generates the necessary voltage at node A to forward bias D_1 and keep the current continuous. During the off time, I_L decays linearly, and reaches zero at t_{off}. Then S_1 closes again, and the cycle repeats itself. While D_1 is conducting, it supplies current both to the load and to the holding capacitor C_O; and when D_1 is nonconducting, the output current is drawn from C_O. Note that at steady state, the average or dc current through D_1 is equal to the output or load current I_o and the net charge supplied to C_O, per cycle of operation, is zero.

The peak current I_{pk} is related to the steady-state output current as

$$I_{pk} = 2I_o \frac{V_D + V_o - V_{sat}}{V_{in} - V_{sat}} \tag{10.28}$$

and the on–off times of the switch, necessary for I_L to ramp from zero to I_{pk} and back to zero, are related as

$$\frac{t_{on}}{t_{off}} = \frac{V_D + V_o - V_{in}}{V_{in} - V_{sat}} \tag{10.29}$$

Solving Eq. (10.29) for V_o, one obtains

$$V_o = V_{in}\frac{T}{t_{off}} - V_{sat}\frac{t_{on}}{t_{off}} - V_D \tag{10.30}$$

where $T(= t_{off} + t_{on})$ is the period of one full cycle of operation. In the idealized

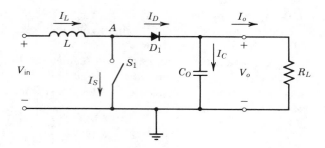

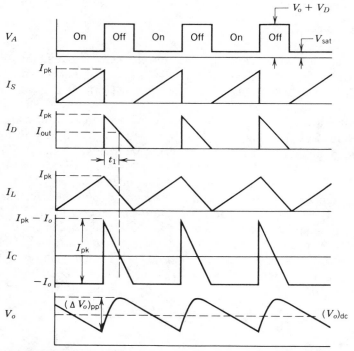

FIGURE 10.29. Voltage and current waveforms in step-up mode.

case, where the diode drop and V_{sat} of the switch are negligible, Eq. (10.30) reduces to

$$(V_o)_{\text{ideal}} = V_{\text{in}}\frac{T}{t_{\text{off}}} \qquad (10.31)$$

or, in other words, the step-up mode of operation results in up-scaling the input voltage by the ratio T/t_{off}.

The peak-to peak output ripple voltage can be expressed as[15]

$$(\Delta V_o)_{pp} \approx \frac{(I_{pk} - I_o)^2}{2I_{pk}} \frac{t_{off}}{C_O} \tag{10.32}$$

with the assumption that $(\Delta V_o)_{pp} \ll V_o$.

Polarity-Inverting Operation

In the polarity-inverting mode of operation, the switching regulator produces an output voltage across the load which has the *opposite* polarity of the input. Figure 10.30 shows the basic circuit configuration for this mode of operation, along with the associated voltage and current waveforms under steady-state operation. The polarity inversion is achieved by making the inductor force a current in the opposite direction through the load.

The operation of the circuit in its inverting mode can be briefly described as follows. The current I_L through the inductor ramps up from zero to I_{pk} during t_{on}, and ramps down to zero during t_{off}, similar to the other two modes of operation described earlier. During t_{off}, with S_1 open, the inductor generates a negative voltage at node A to forward bias D_1 and keep I_L continuous. As in the case of step-up circuits, the average value of the diode current I_D is the steady-state load current I_o since C_O can not carry any dc current. Thus, from the waveforms and timing relations of Figure 10.30, one can express the peak current I_{pk} as

$$I_{pk} = 2I_o \frac{V_{in} + V_D - V_o - V_{sat}}{V_{in} - V_{sat}} \tag{10.33}$$

and the ratio of the on–off times of the switch necessary for I_L to ramp up from zero to I_{pk} and back to zero are

$$\frac{t_{on}}{t_{off}} = \frac{V_D - V_o}{V_{in} - V_{sat}} \tag{10.34}$$

Solving the above equation for V_o, one gets

$$V_o = - \left[\frac{t_{on}}{t_{off}} (V_{in} - V_{sat}) - V_D \right] \tag{10.35}$$

In the idealized case, where V_D and V_{sat} are neglected, Eq. (10.35) will reduce to

$$(V_o)_{ideal} = - \frac{t_{on}}{t_{off}} V_{in} \tag{10.36}$$

or, in other words, the polarity-inverting mode of operation results in the reversal of the polarity of the output voltage as well as its being scaled by the t_{on}/t_{off} ratio.

The peak-to-peak output ripple $(\Delta V_o)_{pp}$ associated with the inverting operation

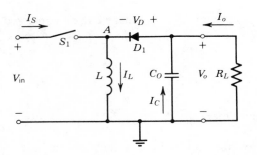

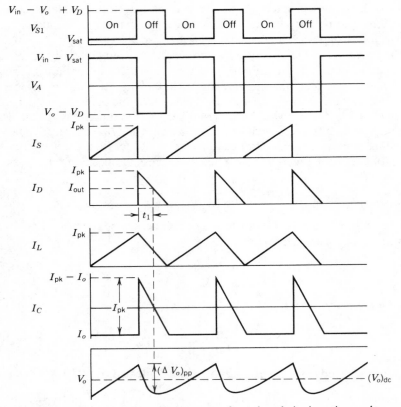

FIGURE 10.30. Voltage and current waveforms in polarity-inverting mode.

can be expressed as[15]

$$(\Delta V_o)_{\text{pp}} \approx \frac{(I_{\text{pk}} - I_o)^2}{2I_{\text{pk}}} \frac{t_{\text{off}}}{C_O} \tag{10.37}$$

which is identical to the case of step-up operation [see Eq. (10.32)].

One word of caution is in order when using the polarity-inverting configuration. Since the output polarity is reversed, the feedback polarity from the sampling resistors to the voltage comparator in the control circuitry (see Fig. 10.25) must be reversed. Normally, this is done by reversing the reference and feedback inputs into the voltage comparator.

10.8. EFFICIENCY CONSIDERATIONS

One of the most important features of switching regulators is efficiency. Using the basic analysis of the previous section, one can readily derive quantitative expressions for the efficiency of each of the three basic switching regulator configurations discussed.

The regulator efficiency is defined by Eq. (10.9) as

$$\eta = \frac{P_o}{P_{in}} = \frac{I_o V_o}{I_{in} V_{in}} \tag{10.38}$$

where I_o and I_{in} are the average (i.e., dc) values of the output and input currents.

For the case of the step-down regulator circuit, from Eq. (10.23) through (10.25), one can derive an expression for efficiency as

$$\eta_{\text{stepdown}} = \frac{V_o}{V_o + V_D} \frac{V_{in} + V_D - V_{\text{sat}}}{V_{in}} \tag{10.39}$$

In a similar manner, the efficiency expression for step-up operation can be derived from Eqs. (10.28) through (10.30) as

$$\eta_{\text{stepup}} = \frac{V_o}{V_o + V_D - V_{\text{sat}}} \frac{V_{in} - V_{\text{sat}}}{V_{in}} \tag{10.40}$$

The efficiency expression for inverted-polarity operation can be derived from Eq. (10.33) through (10.35) as

$$\eta_{\text{inverter}} = \frac{|V_o|}{V_D + |V_o|} \frac{V_{in} - V_{\text{sat}}}{V_{in}} \tag{10.41}$$

The efficiency expressions given above do not take into account the quiescent power dissipation in the control circuitry, which can cause the efficiency to decrease at very low current levels, when the average input current is of the same order of magnitude as the quiescent current. The switching transient losses in the switch transistor and diode, as well as the parasitic resistances associated with the inductor are also not included in the above expressions. In practical regulator systems, these latter losses will cause a small but finite reduction in the observed efficiency, as compared to the theoretical results given in Eqs. (10.39) through (10.41). The exact nature of this reduction in efficiency depends on the specific transistor, diode, and inductor characteristics used, as well as on the selection of the operating frequency.

There are two additional observations which can be made regarding the efficiency expressions:

1. In the ideal case, where both V_{sat} and V_D go to zero, all three regulator configurations provide an ideal efficiency of 100%.
2. The efficiency expressions of Eqs. (10.39) through (10.41) are not sensitive to input–output voltage differential across the regulator. This is very different from the case of a conventional series regulator, where the efficiency varies *inversely* with the input–output voltage differential [see Eq. (10.12)].

As an illustration, it is worthwhile to compare the efficiency of a power supply system with a 20-V input and a 5-V output, with a conventional series regulator, to that with a step-down switching regulator. For simplicity, quiescent power dissipation associated with the control circuitry will be neglected.

In the case of a conventional series regulator, the efficiency is

$$\eta \approx \frac{5}{20} = 25\%$$

For the case of a step-down regulator, assuming typical values of $V_D = 1$ V and $V_{sat} = 1.5$ V, one gets from Eq. (10.39)

$$\eta = \frac{5}{6} \frac{19.5}{20} = 81.25\%$$

which illustrates one of the most important features of switching regulators, namely, the efficient transfer of power.

10.9. PRACTICAL SWITCHING REGULATOR CIRCUITS

The basic control circuitry associated with a switching regulator system is readily suitable for monolithic integration. In terms of circuit implementation, this corresponds to the circuitry within the dashed area of Figure 10.25. Normally, a switching transistor capable of handling several hundred milliamperes of current is also incorporated into the chip. For high-power regulator applications, this internal transistor is used as a switch to activate an external high-current switching transistor.

At present a variety of monolithic switching regulator integrated circuits is available, which are designed to perform the basic control functions shown within the dashed area of Figure 10.25. These circuits often include overload protection circuitry to protect both the monolithic integrated circuit and the entire switching power supply system from permanent damage or burn-out under accidental overload or transient conditions. Many of these circuits also contain two switching transistors, rather than one, which can be used in push–pull switching systems such as the one shown in Figure 10.24a.

In this section, the basic architecture and performance characteristics of some of these monolithic integrated circuits will be reviewed. The design examples are chosen because they represent widely accepted commercial products and because they also provide an excellent demonstration of some of the design considerations discussed in the earlier sections.

The two circuits which will be examined are the SG-1524 switching regulator control circuit introduced by Silicon General Corporation, and the μA-78S40 switching regulator integrated circuit introduced by Fairchild Semiconductor. The first circuit operates on the pulse-width-modulation principle, the second one operates on a modified version of the constant-on-time, variable-frequency principle.

SG-1524 Pulse-Width-Modulating Regulator

The SG-1524 pulse-width-modulating regulator circuit, introduced by Silicon General Corporation in the mid-1970s, contains all of the necessary low-level detection and control circuitry for a switching regulator system.[16] The functional block diagram of the SG-1524 switching regulator circuit is shown in Figure 10.31. The circuit is available in a 16-pin DIP package and, in addition

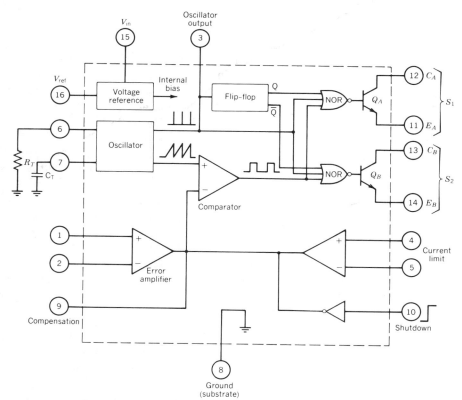

FIGURE 10.31. Functional block diagram of SG-1524 switching regulator.

to the control circuitry, also contains two switching output transistors Q_A and Q_B, each of which is designed to handle up to 100 mA of load current. These two output transistors can be used as independent alternating switches in push–pull systems (see Fig. 10.24a) or can be connected in parallel for driving single-ended output circuits.

As shown in Figure 10.31, the control circuitry is comprised of the following functional blocks:

1. Internal voltage reference.
2. Saw-tooth oscillator.
3. Voltage comparator.
4. Error amplifier.
5. Output driver logic.
6. Output switch transistors.
7. Overload protection and shutdown circuitry.

The operation of the overall circuit can be easily described in terms of the interconnection of these basic blocks. The internal voltage reference section is essentially a temperature-compensated Zener reference with a low-impedance buffered output. It provides a temperature-stabilized 5-V reference voltage and also biases all the rest of the internal oscillator and control circuitry.

The oscillator section is a linear ramp generating R–C oscillator. Its circuit configuration is similar to that shown in Figure 11.19. The Schmitt-trigger section of the oscillator uses a single-comparator-type sensing circuit as shown in Figure 11.15. The oscillator operates by charging an external capacitor C_T with a constant current from a lower threshold voltage V_A to an upper threshold voltage V_B. Once the linear ramp voltage across the capacitor reaches the upper threshold voltage, the capacitor is quickly discharged back to the lower threshold voltage, and the cycle repeats itself. The charging current for the capacitor is set by the external resistor R_T. By proper choice of the internal voltage threshold levels, the frequency of oscillation is set as

$$f_O \approx \frac{1}{R_T C_T} \tag{10.42}$$

The oscillator section produces two output waveforms: a linear ramp voltage across the timing capacitor C_T and a narrow positive pulse (typically on the order of a few microseconds) corresponding to the rapid discharge cycle of the timing capacitor. These two waveforms, which are basic to linear-ramp-type R–C oscillators, are shown as the top two waveforms in the timing diagrams of Figure 10.32 and serve two separate functions.

The linear-ramp, or *saw-tooth*, waveform is compared with the filtered and amplified error voltage at the input of the comparator section, and is used to generate a variable-width pulse output from the comparator. Note that the width of the output pulse from the comparator is directly proportional to the error

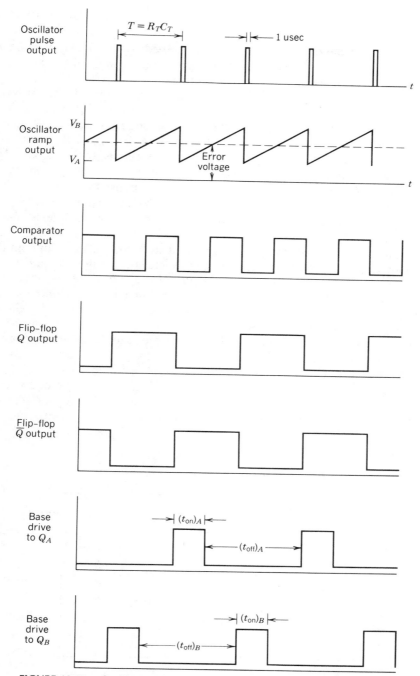

FIGURE 10.32. Oscillator and timing waveforms SG-1524 switching regulator.

voltage level at the comparator input, as indicated by the timing diagram of Figure 10.32. The narrow positive output pulse from the oscillator is used as a timing signal to the output-driver logic section.

The output-driver logic is comprised of an R–S flip-flop and a set of NOR gates which drive the two parallel output-switch transistors Q_A and Q_B. The flip-flop is set and reset alternately by the narrow timing pulse from the oscillator output, and alternately enables one or the other of the two NOR gates. The narrow-pulse output of the oscillator is also directly applied to the output NOR gates, and serves as a blanking pulse to ensure that there is no possibility of having both outputs on simultaneously during transitions. In this manner, the output-driver logic section steers the variable-width output pulse from the voltage comparator alternately to one or the other of the two output transistors, as shown in Figure 10.32.

The two output transistors Q_A and Q_B are identical transistors with both the collector and emitter terminals of each device left uncommitted. Figure 10.33 shows the typical internal circuit configuration for one of these output switch transistors. The output peak-current capability is limited to approximately 100 mA by means of the active current-limit circuit made up of the sensing resistor R_{SC} and the transistor Q_3. The diode D_1, along with the lateral pnp transistor Q_2, serve as antisaturation clamps to keep Q_A out of deep saturation and to improve its switching response. Note that the output transistor Q_A is activated only when all three digital signals appearing at the base of Q_1 are at low state.

The error amplifier section is basically a single-stage transconductance amplifier with a 200-μA full-scale output current capability. Its output is directly connected to the inverting input of the comparator. The output of the error

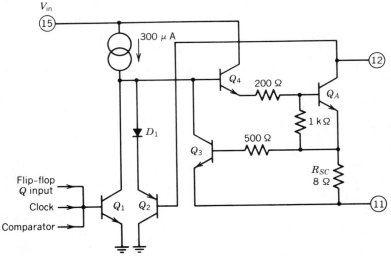

FIGURE 10.33. Current-limited output switch transistor stage of SG-1524 switching regulator. (*Note*: One of two outputs shown only.)

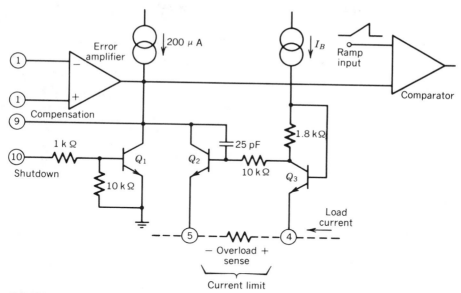

FIGURE 10.34.   Overload current-limit and shutdown circuit of SG-1524 switching regulator.

amplifier provides a convenient point in the circuit for shutdown control and overload protection, as shown in Figure 10.34. By shunting away the limited current drive ($\approx 200 \ \mu$A) available at the output of the error amplifier, this node can be pulled down to ground pontential, which in turn shuts off the comparator output, reduces the output pulse width to zero, and keeps both output transistors in the off state. Both the shutdown amplifier Q_1 and the current-limit circuit, made up of Q_2 and Q_3 of Figure 10.34, operate on this principle. The current-limit amplifier has a temperature-compensated sensing threshold of 200 mV across its inputs, and is used for sensing the total current delivered to the load (see application circuits of Figs. 10.35 and 10.36). If the load current increases beyond a set threshold, due to either transient load changes or an accidental short circuit, the current-limit circuitry causes the output pulse width to decrease and, thus, limits the total current delivered to the load.

Another protection circuit often used in switching regulators is the so-called *soft-start* circuit, which ensures that the output pulse width starts from zero and builds up to its proper level gradually when the regulator is first powered up. This avoids sudden current surges into the output circuitry when the regulator is first activated and lets the output voltage build up to its operating value with a relatively slow transient (typically in several milliseconds or more).

The soft-start circuitry is not directly included in the SG-1524. However, it can be incorporated by connecting an external R–C network to the output of the error amplifier (pin 9) through a disconnect diode.[16]

The SG-1524 is designed to operate with supply voltages of 8–40 V. The entire circuit consumes less than 10 mA of quiescent current. Figures 10.35 and

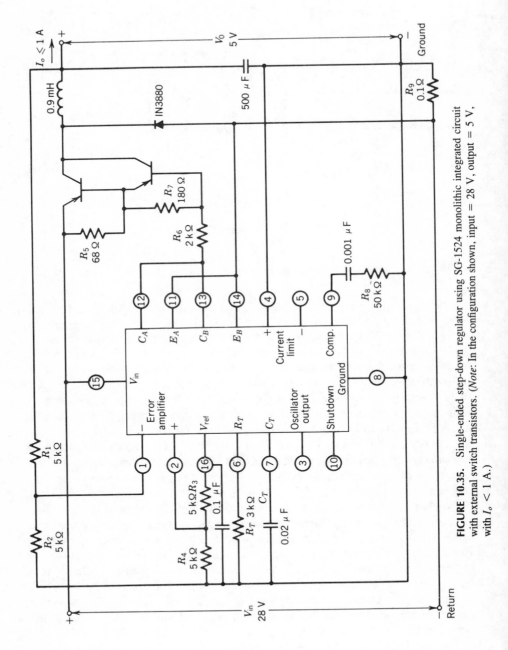

FIGURE 10.35. Single-ended step-down regulator using SG-1524 monolithic integrated circuit with external switch transistors. (*Note:* In the configuration shown, input = 28 V, output = 5 V, with $I_o < 1$ A.)

10.36 show some typical applications of the circuit, both as a single-ended and as push–pull type switching regulators.

Figure 10.35 shows the use of the SG-1524 in a 1-A single-ended switching regulator, with an external Darlington *pnp* transistor as a switch. This is basically a step-down type regulator configuration similar to that shown in Figures 10.25 and 10.28. The appropriate circuit terminals are identified in the figure. The two output transistors of the SG-1524 are connected in parallel by short-circuiting their collectors and emitters, and provide the switching signal to the external switch transistors. The output signal is fed back to the control circuit by the two sampling resistor R_1 and R_2. In this particular configuration, the circuit is designed to generate a 5-V output from a 25-V input. Since the output level is comparable to the level of the internal voltage reference, both V_{ref} and the output are scaled down by resistive divider circuits made up of R_1 through R_4 before they are applied to the inputs of the error amplifier.

Figure 10.36 shows the application of the SG-1524 in a transformer-coupled push–pull dc-to-dc converter application. The basic circuit configuration is similar to that shown in Figure 10.24a and is designed to deliver a 5-V output with a 5-A drive capability from a 28-V input. The circuit uses external *npn* switch transistors, driven in a push–pull manner from the emitters of the on-chip switch transistors. The input current limit to the transformer primary is set at approximately 2 A by the choice of the current-limit resistor. The feedback control loop is closed by sensing the output through the resistors R_1 and R_2 and comparing it to the internal reference voltage, which is divided down by the resistors R_3 and R_4.

μA-78S40 Switching Regulator

The μA-78S40 switching regulator, introduced by Fairchild Semiconductor in 1978, is primarily intended for single-ended load circuits using the inductive switched regulator configurations shown in Figure 10.23. The circuit architecture for the μA-78S40 is shown in Figure 10.37. Unlike the SG-1524 circuit described in the previous section, the μA-78S40 operates in a modified version of the fixed-on-time, variable-frequency mode discussed in Section 10.6.

With reference to the block diagram of Figure 10.37, the principle of operation for the circuit can be described as follows. The oscillator section produces a constant stream of timing signals which are then gated through internal logic circuitry, and are used to turn the internal switch transistor Q_1 on and off. The oscillator frequency is determined by the choice of the external timing capacitor C_T and can be set to be anywhere in the range of 100 Hz to 100 kHz. In most applications, a frequency range of 20–30 kHz is used. The oscillator output is a fixed-duty-cycle pulse train, whose on–off time is fixed internally at a ratio of 6 : 1. This waveform is shown in Figure 10.38, along with other timing signals in the circuit. The high-gain comparator section of the circuit compares the filtered and scaled output voltage level with the 1.3-V internal reference and generates a gating signal which inhibits the oscillator output from reaching the

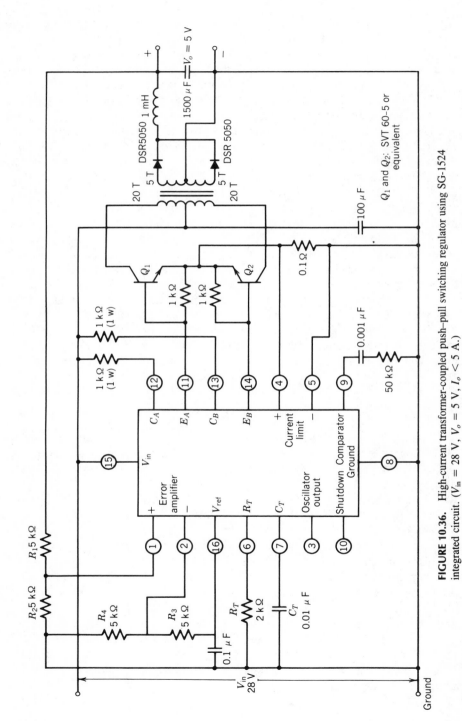

FIGURE 10.36. High-current transformer-coupled push–pull switching regulator using SG-1524 integrated circuit. ($V_{in} = 28$ V, $V_o = 5$ V, $I_o < 5$ A.)

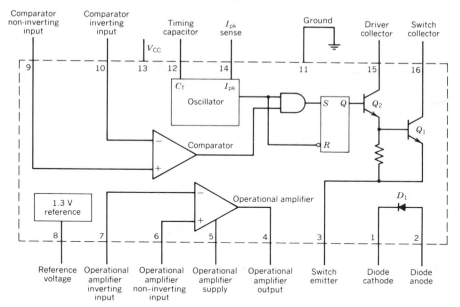

FIGURE 10.37. Functional block diagram of μA-78S40 switching regulator.

output transistors whenever the sampled or scaled output voltage is too high. In this manner, the drive to the output switch transistor is essentially an intermittent pulse train, which is occasionally interrupted or inhibited, as shown in Figure 10.38.

The on-chip Darlington switch transistor and the switch diode D_1 are both rated at 1.5 A peak current capability, with 40-V breakdown voltage, and have typical switching times of 500 nsec.[15]

The internal voltage reference circuit is a temperature-compensated band-gap reference. The operational amplifier section is a general-purpose 741-type operational amplifier with a power output stage which can source up to 150 mA or sink up to 35 mA. Although the operational amplifier is not used in implementing the basic regulator function, it is included on the chip as a general-purpose building block for auxiliary functions.

The overload protection is achieved by a current-sense circuit associated with the oscillator section. This current-sense circuit senses the total current drawn by the output switch transistor or the switching diode. If this current exceeds the rated limit, the current-sense circuit reduces the oscillator duty cycle, which in turn limits the peak current.

Figure 10.39 illustrates the operation of the μA-78S40 as a step-up voltage regulator operated in the circuit configuration of Figure 10.22b. In the circuit of Figure 10.39, the device is interconnected to generate a $+15$ V output voltage from a $+5$ V input. The output is sampled by resistors R_1 and R_2 and compared with the 1.3 V V_{ref} to maintain the regulation. The current-limit function is

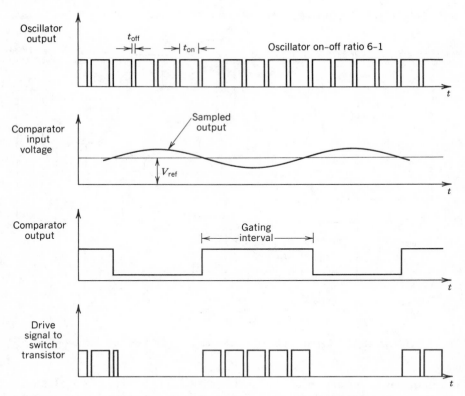

FIGURE 10.38. Timing waveforms for various sections of μA-78S40 switching regulator.

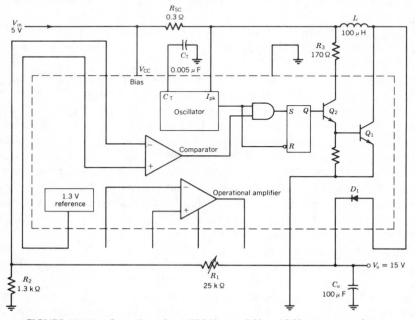

FIGURE 10.39. Operation of μA-78S40 as a 5 V to 15 V step-up regulator.

accomplished by sensing the total supply current delivered to the output circuitry by means of the voltage drop across the external resistor R_{SC} and reducing the oscillator duty cycle, if the rated current limit is exceeded. This protection circuit also functions as a *soft-start* circuit when the regulator is first powered up and limits the transient surge currents.

REFERENCES

1. A. B. Grebene, *Analog Integrated Circuit Design*, Van Nostrand Reinhold, New York, 1972, Chap. 6.

2. W. F. Davis, "A Five-Terminal ±15V Monolithic Voltage Regulator," *IEEE J. Solid-State Circuits*, **SC-6**, 366–376 (December 1971).

3. T. M. Frederiksen, "A Monolithic High-Power Series Voltage Regulator," *IEEE J. Solid-State Circuits*, **SC-3**, 380–387 (December 1968).

4. *National Semiconductor 1982 Linear Databook*, National Semiconductor Corp., Santa Clara, CA, 1982.

5. *Fairchild Voltage Regulator Handbook*, Fairchild Semiconductor Corp., Mt. View, CA, 1978.

6. *The Voltage Regulator Handbook*, Texas Instruments, Inc., Dallas, TX, 1977.

7. R. J. Widlar, "New Developments in IC Voltage Regulators," *IEEE J. Solid-State Circuits*, **SC-6**, 2–7 (February 1971).

8. R. C. Dobkin, "3-Terminal Regulator is Adjustable," National Semiconductor Appl. Note 181, March 1977.

9. A. P. Brokaw, "A Simple Three-Terminal IC Band-gap Reference," *IEEE J. Solid-State Circuits*, **SC-9**, 388–393 (December 1974).

10. R. C. Dobkin, "5 Amp Regulator with Thermal Gradient Controlled Current-Limit," *Digest of Tech. Papers*, IEEE Int. Solid-State Circuits Conf., Vol. 22, pp. 228–229, February 1979.

11. H. A. Schafft and J. C. French, "A Survey of Second Breakdown," *IEEE Trans. Electron Dev.*, **ED-13**, (August 1966).

12. F. Bergmann and D. Gerstner, "Some New Aspects of Thermal Instability of Current Distribution in Power Transistors," *IEEE Trans. Electron Dev.*, **ED-13**, (August 1966).

13. J. R. Black, "Electromigration—A Brief Survey of Some Recent Results," *IEEE Trans. Electron Dev.*, **ED-16**, 338–347 (April 1969).

14. A. I. Pressman, "Switching and Linear Power Supply, Power Converter Design," *Hayden Book Co.*, Rochelle Park, NJ, 1977, Chaps. 1, 2, 8, 9.

15. T. Vaeches, "μA-78S40 Switching Voltage Regulator Applications," Fairchild Appl. Note 344, Fairchild Semiconductor Corp., Mt. View, CA, December 1978.

16. R. Mammano, "Simplifying Converter Design with a New Integrated Regulating Pulse-Width Modulator," Appl. Note, Silicon General Product Catalog, Silicon General Inc., Irvine, CA. 1980.

CHAPTER ELEVEN

INTEGRATED-CIRCUIT OSCILLATORS AND TIMERS

PART I: INTEGRATED-CIRCUIT OSCILLATORS

11.1.   AN OVERVIEW OF OSCILLATOR TYPES

As a general classification, electronic circuits can be divided into two categories: (1) signal-processing circuits which operate on existing signals, to vary their characteristics; and (2) signal-generating circuits which produce or generate their own periodic signals. All of the circuit types discussed so far, such as amplifiers, regulators, modulators, and multipliers, fall into the first category. Oscillators, which will be covered in this chapter, fall into the second category.

The function of an oscillator is to produce a steady and stable, periodic time-varying output waveform which can serve as the information or timing signal for the signal-processing circuits. The ac output signal available from an oscillator can be in any one of a variety of waveforms, such as sinusoidal, triangular, square wave, pulse train, or exponential ramp. Oscillators offer a multitude of applications. They can be used to generate clock signals for system timing, or they may be used to generate the carrier signal which can be modulated in amplitude or frequency to transmit information. In addition, they can produce periodic sweep voltages for display systems or convert analog voltage or current data into frequency. Oscillators are also key building blocks in complex signal-processing systems such as phase-locked loops and frequency synthesizers.

Categorically, oscillator circuits can be broken down into two classes: (1) tuned oscillators which produce nearly sinusoidal outputs, and (2) multivibrator circuits which switch back and forth between two astable states.

The tuned oscillators, which are also called harmonic or near-harmonic oscillators, usually require LC-tank circuits or crystals as the frequency-setting components. Ideally, such oscillator circuits simulate the response of a lossless

LC-tank circuit, that is, they correspond to a pair of complex-conjugate poles exactly on the $j\omega$-axis of the complex frequency plane.

The multivibrators, which are also called relaxation oscillators or charge–discharge oscillators, operate by alternately charging and discharging an energy storage element (normally a capacitor) between two internally set threshold levels within the circuit. In this manner, the circuit *relaxes* into one astable state until it reaches one of the internal threshold levels; then it is switched and relaxes into the other astable state until it reaches the second internal threshold and is switched back to the first astable state; and the cycle continues in this manner. The output of a multivibrator is normally a series of linear or exponential ramp waveforms across the energy storage element and a series of pulse or square-wave signals across the switching components.

At this point, it is also worthwhile to compare the advantages and limitations of tuned oscillators and multivibrators from the point of view of monolithic IC design. The tuned oscillators, particularly those with LC-tuned circuits or crystals, have the following advantages over multivibrator-type oscillators:

1. Higher frequency stability. Since the frequency stability is primarily set by the external components, that is, the Q of the LC-tank circuit and the quality of the crystal, the oscillator frequency stability with regard to temperature is relatively insensitive to the properties of the active components on the integrated circuit, provided that the IC section can supply sufficient gain to maintain the oscillations.

2. High frequency capability. Tuned oscillators are less affected by the switching delays associated with the active devices. Thus, by using a wideband gain block with high f_T transistors, the circuit can be made to oscillate at higher frequencies than is possible by multivibrator-type circuits.

3. Higher spectral purity. The phase noise or jitter associated with the output spectrum is lower in tuned oscillators due to the frequency-selective nature of the external feedback (LC-tank or crystal) circuitry.

On the other hand, they have the following drawbacks:

1. Require expensive and bulky external circuitry. Inductors and crystals are not compatible with monolithic integrated circuits and have to be left external to the monolithic chip.

2. Frequency cannot be varied over a broad range by an external adjustment. Usually, the frequency adjustment by an external control signal is nonlinear, and is limited to a few percent of the center frequency.

Multivibrators or relaxation oscillators also have some unique advantages for IC design. Some of these are:

1. They do not require inductors.

2. They are easy to design and build, and are predictable in their waveforms.

3. Frequency is normally proportional to a current or voltage level and inversely proportional to an external capacitor. Thus, frequency can be set easily by a single external component and can be varied linearly over a wide range (typically over several orders of magnitude).

4. Since output waveforms are predictable and well defined, one can use additional waveshaping networks or filtering to convert waveforms to sinusoids. In the case of emitter-coupled multivibrators, the implicit symmetry of the circuit results in a symmetrical triangle-wave output which is virtually free of even harmonics and can be shaped easily into a sine wave.

5. They inherently provide quadrature output capability, that is, two outputs shifted in phase by 90°. This is particularly advantageous for phase-locked-loop (PLL) applications.

The disadvantages or limitations of multivibrators are:

1. Relatively poor frequency stability, particularly for operation above a few hundred kilohertz.

2. Poor spectral purity. Excessive phase jitter due to random fluctuations in the switching threshold or charging current levels.

In comparing the above features and limitations of both oscillator types, it becomes obvious that the multivibrators, or the relaxation oscillators, are by far the better suited oscillator configurations for monolithic IC design, particularly for applications below the megahertz range. This is because they are easier to design, less affected by parasitics, more predictable in preformance; and they do not require the use of external inductors or crystals.

In the following sections of this chapter, the characteristics of each of these oscillator classes will be examined in more detail to illustrate their features, limitations, and applications, and particular attention will be focused on relaxation oscillators since they are much better suited to monolithic implementation.

11.2. TUNED OSCILLATOR CIRCUITS

A tuned oscillator system is normally made up of two basic circuit sections: (1) an amplifier which provides the necessary signal gain, and (2) a feedback network which feeds a part of the amplifier output back to the input. Figure 11.1 shows a generalized block diagram of such an oscillator system. For simplicity, both the forward and the feedback signal paths are assumed to be unilateral. For

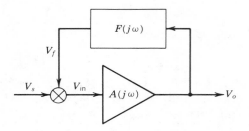

FIGURE 11.1. Simplified block diagram of linear feedback oscillator system.

the circuit to oscillate, the amplifier A must provide sufficient gain and the feedback network $F(j\omega)$ must shift the phase of the output signal a sufficient amount such that it is in phase with the input signal.

Following the signal flow around the feedback loop, one can write

$$V_o = V_{in} A(j\omega) \tag{11.1}$$

and

$$V_f = V_o F(j\omega) = V_{in} A(j\omega) F(j\omega)$$

Then the total signal gain around the feedback loop is

$$\frac{V_f}{V_{in}} = A(j\omega) F(j\omega) \tag{11.3}$$

From Eq. (11.3) one sees that the circuit can sustain oscillations independent of the input signal voltage V_S whenever the conditions are such that

$$\left| \frac{V_f}{V_{in}} \right| = \left| A(j\omega) \right| \left| F(j\omega) \right| \geq 1.0 \tag{11.4}$$

and the phase of the signal, ϕ, around the loop is such that

$$\left(\frac{V_f}{V_{in}} \right) = \phi_A + \phi_F = 0° \tag{11.5}$$

where ϕ_A and ϕ_F are the phase shifts associated with the amplifier and feedback network.

Equations (11.4) and (11.5), which summarize the conditions for the circuit to sustain oscillations are called the *Barkhausen criteria*.[1] For example, if $A(j\omega)$ in Figure 11.1 is an inverting amplifier, such as an operational amplifier which will produce a phase shift of 180° or more (due to phase lag or excess phase shifts within the amplifier), then the system of Figure 11.1 will oscillate provided that $F(j\omega)$ can provide additional phase shift to bring the output phase to zero (or 360°) with $|A(j\omega)F(j\omega)| \geq 1$. Similarly, if $A(j\omega)$ is a noninverting broadband amplifier, a much smaller amount of phase shift would be needed in $F(j\omega)$ to meet the oscillation criteria given by Eq. (11.5).

Tuned Oscillators Using Negative Feedback

In tuned oscillator circuits using negative feedback, the feedback circuit is chosen to introduce approximately 180° of phase shift at the desired frequency of oscillation, to satisfy the oscillation criteria of Eq. (11.5); and the amplifier gain is set sufficiently high to satisfy a loop gain magnitude of ≥ 1.0 [Eq. (11.4)]. There is a wide choice of circuit topologies which can be used for such a configuration.[1-3]

Figure 11.2 shows one of the commonly used circuit configurations for tuned negative-feedback oscillators. By straightforward circuit analysis, one can show that the circuit will oscillate if the gain A_1 is sufficiently high, and if Z_1, Z_2, and Z_3 are reactive elements such that: (1) both Z_1 and Z_3 are capacitors and Z_2 is an inductor, or (2) both Z_1 and Z_3 are inductors and Z_2 is a capacitor.

If circuit configuration (1) is used with Z_1 and Z_3 as capacitors C_1 and C_3, the resulting circuit configuration is called a *Colpitts oscillator*, as shown in Figure 11.3*a*. Assuming pure reactances, the circuit has a frequency of oscillation given by

$$f_{osc} = \frac{1}{2\pi}\left[\frac{1}{L_2}\left(\frac{1}{C_1} + \frac{1}{C_3}\right)\right]^{1/2} \tag{11.6}$$

If both Z_1 and Z_1 are inductors and Z_2 is a capacitor, as shown in Figure 11.3*b*, the resulting circuit configuration is called a *Hartley oscillator* and has a frequency of oscillation given by

$$f_{osc} = \frac{1}{2\pi}\left[\frac{1}{C_2(L_1 + L_2)}\right]^{1/2} \tag{11.7}$$

The Hartley configuration has two practical disadvantages over the Colpitts oscillator: (1) the coils L_1 and L_3 tend to have mutual coupling which causes the

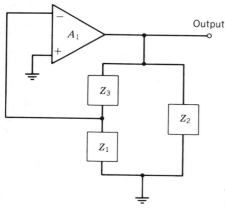

FIGURE 11.2. Commonly used negative-feedback oscillator configuration.

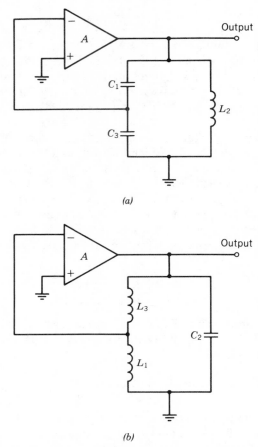

FIGURE 11.3.   Basic oscillator configurations: (a) Colpitts oscillator; (b) Hartley oscillator.

frequency of oscillation to be somewhat different from that given by Eq. (11.7), and (2) the oscillator frequency cannot be readily varied over a wide range.[2] The latter problem arises from the difficulty in changing the value of an inductor. Slug-tuned coils in which the core material of the coil changes its magnetic properties are available, but are too costly for most applications.

The Colpitts oscillator is quite often used in AM and FM receivers where the tuning dial adjusts a "ganged" capacitor which selects both the mixer oscillator frequency as well as the resonant frequency of RF stages. For fine tuning or automatic frequency control (AFC) applications, C_1 and C_3 of a Colpitts oscillator can be replaced by variable-reactance (varactor) diodes.

Figure 11.4 shows another example of a negative-feedback oscillator. In this so-called phase-shift oscillator configuration, a three-stage R-C phase-shift network is added between the output and the input to provide additional phase delay to reduce the total phase shift around the feedback loop to zero (or 360°). At the

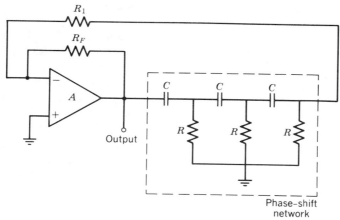

FIGURE 11.4. Typical R–C phase-shift oscillator.

frequency of oscillation, each of the R-C phase shift sections contributes approximately 60° of phase lead, resulting in zero net phase shift around the loop, as required by Eq. (11.5).

Figure 11.5 shows a tuned oscillator configuration using a notch filter in feedback. In this case, a twin-tee network is used which can produce two complex–conjugate zeros on the $j\omega$ axis with the proper choice of component values. At this so-called notch frequency, the twin-tee section behaves as an open circuit, and the positive feedback provided by R_3 and R_4 causes the circuit to oscillate at or very near the notch frequency of the twin-tee. The basic design equations for selecting the components of the twin-tee network for a given notch frequency are well documented in the literature.[1] The twin-tee oscillator is primarily suited for fixed-frequency operation. The adjustment of its frequency requires the change of several component values simultaneously, which is impractical for continuous–tuning applications.

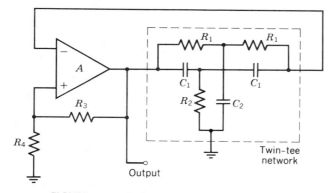

FIGURE 11.5. Twin-tee oscillator configuration.

Positive-Feedback Oscillators

Positive-feedback oscillators often require a much lower value of gain than a negative-feedback circuit. However, in turn, they offer a higher degree of sensitivity to the absolute value of amplifier gain and have relatively poor amplitude stability. One of the most commonly used positive-feedback oscillator configurations is the Wien-bridge circuit shown in Figure 11.6. Assuming that the loop gain is sufficient, the circuit oscillates at a frequency

$$f_{osc} = \frac{1}{2\pi} \frac{1}{\sqrt{R_1 R_2 C_1 C_2}} \tag{11.8}$$

In most cases, one sets

$$R_1 = R_2 = R \quad \text{and} \quad C_1 = C_2 = C \tag{11.9}$$

which simplifies Eq. (11.8) as

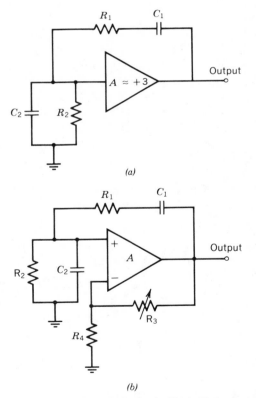

(a)

(b)

FIGURE 11.6. Wien-bridge oscillator: (a) Basic circuit; (b) circuit implementation using an operational amplifier.

$$f_{osc} = \frac{1}{2\pi RC} \qquad (11.10)$$

Under conditions of Eq. (11.9), the amplifier gain A required for oscillation is also minimum and ideally equal to $+3$.

Figure 11.6b shows a typical circuit implementation of a Wien-bridge oscillator, using an operational amplifier or a high-gain differential stage. Resistive negative feedback is used to set the gain of the amplifier. Normally, R_3 is chosen as a variable resistor and is increased until oscillations start. The Wien-bridge oscillator is primarily suited to fixed-frequency applications, since changing its frequency requires the simultaneous adjustment of two or more parameters or component values.[4]

Figure 11.7 shows the block diagram and simplified circuit implementation of an LC-tuned positive-feedback oscillator circuit.[5] Assuming that the loading effects are negligible, the circuit oscillates at the resonant frequency of the parallel LC-tank circuit, where the negative feedback from output to input is negligible and only the unity-gain positive feedback path is present.

Figure 11.8 shows the actual circuit schematic of a monolithic integrated circuit (Motorola MC-1658) which uses the circuit configuration of Figure 11.7. The circuit is intended to be used as a voltage-controlled oscillator (VCO) in high-frequency phase-locked-loop applications. It is designed to interface with emitter-coupled logic (ECL) circuitry, and can operate as a voltage-controlled oscillator for frequencies up to 225 MHz. High-frequency capability is achieved by fabricating the circuit using emitter-coupled logic process technology, which produces very-high-speed transistors with f_T in excess of 2 GHz. In order to enhance high-frequency capability, the circuit is designed to operate with large dc bias currents and very low voltage swings.

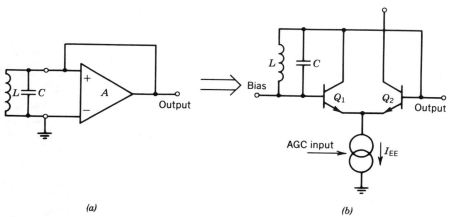

(a)   (b)

FIGURE 11.7. LC-tuned positive-feedback oscillator configuration: (a) Idealized block diagram; (b) simplified circuit implementation.

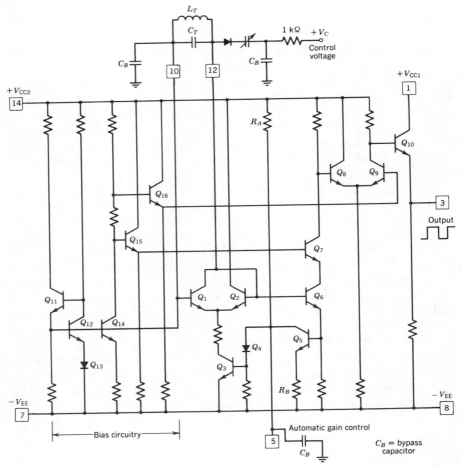

FIGURE 11.8. Circuit diagram of MC-1568 LC-tuned voltage-controlled oscillator.

The oscillator section is formed by the emitter-coupled differential pair Q_1 and Q_2. The positive feedback is obtained by directly short-circuiting the collector of Q_1 to the base of Q_2. The negative feedback from the collector of Q_1 to the base of Q_1 is through the external LC-tank circuit. This configuration forces Q_1 to operate at zero collector–base bias. In order to keep Q_1 out of saturation, an internal automatic-gain-control circuit is used. The output swing at the collector of Q_1 is buffered by Q_6 and amplified by the gain stage formed by Q_5. The negative peak of the voltage swing at the collector of Q_5 is then detected by the peak-detector circuit comprised of the AGC bypass capacitor C_B and the diode Q_4. If the signal swing increases, the AGC voltage decreases and reduces the bias current of the differential pair, to reduce its gain and hence the signal swing. In this manner, a relatively constant-amplitude output signal (typically 500 mV pp) is produced at the collector of Q_1; and Q_1 is kept out of saturation. The oscillator output at the base of Q_2 is then buffered by the cascode stage made up of Q_6 and Q_7, and amplified by the single-ended gain stage of Q_8 and Q_9. The output is then buffered by the emitter follower Q_{10} to make it compatible with the emitter-coupled logic swing and logic levels. The collector of the output transistor is also brought out separately and can be used to drive an LC tank to boost the power output. The voltage control of the frequency of oscillation is achieved either by a varactor diode connected in series with the tank circuit (as shown in Fig. 11.8), or by using two back-to-back varactor diodes to replace C_T in the tank circuit. The frequency stability of the circuit is primarily determined by the quality of the LC-tank circuit.

Crystal Oscillators

Crystal oscillators make use of the stable electromechanical resonance characteristics of a piezoelectric crystal, such as quartz, to set the oscillation frequency. When a quartz crystal is cut along certain crystal axes and is electroplated on opposite faces to form parallel-plate electrodes, it can be made to resonate as a high-Q electromechanical system when it is properly excited. The electromechanical resonance is due to localized deformations within the crystal structure, under applied electrostatic forces. The resonant frequency and Q factor depend on the material and dimensional properties of the crystal, such as the size, the orientation of the crystal faces, and the mounting of the crystal on its mechanical supports. Resonant frequencies from a few kilohertz to several hundred megahertz are possible, depending on crystal design and orientation. Similarly, the Q values can be designed to be in the range of several thousand to several hundred–thousand. These extraordinary high Q values of quartz crystals, and their excellent stability with respect to time and temperature, make them ideal for designing fixed-frequency oscillators, where frequency accuracy and stability are of extreme importance.

Figure 11.9 shows the circuit symbol, the equivalent circuit, and the reactance characteristics of a typical quartz crystal. The equivalent circuit of a crystal can be described by a series RLC circuit with a shunt capacitor C_1, as indicated in Figure 11.9b. Typically, the reactances associated with L_0 and C_0 are much higher than R_0, resulting in an extremely high value of Q. Typical values of Q are in thousands or in tens of thousands.

As an example, it is illustrative to examine some of the component values associated with a typical crystal. In the case of a 90-kHz crystal which would have outside dimensions of $3 \times 4 \times 1.5$ mm, typical values would be $L_0 = 137$ henry (H), $C_0 = 0.0235$ pF, and $R_0 = 15$ kΩ, corresponding to a Q of about 5500.[3] C_1 represents the electrostatic capacitance between the crystal electrodes, and its magnitude is typically several orders of magnitude greater than C_0. In the typical example stated above, C_1 is of the order of 3–5 pF.

If one neglects the finite value of R_0, the impedance of the crystal is a pure reactance jX, which depends on the frequency as

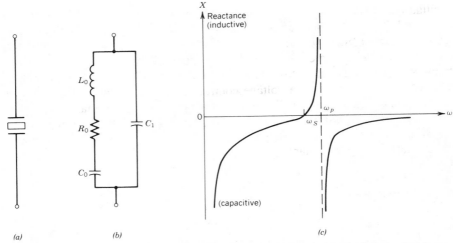

FIGURE 11.9. Piezoelectric crystal: (*a*) Circuit symbol; (*b*) electrical equivalent circuit; (*c*) reactance characteristics, assuming R_o is negligible.

$$Z_{\text{crystal}} = jX = \frac{-j}{\omega C_1}\left[\frac{\omega^2 - \omega_s{}^2}{\omega^2 - \omega_p^2}\right] \tag{11.11}$$

where

$$\omega_s = 2\pi f_s = \left[\frac{1}{L_0 C_0}\right]^{1/2} \tag{11.12}$$

and

$$\omega_p = 2\pi f_p = \left[\frac{1}{L_0}\left(\frac{1}{C_0} + \frac{1}{C_1}\right)\right]^{1/2} \tag{11.13}$$

are the series resonance (zero impedance) and the parallel resonance (infinite impedance) frequencies of the oscillator. Note that ω_p is always higher than ω_s. For almost all crystals, $C_1 \gg C_0$. Therefore, $\omega_p \approx \omega_s$. For example, for the 90-kHz crystal example discussed above, ω_p is only 0.3% higher than ω_s, and this difference can be made even smaller simply by adding a capacitor in parallel with the crystal to increase the value of C_1.

Figure 11.9*c* shows the reactance characteristics of a quartz crystal in the vicinity of its resonant frequencies. For the narrow frequency range, where $\omega_s < \omega < \omega_p$, the crystal looks inductive, and outside of this range it is capacitive. In the design of most crystal oscillators, the crystal is made to operate in its inductive region, which results in an overall frequency of oscillation somewhere between ω_s and ω_p. Since ω_p and ω_s are extremely close in frequency, the resulting oscillator frequency is extremely accurate and stable, determined solely by the characteristics of the crystal and not by the rest of the circuit. The role of the active circuitry is simply to provide enough gain to ensure a loop gain ≥ 1 to sustain the oscillations.

Figure 11.10 shows some of the commonly used crystal oscillator configurations. The circuit of Figure 11.10a is basically the Colpitts oscillator of Figure 11.3a, with L_2 being replaced by the crystal that is operated in its inductive region. The resulting frequency of oscillation is very close to the parallel resonance frequency ω_p of the crystal. The circuit of Figure 11.10b is the basic circuit configuration for a positive-feedback crystal oscillator which operates at or very near the series-resonance frequency ω_s. C_B is a relatively large-value capacitor, which is often included in the circuit to block dc currents through the crystal in order to avoid heating of the crystal.

Figure 11.11 shows another commonly used crystal oscillator configuration, comprised of a common-collector and common-base pair of transistors whose emitters are connected through the crystal. The circuit is a direct derivative of the positive-feedback configuration of Figure 11.10b, and oscillates at or very near ω_s of the crystal. At this frequency, the crystal is replaced by a virtual short circuit, and the circuit configuration of Figure 11.11 reduces to that of a differential pair with an internal positive feedback path from the collector of Q_2 to the base of Q_1.

Crystal oscillators can also be built using MOS transistors, provided that sufficient amounts of loop gain can be obtained to sustain oscillations. Among MOS technologies, CMOS is best suited to crystal oscillator design due to the availability of complementary devices which can be used as active loads to

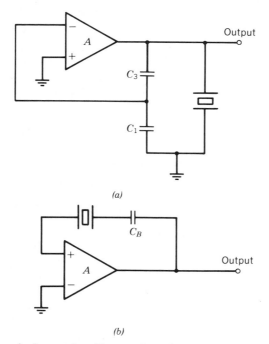

(a)

(b)

FIGURE 11.10.   Some basic crystal oscillator configurations: (a) Colpitts oscillator; (b) positive-feedback oscillator.

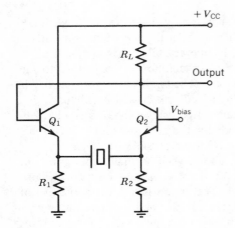

FIGURE 11.11. Crystal-controlled oscillator using common-collector–common-base transistor pair.

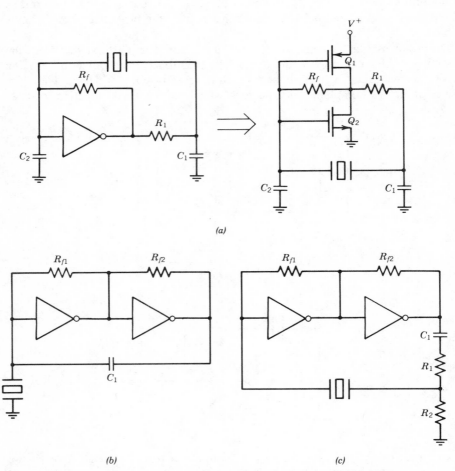

FIGURE 11.12. Various crystal-controlled oscillator circuits using CMOS inverter stages.

obtain high loop gain. The basic CMOS inverter (Fig. 11.32) is often used as the building block for CMOS crystal oscillator circuits. Figure 11.12 illustrates some of the commonly used CMOS crystal oscillator configurations. The circuit of Figure 11.12a, which uses a single inverter stage, is very often used as a configuration for relatively high-frequency applications, since it involves only one gain stage, and thus minimum signal propagation delay within the feedback loop. In the circuit, the crystal operates at or near its series–resonance frequency.

The CMOS oscillator configurations, such as those shown in Figure 11.12b and c, which use multiple inverter stages, are often used for relatively low-frequency applications (i.e., $f < 1$ MHz). Both of these configurations are positive-feedback oscillators. The crystal in Figure 11.12b operates at its parallel resonance, whereas the crystal in Figure 11.12c is operated near its series resonance. In CMOS oscillator design, one normally uses high-value resistors in feedback around inverter stages, as shown in Figure 11.12, in order to linearize the transfer and gain characteristics of CMOS inverter stages.

The biggest advantage of crystal oscillators is their frequency and stability. Typical temperature drift of crystal oscillators is in the 1–2-ppm/°C range. Due to the high-Q of the crystal, the circuit operation is totally determined by the crystal characteristics, provided that the rest of the circuitry can facilitate sufficient gain to sustain oscillations.

The big disadvantage of crystal oscillators is that the key circuit element, the piezoelectric crystal, is not compatible with IC technology and is a relatively expensive circuit component. Another disadvantage of crystal oscillators is that their frequency of operation is restricted to a very narrow range, between ω_s and ω_p, for a given crystal and cannot be easily controlled or "pulled" by external control signals.

Concluding Comments on Tuned Oscillators

The tuned oscillator circuits covered in this section are an important class of circuits and offer a wide range of applications in communication, data transmission, and synchronous logic systems. However, they all have a basic incompatibility with the monolithic IC technology: their frequency-determining components, namely the LC-tank circuits and crystals, are all external to the monolithic integrated circuit and often cost more than the circuit itself. This is also true for tuned RC oscillators such as the Wien-bridge, RC phase–shift, and twin-tee oscillators, which require extremely tightly controlled component absolute-value tolerances. Such a tight control of absolute-value tolerances is not compatible with monolithic IC technology, which is geared to low-cost high-volume production.

Thus, in the design of tuned oscillator circuits, the role of the monolithic integrated circuit is often limited to providing sufficient amounts of signal gain around the loop, while the key circuit function of determining and stabilizing the frequency of oscillation is done by relatively expensive components external to the chip.

11.3. RELAXATION OSCILLATORS

Relaxation oscillators, which are also called multivibrators, are the most commonly used oscillator configurations in monolithic IC design. They operate by alternately charging and discharging a timing capacitor between two internally set threshold voltage levels. This results in a periodic output waveform whose frequency f is inversely proportional to the value of the timing capacitor.

Based on their circuit configurations, relaxation oscillators can be classified into three categories:

1. R-C relaxation oscillators which use resistive charge and discharge paths for the timing capacitor.
2. Constant-current charge and discharge oscillators which use current sources to charge and discharge the timing capacitor.
3. Emitter-coupled multivibrators which use symmetrical charge and discharge paths for a timing capacitor connected across the emitters of a differential gain stage.

These classes of circuits and their derivatives will be discussed in the following sections.

R-C Relaxation Oscillators

The basic circuit configuration of an R-C relaxation oscillator is shown in Figure 11.13. The circuit is made up of three functional sections: (1) a set of timing components R_1, R_2, and C_1, (2) a level-detecting comparator with hysteresis (i.e., a Schmitt trigger), and (3) a grounding switch S_1 which is normally an *npn* transistor driven from cutoff to saturation.

The Schmitt-trigger section is designed to have high input impedance, low input bias current, and nearly ideal switching characteristics, as shown in Figure 11.14; that is, the output changes state when the increasing input level reaches an upper switching threshold V_B and reverts back to its initial state only when the input is reduced to its lower threshold level V_A, where $V_A < V_B$. As will be described later, the net hysteresis voltage $V_B - V_A$ of the Schmitt trigger can be made to be very stable and predictable by proper circuit design.

With reference to Figure 11.13, the principle of operation of the basic R-C relaxation oscillator can be described as follows. With switch S_1 open, the voltage V_{O1} across the timing capacitor rises exponentially toward $+V_{CC}$ with the time constant τ_1, where

$$\tau_1 = R_1 C_1 \tag{11.14}$$

When V_{O1} reaches the upper threshold voltage V_B of the Schmitt trigger, the output changes state, switch S_1 is closed, and the capacitor voltage now decays toward a lower voltage level V_L, where

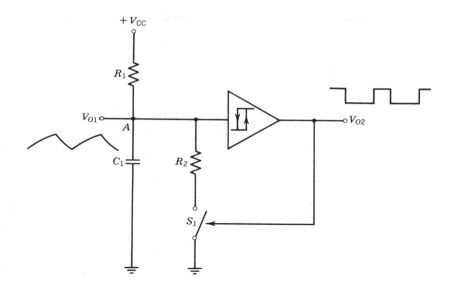

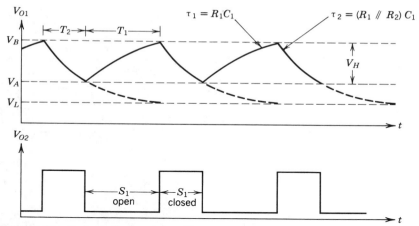

FIGURE 11.13. Basic R–C relaxation oscillator. (a) Generalized circuit configuration; (b) output wave forms.

$$V_L = V_{CC} \frac{R_2}{R_1 + R_2} \qquad (11.15)$$

with a time constant τ_2 such that

$$\tau_2 = (R_1 \| R_2) \, C_1 \qquad (11.16)$$

If the resistor values R_1 and R_2 are chosen such that $V_L < V_A$, then the Schmitt trigger will change state when $V_{O1} = V_A$, causing S_1 to open and the capacitor C_1 to charge up toward $+V_{CC}$ to repeat the cycle. This results in a set of periodic

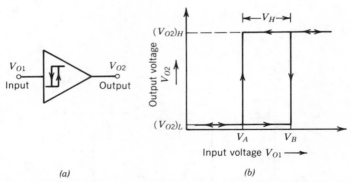

FIGURE 11.14. Circuit symbol and voltage transfer characteristics of a Schmitt trigger: (*a*) Circuit symbol; (*b*) transfer characteristics.

output waveforms, as shown in Figure 11.13*b*, with the assumption that S_1 is closed when the output of the Schmitt trigger is at a high state.

The time T_1 taken for V_{O1} to charge from the lower threshold V_A to the upper threshold V_B can be expressed as

$$T_1 = \tau_1 \ln \left(\frac{V_{CC} - V_A}{V_{CC} - V_B} \right) \tag{11.17}$$

Similarly, the time T_2 for V_{O1} to decay from V_B to V_A is given as

$$T_2 = \tau_2 \ln \left(\frac{V_B - V_L}{V_A - V_L} \right) \tag{11.18}$$

Thus, the total period of oscillation T is equal to $T_1 + T_2$, and the frequency of oscillation f is given as

$$f = \frac{1}{T} = \frac{1}{T_1 + T_2} \tag{11.19}$$

or

$$f = \frac{1}{R_1 C_1} \left[\ln \left(\frac{V_{CC} - V_A}{V_{CC} - V_B} \right) + \frac{R_2}{R_1 + R_2} \ln \left(\frac{V_B - V_L}{V_A - V_L} \right) \right]^{-1} \tag{11.20}$$

A commonly used oscillator configuration derived form Figure 11.13*a* is the case where $R_2 \approx 0$ (i.e., $T_1 \gg T_2$), which produces a series of narrow output pulses. In such applications, one often chooses the voltage levels V_A and V_B to be such that

$$\frac{V_{CC} - V_A}{V_{CC} - V_B} = e = 2.71 \ldots \tag{11.21}$$

which reduces Eq. (11.20) to

$$f = \frac{1}{R_1 C_1} \tag{11.22}$$

and results in a simple frequency-setting expression.

In the design of IC relaxation oscillators, the design of the Schmitt trigger section is very critical since the accuracy and stability of the threshold levels for the Schmitt trigger are directly related to the frequency stability and the frequency-setting accuracy of the oscillator. A wide range of Schmitt trigger circuit configurations have been developed for discrete designs.[6] However, most of these circuits do not meet the threshold stability and accuracy requirements for a high-performance relaxation oscillator. There are two commonly used Schmitt trigger configurations in monolithic relaxation oscillator design. These are the single-comparator and the dual-comparator type Schmitt trigger circuits.

Figure 11.15 shows the general circuit configuration of an R–C relaxation oscillator, using a single-comparator-type Schmitt trigger, which is shown within the dashed area. The operation of the oscillator circuit can be briefly explained as follows: with S_1 and S_2 open, the capacitor C_1 charges exponentially toward $+V_{CC}$. When the voltage level, V_{O1} across C_1 reaches the upper comparator threshold V_B, which is set by the resistor ratio as

$$V_B = V_{CC} \frac{R_B + R_C}{R_A + R_B + R_C} \tag{11.23}$$

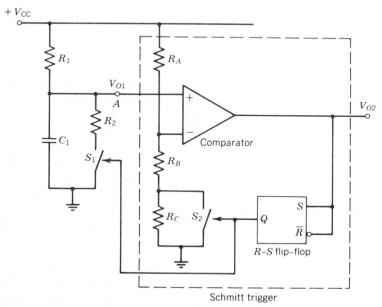

FIGURE 11.15. R–C relaxation oscillator using single-comparator-type Schmitt trigger.

the comparator changes state and sets the flip-flop, causing switches S_1 and S_2 to be closed. The closing of S_2 causes the threshold setting resistor R_C to be short-circuited and generates a new threshold level V_A for the comparator,

$$V_A = V_{CC}\frac{R_B}{R_A + R_B} \qquad (11.24)$$

where $V_A < V_B$. With S_1 closed, the capacitor voltage across C_1 decays exponentially until it reaches the lower threshold level V_A, at which point the comparator again changes state, causing the flip-flop to reset and S_1 and S_2 to open, repeating the oscillation cycle. In this manner, the circuit oscillates by generating exponential ramp voltages across C_1, between the two threshold levels V_A and V_B, as shown in Figure 11.13b. The hysteresis voltage V_H of the Schmitt trigger is primarily determined by the value of R_C, which is switched in and out of the circuit:

$$V_H = V_B - V_A = V_{CC}\frac{R_A R_C}{(R_A + R_B + R_C)\ (R_A + R_B)} \qquad (11.25)$$

It should be noted that both the threshold voltages V_B and V_A as well as the hysteresis voltage V_H are proportional to the supply voltage V_{CC}. However, as long as R_1 is also connected to $+V_{CC}$, the charging and discharging currents available to C_1 are also proportional to V_{CC}. Thus, for example, as V_{CC} is increased, the hysteresis voltage V_H, which determines the amplitude of V_{O1}, increases, and the slope of exponential waveforms will also increase by the same amount. The net result is that the V_{CC} dependence cancels out and the frequency of oscillation is, to first order, independent of the supply voltage, which is given by Eq. (11.20). The output amplitude, however, is proportional to V_{CC}.

Figure 11.16 shows the circuit configuration of a simple R–C relaxation oscillator using the single-comparator-type Schmitt trigger. The comparator is formed by the npn differential pair Q_1, Q_2 with differential current mirror loads. The flip-flop function is performed by Q_8, Q_9, and Q_7. Q_7 also functions as the switch S_2 of Figure 11.15 and short-circuits the resistor R_C when Q_1 and Q_3 are in conduction. The buffered output signal of the comparator at the emitter of Q_{10} is then used as the drive for the switch S_1 made up of transistor Q_5. The timing components R_1, R_2, and C_1 are normally left external to the circuit. The threshold voltages V_A and V_B are determined by the resistor string R_A, R_B, and R_C. The frequency of oscillation can be modulated by applying an external bias voltage to the internal bias string at the base of Q_2.

One basic disadvantage of the single-comparator relaxation configuration of Figure 11.15 is the switching of voltage levels across the bias-setting resistor string, when the voltage swings reach the threshold levels. The parasitic node capacitances across R_A, R_B, and particularly R_C slow down the switching time of the circuit and cause frequency errors for operation of the circuit at frequencies in excess of 50 or 100 kHz. This problem can be partially avoided by using a dual-comparator Schmitt trigger configuration.

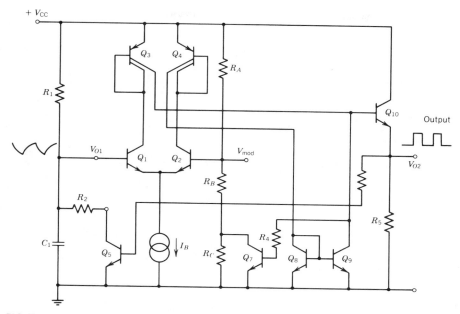

FIGURE 11.16. Simple R–C relaxation oscillator circuit based on single-comparator-type Schmitt trigger.

Figure 11.17 shows the basic circuit configuration for an R–C oscillator using a dual-comparator-type Schmitt trigger. Comparator 1 changes state when the voltage swing across the timing capacitor C_1 reaches the upper threshold V_B, and comparator 2 changes state when V_{O1} reaches the lower threshold voltage V_A. These threshold voltages are set by the resistor ratios as

$$V_B = V_{CC}\frac{R_B + R_C}{R_A + R_B + R_C} \quad \text{and} \quad V_A = V_{CC}\frac{R_C}{R_A + R_B + R_C} \quad (11.26)$$

resulting in a total hysteresis voltage V_H given as

$$V_H = V_B - V_A = V_{CC}\frac{R_B}{R_A + R_B + R_C} \quad (11.27)$$

With S_1 open, V_{O1} rises exponentially toward V_{CC}. When it reaches V_B, comparator 1 changes state, sets the flip-flop which closes S_1, and initiates the discharge cycle. When V_{O1} is reduced to the level of the lower threshold voltage V_A, comparator 2 changes state, resets the flip-flop, and opens S_1, and the cycle repeats itself. As described earlier, although V_A and V_B are proportional to V_{CC}, this dependence is canceled out since charging and discharging currents associated with C_1 are also proportional to V_{CC}. Thus, the frequency of oscillation is, to first order, independent of the supply voltage and is given by Eq. (11.20).

Figure 11.18 shows an example of an R–C oscillator circuit based on the

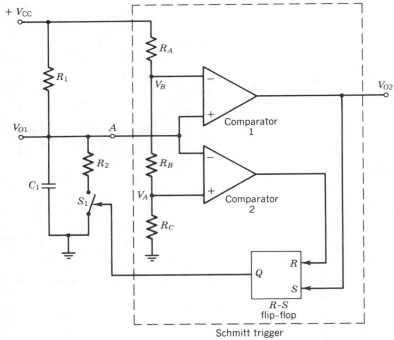

FIGURE 11.17. $R–C$ relaxation oscillator using dual-comparator-type Schmitt trigger.

dual-comparator Schmitt trigger configuration of Figure 11.17. In the circuit, the differential pair made up of Q_1 and Q_2, along with their current–mirror loads, forms comparator 1. Comparator 2 is formed by the Darlington *pnp* differential stage made up of Q_7 through Q_{10}, with the current mirror loads comprised of Q_{11} and Q_{12}. The threshold levels V_B and V_A are set by the resistor chain made up of R_A, R_B, and R_C. The flip-flop section is made up of inverting gain stages Q_{13}, Q_{14}, and Q_{15}, with R_4 providing the positive feedback to ensure latching of the flip-flop. The flip-flop output at the collector of Q_{15} is buffered by the emitter follower Q_{16} and drives the switch transistor Q_5.

During the oscillation of the circuit while V_{O1} is rising, Q_1, Q_3, Q_5, Q_7, Q_8, Q_{13}, and Q_{14} are nonconducting, and the output at V_{O2} is at the low state. When the upper threshold V_B is reached, Q_1 and Q_3 start conducting, which causes Q_{14} to turn on and Q_{15} to turn off; the output goes to the high state, and the discharge cycle is initiated. The feedback resistor R_4 assures that Q_{14} stays on even after Q_1 and Q_3 turn off. At the end of the discharge cycle Q_7 and Q_8 come into conduction and cause Q_{13} to turn on, which in turn resets the flip-flop by causing Q_{14} to turn off and Q_{15} to turn on. Thus, the output again goes to the low state and the charge cycle is initiated.

The frequency of oscillation can be modulated or controlled by means of a control voltage applied to either node A or node B on the bias string. Normally, node B is preferred for modulation, since the voltage level at node A is restricted to be $> V_L$ for the circuit to oscillate, where V_L is given by Eq. (11.15).

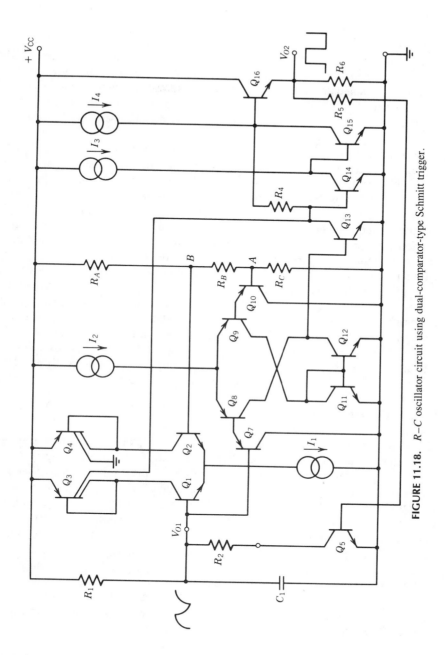

FIGURE 11.18. R–C oscillator circuit using dual-comparator-type Schmitt trigger.

Since the threshold voltage levels V_A and V_B remain stable during the oscillation cycle, the dual-comparator-type R–C oscillator has a somewhat higher frequency capability than the single-comparator relaxation oscillator and is useful for frequencies up to several hundreds kilohertz. At higher frequencies, the accuracy and stability of the oscillations will deteriorate due to excessive switching delays associated with the comparator and flip-flop sections (see Section 11.5).

The bias current drawn by the Schmitt trigger input must be negligible with respect to the charge and discharge currents, in order to assure frequency stability and accuracy. For this purpose, the Schmitt trigger input terminal must have very high impedance. In both the single- and the dual-comparator Schmitt trigger design examples illustrated in Figures 11.16 and 11.18, the high-impedance requirement is met by keeping the input bias current requirements of the comparator stages as low as possible. Furthermore, in the dual-comparator design example of Figure 11.18, the comparator inputs connected to the V_{O1} terminal remain in their off state (i.e., zero input bias current) until the voltage level across C_1 reaches an upper or lower threshold level, and even then the comparator inputs are activated only at the threshold detection state (i.e., at the extremities of V_{O1}) and then revert to the off state. In the case of the circuit configuration of Figure 11.17, this is done by using an *npn* input stage for comparator 1, while comparator 2 has a *pnp* input stage. Because of low β_0 of lateral *pnp* transistors, a Darlington input configuration is normally used for comparator 2, as shown in the design example of Figure 11.18.

Constant-Current Oscillators

Constant-current charge and discharge type oscillators use current sources, rather than resistors, to charge and discharge the timing capacitor. Figure 11.19 shows the generalized circuit configuration for a typical relaxation oscillator of this type. Normally, such an oscillator is made up of two current sources, where one current source, I_1, is always on, and the second current source I_2, is turned on and off intermittently. With I_2 at its off state, the capacitor C_1 is charged with constant current I_1 until the voltage V_{O1} across C_1 reaches the upper threshold V_B of the Schmitt trigger. Then the Schmitt trigger section changes state, turns the intermittent current source I_2 on, and assuming $|I_2| > |I_1|$, discharges C_1 with a net current of $I_2 - I_1$ until V_{O1} reaches the lower threshold voltage V_A. At that point, the Schmitt trigger again changes state and turns I_2 off. In this manner, the circuit oscillates with the linear ramp waveforms shown in Figure 11.19b. The total time T_1 for the charging cycle is

$$T_1 = \frac{(V_B - V_A)C_1}{I_1} \qquad (11.28)$$

Similarly, the total time T_2 for the discharge cycle is

$$T_2 = \frac{(V_B - V_A)C_1}{I_2 - I_1} \qquad (11.29)$$

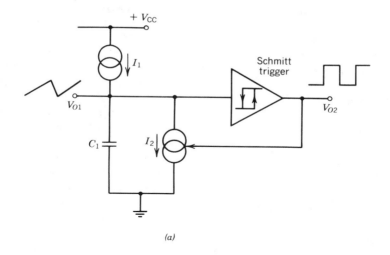

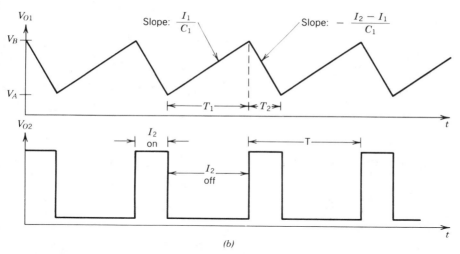

FIGURE 11.19. Constant-current charge and discharge oscillator. (*a*) Basic circuit configuration. (*b*) Output waveform.

The frequency of oscillation f is given as

$$f = \frac{1}{T} = \frac{1}{T_1 + T_2} = \frac{I_1}{(V_B - V_A)C_1}\left(1 - \frac{I_1}{I_2}\right) \qquad (11.30)$$

In the circuit of Figure 11.19, the Schmitt trigger section can be designed using either the single-comparator or the dual-comparator configuration shown in Figures 11.15 and 11.17. In most cases, the dual-comparator configuration using *npn* input and *pnp* input comparator states for comparators 1 and 2, respectively, is preferred, since this configuration minimizes the frequency errors due to finite bias currents at comparator inputs.

In the basic circuit configuration of Figure 11.19a, both I_2 and I_1 have to be varied simultaneously in order to provide linear control of the frequency of oscillation. As shown, this is impractical since I_1 is a *pnp* constant-current stage, whereas I_2 is an *npn* current source. This problem can be avoided by making both current sources to be of the same type and then using a switching current mirror configuration to generate the intermittent current source I_2, as illustrated in Figure 11.20. In the circuit, both I_1 and I_2 are generated by similar *pnp* current sources, and the current I_2 is then subtracted from I_1 at the capacitor charging node, in an intermittent mode, by means of the switched Wilson current mirror comprised of Q_1, Q_2, and Q_3. The switching of the current mirror is done by grounding the base of Q_2 by means of the switch S_1. Normally, S_1 is a grounded-emitter *npn* transistor. Since both current sources I_1 and I_2 have the same polarity, but different current levels (with $I_2 > I_1$), both current sources can be varied with the same control voltage, while maintaining their ratio constant. In this manner, the frequency of oscillation can be controlled linearly by an external control voltage, without affecting the output waveform and the duty cycle.

The voltage-controlled current-source stages can be designed using any one of several circuit configurations described in Chapter 4 (see Fig. 4.20). One of the most common circuit configurations used in oscillator design is the composite *pnp–npn* current-source configuration shown in Figure 11.21. In this circuit, the voltage at the emitter of Q_2 is very closely equal to the control voltage V_C, assuming that the V_{BE} drops of Q_1 and Q_2 are approximately equal. Then,

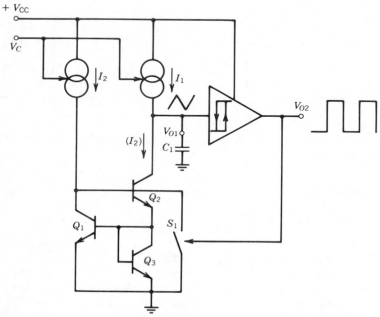

FIGURE 11.20. Constant-current charge and discharge oscillator using switched current mirror.

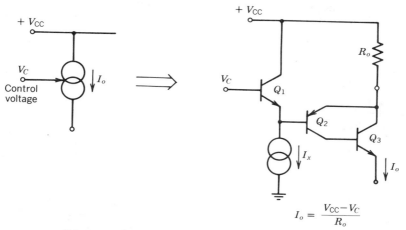

FIGURE 11.21. Typical voltage-controlled current source.

neglecting the base current of Q_2, the output current I_o is directly proportional to the voltage drop across the resistor R_o,

$$I_o = \frac{V_{CC} - V_C}{R_o} \qquad (11.31)$$

Normally, R_o is left external to the circuit. The composite *pnp–npn* transistor configuration of Q_2 and Q_3 functions as a high-β *pnp* transistor with high–current capability.

Triangle-Wave Oscillators

In the circuits of Figures 11.19 and 11.20, if one sets $I_2 = 2I_1$, then both the charge and the discharge currents for the timing capacitor would be equal, and the output waveform V_{O1} becomes a triangle wave with frequency f,

$$f = \frac{I_1}{2C_1(V_B - V_A)} = \frac{I_1}{2C_1V_H} \qquad (11.32)$$

where $V_B - V_A = V_H$ is the hysteresis voltage of the Schmitt trigger. Under this condition, the Schmitt trigger output V_{O2} is a symmetrical (i.e., 50% duty cycle) square wave.

Triangle-wave oscillators are particularly useful for wave-shaping applications, since triangle wave has relatively low harmonic content and can be easily shaped into a low-distortion sine wave (see Section 11.7).

A critical design criterion for triangle-wave output is to maintain the $I_2 = 2I_1$ current ratio under all conditions of circuit operation. Figure 11.22 shows a relatively straightforward method of achieving this 2 : 1 current-ratio condition. In the circuit, Q_2, Q_3 and Q_4, Q_5 form two identical voltage-controlled current sources, of the type shown in Figure 11.21. These current sources are connected

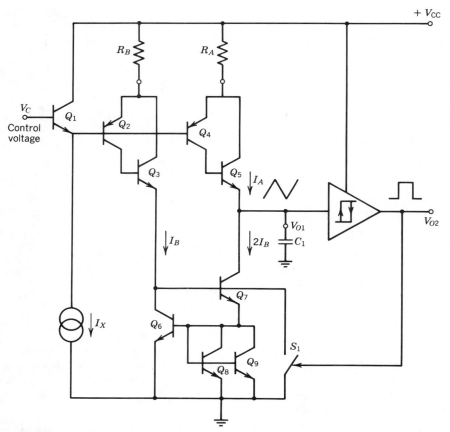

FIGURE 11.22. Triangle-wave oscillator using symmetrical charge and discharge currents. (*Note:* $R_A = R_B$ for triangle-wave output.)

in parallel, and their current levels are set by external resistors R_A and R_B. Normally, $R_A = R_B$, resulting in equal currents I_A and I_B. I_A is used to charge C_1 on a continuous basis. I_B is doubled by the modified Wilson current-mirror and appears as $2I_B$ at the collector of Q_7, which serves as the intermittent discharge current for C_1. In this manner, the matching and tracking accuracy of the two voltage-controlled current sources is optimized by making them identical, and the output symmetry depends primarily on the accuracy of the current doubling of the Wilson current source.

   Figure 11.23 illustrates an alternate circuit configuration for generating symmetrical charge and discharge currents to obtain triangle-wave output.[7] The circuit uses only one constant-current source whose output current I_1 is steered in and out of the timing capacitor C_1 in alternate half-cycles. The operation of the circuit can be briefly described as follows. When the voltage V_{O1} across C_1 is below the Schmitt trigger upper threshold level V_B, the switch S_1 is open, and I_1 flows into C_1 causing V_{O1} to ramp up linearly with time. When $V_{O1} = V_B$, S_1

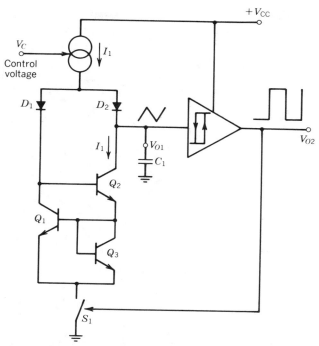

FIGURE 11.23. Alternate circuit configuration for triangle-wave oscillator.

closes. Since voltage across C_1 cannot change instantaneously, the diode D_2 becomes reverse biased, and I_1 is forced to flow through D_1 and Q_1. The Wilson current-mirror formed by Q_1, Q_2, and Q_3 forces an equal amount of current I_1 to flow through Q_2. Since D_2 is reverse biased, this current can only come out of capacitor C_1. Thus, during this period, the capacitor is discharged with constant current, causing V_{O1} to ramp down. When $V_{O1} = V_A$, S_1 is again opened, and the cycle repeats itself. In this manner, C_1 is charged and discharged by equal and opposite currents, resulting in a triangle-wave output. Note that since only one current source is used, the symmetry of the charge and discharge currents is determined primarily by the current transfer accuracy of the Wilson current source.

Both circuit configurations shown in Figures 11.22 and 11.23 have been used in a number of commercially successful IC products. The basic circuit configuration of Figure 11.22 was used as the triangle-wave oscillator section of an ICL-8038 waveform generator integrated circuit, originally introduced by Intersil, Inc., in 1972. The circuit configuration of Figure 11.23 has been used as the oscillator section of the NE-565 phase-locked loop integrated circuit and the NE-566 voltage-controlled oscillator, originally introduced by Signetics Corporation in 1970.

In many applications, particularly for those involving phase-locked loops, an IC oscillator is required to provide two output voltages which are 90° (or $\frac{1}{4}$ cycle)

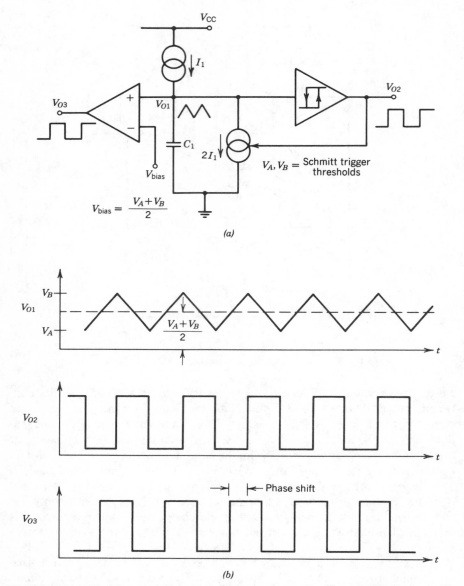

FIGURE 11.24. Obtaining quadrature output from triangle-wave oscillator: (*a*) Basic circuit; (*b*) output waveforms.

out of phase with each other. Outputs having such a phase relationship are called *quadrature*. Such a quadrature–output capability is inherent to the constant-current charge and discharge oscillators, due to the integrating effect of the timing capacitor, as illustrated by the waveforms shown in Figure 11.24. The basic square-wave output of the Schmitt trigger, V_{O2}, changes state at the peaks of the triangle wave. If one were to square up the triangle wave output by passing it through a high-gain voltage comparator, biased at midpoint of the triangle-wave swing, the resulting output V_{O3} will be a square wave which changes state at midpoint of the triangle wave, as shown in Figure 11.23. Thus V_{O3} will be in quadrature with V_{O2}, since it exhibits a net phase shift of 90° or one-quarter cycle.

11.4. EMITTER-COUPLED MULTIVIBRATORS

The emitter-coupled multivibrator circuits are a subclass of the constant-current charge and discharge oscillators described in the previous section. In many monolithic designs, they are often preferred over the conventional Schmitt trigger type oscillator circuits of Figure 11.19 because of the inherent symmetry of output waveforms and the higher frequency capability.

Figure 11.25 shows that basic circuit configuration of the emitter-coupled multivibrator.[7,8] The circuit is comprised of a pair of gain stages made up of Q_1 and Q_2, which are then cross coupled through the emitter-follower buffer stages Q_3 and Q_4. The gain stages have equal load resistor R with clamping diodes D_1 and D_2, which clamp the voltage swing across R to 1 V_{BE}. The emitters of Q_1 and Q_2 are biased with matched constant-current stages and are coupled through the timing capacitor C_1.

The operation of the circuit can be briefly explained as follows. The cross

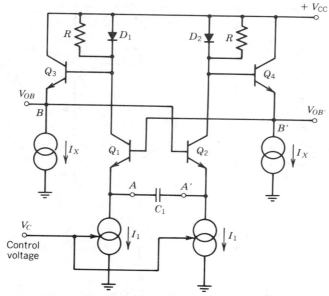

FIGURE 11.25. Basic emitter-coupled multivibrator circuit.

coupling between Q_1 and Q_2 assures that either Q_1 or Q_2 (but not both) is on at any one time. In this manner, the timing capacitor C_1 is alternately charged with equal but opposite currents, first through Q_1 and then through Q_2, in alternate half-cycles of oscillator operation. Assuming for the moment that during one half-cycle, Q_1 is off and Q_2 is conducting, then the equivalent circuit in this state appears as shown in Figure 11.26. Q_2 conducts a total current of $2I_1$, where one-half of the current is forced through the capacitor C_1 in the direction indicated. The current I_1 and the resistor R are chosen such that

$$2I_1R \geq \phi \qquad\qquad (11.33)$$

where $\phi = V_{BE}$ is the diode drop. This condition assures that the diode D_2 would clamp the total voltage swing at the collector of Q_2 to $V_{CC} - \phi$. Under this condition, from Figure 11.26, the base of Q_4 is one diode drop below V_{CC}, and its emitter, which is also the base of Q_1, is two diode drops below V_{CC}. Neglecting the base current of Q_3, its base is at V_{CC}, its emitter is one diode drop below V_{CC}, and the emitter of Q_2 is two diode drops below V_{CC}. Since Q_1 is off, the current I_1, charging the capacitor C_1, is obtained from the emitter of Q_2. This current causes the voltage level V_A at the emitter of Q_1 to ramp down with a constant slope of I_1/C_1 until the voltage level at the emitter of Q_1 becomes equal to three diode drops below V_{CC}. At this point, the transistor Q_1 will turn on. The resulting current through D_1 will cause the base and emitter voltages of Q_3 to shift down by one diode drop, and Q_2 to turn off. When Q_2 is off, the diode drop across D_2 is eliminated, which causes the base of Q_1 to move up by one diode

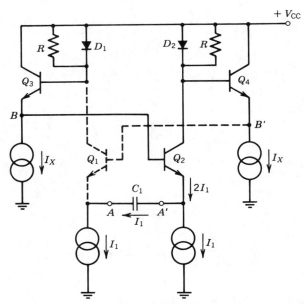

FIGURE 11.26. Equivalent circuit of emitter-coupled multivibrator during one half-cycle of oscillation.

drop. As a result, the circuit has now changed state, with 1 V_{BE} reverse bias across the base–emitter junction of Q_2; and the capacitor C_1 discharges in the opposite direction with constant current I_1. The circuit stays in this state until the voltage level at the emitter of Q_2 ramps downward by an amount equal to $2V_{BE}$, causing Q_2 to turn on and switch the circuit to its prior state, whereby the cycle continues to repeat itself.

Figure 11.27 shows the basic waveforms within the circuit during the transitions between the two astable states. The voltage levels V_A and V_A', correspond

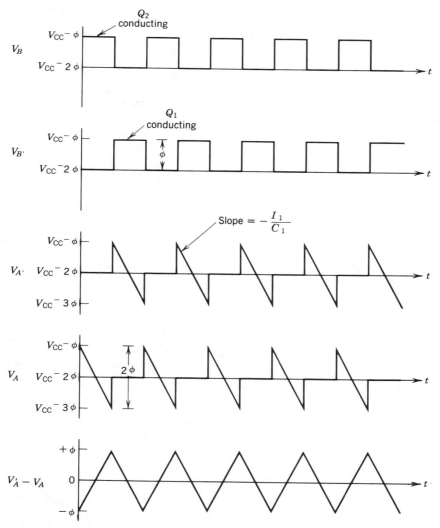

FIGURE 11.27. Voltage waveforms of the emitter-coupled multivibrator circuit of Figure 11.25. (*Note:* $\emptyset = V_{BE}$.)

to the voltages at the emitters of Q_1 and Q_2. Note that the voltage at the emitter of the on–transistor is clamped to two diode drops below V_{CC}, and the voltage level at the emitter of the off–transistor is linearly ramping in a negative direction with a slope of I_1/C_1, for a total voltage swing of $2V_{BE}$ during each half-cycle.

Each half-cycle of the oscillation is equal to the total time Δt for the voltage across C_1 to ramp down by an amount equal to $2V_{BE}$, namely,

$$\Delta t = \frac{C_1 \Delta V}{I_1} = \frac{2C_1 V_{BE}}{I_1} \qquad (11.34)$$

The total period of oscillation T_o is equal to $2\Delta t$, and the frequency of oscillation can be expressed as

$$f_o = \frac{1}{T_o} = \frac{1}{2\Delta t} = \frac{I_1}{4C_1 V_{BE}} \qquad (11.35)$$

Thus, the frequency is proportional to the charging current I_1 and inversely proportional to the timing capacitor and the total voltage excursion across the capacitor.

If the output is taken at the collectors of the switching transistors Q_1 and Q_2, or at the emitters of Q_3 and Q_4, one obtains a set of out-of-phase square-wave signals shown as V_B and $V_{B'}$ in Figure 11.27. Similarly, if the output is taken differentially across the timing capacitor, one obtains a linear triangle wave, corresponding to the voltage difference $V_{A'} - V_A$.

The emitter-coupled multivibrator configuration has several performance advantages:

1. It is an all-*npn* design and is, therefore, capable of high-frequency operation. As described in Section 11.5, it can be made to operate at frequencies in the range of tens of megahertz.
2. It is a symmetrical design, producing symmetrical output waveforms (square wave or triangle), which have low even-harmonic content, as long as the two halves of the circuit are matched.
3. It provides linear control of frequency by means of a control current or voltage.

It can also provide quadrature output (see Fig. 11.24) by squaring the triangle wave available across the capacitor. This can be seen from Figure 11.27 by comparing the timing diagram of the waveforms V_B or $V_{B'}$ with the differential waveform $V_{A'} - V_A$.

The emitter-coupled multivibrator circuit has one major drawback. As indicated in Eq. (11.35), the frequency is inversely proportional to the transistor V_{BE} drop. Since V_{BE} varies at a rate of ≈ -2 mV/°C with temperature, this results in a strong positive temperature coefficient of frequency,

$$\frac{1}{f_o}\frac{df_o}{dT} = -\frac{1}{V_{BE}}\frac{dV_{BE}}{dT} \approx +3300 \text{ ppm/}°\text{C} \qquad (11.36)$$

which is far too large for most applications.

Improving Temperature Stability

The temperature-drift problem associated with the basic emitter-coupled multi-vibrator circuit of Figure 11.25 can be improved by two design techniques:

1. By making I_1 of Eq. (11.35) to be proportional to V_{BE}, so that the ratio I_1/V_{BE} can be made independent of temperature.[9,10]
2. By modifying the basic circuit configuration to eliminate V_{BE} dependence from Eq. (11.35).[11,12]

The first approach usually provides only a partial compensation; the second approach results in somewhat increased circuit complexity and requires the use of *pnp* transistors, which limits the high-frequency capability of the oscillator.

Figure 11.28 shows an all-*npn* emitter-coupled multivibrator circuit which is temperature compensated by making the I_1/V_{BE} term in Eq. (11.35) relatively independent of temperature. With zero applied control voltage, the total bias current I_o is distributed evenly in the circuit so that the current-source transistors Q_5, Q_6, Q_7, and Q_8 each carry an equal amount of current I_1. Q_5 and Q_6 serve as current sources I_1 of the basic emitter-coupled multivibrator section, made up of Q_1 through Q_4 and diodes D_1 and D_2. With $V_C = 0$, the circuit oscillates at frequency f_o given by Eq. (11.35), where $I_1 = I_o/4$.

The frequency control is achieved by the differential control voltage V_C. Increasing V_C in the polarity shown in Figure 11.28 causes the current in Q_5, Q_6 to increase and the currents in Q_7, Q_8 to decrease by the same proportion, while the total current remains constant at I_o, as set by Q_9. Then the frequency of oscillation is given as

$$f = \frac{I_1}{4C_1V_{BE}}\left[1 + \left(\frac{V_C}{R_1}\right)\left(\frac{1}{I_1}\right)\right] \qquad (11.37)$$

where $I_1R_1 >> V_T$ is assumed. Since $I_1 = I_o/4$, Eq. (11.37) can be written as

$$f = \frac{I_o}{16C_1V_{BE}}\left(1 + \frac{V_C}{4I_oR_1}\right) \qquad (11.38)$$

where I_o is the temperature-dependent bias current in Q_9. From the bias configuration used in Figure 11.28, one can show that

$$I_o = \frac{V_{BE}}{R_2} \qquad (11.39)$$

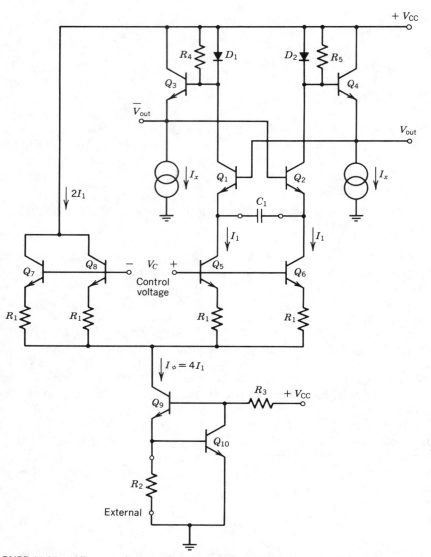

FIGURE 11.28.   All-*npn* emitter-coupled multivibrator circuit with temperature compensation.

where V_{BE} is set by the base–emitter drop of Q_{10}, and R_2 is an external resistor with low temperature coefficient. Then the frequency expression of Eq. (11.38) can be written as

$$f = \frac{1}{16R_2C_1}\left(1 + \frac{V_C}{4V_{BE}}\frac{R_2}{R_1}\right)$$ (11.40)

which indicates that, to a first order, the center frequency (corresponding to the

$V_C = 0$ condition) is insensitive to temperature. In practice, the center frequency stability can be maintained within the ± 300 ppm/°C range, over a 0°C to +70°C range, for frequencies up to 5 MHz. At higher frequencies, the temperature dependence of switching delays in the circuit can cause the frequency stability to deteriorate further, as will be discussed in Section 11.5.

It should be noted that the temperature-compensation technique shown in Figure 11.28 is only a partial–compensation method which works at center frequency only. As shown by the second term in Eq. (11.40), the frequency changes due to the control voltage are not compensated. Thus, for example, a $\pm 10\%$ deviation of the center frequency due to the control voltage would cause an additional ± 300 ppm/°C increase in temperature drift.

The temperature-compensated all-*npn* emitter-coupled multivibrator circuit of Figure 11.28 has been used as the voltage-controlled oscillator section of a number of commercially successful high-frequency phase-locked-loop circuits, such as the NE-560 series circuits introduced by Signetics, or the XR-210 and XR-215 integrated circuits introduced by Exar Integrated Systems.

As mentioned earlier, it is possible to modify the basic emitter-coupled multivibrator circuit to eliminate the V_{BE} dependence of frequency. Figure 11.29 shows a circuit configuration which achieves this objective, at the expense of high-frequency capability.[11] As indicated by the figure, the circuit is fully symmetrical. It operates by alternately charging the external capacitor C_1 first through the transistor Q_1, then through the transistor Q_2, with a constant current I_1 set by an external resistor R_1. The *pnp* transistors Q_{15} through Q_{20} form a balanced current–output gain stage of the type discussed in Section 5.2 (see Fig. 5.19). The differential current outputs of this *pnp* gain stage are cross coupled to opposite halves of the oscillator circuit to assure astable operation. In this manner, one half of the lower portion of the circuit is in the off, or non-conducting, state, while the remaining half is in the on state.

Figure 11.30 shows the equivalent circuit configuration during one half-cycle of oscillation, when Q_1 and Q_4 are on and Q_2 and Q_3 are off. In this state, the capacitor C_1 is charged with the constant current I_1 flowing in the direction shown, from the emitter of Q_1 to the collector of Q_4. The cross-coupled *pnp* gain stage, made up of Q_{15} through Q_{20}, assures that Q_3 along with Q_5, Q_7, Q_9, and Q_{11} are in the off state. During this state, neglecting base currents, the voltage levels at nodes B and B' are clamped at

$$V_B \approx V_{CC} \quad \text{and} \quad V_{B'} = V_{CC} - I_2 R_3 \qquad (11.41)$$

Since the emitters of Q_4 and Q_6 are at virtual–ground potential, the currents I_1 and I_2 are set by the resistors R_1 and R_2 as

$$I_1 = \frac{V_{EE}}{R_1} \quad \text{and} \quad I_2 = \frac{V_{EE}}{R_2} \qquad (11.42)$$

During the half-cycle of oscillator operation shown in Figure 11.30, the emitter of Q_1 is at 1 V_{BE} below V_{CC}, and the collector of Q_4 ramps downward with a slope of I_1/C_1. This state continues until the voltage $V_{A'}$ at the collector of Q_4

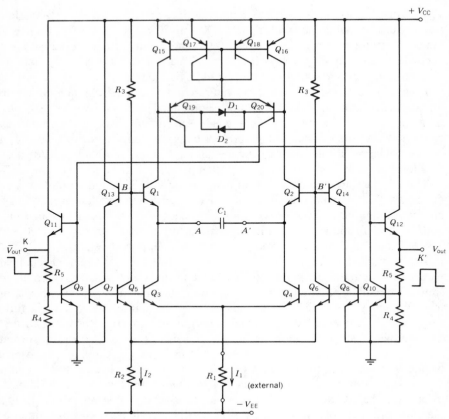

FIGURE 11.29. Stable low-frequency emitter-coupled multivibrator circuit.[11]

reaches 1 V_{BE} below $V_{B'}$ given in Eq. (11.41). At that point, Q_2 is brought into conduction, the circuit changes state; all nonconducting transistors (shown by dashed lines in Figure 11.30) become conducting, and their counterparts are turned off. During the subsequent state, the current through C_1 is reversed; V_A ramps downward with a slope of I_1/C_1 until Q_1 comes into conduction to repeat the cycle. The resulting waveforms available from the circuit at nodes A, A' and B, B' have the same basic forms as those shown in Figure 11.27, except for a scale factor. The waveforms at A and A' are linear half-ramps with a peak-to-peak amplitude $(\Delta V_A)_{pp}$ given as

$$(\Delta V_A)_{pp} = 2I_2R_3 \qquad (11.43)$$

Similarly, the voltage swings at nodes B and B' are out-of-phase square waves with an amplitude of I_2R_3. The square-wave output voltages are also available at low-impedance outputs K and K'.

The half-period of oscillation Δt is again equal to the total voltage swing across the capacitor, divided by the slope of the ramp voltage,

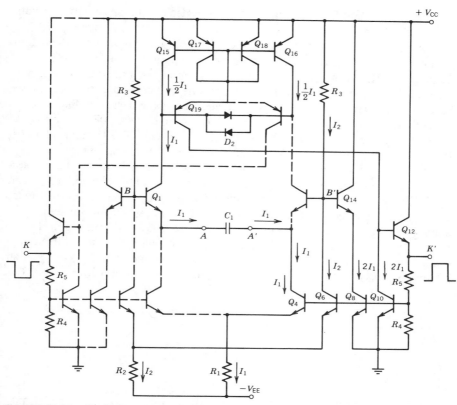

FIGURE 11.30. Equivalent circuit of Figure 11.29 during one half-cycle of oscillation. (*Note:* Nonconducting circuit paths are shown by dashed lines.)

$$\Delta t = \frac{\text{total peak-to-peak swing}}{\text{ramp slope}} = \frac{(\Delta V_A)_{pp}}{I_1/C_1} = \frac{2I_2 R_3 C_1}{I_1} \qquad (11.44)$$

and the frequency of oscillation f_o is given as

$$f_o = \frac{1}{T_o} = \frac{1}{2\Delta t} = \frac{I_1}{4I_2 R_3 C_1} \qquad (11.45)$$

Substituting from Eq. (11.42), one gets

$$f_o = \frac{R_2}{4R_3 R_1 C_1} \qquad (11.46)$$

In most designs it is convenient to set $R_2 = 4R_3$, which reduces Eq. (11.46) to a simple form,

$$f_o = \frac{1}{R_1 C_1} \qquad (11.47)$$

Normally, R_2 and R_3 are internal to the integrated circuit, since only their ratio is critical. R_1 and C_1 are left external to the chip to set the frequency.

The basic design illustrated in Figure 11.29[11] can provide excellent temperature stability (typically on the order of ±20 ppm/°C) over a wide range of frequencies. However, the high-frequency capability is limited to several hundred kilohertz due to the frequency limitations of lateral *pnp* transistors (see Section 11.5). This basic oscillator has been used as the basic building block in a number of commercial IC products, such as the XR-2207 and XR-2206 monolithic waveform generators, and the XR-2211 and XR-2212 phase-locked-loop detector circuits introduced by Exar Integrated Systems, Inc.

An alternate method of improving the temperature stability of the emitter-coupled multivibrator is to clamp the voltage swings at the collectors of Q_1 and Q_2, in Figure 11.25, to a stable voltage level other than the V_{BE} drop. Figure 11.31 shows a practical circuit configuration to achieve this result. In the circuit, unity-gain stages correspond to the emitter followers Q_3 and Q_4 of Figure 11.25. Diodes D_1 through D_4 form a balanced bridge which clamps the total differential swing across the collectors of Q_1 and Q_2 (i.e., voltage $|\,V_B - V_{B'}\,|$) to an internal precision voltage reference V_R. Normally, V_R is a stable reference volt-

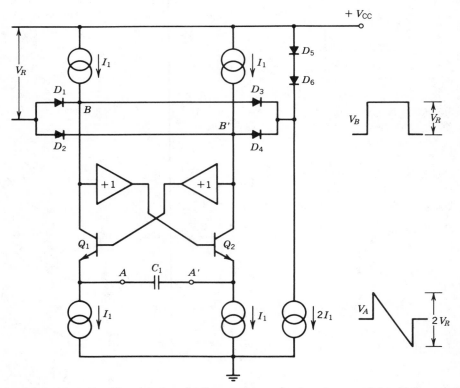

FIGURE 11.31.   Precision clamping of collector voltage swing in emitter-coupled multivibrator to improve temperature stability.[12]

age derived from an internal band-gap reference. The operation of the circuit can be briefly analyzed as follows. Assuming that Q_1 is off and Q_2 is conducting, diodes D_2 and D_3 are conducting and diodes D_1 and D_4 are off. The collector of Q_1 is clamped at 1 V_{BE} below V_{CC}, and the collector of Q_2 is at 1 V_{BE} below V_R. The emitter of Q_2 is at $3V_{\text{BE}}$ below V_{CC}. Since Q_1 is off, the voltage V_A ramps downward with a slope of I_1/C_1 until it reaches $3V_{\text{BE}}$ below V_R and causes Q_1 to turn on. Then Q_2 is turned off and the voltage $V_{A'}$ ramps downward until it reaches $3V_{\text{BE}}$ below V_R to cause Q_2 to turn on and Q_1 to turn off, which will make the circuit revert to its initial astable state. In this manner, the circuit alternates between two astable states and produces the basic waveforms shown in Figure 11.31. Again, these waveforms are the same as those shown in Figure 11.27, except for a scale factor.

Since the half-period is equal to the peak-to-peak voltage swing across C_1, divided by the ramp slope, it can be expressed as

$$\Delta t = \frac{(V_A)_{\text{PP}}}{I_1/C_1} = \frac{2V_R C_1}{I_1} \tag{11.48}$$

and the frequency of oscillation is

$$f_o = \frac{1}{2\Delta t} = \frac{I_1}{4C_1 V_R} \tag{11.49}$$

which is, to a first order, insensitive to temperature.

A detailed analysis of the regenerative switching within the oscillator circuit of Figure 11.31, at the extremities of the voltage swings, indicate that there is an inherent temperature coefficient of frequency, approximately $+250-+300$ ppm/°C, due to the gradual, rather than abrupt, switching of the clamp diodes.[12] For a practical design, this predictable drift can be compensated by introducing an equal drift into V_R.

The basic circuit configuration of Figure 11.31 was utilized in the design of the AD-537, a monolithic precision voltage-to-frequency converter circuit introduced by Analog Devices. The circuit provides a temperature drift of less than ± 100 ppm/°C over a $10,000:1$ linear frequency adjustment range.

In conclusion, the emitter-coupled multivibrator is one of the most versatile building blocks for monolithic designs. However, it requires careful design and temperature-drift compensation techniques in order to optimize its stability; and most of these drift–compensation techniques result in significant circuit complexity and reduction of high-frequency capability.

11.5.   CMOS RELAXATION OSCILLATORS

As described in Chapter 6, MOS technology can also be used in place of bipolar devices to implement a number of analog circuit functions. Among various MOS technologies, complementary MOS (CMOS) technology is by far the best suited for analog design, since complementary devices can be used as active loads to obtain high values of voltage gain.

In this section, basic circuit configurations and design considerations for CMOS relaxation oscillators shall be examined. These oscillator circuits are particularly useful for generating clock or timing signals in monolithic digital systems where the high-density logic requirements would make the use of MOS technology mandatory. In addition, the low power requirements of CMOS circuits make them particularly advantageous for use in battery-powered or portable equipment.

CMOS relaxation oscillators can be built using simple CMOS inverter stages as gain blocks. Figure 11.32 shows that circuit diagram and transfer characteristics of the basic CMOS inverter stage. The transition voltage V_{TR} at which the circuit output changes state is related to the relative geometries of the complementary devices Q_1 and Q_2, as well as to intrinsic process parameters, threshold voltages, and the carrier mobilities associated with each device. Under typical manufacturing conditions and by proper scaling of device geometries, V_{TR} can be maintained within a relatively narrow range of 40–60% of V^+. The slope of the transition characteristics at the switching point, which corresponds to the voltage gain of the stage when it is biased with $V_{in} = V_{TR}$, is normally on the order of 40–50 dB. Thus, the CMOS inverter can be used as a simple comparator circuit which changes state when $V_{in} = V_{TR}$.

Figure 11.33 shows a simple relaxation oscillator using two CMOS inverter stages along with an external R–C network. The operation of the circuit can be briefly described as follows. Since the output of inverter A directly drives the input of inverter B, the two inverter outputs V_{O1} and V_{O2} are at all times in opposite states. Neglecting the finite series resistances of MOS transistors in their on state, when the output of A is low and B is high, $V_{O1} \approx V^+$ and $V_{O2} \approx 0$. Thus, C_1 discharges through R_1 with a time constant $\tau_1 = R_1 C_1$, and the voltage level V_{O3} decreases exponentially toward ground as shown in Figure 11.33b. When $V_{O3} \approx V_{TR}$, inverter A changes state and causes the polarities of V_{O1} and V_{O2} to be reversed. Thus, the capacitor C_1 now charges toward V^+ with a time

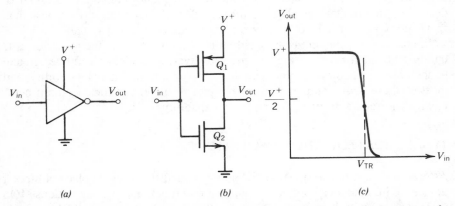

(a)   (b)   (c)

FIGURE 11.32. CMOS inverter as a gain stage: (*a*) Circuit symbol; (*b*) circuit diagram; (*c*) transfer characteristics.

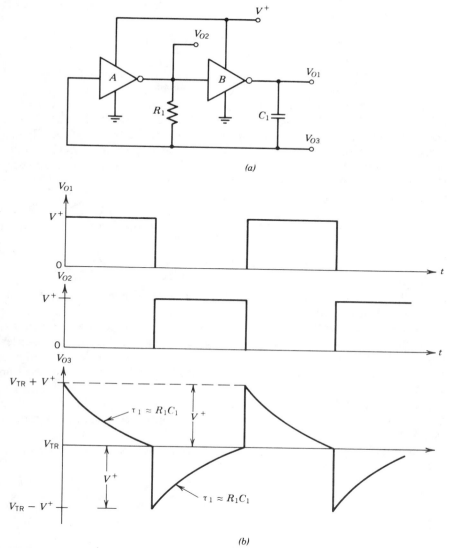

FIGURE 11.33. Simple relaxation oscillator circuit using CMOS inverter stages: (*a*) Circuit diagram; (*b*) output waveforms.

constant τ_1 until V_{O3} again reaches V_{TR} and the circuit reverts to its initial state, continuing to oscillate with the waveforms shown in Figure 11.33*b*. The frequency of oscillation can be calculated from the exponential waveforms as

$$f_o = \frac{1}{R_1 C_1} \left(\frac{1}{\ln \left[\dfrac{(V^+ + V_{TR})(2V^+ - V_{TR})}{V_{TR}(V^+ - V_{TR})} \right]} \right) \qquad (11.50)$$

Assuming that $V_{TR} \approx 0.5V^+$, Eq. (11.50) becomes

$$f_o \approx \frac{1}{R_1 C_1 \ln(9)} = \frac{0.455}{R_1 C_1} \qquad (11.51)$$

For low-frequency operation (i.e., $f_o < 100$ kHz), the value of R_1 can be chosen to be sufficiently large such that $R_1 \gg R_{on}$, where R_{on} is the finite resistance of the MOS transistors at inverter outputs in the "on" condition. In this mode of operation, the circuit is relatively insensitive to the CMOS inverter parameters and exhibits relatively good stability and accuracy characteristics. For operation with supply voltages in the $+5-+15$ V range, with $f_o < 100$ kHz, typical reported circuit performance characteristics are[13]

$$\text{temperature stability} = \frac{1}{f_o} \frac{\partial f_o}{\partial T} \approx -400 \, \text{ppm/°C}$$

$$\text{supply voltage stability} = \frac{1}{f_o} \frac{\partial f_o}{\partial V^+} \approx +2\%/V$$

where the prime reason for the frequency drift is the change of V_{TR} with temperature and supply voltage.

The frequency stability of the simple CMOS oscillator configuration shown in Figure 11.33 can be improved significantly by replacing the simple inverter stages with high-gain voltage comparator stages, similar to that shown in Figure 11.34, where the transition voltage V_{TR} can be derived from an internal resistor string.

The basic CMOS comparator circuit of Figure 11.34 can also be used as a building block to implement the basic R–C relaxation oscillator configurations shown in Figures 11.15 and 11.17 (see Fig. 11.48).

Figure 11.35 gives the simplified circuit schematic of a CMOS relaxation

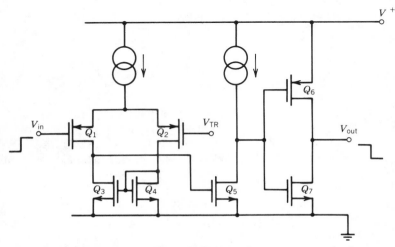

FIGURE 11.34. Typical high-gain CMOS voltage comparator circuit.

oscillator whose frequency can be controlled by a bias current setting.[14] The principle of operation of the circuit is somewhat similar to the case of emitter-coupled multivibrator circuits discussed in the previous section. The digital logic and R–S flip-flop section which cross couples the two inverter stages Q_1, Q_2 and Q_3, Q_4 assures that they are set at opposite states such that either Q_1, Q_4 are on and Q_2, Q_3 are off or vice versa. In the state where Q_1 and Q_4 are on, node B' is clamped to ground; C_1 charges with constant current I_C, and node B ramps up with a slope of I_C/C_1 until the transition voltage V_{TR} of the inverter A_1 is reached. At this point, A_1 changes state and resets the R–S flip-flop formed by the cross-coupled NOR gates. The flip-flop outputs change state and cause Q_2, Q_3 to turn on and Q_1, Q_4 to turn off. Thus, node B is clamped to ground and node B' swings negative. However, the drain–substrate diode of Q_4 clamps the negative excursion of B' to one diode below ground. Then the capacitor charges with constant current I_C through Q_3, and the voltage level at node B' ramps linearly with a slope of I_C/C_1 until it reaches V_{TR} of inverter A_2. When this happens, A_2 changes state, the flip-flop is set, the circuit reverts to its previous state, and the cycle repeats itself. The time Δt for the capacitor C_1 to charge from one diode drop below ground to V_{TR} with constant current I_C is

$$\Delta t = \frac{(V_{TR} + \phi)C_1}{I_C} \qquad (11.52)$$

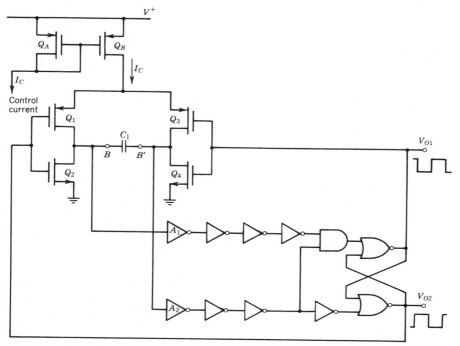

FIGURE 11.35.   A current-controlled CMOS oscillator.[14]

where ϕ is the diode drop. Δt corresponds to one half-period of oscillation. Thus, the frequency of oscillation f_o is

$$f_o = \frac{1}{2\Delta t} = \frac{I_C}{2C_1(V_{\text{TR}} + \phi)} \tag{11.53}$$

which is proportional to the control current I_C.

In the circuit of Figure 11.35, the extra inverters in series with A_1 and A_2 are included to provide additional gain and to shape the slowly–rising ramp waveform across the capacitor to the fast–rising set and reset signals for the R–S flip-flop. They also provide four inverter delays before the removal of the flip-flop set–reset pulses to assure proper toggling action. The AND gate in series with the R–S flip-flop input assures that the flip-flop inputs are of opposite polarity at all times.

For operation with relatively high supply voltages, $V_{\text{TR}} >> \phi$ and Eq. (11.53) reduces to

$$f_o \approx \frac{I_C}{V^+ C_1} \tag{11.54}$$

where $V_{\text{TR}} \approx \frac{1}{2}V^+$ is assumed.

The main advantage of the oscillator circuit of Figure 11.35 is its linear control characteristics which make it suitable for phase-locked-loop applications. In addition, the output waveforms are symmetrical (i.e., 50% duty cycle) due to the symmetrical nature of the circuit. However, its main drawback is relatively poor temperature stability due to the temperature dependence of the inverter transition voltage V_{TR}. The V_{TR} exhibits a negative temperature coefficient of approximately -800 ppm/°C, which would result in a strong positive temperature coefficient of f_o unless additional compensation measures are taken.

11.6. LIMITATIONS OF RELAXATION OSCILLATORS

High-Frequency Capability

The high-frequency capability of relaxation oscillators is limited by the switching delays within the circuit, at the extremities of voltage excursions (i.e., at the switch points of the internal level detector). These switching delays in turn depend on the particular circuit configuration used, the switching speeds of active devices, the internal current levels, and the parasitic capacitances.

The effect of switching delays can be demonstrated considering the case of the constant-current triangular-wave oscillator circuit of Figure 11.36. However, the basic conclusions are applicable to all relaxation oscillator waveforms. Ideally, neglecting the switching delays, the frequency of oscillation f_o is proportional to the charging current I_1, and is inversely proportional to the timing

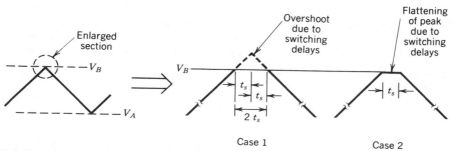

Case 1   Case 2

FIGURE 11.36.   Effects of switching delays on oscillator waveform and frequency. Case 1: Charging current continues to flow until completion of switching cycle. Case 2: Charging current stops at beginning of switching cycle.

capacitor C_1 and the total voltage swing (i.e., hysteresis voltage) V_H, as given by Eq. (11.32). This expression assumes that at the extremity of the voltage swing (i.e., at the peak of the triangle wave) the current direction can change instantaneously. In an actual design, a finite switching–time delay t_s is involved during the current reversal. The effects of this switching delay are illustrated in Figure 11.36, where the circled portion of the waveform at the peak of the triangle wave is enlarged to illustrate the two possible extreme cases:

1.   The charging current continues to flow and is not reversed until the completion of switching time t_s. This results in a total time delay of $2t_s$ at each peak of the triangle.

2.   The charging current stops at the beginning of the switching cycle and is reversed at the completion of switching time. This results in a time delay t_s at each peak of the triangle waveform.

Thus, the total period of oscillation is increased by an amount equal to $4t_s$ for case 1 and by an amount $2t_s$ for case 2. From the above analysis, one can derive an actual frequency expression of the form

$$f_1 = f_0 \frac{1}{1 + f_0 t_s \gamma} \qquad (11.55)$$

The constant γ is in the range of

$$2 \le \gamma \le 4 \qquad (11.56)$$

where the upper and lower limits on γ correspond to cases 1 and 2, respectively. In Eq. (11.55), f_1 is the actual observed frequency, and f_0 is the ideal frequency, assuming zero switching delays.

In deriving Eq. (11.55), we have made the simplifying assumption that the switching delays at the positive and negative extremities of the voltage swing are equal.

In practical circuit configurations, the actual waveform at the point of switching lies somewhere between the two cases illustrated in Figure 11.36 and is often closer to case 1 (i.e., $\gamma = 4$) since several intermediate logic stages have to change state before the charging current is affected. Thus, one can define a hypothetical maximum frequency of oscillation f_{max}, limited solely by the switching delays from Eq. (11.55), as

$$f_{max} = \frac{1}{\gamma t_s} \qquad (11.57)$$

In the relaxation–oscillator circuits of Figures 11.15 through 11.23, which use a long feedback path for the switching signal and also employ lateral pnp transistors in the switching–signal path, typical values of t_s range from 50 to 200 nsec, depending on the circuit configuration and the bias current levels used. The same is also true for the emitter-coupled circuit of Figure 11.29, due to the presence of lateral pnp transistors. Thus, assuming a typical value of $t_s = 100$ nsec and assuming $\gamma = 4$ (i.e., case 1), one gets $f_{max} \approx 2.5$ MHz.

In the case of the all-npn emitter-coupled multivibrator of Figure 11.25, the feedback path for the switching signal is very short, and none of the switching devices are driven into saturation. Thus, typical values of t_s can be kept in the range of 3–5 nsec, giving the circuit an idealized maximum frequency of oscillation in the range of 50–100MHz.

It should be noted that the f_{max} expression of Eq. (11.57) is highly idealized since it assumes zero hysteresis voltage. In practice, frequency accuracy and temperature stability requirements limit the useful range of operation of relaxation oscillators to frequencies that are on the order of 10–20% of f_{max}.

Temperature Stability

The stability of the oscillator frequency f_o with temperature is one of the most important considerations in IC oscillator design. In the case of tuned–oscillator circuits discussed in Section 11.2, frequency stability is primarily determined by the characteristics of the external LC tank or the quartz crystal, and the monolithic integrated circuit does not contribute significantly to the temperature stability. In the case of relaxation–oscillator circuits, the situation is different. The temperature stability of the oscillator depends strongly on current levels and voltage thresholds internal to the IC chip, as well as on the external components.

In specifying the temperature stability of multivibrator-type oscillators, one usually specifies only the temperature drift due to the IC chip, excluding the temperature coefficients associated with the external timing components. This assumption allows IC designers to concentrate their efforts on minimizing or eliminating the temperature drift sources associated with the IC chip itself, independently of the external components which are supplied by the user. In the following discussion, we will examine only those temperature drift sources that are inherent to the IC design, and not those contributed by the external timing components.

There are four contributing factors which affect the temperature stability of a monolithic relaxation oscillator or multivibrator circuit.

1. *Stability of Internal Charge and Discharge Currents with Temperature.* In the case of exponential-ramp R–C oscillators (see Fig. 11.13), charge and discharge current levels are directly set by the external resistors R_1 and R_2. Assuming that the external components are ideal (i.e., zero temperature coefficient), the internal charge and discharge currents will be insensitive to temperature. However, in constant-current oscillator circuits, the charge and discharge currents are set indirectly by converting an external voltage level into an internal current through a voltage-to-current converter stage. This internal current generation contributes an additional drift source. With careful design considerations, the magnitude of this additional drift can be kept well below 100 ppm/°C at low frequencies.

2. *Stability of Schmitt Trigger Thresholds.* The oscillator frequency is inversely proportional to the total hysteresis voltage V_H, associated with the Schmitt trigger. Thus, any temperature drift of V_H results in an equal amount of relative frequency drift. Note that since f_o is inversely proportional to V_H, the polarity is reversed; a positive temperature coefficient of V_H results in a negative temperature coefficient of f_o. In the case of the R–C oscillators described in Section 11.2, the temperature stability of V_H is determined by two separate effects:

(a) Stability of reference voltage settings for V_A and V_B. Typically, these reference voltages are set by a resistor string and are stable. By careful layout of a resistor string and by keeping the bias current in the resistor string much higher than the comparator input bias currents, the magnitude of drift due to reference voltage settings can be kept in the range of 10–20 ppm/°C.

(b) Comparator offset voltages. The switching point of sensing comparators varies with temperature, due to the change of comparator offset voltage with temperature. This is particularly true, if the comparators use Darlington inputs or have relatively high initial offset voltages. The drifts of bipolar comparator offsets can be in the range of 20–50 μV/°C per comparator. In the case of the dual-comparator oscillator circuit of Figure 11.17, with a hysteresis voltage of 2 V, such an offset drift mechanism can contribute approximately ± 50 ppm/°C frequency drift. Since individual comparator offset drifts are relatively constant, their total contribution to oscillator stability varies inversely with the value of V_H. For example, increasing V_H to 4 V in the circuit of Figure 11.17 would reduce the relative drift contribution of comparator offset drifts to approximately 25 ppm/°C. However, this will also have the adverse effect of reducing the high-frequency capability of the circuit due to increased voltage–swing requirements across the timing capacitor. The emitter-coupled multivibrator circuit of Figure 11.25 illustrates an extreme example of frequency drift due to hysteresis voltage change. In this case, V_H is directly proportional to V_{BE}, which varies at a rate of -2 mV/°C, corresponding to an inherent frequency drift of $\approx +3300$ ppm/°C [see Eq. (11.36)].

   3.   *Comparator Bias and Leakage Current.* Any junction leakage current or comparator input bias current at the timing terminal (i.e., node *A* of Figs. 11.15 and 11.17) will add or subtract from the oscillator charge and discharge current and will cause a frequency error. Normally, junction leakage currents are maintained at a low-nanoampere range, and the comparator input bias currents are in the few hundred nanoampere range. Thus, their effects on frequency accuracy and frequency drift are negligible, as long as the timing currents are maintained in the range of ≥ 10 μA. The effects of input bias currents are also minimized by designing the comparator polarities in a dual-comparator circuit, such as that shown in Figure 11.17, so that both comparator inputs connected to the timing terminal are in the off state until the voltage swings reach the upper or lower threshold voltages. The input bias currents flow only during the short time interval corresponding to the switching time t_s at the extremities of the waveform. This is accomplished by making the upper and lower comparators to have *npn* and *pnp* input stages, respectively, as shown in Figure 11.18.

   4.   *Temperature Coefficient of Switching Delays.* As indicated by Eq. (11.55), the actual frequency of operation f_1 becomes strongly influenced by the internal switching delay t_s as the product $\gamma f_0 t_s$ approaches unity. For cases where $\gamma f_0 t_s \ll 1$, Eq. (11.55) can be approximated as

$$f_1 = f_0(1 - \gamma f_0 t_s) \tag{11.58}$$

Differentiating Eq. (11.58) with respect to temperature T, one gets

$$\frac{1}{f_1}\frac{\partial f_1}{\partial T} = -\frac{f_0}{f_{max}}\frac{1}{t_s}\frac{\partial t_s}{\partial T} \tag{11.59}$$

where f_{max} is the hypothetical maximum frequency of oscillation, as defined in Eq. (11.57).

   Typically, switching delays exhibit a strong positive temperature coefficient on the order of $+2000$ ppm/°C, primarily due to parasitic storage time effects and the temperature coefficients of resistances in series with parasitic capacitances at various circuit nodes. This implies that even at frequencies equal to 10% of f_{max}, the switching delays contribute a negative temperature coefficient of approximately -200 ppm/°C to the oscillator frequency. This drift will get progressively worse as f_0 is increased.

   In deriving the approximate equations (11.58) and (11.59), we have made an implicit assumption that switching delays at positive and negative peaks of voltage swings are equal. This is valid for the emitter-coupled multivibrator, but may not be valid for the general charge and discharge oscillators shown in Figures 11.13 through 11.23. In a carefully detailed design, the switching delays associated with each threshold should be analyzed separately and added together to get the total delay, corresponding to the γt_s term in Eqs. (11.55) and (11.58).

Conclusions

From the analysis of the frequency limitation and temperature drift characteristics discussed in this section, the following conclusions can be drawn regarding practical IC oscillator configurations:

1. For low frequencies where switching delays are negligible (i.e., $f_0 \leq 100$ kHz), the temperature stability of R–C relaxation oscillators can be kept to below the ± 100 ppm/°C range by careful design practices, and by using lateral *pnp* transistors in the signal path. At present, several IC oscillator products are available which exhibit frequency drifts of ≤ 50 ppm/°C in this frequency range. These are the XR-2206, XR-2207, and XR-2209 monolithic oscillators by Exar Integrated Systems, which use the circuit configuration of Figure 11.29; the ICL-8038 waveform generator by Intersil Corporation, which uses the circuit configuration of Figure 11.22, and the AD-537 voltage-to-frequency converter integrated circuit by Analog Devices, which uses the circuit topology of Figure 11.31.

2. The frequency stability at high frequencies is determined primarily by switching delays within the oscillator feedback loop. The all-*npn* emitter-coupled multivibrator circuit of Figure 11.25 is the only practical circuit configuration which has sufficiently low switching delays to operate at frequencies in excess of 10 MHz. Unfortunately, this circuit has poor temperature stability due to the V_{BE} dependence of internal threshold levels. Using the temperature compensation techniques shown in Figure 11.28 or Figure 11.31, the frequency stability of such an oscillator can be maintained to within ± 600 ppm/°C for frequencies up to 10 MHz.[15]

11.7. MONOLITHIC WAVE-SHAPING TECHNIQUES

In a wide range of analog signal-processing, telemetry, and data-encoding applications, one requires nearly sinusoidal signals with low harmonic content. The relaxation oscillator and multivibrator circuits discussed in the preceding sections do not provide sinusoidal output signals. In practical applications requiring sinusoidal output signals, this problem can be overcome by using additional wave-shaping circuitry to convert the basic oscillator waveform to a low-distortion sine wave.

In generating a low-distortion sine wave using wave-shaping circuits, one normally starts with a symmetrical triangle wave signal of constant amplitude, and modifies the waveform by passing it through a nonlinear wave-shaping circuit. The symmetry of the initial triangle wave is important in order to eliminate or minimize the even harmonics at the output signal.

There are three basic wave-shaping techniques which are used in monolithic IC design for sine-wave generation:

1. Piecewise linear approximation by introducing breakpoints into a triangle wave.
2. Logarithmic amplification of a triangle wave by an overdriven differential gain stage.
3. Digital staircase signal generation to approximate a sine wave.

In this section, the circuit implementation of these sine-wave-shaping techniques will be examined.

Triangle-to-Sine-Wave Conversion Using Breakpoint Technique

In the so-called *breakpoint* technique of triangle-to-sine conversion, one passes the triangle wave through a nonlinear diode-resistor network which is biased to provide a number of symmetrical breakpoints into the positive- and negative-going sections of a symmetrical triangle wave. Figure 11.37 illustrates the basic principle of converting a triangle wave into a low-distortion sine wave by introducing piecewise-linear breakpoints into the triangle wave. The triangle input signal is supplied from a signal source with a constant-source resistance R_S.

The diodes D_{A1} through D_{An} have their cathodes connected to progressively more positive bias voltages V_{A1} through V_{An}, and they turn on sequentially during the positive swing of the triangle wave input. Depending on the values of the series resistances R_{A1} through R_{An}, associated with them, these diodes introduce n breakpoints into the positive swing of the triangle wave by changing the loading on the source resistance R_S. A symmetrical set of diodes D_{B1} through D_{Bn}, are used to introduce a corresponding set of breakpoints into the negative swing of the triangle-wave signal, where the anodes of the diodes are connected to progressively more negative bias points V_{B1} through V_{Bn}. The breakpoint voltages V_{A1} and V_{B1} have equal magnitudes but opposite polarities referenced to the dc level of the input triangle signal. The piecewise-linear sine wave formed in this manner is then buffered by an internal buffer amplifier to avoid external loading of the wave-shaping circuitry.

In a breakpoint-type triangle-to-sine converter, normally, four breakpoints are required in each half of the triangle wave to obtain a sine wave with a total harmonic distortion (THD) content of $\leq 1\%$.

Figure 11.38 shows a practical circuit implementation of the breakpoint triangle-to-sine conversion technique in a form suitable for monolithic integration. The component values shown in the figure are optimized for an input

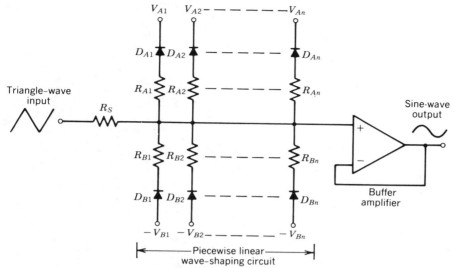

FIGURE 11.37. Triangle-to-sine-wave conversion by piecewise linear wave-shaping circuit.

triangle-wave peak-to-peak amplitude equal to 1/3 of V_{CC} for a range of $10 \text{ V} \leq V_{CC} \leq 30 \text{ V}$.[16] The circuit introduces a total of eight breakpoints (four on each half-cycle) on the input triangle wave to produce a sine-wave output with total harmonic distortion $\leq 1\%$.

In the circuit of Figure 11.38, complementary *pnp–npn* emitter followers are used to perform the function of the breakpoint diodes. These *pnp–npn* emitter-follower pairs are symmetrically biased with respect to the dc level of the input triangle wave which is internally set at $V_{CC}/2$. During the positive swing of the triangle-wave input, the lower half of the wave-shaping circuit is off since the base–emitter junctions of Q_9, Q_{11}, Q_{13}, and Q_{15} are reverse biased. As the input signal swings in the positive direction, *pnp* emitter followers Q_1, Q_3, Q_5, and Q_7, are turned on successively, introducing an increasing amount of signal attenuation near the peak of the triangle swing. For the negative half of the triangle swing, the situation is reversed. The top part of the wave-shaping circuitry is off, and the *npn* emitter followers Q_9, Q_{11}, Q_{13}, and Q_{15} turn on, successively, to attenuate the negative peak of the triangle swing. The component values, and the location of breakpoints in the circuit of Figure 11.38, are chosen to minimize the harmonic content of the output sine wave, using triangle-to-sine conversion algorithms.

Several points regarding the circuit configuration of Figure 11.38 should be noted. The use of complementary *npn–pnp* emitter followers to replace the diodes in the basic wave-shaping circuit of Figure 11.37 simplifies the biasing greatly by buffering the bias string from the input signal swings. The V_{BE} drops, and their temperature drifts associated with the *pnp* and *npn* transistors at each emitter-follower pair, subtract from each other, and result in stable location of

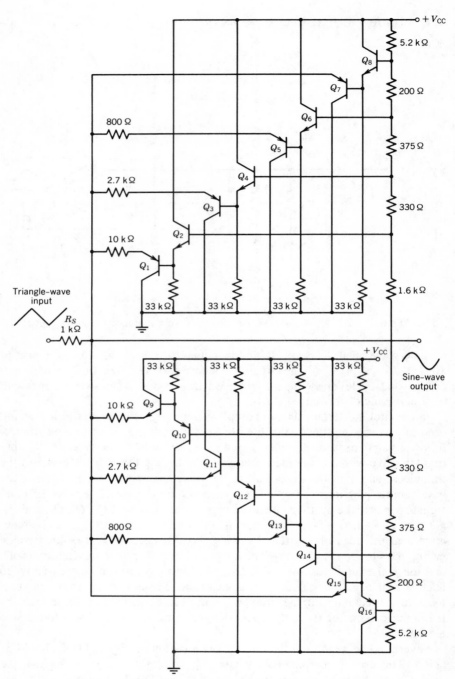

FIGURE 11.38. Piecewise linear sine-shaping circuit using active limiting.[16]

breakpoints, independent of V_{BE} drops. Finally, since all *pnp* and *npn* devices are in a common-collector configuration, the layout can be greatly simplified by combining all *npn* devices in a common isolation pocket, and by using substrate-*pnp* transistors for the *pnp* devices.

Triangle-to-Sine-Wave Conversion by Nonlinear Amplification

Triangle-to-sine conversion can be achieved by the nonlinear amplification of a symmetrical triangle-wave input signal. In this approach, the zero crossing of the triangle wave is amplified linearly, whereas the region near the peaks of the triangle are amplified logarithmically, resulting in a gradual rounding of the peaks of the triangle.

A practical method of obtaining the nonlinear transfer function described above is the use of a differential gain stage with emitter degeneration, as shown in Figure 11.39a.[8,17] The nonlinear signal processing is obtained by slightly overdriving the differential pair such that at the positive and negative extremities of the input triangle-wave swing, Q_2 and Q_1 are driven to the verge of cutoff. As the transistors are driven near cutoff, the transfer characteristics of the circuit tend toward logarithmic behavior, rather than linear, as illustrated in Figure 11.39b. For a given, well-controlled triangle input signal, this causes a gradual rounding of the peaks of the triangle and results in a low-distortion sine-wave output. The output can either be taken differentially across R_1, or across the collector load resistors.

It can be shown, by numerical analysis, that for a given choice of bias current and emitter degeneration setting, the optimum peak-to-peak amplitude $(V_{in})_{pp}$ for minimum output harmonic distortion is obtained when[17]

$$(V_{in})_{pp} = \pi\left(\frac{I_1 R_1}{2} + V_T\right) \tag{11.60}$$

where V_T is the thermal voltage. Both computer simulation and experimental results indicate that the optimum bias setting for minimum total harmonic distortion is when the $I_1 R_1 / V_T$ ratio is chosen to be in the range of

$$2.0 \leq \frac{I_1 R_1}{V_T} \leq 3.3 \tag{11.61}$$

Substituting the above range of values into Eq. (11.60), one obtains the optimum input signal level for minimum output distortion as

$$2\pi \leq \frac{(V_{in})_{pp}}{V_T} \leq 2.65\pi \tag{11.62}$$

Choosing a value in the midrange of Eq. (11.62), with $V_T = 26$ mV at room temperature, one gets

$$(V_{in})_{pp}\bigg|_{\text{optimum}} \approx 2.3\pi V_T = 190 \text{ mV} \tag{11.63}$$

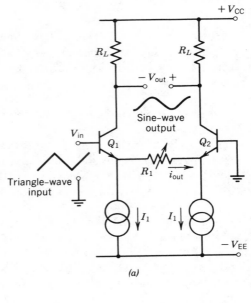

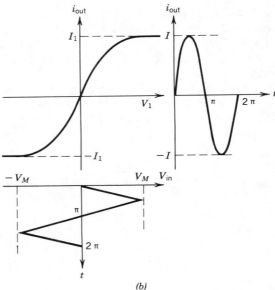

FIGURE 11.39. Differential pair as triangle-to-sine-wave converter. (*a*) Circuit schematic. (*b*) Transfer characteristics.[8, 17]

Experimental results indicate that at optimum setting, the output harmonic content can be maintained under 0.5% at room temperature, and under 1% over the temperature range of $0-+75°C$. However, the output signal amplitude exhibits a relatively strong negative temperature coefficient of approximately -2000 ppm/°C.

The basic advantage of the logarithmic sine-converter circuit of Figure 11.39

over the breakpoint-type sine converter is the low input signal amplitude required. Since a high-amplitude triangle wave is difficult to generate at high frequencies, the logarithmic sine converter circuit using the differential pair offers higher frequency capability and can operate with lower supply voltages than the breakpoint-type converter.

Digital Staircase Sine Converter

In a staircase sine–converter circuit, the output waveform is generated as a series of successive staircase steps. This type of converter is particularly useful for primarily digital systems which operate with internal clock signals and binary data levels, yet which must produce nearly sinusoidal output frequency signals. Examples of such applications are the data MODEMS and tone generators used in telephone systems.

Figure 11.40 shows the typical example of one cycle of a sine wave, approximated by a 14-step staircase waveform. The ideal sine wave is also shown superimposed on the approximate signal. The staircase signal is generated by dividing down a higher frequency input and using decoding logic to activate

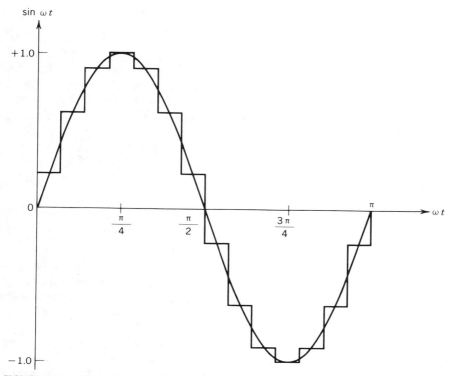

FIGURE 11.40. Approximation of sine wave with digital staircase waveform, using seven current sources in 14 steps.

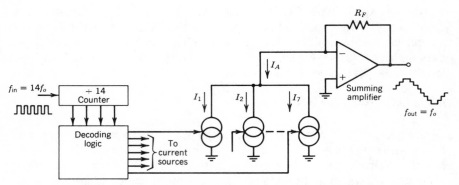

FIGURE 11.41. System for generating digitally synthesized sine wave.

seven current sources in a sequence of 14 steps, as illustrated in the system diagram of Figure 11.41.

The input signal is normally a square wave or a digital pulse train at a frequency Nf_o where N is the number of steps in the staircase, and f_o is the desired frequency of the output waveform. The quality and harmonic content of an output sine wave can be improved by increasing the number of steps used in the staircase. However, this results in additional counting and decoding circuitry and requires higher input frequency.

Assuming a basic staircase algorithm (illustrated in Fig. 11.40), one can show that the harmonic content of the output signal is primarily at frequencies of $(N + 1)f_o$ and $(N - 1)f_o$, and the relative amplitudes of these harmonics are inversely proportional to the harmonic order. In the case of the 14-step waveform shown in Figure 11.40, these correspond to the 13th and 15th harmonics with a total harmonic distortion of approximately 10%.[19]

Since the harmonic content of a digital staircase signal is at higher–order harmonics, it can be easily filtered by a simple one-pole low-pass filter. For example, a one-pole low-pass filter with a bandwidth of f_o would reduce the harmonic content of the waveform in Figure 11.40 to approximately 1%.

Monolithic Function Generators

A monolithic waveform or function–generator integrated circuit is designed to produce a variety of output waveforms which can be modulated in both amplitude and frequency. Figure 11.42 shows the system block diagram of such a monolithic chip. Since all of the blocks shown are readily compatible with monolithic IC design, a number of such commercially available IC designs have been developed.[8,16,18]

The heart of a function–generator integrated circuit is the basic relaxation oscillator capable of producing linear–ramp and triangle waveforms. Such an oscillator can be either an emitter-coupled multivibrator or the charge and discharge type relaxation oscillator of the type shown in Figure 11.19. Its

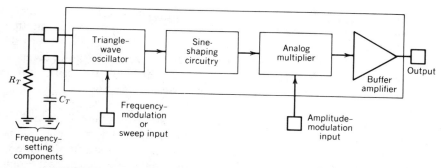

FIGURE 11.42. Block diagram of monolithic waveform generator system.

frequency is set by an external $R–C$ combination and can be modulated by a control voltage. The oscillator output, in a constant-amplitude triangle waveform, is applied to the internal sine-shaping circuitry. This sine-shaping circuit can be of either the breakpoint or the logarithmic amplifier type. After sine shaping, the signal can be applied to a four-quadrant analog multiplier on the IC chip, in order to provide amplitude modulation, output dc level shift and amplitude control capability. Finally, the output is amplified through an on-chip buffer amplifier to provide the desired output impedance level and load–drive capability.

PART II: INTEGRATED-CIRCUIT TIMERS

11.8.   FUNDAMENTALS OF INTEGRATED-CIRCUIT TIMERS

Monolithic timing circuits, or timers, find a wide variety of applications in both linear and digital signal processing. In a large number of industrial control or test sequencing applications, these circuits provide a direct and economical replacement for mechanical or electromechanical timing devices.

Monolithic timers generate precise timing pulses, or time delays, whose length or repetition rate is determined by an external timing resistor R and a timing capacitor C. The timing interval is proportional to the external RC product, and can be varied from microseconds to minutes, or hours by the choice of the external R and C. IC timers can be classified into two categories, based on their principle of operation:

1. *One-Shot or Single-Cycle Timers.*   These timer circuits operate by charging an external capacitor with a current set by an external resistor. Upon triggering, the charging cycle happens only once during the timing interval. The total timing interval T is the time duration necessary for the voltage across the capacitor to reach a threshold value.

2. *Multiple-Cycle Timers or Timer/Counters.* These timer circuits charge and discharge the external timing capacitor not once, but a multiple number of times during the timing interval. The number of times the capacitor is charged and discharged is set by means of a preset count N, stored in a binary counter included on the chip. Thus, the resulting time interval is proportional to N times the external RC product.

Timer circuits are very closely related to relaxation–oscillator circuits. The one-shot timer is essentially a relaxation oscillator which has only one astable state (i.e., once triggered, it operates through only one half-cycle of oscillation). The timer/counter circuit uses an R–C relaxation oscillator as the internal timing component.

Both the one-shot and the timer/counter circuits can operate either in their monostable mode, or in a free-running (i.e., self-triggering) mode. In this latter mode of operation, they function as relaxation oscillators. In addition to generating precise time delays and timing signals, the IC timers can also be used for low-frequency clock generation, sequential timing, and pulse-position and pulse-width modulation.

11.9. ONE-SHOT TIMERS

One-shot or single-cycle timers operate by charging a timing capacitor through an external resistor or a current source. The simplest form of the one-shot-type timer is the exponential-ramp generator circuit shown in Figure 11.43. Normally, all the components, except the R_1 and C_1 shown, are internal to the integrated circuit; and the switch S_1 is a grounded-emitter *npn* transistor included in the IC chip.

The operation of the circuit can be briefly explained as follows. In the rest, or reset, condition, the switch S_1 is closed and the voltage across the capacitor is clamped to ground. The timing cycle is initiated by applying an external trigger pulse to set the flip-flop and to open the switch S_1 across the timing capacitor. The voltage across the capacitor rises exponentially toward the supply voltage $+ V_{CC}$ with a time constant of R_1C_1. When this voltage level reaches an internally set threshold voltage V_{ref}, the voltage comparator changes state, resets the flip-flops, closes the switch S_1, and ends the timing cycle. The output is taken from either the Q or the $\overline{Q}$ terminal of the flip-flop and corresponds to a timing pulse of duration T_1, where

$$T_1 = R_1 C_1 \ln\left(\frac{V_{CC}}{V_{CC} - V_{ref}}\right) \qquad (11.64)$$

Normally, the internal threshold voltage V_{ref} is generated from the supply voltage by means of a resistor divider, as shown in Figure 11.43. Then V_{ref} is directly proportional to the supply voltage as

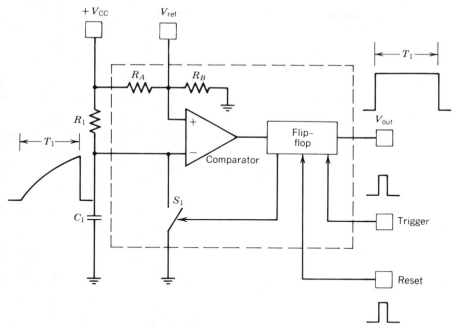

FIGURE 11.43.   Exponential-ramp-type one-shot timer.

$$V_{ref} = V_{CC} \frac{R_B}{R_A + R_B} \tag{11.65}$$

Substituting Eq. (11.65) into Eq. (11.64), the basic timing equation becomes independent of the supply voltage

$$T_1 = R_1 C_1 \ln\left(1 + \frac{R_B}{R_A}\right) \tag{11.66}$$

The resistors R_A and R_B are on the monolithic chip, and their ratio is internally set by the design of the integrated circuit. Their ratio is normally accurate to well within $\pm 1\%$ and is insensitive to temperature. Thus, virtually all the accuracy and stability of the timing interval is primarily determined by the external $R_1 C_1$ product.

An alternate approach to the design of one-shot timers is the linear-ramp generator circuit shown in Figure 11.44. This circuit operates on the same principle as the basic exponential-ramp timer, except that the timing capacitor C_1 is now charged with a constant current I_1, which is in turn set by an external resistor R_1. Once triggered, the circuit generates a linear ramp voltage across C_1 with a constant slope of I_1/C_1. The total timing interval T_1 is given as

$$T_1 = V_{ref} \frac{C_1}{I_1} \tag{11.67}$$

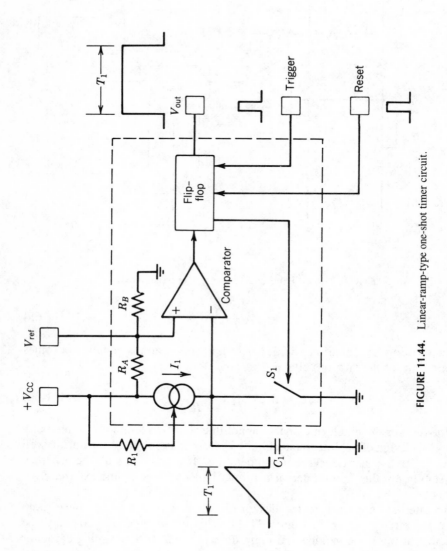

FIGURE 11.44. Linear-ramp-type one-shot timer circuit.

Both V_{ref} and current I_1 are set to be proportional to $+V_{CC}$, and I_1 is also inversely proportional to the external resistor R_1. Under these conditions, the effect of supply voltage variations will cancel, and the basic timing equation for the linear-ramp-type timer becomes

$$T_1 = \alpha R_1 C_1 \tag{11.68}$$

where α is a constant of proportionality set by the internal resistor–divider ratio and by design of the current source I_1.

The exponential-ramp-type timing circuit of Figure 11.43 is inherently simpler and more accurate than the linear-ramp-type timer, since the latter involves the conversion of a voltage drop across a resistor into a current. However, the linear-ramp timer has the advantage of providing a linear voltage ramp across the capacitor C_1, which is proportional to the elapsed time during the timing cycle and can be used as a linear sweep or time-base signal for oscilloscope or $X–Y$ recorder displays.

Normally, the internal threshold reference V_{ref} of one-shot integrated circuits is brought out to a package terminal to permit the user to modulate the timing interval by means of an external control voltage. This feature can also be used for generating pulse-width-modulated or pulse-position-modulated signals.

The NE-555 Timer Integrated Circuit

The versatility of the one-shot-type IC timer has made it one of the basic building blocks of linear or digital system design. Today, a wide range of one-shot IC timer circuits are available.[20] Among these, the circuit which has gained the widest acceptance and has become an industry standard is the NE-555 timer circuit, introduced by Signetics Corporation in 1972. The popularity of the NE-555 timer stems both from its versatility as well as its low cost.

The NE-555 is an exponential-ramp-type timer. Its functional block diagram is shown in Figure 11.45. In addition to the basic blocks necessary for the one-shot timer operation, the circuit also contains an additional comparator (comparator 2 of Fig. 11.45) which is useful for interconnecting the circuit as a relaxation oscillator, similar to that shown in Figure 11.17.

With reference to Figure 11.45, the operation of the circuit can be briefly explained as follows. When the circuit is at reset state, the discharge transistor Q_1 is fully on and the voltage across Q_1 is clamped to ground, the output is at a low logic state. On the leading edge of a negative-going trigger, comparator 2 changes state and sets the flip-flop, turns Q_1 off, and brings the output to a high state. Then the voltage across C_1 rises exponentially toward $+V_{CC}$ with the time constant of $R_1 C_1$. When this exponential ramp voltage reaches the internal threshold voltage V_{ref}, comparator 1 changes state and ends the timing cycle by resetting the flip-flop, which causes the output to go to a low state and Q_1 to be turned on. The result is a positive-going output pulse T_1, which is of the form given by Eq. (11.66). Since, in this case, the bias string is made up of three equal

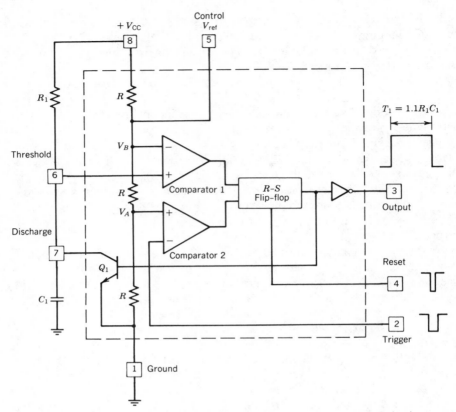

FIGURE 11.45. Functional block diagram of NE-555 timer integrated circuit.

resistors of value R, Eq. (11.66) becomes

$$T_1 = R_1 C_1 \ln(3.0) \approx 1.1 R_1 C_1 \tag{11.69}$$

Figure 11.46 shows the complete circuit schematic of the NE-555 timer. Comparator 1 is made up of the *npn* differential gain stage comprised of Q_1 through Q_4, with *pnp* active loads. Comparator 2 is a *pnp* differential stage made up of Q_{10} through Q_{13}. The R–S flip-flop section is formed by Q_{15} through Q_{17}. The output of the flip-flop drives both the discharge transistor Q_{14} and the high-current output inverter stage. The output is a so-called totem-pole-type push–pull circuit, commonly used in TTL logic. The output can both sink and source 100 mA of load current. However, the push–pull nature of the output circuit can create undesirable current spikes during switching.[20]

Figure 11.47 shows the basic external connection necessary for one-shot (monostable) and free-running (astable) operation in terms of the 8-pin circuit package. Since the output can both source and sink current, the load resistance R_L can be connected to either V_{CC} or ground. The internal reference terminal (pin 5) is normally bypassed to ground by an external capacitor when not in use.

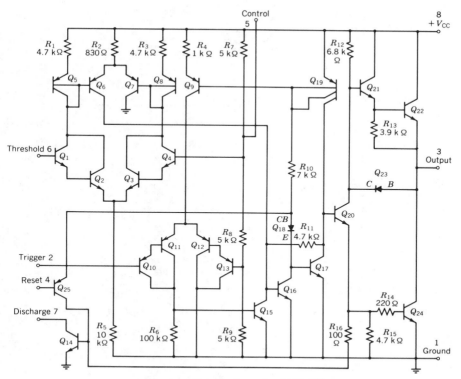

FIGURE 11.46. Circuit schematic of NE-555 timer.

In the free-running mode, the trigger input is connected to the discharge terminal, and an additional timing resistor R_2 is added into the circuit, as shown in Figure 11.47. In this manner, the circuit oscillates as an astable circuit by charging C_1 from the lower threshold level (V_A of Fig. 11.45) with a time constant τ_1,

$$\tau_1 = (R_1 + R_2)C_1 \qquad (11.70)$$

to the upper threshold V_B. Then it discharges it back to the lower threshold with a time constant τ_2, where

$$\tau_2 = R_2 C_1 \qquad (11.71)$$

In this manner, the voltage across C_1 oscillates between the two internal threshold levels set at $\frac{1}{3}V_{CC}$ and $\frac{2}{3}V_{CC}$. During astable operation, the time t_1 necessary for a charging cycle is

$$t_1 = (R_1 + R_2)\ln(2) \qquad (11.72)$$

Similarly, the time t_2 for the discharge cycle is

$$t_2 = R_2 C_1 \ln(2) \qquad (11.73)$$

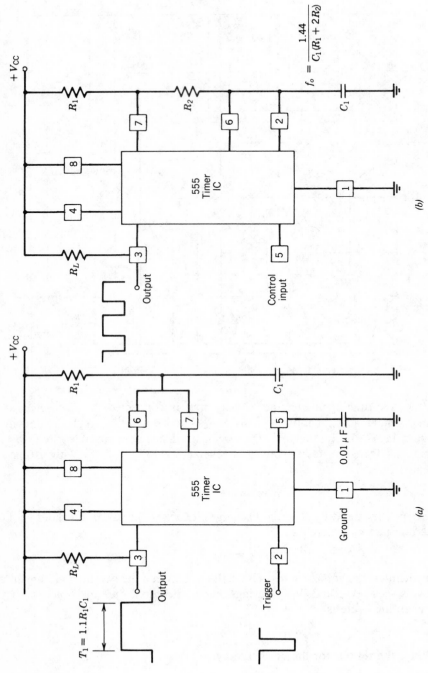

$$f_o = \frac{1.44}{C_1(R_1 + 2R_2)}$$

(b)

$$T_1 = 1.1R_1C_1$$

0.01 μF

(a)

FIGURE 11.47. External circuit connections for monostable and astable operation of NE-555 timer: (a) One-shot operation; (b) astable (free-running) operation.

606

Thus, the frequency of oscillation f_o is given as

$$f_o = \frac{1}{t_1 + t_2} = \frac{1}{C_1(R_1 + 2R_2)\ln 2} \approx \frac{1.44}{C_1(R_1 + 2R_2)} \qquad (11.74)$$

During astable operation, the duty cycle of the output is given as

$$\text{duty cycle} = \frac{t_1}{t_1 + t_2} = \frac{R_1 + R_2}{R_1 + 2R_2} \qquad (11.75)$$

Note that the duty cycle is always >50%, and approaches a square waves as $R_2 \gg R_1$.

The NE-555 timer is designed to operate with a single supply voltage in the range of +4.5–+18 V, with a typical timing error of ≤0.5% and a temperature drift of ±50 ppm/°C in its monostable mode. The accuracy and temperature drift errors are approximately a factor of 3 higher for astable operation. This is because in astable operation, both the upper and lower threshold comparators contribute to the frequency drifts, and the overall voltage swing across C_1 is reduced to $\frac{1}{3}V_{CC}$, rather than $\frac{2}{3}V_{CC}$.

The versatility and the accurate performance characteristics of the NE-555 timer have made it an industry standard in a very broad range of timer applications. As a result, a wide choice of similar timer IC products have been developed as well as the dual, quad, and the low-power versions of the NE-555 timer.

CMOS Timers

The basic timing functions can be implemented using CMOS technology, instead of bipolar. Because of the very-low-power operating capabilities of CMOS circuits, CMOS timers are particularly suited for applications in portable battery-powered instrumentation or remote telemetry and control equipment.

Figure 11.48 shows the functional block diagram and the equivalent circuit schematic of a CMOS one-shot timer (the Intersil ICM-7555), which is designed to be functionally compatible with the NE-555 timer in virtually all timing applications.[21] As shown, the basic circuit architecture is very similar to the original bipolar design (see Figs. 11.45 and 11.47). The internal bias string is made up of three 100 kΩ resistors which are fabricated using the p-well diffusion associated with the CMOS structure (see Section 2.5). The two comparator sections are made up of n-channel and p-channel differential gain stages, with complementary CMOS current–mirror loads. The discharge transistor Q_{16} and the complementary output transistors Q_{14} and Q_{15} are large-area devices, designed to handle higher current levels than the internal small-signal devices.

In the circuit, the npn emitter follower Q_{12} is obtained using the n^+-type source–drain diffusions for n-channel MOS transistors as the emitter and the p^+-type source–drain diffusion for p-channel transistors as the base region, with the n-type substrate as the collector. As described in Chapter 6, such a transistor structure, with the collector tied to the V^+ terminal, is readily compatible with CMOS fabrication and does not require additional processing steps. It is included in the circuit of Figure 11.48, in order to provide active pull-up to the

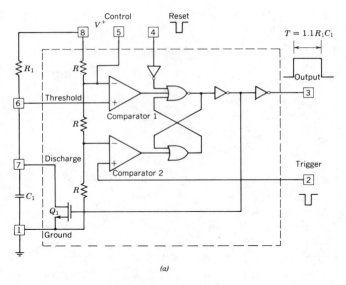

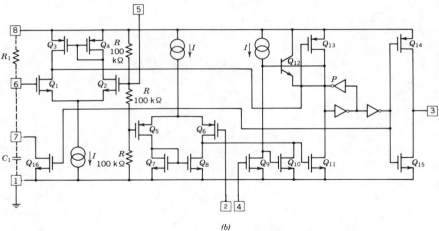

FIGURE 11.48. CMOS functional equivalent of NE-555 timer: (*a*) Functional block diagram; (*b*) Equivalent circuit.

flip-flop input terminal (node P) to assure rapid and reliable resetting of the circuit once Q_4 is turned off.

The basic timing equations and reset and trigger levels of the circuit are the same as for the bipolar NE-555 design. However, the CMOS circuit can operate with supply voltages as low as $+2$ V and dissipates about one-fifteenth as much quiescent power as the conventional NE-555 design.

The output current drive of the ICM-7555 CMOS timer is significantly lower than that of the conventional NE-555 timer, due to the limited current-handling capabilities of CMOS transistors. For operation with $V^+ = 5$ V, the output can source 1 mA, and can sink approximately 5 mA with TTL compatible output levels.

The worst-case initial accuracy and temperature drift characteristics of the

CMOS 555 timer are approximately a factor of 2 poorer than those of the bipolar design, due to higher offset voltages and offset drifts associated with the CMOS comparator stages.

Practical Limitations of One-Shot Timers

The accurate timing intervals which can be obtained from commercially available one-shot-type timer integrated circuits are limited to the range of several microseconds to several minutes. For generating very short timing pulses (in the few microsecond range), the internal time delays associated with the switching speeds of the comparator, the flip-flop, and the discharge transistor may contribute additional timing errors. Similarly, for long time delays (in the several minute range), which require large values of R_1 and C_1, the input bias current of the comparator and the leakage currents associated with the timing capacitor or the internal discharge transistor may limit the timing accuracy of the circuit.

In general, for those timing applications requiring time delays in excess of several minutes, the multiple-cycle or timer/counter type timer circuits provide a more economical and practical solution than the one-shot-type IC timers.

11.10. TIMER/COUNTER CIRCUITS

The timer/counter or multiple-cycle timing circuits use a combination of a time-base oscillator and a binary counter to generate the desired time delay. Figure 11.49 shows a simplified block diagram of a timer/counter circuit which is made up of three basic blocks: (1) a time-base oscillator, (2) a binary counter, and (3) a control flip-flop.

With reference to the simplified block diagram of Figure 11.49, the principle of operation of a timer/counter can be explained as follows. When the circuit is at rest, or in the reset condition, the time-base oscillator is disabled and the counter is reset to zero. Once the circuit is triggered, the time-base oscillator is activated and produces a series of timing pulses whose repetition rate is proportional to the external timing resistor R_1 and the capacitor C_1. These timing pulses are then counted by the binary counter, and when a preprogrammed count is reached, the binary counter resets the control flip-flop, stops the oscillator, and completes the timing cycle. This results in an output timing pulse of duration

$$T_o = NT_1 \tag{11.76}$$

where N is the preprogrammed count in the counter, and T_1 is the period of the time-base oscillator. Normally, the time-base oscillator is an R–C relaxation oscillator with its period T_1 proportional to the external R–C setting as

$$T_1 = \alpha_1 R_1 C_1 \tag{11.77}$$

where α_1 is a constant of proportionality. This results in a total time delay T_o given as

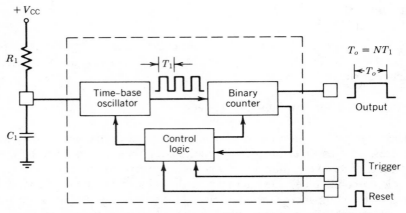

FIGURE 11.49. Simplified block diagram of timer/counter integrated circuit.

$$T_o = \alpha_1 N R_1 C_1 \qquad (11.78)$$

As implied by Eq. (11.78), the major advantage of a timer/counter circuit is its ability to provide N times longer time delays than would be available from a one-shot-type timer with the same external R–C setting.

Time-Base Oscillator

The time-base oscillator is the most important section of the timer/counter circuit since the accuracy of the time interval is determined primarily by the accuracy and stability of the oscillator section. The basic requirements for a time-base oscillator are frequency accuracy, stability, and ease of external setting (i.e., minimum external components). Another important requirement is predictable start-up conditions. In other words, the oscillations should start immediately after the application of the trigger signal, with no start-up transients. This means that the oscillator must be biased, at the reset condition, to start at a given astable state. In practical designs, all of these conditions can be met by using a simple R–C relaxation oscillator of the type shown in Figures 11.15 and 11.16. Normally, the second timing resistor R_2 of Figure 11.15 would be set equal to zero to reduce the external component count, and the internal threshold voltages would be set as given by Eq. (11.21) to produce an output pulse train whose period T_1 is exactly equal to the $R_1 C_1$ product. Such a design approach sets $\alpha = 1$ in Eq. (11.78) and simplifies the external R–C setting.

Binary Counter Section

The binary counter section of a timer/counter circuit is basically a digital design, comprised of a chain of D-type flip-flops. The counter section can be either a fixed-count counter or can be made externally programmable to obtain different settings for the count N in Eq. (11.78). The use of a programmable counter allows one to obtain a multiplicity of time delays from a single R–C setting.

The design of binary counter circuits is well covered in digital circuit design books and will not be repeated here. However, one particular circuit configuration, which is readily compatible with analog circuitry, and commonly used in many frequency divider designs, is worth mentioning. This is the basic ripple counter, or the binary countdown stage shown in Figure 11.50. With reference to the figure, the operation of this countdown stage can be explained as follows.

The circuit is comprised of two R–S flip-flops of the cross-coupled transistors Q_2, Q_3 and Q_4, Q_5. The Q_4, Q_5 pair is referred to as the *master* flip-flop and the

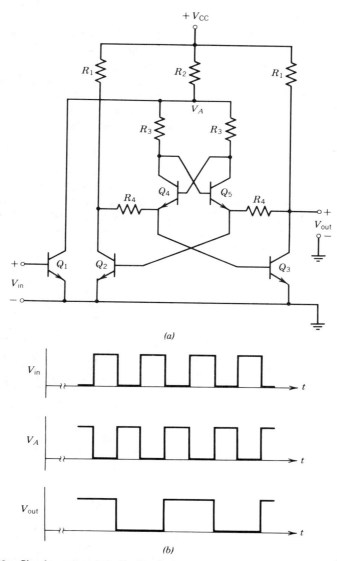

(a)

(b)

FIGURE 11.50. Simple master–slave flip-flop frequency counter stage. (a) Circuit schematic; (b) input and output waveforms.

Q_2, Q_3 as the *slave* flip-flop. If the input makes a low-to-high transition, the collector voltage V_A of Q_1 makes a high-to-low transition and shuts off the master flip-flop, whereas the state of the slave flip-flop remains unchanged. Assume for the moment that the initial state of the circuit is such that Q_3 is off and Q_2 is on and the output V_{out} is high. This state will not be changed by the input making a low-to-high transition.

If the input makes a high-to-low transition (i.e., at the falling edge of the input transition), V_A would go high and cause either Q_4 or Q_5 to turn on. Since the emitter of Q_4 is connected to the base of Q_3, which is at the off stage, Q_4 turns on before Q_5 and the cross coupling between the two assures that Q_5 stays off as long as Q_4 is on. When Q_4 is conducting, Q_3 is turned on and Q_2 is turned off, and the slave flip-flop changes state. In this manner, the cycle repeats itself and the slave flip-flop changes state at every falling edge of V_{in}, resulting in the waveforms shown in Figure 11.50b. Consequently, the output waveform is at one-half the frequency of the input, and the circuit functions as a frequency divider. Many such circuits can be cascaded by simply connecting the output of one into the input of the other to result in long counter chains.

Design Example

At present, a variety of timer/counter integrated circuits are available from several different manufacturers. One of the first timer/counter circuits to be introduced, which has become an industry standard, is the XR-2240 programmable timer/counter introduced by Exar Integrated Systems in 1975. The simplified circuit diagram of the XR-2240 is shown in Figure 11.51. The time-

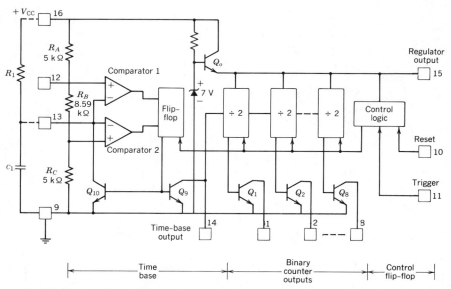

FIGURE 11.51. Simplified circuit diagram of XR-2240 timer/counter integrated circuit.

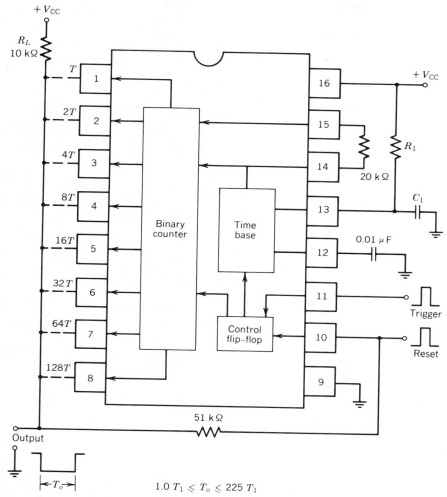

FIGURE 11.52. External connection diagram for generating programmable time delays using XR-2240 timer/counter.

base oscillator section has the same circuit configuration as the two-comparator R–C relaxation oscillator shown in Figure 11.17, with $R_2 = 0$ and with Q_{10} as the discharge transistor.

The internal bias resistors R_A and R_B are chosen such that

$$\frac{R_A + R_B}{R_A} = e = 2.718\ldots \tag{11.79}$$

which sets the oscillator period exactly equal to $1.0\ R_1 C_1$, as indicated by the generalized expressions in Eqs. (11.20) through (11.22). The detailed circuit configuration of the oscillator section is very similar to that shown in Figure

11.18, with the second timing resistor R_2 set to zero. When triggered, the oscillator section produces narrow timing pulses with a period $T_1 = R_1 C_1$. These timing pulses are then counted down by the binary counter section.

The counter section of the XR-2240 is made up of an eight-stage ripple counter, and each counter output is brought out to an open-collector output stage. The basic circuit configuration for the counter stage is similar to that shown in Figure 11.50. The counter section makes extensive use of high-value pinched resistors (see Figs. 3.25 and 3.26) to reduce the chip area required for the counters, as well as to minimize the current drain. In order to avoid voltage breakdown problems associated with a pinched resistor structure, a simple internal Zener regulator is used to bias the counter section.

Figure 11.52 indicates the external circuit connections for generating programmable time delays using the XR-2240. The programming of the total time delay is done by selectively connecting any combination of the eight counter outputs to a common pull-up resistor R_L. Since the counter outputs are open-collector type, connecting them to a common pull-up resistor results in a "wired–or" connection, where the combined output will be at a low state as long as any one of the outputs in the combination is at the low state. For example, if only pin 6 is connected to the output and the rest left open, the total duration of the timing cycle T_o would be $32T_1$. Similarly, if pins 1, 5, and 6 were short-circuited to the output bus, the total time delay would be $T_o = (1 + 16 + 32)T_1 = 49T_1$. In this manner, with the proper choice of counter terminals connected to the output bus, one can program the timing cycle to be in the range of $1T_1 \leq T_o \leq 255T_1$.

Advantages of Timer/Counter Circuits

The combination of a stable time-base oscillator and a programmable binary counter on the same IC chip offers some unique application and performance features. Some of these are:

1. Generating long delays with small capacitors. For a given time-delay setting, the timer/counter would require a timing capacitor C that is N times smaller than that needed for the one-shot-type timer, where N is the count programmed into the binary counter. Since large-value, low-leakage capacitors are quite expensive, this technique may provide substantial cost savings for generating long time delays in excess of several minutes.

2. Generating multiple delays from the same RC setting. By using a programmable binary counter, whose total count can be programmed between a minimum count of 1 and a maximum count of N, one can obtain N different time intervals from the same external RC setting.

3. Generating ultralong delays by cascading. When cascading two timer/counters, one cascades the counter stages of both timers. Since the

second timer/counter further divides down the counter output of the first timer, the total available count is increased geometrically, rather than arithmetically. For example, if one timer/counter gives a time delay of NRC, two such timer/counters, cascaded, will produce a time delay of N^2RC, where N is the count setting of the binary counter. Thus, a cascade of two timer/counter circuits, each with an 8-bit binary counter, can produce a time delay in excess of $32,000RC$. In this manner, accurate time delays in the range of hours, days, months, can be easily accomplished with relatively small values of external timing components.

PART III: FREQUENCY-TO-VOLTAGE AND VOLTAGE-TO-FREQUENCY CONVERTERS

In many telemetry, remote control, and remote data gathering applications it is advantageous to convert the analog information into a frequency, which can be transmitted and decoded, or recorded, more accurately than an analog voltage or current signal. This is particularly true if the signal transmission path is quite long and noisy. Another major category of applications for voltage-to-frequency (V/F) and frequency-to-voltage (F/V) converters is in designing signal isolators, where an analog signal can first be converted to a frequency, then coupled through an optoisolator, and converted back to voltage through an F/V converter to achieve nearly ideal isolation between the system input and output.

A number of circuit techniques have been developed for converting analog data into frequency and vice versa. In this section, some of these circuit techniques will be examined. The V/F and F/V converters generically belong to the class of circuits called *data converters*, where their function is to convert one form of data to another. However, from the point of view of circuit design and architecture, they are closely related to oscillators and one-shot-type timers. Therefore, it is appropriate to cover them in this chapter along with the IC oscillators and timers.

11.11. VOLTAGE-TO-FREQUENCY CONVERTERS

V/F converters are a class of circuits designed to produce an output frequency directly proportional to an analog voltage input signal. These circuits are basically *encoder* circuits, whose function is to encode an analog signal into a stable frequency, which can then be transmitted over long distances or measured and recorded by digital means. Compared to the ordinary voltage-controlled oscillator circuits, V/F converters have some very stringent performance requirements depending on the dynamic range, precision, and resolution requirements of the analog data. Some of the performance requirements commonly encountered in most data encoding applications are:

1. *Dynamic Range.*   A V/F converter is required to have at least 60 dB (i.e., 1000 : 1) dynamic range in terms of its input voltage range. In most designs 80 dB dynamic range is typical, covering input voltages from 1 mV to 10 V.

2. *Frequency Range.*   This defines the range of output frequencies corresponding to the analog input dynamic range. For an 80-dB dynamic range, the output frequency is typically in the range of 1 Hz to 10 kHz or 10 Hz to 100 kHz.

3. *Linearity.*   The V/F conversion characteristics are required to be extremely linear, with a nonlinearity error $\ll$ 1% of full scale. In precision designs, these linearity errors must be kept to $<0.1\%$.

4. *Scale-Factor Accuracy.*   Normally, the scale factor or gain is externally adjusted for a given conversion gain setting. In most applications involving operation over the 1 Hz–10 kHz range, the scale factor is normally set a 1 V/kHz.

5. *Scale-Factor Stability.*   At any given gain setting, the scale factor (and the output frequency) must be stable with temperature and power supply changes. Typical stability requirements are on the order of ≤ 100 ppm/°C drift with temperature and $\leq 0.1\%$/V change with power supply.

6. *Output Waveform.*   The output waveform of the V/F converter is, in general, not a critical factor as long as its levels are compatible with logic signals. Most V/F converters produce constant-width output pulses whose repetition rate, that is, frequency, is proportional to an input voltage. However, such an output can be easily converted to a symmetrical square wave at one-half the normal frequency by passing it through a frequency countdown stage similar to that shown in Figure 11.50.

Today, a variety of monolithic V/F converter products are available which meet and exceed the dynamic range, linearity, and stability requirements mentioned above. The circuit techniques used in these IC designs fall into two categories: (1) ultrawide–sweep multivibrator circuits, and (2) charge-balancing oscillators.

Ultrawide-Sweep Multivibrators

This technique uses the basic current-controlled emitter-coupled multivibrator circuits, such as those shown in Figures 11.29 and 11.31, in conjunction with a voltage-to-current converter stage. Both of these circuits inherently offer linear frequency sweep capability whose dynamic range is primarily limited by the current-handling range of the transistors or diodes in the circuit. With careful design this range can be extended to in excess of 80 dB.

Figure 11.53 shows the simplified block diagram of a V/F converter using the diode-clamped precision multivibrator circuit of Figure 11.31, along with an active voltage-to-current converter.[12] As described in Section 11.4, the frequency of the oscillator can be expressed as

$$f_1 = \frac{I_1}{4C_1V_R} \tag{11.80}$$

where V_R is the voltage level of the clamping voltage generated by the voltage reference circuit internal to the IC chip.

In the block diagram of Figure 11.53, I_1 is derived from the voltage drop across the external resistor R_1 as

$$I_1 = \frac{I_T}{4} = \frac{V_{in}}{4R_1} \tag{11.81}$$

Then the frequency of oscillation becomes

$$f_1 = \frac{V_{in}}{16V_RC_1R_1} = K_1V_{in} \tag{11.82}$$

and the gain constant K_1 is set by the external resistor R_1, and its stability is primarily determined by the stability of the internal voltage reference V_R.

The basic circuit configuration of Figure 11.53, with some modifications, has

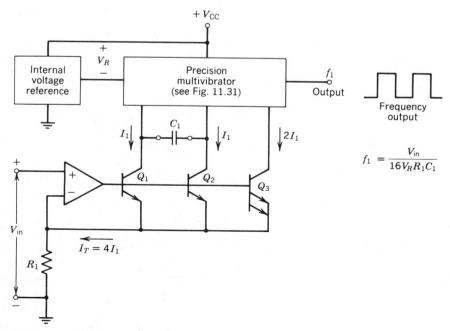

FIGURE 11.53. Block diagram of V/F converter comprised of wide-sweep multivibrator and voltage-to-current converter.

been developed as a successful commercial product which provides 100 dB of dynamic range, with a maximum nonlinearity error of 0.07% and a temperature coefficient of ≤ 50 ppm/°C. [23]

Charge-Balancing V/F Converters

Charge-balancing V/F converters operate on a dynamic charge–balance principle by balancing the charge flow in and out of an integrating capacitor. Figure 11.54 shows the simplified block diagram of a charge-balancing V/F converter. The circuit is comprised of four sections: (1) a precision one-shot timer which produces fixed-duration timing pulses, (2) a switched current source which is turned on only for the duration of the one-shot timing pulse, (3) an integrating R–C network, and (4) a voltage comparator which compares the voltage V_1 across the R–C network with the analog input voltage V_{in} and triggers the one-shot timer if $V_{in} > V_1$.

The operation of the entire circuit can be briefly described as follows. When a positive input voltage V_{in} is applied to the system, the comparator changes state and triggers the one-shot circuit. The output of the one-shot circuit is connected

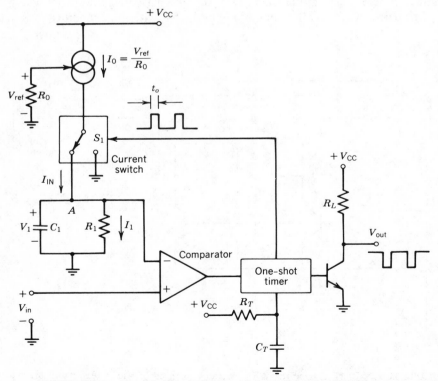

FIGURE 11.54. Simplified block diagram of charge-balancing V/F converter.

to the frequency output of the circuit and also activates a switched current source I_0. When the current source is activated, it injects an amount of charge Q_0, into the R–C integrator, made up of R_1 and C_1, with

$$Q_0 = I_0 t_o \qquad (11.83)$$

where t_o is the one-shot timer output duration. If this charge has not increased the voltage V_1 across C_1 such that $V_1 > V_{in}$, then the comparator triggers the one-shot, again and injects another packet of charge. This process continues until the $V_1 \geq V_{in}$ condition is reached. When this condition is achieved, the current source I_0 remains off until V_1 decays below V_{in} to cause the comparator to trigger the one-shot circuit once more. Thus, at steady state, the one-shot circuit will be triggered at such a rate that the current source I_0 dumps a charge into the capacitor C_1 just at a sufficient rate to balance the charge decaying out of the capacitor, such that the $V_1 \approx V_{in}$ relationship is maintained.

When a packet of charge Q_0 given by Eq. (11.83) is injected into the capacitor C_1, the net voltage across the capacitor changes by an amount ΔV_1 where

$$\Delta V_1 = \frac{Q_0}{C_1} = \frac{I_0 t_o}{C_1} \qquad (11.84)$$

Assuming that C_1 is chosen sufficiently large such that $\Delta V_1 \ll V_1$, then the discharge rate of C_1 at steady state is given by the current I_1 through R_1 as

$$I_1 = \frac{V_1}{R_1} \qquad (11.85)$$

At steady state, this current is balanced by an equal amount of charge flow from the switched current source which delivers Q_0 packets of charge at a frequency f_1, that is,

$$I_{IN} = Q_0 f_1 = I_0 t_o f_1 \qquad (11.86)$$

where f_1 is the steady-state frequency for triggering the one-shot circuit. At steady state, $V_1 = V_{in}$ and $I_1 = I_{IN}$. Thus, equating Eqs. (11.85) and (11.86), one obtains

$$f_1 = \frac{V_{in}}{I_0 t_o R_1} \qquad (11.87)$$

which results in a frequency proportional to V_{in}.

In a practical monolithic design, the one-shot timer section would be designed with a circuit configuration of the form shown in Figure 11.43, with the output pulse duration t_o determined by the choice of timing component R_T and C_T as

$$t_o = K_T R_T C_T \qquad (11.88)$$

where K_T is a constant of proportionality determined by the choice of internal threshold levels within the one-shot circuit as described in Section 11.9. Similarly, the switched current-source current I_0, is set by a resistor R_0 which can be connected to ground or to $+V_{CC}$ such that

$$I_0 = \frac{V_{\text{ref}}}{R_0} \tag{11.89}$$

where V_{ref} is an internal voltage reference level imposed across the current-setting resistor R_0.

Substituting Eqs. (11.88) and (11.89) into Eq. (11.87), one obtains

$$f_1 = V_{\text{in}} \frac{R_0}{R_T} \frac{1}{K_T V_{\text{ref}} C_T R_1} \tag{11.90}$$

In Eq. (11.90), resistors R_0 and R_T appear as a ratio and can be left internal to the IC chip, relying on the very close matching and temperature tracking of IC components. Thus, the circuit requires only two external precision components, C_T and R_1. Normally, R_1 is chosen to set the gain or scale factor.

The internal voltage reference V_{ref} and the one-shot constant of proportionality K_T, along with the R_0/R_T ratio, can be maintained very stable with temperature. Thus, the temperature drift of the entire V/F system can be maintained at <100 ppm/°C range.

Although the basic V/F converter system shown in Figure 11.54 is quite useful in its present form, it has two drawbacks:

1. *Slow Response Time.* For a large step change in input voltage, the capacitor C_1 has to be charged or discharged over a similar voltage step. With a passive integrator network, as shown in Figure 11.54, this may take 50–100 msec for a zero to full-scale step input.

2. *Nonlinearity Due to Finite Output Conductance of Switched Current Source.* This finite output conductance of I_0 causes the value of Q_0 to change with changing voltage levels V_1 at node A.

Both of these problems can be avoided by using an active integrator circuit, placing C_1 in feedback around an operational amplifier as shown in Figure 11.55. This reduces the response time to a few microseconds and virtually eliminates the nonlinearity due to the output conductance of I_0, since node A is now clamped to a virtual ground level. In this manner, the nonlinearity error can be readily reduced to the ≤0.1% level.

The frequency expression for the circuit of Figure 11.55 can be derived by equating I_1 to the net charge flow due to I_{IN} [see Eqs. (11.85) and (11.86)]. The resulting expression for the output frequency is

$$f_1 = -V_{\text{in}} \frac{R_0}{R_T} \frac{1}{K_T V_{\text{ref}} C_T R_1} \tag{11.91}$$

which is identical to Eq. (11.90), except for the negative polarity of the input voltage.

One fundamental drawback of the circuit shown in Figure 11.55 is that the

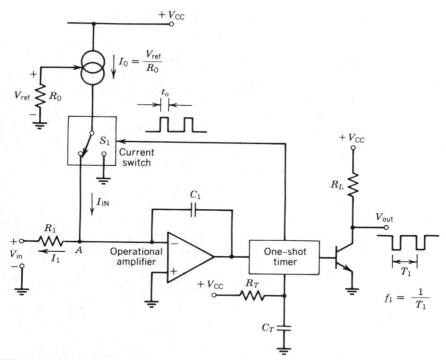

FIGURE 11.55. Functional block diagram of precision V/F converter using charge-balancing technique.

steady-state current I_1, which can be drawn from node A, is unidirectional, and the circuit can operate only with negative values of V_{in}. However, this is a relatively minor inconvenience which can be avoided by putting another operational amplifier gain stage between the input and the scaling resistor R_1.

The frequency accuracy of a V/F converter is also affected by the offset voltages of the comparator or the operational amplifier sections in Figures 11.54 and 11.55. Since the offset voltage V_{OS} directly adds or subtracts from the input voltage, it results in a frequency error Δf_1,

$$\Delta f_1 = \pm K_1 V_{OS} \qquad (11.92)$$

where K_1 is the conversion gain constant of the V/F converter, corresponding to the collection of terms in Eq. (11.90) or Eq. (11.91). Since the frequency error due to V_{OS} is a constant error term, it is particularly significant at low frequencies (i.e., at low-level inputs) in terms of absolute accuracy.

The high-frequency capability of charge-balancing oscillators is primarily determined by the accuracy of the one-shot section. For proper operation, the one-shot period t_o must be less than the minimum period T_{min} of the output frequency. In order to ensure that the one-shot section is fully discharged prior to retriggering, it is customary not to let the one-shot period exceed one-half of

$T_{\min}$. This puts an upper limit on the value of t_o as

$$t_o < \frac{T_{\min}}{2} = \frac{1}{2f_{\max}} \tag{11.93}$$

where $f_{\max}(= 1/T_{\min})$ is the maximum frequency of operation. For $f_{\max} = 10\,\text{kHz}$, as is the case with most V/F converters, the value of t_o is in the 50-μsec range. However, if $f_{\max}$ is extended to 100 kHz, the required value of t_o would be in the range of 5 μsec, which is harder to control accurately over temperature or supply voltage changes.

11.12. FREQUENCY-TO-VOLTAGE CONVERTERS

In many telemetry applications, as well as in opto-coupled isolator circuits, F/V converters find a wide range of applications as a counterpart to V/F converters. They can be used either to decode the frequency data from a V/F converter, or as tachometer circuits for motor speed control or rotation measurements. There are two basic types of F/V converters: (1) pulse-integrating converters, which integrate a fixed-pulse-width, variable-frequency output signal, and (2) phase-locked-loop type F/V converters, which lock on an input frequency and produce an error voltage proportional to the frequency deviations of the input signal. This latter subject will be covered in much greater detail in Chapter 12.

The pulse-integrating F/V converters utilize a precision one-shot timer which, when triggered by a high-level periodic input signal, produces output pulses of fixed amplitude and duration. These pulses can then be integrated to determine their average value over a given time interval, where this average value will be proportional to the repetition rate or the frequency of the output pulse train.

The charge-balancing V/F converter circuits can operate as pulse-integrating F/V converters by disconnecting the feedback loop between the integrator output voltage and the one-shot trigger input. The one-shot section is then triggered directly by the input frequency signal f_{in} and produces output pulses of duration t_o as shown in Figure 11.56. The output of the one-shot circuit is used to switch the current source I_0 on and off and produce the charge packets Q_0 which are supplied to the integrator circuit.

With reference to Figure 11.56a, at steady state the average currents going into and coming out of the R_1–C_1 integrator section are equal, that is,

$$\frac{V_{\text{out}}}{R_1} = Q_0 f_{\text{in}} = I_0\, t_o\, f_{\text{in}} \tag{11.94}$$

or, rearranging terms,

$$V_{\text{out}} = I_0 t_o R_1 f_{\text{in}} \tag{11.95}$$

Using the results of Eqs. (11.88) and (11.89), Eq. (11.95) can be rewritten in terms of circuit component choices and internal design parameters as

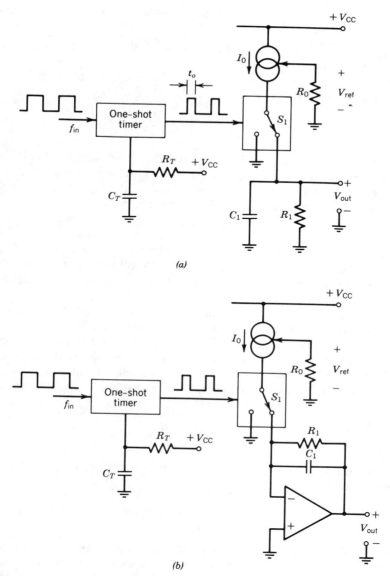

(a)

(b)

FIGURE 11.56. Pulse-integrating F/V converter circuits: (*a*) Basic circuit; (*b*) its improved version using active integrator.

$$V_{out} = f_{in} \frac{R_T}{R_0} V_{ref} K_T R_1 C_T \qquad (11.96)$$

which is the same as Eq. (11.90) now rewritten with f_{in} as the independent variable and V_{in} replaced by V_{out}.

For the charge-balancing F/V circuit to operate properly, it is necessary that the one-shot output pulse width t_o should be less than the period of maximum input frequency. Typically, t_o is chosen to be on the order of 50–80% of the minimum input period T_{min}.

The output voltage has a finite ac ripple riding on it due to the discharge of the holding capacitor C_1, while I_0 is off. The peak-to-peak ripple is

$$(\Delta V_{out})_{pp} = \frac{Q_0}{C_1} = \frac{I_0 t_o}{C_1} \tag{11.97}$$

where the magnitude of the ripple is assumed to be small compared to V_{out}. The ripple can be reduced by additional filtering or by increasing the value of C_1.

The output voltage responds to a step change in the input frequency with a time constant τ_1 associated with the integrator circuit, where

$$\tau_1 = R_1 C_1 \tag{11.98}$$

Thus, increasing C_1 to reduce the output ripple results in a corresponding increase in the response time of the circuit. The trade-off between the output ripple and the system response time is one of the fundamental design considerations in pulse-integrating F/V converters.

The simple F/V converter of Figure 11.56a has the two basic drawbacks discussed earlier, in conjunction with V/F operation, namely, (1) poor linearity due to finite output conductance of I_0, and (2) slow response time due to a passive integrator circuit. Both of these problems can be avoided by using an active integrator circuit, as shown in Figure 11.56b. The output voltage can be calculated by equating the steady-state (i.e., average) current flow in and out of the inverting node of the operational amplifier as

$$\frac{-V_{out}}{R_1} = Q_0 f_{in} = I_0 t_o f_{in} \tag{11.99}$$

or, rearranging terms and substituting from Eqs. (11.88) and (11.89), the output voltage can be written as

$$-V_{out} = f_{in} \frac{R_T}{R_0} V_{ref} K_T R_1 C_T \tag{11.100}$$

Because of the inverting properties of the operational amplifier in the active integrator section, the output voltage becomes more negative as the frequency is increased.

One practical consideration in the design of F/V converters is the input waveform. In order to ensure accurate triggering of the one-shot timer, the input should exhibit a high signal swing with sharp rising and falling edges. In practice, this is accomplished by placing a high-gain voltage comparator ahead of the one-shot trigger input to slice or square up the input signal.

Since the basic F/V circuits of Figure 11.56 are the open-loop versions of the charge-balancing oscillator circuits of Figures 11.54 and 11.55, their accuracy, linearity, and stability characteristics are essentially the same as those obtainable from their V/F counterparts.

REFERENCES

1. L. Strauss, *Wave Generation and Shaping*, 2nd ed., McGraw-Hill, New York, 1970, Chap. 16.

2. T. Young, *Linear Integrated Circuits*, Wiley, New York, 1981, Chap. 6.

3. J. Millman, *Microelectronics: Digital and Analog Circuits and Systems*, McGraw-Hill, New York, 1979, Chap. 17.

4. A. B. Grebene, "A Sinusoidal Voltage-Controlled Oscillator for Integrated Circuits," *IEEE Spectrum*, 79–82 (March 1969).

5. "Phase-Locked Loop Data Book," 2nd ed., Motorola Semiconductor Products, Inc., August 1973, Sec. 3.

6. L. Strauss, *Wave Generation and Shaping*, 2nd ed., McGraw-Hill, New York, 1970, Chap. 10.

7. A. B. Grebene, *Analog Integrated Circuit Design*, Van Nostrand Reinhold, New York, 1972, Chap. 9.

8. A. B. Grebene, "Monolithic Waveform Generation," *IEEE Spectrum*, 34–40 (April 1972).

9. A. B. Grebene, "The Monolithic Phase-Locked Loop—A Versatile Building Block," *IEEE Spectrum*, 38–49 (March 1971).

10. R. R. Cordell and W. G. Garrett, "A Highly Stable VCO for Applications in Monolithic Phase-Locked Loops," *IEEE J. Solid State Circuits* **SC-10**, 480–485 (December 1975).

11. B. Gilbert, "A Stable Second-Generation Phase-Locked Loop," *Digest of Tech. Papers*, IEEE Int. Solid-State Circuits Conf., pp. 16–18, (February 1972.)

12. B. Gilbert, "A Versatile Monolithic Voltage-to-Frequency Converter," *IEEE J. Solid-State Circuits* **SC-11**, 852–864 (December 1976).

13. "CD-4047A Low-Power Multivibrator," in *RCA COS/MOS Integrated Circuits Data Book*, RCA Corp., Camden, NJ, 1978, pp. 493–498.

14. D. K. Morgan and G. Steudel, "The RCA COS/MOS Phase-Locked Loop," App. Note ICAN-6101, RCA Corp., Somerville, NJ, October 1972.

15. "XR-215 Monolithic Phase-Locked Loop," in *Exar Phase-Locked Loop Data Book*, Exar Integrated Systems, Inc., Sunnyvale, California, 1981, pp. 20–27.

16. "ICL-8038 Waveform Generator," in *IC Data Book*, Intersil Corp., Cupertino, CA, July 1979, pp. 5–274.

17. R. G. Meyer, W. M. C. Sansen, S. Lui, and S. Peeters, "The Differential Pair as a Triangle-sine Wave Converter," *IEEE J. Solid-State Circuits (Correspondence)* **SC-11**, 418–420 (June 1976).

18. "XR-2206 Monolithic Function Generator," in *Exar Function Generator Data Book*, Exar Integrated Systems, Inc., Sunnyvale, CA, May 1978, pp. 14–19.

19. D. J. G. Janssen, J. Kaire, and P. Guetin, "An I^2L Circuit for Two-Tone Telephone Dialing," *IEEE J. Solid-State Circuits* **SC-12**, 238–242 (June 1977).

20. W. G. Jung, *IC Timer Cookbook*, Howard Sams and Co., Indianapolis, IN, 1977.

21. "ICM-7555 CMOS Timer," in *Intersil Data Book*, Intersil Corp., Cupertino, CA, 1980, pp. 6.140–6.144.

22. "XR-2240 Programmable Timer/Counter," in *Exar Timer Data Book*, Exar Integrated Systems, Sunnyvale, CA, 1981, pp. 42–49.

23. "AD-537 Voltage-to-Frequency Converter," in *Data Acquisition Products Catalogue*, Analog Devices, Wilmington, MA, 1978, pp. 475–480.

CHAPTER TWELVE

PHASE-LOCKED-LOOP CIRCUITS

The phase-locked loop (PLL) is a frequency-selective feedback system which can synchronize with a selected input signal and track the frequency changes associated with it. Its concept was first introduced in the 1930s. Since then it has been used in many communication, telemetry, and data-synchronization applications.[1,2]

Until the advent of monolithic IC technology, the PLL systems have been too costly and complex. Their applications have been primarily limited to precision measurements or signal-detection systems which require a high degree of noise immunity and very narrow bandwidths. The availability of a complete PLL system as a monolithic integrated circuit, in the late 1960s, has greatly changed this picture.[3] Since the IC PLL systems can be fabricated at a greatly reduced cost compared to discrete (nonintegrated) designs, the economic advantages of monolithic integration has opened up a much wider range of applications for the PLL techniques, particularly in the area of consumer and industrial equipment design. Today, monolithic PLL circuits have become the basic building blocks of FM detectors, stereo demodulators, tone decoders, frequency synthesizers, and television display systems.

Although the basic theory and the principle of operation of a PLL system are straightforward, its rigorous analysis is quite complex. This detailed analysis is well covered in the literature [2] and will not be repeated here. The purpose of this chapter is to examine some of the important characteristics of a PLL system as they relate to monolithic PLL IC designs. This chapter is comprised of two parts: the first part covers the fundamental characteristics, design equations, and applications of a PLL system, and the second part deals with the design and interconnection of various building blocks within a monolithic PLL. In this context, a practical PLL IC design example is presented to illustrate the application of the design equations and to optimize the circuit performance.

627

PART I: FUNDAMENTALS OF PHASE-LOCKED LOOPS

12.1.   PRINCIPLE OF OPERATION OF A PLL SYSTEM

The basic PLL system is comprised of three essential blocks: (1) a phase detector, (2) a loop filter, and (3) a voltage-controlled oscillator (VCO). These three blocks are interconnected to form a feedback system, as shown in Figure 12.1. With reference to the block diagram of Figure 12.1, the principle of operation of a PLL can be qualitatively described as follows.

The phase detector compares the phase of the periodic input signal $V_s(t)$ with the output frequency of the VCO and generates an error voltage $V_d(t)$. This voltage signal is then filtered by the loop filter and is applied to the control terminal of the VCO in the form of the error voltage $V_e(t)$ to control its frequency of oscillation.

Normally, with no input signal applied to the PLL, the filtered error voltage $V_e(t)$ in the feedback loop is equal to zero. This is known as the free-running condition of the PLL, where the VCO operates at a steady-state frequency $\omega_o = 2\pi f_o$, which is called the VCO free-running frequency. If a periodic input signal is applied to the PLL such that the input frequency $\omega_s = 2\pi f_s$ is sufficiently close to the VCO free-running frequency, the feedback nature of the PLL causes an error voltage to be generated, forcing the VCO to synchronize with the input frequency. When this happens, the PLL is said to be *locked* on the input signal frequency.

When the PLL is locked on the input signal, the VCO frequency is *identical* to the input frequency f_s, except for a finite phase difference ϕ_o. This net phase difference, or *phase error*, ϕ_o is necessary to generate the corrective error voltage $V_e(t)$ to shift the VCO frequency from f_o to f_s and thus maintain lock. If the input signal frequency varies slowly, the PLL can still stay locked and track the input signal by generating additional phase error, which would result in a necessary change in the filtered error voltage $V_e(t)$ to maintain lock. The VCO control voltage $V_e(t)$ necessary to maintain lock is proportional to the frequency shift of the incoming signal, relative to the VCO free-running frequency. This self-correcting ability of the system allows the PLL to track the frequency changes of an input signal, once it is locked. The range of frequencies over

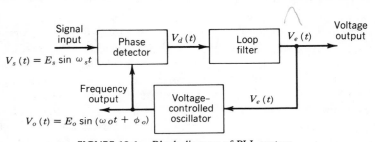

FIGURE 12.1.   Block diagram of PLL system.

which the PLL can maintain lock with an input signal is defined as the *lock range* of the system. This is different from the range of frequencies over which the PLL can acquire lock with the incoming signal. This latter range of frequencies is known as the *capture range,* or acquisition range, of the PLL. For reasons that will be described later, the capture range is always smaller than the lock range in practical PLL circuits, and the difference between the two is related to the loop filter characteristics.

The output from a PLL system can be obtained either as the voltage signal $V_e(t)$ corresponding to the filtered error voltage in the feedback loop, or as a frequency signal at the VCO output terminal. The voltage output is used in frequency discriminator applications, whereas the frequency output is used in signal conditioning, frequency synthesis, or clock recovery applications. Many of these classes of application will be discussed further in Section 12.4.

First, consider the voltage output. When the PLL is locked on an input frequency, the error voltage $V_e(t)$ is proportional to the frequency difference between the input signal at frequency f_s and the original VCO free-running frequency f_o, since $V_e(t)$ corresponds to the corrective voltage applied to the VCO to shift its frequency from f_o to f_s to maintain lock. If the input frequency is varied, as is the case with FM signals, the error voltage also varies proportional to the input frequency changes in order to maintain lock. Thus, the voltage output serves as a frequency discriminator and converts the input frequency changes to voltage changes.

Next, consider the frequency output. When the PLL is locked on an input signal, the VCO output provides a periodic waveform which is at the same exact frequency as the input signal, except for a finite phase difference ϕ_o, which is the phase difference necessary to generate the error voltage to keep the PLL in lock. If the input signal is comprised of many frequency components, along with noise and other disturbances, the PLL can be made to lock, selectively, on one particular frequency component at the input. The output of the VCO would then regenerate that particular frequency while attenuating or eliminating the other undesired frequencies. In other words, the VCO output can be used to regenerate or extract a desired frequency signal out of many other undesirable signals. This property of a PLL makes it particularly attractive for regenerating or reconditioning weak signals buried in noise.

Capture Phenomenon

A very important characteristic of the PLL performance is the capture process by which the PLL acquires lock with an input signal, starting with a free-running condition. This capture or acquisition phenomenon is highly complex and inherently nonlinear; it does not lend itself to simple mathematical analysis. Therefore, it will be described here only in a qualitative way.

First, assume that the PLL feedback loop is opened between the loop filter output and the VCO control input. This would cause the error voltage to be artificially reduced to zero, and the VCO would continue to oscillate at its free-running frequency f_o. Next, assume that an input signal is applied to the loop whose frequency f_s is close but not equal to f_o. The phase detector normally functions as a multiplier or mixer. Thus, coming out of the phase detector will be two frequency components, a sum frequency

$$f_{\text{sum}} = f_o + f_s \qquad (12.1)$$

and a difference frequency

$$\Delta f = |f_o - f_s| \qquad (12.2)$$

Normally, the low-pass loop filter bandwidth is sufficiently narrow such that the sum component is filtered out completely. If f_s is sufficiently close to f_o, then the difference frequency is very small and falls within the passband of the low-pass loop filter; it appears at the output of the loop filter as a sinusoidal *beat note*. This is the waveform shown at the left side of Figure 12.2, where, for illustrative purposes, $f_o > f_s$ is assumed.

Next, assume that the loop is suddenly closed by connecting the low-pass filter output to the VCO control terminal. This would cause the VCO frequency itself to be modulated by the beat note or the difference signal. When this happens, the beat note frequency Δf becomes a function of time. Since the VCO frequency is now a varying function of time, it will alternately move closer to and away from the incoming frequency. Since the filtered error voltage is the difference of the VCO frequency and the input signal, it will also alternately seem to reduce and increase in frequency in its positive and negative half-cycles. Therefore, under this condition, the beat note becomes asymmetrical and looks like a series of cusps, as shown in the middle portion of Figure 12.2. Note that the portion of the beat note that modulates the VCO *closer* to the input signal

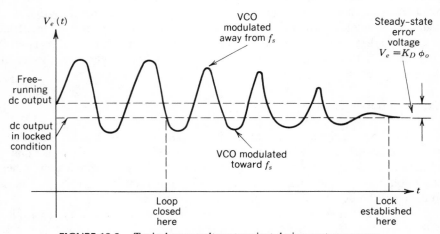

FIGURE 12.2. Typical error voltage transient during capture process.

appears more rounded and the portion that modulates the VCO *away* from the input signal appears more peaked. Because of this asymmetry, the beat note contains a finite dc voltage which steadily pushes the VCO frequency toward the input signal. As the VCO drifts toward f_s, the beat note frequency rapidly decreases, the asymmetry increases, and the transient rapidly converges to a steady-state dc value, corresponding to the lock condition where the VCO frequency is exactly equal to f_s. Once the system is locked, the difference frequency Δf is identically equal to zero, and only a dc voltage remains at the loop filter output. This dc voltage is generated by the phase difference ϕ_o between the VCO output and the input signal. Assuming that the loop filter has unity gain at dc and the phase detector has a conversion gain of K_D V/rad, this steady-state error voltage is

$$V_e = V_e(t)\Big|_{\text{steady state}} = -K_D\phi_o \qquad (12.3)$$

The negative sign in Eq. (12.3) is due to the implicit assumption in the example of Figure 12.2 that $f_o > f_s$ and that a negative voltage had to be generated at the control terminal of the VCO to shift f_o to f_s. This point will be clarified further in the next sections as we study the characteristics of phase detector circuits.

The total time taken by the PLL to establish lock is called *pull-in* time. The pull-in time depends on the initial phase and frequency difference between the two signals, as well as on the overall loop gain and loop filter characteristics. Under certain conditions, the pull-in time can be shorter than the period of beat note, and the loop can lock without an oscillatory error transient of the form illustrated in Figure 12.2.

The main purpose of the loop filter is to filter out the difference components due to undesired signals which are far removed from the VCO free-running frequency. In this manner, it enhances the interference rejection characteristics of the PLL. In other words, qualitatively speaking, the PLL would capture only those signals that are close to the VCO free-running frequency, such that the difference frequency Δf falls approximately within the bandwidth of the loop filter. A second and equally important function of the low-pass filter is that it provides a short-term memory for the PLL and ensures a rapid recapture of the signal, if the system is briefly thrown out of lock due to an interfering transient. In other words, the low-pass loop filter constrains the error voltage $V_e(t)$ to be a slowly varying function of time, such that if the PLL is temporarily thrown out of lock due to a noise or interference transient, the VCO frequency does not change significantly over a short period of time. This condition, then, would facilitate a rapid recapture of the input signal once the transient has passed.

Since the low-pass loop filter attenuates the high-frequency components of the error voltage in the PLL, it has a dominant effect on the capture and transient response characteristics of the system. The reduction of filter bandwidth has the following effects on the system performance:

1. The capture process becomes slower and the pull-in time increases.
2. Capture or acquisition range decreases.
3. Once locked, the interference–rejection characteristics of the PLL improve since the error voltage caused by an interfering frequency is attenuated further by the low-pass filter.
4. The transient response, that is, the response of the PLL to sudden changes of the input frequency within the capture range, becomes underdamped.

This last point also brings about a practical limitation of the loop–filter bandwidth and its frequency rolloff and phase characteristics from the point of view of feedback system stability. These points will be explored further in Section 12.3.

Tracking Characteristics

Once the PLL is locked on an input signal, it can track small frequency changes of the input signal by generating additional phase error ϕ_o between the VCO and the input signal, which is then converted to a dc error voltage V_e by the phase detector. This error voltage, then, keeps the VCO frequency in step with the input. While the PLL is tracking an input signal, the loop error voltage V_e is a direct measure of the *frequency difference* of the input signal from the VCO free-running frequency f_o. In other words, while the PLL is tracking an input signal, the voltage output of the loop functions as a frequency-to-voltage converter.

The tracking range of a PLL is determined by how much corrective error voltage V_e can be generated internally. Assuming that no other amplification or gain stage is present in the loop, the maximum amount of error voltage $(V_e)_{max}$ that can be generated depends on the phase detector gain K_D. Normally, $(V_e)_{max}$ is generated when the phase difference ϕ_o has reached its limiting value of $\pm \pi/2$ rad, about its equilibrium (zero error) point. Then the tracking range of the PLL is given as

$$\pm \Delta f_L = \pm (V_e)_{max} K_0 \qquad (12.4)$$

where K_0 (Hz/V) is the voltage-to-frequency conversion gain of the VCO.

Figure 12.3 illustrates the typical frequency-to-voltage transfer characteristics of a PLL. The input is assumed to be a sine wave whose frequency is swept slowly over a broad frequency range, first from low frequencies across the PLL capture and lock ranges to high frequencies and then back down to low frequencies. The vertical scale is the filtered loop error voltage V_e, and the VCO is assumed to have linear control characteristics where increasing the control voltage causes the frequency to increase. With reference to Figure 12.3a, the transfer characteristics of the PLL can be described as follows. The PLL does not respond to the input signal until its frequency reaches f_1, corresponding to

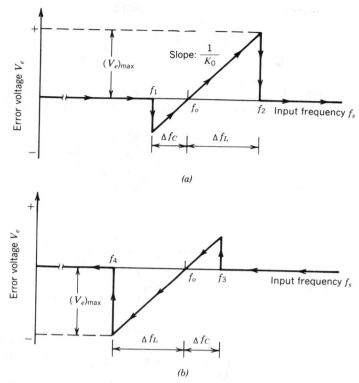

FIGURE 12.3.   Typical PLL frequency-to-voltage transfer characteristics: (*a*) Slowly *increasing* input frequency; (*b*) *decreasing* input frequency.

the lower edge of the capture range. Then the loop suddenly locks on the input signal, causing a negative jump of the loop error voltage. As the input frequency continues to increase, the loop tracks the input signal and V_e continues to increase with a slope equal to the reciprocal of the VCO conversion gain $1/K_0$, and goes through zero at $f_s = f_o$. The loop tracks f_s until the input frequency reaches f_2, corresponding to the upper edge of the tracking range. As the PLL loses lock, the error voltage drops to zero and the VCO frequency returns to f_o.

If the input frequency f_s is now slowly swept back toward low frequencies, the cycle repeats itself, as shown in Figure 12.3*b*, where the loop recaptures the signal at $f_s = f_3$ and tracks it down to $f_s = f_4$. The frequency spread between f_1, f_3 and f_2, f_4 corresponds to the total capture range and the total tracking range of the system, that is,*

$$f_3 - f_1 = 2\Delta f_C \qquad (12.5)$$

---

*In describing PLL characteristics, the term *acquisition range* is used interchangeably with *capture range*. Similarly, the term *tracking range* is used interchangeably with *lock range*.

and

$$f_2 - f_4 = 2\Delta f_L \qquad (12.6)$$

Figure 12.4 shows a composite set of frequency-to-voltage transfer character-istics for a PLL. This is essentially a superposition of the capture and tracking range characteristics shown in Figure 12.3. With reference to Figure 12.4, the basic response characteristics of the PLL can be summarized as follows:

1. The PLL exhibits a frequency-selective frequency-to-voltage conversion characteristic, centered around the VCO free-running frequency f_o.
2. It can acquire lock (capture) with only those signals that fall within the total capture range of $2\Delta f_C$, centered about f_o.
3. Once locked, it can track an input signal over a total tracking range of $2\Delta f_L$, centered about f_o.
4. The slope of frequency-to-voltage conversion characteristics is the recip-rocal of the VCO voltage-to-frequency conversion gain.

Figures 12.3 and 12.4 also indicate the significance of some of the PLL design parameters. The total lock range is primarily determined by the maximum available error voltage $(V_e)_{max}$ and the VCO conversion gain as given by Eq. (12.4). Thus, it can be increased by increasing $(V_e)_{max}$ and adding an amplifier into the basic PLL to provide additional voltage gain in the feedback loop, as shown in Figure 12.5. When the PLL is in lock, V_e is primarily a dc voltage, and the loop filter does not affect the tracking range.

The capture range is always equal to or smaller than the lock range, if a low-pass loop filter is present. As will be described in later sections [see Eq. (12.41)], the capture range decreases as the loop filter bandwidth is reduced.

The VCO free-running frequency f_o determines the nominal center frequency of the capture and lock ranges. Therefore, its accuracy and stability are ex-

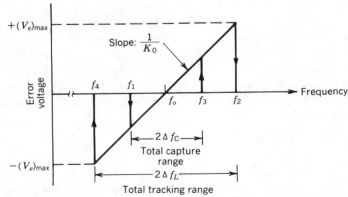

FIGURE 12.4.   Composite voltage-to-frequency transfer characteristics of a PLL.

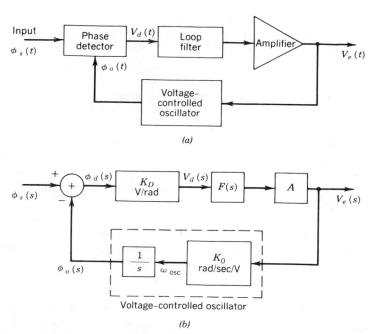

FIGURE 12.5. Block diagram of PLL system in locked condition: (a) Generalized block diagram; (b) its representation as linear feedback system in frequency domain.

tremely important. As the capture and/or tracking range of the PLL are made very narrow by design, these accuracy and stability requirements become more critical.

The VCO control characteristics are particularly important, as shown in Figure 12.4, because the PLL frequency-to-voltage conversion characteristics are solely determined by the VCO control characteristics. The slope of the PLL voltage output, that is, the frequency-to-voltage conversion characteristics, is determined by the VCO conversion gain constant K_0. Similarly, the linearity of the frequency-to-voltage conversion characteristics is solely determined by the linearity of the VCO control characteristics.

Thus, in summary, the overall dc loop gain, loop filter characteristics, VCO stability, and VCO control characteristics make up the four basic parameters in the design of monolithic PLL circuits.

12.2. PLL IN LOCKED CONDITION

When the PLL is in the locked condition, the nonlinear capture transients are no longer present. Instead, a linear relationship exists between the output of the phase detector and the phase difference between the input signal and the VCO output. Under this condition, the PLL can be analyzed as a linear feedback system using conventional feedback system analysis techniques. [4,5] Figure 12.5a shows the block diagram of a PLL. This is essentially the same as the basic PLL system shown in Figure 12.1, except that in this case an additional gain block (voltage amplifier) is added into the system to keep the analysis more general. The presence of a separate amplifier block in the voltage loop of a PLL is often preferred, since this provides a means of controlling the overall feedback loop gain, independent of the phase detector or the VCO gain characteristics.

Figure 12.5b shows the block diagram of the PLL as a linear feedback system in the complex frequency domain. In the figure, s ($= \sigma + j\omega$) is the complex frequency variable. The phase detector provides a voltage output proportional to the phase difference ϕ_d between the input signal and the VCO output, with a conversion gain of K_D V/rad. The loop filter transfer function is $F(s)$, and the amplifier gain is A. $F(s)$ is assumed to have unity gain at dc. The VCO voltage-to-frequency gain is K_0 rad/sec per volt of control voltage.*

The oscillator frequency ω_{osc} is proportional to the error voltage V_e and is given as

$$\omega_{osc} = \omega_o + K_0 V_e \tag{12.7}$$

where ω_o is the VCO free-running frequency ($\omega_o = 2\pi f_o$). With a constant voltage V_e applied to the VCO control terminal, the oscillator frequency ω_{osc} is constant. However, the phase detector is sensitive to the phase difference between the VCO output and the signal input. Thus, for the consistency of analysis, the VCO output should be expressed in phase rather than frequency, since phase is the integral of frequency:

$$\phi_o(t) = \phi_o \bigg|_{t=0} + \int_0^t \omega_{osc}(t) \, dt \tag{12.8}$$

The VCO section actually functions as an integrator. In other words, a step change of error voltage V_e corresponds to a step change of ω_{osc} which corresponds to a ramp change in phase. This process of integration is shown as the $1/s$ block in the diagram. Thus, the transfer function of the VCO section can be expressed as

$$\phi_o(s) = \frac{\omega_{osc}}{s} = \frac{K_0 V_e}{s} \tag{12.9}$$

Using classical linear feedback analysis, the closed-loop transfer function of Figure 12.5b can be written as

$$\frac{V_e}{\phi_s} = \frac{s K_D A F(s)}{s + K_D K_0 A F(s)} \tag{12.10}$$

where $\phi_s(s)$ is the phase of the input signal, relative to the VCO output. For

---

* Note that the units of K_0 must be consistent with the units of output frequency. If the oscillator frequency is expressed in radians per second, then K_0 is in radians per second per volt; if the frequency is expressed in hertz, K_0 must be expressed in hertz per volt.

convenience, it is practical to combine all the gain constants in the loop by defining a loop-gain constant K_L as

$$K_L = K_D K_0 A \qquad (12.11)$$

where K_L represents the total dc loop gain. Note that since K_D and K_0 are conversion gains and A is a dimensionless gain constant, K_L has the units of seconds^{-1}.

In terms of the loop gain constant K_L, Eq. (12.10) can be written as

$$\frac{V_e}{\phi_s} = \frac{K_L}{K_0} \frac{sF(s)}{s + K_L F(s)} \qquad (12.12)$$

In practical applications, one is normally interested in the response of the PLL to variations of input frequency rather than input phase. Since frequency is the time derivative of phase, the frequency deviation $\Delta \omega_s$ of the input signal can be expressed as

$$\Delta \omega_s(t) = \frac{d\phi_s}{dt} \qquad (12.13)$$

and its transformation in the frequency domain is

$$\Delta \omega_s(s) = s\phi_s(s) \qquad (12.14)$$

Thus, in terms of the input frequency changes, the PLL loop transfer function of Eq. (12.12) can be rewritten as

$$\frac{V_e}{\Delta \omega_s} = \frac{1}{s} \frac{V_e}{\phi_s} = \frac{K_L}{K_0} \frac{F(s)}{s + K_L F(s)} \qquad (12.15)$$

Equation (12.15) describes the frequency-to-voltage transfer characteristics of the PLL in the locked condition.

12.3.   EFFECTS OF LOOP FILTER AND LOOP GAIN ON PLL PERFORMANCE

The loop filter function $F(s)$ has a strong effect on the PLL performance characteristics. When the PLL is in the locked condition, the loop filter function along with the loop gain K_L determine both the transient response and the frequency response characteristics of the system. When the PLL is not locked, the loop filter has the dominant effect in controlling the capture characteristics of the loop. In this section, the effect of the loop filter on the overall PLL characteristics will be examined.

First-Order Loop

The simplest case of PLL operation is where the loop filter is removed entirely, which corresponds to setting $F(s) = 1.0$. This type of PLL is called a *first-order loop*, since the transfer function of Eq. (12.15) now reduces to a simple one-pole low-pass function:

$$\frac{V_e}{\Delta\omega_s} = \frac{1}{K_0} \frac{1}{1 + s/K_L} \qquad (12.16)$$

Figure 12.6 shows the root locus (i.e., the pole location) and the frequency response of a first-order PLL. Note that the system behaves as a one-pole low-pass filter with a low-frequency conversion gain of $1/K_0$ and a 3-dB bandwidth of K_L. Also note that at dc or very low frequencies where $s \to 0$ and $F(s) \approx 1.0$, Eq. (12.15) reduces to

$$\left.\frac{V_e}{\Delta\omega_s}\right|_{dc} = \frac{1}{K_0} \qquad (12.17)$$

which is the linear frequency-to-voltage transfer characteristic shown in Figures 12.3 and 12.4.

As indicated by Eq. (12.15), the response of the PLL in the locked condition to changes of input frequency is strongly influenced by two parameters: the loop filter transfer function $F(s)$ and the loop gain factor K_L. In order to investigate both the frequency response and the transient response (i.e., the response of the PLL to step changes of the input frequency), it is instructive to consider various loop filter configurations. At this point, it is instructive to examine the PLL response characteristics by means of an illustrative example. Consider, as an example, the case of a first-order PLL locked on a 100-kHz input signal. The loop is assumed to have the following parameter values:

$$K_D = 0.1 \text{ V/rad}$$
$$K_0 = 10 \text{ kHz/V} = 2\pi \times 10^4 \text{rad/sec.V}$$
$$A = 1$$
$$K_L = K_D K_0 A = 2\pi \times 10^3 \text{ sec}^{-1}$$

Assume that the loop input signal is frequency modulated sinusoidally with $\Delta f_s = \pm 10$ kHz, centered around 100 kHz, with a modulation frequency $\omega_{mod} = 2\pi f_{mod}$. Then, as the PLL is tracking the modulated input signal, the loop error voltage output will be a sinusoid at frequency ω_m. At low modulating frequencies, the output amplitude will be equal to $1/K_0$ for frequencies up to 1 kHz, and then decrease at 6 dB per octave for increasing values of ω_m.

Similarly, in the same example, if the input frequency was changed stepwise from 100 to 110 kHz, the output voltage would be an exponential step change of $+ 1$ V, with a time constant τ_L equal to

$$\tau_L = \frac{1}{K_L} = \frac{1}{2\pi \times 10^3} = 0.16 \text{ msec}$$

The first-order loop finds relatively limited use in practice due to its poor selectivity and interference rejection properties caused by the lack of a loop filter. Without the loop filter, all the high-frequency components coming out of

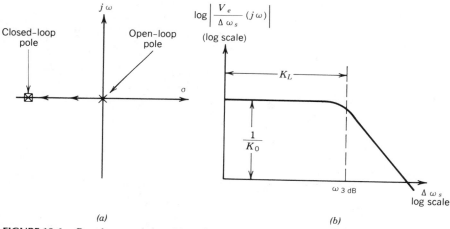

FIGURE 12.6. Root locus and closed-loop frequency response of first-order PLL system: (*a*) Root locus; (*b*) frequency response.

the phase detector appear directly at the output. These include components due to noise as well as undesired signals at the input near the desired signals frequency. Thus, the selectivity characteristics of the PLL become degraded.

Second-Order Loop

In most applications, a one-pole low-pass filter is used as the PLL filter. This results in a quadratic or second-order equation in the denominator of Eq. (12.15), and the PLL response is described by a two-pole transfer function. As a result, such a PLL system is commonly called a *second-order loop*. The most commonly used low-pass filter configuration is the simple one-pole filter shown in Figure 12.7*a*, which has the transfer function

$$F(s) = \frac{1}{1 + s/\omega_1} \tag{12.18}$$

where $\omega_1 = 1/R_1C_1$ is the low-pass filter bandwidth. Since this filter produces a net phase lag of 90° at high frequencies, it is often referred to as a *lag filter*. Substituting Eq. (12.18) into Eq. (12.15) and solving the system poles for the PLL, one obtains

$$p_1, p_2 = -\frac{\omega_1}{2}\left(1 \pm \sqrt{1 - \frac{4K_L}{\omega_1}}\right) \tag{12.19}$$

Figure 12.7*b* shows the locus of the closed-loop poles p_1 and p_2 as a function of the loop gain K_L. For increasing values of K_L, the poles become complex conjugate and the system becomes underdamped.

Using standard feedback servo terminology, substituting Eq. (12.18) into Eq.

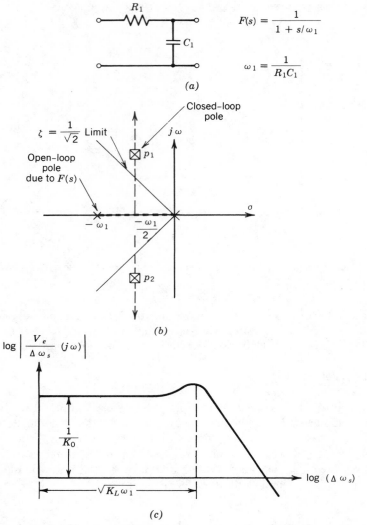

FIGURE 12.7. Root locus and closed-loop frequency response of second-order PLL with single-pole filter: (a) Single-pole loop filter; (b) root locus for increasing K_L; (c) frequency response.

(12.15) and redefining terms, the transfer function can be written as

$$\frac{V_e}{\Delta\omega_s} = \frac{1}{K_0} \frac{1}{s^2/\omega_n{}^2 + (2\zeta/\omega_n)s + 1} \tag{12.20}$$

where

$$\omega_n = \sqrt{K_L\omega_1} \tag{12.21}$$

and

$$\zeta = \frac{1}{2}\sqrt{\frac{\omega_1}{K_L}} \tag{12.22}$$

In classical feedback terminology, ω_n is called the loop natural frequency, and ζ is loop damping. Note that as the loop filter bandwidth ω_1 is reduced, or as K_L is increased, the loop damping decreases, the loop frequency response starts exhibiting peaking, and the loop step response shows oscillatory transients. Figure 12.8 illustrates a typical response of a second-order PLL to a step change in input frequency, for both overdamped ($\zeta > 1$) and underdamped ($\zeta < 1$) conditions.

In general, one tries to minimize or avoid the peaking in the loop frequency response since this results in the distortion of frequency discrimination or FM detection characteristics. A commonly used compromise is to adjust the loop gain to set the closed-loop poles on 45° radials, angled from the origin, as shown in Figure 12.7. This is known as the *maximally flat response* and corresponds to the $\zeta = 1/\sqrt{2}$ condition. It results in a flat frequency response with a slight

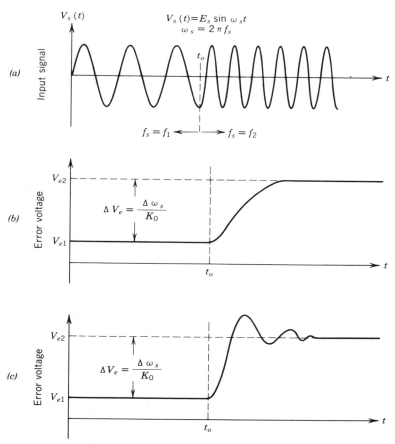

FIGURE 12.8.   Transient response of PLL to step change of input frequency: (*a*) Input signal; (*b*) step response of overdamped loop ($\zeta > 1.0$); (*c*) step response of underdamped loop ($\zeta < 1.0$).

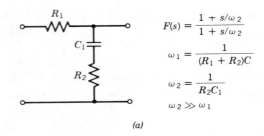

$$F(s) = \frac{1 + s/\omega_2}{1 + s/\omega_2}$$

$$\omega_1 = \frac{1}{(R_1 + R_2)C}$$

$$\omega_2 = \frac{1}{R_2 C_1}$$

$$\omega_2 \gg \omega_1$$

(a)

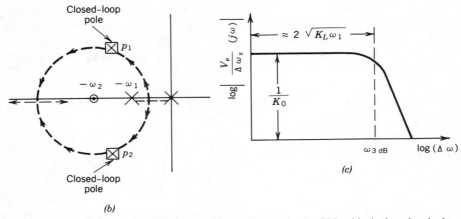

(b)

(c)

FIGURE 12.9. Root locus and frequency response of second-order PLL with single-pole, single-zero (lag–lead) filter: *(a)* Single-pole, single-zero loop filter; *(b)* root locus for increasing K_L; *(c)* closed-loop frequency response.

($\approx 10\%$) overshot in the step response. This condition can be obtained from Eq. (12.22) by setting

$$\omega_1 = 2K_L \tag{12.23}$$

The resulting loop bandwidth is

$$\omega_{3\,dB} = \omega_n = \sqrt{K_L \omega_1} = \sqrt{2}\,K_L \tag{12.24}$$

In many applications, the PLL is required to have both a wide lock range and a narrow bandwidth. The wide lock range is desirable in order to track an input signal over wide frequency deviations; the narrow bandwidth is required for the proper rejection of out-of-band signals. In a second-order PLL, this leads to a serious compromise. To increase the lock range, K_L must be high; yet to reduce the loop bandwidth, ω_1 must be low, which results in the loop being under-damped [see Eq. (12.22)]. This problem can be avoided if a so-called lag–lead filter is used in the loop, as shown in Figure 12.9. Such a loop filter has the transfer function

$$F(s) = \frac{1 + s/\omega_2}{1 + s/\omega_1} \tag{12.25}$$

where

$$\omega_1 = \frac{1}{(R_1 + R_2)C_1} \quad \text{and} \quad \omega_2 = \frac{1}{R_2 C_1} \qquad (12.26)$$

As indicated by the root-locus plot of Figure 12.9b, the effect of a lag–lead filter is to pull the root locus into a nearly circular pattern around the open-loop filter zero at $-\omega_2$ and keep the loop from being underdamped, even at large values of K_L.

By substituting Eq. (12.25) into Eq. (12.15) and rearranging the denominator terms to match those of Eq. (12.20), one can show that the damping factor ζ is

$$\zeta = \frac{1}{2} \sqrt{\omega_1/K_L} + \frac{1}{2} \frac{\sqrt{K_L \omega_1}}{\omega_2} \qquad (12.27)$$

TABLE 12.1. Summary of Design Equations for a Second-Order PLL with Lag and Lag–Lead Type Loop Filters

Lag Filter	Lag–Lead Filter
$F(s) = \dfrac{1}{1 + s/\omega_1}$ $\omega_1 = \dfrac{1}{R_1 C_1}$	$F(s) = \dfrac{1 + s/\omega_2}{1 + s/\omega_1}$ $\omega_1 = \dfrac{1}{(R_1 + R_2)C_1} \qquad \omega_2 = \dfrac{1}{R_2 C_2}$
Natural frequency ω_n	
$\omega_n = \sqrt{K_L \omega_1}$	$\omega_n = \sqrt{K_L \omega_1}$
Damping factor ζ	
$\zeta = \frac{1}{2} \sqrt{\dfrac{\omega_1}{K_L}}$	$\zeta = \frac{1}{2} \sqrt{\dfrac{\omega_1}{K_L}} + \frac{1}{2} \dfrac{\sqrt{K_L \omega_1}}{\omega_2}$
3-dB closed-loop bandwidth for $\zeta = 1/\sqrt{2}$	
$\omega_{3\,\text{dB}} = \sqrt{2} K_L$	$\omega_{3\,\text{dB}} \approx 2 \sqrt{K_L \omega_1}$
	Lag–lead filter zero location for $\zeta = 1/\sqrt{2}$, and $K_L \gg \omega_1$ $\omega_2 \approx \dfrac{\omega_{3\,\text{dB}}}{2\sqrt{2}}$

Comparing Eq. (12.27) with Eq. (12.22), one sees that ζ is increased in the lag–lead filter case by the second additive term in Eq. (12.27). As a result, the loop will not be underdamped as K_L is increased to increase the tracking range.

The natural frequency ω_n for a second-order PLL with a lag–lead filter is still the same as that given by Eq. (12.21), with ω_1 as defined in Eq. (12.26). To obtain a flat frequency response under closed-loop conditions, one can calculate the relations between the breakpoint frequencies ω_1 and ω_2 of the lag–lead filter by setting $\zeta = 1/\sqrt{2}$ in Eq. (12.27) and solving for ω_2 as a function of ω_1. For the condition $K_L \gg \omega_1$, this reduces to

$$\omega_2 \approx \sqrt{\frac{K_L \omega_1}{2}} \tag{12.28}$$

For the $\zeta = 1/\sqrt{2}$ condition, the 3-dB bandwidth of the PLL with a lag–lead filter is given as

$$\omega_{3\,dB} \approx 2\omega_n = 2\sqrt{K_L \omega_1} \tag{12.29}$$

Equations (12.27) through (12.29) illustrate the key advantage of a lag–lead filter over a simple lag filter. It allows the PLL to have a large loop gain, while maintaining a narrow loop bandwidth, without causing the system to be underdamped.

The basic design equations covering the system parameters of second order PLLs with lag or lag–lead filters are summarized in Table 12.1 for easy reference.

The Lock Range

The lock range of a PLL is defined as that range of frequencies, centered about f_o, over which the PLL can track an input signal once it is locked. It is shown as the frequency range $\pm \Delta f_L$ in Figure 12.4. As described in Eq. (12.4), Δf_L is directly proportional to the maximum error voltage $(V_e)_{max}$ that can be generated within the PLL. The maximum error voltage which can be generated is, in turn, related to the maximum phase error $(\phi_s)_{max}$ and phase detector gain K_D. If a generalized PLL configuration, such as the one shown in Figure 12.5, is used, the amplifier gain A also enters into the equation.

In the phase detector circuits most commonly used in IC PLLs (see Section 12.5), the maximum phase error $(\phi_s)_{max}$ which can be detected is

$$\pm (\phi_s)_{max} = \pm \frac{\pi}{2} \text{ rad} \tag{12.30}$$

Thus, assuming a fully switched, balanced-modulator-type phase detector such as the one shown in Figure 12.19, the maximum error voltage that can be generated in the loop is

$$\pm (V_e)_{max} = \pm (\phi_s)_{max} K_D A \tag{12.31}$$

Note that in the calculation of the tracking range, we are talking about steady-state operation with a dc error signal in the loop. Thus, the loop filter characteristics do not enter into the equation. The loop filter is assumed to have unity gain at dc or very low frequencies.

Then, from Eqs. (12.4) and (12.31), one can express the lock range in radians per second as

$$\Delta\omega_L = 2\pi\Delta f_L = \frac{\pi}{2}K_D A K_0 = \frac{\pi}{2}K_L \qquad (12.32)$$

In terms of frequency, this simplifies to

$$\Delta f_L = \frac{\Delta\omega_L}{2\pi} = \frac{K_L}{4} \qquad (12.33)$$

where, for consistency, K_0 is expressed in radians per second per volt. Thus, the tracking characteristics, or lock range, is directly proportional to the total dc loop gain K_L.

An exception to the above condition can arise when a nonlinearity is present in the feedback loop to limit either $(V_e)_{max}$ or the pulling range of the VCO. In that case, the lock range will be set by the *clipping limit* of the nonlinearity rather than the loop gain, provided that the clipping limit is reached before the gain limit. In some IC designs, such a nonlinearity is introduced into the PLL feedback loop intentionally in order to limit the tracking range or make it asymmetrical about f_o. This technique is particularly useful in the presence of strong out-of-band interference signals near the PLL center frequency.

Capture Range

The capture range is the range of input frequencies in the vicinity of f_o over which the PLL can acquire lock on an input, starting with the no-lock condition. As described briefly in Section 12.1 (see Figs. 12.3 and 12.4), the capture range Δf_C or $\Delta\omega_C$ ($=2\pi\Delta f_C$) is always less than the lock range if a low-pass loop filter is present. As a result, the PLL exhibits a hysteresis characteristic, as shown in Figure 12.4, where it can track a signal over a wider frequency range than it can capture. The qualitative reasons for this effect will be discussed in this section. Since the rigorous analysis of the capture phenomenon is cumbersome and difficult, the following analysis is primarily intended as a rule-of-thumb estimate.[5]

With reference to Figure 12.5a, assume that the loop is opened at the output of the amplifier, and a signal frequency f_s, which is close but not equal to f_o, is applied to the input. Then a sinusoidal beat note, similar to that shown on the left-hand side of Figure 12.2, would appear at the phase detector output as

$$V_d(t) = \frac{\pi}{2}K_D \quad \cos\,(\Delta\omega_i t) \qquad (12.34)$$

where

$$\Delta\omega_i = \left| \omega_s - \omega_o \right| \tag{12.35}$$

is the beat note frequency. The amplitude of $V_d(t)$ is

$$\left. V_d(t) \right|_{\text{peak}} = \frac{\pi}{2} K_D \tag{12.36}$$

The amplitude of the corresponding filtered and amplified error voltage appearing at the loop voltage output is

$$\left. V_e(t) \right|_{\text{peak}} = \frac{\pi}{2} K_D A \left| F(j\Delta\omega_i) \right| \tag{12.37}$$

where $\left| F(j\Delta\omega_i) \right|$ is the magnitude of loop filter response at the beat note frequency. Due to low-pass characteristics, $\left| F(j\Delta\omega_i) \right|$ is always less than unity.

In order for capture to occur, the magnitude of the voltage that must be applied to the VCO control terminal is

$$\left| V_{\text{osc}} \right| = \frac{\Delta\omega_i}{K_0} = \frac{\left| \omega_s - \omega_o \right|}{K_0} \tag{12.38}$$

As a rough estimate, the capture effect can take place only if the peak error voltage given by Eq. (12.37) is equal to or greater than that required by Eq. (12.38) to shift the oscillator frequency. Thus, the condition for capture can be estimated by setting Eqs. (12.37) and (12.38) to be approximately equal at the point of capture, that is,

$$\frac{\Delta\omega_i}{K_0} \approx \frac{\pi}{2} K_D A \left| F(j\Delta\omega_i) \right| \tag{12.39}$$

However, at the frequency where the capture occurs, $\Delta\omega_i = \Delta\omega_C$ by the definition of capture range. Substituting this into Eq. (12.39) and rearranging terms, one gets

$$\Delta\omega_C \approx \frac{\pi}{2} K_L \left| F(j\Delta\omega_C) \right| \tag{12.40}$$

or in terms of the lock range, from Eq. (12.32),

$$\Delta\omega_C \approx \Delta\omega_L \left| F(j\Delta\omega_C) \right| \tag{12.41}$$

Since the magnitude of the low-pass filter response is always less than unity, the capture range estimated from Eq. (12.40) or Eq. (12.41) is always less than the lock range. The difference becomes more significant as the filter bandwidth is reduced.

Equation (12.41) is a parametric equation which can be solved by numerical techniques for any generalized filter function. However, in the case of a high-gain second-order loop with a narrowband one-pole low-pass filter of the form shown in Figure 12.7, it can be approximated in a closed form. In such a case,

$$\left| F(j\Delta\omega_C) \right| = \left| \frac{1}{1 + j\Delta\omega_C/\omega_1} \right| = \frac{1}{\sqrt{1 + (\Delta\omega_C/\omega_1)^2}} \qquad (12.42)$$

Assuming that the loop gain K_L is high and the filter bandwidth is narrow, such that $\Delta\omega_C \gg \omega_1$, Eq. (12.42) simplifies to

$$\left| F(j\Delta\omega_C) \right| \approx \frac{\omega_1}{\Delta\omega_C} \qquad (12.43)$$

Substituting Eq. (12.43) into Eq. (12.41) and rearranging terms, one gets

$$\Delta\omega_C \Big|_{\text{lag filter}} \approx \sqrt{\Delta\omega_L\omega_1} = \sqrt{\frac{\Delta\omega_L}{R_1C_1}} \qquad (12.44)$$

where $R_1C_1 = 1/\omega_1$ is the low-pass filter time constant. In using Eq. (12.44), one word of caution is in order. It is derived with the implicit assumption that $\Delta\omega_C \gg \omega_1$. Thus, it should be used only if the value of $\Delta\omega_C$ given by it indeed justifies this assumption.

12.4. APPLICATIONS OF PHASE-LOCKED LOOPS

The PLL is a very versatile building block suitable for a variety of frequency-selective signal demodulation, signal conditioning, synchronization, and frequency synthesis applications. In this section, some of these applications will be reviewed. Particular attention is given to applications which relate to monolithic PLL circuits.

In the applications illustrated in this section, the PLL is shown in its simplest form without an amplifier block within the loop. However, since such an amplifier can always be added into the PLL system when needed, leaving it out does not detract from the general nature of the conclusions.

FM Demodulation

If the PLL is locked on a frequency modulated signal, the VCO tracks the instantaneous frequency deviation of the input. Then the filtered error voltage $V_e(t)$, which constrains the VCO to maintain lock with the input signal, corresponds to the demodulated output.

Figure 12.10a shows the basic block diagram of a phase-locked loop FM demodulator. In this case, the linearity of the demodulated output characteristics is determined by the VCO voltage-to-frequency conversion characteristics, since the reciprocal of the VCO conversion gain $1/K_0$ determines the output voltage swing for a given input frequency change (see Figs. 12.3 and 12.4).

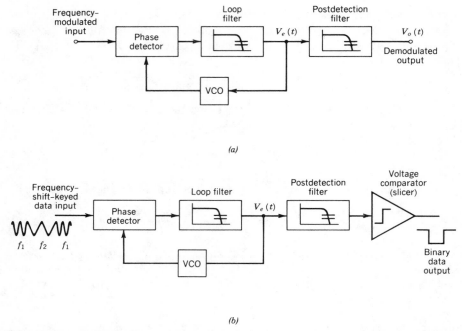

FIGURE 12.10. PLL as FM and FSK demodulator: (*a*) FM demodulator; (*b*) FSK demodulator (decoder).

In many applications, the demodulated signal $V_e(t)$ is filtered further by a so-called *postdetection filter* outside the PLL feedback loop. The function of the postdetection filter is to filter out any carrier feedthrough which may be present in the loop error voltage.

The PLL can be used to detect either wideband (high-deviation) or narrowband FM signals with a higher degree of linearity than that what can be obtained from other FM detection techniques. For this application, the linearity of the VCO control characteristics is critical since any nonlinearity in VCO conversion gain directly affects the distortion of the output signal.

In FM demodulation applications, the PLL loop gain and loop filter are chosen to provide a flat frequency response (i.e., $\zeta \geq 1/\sqrt{2}$), and both the loop lock and the capture range are made significantly larger than the input FM signal frequency deviation. In commercial FM detection, the main drawback of monolithic IC PLLs is the relatively poor signal-to-noise (S/N) ratio at the output due to the VCO phase noise associated with the multivibrator-type oscillators used. This limits the practical signal-to-noise ratio of demodulated output to approximately 65 dB for $\pm$ 1% deviation FM signals.[6]

Frequency-shift keying (FSK) is a special form of frequency modulation used in binary data transmission. The digital information is transmitted by switching the input frequency between two discrete frequencies called *mark* and *space* frequencies, which correspond to digital 1 and digital 0, respectively. Figure

12.10*b* shows the basic PLL system used as an FSK decoder. When the PLL is locked on the input FSK signal and follows the step changes of the input frequency, the error voltage $V_e(t)$ is in the form of discrete voltage steps corresponding to input frequency changes (see Fig. 12.8). These voltage steps correspond to the demodulated data output. In a conventional FSK detector system, as shown in Figure 12.10*b*, this output is filtered further by a postdetection filter to eliminate carrier feedthrough and then converted to logic-level swings by means of an output voltage comparator or *slicer*.

In optimizing a PLL's performance for FSK decoding, one normally chooses the VCO free-running frequency f_o to be midway between mark and space frequencies, that is,

$$f_o = \tfrac{1}{2} \left(f_{\text{mark}} + f_{\text{space}} \right) \tag{12.45}$$

so that the error voltage within the loop would change polarity when the input signal frequency makes a transition. Typically, the loop capture range is chosen to be well in excess of the input frequency step size, and the loop is made slightly underdamped ($\zeta \approx 0.5$) to obtain a rapid rise time of the error voltage, without excessive overshoots (see Fig. 12.8*c*).

Frequency Synchronization

Using the PLL system, the frequency of a relatively poor oscillator, such as an emitter-coupled multivibrator, can be phase locked to a low-level but highly stable frequency reference signal. This can be done by using the oscillator to be synchronized as the VCO portion of a PLL. Then the VCO output reproduces the input signal frequency at the same relative accuracy as the input reference, but at a much higher power level. In some applications, the synchronizing signal can be a low-duty-cycle tone burst at a specific frequency. The PLL can then be used to regenerate a coherent continuous-wave (CW) signal, synchronized to such a short tone burst. A typical example of such an application is the phase-locked chroma reference generator integrated circuit in color TV receivers. The horizontal *raster scan* and the vertical scan information in conventional color or black-and-white TV receivers are also synchronized using PLL systems in a similar manner.

In digital systems, the PLL can be used for a variety of data or clock synchronization functions. For example, two system clocks can be phase locked to each other so that one can function as a backup for the other, or it can be used for synchronizing disk or tape–drive mechanisms in information storage and retrieval systems.

Signal Conditioning

If the input signal spectrum contains several out-of-band signals, interfering channels, and noise, in addition to the desired signal frequency, the PLL system can be used to recreate or "condition" the desired signal and eliminate or

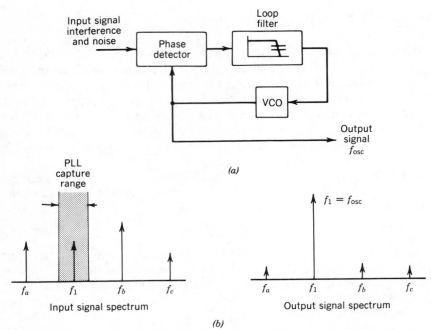

FIGURE 12.11. PLL as signal conditioner: (*a*) System block diagram; (*b*) typical input and output signal spectra. (*Note:* f_1 is the desired signal; $f_a, f_b,$ and f_c represent out-of-band interference signals.)

attenuate the undesired signals. Figure 12.11 illustrates the use of a PLL system in such an application.

By proper choice of the VCO free-running frequency, loop gain, and loop filter function, the selectivity characteristics of the PLL can be centered around the desired signal frequency present at the input; the PLL can be made to lock onto that signal. This is illustrated in Figure 12.11*b* in terms of the desired input signal f_1, which is present at the input along with the undesired signals at frequencies f_a, f_b, and f_c. If the PLL is locked on f_1 and the loop bandwidth is maintained relatively narrow, then the VCO output in Figure 12.11(a) reproduces the frequency of the desired signal while greatly attenuating the out-of-band signals. Note that a small but finite amount of interfering signals still appear at the VCO output spectrum because the difference frequencies between these frequencies and the VCO frequency can still pass through the loop filter, causing a slight modulation or *jitter* of the VCO frequency.

If the loop bandwidth is sufficiently narrow, the signal-to-noise ratio of the VCO output can be much higher than that at the input. Thus, the PLL can be used as a noise filter to regenerate weak periodic signals buried in noise.

A typical signal-conditioning application for monolithic PLL systems is in pulse-code-modulation (PCM) telemetry and telephone systems where the distorted and dispersed pulse-code modulation data have to be accurately regenerated at periodic intervals by means of synchronous *repeater* systems. Often, these repeater systems involve a PLL to extract the synchronous clock signal and then regenerate or recondition the weak input signal into a binary data stream.

Frequency Synthesis

By inserting a frequency divider circuit into the feedback loop of a PLL, between the VCO output and the phase detector input, the PLL can be made to function as a frequency multiplier. A block diagram of this circuit configuration is shown in Figure 12.12a, where N is the frequency-divider modulus. Thus, the actual frequency signal supplied by the loop into the phase detector is $1/N$ times the oscillator frequency. When the PLL is locked on an external reference frequency, the two inputs of the phase detector are at the same frequency, or

$$f_{\text{ref}} = \frac{f_{\text{osc}}}{N} \tag{12.46}$$

Then, taking the signal at the output of the VCO, one obtains

$$f_{\text{osc}} = N f_{\text{ref}} \tag{12.47}$$

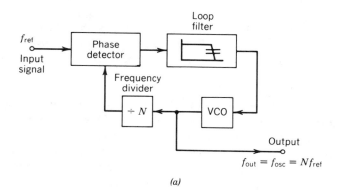

(a)

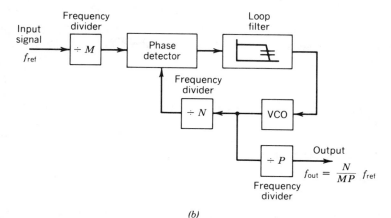

(b)

FIGURE 12.12   PLL as frequency synthesizer: (a) Basic frequency synthesizer; (b) synthesizer with pre- and post-scaling dividers.

By selectively changing the divider modulus N, one can obtain a multiplicity of frequencies which are integer multiples of the input reference frequency.

If a noninteger relationship is desired between the input and output frequencies, additional frequency dividers, such as those shown in Figure 12.12b, can be added into the system. In such a configuration, the divider with modulus M is called a *prescaler,* and the divider with modulus P is called a *postscaler.* By equating the frequency of the signals entering the phase detector, one can show that

$$f_{\text{out}} = \frac{N}{MP} f_{\text{ref}} \qquad (12.48)$$

PLL frequency synthesizers are very widely used in transmitters, receivers, and transponders, as well as in laboratory instrumentation. The examples shown in Figure 12.12 are some of the simplest configurations. In practice, frequency synthesizer circuitry can get quite complex and often involves several inter-locked PLL systems.[7]

It should be noted that when a frequency divider is used between the VCO and the phase detector, the effective conversion gain of the VCO becomes K_0/N, where N is the divider modulus. Since the loop gain K_L is directly proportional to the VCO conversion gain, it is also reduced by the factor $1/N$.

Harmonic Locking. Under certain conditions, frequency multiplication can also be achieved without the use of a frequency divider network, by operating the PLL in its *harmonic-locking* mode. The principle of harmonic lock can be briefly explained as follows. If the VCO output is nonsinusoidal, it will contain a number of harmonics in addition to its fundamental frequency. In other words, the output of the VCO is essentially a composite signal containing frequency components at integral multiples of the VCO fundamental frequency. The same is also true for a nonsinusoidal input signal, such as a pulse-train or a square-wave input. If the VCO free-running frequency is set close to the nth harmonic of the input signal, the VCO fundamental frequency can be made to synchronize with the nth harmonic of the input. In this case, the PLL operates in its harmonic-locking mode with

$$f_{\text{osc}} = n f_s \qquad (12.49)$$

Similarly, if the VCO produces a harmonic-rich output waveform, the mth harmonic of the VCO output can be synchronized with the fundamental of the input frequency. Under this condition, the VCO fundamental is a subharmonic of the input signal frequency,

$$f_{\text{osc}} = \frac{1}{m} f_s \qquad (12.50)$$

When the PLL is operated in its harmonic-locking mode, the spacing between the adjacent harmonics in the frequency spectrum decreases rapidly as the harmonic order n or m is increased. This increases frequency stability and narrow-bandwidth requirements associated with the PLL to enable the system to differentiate between adjacent harmonics. In monolithic PLL integrated circuits which use multivibrator-type oscillators, thermal drifts associated with the VCO frequency usually restrict the harmonic-lock operation of a PLL system to values of n or $m \leq 10$. An additional disadvantage of harmonic locking is that the phase detector gain K_L and thus the PLL loop gain decreases inversely, with the harmonic order n or m, decreasing both the capture and the lock range of the system at higher harmonics.

In most PLL IC designs, one uses a chopper-type phase detector, such as a balanced modulator or an exclusive–OR gate (see Section 12.5), which is driven by square-wave signals. In such systems, the phase detector output is rich in odd harmonics, and the PLL tends to lock on values of n or m equal to 3, 5, 7, 9, and so on.

Although harmonic locking is sometimes used as a special feature, it is often considered to be a parasitic effect since it may interfere with the wide tracking range of a PLL. It can be avoided by using more complex phase detector circuits such as the one shown in Figure 12.22.

Frequency Translation

The PLL can be used to translate the frequency of a highly stable but fixed frequency oscillator by a small amount in frequency. This technique is also called a *frequency-offset loop* since the VCO output differs from the reference frequency by an amount equal to the offset frequency. Figure 12.13 shows the basic block diagram of a frequency-translation loop. In this case, the reference input at frequency f_{ref} is applied to a multiplier, or *mixer*, stage along with the VCO output signal. The mixer is essentially an analog multiplier or a balanced-modulator stage similar to those described in Chapter 9. The mixer output is made up of the *sum* and *difference* components of f_{osc} and f_{ref}. The sum component is filtered out by the low-pass filter, ahead of the phase detector. The translation, or offset, frequency is applied to the phase detector along with the

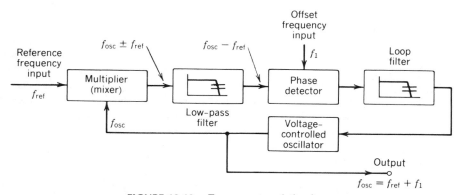

FIGURE 12.13.   Frequency translation loop.

filtered difference component of the mixer output. When the system is in lock, both inputs of the phase detector are identical in frequency, that is,

$$f_{osc} - f_{ref} = f_1 \qquad (12.51)$$

or taking the output from the VCO output,

$$f_{osc} = f_{ref} + f_1 \qquad (12.52)$$

In the above analysis, it was assumed that the VCO frequency was maintained slightly higher than f_{ref}. However, if the VCO frequency is slightly below f_{ref}, the lock can still be obtained at

$$f_{osc} = f_{ref} - f_1 \qquad (12.53)$$

This is known as the imaging problem associated with the frequency translation loop and becomes more critical as the offset frequency is reduced, relative to the reference frequency. The imaging problem can be avoided at the expense of system complexity, by using a so-called *image-rejecting phase detector* configuration.[1]

AM Detection

The PLL can be used as a coherent detector for AM signals.[8] The principle of coherent AM detection was discussed in Section 9.8. It basically consists of multiplying an AM signal with a coherent, unmodulated carrier signal at the same frequency. Then, after low-pass filtering, the residual low-frequency signal corresponds to the demodulated information (see Fig. 9.6c). In such a detection system, the PLL can be used to lock on the carrier of the AM signal and produce an unmodulated coherent reference signal at the output of the VCO. This reference signal has the same frequency as the AM carrier, but no amplitude modulation. Then, by multiplying this coherent reference signal with the AM input and low-pass filtering the output, one can obtain the demodulated information as shown in Figure 9.16c.

In actual implementation of phase-locked AM detection, a practical problem arises. As indicated by Eq. (9.62), the output of the synchronous demodulator is

$$V_o(t)\Big|_{low\ frequency} = K_X E_m(t) \ \cos \phi_o \qquad (12.54)$$

where K_X is the demodulator gain factor, $E_m(t)$ is the amplitude of the AM input, and ϕ_o is the phase difference between the carrier of the AM input and the unmodulated coherent carrier. In most IC PLLs, one uses a chopper-type phase detector which will be described in the next section. Such a phase detector inherently produces a $\pm \pi/2$ rad ($\pm 90°$) phase offset in the PLL. Therefore, when the loop is perfectly locked on an input signal, the VCO frequency has a net steady-state phase difference ϕ_o of $90°$ with the input frequency. The loop stays locked by introducing additional small amounts of phase error, over and above this $90°$ phase offset. Thus, if the VCO output which is fed into the phase

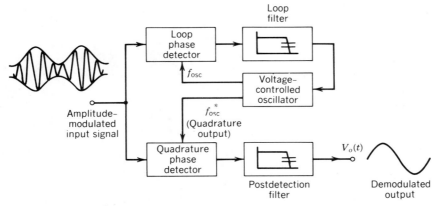

FIGURE 12.14. PLL as synchronous AM detector.

detector were to be used as the coherent carrier signal, the cos ϕ_o factor of Eq. (12.54) will be equal to zero and the demodulation technique will not work. This problem is avoided if the VCO has another output signal which is in quadrature phase (shifted by 90°) from the output signal supplied to the phase detector. In that case, this quadrature output signal would be used as the coherent carrier signal, the cos ϕ_o would be equal to unity, and the detection technique would work.

As described in Chapter 11, many of the monolithic IC oscillators have the quadrature output capability inherent to the basic design of the oscillator. If the quadrature output is not readily available, as is the case with LC-tuned oscillators, it can be obtained by dividing down the oscillator frequency, as shown in Figure 12.26. Thus, in practice, obtaining quadrature-phase output from a VCO does not present a problem.

Figure 12.14 shows the block diagram of a PLL synchronous AM demodulator based on this principle. Assuming that a true quadrature output is available from the VCO, the filtered output signal $V_o(t)$ is

$$V_o(t)\bigg|_{\text{low frequency}} = K_X E_m(t) \qquad (12.55)$$

where K_X is the gain factor of the quadrature phase detector, and $E_m(t)$ is the time-varying amplitude of the AM signal.

Note that the frequency selectivity characteristics of the AM demodulator circuit shown in Figure 12.14 are determined by the capture bandwidth of the PLL section. If the PLL is not locked on the AM signal, the VCO output is no longer coherent with the input signal, and there will be no low-frequency output signal coming out of the quadrature phase detector.

Tone or Carrier Detection

With minor modifications, the basic phase-locked AM detector system can be converted to a tone or carrier detection system as shown in Figure 12.15. The system is basically an amplitude detector, followed by a high-gain voltage comparator or slicer. When there is a carrier or a tone present at the input which

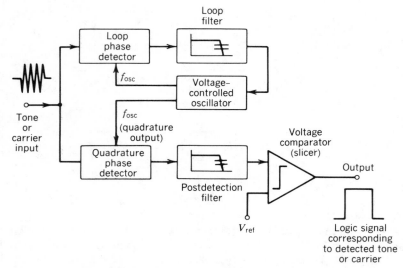

FIGURE 12.15.   Phase-locked tone or carrier detection system.

falls within the capture bandwidth of the PLL, the filtered output of the quadrature phase detector will be a dc level, proportional to the amplitude of the input tone signal. The tone detector system compares this dc level to a preset internal voltage reference, and produces a logic level signal when the input tone amplitude reaches a desired detection threshold.

Note that the tone detector is basically a *lock detector* system, which indicates that the PLL is locked onto a signal whose signal strength (amplitude) is above a given threshold level set by the comparator threshold. Thus, it can perform a number of auxiliary functions for the basic PLL. For example, if the PLL is externally swept over a frequency band, scanning for a given signal, the lock detect section can be used to stop the search once the PLL is locked. It can also be used to switch in a different loop filter, once the PLL is locked, in order to facilitate the rapid capture of an input signal and to enhance interference rejection properties once the lock is established.

The tone or carrier detection principle described above is widely used in commercial IC tone decoder and stereo decoder circuits.[9]

Phase-Angle Modulation

When a PLL is locked on a fixed-frequency signal, a steady-state dc error voltage would exist within the feedback loop to keep the PLL in the locked condition. If an additional slowly-varying ac signal is introduced into the error loop, this would cause the VCO output phase to shift accordingly in order to cancel the effect of the modulating signal and still maintain lock. In other words, the VCO output phase will be modulated by the modulating signal $V_{mod}(t)$ introduced into the error loop. This is illustrated in Figure 12.16.

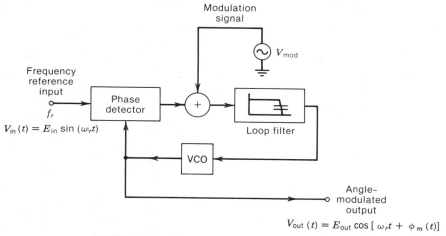

FIGURE 12.16. Phase modulation using PLL system.

The resulting output signal from the VCO is

$$V_o(t) = E_{out} \quad \cos[\omega_r t + \phi_m(t)] \tag{12.56}$$

where $\phi_m(t)$ corresponds to the phase-angle modulation and is equal to

$$\phi_m(t) = \frac{1}{K_D} V_m(t) \tag{12.57}$$

and where K_D is the phase detector conversion gain, and $V_m(t)$ is the modulation voltage.

PART II: BUILDING BLOCKS OF MONOLITHIC PHASE-LOCKED LOOP CIRCUITS

12.5. PHASE DETECTORS

The function of the phase detector is to provide an output voltage proportional to the phase difference of two periodic input signals at the same frequency. There are basically two classes of phase detectors which are readily suited to monolithic IC design. One of these is the so-called *switch* or *chopper* type phase detector which effectively operates as a synchronous switch and "chops" the input signal at the same repetition rate as the reference frequency. The other is the so-called *sequential* phase detectors, comprised of a number of cross-coupled flip-flops which are triggered by the edges of the input pulse signals. In this section, the circuit configurations and performance characteristics of these different classes of phase detectors shall be investigated.

Switch-Type Phase Detectors

The switch-type phase detectors are by far the most widely used phase detector circuits in monolithic PLL design because of their simplicity and ease of implementation in the form of a simple balanced-modulator circuit. These types of phase detectors essentially function as synchronous switches, chopping the input signal in synchronization with the VCO signal. They are also referred to as *multiplier-type* phase detectors since they basically function as mixers, where the output frequency is made up of the sum and difference components of the input frequencies.

Figure 12.17 shows the simplest form of a switch-type phase detector operating with a sinusoidal input signal. For illustrative purposes, the switch is assumed to be conducting only during the positive half-cycles of the VCO drive signal. The filtered error voltage corresponds to the average value of the output waveform, which is comprised of the shaded areas in the waveforms. Note that the error voltage is maximum and positive when both signals are in phase; it is zero when the net phase shift is $\pm$ $\pi/2$ rad (90°), and maximum negative when both signals are out of phase by $\pm$ πrad (180°). This 90° phase shift at zero error voltage is a common property of all switch-type phase detectors and results in an output dc voltage expression of the form

$$\left(V_d\right)_{av} = \frac{E_s \cos \phi}{2\pi} = K_D \cos \phi \qquad (12.58)$$

where E_s is the amplitude of the input signal.

The simple switch-type phase detector shown in Figure 12.17 is called a *half-wave* detector, since the phase information on only one-half of the input waveform is detected and averaged. In many applications, it is more practical to use a balanced modulator, such as the one shown in Figure 9.11, as a phase detector. Assuming that the linear input $V_2(t)$ is used for the sinusoidal input signal and $V_1(t)$ is driven by a high-level drive signal from the VCO, the circuit would produce the output waveform shown in Figure 12.18. Note that the balanced modulator produces an output signal on each half of the VCO drive signal; thus it is called a *full-wave* detector. After low-pass filtering, the average value of the output voltage is

$$\left(V_d\right)_{av} = \frac{2R_L E_s \cos \phi}{\pi R_E} \qquad (12.59)$$

There are two basic problems associated with the switch-type phase detectors which process a sinusoidal input signal linearly. First, the output voltage level is proportional to the input signal amplitude. This is undesirable since it makes the phase detector gain K_0 and the loop gain K_L dependent on the input signal amplitude. Second, the output is not proportional to ϕ but proportional to $\cos \phi$ instead. Both of these problems can be eliminated by amplitude limiting the input signal, that is, converting the input to a constant-amplitude square wave.

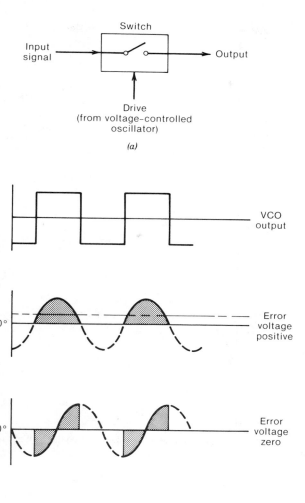

FIGURE 12.17. Operation of switch-type phase detector: (a) Block diagram; (b) typical waveforms for sinusoidal input signal.

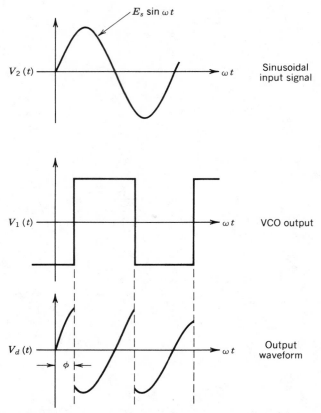

FIGURE 12.18. Input and output waveforms of balanced-modulator circuit of Figure 9.11 when used as phase detector.

In the case of the balanced modulator circuit of Figure 9.11, this can be accomplished easily by eliminating the emitter degeneration resistors R_E and driving the signal input $V_2(t)$ with a high-level signal input such that signal amplitude $E_s \gg V_T$, where $V_T = kT/q$. This circuit configuration is shown in Figure 12.19a. Its phase-to-voltage transfer characteristics can be calculated from the waveform shown in Figure 12.20. Both input and VCO signals are assumed to be high enough to switch all of the transistors in Figure 12.19 fully on and off, resulting in the output waveform of Figure 12.20c.

The average value of the output can be calculated from Figure 12.20c as the difference in areas A_1 and A_2, namely,[5]

$$\left(V_d\right)_{av} = \frac{-1}{\pi} \left[(\text{area } A_1) - (\text{area } A_2)\right] \qquad (12.60)$$

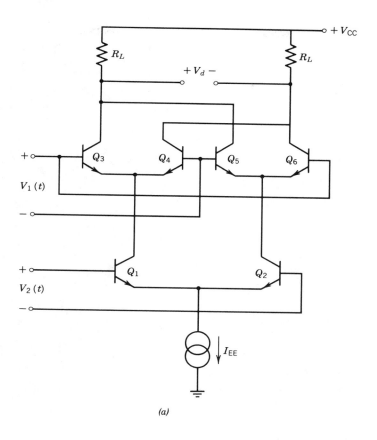

(a)

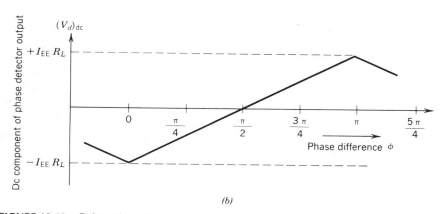

(b)

FIGURE 12.19. Balanced modulator as full-wave switching phase detector: (a) Circuit diagram; (b) output dc voltage versus input phase difference.

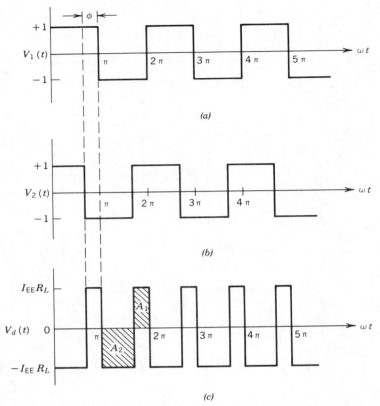

FIGURE 12.20. Timing diagram of input and output waveforms for balanced-modulator circuit of Figure 12.19.

or

$$\left(V_d\right)_{av} = I_{EE}R_L\left(\frac{2\phi}{\pi} - 1\right) \tag{12.61}$$

which has the characteristics shown in Figure 12.19b. Note that $(V_d)_{av}$ given by Eq. (12.61) is both linearly proportional to ϕ and insensitive to input amplitude as long as the input signal is sufficiently large to switch the transistors Q_1 and Q_2 of Figure 12.19 fully on and off. Because of these properties, the basic balanced-modulator circuit of Figure 12.19a is by far the most commonly used phase detector circuit in monolithic IC designs.

By examining the waveforms in Figure 12.20, one can see that the fully switched balanced-modulator circuit has, essentially, an output level that is low when both inputs have the same polarity, going high when both input signals do not have the same polarity. In ordinary combinatorial-logic terms, this corresponds to an exclusive-OR logic function where the symbol and the truth table are given in Figure 12.21. Thus, an exclusive-OR gate can also be used as a

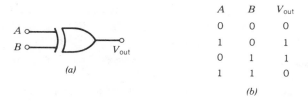

A	B	V_{out}
0	0	0
1	0	1
0	1	1
1	1	0

(b)

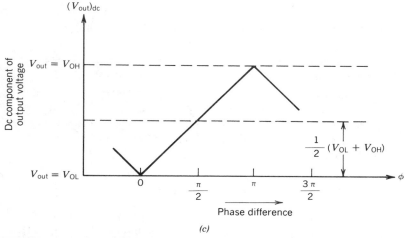

(c)

FIGURE 12.21. Exclusive-OR gate as full-wave switching phase detector: (a) Logic symbol; (b) truth table; (c) output dc voltage versus input phase difference.

phase detector and provides the output transfer characteristics shown in Figure 12.21c,

$$\left(V_{out}\right)_{dc} = V_{OL} + \frac{V_{OH} - V_{OL}}{\pi}\phi \qquad (12.62)$$

where V_{OL} and V_{OH} are the output low and high logic levels. In the case of CMOS logic, V_{OH} and V_{OL} are nominally equal to the supply voltage V^+ and the ground, respectively. Thus, the exclusive-OR gate makes an excellent phase detector for CMOS PLL designs and can provide a nominal output swing from ground to V^+ supply for the range $0 \le \phi \le \pi$.

By examining the input and output waveforms associated with switch-type phase detectors, one can make the following general statements about their electrical characteristics:

1. The phase detector gain is sensitive to the duty cycle of the two signals being compared. It is maximum and equal to that given by Eq. (12.61) if both signals are 50% duty-cycle square wave, as shown in Figure 12.20. The conversion gain decreases as the duty cycle of either input is reduced.

2. There is an inherent *phase offset* of 90° associated with a switch-type phase detector, where the midpoint of the peak-to-peak voltage output associated with the detector occurs at $\phi = \pi/2$.

3. Switch-type phase detectors also respond to odd harmonics in multiples of the input frequency. This can be seen by examining the output waveform available form a full-wave switch-type regulator, when the input frequency f_c is at three times the VCO frequency. However, since only a small portion of the output waveform contributes a net dc voltage, the effective gain of the phase detector is correspondingly reduced.

4. The feedthrough or high-frequency output from a full-wave switch-type phase detector is at twice the signal frequency, which makes it easier to filter with a simple loop filter.

5. With no input signal, the output dc voltage level of switch-type phase detectors stays at midscale (i.e., at or near the $\phi = \pi/2$ point), which corresponds to a perfect-lock condition. Thus, either momentary or long-term loss of signal does not create large changes of the loop error voltage or the VCO free-running frequency.

Sequential Phase Detectors

Sequential phase detectors generate error voltages proportional to the phase difference by detecting the zero crossings of the input and VCO signals. They are edge-triggered circuits, activated on rising or falling edges of signal pulses. Thus, they require high-level input signals which must have sharp edges. The sequential phase detectors differ fundamentally from the simple exclusive-OR-type phase detector circuit of Figure 12.21 which is a combinatorial logic circuit and has a definite input–output relationship in the form of a truth table. Instead, sequential phase detectors are made up of cross-coupled flip-flops which get set and reset in one of several predetermined sequences.

Figure 12.22 shows the logic diagram of one of the commonly used sequential phase detector circuits made up of four R–S flip-flops.[10,11] Because of the sequential nature of the circuit when it is powered up, the internal blocks of the circuit can be in any one of several stable states. Therefore, the analysis of the circuit by means of a timing diagram is too cumbersome. Instead, only the general shape of its transfer characteristics will be described.

The sequential phase detector circuit of Figure 12.21 is triggered only on the negative transitions of the input pulses. Thus the circuit is not sensitive to the duty cycle or the exact waveform of the input as long as each signal has well-defined negative transitions. The two logic outputs, designated as U and D in Figure 12.22, can both be high, but only one can be low at any one time.

The phase-to-voltage transfer characteristics shown in Figure 12.22b have five regions of operation:

1. Both input signals f_s and f_{osc} are at the same frequency with zero phase error.

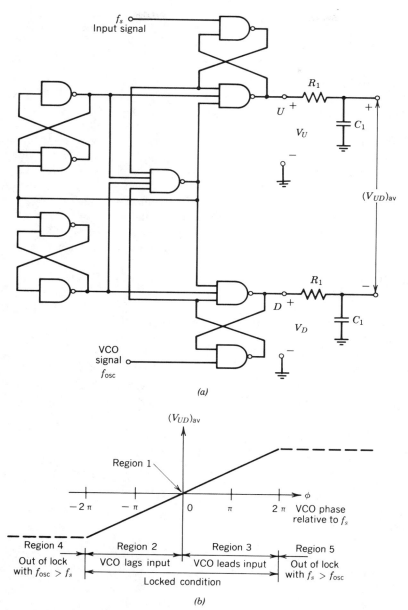

(a)

(b)

FIGURE 12.22. Sequential phase detector circuit with differential input voltage: (a) Logic diagram; (b) frequency and phase-detection characteristics.

2. Both signals are at the same frequency with f_{osc} lagging in phase.

3. Both signals are at the same frequency with f_{osc} leading in phase.

4. Both frequencies are different, with $f_s > f_{\text{osc}}$.

5. Both frequencies are different, with $f_{\text{osc}} > f_s$.

First, consider operation in region 1, that is, with $f_s = f_{\text{osc}}$ with zero phase difference. This is a single point at the center of the transfer characteristics and corresponds to the perfect-lock condition. In this condition, both the U and D outputs are at the high state and the average differential voltage after filtering, which is designated as $(V_{UD})_{\text{av}}$, is equal to zero. If the phase of f_{osc} lags the input, that is, $\phi < 0$, then D would remain high but U would exhibit negative-going pulses whose average duty cycle will increase linearly as the phase lag of f_{osc} increases. This causes a negative differential average voltage to appear at the filtered output of the phase detector and corresponds to operation in region 2 of Figure 12.22b. The negative output increases until the phase lag reaches -2π rad.

Similarly, if both frequencies are equal, but f_{osc} leads f_s, U would remain high but D could produce negative pulses whose increasing duty cycle is proportional to ϕ. This corresponds to operation in region 3 of the transfer characteristics, with increasingly positive output voltage until $\phi = 2\pi$ rad is reached.

When the two input frequencies are different such that $f_s > f_{\text{osc}}$, D stays high, but U continues to oscillate with approximately 50% average duty cycle, and the average differential output voltage remains negative and constant. This corresponds to operation in region 4 of the transfer characteristics, shown as the dashed line on the left-hand side of Figure 12.22b. Conversely, with $f_{\text{osc}} > f_s$, the roles of D and U are reversed; U is constantly high and D oscillates with negative output pulses of approximately 50% duty cycle. This corresponds to operation in region 5 of the transfer characteristics.

In summary, when the PLL is in lock, the sequential phase detector output is ideally at the zero-phase condition at the boundary of regions 2 and 3, while the phase detector output generates the necessary error voltage by shifting back and forth between these two regions. If the signal frequency is varied, the lock is maintained by moving deeper into region 2 or region 3, corresponding to the polarity of the frequency difference. The PLL can track the input signal up to a maximum phase difference of $\pm 2\pi$, which corresponds to one complete cycle of the periodic waveform. If the phase error is increased further, then the loop loses lock and the phase detector moves into region 4 or 5 where the output is a constant voltage level whose polarity indicates the polarity of the frequency difference. This polarity information, which indicates whether f_{osc} is above or below the signal frequency, is useful if the PLL is operated in a search mode, looking for a given signal.

Figure 12.23 shows a modified version of the sequential phase detector for CMOS designs.[11] The basic logic diagram is the same, except that the outputs U and D are now used to drive the upper and lower portions of a CMOS inverter stage which works as a three-state gate. When both U and D are low, the gate

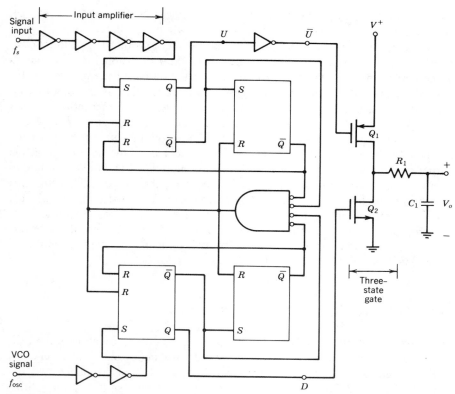

FIGURE 12.23. CMOS logic implementation of sequential phase detector circuit of Figure 12.22.[11]

output is in the off, or open-circuit, state, corresponding to operation in region 1. As the phase detector moves into region 2 or 3, the n-channel or p-channel output transistors turn on with increasing duty cycle. Thus, the low-pass filtered output $(V_o)_{av}$ exhibits the same transfer characteristics as Figure 12.22b with the $\phi = 0$ condition centered at approximately one-half of the supply voltage. In the circuit, additional inverter stages are used at the signal and VCO inputs in order to sharpen the zero-crossing transitions of the input signals and to improve edge definition of the input pulses.

Compared to switch-type phase detectors, the sequential phase detector has the following advantages:

1. It is not sensitive to the duty cycle of the input signal.
2. It does not lock on harmonics.
3. It has a wider phase detection range ($\pm 2\pi$ versus $\pm \pi/2$ rad).
4. It can indicate the polarity of the frequency difference when the loop is out of lock.

Another characteristic of the sequential phase detector is that it establishes lock with zero phase offset versus $\pm 90°$ offset for switch-type detectors.

The main drawback of a sequential phase detector is that it is intolerant to missing pulses or extra transitions at the input. Thus, occasional missing pulses at the input can create large error transients at the output of the phase detector which interprets them as large changes in frequency. This is in contrast to a switch-type phase detector where each transition has little direct influence; and the total integrated waveform determines the dc error voltage output. As an example, consider the case of the input signal dropping out or disappearing completely. In the case of a switch-type phase detector, the dc voltage at the output of the phase detector would go to zero and the VCO would continue to operate at its free-running frequency f_o. In the case of the sequential detector, this would immediately push the output to region 5 of the transfer characteristics, generating maximum error voltage and pushing the VCO to one extreme of its tracking range. Because of these characteristics, sequential phase detector circuits are not suitable for operation with weak signals in a noisy environment or for locking on intermittent signals.

12.6. VOLTAGE-CONTROLLED OSCILLATORS

The voltage-controlled oscillator (VCO) is the most critical block within a PLL system, since the PLL center frequency and the frequency discrimination characteristics are almost entirely determined by the VCO performance. As described in Chapter 11, the oscillator configurations best suited to monolithic design are the relaxation oscillators, or the so-called multivibrator circuits. In all of these circuits, the frequency is controlled by controlling the charge and discharge currents of a timing capacitor. Therefore, in truth, all of these oscillators are current-controlled circuits. However, the term VCO is used to imply, generically, any controlled or variable-frequency oscillator, without specific reference to the particulars of the control mechanism.

The circuit configurations for various multivibrator-type VCO circuits were reviewed in detail in Sections 11.3 through 11.5 and will not be repeated here. In this section, particular attention will be focused on the VCO characteristics as they relate to the overall performance of a PLL system. Some of the important VCO characteristics for various PLL applications will be considered individually.

Frequency Stability

The VCO free-running frequency f_o sets the center frequency of the PLL response characteristics. This makes the VCO frequency stability extremely critical, particularly for the detection of signals over a narrow bandwidth. For example, consider the case of a VCO with a temperature coefficient of ± 200 ppm/°C, operated over a $\pm 50°C$ temperature range. The total frequency drift due to temperature change alone would be approximately $\pm 1\%$ of the f_o setting. Thus, such a VCO circuit would not be suitable for designing a PLL system with a 1 or 2% detection band. As discussed in Section 11.5, the best VCO stability which can be obtained repeatably, under an IC production environment, is approximately 50 ppm/°C at low frequencies (i.e., $f_o < 100$ kHz) and deteriorates rapidly at higher frequencies. Thus, monolithic PLLs with relatively narrow detection bands (i.e., on the order of 2–5% of f_o) can be designed for low-frequency applications, but are not suitable at high frequencies.

In considering the VCO frequency stability with temperature, the effects of additional dc offset drifts, such as the offset drift of the phase detector output or the loop amplifier, must be considered and must be referred to the VCO control terminals as an effective loop-offset drift voltage $(V_e)_{os}$. Thus, the total VCO stability is made up of two components,

$$\text{total VCO drift} = \underbrace{\frac{1}{f_o}\frac{\partial f_o}{\partial T}}_{\text{drift due to VCO}} + \underbrace{\frac{K_0}{f_o}\frac{\partial (V_e)_{os}}{\partial T}}_{\text{drift due to PLL offsets}} \qquad (12.63)$$

where K_0 is in units of hertz per volt.

As indicated by Eq. (12.63), the thermal drifts of PLL offset voltages become a significant contributor to VCO stability when the VCO conversion gain is high. In a well-designed monolithic PLL, the second term in Eq. (12.63) is minimized by careful layout to reduce the dc offsets involved, and by keeping the VCO conversion gain K_0 relatively low.

High-Frequency Capability

The high-frequency capability of a PLL is almost always limited by the VCO frequency capability. Since the phase detectors, particularly the balanced-modulator-type detectors, operate in a current mode and their output is filtered to produce a dc voltage, phase detector response characteristics rarely present a limitation for high-frequency operation. However, the VCO must operate in a stable manner with well-defined control characteristics at a frequency equal to the input signal. As described in Section 11.5, frequency capabilities of most IC relaxation oscillators are limited to 1 MHz. One exception is the all-*npn* emitter-coupled multivibrator circuit of Figure 11.25 or Figure 11.28, which can be operated in tens of megahertz range, provided that the frequency stability requirements can be relaxed.

Frequency Accuracy

From a practical applications point, it is desirable to set the frequency of a monolithic PLL circuit with a minimum number of external components; and that the frequency be *accurately* related to the external component setting. Frequency accuracy is particularly important to minimize the fine adjustments or trimming necessary as different integrated circuits are placed into the same circuit socket.

Most relaxation oscillators require only two precision components, a resistor and a capacitor, to set the frequency. For low-frequency designs, a unit-to-unit accuracy of $\pm 1\%$ can be obtained by careful IC design for the same frequency setting. However, for high-frequency designs, using the all-*npn* emitter-coupled multivibrator, frequency setting accuracy is typically in the $\pm 5\%$ range.

Linear Control Characteristics

In the application of a PLL as a frequency discriminator, the linearity of the demodulated output is directly related to the VCO voltage-to-frequency conversion characteristics. Thus, in those PLL applications requiring conversion of frequency changes to voltage output, the VCO section of the system is required to have linear control characteristics. Any nonlinearity of VCO control characteristics will reflect as distortion in the detected output signal. Fortunately, linear control characteristics are some of the inherent properties of multivibrator-type VCOs which use voltage-controlled current sources. However, the situation is different in the case of LC-tuned VCO circuits, or crystal-controlled VCOs, which require voltage-controlled capacitors (i.e., varactor diodes) to control the frequency. Since varactor characteristics are not linear, these oscillators have relatively poor frequency control characteristics and limited tuning range. As a result, PLL circuits using such oscillators have poorer distortion characteristics for frequency discriminator or FM detector applications.

In the case of PLL applications where the VCO output is used as the loop output, such as in AM and tone-decoding and frequency synthesis, the linearity of VCO control characteristics is not important as long as the control range is large enough to keep the VCO in lock.

Adjustable Tracking Range

In certain applications, it is necessary to limit the tracking range of the VCO by introducing a nonlinear limiting mechanism into the VCO control characteristics. In this manner, the total tracking or lock range of the PLL can be determined by the internal limiting mechanism associated with the VCO control characteristics, rather than by the available loop error voltage. This technique provides an added degree of design freedom to reduce the lock range and yet still maintain a high loop gain.

Figure 12.24 shows a typical VCO control characteristic with limited tracking range. To be effective in PLL operation, the limiting of the control range must come about at a control voltage level $\pm V_L$ that is less than the maximum available error voltage, $(V_e)_{max}$, as shown in Figure 12.24.

Figure 12.25 illustrates how the basic emitter-coupled multivibrator circuit of Figure 11.28 can be modified to exhibit limited tracking characteristics. The tracking range is limited by "offsetting" the center frequency with the addition of two fixed-bias current sources I_1 connected to either side of the timing capacitor. Then, following the derivation of Eq. (11.34), the frequency is

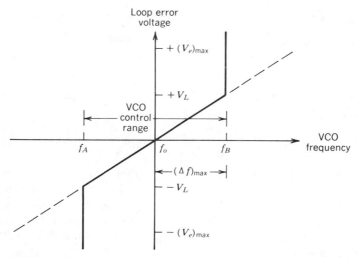

FIGURE 12.24. VCO control characteristics with limited tracking range.

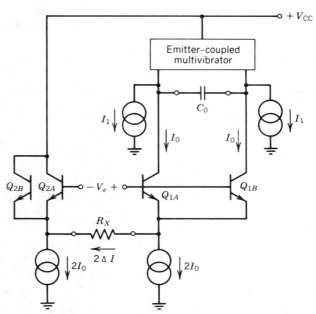

FIGURE 12.25. VCO circuit with adjustable tracking range.

proportional to the total current $I_1 + I_0$ available to charge and discharge C_0, that is,

$$f_o = \frac{I_1 + I_0}{4C_0V_{BE}} \tag{12.64}$$

where I_0 is the current that is controlled by the error voltage input V_e, and I_1 is constant. The conversion gain of the VCO is

$$K_0 = \frac{df_o}{dV_e} = \left(\frac{df_o}{dI_0}\right)\left(\frac{dI_0}{dV_e}\right) = \frac{1}{8C_0V_{BE}R_X} \text{ Hz/V} \tag{12.65}$$

Thus, as the error voltage is varied, the VCO would respond to V_e until the differential current $2\Delta I$ in the resistor R_X is equal to $\pm 2I_0$. When this limit is reached, either the Q_{1A}, Q_{1B} pair or the Q_{2A}, Q_{2B} pair of transistors would be completely cut off, and the VCO frequency will be insensitive to additional changes of V_e. Thus, the total tracking range is limited to frequencies $f_o \pm (\Delta f)_{max}$, where

$$(\Delta f)_{max} = \frac{I_0}{4C_0V_{BE}} \tag{12.66}$$

and it would be reached at the limiting value of error voltage,

$$V_e\bigg|_{limit} = V_L = 2R_XI_0 \tag{12.67}$$

In this manner, by keeping $V_L \ll (V_e)_{max}$, very narrow bandwidths can be obtained and externally adjusted by the choice of the range-setting resistor R_X.

VCO Output Waveform

In most applications, the exact shape of the VCO output waveform is not critical. However, in PLL systems, with switch-type phase detectors, a square-wave or 50% duty-cycle output from the VCO is preferred since this provides the optimum conversion gain for the phase detector.

In PLL applications where a sinusoidal VCO signal is desired as the system output, this can be obtained by either filtering the VCO output or putting the VCO waveform through a wave-shaping circuit (see Section 11.6). In the case of emitter-coupled multivibrator or constant-current charge and discharge oscillators, the triangular voltage waveform across the timing capacitor can be used for wave-shaping purposes.

Quadrature output from the VCO is desirable for AM and tone-detection applications. Usually, with charge and discharge type oscillators, such an output can be available by squaring the triangle-wave output across the timing capacitor (see Fig. 11.24). Another technique commonly used for obtaining 50% duty-cycle output waveforms with quadrature phase relationship is to divide down the

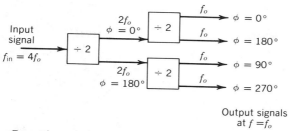

Output signals
at $f = f_o$

FIGURE 12.26. Generating square-wave quadrature outputs by binary countdown technique.

VCO waveform through a series of D-type flip-flops, as illustrated in Figure 12.26. This technique is particularly useful when the VCO output waveform has either a very low duty cycle or an unconventional wave shape.

12.7. MONOLITHIC PLL DESIGN EXAMPLE

In designing a monolithic PLL system, one initially starts with the basic circuit configurations associated with the two key blocks within the system: the VCO and the phase detector sections. Usually, the VCO is the most critical portion of the design. Its circuit configuration is often dictated by the frequency stability and high-frequency capability requirements. In bipolar analog designs, the phase detector is almost always realized as a balanced modulator since this circuit block has a differential configuration which is readily suited to monolithic fabrication. In CMOS analog designs, an exclusive-OR gate is often used to perform the same function.

The loop filter is normally external to the IC chip. In most designs, the filter function is performed by simply bringing the phase detector output to a package pin where an external capacitor, or a series R–C combination, can perform the filter function in conjunction with the internal load resistance on the chip.

Many monolithic PLL designs, particularly those designed for general-purpose applications, have provisions for an external adjustment of loop gain. This is done by means of an external gain-setting resistor which controls the VCO conversion gain or the gain of a separate amplifier block in the loop, if there is one.

Normally, the key system parameters, such as the loop gain, VCO frequency, PLL lock and capture range, loop damping, and closed-loop bandwidth are set by external components in order to optimize the circuit performance for a specific application and a particular input signal level.

The basic characteristics and parameters of a PLL system, covered in preceding sections, can be illustrated best by means of a specific design example which also demonstrates some of the performance trade-offs necessary. The following design example is prepared with this aim in mind.

Figure 12.27 gives a simplified diagram of the PLL design example. The circuit is intended as a monolithic FM demodulator which will lock on an FM input and generate a demodulated audio output signal. The circuit is made up of a multivibrator-type VCO which drives a balanced-modulator-type phase detector. The phase detector output is then low-pass filtered, buffered, and level shifted, and connected back to the VCO control terminals. For the purpose of minimizing internal drift and offset voltage, a fully differential design is used. The demodulated audio output is picked off and amplified from the loop error voltage.

Figure 12.28 shows the detailed circuit schematic of the monolithic FM demodulator PLL circuit. For illustrative purposes, the circuit configuration is kept as simple as possible and unnecessary circuit detail is left out.

With the component values shown, the circuit is designated to operate with nominal ± 6-V supply voltages. The oscillator section is basically the all-*npn* emitter-coupled multivibrator circuit of Figure 11.28. The VCO output is directly connected into the emitters of the cross-coupled transistors Q_1, Q_2 and Q_3, Q_4 which serve as the phase detector. The base–emitter junctions of these transistors, along with resistors R_4 and R_5, serve as diodes D_1, D_2 and resistors R_4, R_5 of Figure 11.28.

The phase detector outputs are filtered by external capacitors and buffered by emitter followers Q_5 and Q_6. Then they are level shifted by the combination of Zener diodes D_{Z1} and D_{Z2} and the current source I_B (see Fig. 4.29) and directly applied to the VCO terminals as a differential control signal.

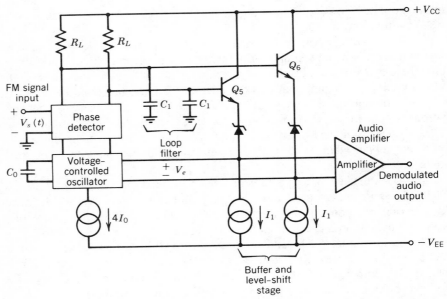

FIGURE 12.27. Simplified diagram of PLL FM demodulator circuit of Figure 12.28.

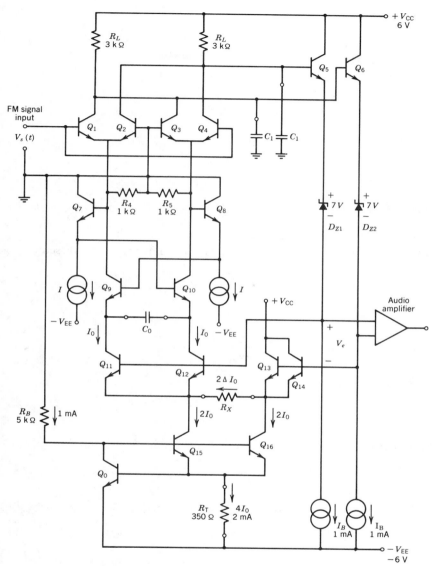

FIGURE 12.28. Design example for PLL FM demodulator circuit.

The VCO section is comprised of cross-coupled transistors Q_9 and Q_{10}, which are connected through emitter-follower buffer stages Q_7 and Q_8. At steady state (i.e., with zero error voltage) current-source transistors Q_{11}, Q_{12}, Q_{13}, and Q_{14} all carry identical amounts of current I_0.

The bias resistor R_B along with transistors Q_0, Q_{15} and Q_{16} and the external resistor R_T form a pair of supply-independent current sources (see Fig. 4.21). The current level is set by forcing 1 V_{BE} drop (due to V_{BE} of Q_0) across the external resistor R_T, such that

$$I_0 = \frac{V_{BE}}{4R_T} \qquad (12.68)$$

The VCO free-running frequency f_o from Eq. (11.34) is

$$f_o = \frac{I_0}{4C_0 V_{BE}} \qquad (12.69)$$

Substituting Eq. (12.68), one obtains

$$f_o = \frac{1}{16C_0 R_T} \qquad (12.70)$$

which, to first order, is insensitive to frequency.

In the present example, assuming a nominal value of $V_{BE} = 0.7$ V, R_T is chosen to be 350 Ω to give set, $I_0 = 0.5$ mA.

The VCO and the phase detector conversion gains can be expressed as

$$K_D = \frac{2R_L I_0}{\pi} \text{ V/rad} \qquad (12.71)$$

and

$$K_0 = \frac{\pi}{2C_0 V_{BE} R_X} \text{ rad/sec. V} \qquad (12.72)$$

and the total loop gain is

$$K_L = K_D K_0 = \frac{R_L I_0}{C_0 V_{BE} R_X} \qquad (12.73)$$

Substituting Eq. (12.69) into Eq. (12.73), one gets

$$K_L = \frac{4f_o R_L}{R_X} \qquad (12.74)$$

The total lock range for the PLL is proportional to K_L, as given by Eq. (12.32),

$$\text{lock range} = \Delta f_L = \frac{K_L}{4} = \frac{f_o R_L}{R_X} \qquad (12.75)$$

Now that the system parameters are well defined, we can proceed to calculate the component values necessary for its application as an FM demodulator, operating with 1-MHz FM carrier, and having $\pm$ 20-kHz deviation range. We will consider the circuit operation for both wideband and narrowband FM detection applications.

Wideband FM Detection. Assume that the information band is 20 kHz and that a flat audio response is required.

Step 1: Set oscillator frequency. f_o is set at 1 MHz by calculating C_0 from Eq. (12.70),

$$C_0 = \frac{1}{16 f_o R_T} = 178 \text{ pF} \tag{12.76}$$

Step 2: Select R_X to set lock range. As a general rule of thumb, Δf_L is chosen to be approximately twice as wide as the FM signal frequency deviation. Choosing $\Delta f_L = 40$ kHz for a ± 20-kHz deviation FM signal, one calculates the gain-setting resistor R_X from Eq. (12.75),

$$R_X = R_L \frac{f_o}{\Delta f_L} = 75 \text{ k}\Omega \tag{12.77}$$

where $R_L = 3$ kΩ is the phase detector load resistance. The corresponding value of the loop gain is

$$K_L = \frac{4 f_o R_L}{R_X} = 1.6 \times 10^5 \quad \text{sec}^{-1} \tag{12.78}$$

Step 3: Select C_1 to determine the loop bandwidth. In order to detect the FM signal with a flat audio response up to 20 kHz, the loop damping factor ζ is set to 0.707, which requires a low-pass filter cutoff frequency ω_1 given as [see Eq. (12.21) and (12.22)]

$$\omega_1 \Big|_{\zeta = \frac{1}{\sqrt{2}}} = \frac{1}{R_L C_1} = 2 K_L \tag{12.79}$$

Solving for C_1, one gets

$$C_1 \approx 1000 \quad \text{pF} \tag{12.80}$$

The corresponding closed-loop bandwidth from Eq. (12.23) is

$$f_{3 \text{ dB}} = \frac{\omega_{3 \text{ dB}}}{2\pi} = \frac{\sqrt{2} K_L}{2\pi} = 35.9 \text{ kHz} \tag{12.81}$$

which is sufficient for detecting the information over a 20-kHz band.

Narrowband FM Detection. Assume that the input FM frequency and deviation are the same, but the information band is 3 kHz. Thus, a narrow loop bandwidth is desired to filter out frequencies beyond 3 kHz.

To reduce the loop bandwidth, one would normally reduce K_L, which in turn causes a proportional reduction in the lock range Δf_L. However, since the input signal deviation is still ± 20 kHz, Δf_L cannot be reduced. Simply increasing C_1 to reduce ω_1 would not be acceptable, since this would cause the loop to be greatly underdamped [see Eq. (12.21)]. The only solution out of this dilemma is to use a lag-lead filter, placing a resistor R_2 in series with each of the filter capacitances. This allows the loop-filter capacitance to be increased without greatly reducing the loop damping. Thus, we reduce the loop bandwidth by increasing C_1 and then correct for loop damping by adding R_2 in series with C_1 into the low-pass filter. With reference to the design equations (12.27) through (12.29), or those shown in Table 12.1, one proceeds as follows.

Step 1: Calculate C_1 to set a 3-dB bandwidth.   From Eq. (12.29), the low-pass filter bandwidth ω_1 can be calculated as

$$\omega_1 = \frac{1}{(R_1 + R_2)C_1} \approx \frac{(\omega_{3\,dB})^2}{4K_L} = 555 \text{ rad/sec} \tag{12.82}$$

Assuming $R_1 \gg R_2$,

$$C_1 = \frac{1}{(R_1 + R_2)\omega_1} \approx \frac{1}{R_1\omega_1} = 0.6 \ \mu\text{F} \tag{12.83}$$

Step 2: Calculate the value of R_2 to keep $\zeta \approx 1/\sqrt{2}$.   From Eq. (12.28), one gets

$$\omega_2 = \frac{1}{R_2C_1} \approx \sqrt{\frac{K_L\omega_1}{2}} = 6.66 \times 10^3 \text{ rad/sec} \tag{12.84}$$

or

$$R_2 = \frac{1}{\omega_2C_1} = 250 \ \Omega \tag{12.85}$$

Note that both the $R_1 \gg R_2$ assumption used in Eq. (12.83) and the $K_L \gg \omega_1$ assumption used in Eq. (12.28) are satisfied.

REFERENCES

1. F. M. Gardner, *Phaselock Techniques,* 2nd ed., Wiley, New York, 1979.
2. W. C. Lindsey and M. K. Simon, Eds. *Phase-Locked Loops and Their Applications,* IEEE Press, New York, 1978.
3. A. B. Grebene, and H. R. Camenzind, "Frequency-Selective Integrated Circuits Using Phase-Lock Techniques," *IEEE J. Solid-State Circuits* **SC-4,** 216–225 (August 1969).
4. A. B. Grebene, *Analog Integrated Circuit Design,* Van Nostrand Reinhold, New York, 1972, Chap. 9.
5. P. R. Gray and R.G. Meyer, *Analysis and Design of Analog Integrated Circuits,* Wiley, New York, 1977, Chap. 10.
6. *Phase-Locked Loop Data Book,* Exar Integrated Systems, Sunnyvale, CA, 1981.
7. W. F. Egan, *Frequency Synthesis by Phase-Lock,* Wiley Interscience, New York, 1981.
8. A. Blanchard, *Phase-Locked Loops: Application to Coherent Receiver Design,* Wiley, New York, 1976.
9. A. B. Grebene, "The Monolithic Phase-Locked Loop—A Versatile Building Block," *IEEE Spectrum,* 38–49 (March 1971).
10. *Phase-Locked Loop Data Book,* 2nd Ed., Motorola, Phoenix, AZ, August 1973.
11. D. K. Morgan and G. Steudel, "The RCA COS/MOS Phase-Locked Loop," App. Note ICAN-6101, RCA Corp., Somerville, NJ, October 1972.

CHAPTER THIRTEEN

INTEGRATED-CIRCUIT FILTERS

In the design of frequency-selective integrated circuits, the lack of inductors is a serious disadvantage. In many applications, this drawback can be overcome by using conventional active-*RC* filter techniques. In designing active-*RC* filters, one uses a combination of resistors, capacitors, and gain blocks in linear feedback loops to obtain the desired frequency selectivity without the need for inductors. Active-*RC* filter design and synthesis techniques have been extensively investigated and are well covered in the literature.[1-4] Since the basic blocks of active filters, namely resistors, capacitors, and amplifiers, are available in monolithic form, active-*RC* filter techniques are, in principle, readily compatible with monolithic integrated circuits. However, in practice, direct application of active *RC* techniques to monolithic IC filters has been somewhat limited because of the following inherent limitations of monolithic IC technology:

1. Limited size of monolithic capacitors.
2. Poor absolute-value tolerances of monolithic components, particularly the resistor and capacitor values.
3. Poor temperature stability of monolithic resistors.

In almost all cases, the performance characteristics of active-*RC* filters depend very strongly on the absolute values of the circuit components and gain parameters. Thus, practical implementation of active-*RC* filters often requires the use of hybrid IC technology, where the resistors and capacitors are left external to the IC chip which normally contains only the amplifier sections. In some cases requiring small capacitor values, active-*RC* filters can be fabricated in single-chip form by depositing thin-film resistors on the monolithic chip. However, in all these cases, the circuit still requires precision trimming of one or more of the circuit components, which makes it incompatible with the low-cost, high-volume production capabilities of basic monolithic IC fabrication technologies. As a result, active-*RC* filters have developed into an excellent application area for monolithic IC products, where low-cost IC gain blocks are used with external precision components, such as single or multiple operational amplifier arrays, to implement the filter function.

Switched-capacitor filters provide an excellent approach to the design of monolithic IC filters, which avoids the basic shortcomings of conventional active-*RC* filters.[4-6] Switched-capacitor filters belong to a class of circuits known as *discrete-time* or *sampled-data* systems, where the information is processed not in a continuous form, but at discrete time intervals by means of periodically operated sampling switches. Switched-capacitor filters are comprised of arrays of capacitors, analog switches, and operational amplifiers which are connected in both the continuous and the sampled-data feedback paths. The analog switches are driven by periodic clock signals. They sample the charge stored in the capacitors, which is then sensed and integrated by operational amplifier integrators. Although the basic concepts of switched-capacitor filters were known for many years, their monolithic implementation has become practical with the advent of MOS analog circuits. In the design of monolithic switched-capacitor filters, one relies on the unique property of MOS circuits to sample and hold a charge in a given circuit node for prolonged periods of time; and be able to sense this charge continuously and nondestructively. In this manner, it is possible to simulate a filter transfer function solely in terms of capacitor *ratios* and the periodic clock frequency which activates the analog switches. Since the capacitor ratios can be held to very accurate and temperature-stable tolerances, and an accurate and stable clock signal is easily derived from a crystal oscillator or system clock, switched-capacitor filters avoid virtually all of the three limitations associated with monolithic technology listed at the beginning of this section, making filter design directly compatible with monolithic technology. However, because of the sampled-data nature of their operation, switched-capacitor filters differ from conventional continuous-time (i.e., nonsampled) filters, and have some unique properties of their own which limit their usage primarily to audio-frequency applications.

The purpose of this chapter is to focus on the monolithic IC implementation of filter functions from the point of view of a monolithic IC designer, rather than a network theorist of a filter design engineer. Therefore, in the course of the discussion, we will avoid the lengthy derivations related to network or transform theory, and concentrate on practical implementation in IC form. Whenever possible, the relative merits of various filter techniques will be viewed with respect to: (1) ease of IC fabrication, (2) sensitivity to component tolerances, and (3) frequency range of operation and selectivity characteristics achievable in integrated design.

Since the switched-capacitor filter is the most readily compatible filter technique for monolithic integrated circuits, the focus of this chapter is directed to switched-capacitor filters. The chapter is comprised of two parts. The first part presents a brief review of some of the basic filter specifications, characteristics, and terminology. Some of the basic differences between continuous-time and discrete-time (i.e., sampled-data) filters are also discussed in this part. The second part of the chapter is devoted to switched-capacitor filters, their principle of operation and methods of synthesis. Special emphasis is given to practical design considerations, particularly with regard to inherent parasitic capacitances associated with monolithic MOS structures.

PART I: A REVIEW OF FILTER CHARACTERISTICS

13.1. BASIC FILTER SPECIFICATIONS

The function of a filter is to provide a frequency-weighted transmission of analog signals. In other words, signals falling within selected regions of the frequency spectrum are attenuated or amplified selectively, relative to those in other parts of the frequency spectrum. Filters are typically categorized according to the parts of the frequency spectrum they emphasize, such as low-pass, bandpass, high-pass, and band-reject filters. The manner of specifying these filter types and their characteristics is illustrated in Figures 13.1 and 13.2. Filter types are normally specified in terms of passband (PB) gain and ripple, stopband (SB) attenuation, passband cutoff frequency ω_C, and stopband corner frequency ω_S. The band of frequencies between the passband and the stopband is called the transition band (TB). As illustrated in Figure 13.1 and 13.2, a desired filter magnitude response $|H(j\omega)|$ must lie within the unshaded region of the filter specifications.

Low-Pass Filters

The function of a low-pass filter is to pass low frequencies from dc to some specified cutoff frequency, and to attenuate high frequencies. As shown in Figure 13.1a, such a filter is specified by its dc gain, passband ripple, and stopband attenuation. The filter passband is defined as the frequencies $0 \le \omega \le \omega_C$, and the stopband corresponds to the frequencies where $\omega > \omega_S$; the transition band covers the frequencies $\omega_C < \omega < \omega_S$.

High-Pass Filters

The function of the high-pass filter is to pass frequencies in excess of a cutoff frequency ω_C and attenuate low frequencies from dc to some specified stopband frequency ω_S. A typical example of such a filter specification is shown in Figure 13.1b. In theory, the passband of a high-pass filter extends to $\omega = \infty$. However,

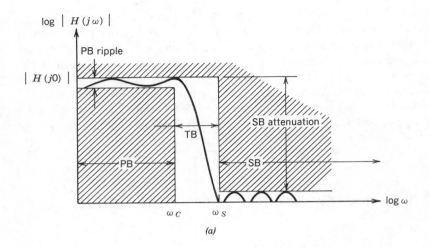

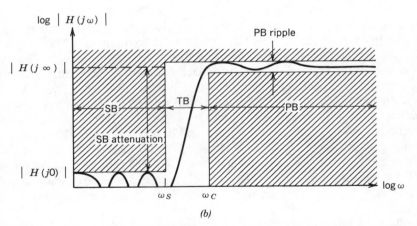

FIGURE 13.1. Basic specifications for low-pass and high-pass filters: (*a*) Low-pass filter; (*b*) high-pass filter.

in practice the passband is limited to finite frequencies due to the finite bandwidth of active devices.

Bandpass Filters

The bandpass filter is designed to pass a finite band of frequencies falling between the lower and upper cutoff frequencies ω_{CL} and ω_{CU}, and to attenuate those that fall outside the lower and upper stopband frequencies ω_{SL} and ω_{SU}, as illustrated in Figure 13.2a. In general, upper and lower stopband attenuations and transition bands do not need to be symmetrical.

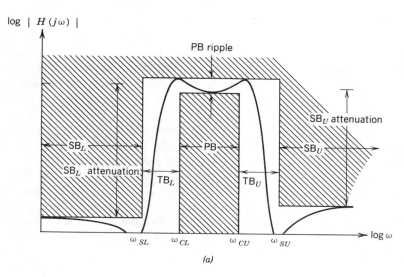

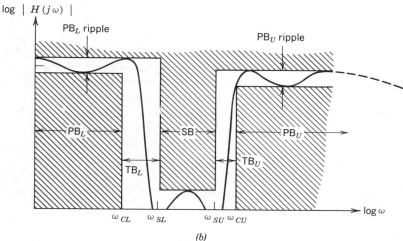

FIGURE 13.2. Basic specifications for bandpass and band-reject filters: (*a*) Bandpass filter; (*b*) band-reject filter.

Band-Reject Filters

The function of a band-reject filter is to attenuate a finite band of frequencies which fall between the lower and upper stopband frequencies ω_{SL} and ω_{SU}, while passing both lower and higher frequencies, as illustrated in Figure 13.2*b*. Such a filter has both a high- and a low-frequency passband, shown as PB_U and PB_L. The gains of the upper and lower passband regions do not need to be symmetrical. Similar to the case of the high-pass filter, the upper passband is normally limited to finite frequencies due to the limited bandwidth of the active components involved.

13.2. A REVIEW OF BASIC FILTER TYPES

An ideal filter would exhibit a perfectly flat response in its passband and infinite attenuation in its stopband, with a very rapid transition from passband to stopband. Furthermore, an ideal filter is assumed ot have *linear phase* characteristics, where the phase shift through the filter increases *linearly* with increasing frequency. This highly idealized model of filter response can only be approximated in practice. Practical filter designs approximate the ideal filter response as a ratio of two rational polynomials in the complex frequency domain, where the highest power in the polynomials determines the order of the filter. The response characteristics and polynomial coefficients associated with these basic filter types are well tabulated in active filter handbooks.[7] The most commonly used filter approximations, and the general features of their response characteristics, are briefly reviewed in this section. For illustrative purposes, the discussion is limited to low-pass filters. However, the basic conclusions are equally applicable to both high-pass and bandpass filters.

Maximally-Flat or Butterworth Filters

The Butterworth filter is designed with emphasis on the flatness of the response within the passband. This is achieved at the expense of phase linearity, and to a lesser degree, at the expense of the steepness of the attenuation slope in the transition band. Figure 13.3*a* shows the typical magnitude response of a low-pass Butterworth filter. Its phase response relative to the other filter types is shown in Figure 13.4. The Butterworth filter has an all-pole transfer function. Its magnitude response decreases monotonically with increasing frequency, and it achieves infinite attenuation only at infinite frequency. Since the attenuation slope and the impulse response of the Butterworth filter are quite good compared to other filter approximations, it is one of the most commonly used general-purpose filter types.

Equiripple or Chebyshev Filters

The Chebyshev filter approximation emphasizes the steepness of the attenuation slope outside the passband. This is achieved at the expense of phase linearity and the passband flatness. It is normally designed with a prescribed ripple in the passband. In return for a finite ripple in the passband, it offers a sharp cutoff (i.e., a high attenuation rate) around the passband edge. The attenuation characteristics within the stopband are similar to those of a Butterworth filter of the same order. The Chebyshev filter exhibits slight overshoot when driven by a step input. Similar to a Butterworth filter, the Chebyshev filter approximation is an all-pole transfer function. Typical magnitude response characteristics of a Chebyshev filter are shown in Figure 13.3*b*.

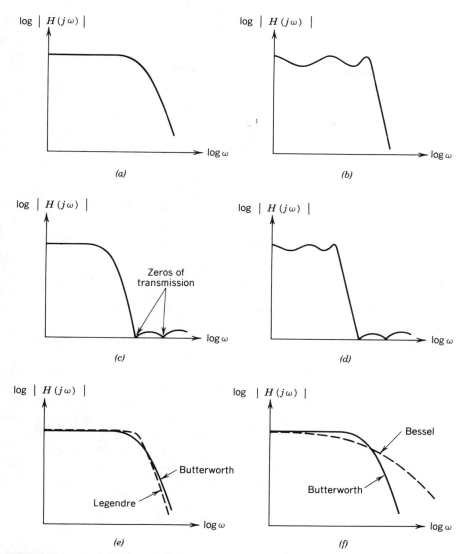

FIGURE 13.3. Typical magnitude response characteristics of various filter types: (a) Butterworth; (b) Chebyshev; (c) inverse Chebyshev; (d) elliptic; (e) Legendre; (f) Bessel (linear phase).

Inverse Chebyshev Filters

The inverse Chebyshev filter approximation provides a monotonic response within the passband with infinite attenuation within parts of the stopband. Thus, it exhibits a flat response in the passband, but finite ripple in the stopband. Its attenuation slope outside the band edge is somewhat less steep than for the

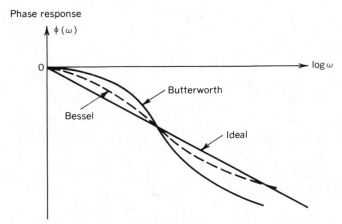

FIGURE 13.4. Relative phase response characteristics of various filter types.

normal Chebyshev approximation. Figure 13.3c shows the typical response characteristics of an inverse Chebyshev filter. It is particularly useful when certain frequencies within the stopband need to be heavily attenuated.

Elliptic Filters

The elliptic filters have a finite ripple in both the passband and the stopband response, as shown in Figure 13.3d. Similar to Chebyshev filters, they exhibit a very sharp attenuation slope (i.e., narrow transition band) outside the passband. Similar to the case of inverse Chebyshev filters, elliptic filters can also provide finite transmission zeros (i.e., infinite attenuation) at prescribed frequencies within the passband. In general, elliptic filters are the most efficient filter type in terms of the component count necessary to implement the filter function.

Optimum Monotonic or Legendre Filters

The Legendre filter is optimized to have a maximum attenuation rate in the transition band, while remaining monotonic in both the passband and the stopband. As a result, its passband response is not as flat as that of a Butterworth filter, but its band-edge attenuation is steeper, as illustrated in Figure 13.3e. It is often used in those applications where the high attenuation rate is required, but the passband or stopband ripple cannot be tolerated.

Linear Phase or Bessel Filters

The Bessel filter is designed primarily to simulate the linear phase response characteristics of an ideal filter. This compromise is made at the expense of attenuation characteristics, as illustrated in Figures 13.3f and 13.4. The linear phase characteristics are particularly important if the filter is designed to handle square-wave or pulse signals. Unlike the other filter types described in this section, Bessel filters do not produce ringing or overshoot when driven with pulsed signals. The Bessel filters also have all-pole transfer functions, similar to those of Butterworth and Chebyshev filters.

13.3. BIQUADRATIC FILTER FUNCTION

One of the basic building blocks of active filter design is the biquadratic filter function, often referred to as the *biquad*. It is a second-order transfer function which, in its generalized form, can be expressed as

$$H(s) = \frac{a_2 s^2 + a_1 s + a_0}{s^2 + b_1 s + b_0} = \frac{a_2 (s + z_1) (s + z_2)}{(s + p_1) (s + p_2)} \qquad (13.1)$$

where s ($= \sigma + j\omega$) is the complex frequency variable, and z_i and p_i are the respective zeros and poles of the transfer function. In the case where the poles and zeros are complex-conjugate pairs,

$$z_{1,2} = -\sigma_z \pm j\omega_{z0} \qquad (13.2)$$

and

$$p_{1,2} = -\sigma_p \pm j\omega_{p0} \qquad (13.3)$$

where σ_z and σ_p represent the real parts, and ω_{z0} and ω_{p0} are the conjugate imaginary parts of zeros and poles. It is more convenient to express $H(s)$ in the form

$$H(s) = \frac{K[s^2 + (\omega_z/Q_z)s + \omega_z{}^2]}{s^2 + (\omega_p/Q_p)s + \omega_p{}^2} \qquad (13.4)$$

where K is a gain constant and

$$\omega_z = |\sigma_z + j\omega_{z0}| = \sqrt{\sigma_z{}^2 + \omega_{z0}{}^2} \qquad (13.5)$$

$$\omega_p = |\sigma_p + j\omega_{p0}| = \sqrt{\sigma_p{}^2 + \omega_{p0}{}^2} \qquad (13.6)$$

which correspond to the radial distance from the origin to the respective zero or pole location, and the parameter Q is given as

$$Q_z = \frac{\omega_z}{2\sigma_z} \quad \text{and} \quad Q_p = \frac{\omega_p}{2\sigma_p} \qquad (13.7)$$

Q_z and Q_p are known as the zero and pole *quality factors*.

The popularity of the basic biquad transfer function stems from the fact that it can be implemented relatively easily with a single operational amplifier or gain block, and can serve as a building block for more complex filter functions. By the proper choice of circuit topology and component values, the coefficient values in Eq. (13.1) or Eq. (13.4) can be selected to provide the basic low-pass, high-pass, bandpass, and band-reject functions. These are briefly reviewed in the following sections.

Low-Pass Function

The purpose of a low-pass filter is to pass low frequencies from dc to some specific cutoff frequency, and attenuate the high frequencies as illustrated in Figure 13.1a. A simple approximation to such a transfer characteristic can be realized by a second-order transfer function of the form

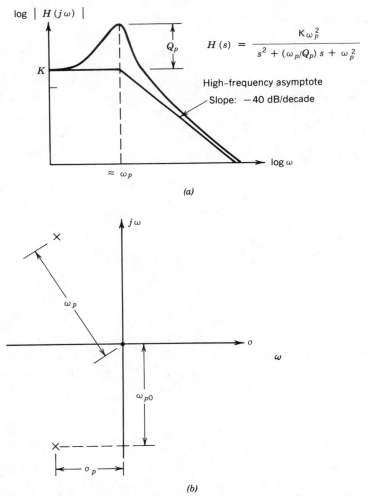

$$H(s) = \frac{K\omega_p^2}{s^2 + (\omega_p/Q_p)s + \omega_p^2}$$

High-frequency asymptote

Slope: -40 dB/decade

(a)

(b)

FIGURE 13.5. Second order (biquadratic) low-pass filter function: (a) Magnitude response; (b) pole locations.

$$H(s) = \frac{K\omega_p^2}{s^2 + (\omega_p/Q_p)s + \omega_p{}^2} \qquad (13.8)$$

In the case of a complex pole pair, this corresponds to the amplitude function shown in Figure 13.5. Note that the dc gain, that is, $|H(j\omega)|$ at $\omega = 0$, is equal to K. The gain decreases at a rate of -40 dB/decade for frequencies where $\omega \gg \omega_p$, and the circuit would exhibit a peaking of the magnitude response at $\omega = \omega_p$ by an amount that is proportional to Q_p. The zeros of the transfer function are located at $\omega = \infty$.

The low-pass biquad transfer function of Eq. (13.8) can be synthesized by using a wide choice of circuit configurations. Figure 13.6 shows two typical examples of obtaining such a transfer function by use of either positive or negative feedback around a gain stage. The transfer functions of these circuits are well covered in the literature, in terms of circuit component values, and will not be repeated here.[4]

The positive-feedback biquad circuits, similar to that shown in Figure 13.6a, have the following advantages:

1. They can be designed with low values of gain. In fact, the circuit can be unstable for $K > 3$.
2. They exhibit a relatively low spread of component values (i.e., between largest and smallest values of Rs and Cs).

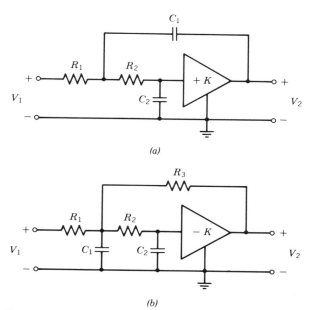

(a)

(b)

FIGURE 13.6. Single-amplifier low-pass biquad filter stages using positive or negative feedback: (a) Positive-feedback circuit; (b) negative-feedback circuit.

However, they have the serious disadvantage that the circuit sensitivity* to component and gain tolerances is extremely high. The circuit is also very sensitive to nonidealities (such as finite input impedance and excess phase shift) associated with the gain block.

The negative-feedback circuits of the type shown in Figure 13.6b have lower sensitivity to passive component values. However, they require high values of gain K and often result in wide component value spreads. As a rule, high Q values normally require very high values of K. Since the practical operational amplifier circuits have finite gain–bandwidth products, the high-gain requirement limits the operation of the negative-feedback circuits to low frequencies.

High-Pass Function

The high-pass filter is designed to attenuate low frequencies, below a cutoff frequency, as illustrated in Figure 13.1b. This can be approximated by a biquad function of the form

$$H(s) = \frac{Ks^2}{s^2 + (\omega_p/Q_p)s + \omega_p{}^2} \tag{13.9}$$

which has the magnitude response characteristics and pole and zero locations shown in Figure 13.7. It differs from the transfer function of Eq. (13.8) in that it contains two zeros at the origin (i.e., $\omega = 0$) and has a low-frequency attenuation characteristic which increases with decreasing values of ω at a rate of 40 dB/decade, for $\omega \ll \omega_p$. For high frequencies where $\omega \gg \omega_p$, the circuit exhibits a transfer gain equal to K.

Similar to the case of the biquad low-pass filter function, the high-pass filter function can be synthesized using a variety of circuit configurations. Figure 13.8 shows two typical circuit examples using either positive or negative feedback, which can generate the biquad high-pass filter transfer function. These circuits are high-pass equivalents of the low-pass configurations given in Figure 13.6, and are derived from them by the low-pass to high-pass transformation technique. Both of these circuits exhibit the same gain and component tolerance sensitivity problems associated with their low-pass equivalents.

Bandpass Function

The bandpass function, whose generalized form is shown in Figure 13.2a, has the property of attenuating both high and low frequencies which are outside the filter passband. It is approximated by a biquadratic transfer function of the form

$$H(s) = \frac{K(\omega_p/Q_p)s}{s^2 + (\omega_p/Q_p)s + \omega_p{}^2} \tag{13.10}$$

---

* See Section 13.4 for a definition of sensitivity terms.

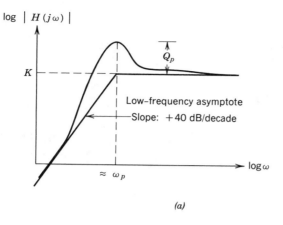

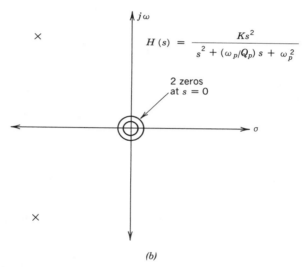

$$H(s) = \frac{Ks^2}{s^2 + (\omega_p/Q_p)s + \omega_p^2}$$

2 zeros
at $s = 0$

(b)

FIGURE 13.7. Second-order high-pass filter function: (*a*) Magnitude response; (*b*) pole and zero locations.

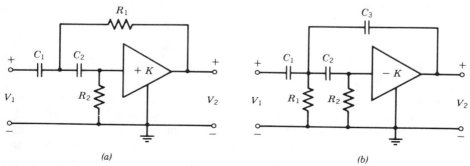

(a) *(b)*

FIGURE 13.8. Examples of single-amplifier high-pass biquad filter stages using positive or negative feedback. (*a*) Positive-feedback circuitry; (*b*) negative-feedback circuit.

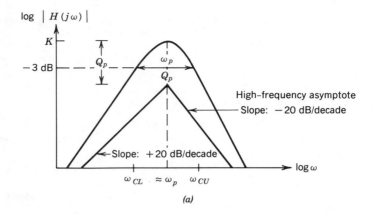

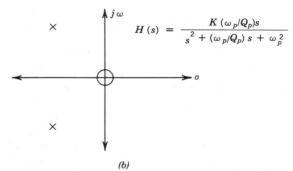

FIGURE 13.9. Second-order bandpass filter function: (*a*) Magnitude response; (*b*) pole and zero locations.

The typical magnitude response and the pole and zero locations of such a transfer function are shown in Figure 13.9. For $Q_p \gg 1$, the filter passband is approximately symmetrical around ω_p, with a total bandwidth $\omega_{3\,\mathrm{dB}}$ given as

$$\omega_{3\,\mathrm{dB}} = \omega_{CU} - \omega_{CL} = \frac{\omega_p}{Q_p} \tag{13.11}$$

At frequencies above and below ω_p, the gain characteristics asymptotically approach a rolloff of 20 dB/decade.

Figure 13.10 shows two typical examples of single-amplifier feedback circuits which can generate the bandpass filter function given in Eq. (13.10). Both circuit configurations exhibit the basic drawbacks described earlier. The positive-feedback circuit is very sensitive to component tolerances and gain accuracy, and the negative-feedback circuit has somewhat lower component sensitivity but requires high values of amplifier gain.

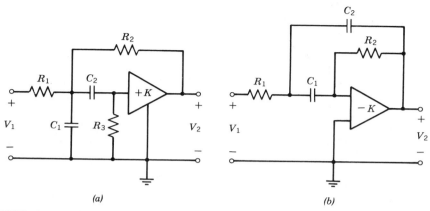

FIGURE 13.10. Single-amplifier bandpass biquad filter configurations using positive or negative feedback: (*a*) Positive-feedback circuit; (*b*) negative-feedback circuit.

Band-Reject Function

The function of a band-reject filter is to attenuate a finite band of frequencies while passing both lower and higher frequencies, as illustrated in Figure 13.2*b*. The biquadratic filter function, which provides such a band-reject characteristic, can be expressed by the transfer function

$$H(s) = \frac{K(s^2 + \omega_z^2)}{s^2 + (\omega_p/Q_p)s + \omega_p^2} \qquad (13.12)$$

The magnitude response characteristics as well as the pole and zero locations associated with the band-reject transfer function of Eq. (13.12) are shown in Figure 13.11. It should be noted that the gain characteristics on either side of the "notch" are not symmetrical, except for the case $\omega_p = \omega_z$. The band-reject filter exhibits low-pass characteristics with higher gain at frequencies where $\omega < \omega_z$, if ω_z is chosen to be larger than ω_p. The filter exhibits zero-transmission (i.e., infinite attenuation) characteristics at $\omega = \omega_z$. The sharpness of the notch in the vicinity of ω_z increases as Q_p increases.

Figure 13.12 shows a typical example of a single-amplifier positive-feedback circuit which can generate the band-stop transfer function given in Eq. (13.12). The total transfer function of the circuit, as shown Figure 13.12, is a third-order transfer function. However, by selecting the component values such that

$$C_3 = C_1 + C_2 \quad \text{and} \quad R_3 = R_1 \| R_2 \qquad (13.13)$$

one real-axis zero of the circuit is made to cancel the third pole and reduce the transfer function to that given by Eq. (13.12). Since the circuit is a positive-feedback stage, it exhibits strong sensitivity to component values and the gain factor K. Improper pole–zero cancellation, due to finite tolerances associated

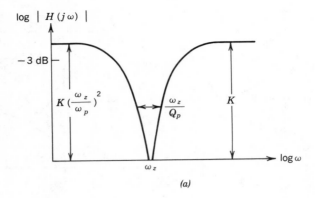

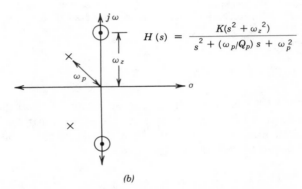

FIGURE 13.11 Second-order band-stop (notch) filter: (*a*) Magnitude response; (*b*) pole and zero locations.

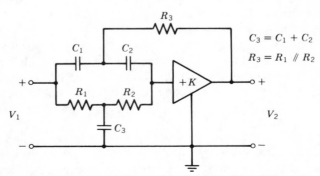

FIGURE 13.12. Single-amplifier band-stop biquad filter.

694

with C_3 and R_3, also contributes to the strong sensitivity characteristics of the circuit.

Multiple Operational Amplifier Biquad Filters

The single-amplifier biquad filter circuits discussed in the previous sections, in general, exhibit very high sensitivity characteristics with respect to component tolerances. Sensitivity to component tolerances and gain parameters increases with increasing values of Q_p, which makes the single-amplifier circuit impractical for high-Q designs with $Q_p > 10$.

In applications requiring high-Q values, it is more practical to use multiple operational amplifier biquad circuit configurations. By introducing additional operational amplifier stages into an active RC circuit, one can obtain any or all of the following benefits:

1. Lower sensitivity of Q and ω_p to both active and passive components.
2. Ease of tuning: allowing ω_p and Q_p to be adjusted independently.
3. Reduced spread of critical component values and critical gain requirements.

In addition, multiple operational amplifier configurations often lead to more versatile filter structures, which can provide a multiplicity of basic filter functions, simultaneously, from a single circuit topology. Figure 13.13 shows such

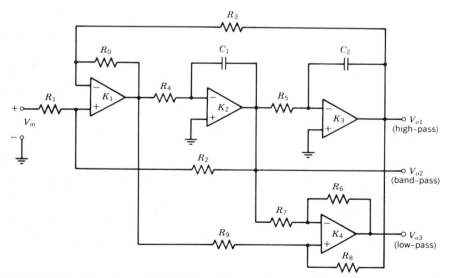

FIGURE 13.13. General-purpose biquadratic filter configuration which can provide low-pass, high-pass, and bandpass outputs.

a "universal" biquad filter configuration[8] which can provide all three basic filter functions (i.e., low-pass, high-pass, and bandpass characteristics) simultaneously. In this case, the operational amplifiers K_1 and K_4 are used as input and output summing amplifiers, K_2 and K_3 are used as integrators. In general, to synthesize an nth-order transfer function, $n + 2$ operational amplifier would be required. Thus, in a discrete design such a synthesis approach will be prohibitively expensive. However, in IC design, where all the operational amplifiers can be fabricated monolithically, the cost and complexity of the design can be greatly reduced. The operational amplifier used in the system are, in general, not critical. They are required to have open-loop voltage gains on the order of 60–70 dB, and can be designed to have low power consumption.

In the circuit of Figure 13.13, the sensitivity of the center frequency ω_0 and Q_p with respect to passive components is equal to or less than unity, which is a big improvement over single-amplifier circuits where Q_p or ω_0 sensitivity factors are on the order of Q_p. If the operational amplifier gains are assumed to be finite, the gain sensitivity of Q_p is on the order of

$$S_{K_1}^Q \approx \frac{Q}{K_1} \quad \text{and} \quad S_{K_2}^Q \approx \frac{Q}{K_2} \tag{13.14}$$

where the sensitivity parameter S is defined in Section 13.4. Thus, the circuit is well suited to synthesizing stable high-Q ($Q > 10$) filters. However, it has the following inherent limitations:

1. It requires large amplifier gains for stable (i.e., low-sensitivity) high-Q circuits. Thus it is mainly limited to applications in the audio frequency range.

2. It is somewhat costly from IC chip area utilization, since it requires $n + 2$ operational amplifiers for synthesizing nth-order transfer functions.

3. An increased number of operational amplifiers degrade the noise characteristics of the circuit.

13.4. SENSITIVITY CONSIDERATIONS

Conventional active-RC systhesis techniques, using resistive and capacitive feedback around the operational amplifier sections, are suited to integration from a conceptual point. However, in practice, the component tolerance and gain stability requirements often impose very significant limitations to their direct application to monolithic integrated circuits. The tight component absolute-value tolerance requirements for linear active-RC filters stems from one fundamental fact. The filter characteristics are solely determined by the poles and zeros of the filter transfer function, which are set by the resistor–capacitor RC products and the overall loop gain in the feedback circuit. The absolute value of the loop gain can be desensitized at low frequencies by using local feedback around individual gain stages. However, the absolute-value control of the *product* of two dissimilar circuit components, such as a resistor and a capacitor, can only be controlled by controlling the absolute value of each R and C. In monolithic IC design, where one can accurately control the *ratios* rather than absolute values of circuit components, this requirement presents a very severe limitation. In other words, RC products that set pole and zero locations do not benefit from the matching and tracking of monolithic IC components and, as a result, require additional trimming of component values subsequent to the fabrication step. This sacrifices the most important feature of monolithic integrated circuits, namely, low-cost batch processing (i.e., high-volume production) capability with no individual trimming.

The required component tolerance or gain accuracy for a given active filter can be directly related to the so-called sensitivity parameters of the circuit. Sensitivity is defined as the fractional or percentage change in a performance parameter of the circuit for a fractional change in the value of any one of the component values, amplifier gains, or any other independent variable in the circuit. The sensitivity S of a given performance parameter, such as the Q factor of an active filter, with respect to a fractional change in an arbitrary circuit parameter x is defined as

$$Q \text{ sensitivity} \triangleq S_x^Q = \frac{\partial Q / Q}{\partial x / x} = \frac{x}{Q} \frac{\partial Q}{\partial x} \qquad (13.15)$$

where x can be either a passive component value or a gain parameter.

The two basic parameters which describe the performance of an active filter are its selectivity Q and the center frequency ω_0. Therefore, the Q sensitivity S_x^Q and the center frequency sensitivity $S_x^{\omega_0}$ are the most commonly used sensitivity terms associated with an active filter. In other words, these two sensitivity parameters indicate the ease of practical implementation of an active filter.

A high value for a given sensitivity parameter indicates that a small change in a component value would result in a large change of the corresponding performance parameter. For example, if $S_x^Q = 5$, then a 1% change in x would cause a 5% variation in Q. Similarly, if $S_x^Q < 1$, then any given fractional change in x would result in a smaller fractional change in Q. In the limiting case, $S_x^Q = 0$ implies that Q is totally insensitive to changes in x. A negative sign associated with the sensitivity parameter implies a polarity reversal in the fractional change. In other words, an increase in x would result in a decrease of Q.

For a given choice of active filter configurations, the sensitivity parameters allow the designer to express component and gain tolerances in terms of filter performance requirements. For this reason, the sensitivity parameters provide an effective means of comparing various filter configurations from the point of view of the practical and economical realization and manufacturability.*

---

* See Chapters 3 and 4 of Ref. (4) for a detailed description of the basic theorems on sensitivity and the calculation of sensitivity factors for various active filter configurations.

13.5. ANALOG SAMPLED-DATA FILTERS

In an analog sampled-data system, the continuous input signal is sampled at periodic intervals. The system transfer function then operates on these discrete samples of the input. The resulting signal is then reconstructed into a continuous analog waveform at the system output. A simplified block diagram of such a system is illustrated in Figure 13.14. For reasons that will be discussed shortly, the frequency range or the bandwidth of the input signal must be limited to frequencies below one-half of the sampling frequency. This is normally achieved by use of a continuous-time low-pass filter, called an *antialiasing* filter, at the system input. Then the band-limited input signal is sampled at periodic intervals, with the information processed through the sampled-data filter, and delivered to an output sample-and-hold stage. The output of the sample-and-hold stage is then converted into a continuous-time output by means of a smoothing filter, called a *reconstruction* filter, which eliminates the high-frequency components introduced into the signal spectrum during the sampling process.

Compared to conventional continuous-time circuits, sampled-data systems offer some unique characteristics. The most important among these is the generation of additional frequency components in the signal spectrum due to the periodic sampling process. Because of this property, the input signal bandwidth has to be limited, relative to the sampling frequency; and special consideration has to be given to characteristics of the output reconstruction filter for the recovery of the analog signal.

The switched-capacitor filters discussed in the second part of this chapter generically belong to the category of sampled-data filters. However, in practice they represent a borderline case between continuous-time and sampled-data systems. As a result, they exhibit some characteristics of both. In the case of switched-capacitor filters, the input and output sample-and-hold functions are performed as an integral part of the filter operation. Thus, no separate sample-and-hold stages are needed. In many cases, the need for separate antialiasing or reconstruction filters is also greatly simplified or eliminated by choosing the sampling frequency to be much higher than the information signal bandwidth.

As a prelude to the discussion of switched-capacitor filters, some of the basic

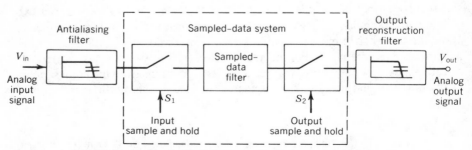

FIGURE 13.14. Basic block diagram of analog sampled-data system.

characteristics of sampled-data systems will be briefly reviewed in this section. It should be noted that the subject of sampled-data systems is a very broad topic, which is well covered in the literature.[9,10] The purpose of this section is to review or point out some of the important properties of such systems, and illustrate under what conditions their characteristics will approach those of continuous-time circuits. For this purpose, mathematical derivations will be avoided, and only the important results will be stated which directly affect their application in monolithic IC filters.

Effect of Sampling on Signal Spectrum

Figure 13.15 illustrates the basic function of a sampling switch S_1 operating on a continuous input signal $V_1(t)$ to produce a sampled output $V_1^*(t)$. The sampling switch is assumed to be closed for the duration t_s during the sampling period T_s. Such a response can be obtained by driving the sampling switch with a series of pulses, as shown in Figure 13.15b. The output of the switch would then be made up of a train of pulses, as shown in Figure 13.15d, whose amplitudes and polarities are equal to those of the continuous input signal at the time of sampling.

Such a sampling step can be considered as a form of the amplitude-modulation process, where the sampling signal $S(t)$ with a unit step height serves as the carrier, and the continuous input signal becomes the modulation. As a result of this modulation process, the frequency spectrum of the output signal is modified and contains additional frequency components centered around integral multiples of the sampling frequency. Figure 13.16 shows the frequency spectrum of the continuous-time input signal $V_1(j\omega)$, and the frequency spectra of the output for different sampling rates. The input signal spectrum contains frequency components from dc to a cutoff frequency ω_c, as shown in Figure 13.16a. The sampling process, similar to the case of amplitude modulation, translates part of this spectrum to other frequency bands, centered around the integral multiples of the sampling frequency ω_s where

$$\omega_s = 2\pi f_s = \frac{2\pi}{T_s} \qquad (13.16)$$

The relative magnitude of each frequency band at integral multiples of the sampling frequency is given by the Fourier coefficients C_k, where

$$C_k\bigg|_{k=0,1,2,\ldots} = \frac{1}{k\pi} \sin\left(\frac{k\pi t_s}{T_s}\right) \qquad (13.17)$$

Normally, only the relative magnitudes of the baseband signal spectrum and the first *lobe* of signal spectra centered around ω_s are important, and these can be expressed from Eq. (13.17) as

$$C_0 = \frac{1}{\pi} \quad \text{and} \quad C_1 = \frac{1}{\pi} \sin\left(\frac{\pi t_s}{T_s}\right) \qquad (13.18)$$

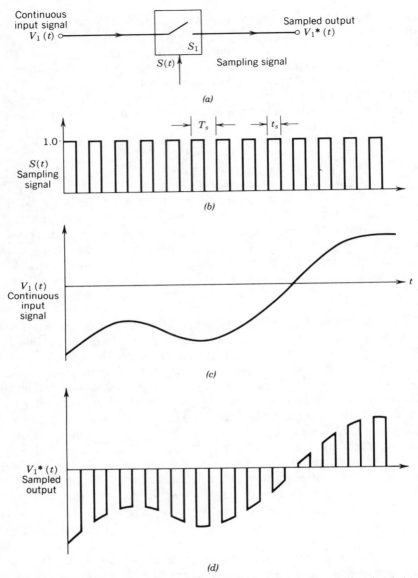

FIGURE 13.15. Operation of sampling switch: (*a*) Block diagram; (*b*) sampling signal; (*c*) continuous input signal; (*d*) sampled output waveform.

First, consider the case where the sampling frequency is sufficiently high such that $\omega_s > 2\omega_c$. This condition is illustrated in Figure 13.16*b*. In addition to the input signal spectrum in the vicinity of $\omega_s = 0$ (i.e., the so-called baseband), the output also contains the replica of the input spectrum, centered on either side of the sampling frequency ω_s, as well as $2\omega_s$, $3\omega_s$ and so on. However, since $\omega_s > 2\omega_s$, these information bands *do not overlap*. Thus, the information con-

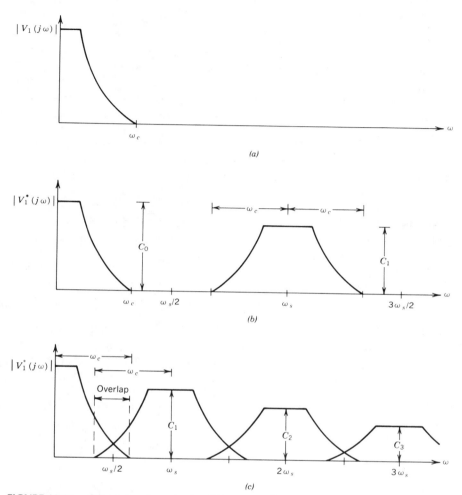

FIGURE 13.16. Output signal spectra for different sampling rates. (*a*) Input signal spectrum; (*b*) sampled output spectra for $\omega_s > 2\omega_c$; (*c*) sampled output spectra for $\omega_s < 2\omega_c$.

tained in the original signal spectrum can still be extracted, or reconstructed, by proper filtering.

Next, consider the case where $\omega_s < 2\omega_s$, as illustrated in Figure 13.16c. In this case, the signal spectrum of the original signal and the additional frequency components created by the sampling process *do overlap*. This overlapping, which is called *aliasing,* introduces an ambiguity into the original signal spectrum and prevents the eventual recovery of the desired baseband signal $V_1(j\omega)$.

The very important conclusion to be gained form Figure 13.16 is that if the sampled signal is eventually to be recovered or reconstructed, the sampling rate must be such that

$$\omega_s > 2\omega_c \qquad (13.19)$$

This can be achieved by either limiting the input signal bandwidth (i.e., reducing ω_c) or increasing the sampling frequency ω_s.

In communication terminology, the frequency $2\omega_c$ is called the *Nyquist rate* and defines the minimum sampling rate necessary in order to reconstruct the input signal without ambiguity.

Figure 13.17 shows the typical frequency response characteristics of a reconstruction filter which can be used to recover the original continuous-time signal spectrum from the sampled signal spectra of Figure 13.16b. Such a reconstruction filter is required to have a flat passband from dc to ω_c and must reject or greatly attenuate signals at frequencies in excess of $\omega_s - \omega_c$. If the sampling frequency is sufficiently high such that $\omega_s \gg \omega_c$, the reconstruction filter becomes very simple to implement. However, as ω_s approaches the Nyquist rate, the attenuation rolloff characteristic of the filter becomes very critical in order to avoid aliasing problems. In such cases, complex filter structures or sample-and-hold circuits are used to recover or reconstruct the baseband signal. In the case of switched-capacitor filters, ω_s is normally chosen to be at least an order of magnitude higher than ω_c in order to simplify, or avoid, the reconstruction filter requirements.

To assure proper operation of a sampled-data filter, it is essential to ensure that the input signal spectrum is band limited such that $\omega_c < \omega_s/2$ to satisfy the Nyquist criteria. In practice, this is normally achieved by preceding the sampling operation with a low-pass band-limiting filter, called an antialiasing filter, as illustrated in Figure 13.14. Such a filter also band-limits the noise present in the input to avoid aliasing distortion due to broadband noise signals, which will get "folded back" into the baseband due to the sampling process. Note that as the sampling frequency is increased relative to ω_c, the antialiasing filter requirements also become simplified.

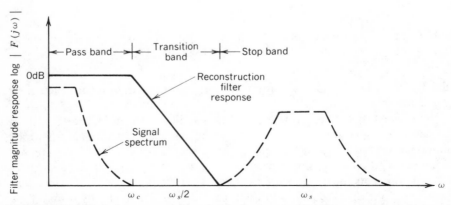

FIGURE 13.17. Typical response characteristics of a reconstruction filter superimposed on sampled signal spectra. (*Note*: Filter passband should extend to ω_c and stopband should start from $\omega_s - \omega_c$.

PART II: SWITCHED-CAPACITOR FILTERS

Switched-capacitor filters are comprised of arrays of capacitors, analog switches, and operational amplifier integrators. Compared to conventional active-*RC* filters, they offer a unique advantage. Under certain conditions, the pole positions of the filter function become determined not by the *RC* products, but by capacitor *ratios*. Since capacitor ratios can be very precisely controlled, and are stable with temperature, very accurate filter transfer functions can be implemented in completely monolithic form. However, there is one very distinct difference between switched-capacitor filters and the conventional active-*RC* circuits. Switched-capacitor filters belong to the category of analog sampled-data filters, discussed briefly in Section 13.5. Thus, switched-capacitor filters provide an equivalent substitute for continuous-time circuits, such as the active-*RC* filters, only over a limited range of the frequency spectrum where the input signal bandwidth is much less than the switch frequency.

The exact and detailed analysis of switched-capacitor filters requires the use of sampled-data techniques and the z-transform theory in place of the conventional Laplace tranforms and s-plane analysis.[4] However, the z-transform analysis is somewhat specialized and is often unfamiliar to an analog IC designer. Therefore, in the following discussions, we will avoid the use of z-transform methods and instead rely on the similarity of the switched-capacitor filters to their active-*RC* counterparts under the condition that the switch frequency is far in excess of the information bandwidth.

13.6. FUNDAMENTALS OF SWITCHED-CAPACITOR CIRCUITS

A simple but very fundamental feature of switched-capacitor circuits is the use of a capacitor and a number of analog switches to simulate the circuit behavior of a resistor, as shown in Figure 13.18. Assuming that the circuit nodes 1 and 2 are voltage sources with negligible impedance, the simulation of a resistor can be described as follows.

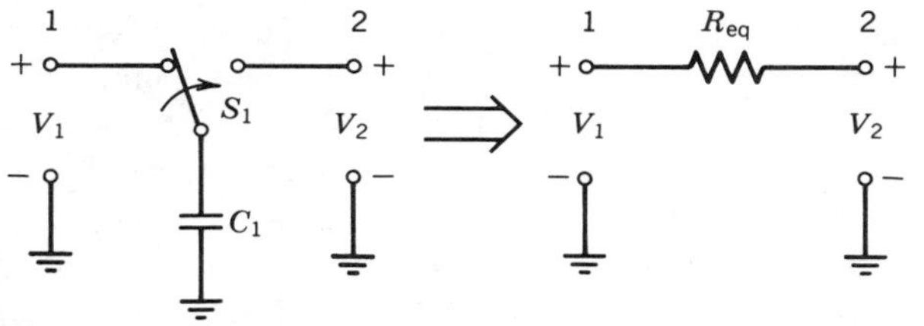

$$i_c = c \frac{dv_c}{dt}$$
$$= c(v_1 - v_2)$$

FIGURE 13.18. Simulation of resistor across two voltage sources by means of a switched capacitor.

With the switch in the left-hand position, as shown in Figure 13.18, the capacitor C_1 is charged to the voltage V_1. Then the switch is thrown to the right, and the capacitor is discharged to the voltage V_2. For each cycle of the switch, a total unit charge Q_0 is transferred from one node to another, where

$$Q_0 = C_1 \Delta V = C_1(V_1 - V_2) \tag{13.20}$$

If the switch is operated at a clock frequency f_s, this amount of charge transfer per switch cycle would correspond to an equivalent current flow I_{eq}, where

$$I_{eq} = Q_0 f_s = C_1 f_s(V_1 - V_2) \tag{13.21}$$

In other words, the switched capacitor C_1 behaves as a *conveyor belt*, delivering discrete packets of charge from one node of the circuit to another, where the total amount of charge delivered per unit of time (i.e., I_{eq}) is proportional to the size of the capacitor and the frequency of switching.

From the above relations, one can define an equivalent resistor R_{eq}, which will give the same equivalent average current as

$$R_{eq} = \frac{V_1 - V_2}{I_{eq}} = \frac{1}{C_1 f_s} \tag{13.22}$$

It should be remembered that the R_{eq} given by Eq. (13.23) can be used to replace a conventional resistor only under special conditions, where the two implicit assumptions used in its derivation are satisfied, namely,

1. The sampling frequency f_s is much higher than the signal frequencies of interest.
2. The voltages V_1 and V_2 at nodes 1 and 2 are unaffected by switch closures (i.e., V_1 and V_2 are voltage sources with virtually zero impedance).

The first assumption stems from the sampled-data nature of the switched-capacitor circuits. The switched capacitor is essentially a sampling element similar to that illustrated in Figure 13.15. Thus, it creates additional frequencies and sidebands at its output spectrum, similar to that shown in Figure 13.16. However, if the sampling or switch frequency f_s is high compared to the input cutoff frequency f_c ($= \omega_c/2\pi$), the output spectrum is similar to that shown in Figure 13.16b, and in the limit where $f_s \gg f_c$, the undesired sidebands and spectral components are located far outside the frequency band of interest. The second assumption is necessary to avoid the circuit transients and the instantaneous signal level changes due to switch closures. In the class of switched-capacitor circuits to be discussed in this chapter, which make use of switched-capacitor integrators, both of these assumptions are satisfied by the proper choice of switch frequencies and circuit configuration.

Figure 13.19 shows a simple implementation of the switched-capacitor circuit of Figure 13.18 by means of two MOS analog switches, driven by non-overlapping out-of-phase clock signals ϕ and $\bar{\phi}$. For illustrative purposes, NMOS enhancement-mode devices are assumed for Q_1 and Q_2 to implement the

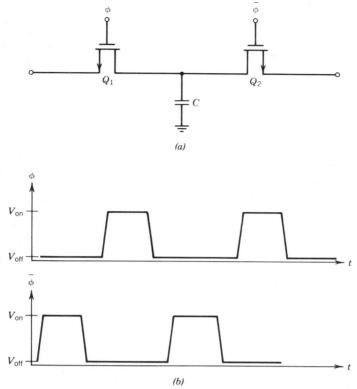

(a)

(b)

FIGURE 13.19.   MOS implementation of switched-capacitor stage using nonoverlapping clock signals. (*a*) Circuit connection; (*b*) nonoverlapping clock signals for switch operation.

switch function. The switches formed by Q_1 and Q_2 are conducting (i.e., closed) only when ϕ or $\bar{\phi}$ are at their high state. The nonoverlapping requirement of clock pulses assures that both switches are never closed at the same time.

The equivalent resistor formed by a switched-capacitor stage, as given in Eq. (13.22), has certain unique features from an IC design point of view:

1.   It requires very little silicon area to implement high-value resistors. For example, a 10-MΩ resistor can be simulated by switching a 1-pF capacitor at 100-kHz clock rate. This makes it possible to generate high-value RC time constants on a monolithic chip, without requiring excessive chip area or external components.

2.   If an RC time constant is formed by combining the effective resistor R_{eq} of Eq. (13.22) with a discrete capacitor C_2, the resulting time constant τ_{eq} would be

$$\tau_{eq} = R_{eq}C_2 = \frac{C_2}{C_1 f_s} \qquad (13.23)$$

which is proportional to the capacitor *ratio* and inversely proportional to the clock frequency f_s. Since monolithic capacitors are extremely stable (see Section 13.7), and their ratio and clock frequency can be very accurately controlled, one can now generate very accurate time constants.

As a consequence of Eq. (13.23), viewed from a conventional active-*RC* filter standpoint, all the *RC* time constants of a switched-capacitor filter are determined by capacitor ratios and the clock frequency that drives the switches. Thus, the problem of controlling *RC* products to a high degree of accuracy, which is the characteristic of active-*RC* filters, is reduced to the problem of maintaining accurate capacitor *ratios* and generating an accurate clock signal. This property makes the switched-capacitor filters very attractive for monolithic integration. However, one must always remember that this simplified behavior of a switched-capacitor stage is valid only for switching frequency $f_s \gg f_c$, and with virtually zero impedance nodes (i.e., voltage sources) at both sides of the circuit. Another significant factor which will be considered later is the effect of inherent circuit parasitics, such as parasitic capacitances and leakage paths.

Switched-Capacitor Integrators

Switched-capacitor integrators are the basic buidling blocks of switched-capacitor filters. Figure 13.20 shows the switched-capacitor equivalent of a conventional lossless *RC* integrator circuit. By substituting Eq. (13.22) for resistor R_1, the transfer function of the switched-capacitor integrator can be written as

$$\frac{V_{\text{out}}}{V_1}(j\omega) = \frac{-f_s C_0 / C_1}{j\omega} \tag{13.24}$$

for frequencies where $\omega \ll \omega_s \,(= 2\pi f_s)$.

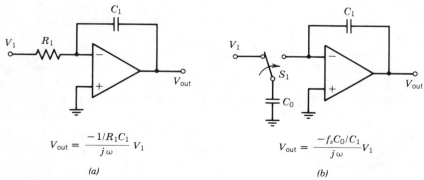

$$V_{\text{out}} = \frac{-1/R_1 C_1}{j\omega} V_1 \qquad\qquad V_{\text{out}} = \frac{-f_s C_0 / C_1}{j\omega} V_1$$

(a)   (b)

FIGURE 13.20. Single-ended integrator circuit: (*a*) Conventional *RC* circuit; (*b*) its switched-capacitor equivalent.

If a number of input signals are simultaneously applied to the switched-capacitor integrator through separate switches and capacitors, as illustrated in Figure 13.21, the circuit can also perform the summing and scaling functions, combined with the integration operation. In this case, the resulting output is

$$V_{\text{out}} = -\frac{f_s}{j\omega C_1}(V_1 C_{01} + V_2 C_{02} + \ldots + V_n C_{0n}) \qquad (13.25)$$

which corresponds to the switched-capacitor equivalent of a multiple-input RC integrator. Note that, for proper operation of the circuit, the switch phasing should be as shown in the figure, where all switched capacitors are connected to the inverting input of the operational amplifier at the same time.

If a lossy integrator circuit is required, as shown in Figure 13.22a, such a circuit can also be realized in a switched-capacitor from by replacing the feedback resistor with its switched-capacitor equivalent, as given in Figure 13.22b. This results in a transfer function of the form

$$\frac{V_{\text{out}}}{V_{\text{in}}} = -\frac{C_0}{C_2}\frac{1}{1 + j\omega C_1/f_s C_2} \qquad (13.26)$$

subject to the frequency restriction that $\omega \ll \omega_s$.

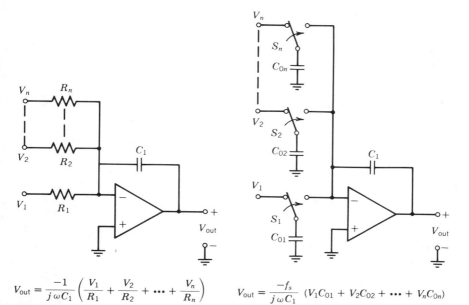

$$V_{\text{out}} = \frac{-1}{j\omega C_1}\left(\frac{V_1}{R_1} + \frac{V_2}{R_2} + \cdots + \frac{V_n}{R_n}\right) \qquad V_{\text{out}} = \frac{-f_s}{j\omega C_1}(V_1 C_{01} + V_2 C_{02} + \cdots + V_n C_{0n})$$

FIGURE 13.21.   Multiple-input integrator circuit combining summing and scaling functions with integration operation: (a) Conventional RC circuit; (b) switched-capacitor equivalent.

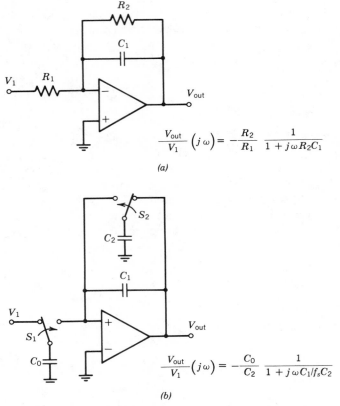

$$\frac{V_{out}}{V_1}(j\omega) = -\frac{R_2}{R_1}\frac{1}{1 + j\omega R_2 C_1}$$

(a)

$$\frac{V_{out}}{V_1}(j\omega) = -\frac{C_0}{C_2}\frac{1}{1 + j\omega C_1/f_s C_2}$$

(b)

FIGURE 13.22. Lossy integrator circuit: (*a*) Conventional *RC* circuit; (*b*) its switched-capacitor equivalent.

Another useful building block of switched-capacitor filters is the differential integrator circuit of Figure 13.23, which has the transfer function

$$V_{out} = -\frac{f_s C_0/C_1}{j\omega}(V_1 - V_2) \qquad (13.27)$$

for frequencies where $\omega \ll \omega_s$. In this circuit, the capacitor C_0, along with two sets of switches, replaces both of the input resistors and the capacitor at the (+) input, since it applies a net change proportional to the input voltage differential $(V_1 - V_2)$ to the inverting input of the operational amplifer. Note that another useful feature of the differential integrator is that it can be used as a single-ended noninverting integrator simply by setting $V_2 = 0$.

Figure 13.24 shows a general case of a differential integrator which can simultaneously sum and scale several inputs. Summing is done by changing C_{01} and C_{02} to various input voltage differentials and then connecting them in parallel across the operational amplifier input during the next clock cycle. The resulting

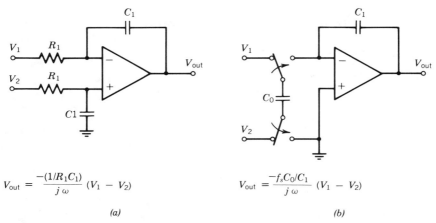

$$V_{out} = \frac{-(1/R_1C_1)}{j\omega}(V_1 - V_2)$$

$$V_{out} = \frac{-f_sC_0/C_1}{j\omega}(V_1 - V_2)$$

(a)   (b)

FIGURE 13.23. Differential integrator: (a) Conventional RC circuit; (b) its switched-capacitor equivalent.

transfer function is

$$V_{out} = [(V_1 - V_2) + \alpha(V_3 - V_4)]\frac{f_s(1 + \alpha)C_{01}}{C_1}\frac{1}{j\omega} \qquad (13.28)$$

where the scaling constant α is the capacitor ratio C_{02}/C_{01}.

One of the commonly used circuit functions in switched-capacitor filter synthesis is the summing and scaling operation, separate from the integration operation. Figure 13.25a shows a simple integrator/summer circuit which can perform such an operation and produce an output voltage,

$$V_{out} = -\frac{f_sC_0}{j\omega C_1}V_1 - \frac{C_x}{C_1}V_x \qquad (13.29)$$

The summing operation comes about because the net charge Q_x induced at the inverting input of the operation amplifier (node A) due to the voltage V_x. applied

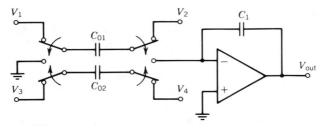

FIGURE 13.24. Differential integrator with scaling and summing capability.

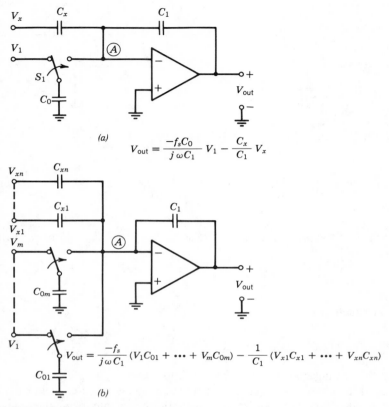

$$V_{out} = \frac{-f_s C_0}{j \omega C_1} V_1 - \frac{C_x}{C_1} V_x$$

$$V_{out} = \frac{-f_s}{j \omega C_1} (V_1 C_{01} + \cdots + V_m C_{0m}) - \frac{1}{C_1} (V_{x1} C_{x1} + \cdots + V_{xn} C_{xn})$$

FIGURE 13.25. Switched-capacitor integrator/summer circuit: (*a*) Basic circuit; (*b*) its generalized configuration.

across C_x, must be canceled by an equal and opposite amount of charge Q_1, induced due to voltage appearing across C_1,

$$Q_x = V_x C_x \quad \text{and} \quad Q_1 = - V_{out} C_1 \qquad (13.30)$$

Equating Q_x and Q_1, and superposing the effect of V_1, one obtains the output voltage expression given by Eq. (13.29). Figure 13.25*b* shows the generalized version of the integrator/summer circuit where any arbitrary number of inputs can be integrated and/or summed and scaled simultaneously.

A special application of the integrator/summer circuit is to use it only as a scaling circuit, without the integration operation. This corresponds to operating the circuit of Figure 13.25*a* with $V_1 = 0$. In such an application, the switched capacitor C_0, and the switch S_1 must still be retained in the circuit in order to avoid parasitic charge accumulation at node A, and to satisfy the practical design constraints given in Section 13.9. In all cases, one side of the switched capacitor is connected to the "virtual ground" formed by the inverting input of the operational amplifier, the other side is driven from a low-impedance signal source or an operational amplifier output.

Although the basic switched-capacitor circuits in Figure 13.20 and 13.25 are fully functional as shown, their performance characteristics are particularly sensitive to parasitic and stray capacitance associated with the ungrounded plates of the switched capacitors. This problem and its solutions will be covered in later sections.

Discrete-Time Effects

Using sampled-data techniques, the exact transfer function of a switched-capacitor integrator, such as the one shown in Figure 13.20b, can be expressed as

$$\frac{V_{out}}{V_1}(j\omega) = -\underbrace{\left[\frac{f_sC_0/C_1}{j\omega}\right]}_{\substack{\text{ideal} \\ \text{response}}}\underbrace{\left[\frac{\omega T_s}{2\sin(\omega T_s/2)}\right]}_{\substack{\text{amplitude} \\ \text{error}}}\underbrace{\left[\exp-(j\omega T_s/2)\right]}_{\substack{\text{phase} \\ \text{error}}} \quad (13.31)$$

where $T_s(=1/f_s)$ is the period of the clock signal.

Comparing Eq. (13.31) with the simplified result given in Eq. (13.24), one sees that the exact expression differs from the approximate response by both an amplitude and a phase error. In the limit, as the switch frequency increases such that ωT_s approaches zero, both of these error terms approach unity and the response reduces to the case of the ideal RC integrator.

The amplitude error is usually of very little concern in the frequency of interest where $\omega T_s \ll 1$. The phase error, which represents a phase delay proportional to frequency, is the most bothersome. This phase shift or delay is due to the one-half clock cycle delay in replenishing the charge on the capacitor C_0 in Figure 13.20b. This phase delay will distort the frequency response of a complete filter in a way that is similar to the effects of excess phase in operational amplifiers used in conventional active-RC filters. Typically, the effect of this excess phase is in the form of Q-enhancement, or undesired peaking, in the filter response. The phase error or the delay term associated with Eq. (13.31) can be compensated by one of two approaches:

1. By using predistortion technique. The filter can be "predistorted" in actual design so that the response is the desired one when the excess phase is present. This approach requires fairly extensive numerical calculations, and has the disadvantage of increasing the component sensitivity and the design complexity in filters requiring transmission zeros in the response.

2. By adjusting the phasing of clock cycles between successive integrator stages, using the so-called lossless digital integrator, or LDI, clocking scheme. [11] In this scheme, the clock phases are arranged so that the delay time through every other integrator is reduced by one-half clock cycle.

The LDI clocking scheme is illustrated in Figure 13.26 in terms of a cascade of two integrators. In order to reduce the delay through the second integrator by one-half clock cycle, the switch S_2 in integrator 2 is phased to sample the output of integrator 1 as the output is ready for sampling (i.e., at the moment the switch S_1 is thrown to the right. Thus, the switches in adjacent integrators face in opposite directions. The LDI clocking technique is particularly effective in eliminating the excess-phase effects in active-ladder filters using switched-capacitor integrators where one integrator always samples the output of the other (see design examples of Section 13.10 and 13.11.)

Although the LDI clocking technique is effective for correcting the phase delay associated with lossless integrators, it does not compensate for the phase delay introduced by the switched capacitor in the feedback path of an integrator. Thus, it is not effective for the lossy integrator stage (see Fig. 13.22*b*). As will be shown in the design examples of Section 13.10 and 13.11, lossy integrator stages are required to simulate the termination resistances in active-ladder filters. In such cases, predistortion techniques have to be used, in addition to proper clock phasing, to compensate for such phase delays.

13.7.   CHARACTERISTICS OF MOS CIRCUIT ELEMENTS

The three essential types of circuit elements in a monolithic switched-capacitor filter are MOS capacitors, analog switches, and operational amplifiers. In order to optimize the performance of a given monolithic switched-capacitor filter design, it is imperative that the IC designer be familiar with the basic capabilities and limitations associated with each of these types of elements. In this section, the basic performance capabilities, inherent parasitics and design tolerances associated with each of these circuit elements will be reviewed.

MOS Capacitors

The important features of the MOS capacitor, which make it nearly ideal for switched-capacitor filter applications, are its linearity, temperature stability, and

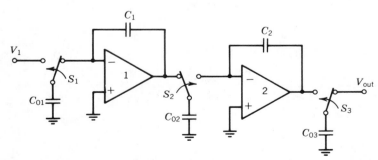

FIGURE 13.26.   Minimizing phase-delay errors by proper phasing of switches between successive integrator stages.

ratio acuracy. There are two basic types of capacitor structures available in MOS circuits, depending on the particular fabrication process used. These are: (1) the conventional MOS capacitor structure, and (2) the double polysilicon capacitor.

Conventional MOS Capacitor. The basic MOS capacitor structure (comprised of a heavily doped silicon substrate, a thin (SiO$_2$) dielectric layer, and an aluminum top plate) was discussed in Chapter 3 (see Section 3.11 and Fig. 3.44) as it related to conventional bipolar IC design. In the case of conventional NMOS or CMOS process technologies, such capacitance structures are directly compatible with the metal-gate MOS process by simply using the low-resistivity source–drain diffusions to form the bottom plate, with the gate oxide as the dielectric and the metal as the top plate. Figure 13.27 shows the top view and the cross section of such a MOS capacitor structure. In this case, for illustrative purposes, a CMOS process is assumed, and the p^+ source–drain diffusion of a PMOS transistor is used to form the low-resistivity bottom plate of the capacitor. The capacitor value is proportional to the area of the capacitor plates and is inversely proportional to the dielectric oxide thickness t_{ox}. The capacitor area can be determined either by the size of the metal top plate or by the size of the thin dielectric oxide area under the top plate. In the example of Figure 13.25, this

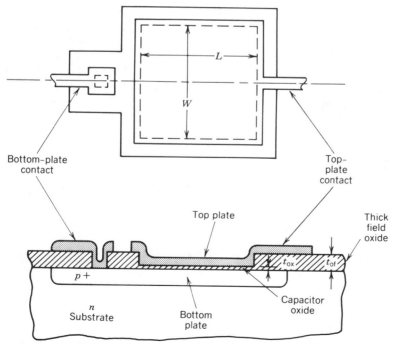

FIGURE 13.27. Top view and cross-section of MOS capacitor. (*Note*: If NMOS process is used, a p-type substrate would be used with n^+ bottom plate.)

latter case is demonstrated, where the metal slightly overlaps the edges of the thin oxide region. In most designs, this is preferred because the edge definition of the thin oxide pattern is better than that of the metal pattern during the photomasking process, thus resulting in improved ratio accuracy. Typically, the thickness of the dielectric SiO_2 layer is approximately 1000 Å, whereas the thick "field oxide," which covers the rest of the chip surface, is in excess of 10,000 Å. Thus, the slight overlap of the top plate metal does not contribute any significant error in the ratio accuracy. In metal-gate NMOS structures, the bottom plate of the MOS capacitor is heavily doped n type, corresponding to the source–drain diffusion. In the case of a CMOS process, it can be either n or p type. In general, the p-type bottom plate is preferred, since the dielectric oxide thickness over p-type regions can be controlled somewhat more accurately during the oxidation process.

The basic MOS capacitor structure shown in Figure 13.27 is directly compatible with the conventional metal-gate CMOS processes, and requires no additional masking or diffusion steps. However, in the case of silicon-gate MOS technology, which uses deposited polycrystalline silicon (or the so-called poly layer) to form the gate electrodes of the MOS devices, the conventional MOS capacitor is not directly compatible with the basic process, and would require additional process steps.

Double-Polysilicon Capacitor.   Since the conventional MOS capacitor structure requires additional process steps in silicon-gate technology, an alternate approach is to use a thin dielectric oxide layer between two polysilicon layers, on the chip surface, to form a capacitor, as shown in Figure 13.28. This layer of thin oxide is formed by oxidizing the top surface of the first polysilicon layer. In this case, the first polysilicon layer serves as the bottom plate, and the second becomes the top plate. The dielectric layer between the two is formed by the

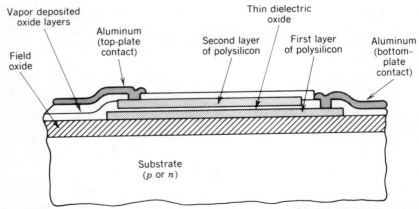

FIGURE 13.28.   Cross section of a polysilicon–oxide–polysilicon (double-polysilicon) capacitor.

oxidation of the first polysilicon layer. The double-polysilicon capacitor is particularly useful in so-called *double-poly* silicon-gate MOS structures, which utilize two separate polysilicon layers. In such device structures, a double-poly capacitor can be incorporated into the circuit without requiring additional process steps or process modification.

The linearity, temperature coefficient, and ratio-accuracy characteristics of the conventional MOS capacitors, and those formed by the double-poly process, are roughly comparable. In general, conventional MOS capacitor structures offer better absolute-value control, whereas the double-poly capacitors have somewhat lower stray parasitics associated with them.

Parasitic Capacitances. In both the conventional MOS and the double-poly capacitors, a sizable parasitic capacitance exists between the bottom plate and the substrate. In the case of the conventional MOS capacitor of the type shown in Figure 13.27, this is the capacitance of the p-n junction surrounding the heavily doped bottom plate region. For the double-poly capacitor of Figure 13.28, this is the capacitance between the first polysilicon layer and the silicon substrate. Typically, the value of this parasitic bottom plate capacitance is in the range of 5–20% of the actual capacitor value, depending on the particular process or layout technique.

The top plates of MOS capacitors also have small but finite amounts of parasitics associated with them. These are primarily due to metal interconnections which connect the top plate to various parts of the circuit. This top plate parasitic capacitance is on the order of 0.1–1% of the nominal capacitance value, depending on the circuit layout and the field-oxide thickness.

By proper choice of circuit configuration, the detrimental effects of the bottom plate parasitics can often be eliminated by using grounded, rather than floating, capacitors in the circuit, where the bottom plate would be connected to an ac ground point. However, the presence of top plate parasitic capacitance is unavoidable. Their finite values must be taken into account, along with other circuit parasitics, in calculating the overall accuracy of the filter response.

Linearity and Temperature Coefficient. MOS capacitors with heavily doped bottom plates exhibit a positive voltage coefficient in the range of +20 to +100 ppm/V.[12] In the case of double-poly capacitors, this value is in the range of +10 to +20 ppm/V. The temperature coefficients of MOS and double-poly capacitors are in the range of +20 to +50 ppm/°C. The above coefficients relate to absolute-value changes. In the case of capacitor ratios, as used in switched-capacitor filters, the linearity and temperature coefficient values associated with the capacitor *ratios* are much lower than the above values. Thus, for most practical designs utilizing capacitor ratios, the errors due to the voltage or temperature coefficients of MOS capacitors are low enough to be insignificant.

Ratio Accuracy. The key factor in switched-capacitor filter design is the accuracy and reproducibility of the capacitor ratios. Assuming that the dielectric thickness for the capacitors within a given IC chip is uniform, the ratio of two capacitors on the same monolithic chip, will depend solely on their area ratio. The area ratio is in turn determined by the lateral dimensions of the capacitor layout and the edge-definition characteristics of the photolithographic masks used in the manufacturing process.

To minimize mismatch errors and improve ratio accuracy, capacitors are often made up of arrays of identical "unit geometry" capacitors, which are interconnected in parallel to give specific capacitor values. Generally speaking, achievable ratio tolerances are on the order of 1% for small capacitors (area $\leq 500\ \mu m^2$) and approach 0.1% for large capacitors (area $> 10,000\ \mu m^2$). The above ratio tolerances are stated for identical capacitors, having a ratio of 1 : 1. The ratio tolerance deteriorates somewhat as very large or small capacitor ratios are needed, since one of the capacitors would be forced to be near minimum size.

MOS Analog Switches

The MOS transistor makes a nearly ideal analog switch. Its basic features as an analog switch were discussed in Section 6.6. Some of these characteristics will be briefly reviewed in this section, particularly as they relate to switched-capacitor filter applications.

In switched-capacitor filter design, the MOS analog switches carry very little current. Therefore, they are normally designed as minimum-geometry devices, with minimum channel width, and a channel width-to-length ratio of unity. In CMOS designs, an n-channel transistor is normally used as an analog switch, since it offers lower resistance (due to higher carrier mobility) than a p-channel device of identical geometry.

For a typical NMOS analog switch with channel length of 5 μm and channel width-to-length ratio of unity, typical on resistance is 5 kΩ, with a gate drive of +5 V with respect to the source. Typical leakage currents to the substrate, at its off state, are about 10^{-14} A at 70°C. The parasitic capacitance from source and drain to the substrate are typically on the order of 0.02 pF each. The gate–source and gate–drain overlap capacitances are about 0.01 pF for metal-gate devices, typically 0.005 pF or less for self-aligned silicon-gate devices. Typical values of induced charge in the channel, for $V_{GS} > 5$ V are approximately 0.03 pC.

As indicated above, the minimum-geometry NMOS transistor makes an excellent, nearly ideal analog switch. However, since the majority of capacitors used in switched-capacitor filters are in the range of 1–10 pF, the parasitic capacitance associated with the analog switches must also be taken into account in calculating capacitor ratios.

MOS Operational Amplifiers

As discussed in Chapters 6 and 7, the basic operational amplifier function can be implemented with MOS technology, with very efficient use of the chip area.

In switched-capacitor filter applications, the performance requirements for the operational amplifier stages are relatively modest for operation in voice-band frequencies. Typically, open-loop gains of 60–80 dB, with a common-mode rejection of 50–60 dB and a unity-gain bandwidth of approximately 2 MHz, are sufficient. In addition, since the operational amplifiers interface only with small capacitors and analog switches, output current drive requirements are minimal.

MOS operational amplifiers meeting the above performance requirements can be designed easily using either NMOS or CMOS technologies. In general, CMOS technology has the added benefit of reduced power dissipation and simplified circuit design, which is important when many operational amplifier circuits may have to be integrated on the same chip.

Figure 13.29 shows a very simple CMOS operational amplifier circuit for a monolithic switched-capacitor filter application,[13] which meets the basic performance requirements stated above, with a power dissipation of $\approx$ 5 mW. Since the operational amplifier output is not required to drive high current loads, the compensation capacitor is not buffered with a source-follower stage, as is customary for conventional operational amplifiers (see Fig. 7.43).

A simple operational amplifier circuit similar to that shown in Figure 13.29 can be integrated within a chip area of approximately 200 mil^2. Thus, several tens of such operational amplifier blocks can be combined on the same monolithic chip to perform a complex switched-capacitor filter function.

An important operational amplifier characteristic for switched-capacitor filter design is the settling time requirement, since this determines the maximum switching frequency for the analog switches. Most MOS operational amplifier

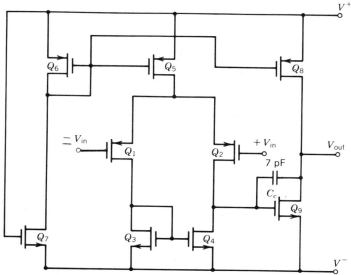

FIGURE 13.29. Simple CMOS operational amplifier for switched-capacitor filter applications.[13]

circuits can settle to 0.1% of final value within 1–2 μsec settling time, with 20-pF load capacitance, which make them suitable for operation with switch frequencies up to 500 kHz.

Noise Characteristics

There are two primary sources of noise in switched-capacitor integrators. The first is due to the thermal noise of the MOS switches, and the second is due to the operational amplifiers. The qualitative effects of each of these noise sources are briefly examined below.

Noise Due to MOS Switches. When a MOS transistor switch is in the on condition, connecting a capacitor C to a voltage source, the thermal noise power in the resistive channel is given by the thermal noise equation

$$(V_n)_{\text{rms}}^2 = 4\ kTR\,(\Delta f)_n \tag{13.32}$$

where k is the Boltzmann's constant, R is the on resistance of the channel, and $(\Delta f)_n$ is the noise-power bandwidth of the RC circuit formed by the sampling capacitor and the switch resistance R. The noise-power bandwidth $(\Delta f)_n$ associated with a single-pole low-pass circuit of resistance R and capacitor C is given as

$$(\Delta f)_n = \frac{1}{4RC}\ \ \text{Hz} \tag{13.33}$$

Substituting Eq. (13.33) into Eq. (13.32), one gets

$$(V_n)_{\text{rms}}^2 = \frac{kT}{C} \tag{13.34}$$

When the switch is turned off, this noise is then sampled and held on the sampling capacitor.

In a switched-capacitor integrator, the noise contributed by each one of the two MOS switches must be considered. The simple analysis leading to Eq. (13.34) is valid for the first MOS switch between the input and the sampling capacitor. The analysis of the noise contributed by the second MOS switch, between the capacitor and the operational amplifier input, is more complicated. However, by a similar analysis one can show that it is of comparable magnitude to that given by Eq. (13.34).[14]

The two important conclusions to be drawn from this qualitative discussion are as follows. To reduce noise problems associated with analog switches, one should (1) keep the sampling capacitor sizes as large as possible, and (2) minimize the number of switches in the circuit. Unfortunately, the sampling capacitor is often chosen as the smallest practical "unit capacitor," against which the other capacitors in a switched-capacitor filter are scaled. Therefore, increasing its size is often not practical since it leads to increased chip area.

Noise Due to Operational Amplifiers. The MOS operational amplifiers are relatively noisy compared to bipolar ones. The primary source of noise in MOS operational amplifiers is the high $1/f$ noise due to the surface recombination effects within the channel region of the MOS transistor. This low-frequency noise falls within the passband of voice-frequency filters, and puts a lower limit on the minimum detectable signal. For conventional MOS processes, the $1/f$ noise associated with operational amplifiers restricts the dynamic range (i.e., the signal-handling capability) of switched-capacitor filters to less than 90 dB in most applications.

Another important noise source related to MOS operational amplifiers is the broadband noise of the amplifiers. Due to the sampled-data nature of the circuit, some of this noise which is normally outside the filter bandwidth gets "aliased," or frequency translated into the passband. To avoid this problem, care must be taken to minimize the broadband operational amplifier noise, even beyond the unity-gain bandwidth frequencies of the operational amplifiers.

13.8. EFFECTS OF PARASITIC CAPACITANCES

The main parasitic associated with the MOS capacitor structure is the parasitic capacitance from the bottom plate to the substrate, that is, to ac ground. As described in the previous section, the value of this parasitic capacitance can be in the range of 5–20% of the actual capacitor value, depending on the particular process and the capacitor structure used. In practical monolithic designs, the problems associated with this particular parasitic can be avoided by using circuit configurations where the bottom plate of the capacitor can be connected to ac ground at all times. Thus, only the top plate of the capacitor is connected to the circuit nodes which are sensitive to parasitic capacitance.

In addition to bottom-plate parasitics, the other remaining circuit parasitics are the stray capacitance associated with the top plate, metal interconnection paths, and the analog switches. In a well-designed and laid out monolithic circuit, the parasitic values due to these sources are much smaller than those due to the bottom-plate stray capacitance. However, they still set a practical limit to the achievable filter precision. The effect of these circuit parasitics can be examined using the simple switched-capacitor integrator circuit of Figure 13.30a, where all the basic parasitics associated with the circuit nodes are shown separately. C_{p1} is the parasitic source–substrate capacitance of the analog switch transistor Q_1, C_{p2} and C_{p3} are the drain–substrate capacitances of Q_1 and Q_2 as well as the top-plate and interconnection parasitics associated with C_0. C_{p4} is the source-substrate capacitance of Q_2 and the input capacitance of the operational amplifier. C_{p5} and C_{p6} are the top- and bottom-plate parasitics associated with the integrator capacitance C_1. The bottom-plate parasitic C_{p6} has no effect on circuit performance since it is driven from the operational amplifier output. C_{p1} is in parallel with the voltage-source input C_{p4} and C_{p5} are in parallel with the virtual

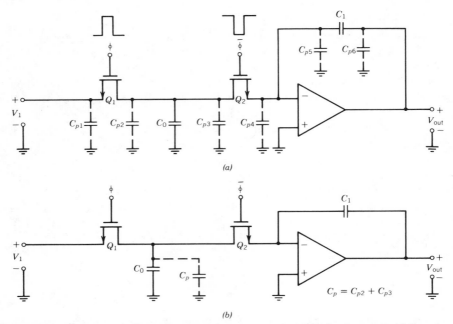

FIGURE 13.30. Effects of a parasitic capacitances in switched-capacitor integrator: (*a*) Locations of parasitic capacitances; (*b*) dominant parasitic capacitance.

ground node (inverting input) of the operational amplifiers. Therefore, they do not affect circuit performance as long as the input signal source impedance is very low and the operational amplifier gain is high. The parasitic capacitances C_{p2} and C_{p3} are in parallel with the switched capacitor C_0; therefore, they directly affect the circuit performance by causing C_0 to have an apparent value of C_0', where

$$C_0' = C_0 + C_p = C_0(1 + \gamma_0) \qquad (13.35)$$

where $C_p = C_{p2} + C_{p3}$ is the total parasitic capacitance associated with the top plate of C_0, and $\gamma_0 = C_p/C_0$ is the fractional parasitic top-plate capacitance associated with C_0.

The effect of the parasitic top-plate capacitor is to modify the integrator gain constant by a factor of $(1 + \gamma_0)$, such that the basic transfer function for the circuit of Figure 13.30*b* becomes

$$\frac{V_{\text{out}}}{V_1} = -\frac{f_s C_0'/C_1}{j\omega} = \frac{(f_s C_0/C_1)(1 + \gamma_0)}{j\omega} \qquad (13.36)$$

In the case of the lossy integrator circuit of Figure 13.22*b*, the parasitic top plate capacitance associated with C_2 would also come into the picture and result in a transfer function of the form

$$\frac{V_{out}}{V_1} = -\frac{C_0(1 + \gamma_0)}{C_1}\left(\frac{1}{1 + \dfrac{j\omega C_1}{1 + \gamma_2 f_s C_2}}\right) \tag{13.37}$$

where γ_2 is the parasitic factor associated with C_2. Note that in this case, the parasitic capacitance results in both gain and pole-position error. In a well-designed and laid out switched–capacitor filter, the top-plate parasitics associated with the key capacitors in the circuit must be kept to within less than 1% of the nominal capacitor value. This can be done by maintaining relatively large capacitor values (at the expense of the chip area) or by using parasitic-insensitive switch circuits as discussed below.

The detrimental effects of top-plate parasitics can be greatly reduced by using *dual-switch* equivalents of the simple switched-capacitor stage, as shown in Figure 13.31.[13,15] By simple analysis, one can show that the equivalent average current carried across the switched capacitor is equal to

$$I_{eq} = Q_0 f_s = f_s C_0(V_1 - V_2) \tag{13.38}$$

for the circuit of Figure 13.31a, which is identical to that obtained from the basic switched-capacitor circuit of Figure 13.19 [see Eq. (13.21)]. In the case of the circuit of Figure 13.31b, the equivalent average current is

$$I_{eq} = -f_s C_0(V_1 + V_2) \tag{13.39}$$

which is equivalent to inverting the polarity of the input voltage V_1.

The dual-switch circuits can be implemented by using a set of four MOS transistors, as shown in Figure 13.32. Note that whether a configuration is

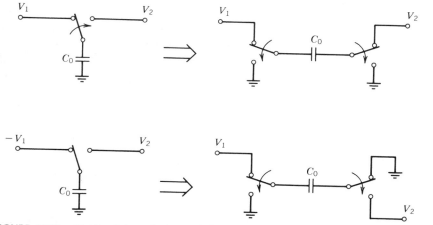

FIGURE 13.31.   Dual-switch equivalents of simple switched-capacitor stage: (*a*) Noninverting configuration; (*b*) polarity-inverting configuration.

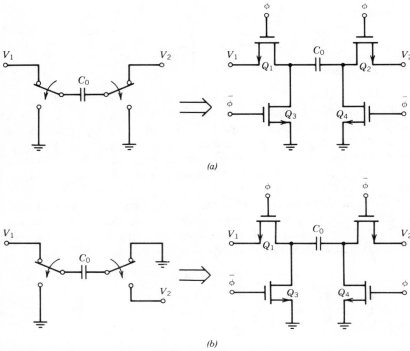

(a)

(b)

FIGURE 13.32. Circuit implementation of dual-switch switched-capacitor-stages: (a) Non-inverting switch; (b) inverting switch.

inverting or noninverting is determined solely by the relative phasing of the clock signals. Figure 13.33 shows the dominant parasitic capacitance associated with an inverting integrator circuit, using dual-switch capacitor configurations. The effect of parasitic capacitance can be determined by examining the circuit behavior during one complete clock cycle. With clock phase ϕ in the high state, Q_1 and Q_2 are on, and Q_3 and Q_4 are off. During this half-cycle, parasitic C_{p1} is

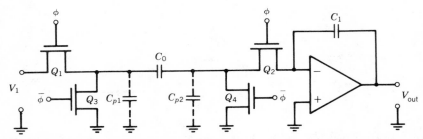

FIGURE 13.33. Dominant parasitic capacitances associated with switched-capacitor integrator using dual-switch configuration.

charged to V_1, but C_{p2} remains uncharged since it is short-circuited to virtual ground through Q_2 at the operational amplifier input. During the next half-cycle, $\bar{\phi}$ is high, Q_1 and Q_2 are off, and Q_3 and Q_4 are on. Thus, C_{p1} is discharged to ground through Q_3 harmlessly, and C_{p2} remains short-circuited to ground through Q_4. Thus, neither C_{p1} or C_{p2} interferes with the charge delivered to C_0.

Figure 13.34 gives the basic circuit configurations for single-ended and differential integrator circuits, using parasitic-insensitive switching. Note that the change of a single switch phasing results in the polarity reversal of the integrator circuit.

The parasitic-insensitive switching technique can also be applied to the lossy integrator circuit of Figure 13.35. By inspecting the circuit of Figure 13.35a one sees that the phasing of switches S_2 and S_3 is such that they effectively operate

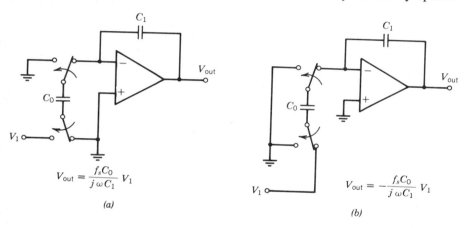

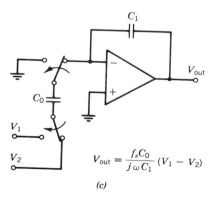

FIGURE 13.34. Switched-capacitor integrator circuits using parasitic-insensitive switching: (a) Noninverting integrator; (b) inverting integrator; (c) differential integrator.

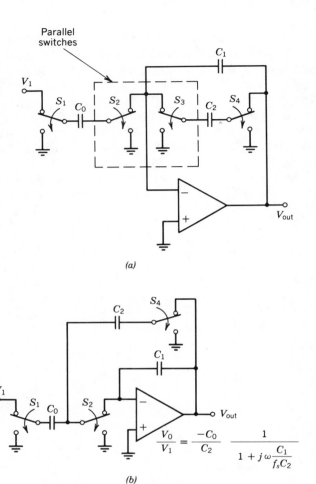

FIGURE 13.35. Parasitic-insensitive lossy integrator: inverting configuration. (*a*) Basic circuit; (*b*) its simplified version with reduced switch count.

in parallel, connecting to the same node and to ground simultaneously. Thus S_3 can be eliminated by rearranging the circuit configuration as shown in Figure 13.35b. The circuit as shown is an inverting integrator. However, one can see by inspection that the reversing of polarity of the switch S_2 will cause the circuit to be noninverting.

Although the parasitic-insensitive switching technique described above is quite effective in improving circuit performance, it also requires a significant increase in circuit complexity in terms of increased switches and corresponding clock signal paths, and increased noise sources. In many cases, its utilization is a design trade-off, depending on the precision requirements.

13.9. PRACTICAL DESIGN CONSTRAINTS

In the actual circuit implementation and interconnection of monolithic switched-capacitor filters, a number of practical design constraints must be kept in mind to assure stable and reliable circuit operation. Some of these fundamental points are listed below:

a. An equivalent resistor generated by switched-capacitor techniques cannot be used to close an operational amplifier feedback path by itself. An operational amplifier requires a continuous-time feedback path to assure stability. Since a switched-capacitor resistor does not provide a continuous-time feedback path, it cannot be used as the only feedback path around an operational amplifier. However, it can be used in conjunction with a capacitor to shape the frequency response of the feedback path, as in the case of the lossy integrator circuit of Figure 13.22. This constraint complicates the design of circuits which require precise control of closed-loop gains which would normally be set using resistor ratios. Alternatively, since closed-loop gains are ratios of resistors, they can be obtained using diffused or ion-implanted resistors at the expense of chip area, additional processing steps, and increased power dissipation.

b. No circuit node must be left floating. All circuit nodes must have a resistive path to ground in order to avoid charge accumulation at capacitor plates. A path formed by a switched-capacitor resistor is acceptable for this purpose.

c. The bottom plates of MOS capacitors must be connected to ground or to a voltage source. The bottom-plate parasitic capacitance is in the range of 5–20% of the nominal capacitor value, and is often nonlinear in its response. It must be connected to an ac ground or be switched between voltage sources so that the charge flow in and out of it will not affect the filter response. This rules out floating switched-capacitor configurations, capacitive voltage dividers comprised of three or more capacitors, or circuits which sequentially switch both ends of a capacitor into an operational amplifier input.

d. The noninverting operational amplifier input should be kept at a constant voltage. If the positive input of the operational amplifier is connected into the signal path, then the inverting input no longer behaves as a virtual ground, and the filter response becomes sensitive to all parasitic capacitance due to substrate capacitance, bus lines, and switches connected to the inverting input.

These basic design constraints impose significant restrictions on switched-capacitor filter configurations which can be integrated in'monolithic form. However, there still remains a wide range of filter configurations which can be implemented in IC form, using switched-capacitor techniques. In this section, some of these filter configurations will be examined, starting with the simplest and working toward more complex circuit configurations.

One general class of filters which are particularly well suited to switched-capacitor techniques are those that use operational amplifiers as integrators in the same way as they would be used in analog computer simulation. Two commonly used filter types which fall into this category are the state-variable and the active-ladder filters. An important advantage of these filters is that they can be organized to simulate the transfer functions of passive-LC ladders. This feature has two practical advantages:(1) it greatly simplifies the design effort by enabling the designer to calculate component values directly from existing filter tables; and (2) it provides low sensitivity to component tolerances, which is an inherent property of passive-LC ladder filters. For switched-capacitor implementation, this implies low sensitivity of the filter response to the accuracy of capacitor ratios.

Circuit configurations which provide a given filter transfer function are not unique. Often, many circuit configurations exist which can provide the same transfer function. In designing switched-capacitor filters based on passive-component filter configurations, one normally progresses through the following sequence of steps:

1. The node voltage and loop current equations of the passive filter are written so that they contain only summations and integrations.

2. Since operational amplifiers used in switched-capacitor circuit implementation are voltage-controlled voltage amplifiers, loop currents are represented as "voltages," by multiplying them with an arbitrary *scaling resistor* R_s. As will be demonstrated in later examples, the choice of this scaling resistor allows one to scale current and impedance levels in the circuit, without affecting the transfer function.

3. The basic circuit configuration is drawn from a set of circuit equations using symbolic analog computer notations, showing the summing and integration operations. The resulting diagram is called a *state-variable* diagram, since it describes the individual voltages and currents at each of the circuit nodes. Instead of analog computer notations, flow-graph techniques can also be used as effectively.

—

4. From the symbolic state-variable diagram, corresponding switched-capacitor circuit building blocks can be determined and interconnected by inspection, to form the final switched-capacitor filter configuration.

5. Finally, capacitor ratios are calculated from R, L and C values given in the passive filter, and from the switched-capacitor filter design parameters. The switched-capacitor filter design parameters that affect the capacitor ratios are the clock frequency f_s and the scaling resistance R_s.

13.10. SECOND-ORDER FILTER CONFIGURATIONS

The second-order or *biquadratic* switched-capacitor filter configurations are often used in monolithic IC designs to simulate the conventional active-RC biquad filters discussed in Section 13.3. Although some of the switched-capacitor equivalents of the conventional active-RC filters can be derived by direct inspection and replacement of resistors with switched capacitors, such a synthesis technique is often not practical because the resulting switched-capacitor circuit equivalents still suffer from the high component sensitivity problems. A more practical approach is to derive the switched-capacitor equivalent circuits directly from passive component networks comprised of R, L, and C elements, using the basic five-step synthesis procedure outlined in the previous section.

One unique advantage of the biquadratic filters derived using the state-variable techniques is the simultaneous availability of several filter functions, such as low-pass, bandpass, or high-pass, from a given filter configuration. This very useful feature makes the switched-capacitor biquad filters particularly useful as general-purpose building blocks in designing complex monolithic LSI systems.[13,16]

Low-Pass Function

The biquadratic low-pass filter function is given by Eq. (13.8) and has the generalized response shape illustrated in Figure 13.5. In terms of passive components, it can be represented by the circuit of Figure 13.36a. The circuit has the basic transfer function

$$\frac{V_{out}}{V_{in}} = H(s) = \frac{1/LC}{s^2 + s(R_T/L) + 1/LC} \tag{13.40}$$

which is the same form as Eq. (13.8), with

$$\omega_p = \frac{1}{\sqrt{LC}} \tag{13.41}$$

and

$$Q_p = \omega_p \frac{L}{R_T} = \frac{1}{R_T} \sqrt{\frac{L}{C}} \tag{13.42}$$

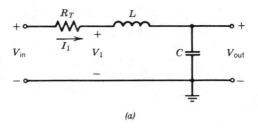

(a)

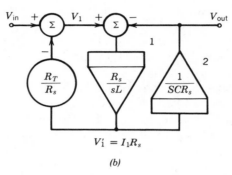

$V_1' = I_1 R_s$

(b)

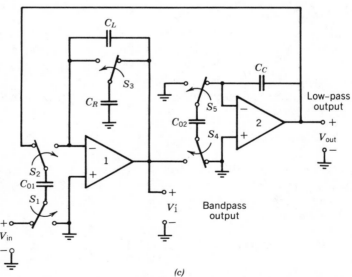

(c)

FIGURE 13.36. Two-pole low-pass filter example: (*a*) *RLC* equivalent circuit; (*b*) state-variable diagram; (*c*) switched-capacitor equivalent.

The filter topology of Figure 13.36a is called a *singly-terminated* filter configuration, since the filter output does not contain a resistive element. In general, *RLC* filters which contain resistive elements both at the input and at the output, are called *doubly-terminated* filters, and offer the lowest component sensitivity. However, for low-order filters, sensitivity degradation due to lack of output termination is relatively small and does not justify the extra circuitry.

In deriving the switched-capacitor equivalent of the second-order low-pass filter of Figure 13.36a, the first step is to list the loop current and node voltage equations so that they contain only summation and integration operations. Thus, from the figure, one obtains

$$V_1 = V_{in} - R_T I_1 \tag{13.43}$$

$$I_1 = \frac{1}{sL}(V_1 - V_{out}) \tag{13.44}$$

$$V_{out} = \frac{1}{sC} I_1 \tag{13.45}$$

Next, the current variable I_1 is converted to a voltage variable V_1' by defining a scaling resistance R_s where

$$I_1 \triangleq \frac{V_1'}{R_s} \tag{13.46}$$

Then Eqs. (13.43) through (13.45) can be rewritten in all-voltage form as

$$V_1 = V_{in} - \frac{R_T}{R_s} V_1' \tag{13.47}$$

$$V_1' = \frac{R_s}{sL}(V_1 - V_{out}) \tag{13.48}$$

$$V_{out} = \frac{1}{sCR_s} V_1' \tag{13.49}$$

These equations can be directly represented by the state-variable diagram of Figure 13.36b in terms of the integration, scaling, and summation operations performed. In the figure, integrator 2 is a standard noninverting integrator, integrator 1 represents a lossy differential integrator.

Figure 13.36c shows the corresponding switched-capacitor circuit equivalent which can be derived by inspection from the state-variable diagram. The switched capacitor C_R across integrator 1 simulates the input termination resistance R_T. Note that when the switches S_1, S_2, and S_3 are in the position shown, the total charge Q_T delivered to the inverting node of operational amplifier 1 is equal to

$$Q_T = C_0\left(V_{in} - V_1'\frac{C_R}{C_0} + V_{out}\right) \tag{13.50}$$

which satisfies the summation operations indicated in the state-variable diagram, with the scale factor

$$\frac{R_T}{R_s} = \frac{C_R}{C_0} \tag{13.51}$$

Note that integrator 1, with its associated switches, is essentially equivalent to that shown in Figure 13.24, with $V_4 = 0$.

By equating the integration coefficients in Figure 13.36b and c, the component values of the passive filter can be related to the switched-capacitor filter component values as

$$R_T = \frac{C_R}{C_{01}} R_s \tag{13.52}$$

$$L = \frac{C_L R_s}{f_s C_{01}} \tag{13.53}$$

$$C = \frac{C_C}{f_s R_s C_{02}} \tag{13.54}$$

where C_{01} and C_{02} are the switched capacitors at the input of the respective integrators. Normally, one would set

$$C_{01} = C_{02} = C_0 \tag{13.55}$$

which then serves as the unit capacitance against which all other capacitors are ratioed.

The capacitor ratios for the switched-capacitor filter can be expressed directly from Eqs. (13.52) through (13.54) as

$$\frac{C_C}{C_0} = C f_s R_s \tag{13.56}$$

$$\frac{C_L}{C_0} = \frac{L f_s}{R_s} \tag{13.57}$$

The value of the scaling resistor R_s is a free variable which can be used to optimize the filter dynamic range or to minimize the capacitance ratios to reduce the monolithic chip area. First, consider the case of the minimum chip area. From Eqs. (13.56) and (13.57), this comes about when $C_C/C_0 = C_L/C_0$, which corresponds to the condition that

$$R_s \bigg|_{\substack{\text{minimum} \\ \text{area}}} = \sqrt{\frac{L}{C}} \tag{13.59}$$

Next, consider the dynamic range. The maximum dynamic range of the filter is achieved when all the operational amplifier outputs have the same peak voltage. If the gain from the filter input to the output of a particular operational amplifier in the filter is greater than the gain to the output of the filter, then that particular internal operational amplifier will saturate before the output amplifier saturates. Thus, for maximum dynamic range, the voltage gain at the first operational amplifier output should be equal to the voltage gain at the filter output. The voltage swings within the circuit are maximum at $\omega = \omega_0$. The voltage gain at the output of the first operational amplifier, at $\omega = \omega_0$ is

$$\frac{V_1'}{V_{in}}(j\omega_0) = \frac{R_s}{R_T} \tag{13.60}$$

The voltage gain of the entire filter, at $\omega = \omega_0$ is

$$\frac{V_2}{V_{in}}(j\omega_0) = \frac{1}{R_T}\sqrt{\frac{L}{C}} \tag{13.61}$$

Equating results of Eqs. (13.60) and (13.61), one sees that for maximum dynamic range,

$$R_s\bigg|_{\substack{\text{maximum}\\\text{dynamic range}}} = \sqrt{\frac{L}{C}} \tag{13.62}$$

Thus, for this case, both the dynamic range and the chip area requirements are optimized simultaneously, when the scaling resistance R_s is chosen to have the value given in Eq. (13.59) or Eq. (13.62). This choice of R_s holds for optimizing most second-order filters. However, for high-order filters, calculating the optimum value of R_s becomes more complex and depends on the particular frequency response requirements.

Design Example. As an example, we shall now consider the design of a two-pole Butterworth filter using the low-pass filter configuration of Figure 13.36. The filter is assumed to have a 3-dB bandwidth of 1 kHz with a termination resistance $R_T = 1.0$ kΩ. The clock frequency for the switched-capacitor filter is chosen to be 100 kHz. For convenience, unit capacitors C_{01} and C_{02} are assumed to be equal, that is, $C_{01} = C_{02} = C_0$.

From the choice of $\omega_{3\,dB}$ and R_T, the values of L and C can be readily found from filter tables[17] as

$$L = 0.1126 \text{ H} \quad \text{and} \quad C = 0.2247 \text{ }\mu\text{F} \tag{13.63}$$

For this choice of L and C values, the value of R_s can be calculated from Eq. (13.62) as

$$R_s = \sqrt{\frac{L}{C}} = 0.707 \text{ k}\Omega \tag{13.64}$$

Then the capacitor ratios of the switched-capacitor filter can be calculated from Eqs. (13.56) through (13.58) as

$$\frac{C_C}{C_0} = \frac{C_L}{C_0} = 15.91 \quad \text{and} \quad \frac{C_R}{C_0} = 1.414 \tag{13.65}$$

Taking the unit capacitance $C_0 = 1$ pF, results in a total capacitance value for the circuit of less than 50 pF, making it very efficient in terms of the total chip area.

Bandpass Function

The biquadratic bandpass filter function is given by Eq. (13.10) and has the response characteristics shown in Figure 13.9. This basic transfer function can be simulated as the admittance function of the RCL circuit given in Figure 13.36a, where

$$\frac{I_1}{V_{in}} = \frac{s(1/L)}{s^2 + s(R_T/L) + 1/LC} \tag{13.66}$$

In terms of the switched-capacitor circuit of Figure 13.36c, the voltage V_1 at the output of integrator 1 is proportional to I_1, as defined in Eq. (13.44). Thus, the V_1 output of the circuit exhibits a bandpass response given as

$$\frac{V_1'}{V_{in}} = \frac{s(R_s/L)}{s^2 + s(R_T/L) + 1/LC} \tag{13.67}$$

which is the same form as Eq. (13.10), with the parameters ω_p, Q_p, and K defined as

$$\omega_p = \frac{1}{\sqrt{LC}} \tag{13.68}$$

$$Q_p = \frac{1}{R_T}\sqrt{\frac{L}{C}} \tag{13.69}$$

$$K = \frac{R_s}{R_T} \tag{13.70}$$

For $Q_p \gg 1$, the filter passband is approximately symmetrical around ω_p, with the 3-dB bandwidth given by Eq. (13.11).

In the case of the bandpass circuit, it is more convenient to relate the capacitor ratios of the switched-capacitor circuit to Q_p and the center frequency ω_p of the circuit. This can be done by substituting Eqs. (13.68) and (13.69) into Eqs. (13.56) and (13.57). The result is

$$\frac{C_C}{C_0} = \frac{R_s}{R_T}\frac{f_s}{\omega_p}\frac{1}{Q_p} \tag{13.71}$$

$$\frac{C_L}{C_0} = \frac{R_T}{R_s}\frac{f_s}{\omega_p}Q_p \tag{13.72}$$

$$\frac{C_R}{C_0} = \frac{R_T}{R_s} \tag{13.73}$$

If the optimum value of R_s given by Eq (13.59) or Eq. (13.62) is substituted into the above equations to optimize both the chip area and the output dynamic range, one obtains

$$\frac{C_C}{C_0} = \frac{C_L}{C_0} = \frac{f_s}{\omega_p} \tag{13.74}$$

$$\frac{C_R}{C_0} = \frac{1}{Q_p} \tag{13.75}$$

Equations (13.71), (13.72), and (13.75) illustrate a fundamental problem associated with high-Q switched-capacitor bandpass filters. As Q_p is increased, the capacitor values for C_L and C_C become widely separated. If the scaling resistor is optimized to avoid this problem, then the termination resistor capacitance C_R becomes too small. Since, normally, C_0 is chosen as the smallest unit capacitance, Eq. (13.75) becomes impractical to realize. For example, if C_0 is chosen as 1 pF, the value of C_R required by Eq. (13.75) would be 0.02 pF for a $Q = 50$ circuit, which is of the same order of magnitude as the parasitic capacitance. If C_0 is increased to avoid this problem, the values of C_C and C_L would be too large, since both of these capacitors are ratioed with C_0.

One practical solution to this problem is the use of a *charge-scaling* circuit to scale the value of C_R to a more realistic level.[18] This charge-scaling circuit can be implemented as a special case of the integrator/summer circuit of Figure 13.25a with $V_1 = 0$, which results in a transfer function

$$\frac{V_0}{V_x} = -\frac{C_x}{C_0} \tag{13.76}$$

In the circuit of Figure 13.36c, the charge Q_R supplied to the inverting input of integrator 1 by C_R is

$$Q = V_1' C_R \tag{13.77}$$

for each cycle. If V_1' can be scaled down along with C_R, then the value of C_R would not have to be made prohibitively small to reduce Q_R. Figure 13.37 shows a modified version of the bandpass filter of Figure 13.36, which uses such a scaling circuit, along with an inverting switch S_3, S_3', to replace the capacitor C_R and the switch S_3 of Figure 13.36c.

The circuit of Figure 13.37 has the same characteristics as that of Figure 13.36c, except that the C_R/C_0 ratio given by Eq. (13.73) is now replaced with

$$\frac{C_R}{C_0} \frac{C_A}{C_B} = \frac{1}{Q_p} \tag{13.78}$$

Thus, by choosing

$$C_R = C_A C_0 \quad \text{and} \quad C_B = Q_p C_0 \tag{13.79}$$

all capacitor values can be kept at realizable levels. However, it should be noted that this technique requires the addition of an operational amplifier and a relatively large capacitor C_B of the order of QC_0 into the circuit, which results in an increase of the chip area and degradation of the noise characteristics.

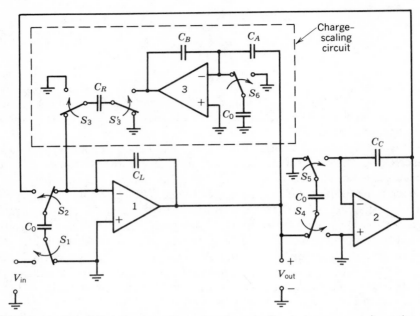

FIGURE 13.37. Biquadratic bandpass filter using charge scaling to improve capacitor ratios.

Figure 13.38 shows an alternate switched-capacitor bandpass filter configuration which is based on the shunt-LC resonant circuit configuration. From Figure 13.38, the discrete circuit transfer function is

$$\frac{V_{\text{out}}}{V_{\text{in}}} = \frac{s(1/R_T C)}{s^2 + s(1/R_T C) + 1/LC} \tag{13.80}$$

which has the Q and center frequency values

$$Q_p = R_T \sqrt{\frac{C}{L}} \tag{13.81}$$

and

$$\omega_p = \frac{1}{\sqrt{LC}} \tag{13.82}$$

From Figure 13.38a, one can write the scaled circuit node and loop equations as

$$V_{\text{in}}\frac{R_s}{R_T} - V_{\text{out}}\frac{R_s}{R_T} = V_0' \tag{13.83}$$

$$V_0 = V_1' + V_2' \tag{13.84}$$

$$V_{\text{out}} = V_1'\frac{1}{sR_s C} \tag{13.85}$$

$$V_2' = V_{\text{out}}\frac{R_s}{sL} \tag{13.86}$$

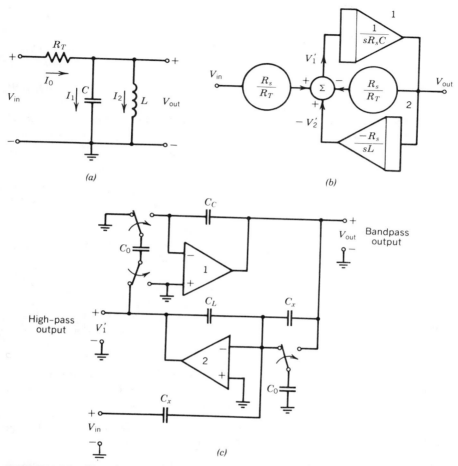

FIGURE 13.38. Biquadratic bandpass filter configuration with improved capacitor ratios: (a) *RLC* equivalent circuit; (b) state-variable diagram; (c) switched-capacitor equivalent circuit.[13]

where R_s is the scaling resistance, and voltages V_0', V_1', and V_2' correspond to the currents I_0, I_1, and I_2, respectively [see Eq. (13.46)]. The resulting state-variable diagram is shown in Figure 13.38b.

Figure 13.38c shows a practical realization of the corresponding switched-capacitor circuit, using the integrator/summer configuration (see Fig. 13.25) for integrator 2. As a result, both the input and the output voltages are scaled by the ratio C_x/C_L at the *output* of integrator 2.[13] Thus, the summing function shown in Figure 13.38b actually takes place at the *output* of integrator 2, rather than at the input of integrator 1. This technique has the benefit of both reducing the number of switches and also improving the capacitor ratios.

By equating the switched-capacitor circuit components to their state-variable equivalents in Figure 13.38, and by substituting from Eqs. (13.81) and (13.82), the capacitor ratios can be expressed as

$$\frac{C_C}{C_0} = \frac{R_s}{R_T} \frac{f_s}{\omega_p} Q_p \qquad (13.87)$$

$$\frac{C_L}{C_0} = \frac{R_T}{R_s} \frac{f_s}{\omega_p} \frac{1}{Q_p} \qquad (13.88)$$

$$\frac{C_x}{C_L} = \frac{R_s}{R_T} \qquad (13.89)$$

One can show that both the chip area and the dynamic range are optimized if R_s is chosen such that

$$\left. \frac{R_T}{R_s} \right|_{\text{optimum}} = Q_p \qquad (13.90)$$

Substituting Eq. (13.90) into Eqs. (13.87) through (13.89), one obtains the optimized capacitor ratios as

$$\frac{C_C}{C_0} = \frac{C_L}{C_0} = \frac{f_s}{\omega_p} \qquad (13.91)$$

$$\frac{C_x}{C_0} = \frac{f_s}{\omega_p Q_p} \qquad (13.92)$$

Note that the C_x/C_0 ratio given by Eq. (13.92) is much more favorable than that of Eq. (13.75) for practical circuit implementation, particularly if a high clock frequency is utilized.

Design Example.　Consider the typical component values associated with a 1-kHz bandpass filter, with a Q of 10, operating with a 200-kHz clock rate. Then from Eq. (13.91) and (13.92), assuming $C_0 = 1$ pF, one obtains

$$C_C = C_L = 31.83 \text{ pF} \quad \text{and} \quad C_x = 3.183 \text{ pF} \qquad (13.93)$$

If the required value of Q is raised to 100, then C_C and C_L would remain unchanged, but C_x would be reduced to

$$C_x = \frac{C_L}{Q} = 0.318 \text{ pF} \qquad (13.94)$$

which is small but still realizable. Note that the total capacitance requirement for the circuit is less than 100 pF in both cases.

Band-Stop Function

The biquadratic band-stop function has the basic response characteristics shown in Figure 13.11. Its generalized transfer function is given in Eq. (13.12). A biquadratic band-stop function can be obtained using the simple RLC circuit shown in Figure 13.39a. The resulting transfer function is

$$\frac{V_{\text{out}}}{V_{\text{in}}} = \frac{s^2 + 1/LC}{s^2 + s(1/R_T C) + 1/LC} \qquad (13.95)$$

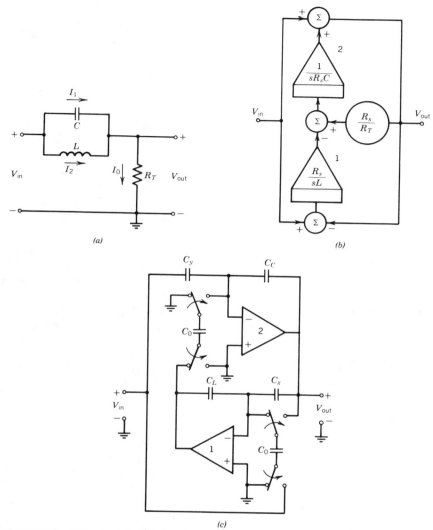

(a)

(b)

(c)

FIGURE 13.39.   Biquadratic band-reject filter configuration: (a) *RLC* equivalent circuit; (b) state-variable diagram; (c) switched-capacitor eqiuvalent circuit.

Equation (13.95) corresponds to a special case of Eq. (13.12), where

$$\omega_z = \omega_p = \frac{1}{\sqrt{LC}} \qquad (13.96)$$

$$Q_p = R_T \sqrt{\frac{C}{L}} \qquad (13.97)$$

Note that the circuit of Figure 13.39a is essentially the same as that of Figure 13.38a, where the output is taken across R_T, rather than across the *LC* tank.

Figure 13.39b shows the state-variable diagram of the circuit, derived from the scaled loop current and node voltage equations

$$(V_{in} - V_{out}) \frac{R_s}{sL} = V_2' \tag{13.98}$$

$$V_0' = \frac{R_s}{R_T} \tag{13.99}$$

$$V_1' = V_0' - V_2' \tag{13.100}$$

$$V_1' \frac{1}{sRC} = V_{in} - V_{out} \tag{13.101}$$

Figure 13.39c shows a switched-capacitor implementation of the circuit, using switched-capacitor integrator/summer circuits similar to those described in Figure 13.38c.[13] In this manner, the summation of signals is performed at the output of integrators 1 and 2, without requiring additional switches. In the switched-capacitor equivalent circuit of Figure 13.39c, integrator 1, along with the scaling capacitor C_x, performs the operations described by Eqs. (13.98) through (13.100). Integrator 2 performs the function indicated by Eq. (13.101). As described in connection with the bandpass filter of Figure 13.37c, the use of an integrator/summer configuration provides a smaller spread of capacitor ratios for high-Q circuits. In addition, the noise characteristics of the circuit are also improved, since the total number of switches can be reduced.

The capacitor ratios for the band-stop circuit are the same as those given in Eqs. (13.87) through (13.89), since both circuits are derived from the same basic RLC circuit (see Figs. 13.38a and 13.19b),

$$\frac{C_C}{C_0} = \frac{R_s}{R_T} \frac{f_s}{\omega_p} Q_p \tag{13.102}$$

$$\frac{C_L}{C_0} = \frac{R_T}{R_s} \frac{f_s}{\omega_p} \frac{1}{Q_p} \tag{13.103}$$

$$\frac{C_x}{C_0} = \frac{R_s}{R_T} \tag{13.104}$$

$$\frac{C_y}{C_C} = 1.0 \tag{13.105}$$

The dynamic range and the capacitor ratios are optimized by setting $R_T/R_s = Q_p$, as in the case of the bandpass circuit.

As shown in Figure 13.39, with $C_C = C_y$, the circuit provides a symmetrical band-stop response, with $\omega_z = \omega_p$ (see Fig. 13.11). However, if an arbitrary capacitor ratio is chosen such that

$$\frac{C_y}{C_C} = \beta \tag{13.106}$$

then the overall transfer function of the filter becomes

$$\frac{V_{out}}{V_{in}} = \frac{s^2 + \omega_p^2/\beta}{s^2 + s(\omega_p/Q_p) + \omega_p^2} \tag{13.107}$$

which is equivalent to the general biquad band-stop transfer function of Eq. (13.12), with

$$K = 1.0 \quad \text{and} \quad \omega_z = \omega_p/\sqrt{\beta} \tag{13.108}$$

Thus, by controlling the C_y/C_C ratio, one can change the symmetry of the response characteristics as well as the notch frequency ω_z. Setting $\beta > 1$, increases the circuit gain in the upper passband, whereas $\beta < 1$ has the opposite effect.

High-Pass Function

The biquadratic high-pass transfer function given by Eq. (13.19) can be implemented as the current I_1 through the capacitor C in Figure 13.38a. The resulting transfer function is

$$\frac{V_{out}}{V_{in}} = \frac{s^2(1/R_TC)^2}{s^2 + s(1/R_TC) + 1/LC} \tag{13.109}$$

In terms of the scaled voltage equivalent, this corresponds to the voltage V_1' in Figure 13.38c. For high-pass applications, similar to the case of low-pass circuits, it is more convenient to express the capacitor ratios directly in terms of L and C values of the original transfer function, rather than ω_p and Q_p. By proper substitutions, one obtains form Eqs. (13.87) through (13.90),

$$C_C = R_s f_s C \tag{13.110}$$

$$\frac{C_L}{C_0} = \frac{f_s L}{R_s} \tag{13.111}$$

$$\frac{C_x}{C_L} = \frac{R_s}{R_T} \tag{13.112}$$

From Eq. (13.90), the optimum value of R_s for minimum chip area (i.e., $C_C = C_L$) and maximum dynamic range is

$$R_s\bigg|_{optimum} = \frac{R_T}{Q_p} = \sqrt{\frac{L}{C}} \tag{13.113}$$

13.11. HIGHER ORDER FILTERS

The basic design and synthesis methods for switched-capacitor filters discussed in the previous sections are also applicable to higher order (i.e., $n > 2$) filters. Higher-order filters can be designed either by cascading a number of second-order (biquad) sections or by direct synthesis methods from a higher order RLC ladder network.

Cascaded Biquad Filters

The cascaded biquad method is straightforward and lends itself to a "building-block" approach where a given filter transfer function can be synthesized as a collection of *pole pairs*. In this approach, the dynamic range of each of the filter sections can also be optimized individually, by the choice of proper terminations in each section.

The cascaded biquad technique also lends itself to the design of general-purpose high-order monolithic filters in semicustom form, where many biquad sections can be prefabricated on an IC chip and customized into a high-order filter by the choice of capacitor and interconnection masks. Figure 13.40 shows the photomicrograph of an NMOS semicustom, or mask-programmable, switched-capacitor filter chip which is capable of implementing up to 22 poles and zeros in the form of cascaded biquad sections.[20] Some of these sections can also be used for continuous-time antialiasing or output reconstruction filters.

The mask-programmable switched-capacitor filter chip shown in Figure 13.40 is designed with a double-polysilicon NMOS process. It uses a double-polysilicon capacitor structure similar to that shown in Figure 13.28, and operates with $\pm$ 5-V supplies. The monolithic chip measures 120 $\times$ 240 mils and is designed to fit into a 16-pin standard DIP IC package. It contains 22 operational amplifiers, 88 analog switches, input/output buffers, on-chip clock generation and busing circuitry, second-order continuous antialiasing, and smoothing filters. The basic architecture of the chip is identified in the figure. This chip

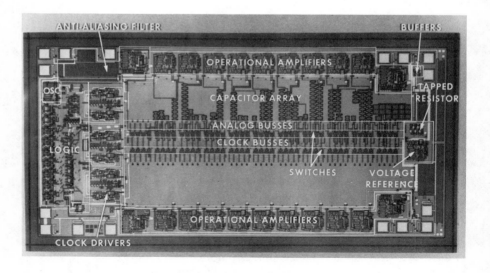

FIGURE 13.40. Photomicrograph of a mask-programmable NMOS chip containing 11 undedicated biquad filter sections. Chip size: 120 mils $\times$ 240 mils. (Photo: Bell Laboratories.)

is customized by mask-programming the first and second layers of poly-crystalline silicon, which define the capacitor areas and the essential inter-connections. All the other masks are held fixed. In this manner, the basic design can be quickly and inexpensively "customized" to fit a particular filter function.

In the particular case illustrated in the photomicrograph of Figure 13.40, the NMOS switched-capacitor filter chip is used for implementing a tenth-order filter function, where only the 10 operational amplifiers along the upper periphery of the chip are used. The empty area in the lower middle portion of the chip is reserved for implementing the capacitor arrays associated with the 10 operational amplifier in the lower part of the chip, which are not utilized in this particular implementation. The parts of the active circuitry, such as the operational amplifiers which are not used in a particular implementation, are "powered down," and only the active part of the chip is energized.

The key advantage of the cascaded biquad design for higher order switched-capacitor filters is its versatility. However, the filter frequency response exhibits relatively high sensitivity to component tolerances (i.e., capacitor ratios and parasitic capacitances). Some of these sensitivity problems can be greatly re-duced by using the direct synthesis methods described in the following sections.

All-Pole Low-Pass Ladder Filters

The doubly terminated low-pass ladder filter configuration shown in Figure 13.41 exhibits very low sensitivity to the absolute-value tolerance of circuit components.[19] Therefore, it is often used as the starting point of many high-order low-pass filter designs. Such a circuit provides an all-pole transfer function of the form

$$\frac{V_{\text{out}}}{V_{\text{in}}} = \frac{K}{s^n + a_{n-1}s^{n-1} + \ldots + a_1 s + a_0} \qquad (13.114)$$

By the proper choice of the pole locations (i.e., the choice of coefficients a_{n-1} through a_0) one can implement Butterworth, Chebyshev, Legendre, and Bessel low-pass filters (see Fig. 13.3), using the all-pole transfer function.

In implementing a particular high-order filter function using the switched-capacitor equivalent of ladder networks, one follows the sequence of basic synthesis steps outlined in Section 13.9, namely, starting with node voltage and

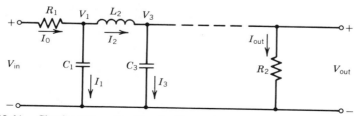

FIGURE 13.41. Circuit configuration for a doubly terminated RLC all-pole low-pass ladder filter.

loop current equations, proceeding to a state-variable diagram, and finally to a switched-capacitor equivalent circuit. However, since the basic structure of the filter is repetitive, such a synthesis procedure can be greatly simplified by examining one section of the ladder and then repeating this structure for the remaining sections. Figure 13.42a shows a simple LC section within the ladder which can be described by the state-variable equations

$$(V_1 - V_3)\frac{R_s}{sL_2} = V_2'$$ (13.115)

$$(V_2' - V_4')\frac{1}{sR_sC_3} = V_3$$ (13.116)

where V_2' and V_4' correspond to the respective currents multiplied with the scaling resistor R_s. The corresponding state-variable diagram and switched-capacitor equivalent circuit of the ladder section is shown in Figure 13.42b and c. Note that the switch phasing between successive integrator stages is alternated to eliminate the excess phase shifts due to the sampling process (see Fig. 13.26).

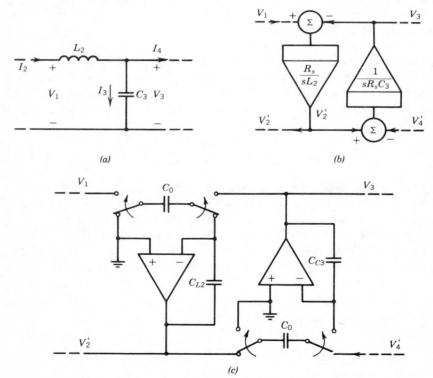

FIGURE 13.42. Switched-capacitor equivalent of one section of RLC ladder filter of Figure 13.41: (a) Circuit schematic; (b) state-variable diagram; (c) switched-capacitor equivalent circuit.

Figure 13.43 shows a complete fifth-order all-pole low-pass filter implemented in this manner. The switched-capacitor filter capacitor ratios can be obtained by equating the integrator time constants in the switched-capacitor circuit with those of the equivalent state-variable diagram, in the same manner as in Eqs. (13.56) through (13.58), namely,

$$\frac{C_{Ci}}{C_0} = C_i f_s R_s \qquad (13.117)$$

$$\frac{C_{Li}}{C_0} = \frac{L_i f_s}{R_s} \qquad (13.118)$$

$$\frac{C_{Ri}}{C_0} = \frac{R_i}{R_s} \qquad (13.119)$$

where the subscript i refers to the particular component designations in Figure 13.43. In designing a particular filter response, one would use the values of L_i and C_i given in the filter tables to calculate the corresponding switched-capacitor circuit capacitor ratios.

Figure 13.44 shows a photomicrograph of an experimental fifth-order Chebyshev low-pass filter chip, using the circuit configuration of Figure 13.43b. The circuit is designed as a 3400-Hz low-pass filter, with a passband ripple of 0.1 dB, and operates with a 128-kHz clock frequency.[19]

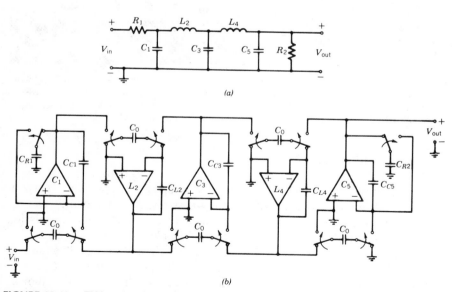

FIGURE 13.43. Fifth-order all-pole low-pass ladder filter configuration: (a) RLC equivalent circuit; (b) switched-capacitor equivalent circuit.

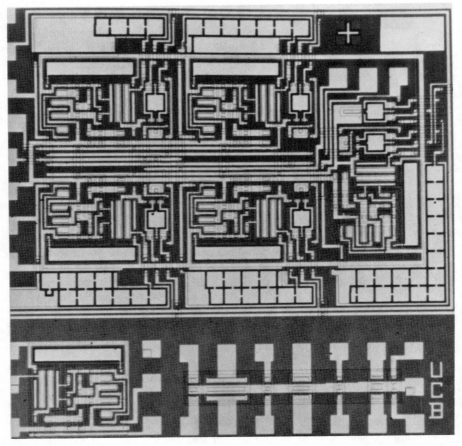

FIGURE 13.44. Photomicrograph of fifth-order Chebyshev low-pass filter using circuit configuration of Figure 13.43. Technology used: metal-gate NMOS with depletion loads, 10-μm feature size, chip size 100 mils × 100 mils, lower half of chip contains test devices for characterization. (*Photo*: University of California, Berkeley, Department of EECS.)

Filters with Transmission Zeros

Filters with finite transmission zeros within the stopband are particularly useful in many applications. Examples of such low-pass filters are the elliptic and the inverse Chebyshev filters illustrated in Figures 13.3c and d. The transmission zero addition is easily implemented in an *RLC* ladder configuration by adding shunt capacitors across the series inductors in the circuit. This is illustrated in Figure 13.45a for the case of a third-order elliptic filter. The capacitor C_2 across L_2 provides a $j\omega$-axis zero at the resonant frequency of the shunt *LC* combination. The state-variable diagram corresponding to the circuit of Figure 13.45a is not practical for switched-capacitor implementation since it contains voltage

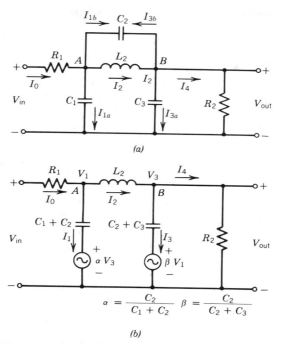

FIGURE 13.45. Representation of third-order elliptic filter section by its Thévenin eqivalent: (*a*) Equivalent *RLC* circuit; (*b*) Thévenin equivalent of actual *RLC* circuit.

scaling (i.e., multiplications) separate from operational amplifier integrators. In switched-capacitor realization, it is desirable to avoid such scaling operations which are not included with the integrators, in order to minimize the number of operational amplifiers and the capacitor sizes used.[11]

In order to avoid the use of additional operational amplifiers in the snythesis of the circuit of Figure 13.45*a*, it is useful to examine in detail the operation performed by the shunt capacitor C_2. It forms a bidirectional path between nodes A and B and feeds some of the signal from A to B, while simultaneously feeding a fraction of the signal from B back to A. In terms of the currents and voltages shown, the function of C_2 can be expressed as

$$V_1 = \frac{I_0 - I_2}{s(C_1 + C_2)} + V_3 \frac{C_2}{C_1 + C_2} \tag{13.120}$$

$$V_3 = \frac{I_2 - I_4}{s(C_2 + C_3)} + V_1 \frac{C_2}{C_2 + C_3} \tag{13.121}$$

From Eqs. (13.120) and (13.121) one can replace C_2 in Figure 13.45*a* with the equivalent circuit of Figure 13.45*b*, which contains two shunt capacitances and two dependent voltage sources where the scaling constants α and β are given as

$$\alpha = \frac{C_2}{C_1 + C_2} \quad \text{and} \quad \beta = \frac{C_2}{C_2 + C_3} \tag{13.122}$$

The shunt branches of the equivalent circuit of Figure 13.45b, which contain the dependent voltage sources, can be implemented by using the integrator/summer circuit of Figure 13.25 with appropriate sign changes.

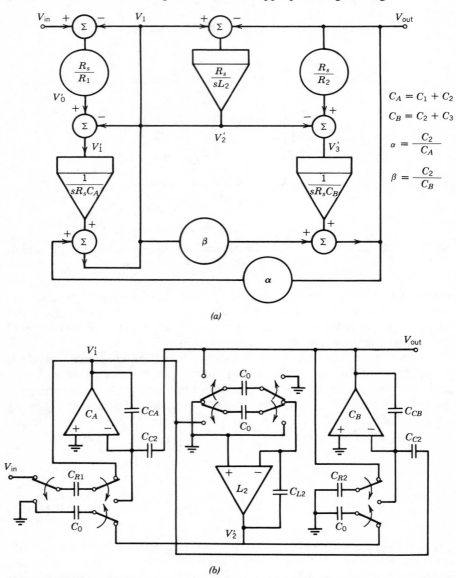

(a)

(b)

FIGURE 13.46. Switched-capacitor equivalent of third-order elliptic low-pass filter of Figure 13.45: (a) State-variable diagram; (b) switched-capacitor equivalent circuit.

Figure 13.46 illustrates the state-variable diagram and the switched-capacitor filter circuit which corresponds to the basic circuit configuration of Figure 13.45. Note that by using the equivalent circuit of Figure 13.45b, the integration, scaling and summing operations associated with capacitors C_A ($=C_1 + C_2$) and C_B ($=C_2 + C_3$) are all performed by the op amps designated as C_A and C_B and the scaling capacitors C_{C2}, in Figure 13.46b.

Bandpass Filters

The high-order bandpass switched-capacitor filters can be directly synthesized from bandpass ladder filter topology. The bandpass ladder configuration can be derived from the low-pass ladder circuit of Figure 13.41 by replacing the shunt capacitors with a shunt-LC combination, and by replacing the series inductor with a series-LC combination. This operation is known as the low-pass-to-bandpass transformation.

Figure 13.47a shows the circuit configuration for a fourth-order bandpass ladder filter derived from the low-pass ladder by means of the low-pass-to-bandpass transformation. The state-variable diagram of the circuit is shown in Figure 13.47b, with its switched-capacitor circuit implementation given in Figure 13.47c. Because of the presence of C_B in series with the termination resistor R_2, the actual output V_{out} appears at the input, rather than the output, of an integrating amplifier. Therefore, it cannot be extracted without using an additional buffer amplifier. However, this problem can be avoided by taking the output from the V_3' terminal, which corresponds to the scaled version of the current I_3 and is related to V_{out} by the scale factor R_s/R_2.

High-Pass Filters

The higher order high-pass filters can be directly synthesized from high-pass ladder networks. The high-order ladders can be derived from the basic low-pass ladder of Figure 13.41 by replacing all the capacitors with inductors and vice versa. Similar to other filter configurations, high-pass filters can be either doubly or singly terminated. Figure 13.48a shows the RLC equivalent circuit of a singly terminated fourth-order high-pass filter. For switched-capacitor high-pass filters, the input termination resistor is inconvenient to implement since it often requires an additional operational amplifier. Therefore, in most switched-capacitor high-pass filters, a singly terminated ladder configuration is used.

Similar to the case of elliptic filters, the high-pass ladders can be most efficiently implemented using the integrator/summer circuit (see Fig. 13.25) as the basic building block, and by performing the summing operation directly at the output, rather than at the input, of the integrators.[21] Figure 13.48b shows a state-variable diagram of the high-pass ladder filter, which is rearranged to transfer the summing operation to integrator outputs. A direct switched-capacitor implementation of this circuit is shown in Figure 13.48c. Note that the summing operations at the outputs of the first three integrators do not require

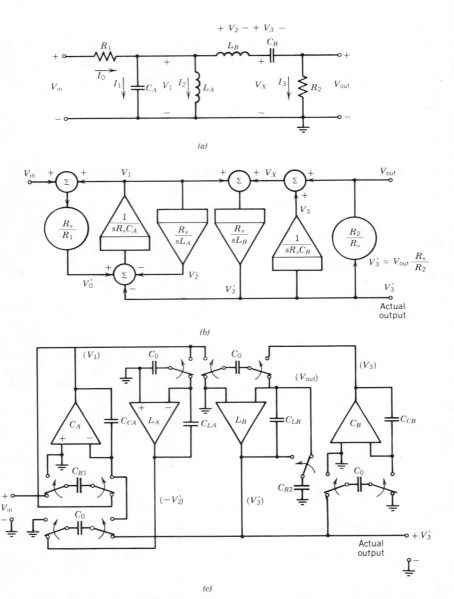

FIGURE 13.47. Fourth-order bandpass filter configuration: (*a*) *RLC* equivalent circuit; (*b*) state-variable diagram; (*c*) switched-capacitor equivalent circuit.

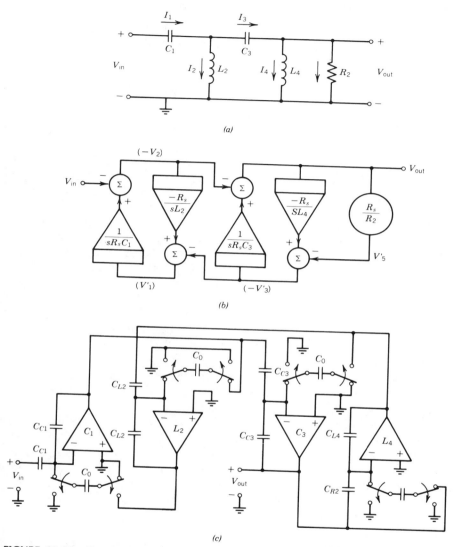

FIGURE 13.48. Fourth-order high-pass filter example: (a) RLC equivalent circuit; (b) state-variable diagram; (c) switched-capacitor equivalent circuit.

scaling. Therefore, the summing capacitor ratio associated with each integrator is unity. The summing operation at the output of the fourth integrator requires scaling by a factor of R_s/R_2, which is achieved by setting

$$\frac{C_{R2}}{C_{L4}} = \frac{R_s}{R_2} \tag{13.123}$$

13.12. APPLICATIONS AND LIMITATIONS OF SWITCHED-CAPACITOR FILTERS

Switched-capacitor circuit design techniques, together with the advances in MOS analog design, have made it possible to implement a wide range of filter functions on the silicon chip. The biggest advantage of switched-capacitor circuits, compared to conventional active-RC filters, is the elimination of the need for precise control of capacitor and resistor absolute values. Instead, the filter performance is determined by capacitor ratios and the choice of the clock or sampling frequency f_s. Normally, the clock signal is either supplied externally or derived from an on-chip oscillator and countdown circuitry. In the latter case, an external crystal would be used to set the oscillator frequency. In most applications, an inexpensive 3.58-MHz color TV chroma-regeneration crystal is used as the external frequency-setting component. Since no external adjustments or trimming is needed, other than a clock signal, the packaging of the IC chip is greatly simplified.

There are two practical limitations to monolithic switched-capacitor filters: high-frequency capability and noise characteristics. The high-frequency capability is limited by the clock frequencies that can be used in practical MOS circuits. As discussed in Section 13.7, the practical choices for f_s are limited to no more than several hundred kilohertz. In theory, the maximum useful filter frequency is limited to one-half the clocking frequency, as given by the Nyquist criteria [Eq. (13.19)]. In practice, to minimize or avoid continuous-time antialiasing and reconstruction filter requirements, f_s is chosen to be at least one order of magnitude higher than the filter cutoff frequency f_c. This limits the useful frequency range for monolithic switched-capacitor filters to the audio band, that is, from dc to approximately 20 kHz.

The noise characteristics of monolithic switched-capacitor filters are somewhat poorer than their active-RC counterparts. This is primarily due to the noise associated with the MOS operational amplifiers and the analog switches. Thus, the dynamic range, and the best signal-to-noise ratio that can be achieved by monolithic switched-capacitor filters using conventional MOS process technologies, is limited to 80–90 dB, which is not suitable for very-high-quality audio signal processing or for the detection of low-level input signals.

Another very important point which should always be remembered is that switched-capacitor filters are discrete-time or sampled-data circuits. They very closely approximate the continuous-time filter response *only* when f_s is much larger than the signal frequency. This is illustrated in Figure 13.49, which shows the actual frequency response of a 1-kHz low-pass filter for various clock frequencies.[5] Note that the aliased response of the switched-capacitor filter, outside the passband, becomes very noticeable as f_s is reduced.

The time delay inherent to the sampled-data nature of switched-capacitor integrators also introduces excess phase shifts into the filter response, which can cause instability or excessive peaking [see Eq. (13.29)]. In the case of simple lossless integrator loops, such as those shown in Figure 13.42, this problem can be avoided by proper phasing of the sampling switches (see Fig. 13.26). However, it is inherently present in the lossy-integrator stages which are used to simulate the termination resistances, and cannot be avoided by clock phasing. Thus, it must be taken into account in calculating the actual filter response.

In spite of their limitations, particularly with regard to frequency range and noise, switched-capacitor filters find a wide range of usage in voice-band and telephone related applications. Some of their primary areas of applications are in pulse-code modulation coder–decoder (CODEC) circuits, tone encoding and decoding, echo cancellation, speech synthesis, and speech recognition.

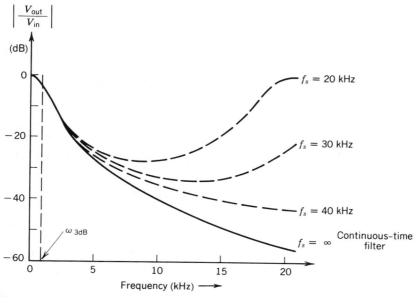

FIGURE 13.49. Magnitude response of 1-kHz low-pass filter example of Figure 13.36 for various clock frequencies.

REFERENCES

1. G. Moschytz, *Linear Integrated Networks Design*, Van Nostrand Reinhold, New York, 1975.

2. L. Huelsman, Ed., *Active Filters*, McGraw-Hill, New York, 1970.

3. R. Schaumann, M. A. Soderstrand, and K. R. Laker, Eds., *Modern Active Filter Design*, IEEE Press, New York, 1981.

4. M. S. Ghausi and K. R. Laker, *Modern Filter Design: Active-RC and Switched Capacitor*, Prentice-Hall, Englewood Cliffs, NJ, 1981.

5. R. W. Brodersen, P. R. Gray, and D. A. Hodges, "MOS Switched Capacitor Filters," *Proc. IEEE*, **67**, 61–75 (January 1979).

6. P. R. Gray, D. A. Hodges, and R. W. Brodersen, Eds., *Analog MOS Integrated Circuits*, IEEE Press, New York, 1980.

7. G. S. Moschytz and P. Horn, *Active Filter Design Handbook*, Wiley, New York, 1981.

8. W. J. Kerwin, L. P. Huelsman, and R.W. Newcomb, "State-Variable Synthesis for Insensitive Integrated Circuit Transfer Functions," *IEEE J. Solid-State Circuits* **SC-2**, 87–92 (September 1967).

9. L. R. Rabiner and B. Gold, *Theory and Application of Digital Signal Processing*, Prentice-Hall, Englewood Cliffs, NJ, 1975.

10. S. A. Tretter, *Introduction to Discrete Time Signal Processing*, Wiley, New York, 1976.

11. G. M. Jacobs, D. J. Allstot, R. W. Brodersen, and P. R. Gray, "Design Techniques for MOS Switched Capacitor Ladder Filters," *IEEE Trans. Circuits and Systems*, **CAS-25**, 1014–1021 (December 1978).

12. J. L. McCreary and P. R. Gray, "All-MOS Charge Redistribution Analog-to-Digital Conversion Techniques—Part I," *IEEE J. Solid-State Circuits* **SC-10**, 371–379 (December 1975).

13. B. J. White, G. M. Jacobs, and G. F. Landsburg, "A Monolithic Dual Tone Multifrequency Receiver," *IEEE J. Solid-State Circuits* **SC-14**, 991–997 (December 1979).

14. G. M. Jacobs, "Practical Design Considerations for MOS Switched Capacitor Ladder Filters," University of California, Berkeley, Electronics Research Lab. Memo. UCB/ERL M77/69, November 1977.

15. K. Martin, "Improved Circuits for Realization of Switched-Capacitor Filters," *IEEE Trans. Circuits and Systems* **CAS-27**, 237–244 (April 1980).

16. P. E. Fleischer and K. R. Laker, "A Family of Active Switched-Capacitor Biquad Building Blocks," *Bell System Tech. J.* **58**, 2235–2269 (December 1979).

17. A. B. Williams, *Electronic Filter Design Handbook*, McGraw-Hill, New York, 1981.

18. D. J. Allstot, R. W. Brodersen, and P. R. Gray, "An Electrically Programmed Switched Capacitor Filter," *IEEE J. Solid-State Circuits* **SC-14**, 1034–1041 (December 1979).

19. D. J. Allstot, R. W. Brodersen, and P. R. Gray, "MOS Switched Capacitor Ladder Filters," *IEEE J. Solid-State Circuits* **SC-13**, 806-814 (December 1978).

20. P. E. Fleischer, K. R. Laker, D. G. Marsh, J. P. Ballantyne, A. A. Yiannoulus, and D. L. Fraser, "An NMOS Building Block for Telecommunications Applications," *IEEE Trans. Circuits and Systems* **CAS-27**, 552–559 (June 1980).

21. D. J. Allstot, "MOS Switched Capacitor Ladder Filters," University of California, Berkeley, Electronics Research Lab. Memo. UCB/ERL M79/30, May 1979.

CHAPTER FOURTEEN

DATA CONVERSION CIRCUITS: DIGITAL-TO-ANALOG CONVERTERS

Digital and analog signal processing correspond to two fundamental, yet distinctly different, modes of information handling. In the analog case, the signals are handled in a nonquantized, continuously variable manner; in digital processing, they are quantized in binary *bits*, that is, in 1's and 0's. In many cases, it is necessary to interface these two fundamental means of signal processing and to convert the data from digital to analog or vice versa. This is accomplished by the use of digital-to-analog (D/A) and analog-to-digital (A/D) converter circuits.

In their natural state, all variables (such as current, voltage, pressure, distance, time) appear in analog form. However, for signal transmission and computation purposes, they are often handled in a digital manner. Therefore, the A/D and D/A converters can be considered to be a class of *coding* and *decoding* devices, respectively. The input to a D/A converter is a digital *word* of a prescribed number of bits, and the output is an analog voltage level uniquely corresponding to the input word. Conversely, the analog input applied to the A/D converter results in a digital word of a prescribed number of bits.

The performance of data converter circuits will depend very strongly on the matching and thermal tracking of component values in the circuit. This characteristic makes them ideally suited for monolithic integration. The subject of monolithic data converters can be covered best by dividing it into its two basic categories, D/A and A/D converters, and examining each one separately. The subject of D/A converters will be covered in this chapter, and the subject of A/D converters will be examined in the next chapter.

14.1. PRINCIPLES OF D/A CONVERSION

The digital-to-analog conversion circuits, which are also called D/A converters, or DACs, can be considered as decoding devices that accept digitally coded signals and provide analog outputs in the form of currents or voltages. In this manner, they provide an interface between the digital signals of the computer systems and continuous signals of the analog world.

Figure 14.1a shows the functional block diagram of a basic D/A converter system. The input to the D/A converter is a digital word D, made up of a stream of binary bits comprised of 1's and 0's. The output analog quantity A, which can be a voltage or a current, is related to the input as

$$A = KV_{ref}D \tag{14.1}$$

where K is a scale factor, V_{ref} is a reference voltage, and D is the digital word of a given number of bits. D can be represented as

$$D = \frac{b_1}{2^1} + \frac{b_2}{2^2} + \frac{b_3}{2^3} + \ldots + \frac{b_N}{2^N} \tag{14.2}$$

where N is the total number of bits, and $b_1, b_2, \ldots$ are the bit coefficients, which

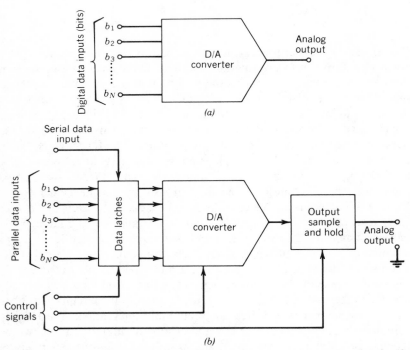

FIGURE 14.1. Functional block diagram of a D/A converter: (a) Basic D/A converter; (b) microprocessor-compatible D/A converter with interface circuitry.

are quantized to be either a 1 or a 0. Thus, the transfer function of an N-bit D/A converter can be written as

$$A = KV_{\text{ref}}\left[b_1 2^{-1} + b_2 2^{-2} + \ldots + b_N 2^{-N} \right] \qquad (14.3)$$

In many practical applications, particularly those requiring direct interface with data buses or microprocessors, the basic D/A system also requires some additional support circuitry, such as data latches and an output sample-and-hold circuit, as shown in Figure 14.1b. The function of input data latches would be to store the digital input data and prevent them from changing until the conversion step is completed. Similarly, the output sample-and-hold circuit would hold the output level constant, corresponding to the last conversion step, while the new digital data are being loaded into the input latches for the next conversion step.

In its most commonly used form, the basic D/A converter system of Figure 14.1a consists of four separate blocks, as shown in Figure 14.2: a reference voltage source which generates V_{ref}, a set of binary switches which set the binary bit coefficients (b_1 through b_N), a resistive current-scaling circuit, and an output current-summing amplifier which essentially converts the current output to an analog voltage level. In certain cases, a current rather than a voltage output would be sufficient, and the summing amplifier can be left out of the system.

The voltage reference circuitry can be either internal or external to the monolithic D/A converter circuit, depending on system accuracy, stability, and resolution requirements. As indicated by Eq. (14.3), the analog output is a product of the reference voltage and the digital word. Thus, the analog output can be modulated by varying the value of V_{ref}. The D/A converter circuit which has this

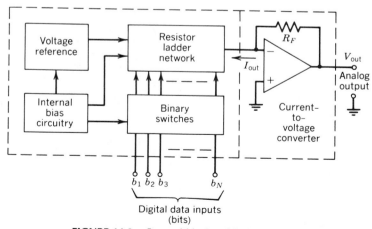

FIGURE 14.2. Internal blocks of D/A converter.

property, or is designed to operate with variable values of V_{ref}, is called *multiplying* D/A converter. Normally, all D/A converters that operate with an external reference fall into this category.

Assuming a voltage output, the transfer function of an N-bit D/A converter can be written as

$$V_O = V_{FS}(b_1 2^{-1} + b_2 2^{-2} + \ldots + b_N^{-N}) \tag{14.4}$$

where V_{FS} is the full-scale output voltage, which is equal to KV_{ref} of Eq. (14.3). As a function of the input binary word which determines the bit coefficients, the output exhibits 2^N discrete voltage levels ranging from zero to a maximum value of

$$(V_O)_{max} = V_{FS} \frac{2^N - 1}{2^N} \tag{14.5}$$

with a minimum step change ΔV_O given as

$$\Delta V_O = \frac{V_{FS}}{2^N} \tag{14.6}$$

Figure 14.3 shows the transfer characteristics of a 3-bit D/A converter as a function of the binary input code. In practical implementations of the circuit, as shown in the block diagram of Figure 14.2, the digital input signal activates the binary switches within the converter, which in turn set the values of the bit coefficients b_1 through b_N. The bit coefficient b_1 is called the *most significant bit* (MSB) since it carries the highest numerical weight. In a binary-weighted

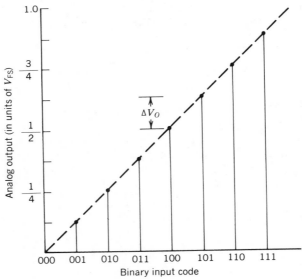

FIGURE 14.3. Transfer characteristics of 3-bit D/A converter.

converter whose transfer function is given in Eq. (14.4), 1 MSB change creates an analog output level shift equal to $V_{FS}/2$.

The bit coefficient b_N associated with the last bit of an N-bit input word carries the least numerical weight and is called the *least significant bit* (LSB). It designates the smallest analog step size available at the output [see Eq. (14.6)] and is equal to $V_{FS}/2^N$.

Note that in a binary-weighted D/A converter circuit, as described by Eq. (14.4), the output is zero when all binary bit coefficients are equal to zero, and is 1 LSB less than of the full-scale output when all bit coefficients are equal to 1 (see Fig. 14.3).

In the process of D/A conversion, the digital input can be applied either in parallel format (as shown in Fig. 14.2) or serially, that is, one bit at a time, starting with the MSB. In almost all cases, parallel input is preferred since it greatly speeds up the conversion process.

14.2. BASIC D/A CONVERTER CIRCUITS

A large number of D/A conversion techniques have been developed and are well covered in the literature.[1-4] In this section, some of the very basic converter configurations will be discussed to illustrate the basic principles of operation. These circuit configurations form the starting point of more complex designs which will be described in later sections.

The basic D/A converter circuits suitable for monolithic IC designs fall into three basic categories, based on their principles of operation: (1) current-scaling circuits, (2) voltage-scaling circuits, and (3) charge-scaling circuits. Among these, the first group of circuits are by far the most commonly used approach in bipolar D/A converter designs. The voltage-scaling and charge-scaling type D/A converter circuits are particularly suitable for MOS designs.

Current-Scaling D/A Circuits

In D/A converters operating on the current-scaling principle, the conversion is achieved by generating a set of binarily weighted currents within the circuit, which are then selectively summed to provide an analog output. Figure 14.4 shows two basic circuit configurations for generating and summing a set of binary-weighted currents $I_1, I_2, \ldots, I_N$. The currents are generated by connecting a binary-weighted resistor network across the voltage reference V_{ref}. The positions of switches $S_1, S_2, \ldots, S_N$ simulate the values of the bit coefficients $b_1, b_2, \ldots, b_N$ in Eq. (14.4). The coefficients are either 0 or 1, depending on the corresponding switch being in position 1 or 2, respectively. The output current I_O is then summed at the virtual ground point, that is, at the inverting input of the operational amplifier A_1; and results in an output voltage,

$$V_O = -I_O R_O = -V_{ref}(b_1 2^{-1} + \ldots + b_N 2^{-N}) \qquad (14.7)$$

where the operational amplifier feedback resistor R_O, which sets the scale factor, is chosen to be equal to $R/2$ for convenience.

The binary bit coefficients are determined by the positions of the corresponding switches in the figure. One has the option of switching either a voltage or a current in the circuit, as a function of the digital input. In the circuit of Figure 14.4a, voltage switching is employed where the net voltage across any one of the weighting resistors is switched either to ground or to V_{ref}. Figure 14.4b shows an alternate switching arrangement for the same circuit. In this case, one terminal of each of the resistors remains connected to V_{ref}, the other terminal is switched between the actual ground (position 1) and the "virtual" ground formed at the operational amplifier input. This method of switching is called *current switching*.

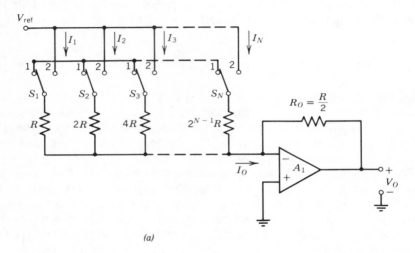

(a)

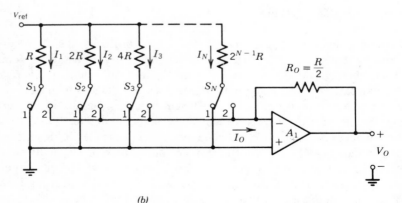

(b)

FIGURE 14.4. Current-scaling D/A converter circuits using binary-weighted resistor ladders: (a) Voltage switching; (b) current switchings.

In most applications, and particularly in integrated circuits, current switching is normally preferred over voltage switching because it offers significant speed advantages. The network nodes have parasitic stray capacitances associated with them. Therefore, the sudden changes of node voltages during voltage switching create voltage transients that have to die down before the circuit settles to its final state. On the other hand, during current switching, the node voltages remain unchanged. This minimizes the switching transients and the corresponding settling time.

In the D/A converter configurations shown in Figure 14.4, the current-weighting function is achieved by using N parallel, independent branches in the weighting network, which have respective impedance levels of R, $2R$, $4R$, $8R$, ... In this type of resistor network, the spread of resistor values increases very rapidly as the number of bits increases, such that the resistance of the MSB branch is related to that of the LSB branch as

$$\frac{R_{\text{MSB}}}{R_{\text{LSB}}} = \frac{1}{2^{(N-1)}} \tag{14.8}$$

Thus, for example, for 8-bit resolution one needs a set of precision resistors covering a range of resistor values from R to $128R$. In monolithic or thin-film circuits, such a wide range of resistor values is difficult to obtain with sufficient precision without resorting to expensive trimming processes.

An alternate resistor array configuration which eliminates the large component spread of the binary-weighted resistor networks is the R-$2R$ ladder network shown in Figure 14.5. In this type of network, the binary division of the currents I_1, I_2, ... is achieved by successive partitioning of current between

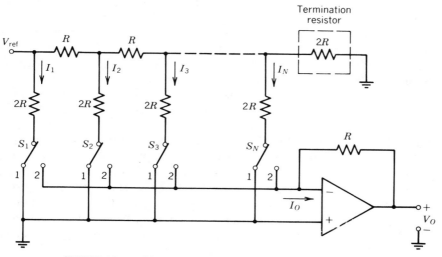

FIGURE 14.5. Current-scaling D/A converter using R-$2R$ ladder.

the shunt ($2R$) and series (R) branches. Thus, the branch currents still satisfy the binary relationship

$$I_1 = 2I_2 = 4I_3 = \ldots = 2^{N-1}I_N \qquad (14.9)$$

while maintaining the resistor values within an easily attainable 2 : 1 ratio. The R-$2R$ ladder requires twice as many resistors as the binary-weighted resistor network and must be properly terminated by a termination resistor (see Fig. 14.5). A particular advantage of the R-$2R$ ladder is that it can be used in conjunction with equal-value current sources to perform the D/A conversion, as shown in Figure 14.13.

For monolithic D/A converters, the R-$2R$ ladder is generally preferred over the binary-weighted ladder for circuits of 6-bit or higher complexity, because it offers a lesser spread of component values. However, the weighted resistor ladders are better suited for the design of 4-bit converter sections which can then be cascaded as building blocks to form higher-order converters (see Fig. 14.12).

Voltage-Scaling D/A Converters

Voltage-scaling D/A converters produce an analog output voltage by selectively "tapping" a voltage-divider resistor string connected between the reference voltage and ground.[5,6] For an N-bit converter circuit, the resistor string is made up of 2^N identical segments connected in series, and it is used as a potentiometer where the voltage levels between the resistor segments are sampled by means of binary switches. For this reason, these types of converters are also called *potentiometric* D/A converters.

Figure 14.6 shows the conceptual diagram of a 3-bit D/A converter operating on the voltage-scaling principle. The resistor string is comprised of eight identical resistors, connected between V_{ref} and ground. The voltage drop across each resistor section is equal to 1 LSB of output voltage change, or $V_{\text{FS}}/2^N$. The output is sampled by means of a decoding switch matrix, and is sensed by a high-impedance buffer amplifier or voltage follower. In order to assure accuracy of the D/A conversion, the buffer amplifier should draw negligible dc bias current, compared to the bias current within the resistor string.

With reference to Figure 14.6, the operation of the switch matrix which decodes the input logic signal into an analog voltage can be described as follows. The analog switches marked A, B, and C are driven by the input logic lines corresponding to the input bits b_1, b_2, and b_3, where b_1 corresponds to the MSB and b_3 corresponds to the LSB. The switches designated $\overline{A}$, $\overline{B}$, and $\overline{C}$ are driven by the complements of the input logic levels. Assuming that a given switch is *closed* when the corresponding logic level is 1, the output corresponding to an input code 000 will be one that corresponds to switches A, B, and C all open, and $\overline{A}$, $\overline{B}$, and $\overline{C}$ closed, resulting in an output level of 0 V. Similarly, if an input code of 100 is applied, switches A, $\overline{B}$, and $\overline{C}$ will be closed, resulting in an output level of $V_{\text{ref}}/2$, which corresponds to 1 MSB step.

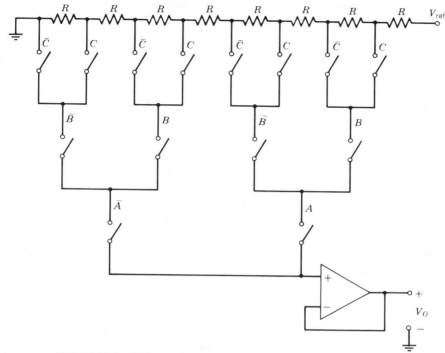

FIGURE 14.6. 3-bit converter example using voltage-scaling principle.

The voltage dividing the D/A converter configuration of Figure 14.6 is particularly suited to MOS technology, where the analog switches can be implemented very efficiently, and the dc bias current of the buffer amplifier can be kept negligibly small.[6]

One main drawback of the circuit for high-bit-count D/A conversion is the excessive number of components required. For N-bit conversion, 2^N resistors and approximately 2^{N+1} analog switches and $2N$ logic drive lines would be required. However, several multiplexing techniques have been reported which can reduce this complexity by cascading and time-sharing several 2-bit ladders to obtain conversion capabilities up to 12 bits.[7] Some of these multiplexing techniques are discussed further in Chapter 15.

The voltage-scaling D/A converter finds its widest application as a building block in MOS A/D conversion systems, where it is used as the D/A converter subsection of a successive-approximation-type A/D converter.

Charge-Scaling D/A Converter

Charge-scaling D/A converters generate an analog voltage by scaling the total charge applied to a capacitor array. Their principle of operation can be illustrated

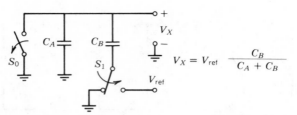

FIGURE 14.7.   Illustration of charge-scaling principle.

by the simple circuit example of Figure 14.7. In the circuit, C_A is connected to ground, and C_B is periodically switched between ground and an internal reference voltage V_{ref}. Assume that, initially, both switches S_0 and S_1 are connected to ground. In this so-called *reset* mode, both capacitors are discharged, and the output voltage V_X is equal to zero. Next, assume that S_0 is opened and S_1 is connected to a reference voltage V_{ref}. If the output is measured during this so-called *sample* mode, one obtains an output voltage V_X

$$V_X = V_{ref}\frac{C_B}{C_A + C_B} \tag{14.10}$$

This same principle can be applied to the binary-weighted capacitor array shown in Figure 14.8, where during the reset mode, all switches, including S_0, would be connected to ground; during the sample mode, S_0 would be open and S_1 through S_N would be controlled by the binary bit coefficients associated with an N-bit digital input signal. A logic 1 would cause the corresponding switch to be connected to V_{ref}, whereas a logic 0 would cause the switch to remain connected to ground. Under this condition, the output voltage would be given from Eq. (14.10) as

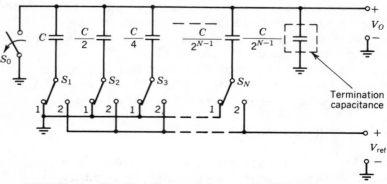

FIGURE 14.8.   Simplified diagram of charge-scaling D/A converter.

$$V_O = V_{ref}\frac{C_{eq}}{C_{tot}} \qquad (14.11)$$

where C_{eq} is the sum of all capacitors connected to V_{ref} and C_{tot} is the total capacitance in the array.

C_{eq} is determined by the choice of individual bit coefficients $b_1, b_2, \ldots, b_N$, which set the switch positions in Figure 14.8. Thus, it can be expressed as

$$C_{eq} = b_1 C + \frac{b_2 C}{2} + \frac{b_3 C}{2^2} + \ldots + \frac{b_N C}{2^{N-1}} \qquad (14.12)$$

where C is the largest capacitor in the binary-weighted array, corresponding to the MSB data input. Similarly, the total capacitance C_{tot} in the array is

$$C_{tot} = C + \frac{C}{2} + \frac{C}{2^2} + \ldots + \frac{C}{2^{N-1}} + \frac{C}{2^{N-1}} = 2C \qquad (14.13)$$

Note that the convergence of the summation in Eq. (14.13) to $2C$ is due to the presence of a termination capacitance in the capacitor array, equal to the LSB capacitance. This termination capacitance is shown on the extreme right-hand side of the capacitor array in Figure 14.8.

Combining Eqs. (14.11) through (14.13), one can express the output voltage during the sample mode as

$$V_O = V_{ref}(b_1 2^{-1} + b_2 2^{-2} + \ldots + b_N 2^N) \qquad (14.14)$$

which is the desired analog output corresponding to the N-bit binary input word.

The charge-scaling D/A converter, which requires precisely ratioed capacitor arrays and nearly ideal analog switches with very-high-impedance sample-and-hold circuitry, is ideally suited for MOS integrated circuits. It is essentially the capacitor equivalent of the weighted resistor ladder discussed in Figure 14.4, where the charge rather than the current is used as the circuit variable to be measured.

As described in Chapter 13, MOS capacitors can be fabricated to very accurate and stable ratio tolerances. Furthermore, the effects of the bottom-plate parasitic capacitance associated with a MOS capacitor is eliminated in the circuit of Figure 14.8 by connecting the bottom plates to the bit switches S_1 through S_N. This causes the bottom-plate parasitics to be switched between low-impedance reference voltage and ground, while the parasitic-free top plates are connected to the output terminal.

The only drawback of the circuit is the large capacitor ratios required for high-bit-count conversion. Similar to the case of binary-weighted resistor ladders, the capacitor ratio between MSB and LSB capacitors is

$$\frac{C_{MSB}}{C_{LSB}} = 2^{N-1} \qquad (14.15)$$

which, for an 8-bit converter, requires a ratio of 128. Assuming that the lowest practical value of the capacitor is approximately 1 pF due to parasitic considerations, an 8-bit converter would require an MSB capacitor of 128 pF and a total capacitance of 256 pF. Because of these practical considerations, monolithic D/A converters using the charge-scaling principle with simple binary ladders are generally limited to 8-bits.

It is possible to reduce the capacitor ratios required by using a capacitor equivalent of the R-$2R$ ladder (see Fig. 14.5), rather than the binary-weighted ladder. However, this results in *floating* capacitors, where one terminal of the capacitor cannot be connected to a low-impedance node. Thus, it makes the circuit performance sensitive to bottom-plate parasitic capacitances of MOS capacitors.

Although the charge-scaling D/A converter is very useful on its own, it is mostly used as a building block in the design of MOS A/D converters operating on the so-called *charge-redistribution* principle.[8,9]

Overview of Basic D/A Conversion Circuits

Although a wide range of D/A conversion techniques is available,[5] the three basic D/A conversion methods outlined in this section are by far the most commonly used in monolithic IC design. In practice, the choice of which converter technique to use is divided along the lines of available technology. The current-scaling technique is almost always used in conjunction with bipolar technology, which is suitable for building precision resistor arrays and high-speed current switches. The voltage-scaling and charge-scaling techniques are especially suited to analog MOS designs where MOS capacitors, analog switches, and charge-sensing circuitry can be easily fabricated.

In terms of precision designs where 10-bit or higher resolution is required, bipolar data conversion tecniques are in general preferred over MOS approaches.

14.3. DEFINITIONS OF D/A CONVERTER TERMS

The performance criteria and design parameters for D/A converters are significantly different from those for other classes of circuits discussed so far. Therefore, it is useful at this point to review some of the basic parameters, terminology, and typical system errors associated with the performance of practical D/A converter circuits.

Resolution

Resolution of a D/A converter is defined as the smallest distinct change that can be produced at the analog output in response to a change in the digital input code. It also indicates the total possible discrete analog levels available at the output, and is normally expressed in bits. For N-bit resolution, the converter output must be capable of producing 2^N discrete analog levels.

Converter resolution can also be expressed in percentage of full-scale output V_{FS}. If the full-scale output is fixed, then resolution may also be expressed in volts. Since the smallest analog step size is given as ΔV_O in Eq. (14.6) (see Fig. 14.3), the D/A converter resolution, expressed in volts, becomes

$$\text{converter resolution} = \Delta V_O = \frac{V_{FS}}{2^N} \qquad (14.16)$$

for an N-bit converter. As an example, a D/A converter with 5-V full-scale output and 12-bit resolution must be able to resolve one part in 2^{12} or 1.221 mV, at its analog output, for 1 LSB change of the input code.

Accuracy

Accuracy is the measure of the deviation of the analog output from its "ideal" or predicted value. It can be described in terms of either absolute or relative terms.

Absolute Accuracy.   Absolute accuracy is the measure of worst-case error (i.e., inaccuracy) between the actual and the ideal converter output. Absolute accuracy implies the use of a standard traceable to the National Bureau of Standards.

Relative Accuracy.   The relative accuracy of a D/A converter is the worst-case error (i.e., inaccuracy) between the actual and the ideal converter output *after* the gain and offset errors have been removed (see Fig. 14.9). It can be expressed as a percentage of full-scale output, as the number of bits, or as a fraction of the LSB. N-bit relative accuracy means a possible error ΔV_e of

$$\Delta V_e \leq \frac{V_{FS}}{2^N} \qquad (14.17)$$

Similarly, for an N-bit converter, $\frac{1}{2}$ LSB relative accuracy means

$$\Delta V_e \leq V_{FS} \frac{1}{2} \frac{1}{2^N} = V_{FS} \frac{1}{2^{N+1}} \qquad (14.18)$$

In practice, the term accuracy is used to imply relative rather than absolute accuracy. Note that resolution and accuracy are not necessarily equal in bits. For example, one can have a D/A converter with 10-bit accuracy, but with 12-bit resolution, and vice versa.

Converter Error Terms

As indicated in Figure 14.3, the ideal analog output of a D/A converter is made up of 2^N distinct output levels, corresponding to the height of the "bars" in the figure, for each given digital input. The types of errors or deviations from this ideal condition which may be encountered in practice are graphically illustrated in Figure 14.9. They are briefly explained below.

Offset Error.   This is the deviation of the actual output from the ideal output when the ideal output should be zero. It is shown as ΔV_{OS} in Figure 14.9a.

Gain Error.   This is the relative error at the analog output, due to the inaccuracy of the scale factor or the reference voltage. It corresponds to a change in the slope of transfer characteristics, as shown in Figure 14.9b.

Normally, both the offset and the gain errors can be reduced to zero by conventional trimming techniques.

Linearity

This is a measure of the nonlinearity error at the output, after the gain and the offset errors have been trimmed. It is essentially a measure of the uniformity of the relative step size. Normally, there are two types of nonlinearity errors.

Integral Nonlinearity.   Integral nonlinearity is the measure of the deviation of the D/A converter transfer function from an ideal straight line. It is normally defined as the worst-case deviation of the transfer function from a straight line between zero and full-scale readings (i.e., the end points of the transfer characteristics). Note that this is somewhat different from the conventional definition of nonlinearity as the deviation from a *best-fit* straight line, and it is essentially a measure of the *curvature* of the transfer characteristics (see Fig. 14.9c). This definition of nonlinearity is preferred by IC manufactures since it is easier to measure under production conditions.

Differential Nonlinearity.   Differential nonlinearity is the measure of nonuniform step sizes between adjacent transitions. Ideally, the step size is 1 LSB. The differential nonlinearity is a measure of the deviation of each differential step size from this ideal value. It is normally specified as a fraction of LSB. For example, if a D/A converter is specified to have a maximum differential nonlinearity of $\pm\frac{1}{2}$ LSB, then the minimum and maximum allowable step sizes between adjacent transitions would be $\frac{1}{2}$ LSB and $1\frac{1}{2}$ LSB, respectively.

Unlike the gain and offset errors, the nonlinearity errors depend on the input digital code. Therefore, they cannot be corrected by a simple trimming operation. The only way of minimizing the nonlinearity errors is by improving the matching and tracking of the circuit components (i.e., resistor or capacitor ladders).

Monotonicity

Monotonicity in a D/A converter implies that the analog output is continuously increasing as the digital code is continuously increased. The D/A converter

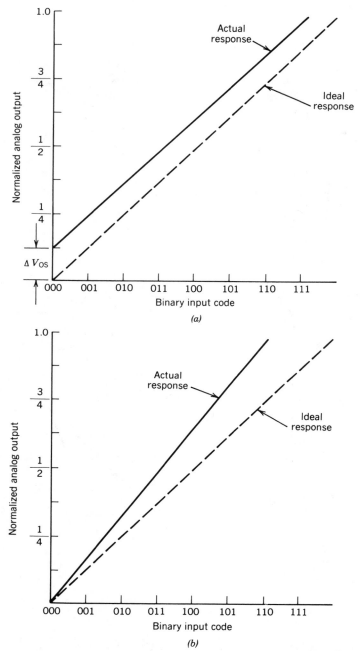

FIGURE 14.9. Typical sources of error in 3-bit D/A converter: (*a*) Offset error; (*b*) scale factor (gain) error; (*c*) nonlinearity error; (*d*) nonmonotonicity (due to excessive differential nonlinearity).

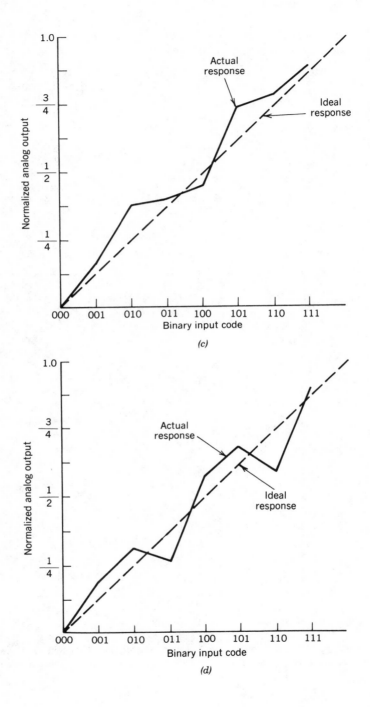

(c)

(d)

would be nonmonotonic if the analog output *decreased* at any point in its dynamic range for increasing input codes. Nonmonotonicity comes about as a result of excessive differential nonlinearity, as illustrated in Figure 14.9*d*. Monotonicity can be assumed if the differential nonlinearity error is at all times less than ±1 LSB. Figure 14.10 gives a graphic illustration of monotonic and nonmonotonic nonlinearities present at the analog output.

Monotonicity is an important requirement for D/A converters, particularly if the converter is used in the feedback path of an A/D conversion system. In such a case, lack of monotonicity would result in missing digital codes at the converter output, which will make it unusable in most applications.

Monotonicity can be a problem with current-scaling or charge-scaling D/A converters described in Section 14.2. However, it is inherent in voltage-scaling converter circuits, such as the one shown in Figure 14.6.

Stability

The performance of a D/A converter can change with time, temperature, and supply voltage. Therefore, the gain, offset, integral and differential nonlinearity

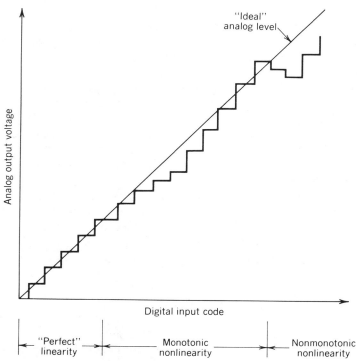

FIGURE 14.10. Graphical illustration of various nonlinearities present at the analog output.

must all be specified over an operating temperature and power supply range to be generally useful. It is also important to know whether the converter is monotonic over its full temperature and supply voltage range.

Settling Time

Settling time is the time taken by the D/A converter output to settle within some specified error band (typically $\pm\frac{1}{2}$ LSB) about its final value, subsequent to a change of the input digital code. It is determined by the switching time of the internal logic circuitry and the settling of the circuit transients due to parasitic node capacitances.

The fastest settling D/A converters are the current-switching current-output converters which do not include a current summing amplifier. Such circuits can settle to $\pm\frac{1}{2}$ LSB accuracy in less than 300 nsec. In voltage output D/A converters, which use a summing amplifier to convert the internal current to a voltage level, the settling time of the system is normally limited to several microseconds due to the slew rate and settling time of the amplifier section. The settling time is primarily dominated by the settling of the MSB contribution; the settling times associated with the lower order bits are usually negligible.

Glitches

One of the problems in high-speed D/A converters is the presence of output transient spikes, called *glitches*, during the conversion process. Glitches are associated with current-scaling D/A converters, and are caused by the small time differences, or unequal delays, in switching various current sources within the converter. Glitches occur at major transitions, at $\frac{1}{2} V_{FS}$, as well as at $\frac{1}{4}$ and $\frac{3}{4} V_{FS}$, where one major current source turns on while a group of minor current sources turn off. An example of this is the so-called *major carry* transition at $\frac{1}{2} V_{FS}$, where the input code changes from 011 ... 111 to 100 ... 000, causing the MSB current source to turn on, while $N - 1$ lower bits are turned off. The presence of output glitches is particularly bothersome when a video or CRT display of the analog output is produced.

The presence of glitches can be minimized by careful circuit design to equalize the switching delays associated with individual bit signals, or by using a sample-and-hold circuit at the D/A output, as shown in Figure 14.1b. In this latter case, the glitch problem is eliminated by sampling the D/A converter output *after* the conversion is completed and the transients have died down.

14.4. D/A CONVERTER ARCHITECTURE

Among the three fundamental classes of D/A converters discussed in Section 14.2, the current-scaling D/A converter is by far the most widely used because of its high conversion speed and high accuracy capability.[3,4] In this section, some of the basic circuit configurations used in the design of current-scaling monolithic D/A converters will be examined.

Circuits Using Binary-Weighted Current Sources

The D/A converters which use binary-weighted current sources are essentially derived from the basic circuit configurations of Figure 14.4 by replacing the resistors with constant-current stages. Figure 14.11 shows the simple circuit configuration for a 4-bit D/A converter using binary-weighted current sources and a R-$2R$ resistor ladder. In the monolithic implementation of the circuit, it is necessary to maintain equal current density in the emitters of binary-weighted current-source transistors, in order to avoid errors due to V_{BE} mismatches. In practice, this is achieved by scaling the emitter areas of the current-source transistors. However, complete scaling of emitter areas for all current sources is impractical in high-bit-count converters. For example, for a 10-bit converter, emitter scaling would require a $512:1$ ratio between the emitters of the MSB and LSB current sources. Therefore, in general, emitter-area scaling is confined to the first 4 or 6 significant bits, and sufficiently large resistor values are used in the resistor ladder so that the ohmic drops in the emitter resistors would dominate the V_{BE} mismatches of the less significant bits, and provide sufficiently accurate current scaling.

There are two inherent drawbacks in using binary-weighted current sources. First, the transient response of binary-weighted current sources is unequal due to unequal currents. This may result in faster settling times for low-order bits, resulting in output glitches. Second, since each bit section carries a different amount of current, thermal drifts and temperature gradients associated with each bit due to self-heating would be different, and may cause local mismatch errors.

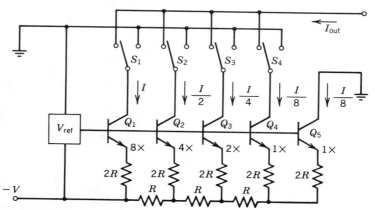

FIGURE 14.11.   4-bit D/A converter using R-$2R$ ladder and binary-weighted current sources.

This is particularly true if high supply voltages and bit currents are used, resulting in high power dissipation. However, both of these effects can be minimized by careful circuit design and chip layout.

Binary-weighted current sources can also be used in groups of four as building blocks for high-accuracy D/A converters. Figure 14.12 shows an example of cascading two identical 4-bit binary-weighted current sources, along with a 16 : 1 attenuator, to form an 8-bit D/A converter.[9] The transfer function of the converter can be written as

$$I_{\text{out}} = 2I_1\left(\frac{b_1}{2} + \frac{b_2}{4} + \frac{b_3}{8} + \frac{b_4}{16}\right) + I_A \qquad (14.19)$$

where I_A is the residual current drawn from the first four current sources. The resistive divider scales this current such that it contributes only one-sixteenth of the total current drawn by the second set of four current sources, that is,

$$I_B = \frac{I_A}{16} = \frac{2I_1}{16}\left(\frac{b_5}{2} + \frac{b_6}{4} + \frac{b_7}{8} + \frac{b_8}{16}\right) \qquad (14.20)$$

where b_i are the particular binary bit coefficients. Combining Eqs. (14.19) and (14.20), one obtains

$$I_{\text{out}} = 2I_1\left(\frac{b_1}{2} + \frac{b_2}{4} + \ldots + \frac{b_8}{256}\right) \qquad (14.21)$$

which is the transfer function of an 8-bit D/A converter.

The concept outlined above can be used to cascade three or four binary-weighted quad current sources with 16 : 1 current dividers to obtain 12- or 14-bit converters. However, as the bit count is increased, the relative accuracy and precision requirements on current sources and the scaling resistors become very stringent.

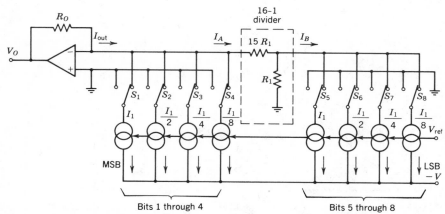

FIGURE 14.12.   Cascading two 4-bit binary-weighted current sources.

Circuits Using Equal-Value Current Sources

Some of the inherent problems, such as emitter-area scaling or unequal switching associated with binary-weighted current sources, can be avoided by using equal-value current sources in conjunction with an R-$2R$ ladder, which is used as an *attenuator* to scale the currents in a binary manner. Figure 14.13 shows the circuit configuration of a 4-bit D/A converter using this principle. Transistors Q_1 through Q_4 function as identical current sources, each carrying an equal current I_1. The emitter resistors are chosen to be sufficiently large to eliminate V_{BE} mismatch effects. The R-$2R$ ladder attenuates the contributions of each of the current sources to the output current I_{out} in a binary manner, resulting in the desired transfer function

$$I_{out} = 2I_1\left(\frac{b_1}{2} + \frac{b_2}{4} + \frac{b_3}{8} + \frac{b_4}{16}\right) \tag{14.22}$$

The circuit can be extended to higher bits by adding additional current sources and extending the R-$2R$ ladder, to result in the general transfer function

$$I_{out} = 2I_1\left(\frac{b_1}{2} + \frac{b_2}{2^2} + \ldots + \frac{b_N}{2^N}\right) \tag{14.23}$$

The equal-value current source configuration of Figure 14.13 is an inherently faster circuit than the binary-weighted current source configuring, due to a

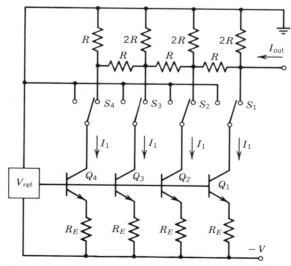

FIGURE 14.13. 4-bit D/A converter using equal-value current sources.

higher current level of operation. Furthermore, the conversion speed can be enhanced by using relatively low values for the resistors in the R-$2R$ ladder.[10] Since the R-$2R$ ladder is in the collector, rather than in the emitter, of the current source transistors, the voltage drops across resistor-ladder elements do not need to be comparable to the transistor V_{BE} drops to equalize the emitter current densities. Instead, emitter currents are set by the choice of emitter resistors R_E, which do not affect switching speeds.

An added advantage of the D/A converter configuration, using equal-value current sources, is its reduced sensitivity to the voltage coefficient of resistors (VCRs), which can affect accuracy and resolution for 10-bit or higher converters.

Master–Slave Ladders

Some of the emitter-area scaling problems associated with conventional binary-weighted current sources can be reduced significantly by using a so-called master–slave ladder arrangement, as shown in Figure 14.14. In this approach, a primary or *master* ladder is used for the major bits, and a secondary or *slave* ladder is used for the minor bits.

The operation of the circuit can be explained as follows, with reference to Figure 14.14. The last transistor Q_{4B} in the master ladder generates a current I_1 equal to that of the highest bit (in this case, bit 4) in the ladder. This current is then used to drive the slave ladder, and is then partitioned further by the slave ladder to form the remaining bit currents.

The master–slave ladder configuration is very often used in the design of D/A converters having 8-bit or higher resolution since it greatly simplifies the emitter-area scaling requirements and thus makes more efficient use of the chip area.[11,12]

Segmented Ladders

One of the main performance requirements in a D/A converter design is the monotonicity of transfer characteristics. This is a particularly serious problem for precision D/A converters having 12-bit or more resolution. Monotonicity is hardest to realize at the points of *major carry*, that is, when the MSB or the second bit changes state. This problem can be avoided, to a large extent, by using a so-called *segmented* converter structure. An illustrative example of such a segmented D/A converter architecture is given in Figure 14.15.

The segmented converter is made up of three sections: a step generator, a segment generator, and a segment decoder. Using the illustrative example of a 6-bit segmented converter in Figure 14.15, the principle of operation of the segmented D/A converter can be explained as follows. The segment currents I_A, I_B, I_C, and I_D are generated by the four equal-value current sources, and are each equal to one-quarter of the full-scale output current I_{FS}. Particular combinations of segment currents are selected by the output of the segment decoder. In the

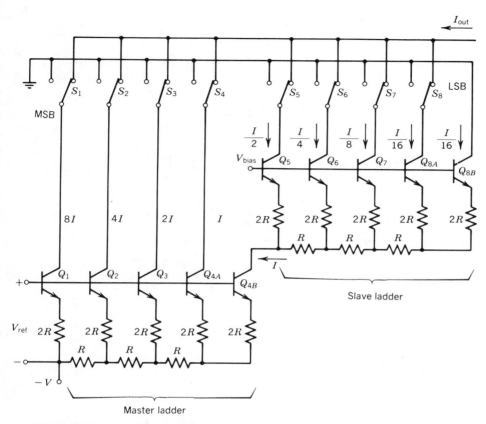

FIGURE 14.14. 8-bit D/A converter using two 4-bit ladders in master–slave configuration.

example of Figure 14.15, the first 2 bits of digital input drive the segment decoder whose four output states determine the status of segment switches S_A through S_D. The segment currents can be connected either to the step generator or directly to the output bus I_{out}.

The step generator is a self-contained D/A converter driven by the output current of one of the segment generators, and its output current is partitioned in accordance with the remaining 4 LSBs of the digital input.

As an example, consider the operation of the circuit with input codes successively increasing from 000000 to 111111, resulting in an analog output which starts from zero and moves toward I_{FS}, as shown in Figure 14.16. At the input code 000000 segment currents I_B, I_C, and I_D are connected to ground, and I_A is connected to the step generator ladder. As the code modes toward 001111, the output will be a staircase of $2^4 = 16$ steps, due to I_A reaching the output bus I_{out} after being partitioned by the 4-bit converter section.

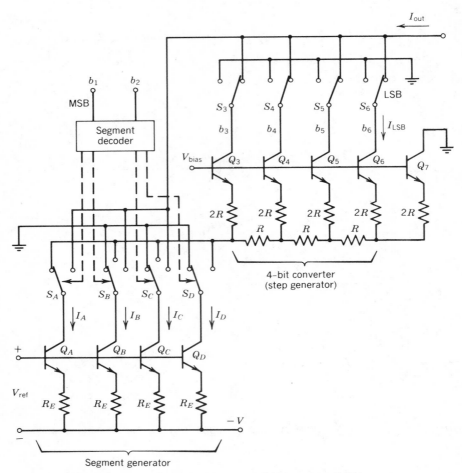

FIGURE 14.15.   Functional block diagram of 6-bit segmented D/A converter.

At input code 001111, I_{out} is equal to

$$(I_{out})_{001111} = I_A \frac{15}{16} \tag{14.24}$$

At the next successive input code, 010000, I_A is switched directly to the output, I_B is switched into the step generator, and

$$(I_{out})_{010000} = I_A \tag{14.25}$$

During the next sequence of 15 steps, I_A of segment 1 serves as a *pedestal* to which steps generated from the second-segment current I_B are added. At code 011111, the output is equal to

$$(I_{out})_{011111} = I_A + \frac{15}{16}I_B \tag{14.26}$$

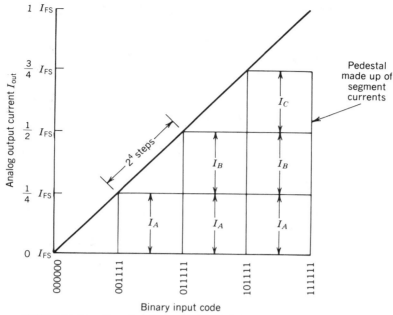

FIGURE 14.16. Transfer characteristics of 6-bit segmented D/A converter.

The subsequent input code, 100000, causes I_B to be switched to the output, with

$$(I_{out})_{100000} = I_A + I_B \tag{14.27}$$

and I_C is switched into the step generator. The next 15 steps are then generated by partitioning I_C in the step generator, where $I_A + I_B$ serve as a pedestal. The sequence progresses in this manner until the final code, 111111, where

$$(I_{out})_{111111} = I_{FS} \frac{2^{N-1}}{2^N} = I_{FS} \frac{63}{64} \tag{14.28}$$

A very important feature of the segmented converter is its monotonicity. Provided that the internal D/A converter which serves as a step generator is monotonic, overall monotonicity is assured.

As a general case of the simple example illustrated in Figures 14.15 and 14.16, one can form a monotonic N-bit D/A converter by combining a simpler monotonic M-bit converter (where $M < N$) and 2^{N-M} segment generators. For example, a 12-bit monotonic D/A converter can be built by using a 9-bit converter in conjunction with eight segment generators.[13]

Companding D/A Converters

In particular applications, such as pulse-code-modulated telephone systems or wide-dynamic-range audio signal processing, nonlinear conversion characteristics are desirable to enhance the dynamic range (i.e., the signal-handling and resolution capability) of the system. In such cases, the analog output is required to have a high degree of resolution at low values of the input digital code, whereas a coarser resolution (i.e., larger output steps) can be permitted at high values of the digital code. If a D/A converter which has such characteristics is used in the feedback path of an A/D converter system, it can "compress" an analog signal into a relatively compact digital code. Conversely, if it is used as a D/A converter, it can, in turn, "expand" the digital code into a wide-dynamic-range analog signal. Because of this property, such a converter is called a compressing/expanding or, simply, a *companding* converter.

A very commonly used format of the companding D/A converter characteristic is the so-called "μ-255 companding law," which has the basic transfer characteristics that can be approximated by the D/A conversion characteristics illustrated in Figure 14.17.[14] The approximation is obtained by using an eight-segment conversion characteristic where each segment, called a *chord*, is made up of 16 steps. The analog output is selected by an 8-bit code, where the first bit denotes the sign (i.e., polarity) of the output, the next 3 bits denote the

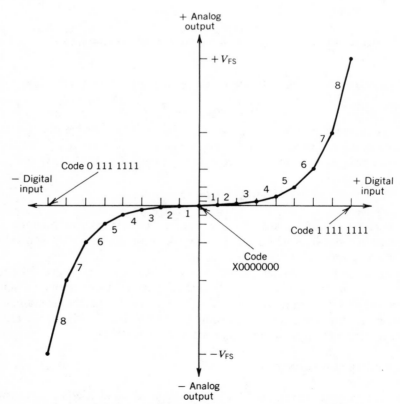

FIGURE 14.17. Transfer characteristics of companding D/A converter.

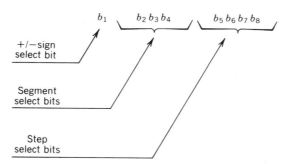

FIGURE 14.18. Typical 8-bit (7-bit plus sign) binary coding format used in companding data converters.

particular segment, and the last 4 bits denote the particular step within the segment. This particular coding format is illustrated in Figure 14.18.

The companding D/A conversion characteristics shown in Figure 14.17 have binarily increasing segment heights and step sizes in each successive segment. The step sizes within a given segment are equal. However, the absolute step sizes in each successive segment is twice as high as those in the preceeding segment. Thus, for example, it the step size within the fist segment is 5 mV, it becomes 10 mV in segment 2, 20 mV in segment 3, and so on.

Figure 14.19 shows the functional block diagram of a current-scaling companding D/A converter which has the conversion transfer characteristics given by Figure 14.17. It is made up of eight segment-generator current sources I_1 through I_8, which are controlled by the output of the segment decoder in response to the 3-bit segment-select code. The step generator is made up of a 4-bit D/A converter, using emitter-area scaling to set the current ratios.

With some minor differences, the operational concept of the circuit is similar to the case of the segmented ladder D/A converter shown in Figure 14.15. The main difference is that the segment currents are no longer equal, but are binary-weighted. This results in doubling the height of the pedestal currents and the converter step size when proceeding from a lower segment to the next higher one. The polarity conversion is done by reversing the polarity of the current-to-voltage conversion stage at the output by the position of the sign-bit switch S_0.

The basic circuit architecture of Figure 14.19 is also suitable for CMOS analog designs, where the basic current sources can be implemented with scaled MOS current mirrors (see Figs. 6.11 and 6.12) and the current switches can be implemented as transmission gates.[15] With minor modifications, the concept of a segmented D/A converter can be adopted to charge-scaling techniques using MOS technology and binary-weighted capacitor ladders. Several monolithic telecommunication IC products are commercially available at the present time, which utilize companding D/A converter structures implemented with NMOS and CMOS technology using charge-scaling techniques.[16–18]

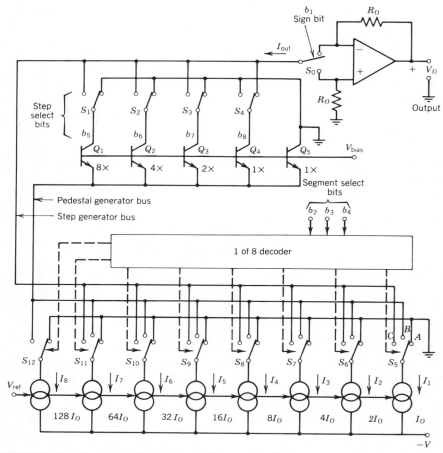

FIGURE 14.19. Functional block diagram of companding D/A converter using current scaling.

14.5. CURRENT SWITCHES

The performance characteristics and the response speed of a current-scaling D/A converter depend strongly on the characteristics of the current switches used in the circuit. A number of current-switching schemes have been developed, which are readily suitable for integration, using bipolar or MOS technologies. Some of these circuits will be examined in this section.

To be suitable for D/A converter applications, a current switch must have the following properties:

1. *High Speed.* Rapid switching is desired with minimum settling time or transients. In order to reduce the effects of parasitic capacitances, the voltage swings at the switching node must be kept to a minimum.

2. *Good Buffering.* It should provide good isolation between the digital switching signals and the analog portion of the circuitry.

3. *Low Reverse Leakage.* Leakage current through the switch in the off state should be negligibly small.

4. *Logic Compatibility.* The logic control signal levels and amplitudes required for switch operation must be compatible with conventional logic levels.

In bipolar circuits, the current-switching action is accomplished by forward or reverse-biasing a diode or a transistor junction. Since the reverse leakages associated with silicon *p-n* junctions under normal operating conditions can be made negligibly small compared to bit currents, the low leakage requirement normally does not present a design problem, except at elevated temperatures.

Logic level compatibility requirements often necessitate the inclusion of dc level-shifting functions within the switch. This is usually done by using lateral *pnp* transistors, to provide both buffering and level shifting within the switch circuit.

The current switches used in D/A converter design can be either single-ended or differential. In general, differential current switching is preferred over the single-ended approach since it can provide faster switching speeds by avoiding large voltage swings at the switching nodes.

Single-ended switches are used in modular D/A converter designs using hybrid technology. Figure 14.20 illustrates a simple current switch configuration suitable for use in conjunction with binary-weighted resistor ladders. The lateral *pnp* transistor Q_1 is used to provide both the buffering and the level shifting of the digital control signal. The base of the *pnp* transistor is biased from an internal

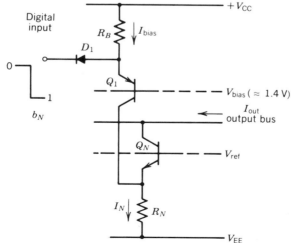

FIGURE 14.20. Single-ended current switch for binary-weighted resistor ladders.

threshold bias source V_{bias}. For compatibility with TTL logic levels, V_{bias} is normally chosen to be approximately 2 diode drops above ground level (i.e., ≈ 1.4 V). With the input logic at its low state, diode D_1 is forward biased and Q_1 is turned off. In this condition, the bit transistor Q_N conducts the normal bit current I_N, which is diverted to the output bus. Conversely, with the input logic at its high state, diode D_1 is reverse biased, decoupling the logic signal from the circuit. Under this condition, the *pnp* transistor Q_1 conducts the bias current I_{bias} set by the choice of its bias resistor R_B. I_{bias} is chosen to be bigger than I_N. Thus, the voltage drop across the bit resistor R_N due to I_{bias} causes the bit transistor Q_N to be turned off. Typically, I_{bias} is chosen to be $\approx 1.1\,I_N$, to minimize the voltage swing at the emitter of Q_N but still assure proper switching.

In the type of current switch shown in Figure 14.20, the current level within the bit resistor R_N is *not* equal to the bit current when Q_N is turned off and Q_1 is on. Therefore, this type of current switch is useful only with binary-weighted resistor ladders (see Fig. 14.12) or with equal-value current source D/A converters (see Fig. 14.11).

The differential current switch is by far the most commonly used switch configuration in high-speed, current-scaling D/A converters. This switch is essentially a fully-switched differential pair, as shown in the simplified schematic of Figure 14.21. The bit current I_m is steered either to ac ground or to the output bus, depending on the polarity of the digital input signal applied to the base of Q_1. The switch threshold level is set by an internal bias voltage V_{bias}. As will be shown in later examples, the digital input swing is limited, by circuit design, to several hundred millivolts around the threshold bias level, in order to

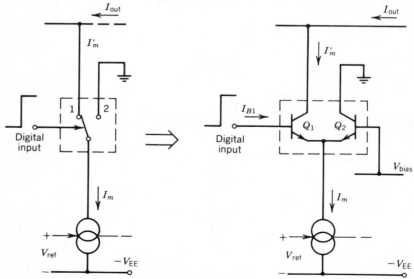

FIGURE 14.21. Simplified schematic of differential current switch.

keep the voltage level at the emitters of Q_1 and Q_2 relatively constant during switching. This minimizes the effects of parasitic capacitances, and thus both speeds up the switching time and minimizes the switching transients reaching the output bus.

A very important advantage of the differential switch, in addition to its switching speed, is that it leaves the bit currents in the resistor ladder network unchanged during the switching operation, and thus avoids the propagation of transients within the ladder network. As a result, it can be used with both the R-$2R$ and the binary-weighted resistor ladders.

The only practical problem associated with the differential current switch is the error in the output current due to the finite β of the switch transistors. For example, in the circuit of Figure 14.21, the actual current I_m' delivered to the output bus differs from the actual bit current I_m by an amount equal to the base current I_{B1} of the transistor Q_1

$$I_m' = I_m - I_{B1} = \frac{\beta}{\beta+1} I_m \qquad (14.29)$$

where β is the forward-current gain of Q_1. A simple way of minimizing this error is to use a Darlington-connected pair of transistors for Q_1 and Q_2. Although this is used in some designs,[10] it tends to degrade the switching speed of the circuit significantly. A more effective method of avoiding this base current error is to use a feedback bias arrangement which automatically corrects for base current errors. This technique will be discussed in Section 14.8.

In the actual circuit implementation of the differential switch, one has to consider two problems: (1) provide appropriate dc level shifting and buffering between the input logic signal and the switch transistor, and (2) limit the signal swing at the bases of switch transistors to a fraction of a volt to provide fast and transient-free switching. Figures 14.22 and 14.23 show two circuit configurations that overcome these problems.

Figure 14.22 shows a two-stage differential current switch which uses lateral *pnp* transistors to provide level shifting at the input.[19] The logic threshold is set by the internal bias setting V_{B1}. The switch bias voltage V_{B2} is internally generated and tracks the reference voltage V_{ref}. With the use of Schottky diode clamps D_1 and D_2, the voltage swing at the input of Q_2 is ± 500 mV, independent of the input logic swing.

An alternate configuration for a high-speed differential current switch is given in Figure 14.23, which again uses a lateral *pnp* voltage comparator stage driving an *npn* differential pair.[20] The internal bias voltage V_{B2} sets the logic threshold of the digital input. The bias voltage V_{B1} is normally set approximately 2 or 3 V_{BE} above V_{ref}, and maintains a constant collector–base voltage across the bit transistor Q_N under all switching conditions. The bias current I_B in the *pnp* comparator stage is set to maintain a voltage drop of approximately 1 V_{BE} across R_B when it is fully switched, to limit the maximum voltage swing at the bases of Q_3 and Q_4.

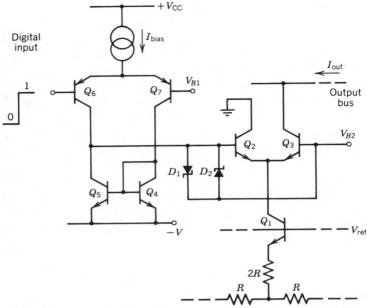

FIGURE 14.22. Differential current switch using *pnp* level-shift stage.[19]

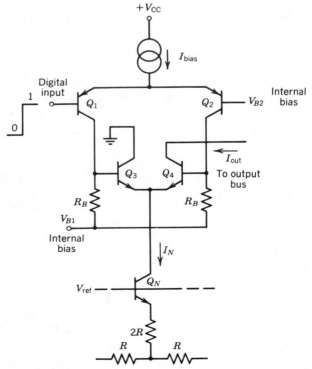

FIGURE 14.23. High-speed differential current switch circuit.[20]

Both differential switch configurations provide very rapid switching times, with propagation delays on the order of 100 nsec or less. Current output D/A converters using these types of switches exhibit very rapid settling times, typically on the order of 300–500 nsec for settling to within $\pm 0.05\%$ of I_{FS}.

14.6. RESISTOR AND CAPACITOR NETWORKS

In current-scaling D/A converters, the generation of binary-weighted currents is achieved by the use of a resistive ladder network. In the case of charge-scaling D/A converters, a binary-weighted capacitor array is used to perform the charge-scaling function. The accuracy, matching, and tracking characteristics of these resistor and capacitor ladder networks, or arrays, are dominant factors in determining the accuracy, resolution, and stability of the overall converter system. In this section, some of the practical component tolerances and precision requirements associated with these ladder networks will be examined.

A major source of error in the matching and scaling of monolithic components is the uncertainty in the photolithographic edge definition during the masking step. In a well-controlled manufacturing process, the probable amount of edge-definition uncertainty can be reasonably predicted. With the present photo-masking technology, the standard deviation (sigma limit) of an edge definition error is in the range of 0.1–0.05 μm under careful processing conditions. Thus, for example, in the case of two identical resistor strips, each having a width of 50 μm, the mask-resolution and edge-definition tolerances can account for a matching error of 0.1–0.2%.

In general, the effects of mask-resolution or edge-definition errors can be minimized by using large-geometry patterns, such as wide-geometry resistors and large-area capacitors. Unfortunately, increasing the device geometries has practical limitations. Chip-area requirements become excessive, and device matching and tracking become more sensitive to the process-related parameter gradients and the temperature variations across the chip. As a compromise between these opposing effects, the ladder components are designed to have relatively large geometries, and are located in close proximity at or near the center of the chip layout, in order to optimize their matching properties.

Resistor Ladders

The resistor ladders used in current-scaling converter circuits can be either binary-weighted or R-$2R$ type networks, as shown in Figures 14.4 and 14.5. In general, R-$2R$ ladders are preferred for most designs, since all of the resistor values in the ladder conform to a $2:1$ ratio.

The resistor structures used in the ladder networks of monolithic D/A converters can be fabricated with any one of the three basic technologies:

1. Diffused resistors.
2. Ion-implanted resistors.
3. Thin-film (deposited) resistors.

The basic characteristics of these resistor structures were covered in Chapter 3. Their salient features will be reviewed briefly for their application to ladder networks.

Diffused Resistors (see Fig. 3.15). Diffused resistors represent the least complicated device structure from the processing standpoint since they require no additional fabrication steps beyond the standard bipolar process sequence. However, the available sheet resistance range is limited to under 200 $\Omega/\square$. They also exhibit relatively high voltage coefficients of resistance, typically on the order of +200–300 ppm/V. Their temperature coefficient is relatively poor (≈ 1500 ppm/°C); however, temperature drift tracking characteristics are good (typically $\approx \pm 15$ ppm/°C for closely spaced resistors carrying equal current). They have excellent long-term stability.

Ion-Implanted Resistors (see Fig. 3.31). These resistors are normally formed by a p-type (boron) ion-implant step on contact beds formed by p-type diffused regions. The ion-implantation step necessary for their fabrication can be readily incorporated into the basic bipolar process sequence, with a very minor increase in fabrication cost. Matching characteristics of ion-implanted resistors are somewhat better than those of diffused resistors.[10] The sheet resistance of ion-implanted resistors is in the range of 1–2 k$\Omega/\square$. Their temperature coefficient is lower than that of diffused resistors. Due to the lightly doped structure of the resistor, ion-implanted resistors have relatively high voltage coefficients of resistance (typically of the order of +300–800 ppm/V, see Fig. 3.33). Since the ion-implanted resistors are formed in the bulk silicon, similar to the case of diffused resistors, they exhibit excellent long-term stability characteristics.

Thin-Film Resistors. Thin-film resistors are fabricated by depositing a thin layer of resistive material on the oxidized silicon surface. Normally, tantalum, nickel-chromium, or silicon-chromium is used as the resistive film. Being surface devices, these thin-film resistors require a passivation layer over the resistor to maintain long-term stability. They offer a wide range of resistivity values and lower temperature coefficients than diffused or ion-implanted resistors and exhibit no voltage coefficient (see Table 3.2). The biggest advantage of thin-film resistors is that they can be trimmed, usually with the use of a laser beam, after completion of the fabrication process and during the wafer probing stage. The main disadvantage of thin-film resistors is the extra processing steps required for their fabrication. Their long-term drift characteristics are also somewhat poorer than those of ion-implanted of diffused resistors, due to long term aging or surface oxidation effects.

    All three resistor types can be used in the fabrication of resistor ladders for D/A converters. In general, diffused resistors are used for all designs up to 10-bit accuracy. Ion-implanted resistors are utilized for converter designs in the 10–12-bit range, and thin-film resistors are used for converters of 12 or more bits.

Layout of Resistor Ladders

In the layout of resistor ladders, care is taken to minimize the matching errors. This is done by laying out each resistor in straight-line segments, with no bends. In general, R-$2R$ ladders are preferred for most designs since all the resistor values in the ladder conform to a $2:1$ ratio. The resistor ladder is constructed of identical unit-resistor segments which are laid out parallel to each other and then interconnected to form the ladder structure. The ladder is normally located near the center of the chip, and is situated along isothermal lines on the chip surface to minimize the temperature gradients across it. Typical examples of ladder layouts can be seen in chip photographs of Figures 14.37 and 14.41. Note that in each case, the ladder network takes up a substantial portion of the chip area. Normally, resistor widths are maintained relatively wide (i.e., ≥ 40 μm) in order to minimize the edge-definition errors due to the photomasking process.

Capacitor Arrays

Binary-weighted capacitor ladders, or arrays, are used in charge-scaling D/A converters, similar to that shown in Figure 14.8. The ratio accuracy, matching, and tracking requirements for capacitor ladders are similar to those associated with resistor ladders of comparable bit resolution. Capacitor ladders are made up of MOS capacitors, arranged in a parallel combination of binary-weighted component values. Because of their poorer matching characteristics and voltage dependence, junction capacitors are not suitable for charge-scaling applications.

The capacitor structures used for capacitor ladders are either conventional MOS structures or double-polysilicon (i.e., polysilicon–oxide–polysilicon) structures which are the same as those discussed in Chapter 13 in connection with switched-capacitor filters. The cross sections of these devices are shown in Figures 13.27 and 13.28, and their electrical characteristics are reviewed in Section 13.7.

The requirements for D/A converter capacitor ladders are virtually identical to those for switched-capacitor filters, namely, the accurate control of capacitor ratios and minimizing the effects of bottom-plate parasitics. The ratio accuracy is optimized by forming arrays made up of identical unit-geometry capacitors, which are then interconnected to obtain the desired ratios. The effect of bottom-plate parasitics are minimized by making the bottom plate connect to a voltage source at all times. For example, in the capacitor ladder of Figure 14.8, the bottom plates of the capacitors are switched between ground and V_{ref}, whereas the top plates are connected together to the common output bus.

The MOS capacitors have extremely low temperature and voltage coefficients. Typically, the temperature coefficients are on the order of $+20$ to

+50 ppm/°C and the voltage coefficient is on the order of -10 to -20 ppm/V. These characteristics are significantly better than those associated with resistor structures.

Layout of Capacitor Arrays

The capacitors forming the ladder array are designed to optimize the ratio matching. The capacitor ratio error stems from three sources:

1. Edge definition of masking process.
2. Top-plate parasitics due to metal interconnections and metal overlap over dielectric oxide step.
3. Oxide thickness gradients across the chip.

The edge-definition limitation associated with the photomasking process contributes an error of approximately 0.1–0.2 μm uncertainty to the capacitor length and width dimensions. If slight metal overlap of the top plate over the field dielectric is used (see Fig. 13.27), this may also contribute to an edge parasitic. Normally, the relative uncertainty ΔX in any edge dimension is relatively well fixed by process tolerances. To minimize the effect of this uncertainty, the capacitor length and width ratios are normally chosen to be equal, and the capacitors are laid out as square sections, as shown in Figure 14.24.

If direct area scaling is used to set the capacitor ratios, as shown in Figure 14.24a, the capacitor ratios become increasingly sensitive to the edge-definition error ΔX as the capacitor area is reduced. This can be avoided by using the capacitor array approach shown in Figure 14.24b, where the capacitor ratios are scaled by interconnecting a number of identical unit capacitors. In this approach, the edge-definition errors have a negligible effect on the ratio accuracy, since both the area and the periphery ratios are scaled simultaneously. The effect of top-plate parasitics due to interconnection lines can also be accounted by proper scaling of the interconnection line lengths and widths between the capacitors.

Another source of error in capacitor ratio accuracy is the presence of long-range thickness gradients in the thin dielectric oxide. The oxide thickness gradients can be on the order of ± 10–± 100 ppm per mil of dimensional length along the chip surface.[8] The effects of these long-range gradients can be minimized by using a common-centriod geometry in the layout and the interconnection of the capacitor array. This is illustrated in Figure 14.25. In this manner, the capacitor ratio accuracy can still be maintained in spite of first-order gradients in the capacitor oxide thickness.

Common-centriod layout becomes cumbersome and wasteful of the chip area if a higher order of symmetry is required. Therefore, it is normally limited to the first several bits of the capacitor ladder.

In the case of D/A converters, the capacitor ratios are in integral multiples, which makes them easy to generate using a unit-capacitor array. In the case of

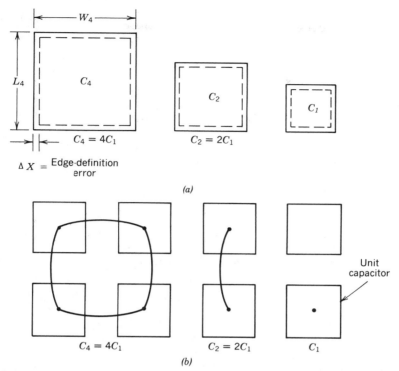

FIGURE 14.24. Layouts for scaling capacitor values: (*a*) Area scaled layout—sensitive to edge definition errors; (*b*) array layout—insensitive to edge definition errors for ratio accuracy.

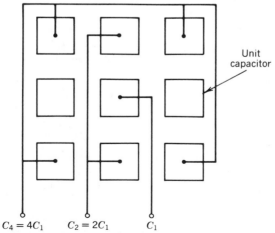

FIGURE 14.25. Common-centroid layout of capacitor array to minimize ratio errors due to oxide thickness gradients across chip.

switched-capacitor filters where noninteger capacitor ratios are used, the scaling of capacitors becomes more complex.

14.7. VOLTAGE REFERENCES

As described in Section 14.1, the analog output of a D/A converter is proportional to the product of an analog reference and the binary digital word. In the case of a voltage output D/A converter, the analog output is expressed as

$$V_{out} = KV_{ref}D \qquad (14.30)$$

where K is an arbitrary scale factor, and D is the digital word expressed in terms of weighted binary bit coefficients b_1 through b_N, as given in Eq. (14.2).

As indicated by Eq. (14.30), the voltage reference, along with the scale factor K, sets the full-scale output level. Thus, the voltage reference V_{ref} has a direct effect on the accuracy and stability of the output level.

The voltage reference circuits used in monolithic D/A converter designs can be either the Zener-breakdown or the band-gap type. The basic characteristics and design principles for these reference circuits were covered in Chapter 4. Typical voltage reference circuit configurations are illustrated in Figures 4.33 through 4.36. For Zener-breakdown-type references, normally a buried-Zener structure is used to keep the localized voltage breakdown point away from the device surface, to enhance the long-term stability of the breakdown voltage. Typical buried-Zener structures are illustrated in Figures 3.11 through 3.13.

For MOS designs, the voltage reference function can be implemented using the circuit configurations discussed in Section 6.5. However, the temperature stability and reproducibility characteristics of MOS voltage references are still somewhat poorer than those of bipolar designs.

The voltage reference circuitry is normally designed to have a buffered low-impedance output so that it can provide some amount of current drive.

Typical temperature coefficients obtainable with the monolithic reference circuits are in the range of ±20–±50 ppm/°C. If a higher degree of stability is required, an external precision voltage reference is used. For such applications, a number of monolithic precision voltage reference circuits have been developed using substrate-temperature stabilization techniques,[21,22] which provide a voltage stability of approximately ±1 ppm/°C.

The initial accuracy of a D/A converter depends on the absolute value of V_{ref}. Typical absolute-value tolerance of monolithic voltage references are on the order of $\pm3\%$ under normal manufacturing conditions. Without trimming or external adjustments, this is not accurate enough for D/A converter applications in excess of 4 or 5 bits. Thus, in most applications, either the value of V_{ref} or the scale factor K in Eq. (14.30) has to be adjusted to meet the initial accuracy specifications. This can be done either by external adjustment, or performed internally to the chip by using any one of the trimming techniques discussed in Chapter 3 (see Section 3.9).

As will be described in Section 14.10, the temperature stability of the voltage reference is extremely important to assure a given relative accuracy specification for the D/A converter over a wide operating temperature range. For example, assuming an rms distribution of errors [see Eq. (14.42)], to assure an upper error limit of less than $\pm\frac{1}{2}$ LSB over an operating temperature range of $\pm50°C$, a V_{ref} temperature coefficient of less than ±20 ppm/°C would be needed for an 8-bit D/A converter. If a 10-bit or 12-bit accuracy is needed, with $\pm\frac{1}{2}$ LSB tolerance, this specification has to be tightened to ±5 and ±1 ppm/°C, respectively.

14.8.   BIASING OF CURRENT SOURCES

Current-scaling D/A converters operate by first generating a reference current I_{ref} from a voltage reference source, then scaling and summing this current in accordance with the binary bit coefficients, to form an output current I_{out}. If a voltage output is desired, a summing amplifier is used, as shown in Figure 14.2, to convert the output current to an output voltage.

Generating an accurate and stable set of current levels in binary current sources requires setting up an accurate bias source. The output of the voltage reference section V_{ref} is used to establish this bias level. The simple biasing scheme shown in Figures 14.11 through 14.15, where V_{ref} is directly connected to the bases of the bit transistors, has two practical limitations associated with it: (1) it is sensitive to transistor V_{BE} drops; and (2) the output currents which are taken at the collectors of bit transistors are sensitive to the β of bit transistors.

The first of the problems described above can be avoided by using a feedback-bias technique, as shown in Figure 14.26. In this circuit, first a reference current I_{ref} is generated from the voltage reference V_{ref} as

$$I_{ref} = \frac{V_{ref}}{R_{ref}} \tag{14.31}$$

and a feedback loop around the operational amplifier A_1 forces the bit currents I_1 through I_N to be proportional to I_{ref}. I_{ref} is set equal to one of the major bit currents I_1, I_2, or I_3 by proper choice of the emitter bias resistor R_X associated with Q_0, and the emitter area of Q_0 is scaled accordingly.

The current output I_{out} is given as

$$I_{out} = I_{FS}D = 2I_1D \tag{14.32}$$

where D is the sequence of binary-weighted bit coefficients corresponding to the digital input [see Eq. (14.2)].

Since I_1 is related to the reference current as

$$I_1 = I_{ref}\frac{R_X}{2R} = \frac{V_{ref}}{R_{ref}}\frac{R_X}{2R} \tag{14.33}$$

the current output transfer function can be written as

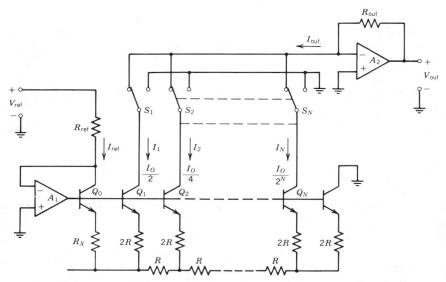

FIGURE 14.26. Feedback-bias circuit for setting bit-current levels.

$$I_{out} = \frac{V_{ref} R_X}{R_{ref} R}\left(\frac{b_1}{2} + \frac{b_2}{2^2} + \ldots + \frac{b_N}{2^N}\right) \tag{14.34}$$

If a voltage output is desired, a current-to-voltage conversion can be obtained at the output by means of the summing operational amplifier A_2 as

$$V_{out} = I_{out} R_{out} = V_{ref}\frac{R_{out}}{R_{ref}}\frac{R_X}{R}\left(\frac{b_1}{2} + \ldots\right) \tag{14.35}$$

In order to optimize the accuracy and stability of the output, the resistor pairs R_X and R, as well as R_{ref} and R_{out}, are designed for optimum matching. The resistors in each group are made to have identical geometries, and are located in close proximity to each other in the chip layout.

In current output D/A converters, where the transfer function is given by Eq. (14.33), R_{ref} is either external to the chip or is made up of a low temperature-coefficient thin-film resistor.

Base Current Compensation

One very important feature of the biasing arrangement shown in Figure 14.26 is that the base bias to all current source transistors is no longer a constant bias voltage equal to V_{ref} as shown in the simplified diagrams of Figures 14.11 through 14.15. Instead, due to the feedback path around the biasing operational amplifier A_1, the voltage at the base bias line is automatically adjusted to keep the current I_{ref} constant, as given by Eq. (14.31). This bias adjustment has the advantage that the output bit currents $I_1, I_2, \ldots, I_N$ do not depend on the β of the respective bit transistors. However, due to the presence of current switches S_1, $S_2, \ldots$, the actual bit currents reaching the output bus still differ from the true bit currents by an amount equal to the base current of the switch transistors [see Eq. (14.29)].

The base current associated with the current switch transistors can be compensated by including an equivalent transistor within the bias feedback loop, as shown in Figure 14.27. In this circuit, the transistor Q_A is added into the feedback loop around operational amplifier A_1 so that the current I_A, against which all bit currents are scaled, is higher than I_{ref} by an amount equal to the base current I_{BA} of Q_A

$$I_A = I_{ref} + I_{BA} = \frac{\beta + 1}{\beta} I_{ref} \tag{14.36}$$

and the actual bit current I_1' reaching the output bus is then equal to

$$I_1' = \frac{\beta}{\beta + 1} \frac{R_X}{2R} I_A = \frac{R_X}{2R} I_{ref} \tag{14.37}$$

which is free of base current errors. In other words, compensation of base current errors associated with current switches is achieved by introducing an equal amount of offset current into the bias feedback loop via the base current of the transistor Q_A in Figure 14.27.

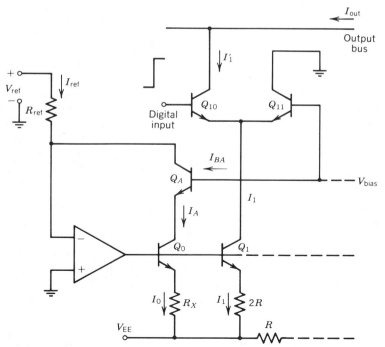

FIGURE 14.27. Feedback-bias circuit for canceling base current errors.

In precision designs, the dependence of β on current density in a transistor must also be taken into consideration. Normally, this is done by scaling the emitter areas of the current switches associated with the major bits.

The use of master–slave ladder circuits (see Fig. 14.14) also introduces a base current error. The output currents from the slave ladder will be less than the ideal currents due to the base current lost in the slave ladder. This effect can be compensated by introducing an equal amount of additional current into the slave ladder to make up the difference. As illustrated in Figure 14.28 this can be implemented through a minor modification of the circuit of Figure 14.14 by taking out a base current from the collector current of Q_{4B}. The compensation is provided by extracting an amount of current I_{BX} equal to the base current of Q_{X1} from the collector current of Q_{4B}. By proper choice of the resistor R_Y, the current mirror transistor Q_{X2} and the compensation transistor Q_{X1} are both operated at a current level $I_X \approx I_1$. In this manner, the total current available to the slave ladder is equal to

$$I_{\text{slave}} = I + I_{BX} = \frac{\beta + 1}{\beta} I \qquad (14.38)$$

which makes up for the base current loss in the ladder.

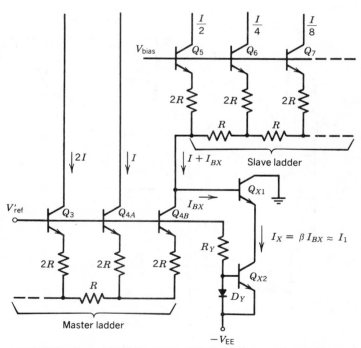

FIGURE 14.28. Base current compensation in master–slave ladders.

It should be pointed out that the feedback bias circuit of Figure 14.26 and the base current compensation techniques shown in Figures 14.27 and 14.28 only work if all transistor β's are well matched. The do *not* compensate for the errors introduced due to β mismatches. The β mismatch effects still result in non-linearity errors, as will be discussed in the next section.

14.9. EFFECTS OF DEVICE MISMATCHES

Device mismatches in a D/A converter result in an uneven distribution of bit currents, which in turn appears as a nonlinearity error in the converter output. In the extreme case, when this nonlinearity exceeds ± 1 LSB, it can lead to nonmonotonic operation, as illustrated in Figure 14.9c and d. Since the non-linearity errors due to device mismatches are input-code dependent (i.e., they depend on which particular currents are summed), they cannot be corrected by trimming. Therefore, they have to be minimized by design. Three basic sources of nonlinearity errors are the transistor β mismatches, V_{BE} mismatches, and the resistor mismatch errors.

The effects of device mismatches can be investigated by examining the current distribution in a typical bit-transistor Q_m in an N-bit binary-weighted current source array. This is illustrated in Figure 14.29 for the case of an R-$2R$ ladder network. The bit current I_m at the collector of Q_m can be expressed as

$$I_m = \frac{I_{FS}}{2^m} = \frac{\beta}{\beta + 1} \frac{1}{2^m} \frac{V_{ref} - V_{BE}}{R} \tag{14.39}$$

Assuming that R, V_{BE}, and β are independent variables, and the mismatches associated with them are small, one can approximate the change ΔI_m in the bit current due to a change ΔR, ΔV_{BE}, or $\Delta \beta$ by means of the partial derivative of the particular parameter from Eq. (14.39). In other words, the change ΔI_m in the bit current I_m due to a resistor mismatch error ΔR can be approximated as

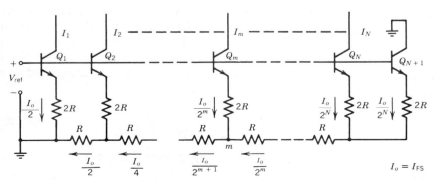

FIGURE 14.29. Current distribution in R-$2R$ ladder.

$$\Delta I_m\bigg|_{\Delta R} \approx \frac{\partial I_m}{\partial R}\Delta R = -\frac{I_m\Delta R}{R} \tag{14.40}$$

From Eq. (14.40) one can define the fractional error term $(\epsilon_{\Delta R})_m$, associated with the resistor mismatches of the mth bit, as

$$(\epsilon_{\Delta R})_m \overset{\Delta}{=} \left|\frac{\Delta I_m}{I_{\mathrm{FS}}}\right|_{\Delta R} \approx \frac{\Delta R}{2^m R} \tag{14.41}$$

Assuming that the resistor mismatches are distributed randomly between the bits, the total resistor mismatch error is the rms sum of the individual bit errors

$$\epsilon_{\Delta R} = \left|\frac{\Delta I_{\mathrm{FS}}}{I_{\mathrm{FS}}}\right| = \sqrt{(\epsilon_{\Delta R})_1^2 + (\epsilon_{\Delta R})_2^2 + \ldots + (\epsilon_{\Delta R})_N^2} \tag{14.42}$$

or

$$\epsilon_{\Delta R} = \frac{\Delta R}{R}\sqrt{\left(\frac{1}{2}\right)^2 + \left(\frac{1}{4}\right)^2 + \ldots + \left(\frac{1}{2^N}\right)^2} \approx \frac{\Delta R}{2R} \tag{14.43}$$

Equation (14.43) implies that due to the binary weighting of the bit coefficients in the rms adding of bit errors, only the error contributed by the MSB is significant, and the effects of the higher bit errors are negligible and diminish very quickly as m increases.

In the same manner, the particular bit current error due to the V_{BE} mismatch of the mth bit can be written from Eq. (14.37) as

$$\Delta I_m\bigg|_{\Delta V_{\mathrm{BE}}} \approx \frac{\partial I_m}{\partial V_{\mathrm{BE}}}\Delta V_{\mathrm{BE}} = \frac{I_m\Delta V_{\mathrm{BE}}}{V_{\mathrm{ref}} - V_{\mathrm{BE}}} \tag{14.44}$$

Then the fractional error term $(\epsilon_{\Delta\mathrm{VBE}})_m$ associated with the V_{BE} mismatch of the mth bit, can be expressed as

$$(\epsilon_{\Delta V_{\mathrm{BE}}})_m = \left|\frac{I_m}{I_{\mathrm{FS}}}\right|_{\Delta V_{\mathrm{BE}}} \approx \frac{V_{\mathrm{BE}}}{(V_{\mathrm{ref}} - V_{\mathrm{BE}})\,2^m} \tag{14.45}$$

Assuming that the transistor emitters are properly scaled so that all bit transistors operate at the same current density, the ΔV_{BE} errors would be randomly distributed through all the bits. As a result, the total expected ΔV_{BE} error can be approximated as the rms sum of the individual bit errors, similar to the case of the resistor mismatch error in Eq. (14.42). Due to the binary weighting of bit coefficients, only the error contributed by the MSB is significant, and one obtains

$$\epsilon_{\Delta V_{\mathrm{BE}}} = \left|\frac{\Delta I_{\mathrm{FS}}}{I_{\mathrm{FS}}}\right|_{\Delta V_{\mathrm{BE}}} \approx \frac{\Delta V_{\mathrm{BE}}}{2(V_{\mathrm{ref}} - V_{\mathrm{BE}})} \tag{14.46}$$

In the case of β mismatches, the resulting change in bit current ΔI_m, can be written as

$$\Delta I_m \bigg|_{\Delta\beta} \approx \frac{\partial I_m}{\partial \beta} \Delta\beta = -\frac{I_m \Delta\beta}{\beta^2} \qquad (14.47)$$

or

$$\left| \frac{\Delta I_m}{I_m} \right| \approx \frac{\Delta\beta}{\beta^2} \qquad (14.48)$$

The case of base current mismatch contribution requires additional attention. The bit current I_m has to go through a differential current switch, similar to those shown in Figures 14.21 through 14.23, before it reaches the output bus. Thus, the finite β variation of the current switch also contributes a potential mismatch error, similar to that given by Eq. (14.48). Assuming the β variations are random, these two equal error contributions add in an rms manner [see Eq. (14.42)] so that the total β mismatch error is $\sqrt{2}$ times greater than that given by Eq. (14.48). Following this arguement, on can express the β mismatch error contribution as

$$(\epsilon_{\Delta\beta})_m = \sqrt{2} \left| \frac{\Delta I_m}{I_{FS}} \right| \approx \frac{\sqrt{2}\,\Delta\beta}{2^m \beta^2} \qquad (14.49)$$

Assuming the β mismatches are randomly distributed through all the bits, the total β mismatch error would be the rms sum of the individual bit errors, which will be dominated by the MSB error, that is,

$$\epsilon_{\Delta\beta} = \left| \frac{\Delta I_{FS}}{I_{FS}} \right|_{\Delta\beta} \approx \frac{\Delta\beta}{\sqrt{2}\,\beta^2} \qquad (14.50)$$

Assuming that R, V_{BE}, and β mismatch errors are uncorrelated, the total error ϵ_{tot} can be approximated as the rms sum of the individual error terms in Eqs. (14.43), (14.46), and (14.50),

$$\epsilon_{tot} = \left| \frac{\Delta I_{FS}}{I_{FS}} \right| = \sqrt{(\epsilon_{\Delta R})^2 + (\epsilon_{\Delta V_{BE}})^2 + (\epsilon_{\Delta\beta})^2} \qquad (14.51)$$

The total error given by Eq. (14.51) is a direct measure of the differential nonlinearity that can be expected in converter operation. To assure monotonicity, the differential nonlinearity error cannot exceed ± 1 LSB. Then to guarantee monotonicity

$$\epsilon_{tot} \leq \pm 1 \text{ LSB} = \frac{1}{2^N} \qquad (14.52)$$

If a complete N-bit resolution is required, with 2^N distinct discrete steps, then the differential nonlinearity error has to be limited to $\pm\frac{1}{2}$ LSB, and the total allowable error becomes

$$\epsilon_{tot} \leq \frac{1}{2} \text{ LSB} = \frac{1}{2^{N+1}} \qquad (14.53)$$

At this point, a quantitative analysis of the error terms is instructive to judge their relative magnitudes. Consider, as an example, the case of a converter circuit having a nominal resistor matching error of 0.1%, a V_{BE} matching error of ± 1 mV, and a β matching error of $\pm 5\%$. Assume that the nominal value of β is 200, with $V_{BE} = 0.7$ V and $V_{ref} = 3$ V. Then the estimated values of the error terms are

$$\epsilon_{\Delta R} = \frac{\Delta R}{2R} = 0.05\% \tag{14.54}$$

$$\epsilon_{\Delta V_{BE}} \approx \frac{\Delta V_{BE}}{2(V_{ref} - V_{BE})} = 0.022\% \quad \text{and} \quad \epsilon_{\Delta \beta} \approx \frac{\Delta \beta}{\sqrt{2} \, \beta^2} = 0.018\% \tag{14.55}$$

and the total error is

$$\epsilon_{tot} \approx \sqrt{(\epsilon_{\Delta R})^2 + (\epsilon_{\Delta V_{BE}})^2 + (\epsilon_{\Delta \beta})^2} \approx 0.06\% \tag{14.56}$$

Comparing the value of ϵ_{tot} given by Eq. (14.56) with those required by Eq. (14.52) and (14.53), one sees that the present example can meet the monotonicity requirement for 10-bit operation. However, it will not be able to meet the $\pm\frac{1}{2}$ LSB nonlinearity specification.

In conclusion, it should be pointed out that among the three types of device mismatches, the ladder resistor mismatch is usually the dominant source of nonlinearity error. Provided that the emitter areas of the bit transistors and the major-bit current switches are properly scaled, the β and V_{BE} mismatch effects are usually lesser contributors to the overall nonlinearity error.

The β mismatches become significant if the value of β is low (i.e., $\beta < 100$). V_{BE} mismatch errors become important if the voltage drop across the resistor ladder is low, and if proper emitter-area scaling is not used.

In practice, scaling of the emitter areas of bit transistors across the entire ladder presents some practical problems in high-bit converters. To maintain equal emitter current density, the emitter areas of the MSB and LSB transistors must be related as

$$\frac{(\text{area})_{MSB}}{(\text{area})_{LSB}} = \frac{I_{MSB}}{I_{LSB}} = 2^{N-1} \tag{14.57}$$

The lower bound on the emitter area is set by the photomasking tolerances. As the bit count N is increased, the required emitter area of the MSB transistor gets prohibitively large. For example, for a 10-bit converter, an area mismatch of 512:1 would be required if simple area scaling is used. A practical solution to the area-scaling problem is to use a master–slave ladder arrangement similar to that shown in Figure 14.14. In this case, the more significant bits grouped into the master ladder can be area-scaled separately from the less significant bits grouped into the slave ladder. Thus, for example, in the case of the 8-bit converter circuit of Figure 14.14, only an 8:1 area scaling would be required between the transistors in each ladder section.

An alternate approach to the problem of emitter-area scaling is the use of active current scaling in the less significant bits of the circuit by means of an area-scaled current mirror configuration. An example of such a circuit configuration for an 8-bit D/A converter is illustrated in Figure 14.30, where the last 4 bits are formed by active current scaling.[13] The active current scaling method illustrated in Figure 14.30 is essentially a special case of the master–slave ladder, where the current splitting in the slave ladder is done entirely by area-scaled current mirrors.

In the case of the area-scaled current mirror ladder, the error coefficient due to V_{BE} mismatch of the mth bit is

$$(\epsilon_{\Delta V_{BE}})_m = \frac{\Delta V_{BE}}{2^m V_T} \tag{14.58}$$

where $V_T (= kT/q)$ is the thermal voltage. In the example shown, the main error contribution of active scaling is in the first bit of the active ladder portion (i.e., bit 5 in Figure 14.30). For an anticipated V_{BE} mismatch of ± 1 mV, the estimated error contribution from Eq. (14.58) (for $m = 5$) is approximately 0.12% of full scale, which is well within the $\pm\frac{1}{2}$ LSB nonlinearity tolerance of the 8-bit converter.

14.10. ACCURACY CONSIDERATIONS

The relative accuracy of a D/A converter is stated in terms of the maximum allowable error in the analog output level. The output error is the difference between measured and predicted output level, and is normally expressed either as a percentage of the full-scale output or as a fraction of the smallest output step

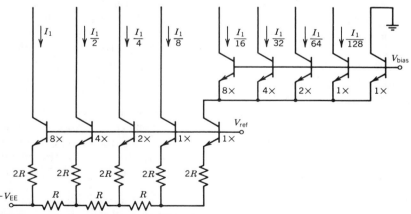

FIGURE 14.30. Example of active current splitting in lesser bits to minimize emitter-area scaling requirements.

LSB. In the latter case, typical accuracy requirements are usually on the order of $\pm\frac{1}{2}$ LSB. This implies that the total allowable error voltage ΔV_ϵ which is defined as the difference between the ideal and the actual analog output level anywhere within zero to full-scale reading, cannot exceed

$$\Delta V_E \leq \tfrac{1}{2} \, \text{LSB} = \frac{V_{\text{FS}}}{2^{N+1}} \tag{14.59}$$

Thus, the total percentage of error ϵ_T allowable at the output is

$$\epsilon_T = \pm \frac{100}{2^{N+1}} \tag{14.60}$$

for $\frac{1}{2}$ LSB accuracy. Thus, the accuracy requirement increases very rapidly as the bit count is increased. For example, a 4-bit converter with $\pm\frac{1}{2}$ LSB accuracy can allow a total error of $\pm 3.2\%$ at the output, but for an 8-bit converter it decreases to 0.195%, and for the case of a 12-bit converter, the maximum allowable error is on the order of 0.012%.

Accuracy is a much more comprehensive specification than resolution or nonlinearity, since it includes all the error sources within the converter system. A complete current-scaling D/A converter system, as shown in Figure 14.2, contains four basic sections: (1) voltage reference, (2) resistor ladder, (3) current switches, and (4) output amplifier. The resistor ladder and current switches will primarily contribute to the nonlinearity errors, as discussed in Section 14.9. The voltage reference and output amplifier are the main contributors to the gain (i.e., scale factor) and dc offset errors.

The total error ϵ_T determining the accuracy can be separated into three parts

$$\epsilon_T = \epsilon_{T0} + \epsilon_{T1} + \epsilon_T(T) \tag{14.61}$$

where ϵ_{T0} and ϵ_{T1} are the initial errors independent of temperature, and $\epsilon_T(T)$ is the temperature-dependent portion of the total error arising from thermal drifts and imperfect tracking of component values.

ϵ_{T0} is an initial offset and scale-factor error due to absolute-value tolerances associated with the voltage reference, the resistor values, and the operational amplifier offset voltage. In converters with 8-bit or higher resolution, an offset adjustment is usually provided to null out ϵ_{T0}.

ϵ_{T1} is the collection of initial nonlinearity error terms due to device mismatches, as described in Section 14.9. It corresponds to ϵ_{tot} of Eq. (14.51), and since it is dependent on a digial input code (i.e., a particular choice of switch positions) it cannot be nulled out. However, since the major contributor to ϵ_{T1} is the resistor ladder error, it can be minimized by trimming the resistors in the most significant bits to reduce their mismatches.

At this point it is instructive to examine the degree of precision required to meet a specific accuracy requirement. For illustrative purposes, consider the case of a complete 8-bit monolithic D/A system, which is required to meet $\pm\frac{1}{2}$ LSB relative accuracy specification over a temperature range of $-55\degree$C to

+125°C. It is assumed that the offset error term ϵ_{T0} can be nulled out at room temperature. From Eq. (14.60), the total allowable error is approximately 0.2%. As a somewhat arbitrary but reasonable choice, this error budget can be allocated approximately equally to ϵ_{T1} and $\epsilon_T(T)$, that is,

$$\epsilon_{T1} \approx 0.1\% \quad \text{and} \quad \epsilon_T(T) \approx 0.1\%$$

ϵ_{T1} is essentially the total nonlinearity error described in Eq. (14.51) involving R, V_{BE}, and β mismatches. Following the example given in Section 14.9, for a resistor mismatch of 0.1%, a β mismatch of $\pm5\%$, and V_{BE} mismatch of ±1 mV, the value of ϵ_{T1} is approximately 0.06% [see Eq. (14.56)].

The temperature drift error $\epsilon_T(T)$ is made up of three contributors:

$\epsilon_{ref}(T)$ = error due to temperature drift of V_{ref}
$\epsilon_{lad}(T)$ = error due to temperature drift of ladder mismatches
$\epsilon_{amp}(T)$ = error due to drift of operational amplifier offset

In general, assuming that the operational amplifier bias current is $\leq5\%$ of the LSB current in the ladder, the operational amplifier bias current introduces negligible error. Typical operational amplifier offset drifts are in the range of $\pm10\ \mu V/°C$, which translates to ±1 mV output level change over the operating temperature range. If the full-scale output swing V_{FS} is 10 V, this represents an error of $\pm0.01\%$, which is negligible compared to the other two error sources. Assuming that the error sources making up $\epsilon_T(T)$ are uncorrelated, and thus add in an rms manner, one can write

$$\epsilon_T(T) = \sqrt{\epsilon_{ref}(T)^2 + \epsilon_{lad}(T)^2} \tag{14.62}$$

where the operational amplifier error has been neglected. A reasonable (but somewhat arbitrary) assumption is to consider the two error contributions in Eq. (14.62) to be approximately equal, that is,

$$\epsilon_{ref}(T) \approx \epsilon_{lad}(T) = \frac{\epsilon_T(T)}{\sqrt{2}} = 0.07\% \tag{14.63}$$

which can be met only if the voltage reference has a temperature coefficient of $\leq \pm7$ ppm/°C and if the resistor ratios track to better than ±7 ppm/°C.

The above examples illustrates the degree of precision and temperature stability required to meet a relatively modest 8-bit $\pm\frac{1}{2}$ LSB accuracy requirement over the full (-55 to $+125°C$) operating temperature.

Monolithic Converter Specifications

Overall accuracy is a very comprehensive specification, which includes many different error sources. It is very difficult to measure and guarantee under a production environment. Furthermore, many monolithic D/A converters have different architectures; some use external reference, and some have only current output. Because of this diversity of options in circuit configurations and specifications, a certain amount of confusion exists in the D/A converter specifications available from the monolithic converter manufacturers. Instead of accuracy, the IC converter manufacturers normally give the following set of specifications:

1. Resolution (bits).
2. Linearity error (integral nonlinearity).
3. Monotonicity (differential nonlinearity).
4. Gain (scale factor) drift (ppm/°C).
5. Initial offset errors.
6. Operating temperature range over which all of the above specifications are guaranteed.

It is left to the users to relate all of these separate specifications to an ultimate accuracy specification for their complete system.

14.11. MONOLITHIC DESIGN EXAMPLES

At present, a wide choice of monolithic D/A converters are available from many different IC manufacturers. These designs vary from the medium-accuracy units in the 8- to 10-bit range to precision circuits in the 10- to 12-bit range. Commercially available D/A converters also vary greatly in their architecture. In one extreme are the simple building-block-type D/A converter subsections, such as binary current source or resistor arrays, which can be externally interconnected to form a complete converter system. In the other extreme are the complete systems that contain all the subsections of a D/A converter, including the voltage reference and the output amplifier, on the monolithic chip. The majority of the commercially available monolithic D/A converters fall somewhere between these two extremes.

Most of the commercially available D/A converter chips are designed for maximum versatility. Therefore, they tend to be at the subsystem level, rather than at a complete system level of integration. For this purpose, the voltage reference and/or the output amplifier are often left external to the chip.

In this section, several practical design examples will be examined to illustrate the capabilities of monolithic D/A converters, and to demonstrate the practical implementation of some of the design techniques described in the preceding sections. The design examples to be considered are the 8-bit DAC-08, the 10-bit AD-7520, and the 12-bit Am6012 D/A converters. These particular examples are chosen because they represent different design approaches and IC technologies, yet their overall architectures are similar. They are all current output D/A converters and operate with external reference and output amplifiers, and they all represent successful commercial products.

DAC-08 8-Bit D/A Converter

The DAC-08 circuit, introduced by Precision Monolithics, Inc., is an 8-bit current output converter circuit fabricated with conventional bipolar technology. It is a so-called multiplying D/A converter circuit, which is designed to operate with an external voltage reference. The circuit is packaged in a 16-pin standard IC package and has the basic architecture shown in Figure 14.31.

The main features of the circuit can be identified from the figure. It is comprised of eight binary-weighted current sources and their corresponding differential current switches. The converter output is in the form of a pair of complementary currents I_{out} and $\overline{I}_{out}$, which vary in opposite directions with the changing binary input code.

The current scaling is obtained by means of a master–slave ladder configuration, and active current scaling is used in the last bits of the slave ladder to reduce circuit complexity and device scaling requirements (see Fig. 14.30).

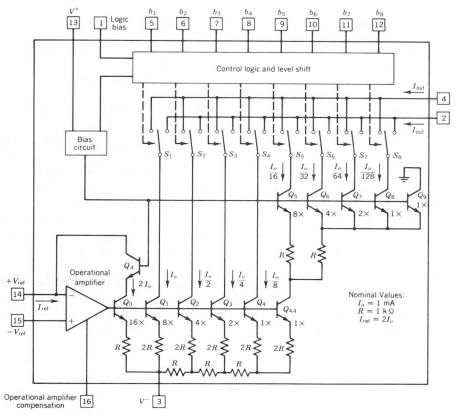

FIGURE 14.31. Simplified circuit diagram of DAC-08 8-bit D/A converter.

The relative sizes of the key devices in the circuit are indicated in the block diagram. The largest active area scaling ratio is $16:1$ between the current source transistors, and the largest device area is associated with the current-setting transistor Q_0 which carries twice the MSB current. The R-$2R$ ladder section of the converter is made up of approximately 40-μm-wide unit resistor segments of 1-kΩ nominal value, and the nominal values of the bit currents are in the range of 1 mA for the MSB down to 7.8 μA for the LSB, with nominal bias setting current $I_{ref} = 2$ mA.

The bit current levels are set from an external reference voltage V_{ref} and a current setting resistor R_{ref}, as described in Section 14.8, by means of a feedback bias circuit around the internal operational amplifier section [see Eq. (14.31)]. The transistor Q_A is added into the feedback loop of the bias circuit to cancel the base current errors due to finite β of the bit transistors (see Fig. 14.27). This feedback bias circuit results in a nearly ideal constant-current output at the D/A converter output terminals, with an output impedance in excess of 20 MΩ, and with a very high degree of voltage compliance* nearly equal to the total supply voltage.

The reference amplifier is basically a single-stage operational amplifier circuit with *pnp* inputs and *npn* active loads. The value of the operational amplifier compensation capacitor depends on the value of the reference current setting and increases with increasing I_{ref}. Therefore, it is left external to the IC chip for maximum versatility. For $I_{ref} = 2$ mA, the value of the compensation capacitor is approximately 75 pF.[23]

The differential current switches utilize the basic circuit configuration shown in Figure 14.23, operating with a typical bias current of ≈ 0.1 mA per switch. The propagation delay through the input control logic to the output of the current switch is approximately 35 nsec, and the settling time of the output current to within $\pm\frac{1}{2}$ LSB is less than 135 nsec. The DAC-08 circuit is designed to operate with supply voltages of ± 4.5–± 18 V. Other electrical characteristics of the circuit are listed in Table 14.1.

The DAC-08 is integrated on an 82- $\times$ 62-mil chip (2.05 $\times$ 1.55 mm), using conventional linear bipolar process technology. The photograph of the chip is shown in Figure 14.32. The main features of the circuit are identified on the photomicrograph.

Figure 14.33 shows the basic external circuit connections of the DAC-08 converter for obtaining either unipolar (i.e., single-polarity) or bipolar outputs (i.e., swinging both positive and negative with respect to ground). The full-scale output current I_{FS} is set by the external reference resistor R_{ref} as

$$I_{FS} = I_{ref} = \frac{V_{ref}}{R_{ref}} \qquad (14.63)$$

---

*Voltage compliance is the output voltage range over which the current output of the converter can maintain $\pm\frac{1}{2}$ LSB accuracy. True compliance requires extremely high output impedance.

TABLE 14.1. Comparison of Electrical Characteristics of DAC-08, AD-7520, and Am-6012 Current Output D/A Converter Circuits.[a]

Parameter	DAC-08	AD-7520	Am-6012
Resolution	8 bits	10 bits	12 bits
Differential nonlinearity	$\pm\frac{1}{2}$ LSB (M)[b]	$\pm\frac{1}{2}$ LSB (M)	$\pm\frac{1}{2}$ LSB (M)
Integral nonlinearity (% of full scale)	$\pm 0.1\%$ (M)	$\pm 0.05\%$ (M)	$\pm 0.05\%$ (M)
Settling time to $\pm\frac{1}{2}$ LSB	135 nsec (M)	500 nsec (T)	500 nsec (M)
Full-scale current I_{FS} (nominal value)	2 mA	2 mA	4 mA
Temperature Coefficient of I_{FS} (with external reference)	10 ppm/°C (T)	10 ppm/°C (M)	5 ppm/°C (T)
Output impedance R_{out}	>20 MΩ (T)	>10 kΩ (T)	>10 MΩ (M)
Output voltage compliance range (± 15 V operation)	-10 to $+18$ V	—	-5 to $+10$ V
Power supply range			
Nominal supply	± 15 V	$+15$ V and ground	± 15 V
Operating range	$+4.5$ to $+18$ V -18 to -4.5 V	$+5$ to $+15$ V	$+4.5$ to $+18$ V -18 to -11 V
Power dissipation (at nominal supply and I_{FS} values)	174 mW (M)	20 mW (T)	397 mW (M)
Technology used	Bipolar with diffused R's	CMOS with thin-film R's	Bipolar with diffused R's
Chip size (mil)	82×62	96×74	133×93
Package type	16-pin DIP	16-pin DIP	20-pin DIP

[a]Specifications indicated correspond to prime grade units, that is, the DAC-08A, AD-7520L, and Am-6012A.
[b](M) = minimum or maximum specification; (T) = typical specification

The maximum value of the output current is 1 LSB less than I_{FS}, as given by Eq. (14.5), for $N = 8$, that is,

$$(I_{out})_{max} = I_{FS}\frac{2^N - 1}{2^N} = \frac{255}{256}I_{FS} \qquad (14.64)$$

The complementary output currents I_{out} and $\overline{I_{out}}$ vary in opposite direction as a function of the digital input code. However, their sum is constant at all times,

$$I_{out} + \overline{I_{out}} = (I_{out})_{max} = \frac{255}{256}I_{FS} \qquad (14.65)$$

The resistor R_X connected between the noninverting input of the operational amplifier (pin 15) and ground is chosen to be approximately equal to R_{ref} in order to avoid offset voltage due to bias currents of the operational amplifier input stage.

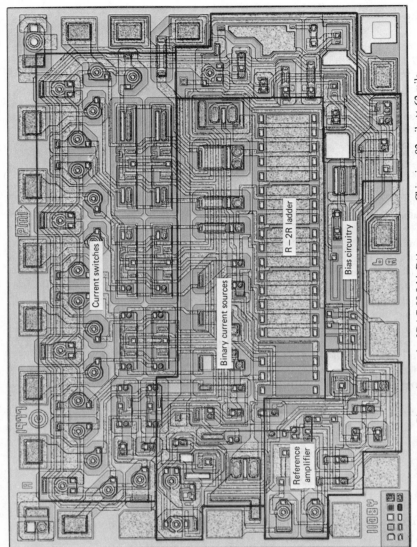

FIGURE 14.32. Photomicrograph of DAC-08 8-bit D/A converter. Chip size: 82 mils × 62 mils. (*Photo*: Precision Monolithics, Inc.)

Current switches

Binary current sources

R − 2R ladder

Bias circuitry

Reference amplifier

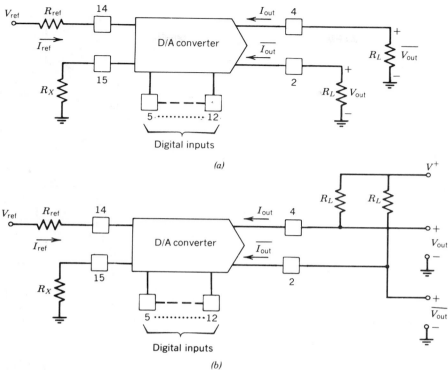

FIGURE 14.33. Basic external circuit connections for current output D/A converter circuit of Figure 14.31. (*a*) unipolar operation; (*b*) bipolar operation.

In the simple unipolar output configuration of Figure 14.33*a*, the output voltage swings from zero toward V^-, with a full-scale output swing

$$V_{FS} = -I_{FS}R_L = -V_{ref}\frac{R_L}{R_{ref}} \qquad (14.66)$$

To assure temperature stability of the output, both R_L and R_{ref} must track with temperature.

Figure 14.33*b* shows the circuit connection for obtaining bipolar outputs. In this case, the output voltage swings from V^+ to $V^+ - V_{FS}$. In the circuits shown in Figure 14.33, the load resistance R_L must be chosen to maintain the output voltage swing within the compliance limits of the output stage. The response speed of the circuit is sensitive to the parasitic capacitances associated with the output load resistors.

Figure 14.34 shows the methods of obtaining low-impedance voltage output by adding an external operational amplifier into the circuit. The circuit of Figure 14.34*c* is similar to that of Figure 14.34*a*, except that an offset current equal to $I_{FS}/2$ is added to the output current through the feedback resistor $2R_{ref}$ to set the

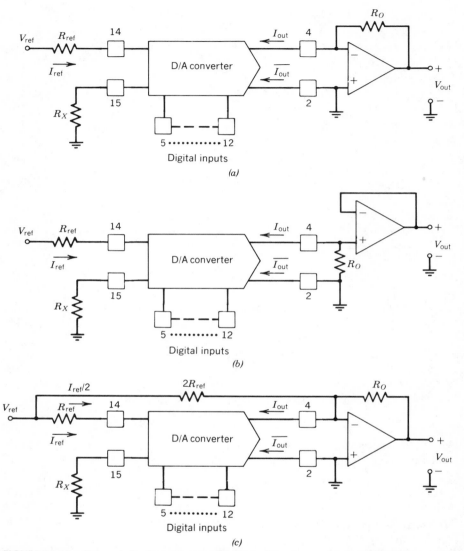

FIGURE 14.34. External circuit connections for converting current output of DAC-08 D/A converter to low-impedance voltage output: (*a*) Positive output; (*b*) negative output; (*c*) offset binary voltage output.

output equal to zero with only the MSB turned on. In this manner, the output can swing to within 1 LSB of $V_{FS}/2$ in the positive direction, and to $-V_{FS}/2$ in the negative direction.

When used with an external operational amplifier, the settling time of the voltage output is normally limited by the operational amplifier slew rate and the phase-margin characteristics.

AD-7520 10-Bit CMOS D/A Converter

The AD-7520, developed by Analog Devices, Inc., is essentially a precision resistor array with CMOS binary switches.[24] The basic circuit configuration is shown in Figure 14.35. It is made up of an R-$2R$ ladder, fabricated using deposited silicon-chromium resistors. These resistors have a nominal sheet resistivity of 2 kΩ/$\square$. They exhibit a nominal temperature coefficient of $+150$ ppm/°C and track to within ± 1 ppm/°C. Similar to the DAC-08, the circuit operates with an external voltage reference and provides a current output. Since the output impedence is determined primarily by the impedence of the R-$2R$ ladder, the output impedance is low (≈ 10 KΩ), and the output voltage compliance characteristics are poor. Therefore, in normal operation, an external operational amplifier is used to convert the output current into an output voltage. An on-chip feedback resistor R_{FB} is provided for this purpose.

The digital-input buffer stages and the differential bit switches are implemented using CMOS transistors. Figure 14.36 shows the basic circuit diagram of one of the current switch stages, along with its driver. The driver stage is essentially a three-stage inverter chain. A small amount of internal positive feedback is provided from the output of the second stage to the first stage through the transistor Q_3, which increases the switching speed of the driver stage and adds a small amount of hysteresis to the logic threshold. The logic threshold is set to be at the TTL level (i.e., ≈ 1.4 V above ground) by proper scaling of the Z/L ratios of Q_1 and Q_2.

The use of CMOS devices as current switches has the advantage of avoiding nonlinearity due to base current mismatches in bipolar transistor switches. However, since the switches are in series with the ladder network and ground, their series resistance is critical. The effect of this series resistance is minimized by choosing a high value for the ladder resistors (i.e., $R = 10$ kΩ) and by scaling the area of the current switches in a binary manner for the major bits.

Figure 14.37 shows the chip photograph of the CMOS D/A converter circuit. The locations of the key circuit components are identified. The R-$2R$ ladder occupies the center section of the chip. The ladder is comprised of an array of unit resistors of 10-kΩ nominal value, which are laid out in close proximity to each other. Large resistor geometries are used to minimize the photomasking tolerances. Resistor strips are approximately 70 μm wide by 350 μm long.

The locations of the CMOS switches are identified in the chip photograph.

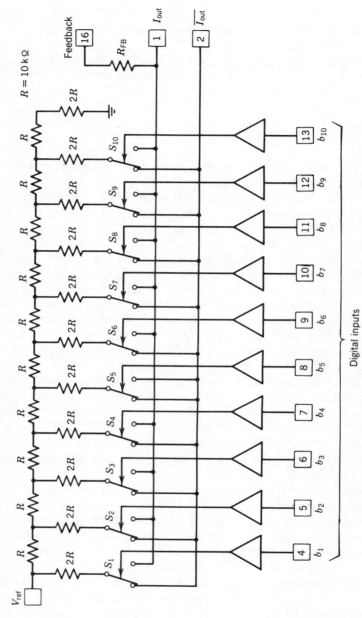

FIGURE 14.35. Simplified circuit diagram of AD-7520 10-bit CMOS A/D converter.[24]

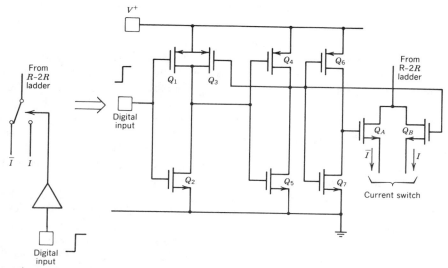

FIGURE 14.36. Circuit schematic of CMOS drivers and current switches in AD-7520 A/D converter.

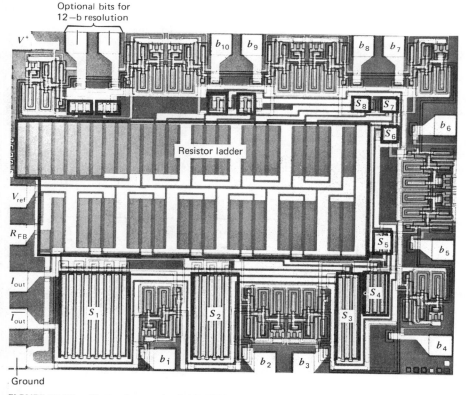

FIGURE 14.37. Photomicrograph of AD-7520 CMOS A/D converter. Chip size: 74 mils × 96 mils. Technology used: metal-gate CMOS process with deposited silicon-chromium resistors. (*Photo:* Analog Devices, Inc.)

These switches have their areas scaled in a binary manner for the first 7 bits, with the switch on resistance varying from 20 Ω for S_1 to 640 Ω for S_7 and higher bits.

Figure 14.38 shows the operation of the circuit with external operational amplifier to provide voltage output. The use of operational amplifiers, rather than resistive loads, is required to obtain a voltage output due to the poor compliance and relatively low impedance of current outputs.

If unipolar operation is desired, where the output varies from 0 to $+V_{FS}$, then only one of the output currents is used, as shown in Figure 14.38a. The feedback resistor for the output operational amplifier is internal to the chip. It matches and tracks the ladder resistors.

If a bipolar output is desired, which can swing $\pm\frac{1}{2} V_{FS}$ around zero, the difference of the two output currents can be used to produce an output voltage, as shown in Figure 14.38b. In this case, the second operational amplifier A_2 inverts $\overline{I_{out}}$ and sums it with I_{out} at the input of operational amplifier A_1.

Since the circuit is fabricated using high-value resistors and CMOS switches, it operates with very low power dissipation ($\approx$ 30 mW). This reduces the thermal gradients on the chip and greatly simplifies the chip layout.

Basic electrical characteristics of the AD-7520 D/A converter are listed in Table 14.1.

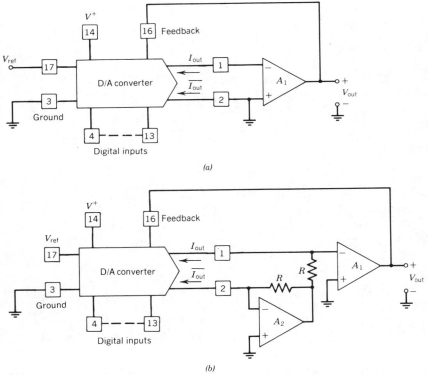

(a)

(b)

FIGURE 14.38. Typical external circuit connections for AD-7520 CMOS D/A converter: (*a*) Unipolar operation; (*b*) bipolar operation.

Am-6012 12-Bit D/A Converter

The Am-6012, developed by Advanced Micro Devices, Inc., is a 12-bit mono-lithic D/A converter fabricated using conventional linear bipolar technology and diffused resistors. A unique feature of the circuit is that it provides inherent monotonicity and a differential nonlinearity of less than $\frac{1}{2}$ LSB, without requiring any trimming of internal circuitry. This performance characteristic is achieved by using the segmented-ladder architecture discussed in Section 14.4 (see Fig. 14.15).[13]

Figure 14.39 shows the functional block diagram of the Am-6012 D/A con-verter. The circuit is a current-scaling converter, providing complementary current outputs. It is comprised of four major sections: (1) internal 9-bit D/A converter subsection, (2) segment current generators, (3) segment decoder switches, and (4) reference amplifier and biasing circuitry.

The reference amplifier and internal biasing circuitry set the current levels in each of the eight segment generator stages, using a feedback-bias configuration similar to that shown in Figures 14.26 and 14.27. Each of the eight segment currents is set to be equal to $\frac{1}{8} I_{FS}$.

The principle of operation of the circuit is identical to that of the 6-bit segmented D/A converter example illustrated in Figures 14.15 and 14.16 except that eight rather than four segment generators are used. Similarly, the 4-bit step generator section of Figure 14.15 is now extended to a 9-bit D/A converter capable of producing 2^9 steps. Thus, the output transfer characteristics are similar to those shown in Figure 14.16; there are up to eight segment currents or pedestals, with 512 steps per pedestal. As described in Section 14.4, a very important feature of the segmented D/A converter architecture is its inherent monotonicity. Provided that the internal 9-bit D/A converter which serves as the step generator is monotonic, overall monotonicity is assured.

The 9-bit step generator D/A converter section uses a current-scaling master–slave ladder, similar to that discussed in connection with the DAC-08 design (see Fig. 14.31), where active current scaling is used in the last 4 bits. The bit-current switches S_8 through S_{16} are designed as differential current switches with the basic circuit configuration given in Figure 14.23.

The function of the segment decoder and the switches S_0 through S_7 is to connect various combinations of segment current generators to three analog buses in proper priority, in accordance with the binary stages of bits b_1, b_2, and b_3. These essentially correspond to priority-encoded three-way analog current switches.

Figure 14.40 shows the circuit configuration used in implementing the seg-ment decoder and switch section of the Am-6012 converter. The three segment switch drivers, which are activated by the input bits b_1, b_2, and b_3, are on the left-hand side of the figure, and the decoding switches and segment current sources are on the right-hand portion. The circuit uses multiple logic levels V_{B1}, V_{B2}, and V_{B3}, which are spaced approximately 0.5 V apart. This stacking of switch stages results in an inherent "priority" arrangement. For example, the transistors Q_{10} through Q_{13} would override the transistors below them. Thus,

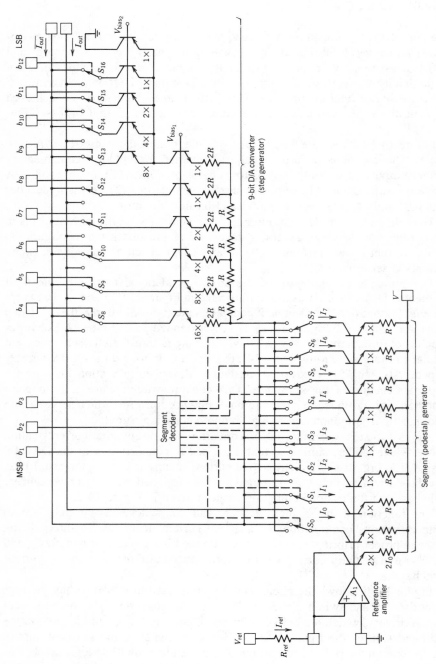

FIGURE 14.39. Simplified circuit diagram of Am-6012 12-bit D/A converter.[13]

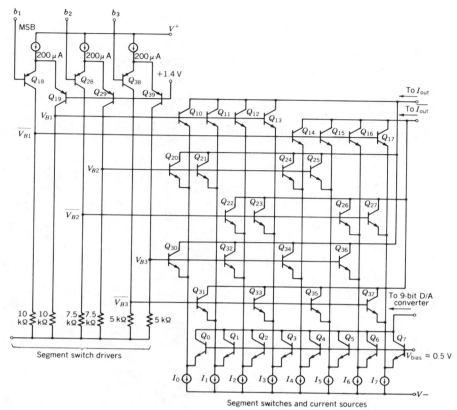

FIGURE 14.40. Circuit configuration for segment decoder section of Am-6012 D/A converter.

when the MSB is high, I_0 through I_3 will be routed to I_{out}, irrespective of the state of the lesser bits. For any given input code, seven of the eight segment currents will be switched to either the I_{out} or the $\overline{I_{out}}$ bus, and the eighth will be switched into the 9-bit D/A converter (i.e., the step generator).

The segment decoder circuit of Figure 14.40 results in a very efficient circuit layout using only three isolation pockets, since most of the transistors are in the common-collector configuration; and the switching speed of the decoder is comparable to that of the conventional differential current-switch stages used in the step generator section.

Figure 14.41 shows the chip photograph of the Am-6012 12-bit D/A converter. The locations of various circuit blocks on the chip are identified. The R-$2R$ and the segment generator ladder networks are made up of identical resistor segments, located in the central section of the chip. Each unit resistor strip R is approximately 30 mils long and 2 mils wide, and is designed for an initial matching tolerance of approximately $\pm 0.2\%$. The monotonicity (i.e., differential nonlinearity) characteristics are determined by the differential non-

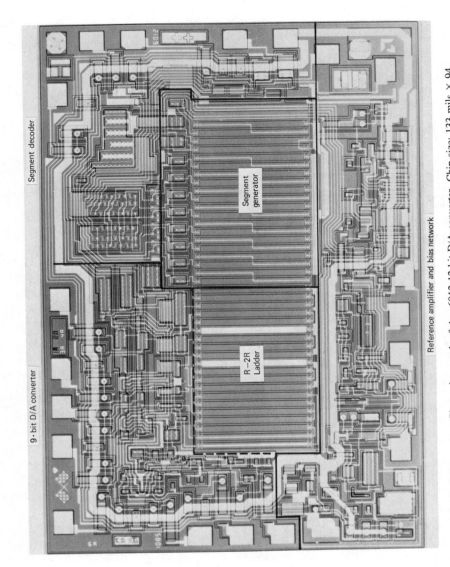

FIGURE 14.41. Photomicrograph of Am-6012 12-bit D/A converter. Chip size: 133 mils × 94 mils. (*Photo:* Advanced Micro Devices, Inc.)

Segment decoder

9 · bit D/A converter

Reference amplifier and bias network

Segment generator

R – 2R Ladder

linearity of the 9-bit step generator D/A converter (i.e., the R-$2R$ ladder), whereas the integral nonlinearity (i.e., the straight-line conformance) is primarily determined by the matching of segment currents.

Table 14.1 lists some of the performance characteristics of the Am-6012 converter integrated circuit. Note that its output impedance characteristics are similar to those of the DAC-08. However, the output compliance range is somewhat lower due to the stacking of multiple current source stages within the D/A converter architecture (see Fig. 14.39).

14.12. ULTRAPRECISION D/A CONVERTER CIRCUITS

The matching and tracking tolerances associated with IC components limit the resolution and accuracy specifications of commercially available monolithic D/A converters to 12 bits. However, there is a very rapidly developing market for precision converter circuits with resolution and linearity characteristics in excess of 12 bits. Traditionally, the main applications for ultraprecision D/A converters have been in the areas of precision instrumentation, test, and calibration systems. However, a new and rapidly expanding market is developing in the fields of digital audio recording and sound reproduction, which require D/A converters having resolution and linearity characteristics in the 14- to 16-bit range.

A number of circuit design and fabrication techniques have been developed to extend the capabilities of D/A converters to higher levels of precision. These techniques fall into two major categories: (1) trimming techniques, and (2) self-correction techniques. In the former technique, the level of precision is improved by individual adjustment of component values, subsequent to IC fabrication. In the latter case, code-dependent error-correction methods using digital data storage techniques are utilized.

Trimming Techniques

Two basic trimming techniques commonly used in designing precision D/A converts are the Zener-zapping and the laser-trimming methods, which are discussed in Section 3.9.

The Zener-zapping technique can be used to add or subtract corrective currents from the converter bits in order to improve linearity and resolution. Normally, the corrections are confined to the first several bits. Figure 14.42 shows the schematic block diagram of a 12-bit D/A converter circuit using the Zener-zapping technique.[11] In the circuit, trimming is done by adding or subtracting currents from the reference current and the first 6 bits of the circuit by means of a trim network, which generates additive or subtractive trim currents. These trim currents are proportional to the LSB current I_{LSB}, and are generated by scaling the residual current I_{res} in the ladder termination transistor Q_{13}.

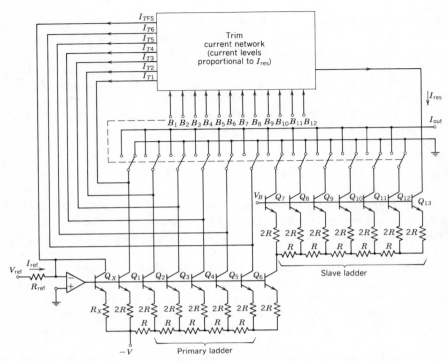

FIGURE 14.42. Functional block diagram of 12-bit D/A converter trimmed by selective on-chip Zener zapping after packaging.[11]

Figure 14.43 shows a simplified schematic of the trim circuit used for generating the correction current I_{T1} for bit 1 of the converter circuit of Figure 14.43. Binary-weighted current mirrors made up of split-collector *pnp* transistors are used in series with zapping diodes Z_1 through Z_4 to generate four binarily weighted currents, proportional to I_{LSB}. The total trim current is selected by the selective short-circuiting of zapping diodes; and the polarity of the current is determined by the fifth diode Z_5, which determines whether the current mirror made up of D_1 and Q_1 is included in the output. The same or a similar trim circuit is used for each of the other individual bits to be trimmed. The trimming operation is performed after the circuit has been packaged, by comparing the actual output to an ideal output at each of the first set of significant bits, and by selectively short-circuiting internal Zener diodes until the bit errors are reduced to below a prescribed level. Trimming is performed by applying current through the proper combination of digital input terminals. Once the trimming is complete, a "lock-out" diode is short-circuited to prevent any further changes.

The main advantage of the Zener-zapping (or fusible-link) technique is that it can be implemented after the device has been packaged and burned in, and is free of any initial component drifts. Its main disadvantage is that it requires a fairly complex decoding and selection matrix on the chip, which takes up an

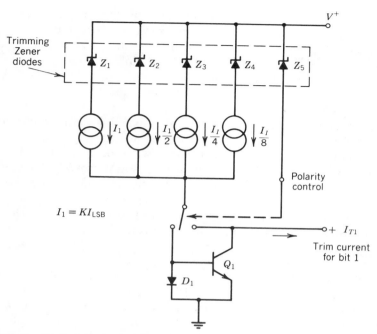

FIGURE 14.43. Block diagram of trim circuit for MSB (bit 1) of converter circuit of Figure 14.44.

inordinate amount of chip area for resolution requirements in excess of 12 bits. For example, in the case of the circuit shown in Figure 14.41, approximately 50% of the chip area is taken up by the trim network.

Laser-trimming techniques are quite feasible for D/A converter design, and have been used extensively.[20,25] The advantage of laser trimming is that it can be implemented at the wafer probe stage. Since each component is trimmed individually, it does not require additional circuitry on the chip (other than special trimming pads) to complete the trimming operation. However, since laser trimming is performed at the wafer probe stage, it is not immune to component drifts during the assembly and burn-in stages. Monolithic D/A converter designs having 16-bit resolution have been reported[26] using Ni-Cr resistors and dielectric isolation. However, the long-term stability of trimmed thin-film resistors is still not fully characterized, and the long-term accuracy may be a problem at such precision levels.

Self-Correcting Circuits

Self-correcting circuits represent a novel and powerful technique for achieving very high levels of precision in data converters, without resorting to costly trimming operations.[27,28] In this technique, the nonlinearity errors at key points of the D/A converter transfer characteristics (i.e., at so-called major carry points) are measured, and the error is digitally stored in a memory bank. Later it is used to provide code-dependent error correction. This type of error correction is particularly effective in eliminating linearity errors without requiring extremely high degrees of initial component-matching tolerances.

Figure 14.44 shows the conceptual example of a self-correcting D/A converter system. The system is made up of three subsections: (1) a main D/A converter section, (2) a correction converter, and (3) a memory element. The operation of the entire system can be described as follows. The main converter converts the N-bit digital input $b_1, b_2, \ldots, b_N$ into an output current I_o. The actual output is measured and compared against the ideal output level, and the difference is stored in the memory section in the form of an M-bit digital word for each of the selected points on the transfer characteristics during the initial calibration stage at the time of fabrication. These data are later applied to the M-bit correction converter to generate a correction output ΔI, which can be added to or subtracted from I_o to give the corrected output

$$I_{\text{out}} = I_o \pm \Delta I \tag{14.67}$$

Normally, the resolution of the correction converter is significantly less than that of the main converter (i.e., $N > M$). The correction converter is required to have output current ΔI that can be either additive or subtractive to the main converter output, as shown by Eq. (14.67). In practice, this can be implemented by adding a current mirror to the output of the correction converter, in a manner similar to that illustrated in the trim network of Figure 14.43.

In the simplest case, an electrically programmable read-only memory (EPROM) is used to store the correction data on the chip. These data are written into the memory during the initial "calibration" operation, and retained there indefinitely. The memory is then addressed at the same time as the main converter and provides the corrective data to the M-bit correction converter in the form of an M-bit word ($\gamma_1, \gamma_2, \ldots, \gamma_M$). In the ideal case, if each of the 2^N points on the transfer characteristics is to be corrected, a memory capacity of 2^N words $\times$ M bits would be needed. In practice, the correction is only applied to

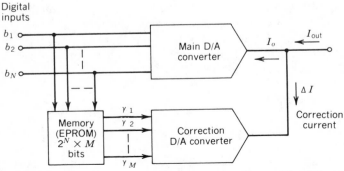

FIGURE 14.44.   Conceptual block diagram of self-correcting precision D/A converter.

the first several major bits, say bits 1 through K. This would reduce the memory capacity requirement to 2^K words $\times$ M bits. As an example, consider the case of a 14-bit main converter with an 8-bit (i.e., 7-bit plus sign) correction converter, using a correction applied to the first 5 major bits of the input data (i.e., $K = 5$). Then the total required memory capacity would be 2^5 words $\times$ 8-bits, or a total of 256 bits. Since an EPROM structure is feasible with MOS technology, a circuit similar to that illustrated in the block diagram of Figure 14.44 would be suitable for fabrication using CMOS technology. A somewhat different implementation of the concept shown in Figure 14.44 has been developed as a commercial product.[27]

The concept of self-correction of errors can be extended one step further, to "self-calibration," where instead of a fixed read-only memory (ROM), a random-access memory (RAM) block can be used in the system. Then, provided that a sufficiently accurate reference source is available to the system, the circuit can calibrate itself periodically and eliminate the long-term component drift problems. Figure 14.45 shows the conceptual block diagram of such a self-calibrating D/A converter system.[28] With reference to the figure, the operation of the circuit can be explained as follows. In the conversion mode, the circuit operates in a manner similar to that shown in Figure 14.44. The main converter performs the conversion, and the correction converter provides the necessary correction current $\pm \Delta I$, in accordance with the error data stored in the memory,

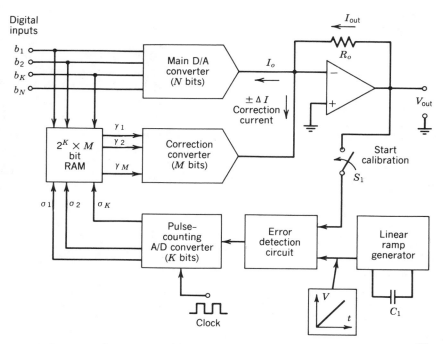

FIGURE 14.45. Conceptual block diagram of self-calibrating D/A converter system.[28]

which determine the correction coefficients γ_1, γ_2, ..., γ_M. During the conversion cycle, switch S_1 is open, and the calibration section is inactive.

During the calibration cycle, the output is compared against a linear ramp signal generated by an on-chip ramp-generator or integrator circuit. The error-detection circuitry generates an output pulse whose duration is proportional to the error (i.e., the difference between the linear ramp level and the converter output). These pulses are then converted into digital data by means of a simple pulse-counting A/D converter (see Section 15.3) and are stored in the memory to be used as the input to the correction converter during the subsequent conversion cycles. Normally, the correction would apply only to the first K bits b_1, b_2, ..., b_K of the converter ($K < N$). In that case, only these first K bits would be in sequence during the calibration cycle, generating a 2^K step staircase, where each output level would then be compared against the linear ramp waveform to generate the correction data.

This calibration technique is particularly effective for correcting linearity errors, rather than absolute accuracy. In the case of digital audio and sound processing, linearity rather than accuracy is critical. Therefore, the accuracy limitation does not present a problem.

A 14-bit version of the self-calibrating D/A converter system shown in Figure 14.45 has been integrated in monolithic form on a 4.1×5.2-mm^2 chip, using bipolar and integrated injection logic (I^2L) technology.[28] The circuit provides untrimmed resolution of 14-bits with $\pm\frac{1}{2}$ LSB linearity, and a conversion time of less than 2 μsec. In this particular implementation, an 8-bit correction converter was used, and the correction was applied to the first 5 major bits (i.e., $K = 5$) of the input data. This required the use of a random access memory of 2^5 words $\times$ 8 bits, or a total of 256 bits. The self-calibration cycle takes approximately 120 msec with 1-MHz external clock. A calibration cycle is normally performed automatically at prescribed intervals, such as once every hour.

Self-correcting precision circuits represent an added level of sophistication in monolithic IC design. As the level of required precision is increased, the circuit complexity tends to move from a circuit to a subsystem and finally to a system level, where both analog and digital design technologies "merge" together on the IC chip. The self-correcting circuits illustrate a clear trend in the future of VLSI design as a means of scaling the limits of precision.

REFERENCES

1. D. F. Hoeschele, *Analog-to-Digital and Digital-to-Analog Conversion Techniques*, Wiley, New York, 1968.

2. H. Schmid, *Electronic Analog/Digital Conversions*, Van Nostrand Reinhold, New York, 1970.

3. D. J. Dooley, Ed. *Data Conversion Integrated Circuits*, IEEE Press, New York, 1980.

4. E. R. Hnatek, *A User's Handbook of D/A and A/D Converters*, Wiley, New York, 1976.

5. B. M. Gordon, "Linear Electronic Analog/Digital Conversion Architectures, Their Origins, Parameters, Limitations and Applications," *IEEE Trans. Circuits and Systems* **CAS-25,** 391–418 (July 1978).

6. A. R. Hamadé, "A Single-Chip All-MOS 8-Bit A/D Converter," *IEEE J. Solid-State Circuits* **SC-13,** 785–791 (December 1978).

7. T. Redfern, J. J. Connoly, S. W. Chin, and T. M. Frederiksen, "A Monolithic Charge-Balancing Successive Approximation A/D Technique," *IEEE J. Solid State Circuits* **SC-14,** 912–919 (December 1979).

8. J. L. McCreary and P. R. Gray, "All-MOS Charge Redistribution Analog-to-Digital Conversion Techniques—Part I," *IEEE J. Solid-State Circuits* **SC-10,** 371–379 (December 1975).

9. D. H. Sheingold, Ed., *Analog-Digital Conversion Notes*, Analog Devices, Norwood, MA, 1977.

10. G. Kelson, H. Stellrecht, and D. S. Perloff, "A Monolithic 10-Bit Digital-to-Analog Converter Using Ion Implantation," *IEEE J. Solid-State Circuits* **SC-8,** 396–403 (December 1973).

11. D. T. Comer, "A Monolithic 12-Bit DAC," *IEEE Trans. Circuits and Systems* **CAS-25,** 504–509 (July 1978).

12. J. A. Schoeff, "Microprocessor Compatible High-Speed 8-Bit DAC," *IEEE J. Solid-State Circuits* **SC-13,** 746–753 (December 1978).

13. J. A. Schoeff, "An Inherently Monotonic 12-Bit DAC," *IEEE J. Solid-State Circuits* **SC-14,** 904–911 (December 1979).

14. J. A. Schoeff, "A Monolithic Companding D/A Converter," *Digest of Tech. Papers*, IEEE Int. Solid-State Circuits Conf., pp. 58–59, February 1977.

15. S. Kelley and R. Ulmer, "A Single-Chip CMOS PCM CODEC," *IEEE J. Solid-State Circuits* **SC-14,** 54–59 (February 1979).

16. Y. P. Tsividis, P. R. Gray, D. A. Hodges, and J. Chacko, "A Segmented μ-255 Law PCM Voice Encoder Utilizing NMOS Technology," *IEEE J. Solid-State Circuits* **SC-11,** 740–747 (December 1976).

17. K. B. Ohri and M. J. Callahan, "Integrated PCM CODEC," *IEEE J. Solid-State Circuits* **SC-14,** 38–46 (February 1979).

18. M. E. Hoff, J. Huggins, and B. M. Warren, "An NMOS Telephone CODEC for Transmission and Switching Applications," *IEEE J. Solid-State Circuits* **SC-14,** 47–53 (February 1979).

19. D. J. Dooley, "A Complete Monolithic 10-Bit D/A Converter," *IEEE J. Solid-State Circuits* **SC-8,** 404–408 (December 1973).

20. R. B. Craven, "An Integrated Circuit 12-Bit D/A Converter," *Digest of Tech. Papers*, IEEE Int. Solid-State Circuits Conf., pp. 40–41, February 1975.

21. R. Dobkin, "IC Reference has 1 ppm/°C Drift," National Semiconductor Appl. Note 181, 1977.

22. D. P. Laude and J. D. Beason, "5 Volt Temperature-Regulated Voltage Reference," *IEEE J. Solid-State Circuits* **SC-15,** 1070–1076 (December 1980).

23. *1981 Full Line Catalog*, Precision Monolithics, Santa Clara, CA, 1980, Sec. 11.

24. J. B. Cecil, "A CMOS 10-Bit D/A Converter," *Digest of Tech. Papers*, IEEE Int. Solid-State Circuits Conf., pp. 196–197, February 1974.

25. P. R. Holloway and M. P. Timko, "Circuit Techniques for Achieving High-Speed, High-Resolution A/D Conversion," *IEEE J. Solid-State Circuits* **SC-15,** 1040–1051 (December 1980).

26. T. S. Guy and L. M. Trythall, "A 16-Bit Monolithic Bipolar DAC," *Digest of Tech. Papers*, IEEE Int. Solid-State Circuits Conf., pp. 88–89, February 1982.

27. Z. Boyacigiller, B. Weir, and P. Bradshaw, "An Error-Correcting 14-b/20 μs CMOS A/D Converter," *Digest of Tech. Papers*, IEEE Int. Solid-State Circuits Conf., pp. 62–63, February 1981.

28. K. Maio, et al., "An Untrimmed D/A Converter with 14-Bit Resolution," *IEEE J. Solid-State Circuits* **SC-16,** 616–621 (December 1981).

CHAPTER FIFTEEN

DATA CONVERSION CIRCUITS: ANALOG-TO-DIGITAL CONVERTERS

The function of an A/D converter is to convert any analog quantity such as a voltage or a current into a digital word. Figure 15.1 shows the very simplistic block diagram of an A/D converter circuit. The A/D converters are a logical counterpart to the D/A converters discussed in the previous chapter, and they perform the opposite function. They can be thought of as encoding devices, where an analog quantity is converted into a digital word of prescribed bit count. The output of an A/D converter can be in either serial or parallel format. In the serial format, the digital data are delivered to the output serially, that is, 1 bit at a time, starting with the MSB. In the parallel format, the digital output is presented as a binary word available from N parallel terminals, each corresponding to 1 bit of the output digital word. In almost all applications, parallel output is preferred over serial output for speed and efficiency of data handling.

In many data-conversion applications, A/D converter circuits are used in conjunction with additional support or interface circuitry to enhance their capabilities. Figure 15.2 illustrates the functional block diagram of such an A/D conversion system. A single converter can be made to serve a number of separate analog channels by use of an analog multiplexer at its input. To avoid changing the analog input signal during the conversion process, a sample-and-hold, or track-and-hold, circuit is often used at the input of an A/D converter.

In many applications, the output of an A/D converter has to interface with a microprocessor data bus. This can be implemented by including a set of data latches at the output of the A/D converter, as shown in Figure 15.2, both to

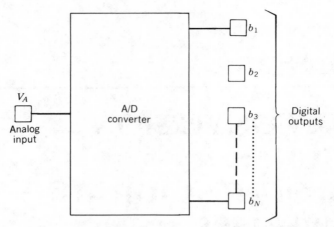

FIGURE 15.1. Symbolic block diagram of A/D converter.

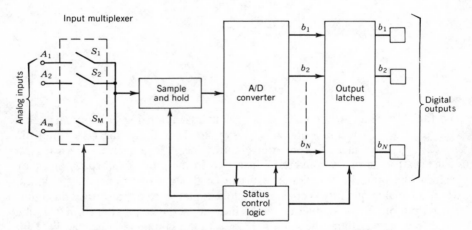

FIGURE 15.2. Complete A/D converter system with support circuitry.

buffer the outputs and to store the results of the previous conversion step, while a new conversion cycle is in progress. A complete A/D converter system also includes a significant amount of control logic, which would sense the end of the conversion cycle, and activate the output data latches, the sample-and-hold, and the input multiplexer circuitry.

The A/D conversion process is a natural evolution of the basic D/A converter techniques described in the preceding chapter. In fact, many of the A/D converter circuits employ a D/A converter as a subblock within the circuit. A wide range of A/D conversion techniques have been developed for a wide diversity of applications.[1-3] The choice of a particular conversion technique depends very strongly on a particular application.[4-6] In certain cases, the dominant parameters are precision and stability of conversion; in others, conversion speed is of importance. These diverse and often conflicting requirements very strongly influence the choice of the particular conversion technique used. As a general rule, precision is obtained at the expense of conversion rate.

Although a wide range of circuit techniques are known for A/D conversion, the great majority of monolithic IC A/D converters fall into one of the following categories:

1. Integrating A/D converters which charge or discharge a timing capacitor during the conversion cycle.

2. *Digital-ramp*, or *servo*, type converters which use a binary counter and D/A converter in a feedback loop.

3. Successive-approximation-type A/D converters which produce a digital output by a succession of trial-and-error steps.

4. Parallel, or *flash*, A/D converters which perform the conversion operation in a single step.

In the following sections, each of these classes of A/D converters will be examined separately, regarding their capabilities and limitations, and their applications. Particular emphasis will be placed on their implementation in monolithic IC form using both bipolar and MOS technologies.

15.1. FUNDAMENTALS OF A/D CONVERSION

The A/D converter encodes an analog input into a digital output of predetermined bit length. The analog input voltage V_A of Figure 15.1 is approximated as a binary fraction of a full-scale output voltage V_{FS}. Thus, the output of the converter corresponds to an N-bit digital word D given as

$$D = \frac{V_A}{V_{FS}} = \frac{b_1}{2^1} + \frac{b_2}{2^2} + \cdots + \frac{b_N}{2^N} \tag{15.1}$$

where b_1, b_2, ..., b_N are the binary bit coefficients having a value of either a 1 or a 0. The bit coefficients which form the digital data can be obtained from the output of the A/D converter, either simultaneously in the form of N parallel outputs, or serially with one bit coefficient at a time. In the latter case, bit coefficient b_1, corresponding to the MSB, is calculated and displayed first, then followed by bits of successively decreasing importance.

Figure 15.3 shows the ideal transfer characteristics of an A/D converter. For illustrative purposes, a simple 3-bit converter is assumed. However, all the definitions and characteristics are applicable to any A/D converter transfer function, irrespective of the bit count. The transfer function is discontinuous, and does *not* have a one-to-one correspondence between the analog input and the digital output. Instead, the output is a "quantized" version of the continuously variable analog input. As a result, each output *code* corresponds to a small range

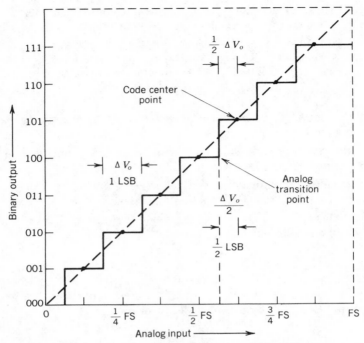

FIGURE 15.3. Ideal transfer characteristics of 3-bit A/D converter.

ΔV_O of analog input values. An N-bit A/D converter has 2^N output states and $2^N - 1$ transitions between states, as the analog input varies from zero to full-scale (FS) reading.

The smallest quantizing step, ΔV_O, between the discrete output levels corresponds to 1 LSB,

$$\Delta V_O = 1 \text{ LSB} = \frac{V_{\text{FS}}}{2^N} \tag{15.2}$$

and is the smallest code change that the converter can reproduce.

Similar to the D/A conversion described in Chapter 14 (see Fig. 14.3), the transfer function of an A/D converter stops 1 LSB short of its full-scale value, irrespective of its resolution. This is because the analog zero is one of the 2^N output states, leaving only $2^N - 1$ states for the converter output above zero. However, for simplicity and convenience, the analog range is always defined as the nominal full-scale voltage V_{FS}, rather than the actual one, which is 1 LSB less.

In an ideal A/D converter transfer characteristic, such as the one shown in Figure 15.3, the code center point is located at the analog level ideally corresponding to that particular code, and the code transitions take place $\pm\frac{1}{2}$ LSB away from this center point.

Effects of Quantization

An ideal A/D converter has an irreducible error associated with the conversion process, due to the quantization uncertainty. For example, in the ideal transfer characteristics of Figure 15.3, the converter cannot distinguish an analog difference less than ΔV_O at its output, and its output at any point can be in error by as much as $\pm\frac{1}{2}\Delta V_O$ (i.e., $\pm\frac{1}{2}$ LSB). This effect is referred to as *quantization error* or *quantization noise*, and is an inherent part of the A/D conversion process.

Figure 15.4 illustrates the presence of the quantization noise in the A/D conversion process. Figure 15.4*a* shows the block diagram of an ideal experiment where an analog signal is first quantized by an ideal N-bit A/D converter and then converted back to analog by an ideal N-bit D/A converter. The uncertainty in the A/D conversion process leads to a finite error voltage $V_\epsilon(t)$ to appear as the difference of the actual and the reconstructed analog signals. If the input

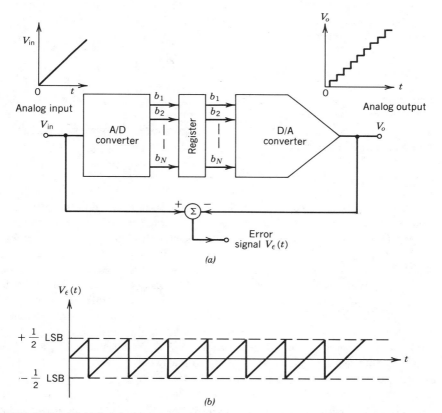

FIGURE 15.4. Digitizing and reconstructing analog signal to illustrate effect of quantization noise: (*a*) Digitizing and reconstructing analog signal; (*b*) quantization noise in reconstructed analog signal.

is a ramping analog signal, then the output would be a staircase voltage with 2^N steps. The difference of the two signals would then correspond to the triangular quantization uncertainty, or noise signal, shown in Figure 15.4b, which in the case of an ideal converter has a peak-to-peak value of 1 LSB.

As with most noise sources, the average value of the quantization noise is zero. However, its rms value can be determined from the triangular wave shape of Figure 15.4b as[7]

$$E_n(\text{rms}) \approx \frac{\Delta V_O}{\sqrt{12}} = \frac{V_{FS}}{2^N \sqrt{12}} \qquad (15.3)$$

Note that since quantization noise is proportional to ΔV_O, it decreases by a factor of 2 (i.e. 6 dB) for each additional bit of resolution.

Dynamic Range

The dynamic range of an A/D converter is the measure of the largest and smallest analog signals that can be handled by it. Since the largest signal is equal to V_{FS} and the smallest signal that can be quantized is equal to 1 LSB, the dynamic range can be expressed as

$$\text{Dynamic range} = \frac{V_{FS}}{\Delta V_O} = 2^N \qquad (15.4)$$

and is normally stated in decibels. Note that dynamic range increases by 6 dB for each bit of added resolution. For example, an 8-bit converter provides a dynamic range of 48 dB, whereas a 10-bit converter gives a dynamic range of approximately 60 dB.

Effects of Finite Conversion Time

The A/D conversion process requires a finite amount of time due to switching delays, integration cycles, and settling times associated with the converter operation. Thus, the conversion rate of the system in terms of the number of conversions per second is fixed by system design. It may vary from extremely slow systems with less than one conversion per second to very-high-speed converters which can perform $\approx 10^7$ conversions per second. The particular choice depends very strongly on the system application, and this choice in turn determines the particular converter architecture and conversion technique used. As a general rule, the greater the required accuracy and resolution, the slower is the speed of conversion, since more time is required for system transients to settle to within the prescribed accuracy limits.

The conversion time T_X required for the completion of one conversion cycle is called the *aperture time* of the converter. Thus, for example, an A/D converter capable of performing 2000 conversions per second will have an aperture time of 500 μsec. If the input analog signal varies as a function of time, the presence of the finite aperture time leads to an additional uncertainty in the encoded signal. This uncertainty is called the *aperture error*.

For example, if the input is a linearly varying function of time, the aperture error ΔV_X can be related to the analog input as

$$\Delta V_X = \frac{dV_A}{dt} T_X \qquad (15.5)$$

where T_X denotes the aperture time. Thus, as the frequency content of the input increases, the aperture error due to the finite conversion rate increases quite rapidly.

Figure 15.5 illustrates the effect of amplitude uncertainty in the analog input level due to aperture error. As an example, it is illustrative to examine the possible aperture error in converting a sinusoidal input signal at frequency f and amplitude E_A,

$$V_A(t) = E_A \sin(2\pi ft) \qquad (15.6)$$

which would have its maximum slope at its zero crossing point, where

$$\left. \frac{dV_A}{dt} \right|_{max} = 2E_A \pi f \qquad (15.7)$$

From Eq. (15.7), the corresponding aperture error would be

$$\Delta V_X = 2\pi f E_A T_X \qquad (15.8)$$

If an A/D conversion with N-bit resolution is required, then the maximum aperture error ΔV_X cannot exceed 1 LSB (or 1 part in 2^N) of the full-scale swing. Assuming a sinusoidal signal with full-scale peak-to-peak swing (i.e., $V_{FS} = 2E_A$), then the maximum allowable aperture time T_X for N-bit resolution would be, from Eq. (15.8),

$$T_X = \frac{\Delta V_X}{2E_A} \frac{1}{\pi f} = \frac{1}{2^N} \frac{1}{\pi f} \qquad (15.9)$$

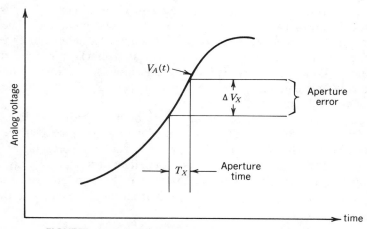

FIGURE 15.5. Amplitude uncertainty due to aperture error.

From this result, the aperture time required to digitize a 1-kHz sine wave, with 10-bit resolution, would be $\approx$320 nsec. Although 1 kHz is a relatively modest signal frequency, a 10-bit A/D converter with 320-nsec response falls outside the state-of-the-art in A/D converters. In practice, this problem is avoided by using a sample-and-hold or track-and-hold circuit at the input of the converter for almost all signals except those with very low frequency content.

Even when a sample-and-hold circuit is used ahead of an A/D converter, the necessary conversion rate is still determined by the frequency content of the input analog signal. Assuming that a sample-and-hold circuit is used at the input to sample the analog signal and hold its level constant during the conversion process, the sampling frequency of the input signal must be at least twice as high as its highest frequency component. This minimum sampling rate, called the *Nyquist rate*, is an outcome of the sampling theorem, as described in Chapter 13 (see Section 13.5). Therefore, as an example, if an A/D converter is to be used for quantizing a signal with an information bandwidth from dc to 1 kHz, it must be able to perform a minimum of 2000 conversions per second.

Nonideal A/D Converters

Similar to the case of D/A converters discussed in Section 14.3, practical A/D converters also exhibit a number of additional error sources due to the nonideal device characteristics and component mismatches. The most common among these are the offset, gain, linearity, and differential nonlinearity errors. The effects of these error sources on the A/D converter characteristics are illustrated in Figures 15.6 and 15.7 in terms of a simple 3-bit converter characteristic. In a practical A/D converter, all of these errors are present simultaneously.

The offset error is defined as the amount by which the actual code center line misses the origin of the transfer characteristics (see Fig. 15.6a). The code center line is a hypothetical line connecting the center points of each of the code transitions. Similarly, the gain error is defined as the difference in the slope of the actual and the ideal code center lines (see Fig. 15.6b). In practice, it is easier to measure the locations of code transitions, rather than center points. Therefore, the offset error is normally measured as the amount by which the first transition differs from $+\frac{1}{2}$ LSB. Similarly, the gain (or scale-factor) error is measured as the difference between the first and the last transitions in the transfer characteristics. Ideally, this separation would be equal to $V_{FS} - 2$ LSB, and any deviation from that would be the gain error.

The linearity error indicates the curvature of the code center line from a straight line drawn through the end points of the transfer characteristics (i.e., between zero and V_{FS} readings). It is expressed as either a percentage of V_{FS} or a fraction of LSB. Figure 15.7a illustrates the effect of the linearity error on an A/D converter transfer characteristic. This type of linearity error is also called an *integral nonlinearity* error, to separate it from the differential nonlinearity error.

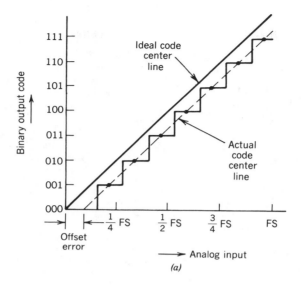

(a)

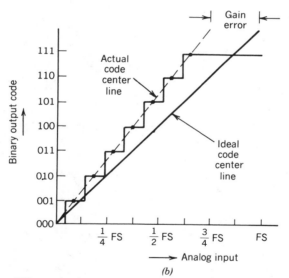

(b)

FIGURE 15.6. Offset and gain errors in A/D converters: (*a*) offset error; (*b*) gain error.

Differential nonlinearity is the measure of nonuniform step size between adjacent code transitions. Ideally, this step size is 1 LSB. Differential nonlinearity is the deviation of each differential step size from this value. For example, if an A/D converter is specified to have a differential nonlinearity of $\pm\frac{1}{2}$ LSB, then the minimum and maximum step sizes between adjacent transitions would be $\frac{1}{2}$ LSB and $1\frac{1}{2}$ LSB, respectively. If the differential nonlinearity

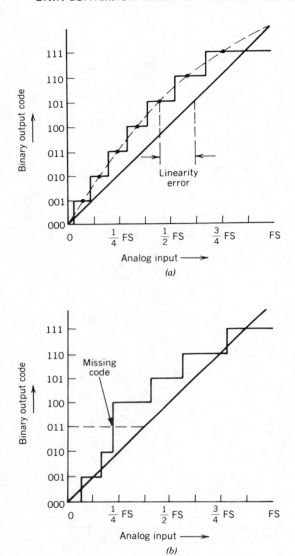

FIGURE 15.7.   Linearity and differential nonlinearity errors in A/D converters: (*a*) Linearity error; (*b*) differential nonlinearity error.

exceeds ± 1 LSB, then one or more of the digital codes may be skipped at the output, as illustrated in Figure 15.7*b*. Such missing codes cannot be allowed in most converter applications, since they lead to erroneous results or system instability.

Differential nonlinearity does not present a problem in integrating-type A/D converters. However, it requires special attention in successive-approximation and digital-ramp or staircase type A/D converters, which use a parallel input D/A converter in a feedback path (see Figs. 15.14 and 15.15). In such a case, potential nonmonotonic behavior of the D/A converter can result in missing codes at the digital output. The missing codes are most likely to occur at the so-called major carry points in the transfer characteristics, where the MSB or the next significant bit is changing sign.

The gain and offset errors can be trimmed externally. However, linearity and differential nonlinearity errors cannot be eliminated by adjustment. They can be minimized by improving the matching and tracking of circuit components, or by the proper choice of the circuit configuration.

15.2. INTEGRATING-TYPE A/D CONVERTERS

Integrating A/D converters perform the A/D conversion in an indirect manner. The analog input is first converted to a timing pulse whose duration is proportional to the analog voltage V_A. The duration of the timing pulse is then measured in a digital format by counting the number of cycles of a stable reference frequency (i.e., the clock signal) between the beginning and the end of the timing pulse. Because of this basic principle of operation, such converters are often called *indirect* or *pulse-width-modulating* converters.

Figure 15.8 shows an illustrative example of a simple integrating A/D converter system. The operation of the converter is as follows. Before the start of

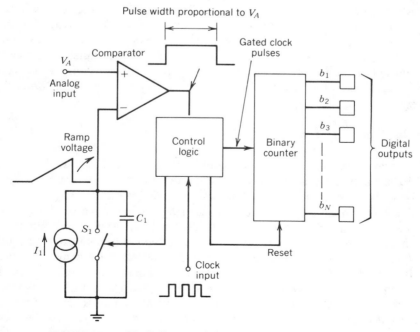

FIGURE 15.8.   Block diagram of simple integrating A/D converter.

the conversion cycle, the counter is reset to zero and the switch S_1 is closed. When the conversion cycle is started, the switch S_1 is opened, and the current source I_1 generates a linearly rising ramp voltage across the integrating capacitor C_1. During this time, the counter starts counting the input clock cycles. When the linear ramp across C_1 reaches the level of the analog input V_A, the comparator changes state, stops the counter, and ends the conversion cycle. The final count in the counter is then the digital equivalent of the analog signal V_A. For example, if it takes a total count N_T to reach the full-scale output V_{FS}, then the count n in the counter when the ramp reaches V_A is

$$n = N_T \frac{V_A}{V_{FS}} \qquad (15.10)$$

For N-bit resolution, $N_T = 2^N$, and the accumulated count n in the counter is

$$n = \frac{2^N}{V_{FS}} V_A \qquad (15.11)$$

The accumulated count n in the counter is quantized in increments of unit clock cycles. Therefore, each additional clock cycle corresponds to 1 LSB increment in V_A. The accumulated count is normally displayed as the digital states of the N-stage binary counter. The first counter stage output which changes state with every clock input corresponds to the LSB, and the last counter stage which changes state with every 2^{N-1} clock cycles corresponds to the MSB output.

Although the simple counter circuit of Figure 15.8 illustrates the basic principle of operation of integrating A/D converters, it is seldom used in practice because of inaccuracies associated with the initial ramp startup point, as well as with the control of the absolute values of I_1 and C_1. Some of these errors can be eliminated to a large degree by using more complex circuit techniques. Two such circuit techniques used for this purpose are the *single-slope* and the *dual-slope* integration methods.

Single-Slope A/D Converters

The single-slope converter is essentially identical to the basic circuit of Figure 15.8, except that the ramp voltage does not start exactly at zero, but at a slightly lower voltage level. Figure 15.9 shows the functional block diagram and the timing waveforms associated with a typical single-slope converter system. Prior to the start of the conversion cycle, the switch S_1 across the integrating operational amplifier A_1 is closed, and the integrator output is clamped to a negative voltage $-V_1$ at a level slightly below zero. At the start of the conversion cycle, S_1 is opened, and the integrator output ramps in the positive direction with a slope equal to $1/R_1C_1$. When this ramp reaches zero, Comparator 1 changes state and gates the clock signal into the counter. The clock pulses are counted, and the count is accumulated in the counter until the ramp level reaches the analog

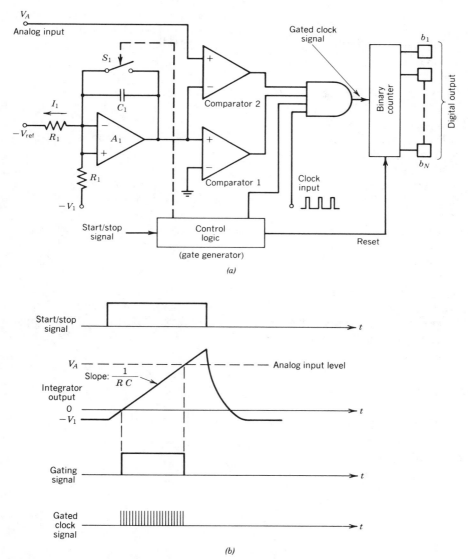

FIGURE 15.9. *Ramp-pick-off* or single-slope-type integrating A/D converter. (*a*) Basic block diagram. (*b*) Timing waveforms.

input level V_A. At that point, Comparator 2 changes state and gates off, the clock signal going into the counter. The accumulated count n in the counter is the digital equivalent of the analog signal V_A, as given by Eq. (15.11).

The single-ramp converter has one basic drawback. Assuming that the comparator, the integrator, and the voltage reference characteristics are ideal, the system accuracy depends strongly on the absolute value of the R_1C_1 product which determines the scale-factor accuracy and stability. This is inherent in all single-slope-type converters, and can be avoided by using the dual-slope conversion techniques described in the next section.

Dual-Slope A/D Converters

The dual-slope-type converter is one of the most popular types of integrating A/D converters. It eliminates most of the error sources and component tolerance requirements associated with the single-slope circuits. Figure 15.10 shows the functional block diagram of a dual-slope converter. The system operates by integrating the unknown analog signal V_A for a fixed period of time (i.e., for a fixed count). The resulting integrator output level is then returned to zero, by integrating a known reference voltage of opposite polarity. The length of time required for the integrator output to return to zero, as measured by the number of clock cycles gated into a counter, is proportional to the value of the input signal, averaged over the integration period.

With reference to the block diagram of Figure 15.10, the principle of operation can be illustrated as follows. Prior to the start of conversion, the switch S_2 is closed, the integrator output voltage V_X is clamped to ground, and S_1 is connected to the analog voltage $-V_A$. For illustrative purposes, the analog input is assumed to be negative and varying between zero and $-V_{FS}$. At the start of

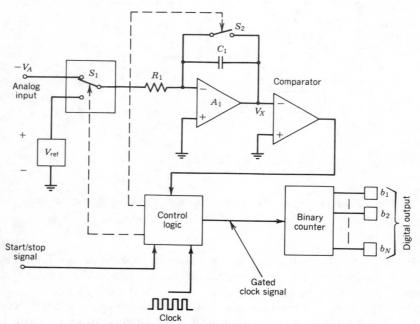

FIGURE 15.10. Block diagram of dual-slope A/D converter.

the first phase of the conversion cycle, the switch S_2 is opened, and the input signal is integrated for a predetermined time period, normally equal to 2^N clock cycles. During this time period, the integrator output voltage V_X ramps up with the slope,

$$\left(\frac{dV_X}{dt}\right)_{\text{phase I}} = \frac{+V_A}{R_1 C_1} \qquad (15.12)$$

At the end of 2^N clock cycles, the counter is reset to zero, switch S_1 is connected to V_{ref}, and the integrator output ramps down with the slope

$$\left(\frac{dV_X}{dt}\right)_{\text{phase II}} = \frac{-V_{\text{ref}}}{R_1 C_1} \qquad (15.13)$$

During this phase, the clock cycles are accumulated in the counter until the ramp reaches back to zero, and the comparator changes state to stop the conversion. Then the count n stored in the counter during this reference integration phase is the digital equivalent of the analog voltage

$$n = -V_A \frac{2^N}{V_{\text{ref}}} \qquad (15.14)$$

Figure 15.11 shows the integrator output waveforms during the two phases of the converter operation for various values of analog input voltage. Note that the signal integration phase is associated with a fixed count and variable ramp slope depending on the value of the analog input. The reference integration phase is associated with a fixed ramp slope, but a variable count n. Since all ramp-down integration paths have the same slope during phase II, longer ramp-up signals due to large values of analog input take proportionally longer to return to zero during the ramp-down phase. Thus, the duration of the ramp-down phase, and the count accumulation during it, are both proportional to the value of the analog input.

The dual-slope conversion technique has some unique advantages for high-resolution converter design. Some of these are the following:

1. The conversion accuracy is independent of both the value of the integrator time constant (i.e., the RC product) and the clock frequency, since these parameters affect the ramp-up and ramp-down times equally, as long as their values remain stable during the integration cycle. In this manner, long-term drifts due to time or temperature effects are largely avoided.

2. The linearity is excellent, determined primarily by the quality of the integrator ramp waveform. Differential nonlinearity is virtually eliminated since the analog function and the ramp waveform remain continuous during the conversion process. The only contribution to differential nonlinearity is the clock jitter during the counting interval.

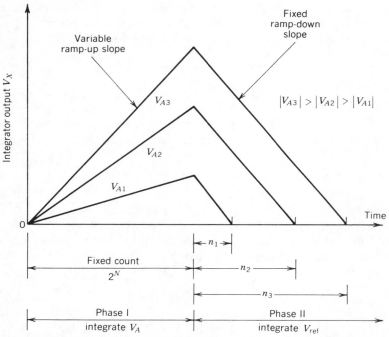

FIGURE 15.11. Integrator output waveforms in dual-slope A/D converter.

3. The power-line noise rejection characteristics are excellent. One of the limitations to high resolution in line-operated instruments is the presence of noise at the power-line frequency or its integer multiples. This noise normally appears superimposed on the input signal, and has zero average value. During the conversion process, the input signal is integrated for a fixed period of time (i.e., for the duration of phase I of Fig. 15.11). If this period is made equal to the period of the power-line signal (i.e., 16.67 msec for 60-Hz line) or its integer multiples, then the power-line noise is greatly attenuated during the integration process. In practical designs, power-line noise can be attenuated by as much as 70 dB using this technique.[7]

Because of the advantages listed above, the dual-slope conversion technique has become one of the most commonly used high-resolution A/D conversion methods for low-speed converter applications. Using auto-zero techniques (see Figs. 7.46 and 7.47) to reduce operational amplifier and comparator offsets to well below 1 mV, it is possible to build monolithic A/D converters in the 12- to 14-bit resolution and linearity range. At present, a number of monolithic A/D converter products are available which use the dual-slope technique or its derivatives[7, 8] which are fabricated with CMOS technology.

Charge-Balancing Converters

Although the charge-balancing A/D converter belongs to the family of integrating converters, its basic principle of operation is quite different from that of single- or dual-slope-type converters. It operates on the principle of balancing or "canceling" a charge applied to the input terminal of an integrator by generating an equal amount of charge of opposite polarity which is supplied in discrete *quantum* packets.

The principle of operation of a charge-balancing A/D converter is conceptually very similar to the case of the voltage-to-frequency (V/F) converter configurations discussed in Chapter 11 (see Fig. 11.55). First, an analog signal is converted to a periodic pulse train whose pulse-width is fixed and whose repetition rate is proportional to the analog signal. Next, the number of such pulses per unit of time interval are counted by a binary counter to form a digital output.

Figure 15.12 shows the conceptual block diagram of a charge-balancing A/D converter. Prior to the start of the conversion cycle, the switch S_2 is closed, and S_1 is switched to ground. During the conversion cycle, S_2 is opened, and the integrator starts generating a negative-going ramp whose slope is proportional to the input current, $I_1(=V_A/R_1)$. When the negative ramp at the integrator output (node B) causes the comparator to change state, the switch S_1 is activated for a time duration of t_1 corresponding to one-half clock cycle, and an amount of charge Q_0 is extracted from the integrator summing node (node A), where

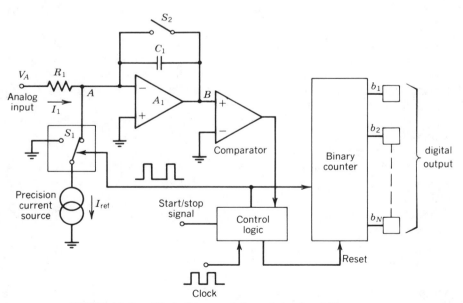

FIGURE 15.12. Block diagram of charge-balancing A/D converter.

$$Q_0 = I_{\text{ref}} t_1 \qquad (15.15)$$

This charge packet momentarily subtracts from the analog input current I_1 and causes the integrator output to ramp up again. In this manner, the integrator output is constantly forced down with a steady input current I_1, and is intermittently forced up with the internal current I_{ref} applied for discrete intervals of duration t_1. At the equilibrium condition, the net charge accumulated at node A is identically equal to zero. Over a conversion period of N_t clock cycles, the total charge Q_{in} supplied by the current I_1 is

$$Q_{\text{in}} = I_1 N_t 2 t_1 = \frac{V_A}{R_1} 2 N_t t_1 \qquad (15.16)$$

During the same time interval, a charge Q_{out} is extracted from the same node by n pulses of the intermittent current source I_{ref}, where

$$Q_{\text{out}} = n Q_0 = I_{\text{ref}} n t_1 \qquad (15.17)$$

Equating the net input charge to the outgoing charge, from Eqs. (15.16) and (15.17), one obtains

$$n = \frac{V_A}{R_1 I_{\text{ref}}} 2 N_t \qquad (15.18)$$

Setting $I_{\text{ref}} R_1$ equal to $V_{\text{FS}}/2$, and choosing the total number of clock pulses N_t available during the conversion step equal to 2^N for N-bit resolution, one can rewrite Eq. (15.18) as

$$n = \frac{2^N}{V_{\text{FS}}} V_A \qquad (15.19)$$

which is of the same form as Eq. (15.11). Then, accumulating the total count n of the clock cycles which activate the switch S_1 in the binary counter, one obtains the digital equivalent of the analog input. Note that each unit packet of charge Q_0, delivered from the reference current source, corresponds to 1 LSB increment of the digital output.

To control the accuracy of the charge packet Q_0, symmetrical clock pulses with 50% duty cycle are required. In practice, this is achieved by dividing down the clock frequency by an R–S flip-flop.

Figure 15.13 illustrates a practical method of implementing the switching current reference of Figure 15.12, using a voltage follower and resistor R_2, which is switched between $-V_{\text{ref}}$ and ground. Equating the incoming and outgoing charge at the integrator input node, under equilibrium conditions, the number of charge packets n can be expressed as

$$n = \frac{2 R_2}{R_1} \frac{2^N}{V_{\text{ref}}} V_A \qquad (15.20)$$

Setting the full-scale output V_{FS} to be equal to

$$V_{\text{FS}} = V_{\text{ref}} \frac{R_1}{2 R_2} \qquad (15.21)$$

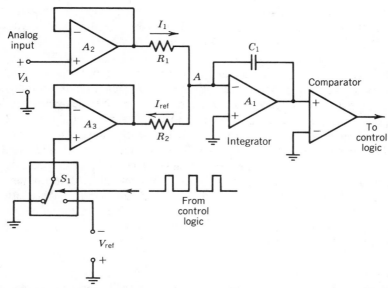

FIGURE 15.13. Method of implementing the switching current reference of Figure 15.12.

Eq. (15.20) becomes the same as Eq. (15.19). Note that since R_1 and R_2 appear as a ratio in the transfer function, only their matching and tracking is critical.

The basic circuit configuration of Figure 15.13 is particularly suited to circuit design with CMOS technology, using internal offset correction (i.e., auto-zero techniques) to eliminate operational amplifier and comparator offsets.[9]

The charge-balancing A/D conversion technique has some of the advantages of the dual-slope conversion. The dependence on absolute values of R and C is avoided, and due to a fixed-interval integration process, the effects of periodic power-line noise can be significantly reduced. However, the charge-balance conversion technique is inherently slower and requires careful control of clock signal symmetry. Therefore, it does not lend itself to the same degree of precision as the dual-slope technique.

Error Sources in Integrating A/D Converters

Major error sources in integrating A/D converters are the offset voltages associated with the detection and integration circuitry, nonlinearity of the integrator output, and comparator response time. In addition to these inherent errors, other error sources, such as the stability and accuracy of the voltage reference and the external R and C components for the integrator section also affect the accuracy of conversion.

The effect of the reference voltage inaccuracy is an additive error term, present in every A/D and D/A converter. The effect of external R and C values on circuit accuracy can be greatly reduced by using self-compensating techniques, such as the dual-slope conversion. In this discussion, we will briefly examine the effects and magnitudes of the remaining error sources, which are inherent to the integrating-type A/D converter operation.

Offset Errors.   The offset voltages of operational amplifier and comparator circuits would appear superimposed on the analog signal input, and thus cause errors in the detected or integrated voltage levels. In MOS designs where offset voltage levels are in the range of ± 10–± 30 mV, these errors would correspond to approximately 0.3% of V_{FS} for a 10-V full-scale signal. This would correspond to greater than $\pm\frac{1}{2}$ LSB error for 8-bit or higher resolution levels. Fortunately, in MOS technology where sampling switches and high-impedance levels are available, one can use self-correcting sample-and-hold techniques (i.e., so-called auto-zero methods) to reduce the apparent level of these offsets to approximately 1 mV (see Figs. 7.46 and 7.47). Thus, using auto-zero techniques, offset errors can be maintained to within $\pm 0.01\%$ of V_{FS}, which is suitable for 12- to 14-bit resolution.

Integrator Linearity Error.   The nonlinearity in the integrator output is primarily due to the finite open-loop gain A_o of the integrating operational amplifier. The error voltage ΔV_I between the ideal and the actual integrator outputs can be approximated as[10]

$$\Delta V_I = (V_{\text{out}})_{\text{ideal}} - (V_{\text{out}})_{\text{actual}} \approx V_{FS}\frac{t_I}{2A_oR_1C_1} \tag{15.22}$$

where t_I is the total integration time, and R_1 and C_1 are the integrator time constants. Normally, (R_1C_1) product is set equal to t_I, resulting in

$$\Delta V_I = \frac{V_{FS}}{2A_o} \tag{15.23}$$

To keep the linearity error under $\pm\frac{1}{2}$ LSB in a 12-bit integrating converter, a ramp nonlinearity of less than 0.01% of V_{FS} is required. Equation (15.23) implies that under such circumstances, the minimum operational amplifier gain A_o must be in excess of 5000 (i.e., ≈ 74 dB).

Comparator Response Time Errors.   The comparator section must have enough gain to switch fully with an input level of 1 LSB. For a 12-bit converter with $V_{FS} = 10$ V, this implies that the comparator must fully switch with ± 2.5 mV input differential, and that such switching must be accomplished within one clock cycle. Any finite delay t_d in comparator switching causes a resolution error ϵ_d in units of LSB,

$$\epsilon_d = \frac{t_d}{T_{ck}} \quad \text{LSB} \tag{15.24}$$

where T_{ck} is the period of the clock frequency. Typical CMOS comparator stages have response times of approximately 0.5–1 μsec, with 2.5-mV input drive.

Thus, from Eq. (15.24), to maintain 12-bit resolution, the maximum allowable clock rate must be limited to under 1 MHz.

Characteristics and Applications

From the previous discussions, the basic characteristics of integrating A/D converters can be summarized as follows:

1. *Low Conversion Rate.*   A total of 2^N clock cycles would be required for a full-scale conversion. For a 10-bit converter operating with 1-MHz clock frequency, this would result in a conversion time of approximately 1 msec, or a rate of about 1000 conversions per second.

2. *No Missing Codes.*   Since all clock cycles are accumulated in the counter during the conversion process, all digital codes between zero and full scale will be present at the output.

3. *High Linearity.*   Linearity is primarily determined by the linearity of the ramp signal. By proper design, the ramp nonlinearity error can be reduced to less than 0.01%.

4. *High Resolution.*   Within the limits of ramp linearity and comparator-sensing capability, the resolution can be increased by increasing the total clock cycles N_T. Since the maximum clock frequency is fixed by the comparator response time and the control logic capability, the only way to increase N_T is by increasing the ramp time, which decreases the conversion rate. Thus, in general, the conversion rate varies inversely with the resolution.

5. *Noise Rejection.*   Integrating converters have excellent noise-rejection characteristics for periodic noise sources whose frequency is an integral multiple of the conversion rate. For example, 60- and 120-Hz power-line noise can be greatly attenuated by choosing the conversion rate to be 30 and 60 conversions per second, respectively.

Because of these characteristics, integrating A/D converters are primarily suited for high-accuracy and low-speed measurements of slowly varying signals. Their prime areas of application are in digital panel meters and multimeters, as well as in remote-area data-acquisition and monitoring systems. In digital panel-meter applications, the counter outputs are normally brought out in binary-coded decimal (BCD) format, rather than as a simple binary code, so that they can directly interface with numerical displays.

The circuit configuration of integrating A/D converters is best suited to the fabrication with CMOS process technology. This is because the high-impedance nature of CMOS devices and the availability of analog switches greatly simplify the system architecture and permit on-chip offset correction and sample-and-hold capability. The design of digital control logic and the counter circuitry is also simplified by using CMOS technology, due to the high functional density of MOS devices on the IC chip. An important added benefit of the use of CMOS technology is the significant reduction in power dissipation. As a result, almost all of the commercially available integrating-type A/D converters are fabricated with CMOS technology.

15.3. DIGITAL-RAMP-TYPE A/D CONVERTERS

Digital-ramp-type A/D converters operate by using a combination of a binary counter and a D/A converter in a feedback loop around a voltage comparator. The binary counter and D/A converter combination is used to generate a digital ramp or staircase output, which is compared to the level of the analog input signal. There are two main classes of digital-ramp-type converters: (1) tracking or servo type converters; and (2) staircase converters. The principles of operation of both converter types are very similar; they only differ in circuit complexity.

Tracking Converter

Figure 15.14 shows the functional block diagram of a tracking or servo type A/D converter. It operates with a clock signal which is counted by the binary up-down counter. Whether the counter is counting up or down is set by the polarity of the control input from the voltage comparator. The binary outputs of the counter drive a D/A converter whose analog output V_O is connected to one of the comparator inputs. At the start of the initial conversion cycle, the system counts

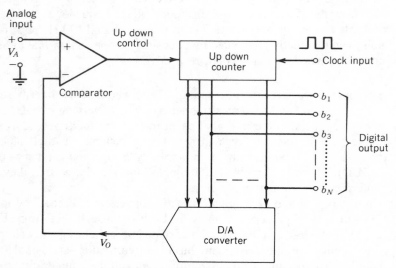

FIGURE 15.14. Functional block diagram of tracking A/D converter.

the clock pulses. At the output of the D/A converter, the binary count is converted to a staircase output, corresponding to the uniformly increasing sequence of binary codes. This analog level V_O is continuously compared with the analog input level. When it reaches the analog input level V_A, the comparator changes state and stops the count. The counter output code necessary to set $V_O = V_A$ is the digital word corresponding to the analog input V_A. When the next cycle of A/D conversion is initiated, the counter does not start from zero. Instead, it retains the previous count and starts incrementing or decrementing the count, based on the polarity of the error signal from the comparator. The conversion is again completed when V_O reaches V_A and causes the comparator to change state.

For full-scale digital output this type of converter is quite slow, since $2^N - 1$ clock cycles need to be counted. However, it has a very rapid small-signal response. For example, if the input varies only a small amount, the output can follow it within a few clock cycles. The term *tracking* or *servo* type converter stems from this property. Note that in Figure 15.14, the reference voltage V_{ref} is not shown separately, but is assumed to be an integral part of the D/A converter.

Staircase Converter

This is a simplified version of the tracking converter. Instead of an up–down counter, a simple one-directional binary counter is used. In each conversion step, the counter is reset to zero and starts counting the clock pulses. This creates a staircase output from the D/A converter. The count is stopped when the D/A converter output equals the analog input, and the last count present in the counter corresponds to the digital word for the analog input. Unlike the tracking converter, the staircase-type A/D converter resets itself to zero after every conversion step. Thus, it does not have the rapid small-signal response properties of a tracking converter. However, since only unidirectional count is used, the counter circuitry is somewhat simplified.

15.4. SUCCESSIVE-APPROXIMATION A/D CONVERTERS

The successive-approximation converter is a feedback system which operates on a trial-and-error technique to approximate an analog input with a corresponding digital code. The system is comprised of a so-called successive-approximation register (SAR) and a D/A converter in feedback around a voltage comparator, as shown in the functional block diagram of Figure 15.15. With reference to the block diagram, the operation of the circuit can be described as follows. Prior to the start of the conversion process, the N-bit shift register and the holding register which form the successive-approximation register section are cleared. In the first step of conversion, a 1 is inserted as a trial bit for the MSB in the holding register, with the rest of the bits remaining at 0. If the resulting analog output

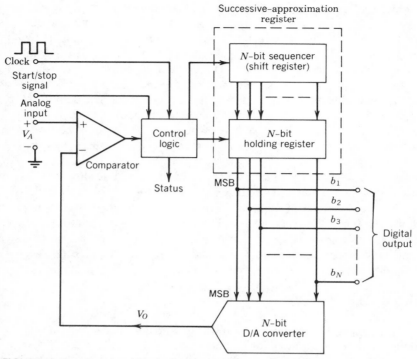

FIGURE 15.15. Functional block diagram of successive-approximation A/D converter.

V_O of the D/A converter is $\leq V_A$, the output state of the comparator remains unchanged, and the 1 is retained for the MSB; otherwise it is replaced by a 0. Then, in the next cycle, a 1 is tried for the second significant bit. If the comparator output does not change state, it is retained; otherwise it is replaced by a 0. In this manner, the approximation process continues until all bits are calculated in N successive cycles.

The decision-making logic, whether to leave the trial 1 or replace it with 0 during each trial cycle, is performed by the comparator and the successive-approximation register logic. The trial bit is shifted sequentially from the MSB toward the LSB by means of the N-bit sequencer section of the successive-approximation register, and the results of each successive approximation are retained in the holding register. The control logic performs the start/stop function for each approximation step, as well as for the entire sequence, in synchronization with a clock signal. The digital output, which corresponds to the data stored in the holding register, is not valid until the entire conversion step is complete and all bits, MSB through LSB, have been evaluated. When this is complete, the control logic provides a "status" signal, which allows the digital output to be read as valid data.

Figure 15.16 illustrates the decision sequence for a 3-bit successive-approximation A/D converter.[11] The heavy dashed line shows the decision steps

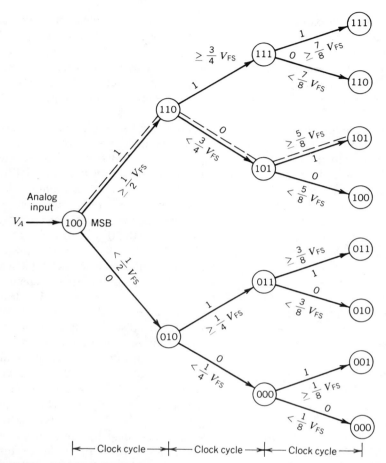

FIGURE 15.16. Decision diagram for 3-bit successive-approximation A/D converter. (*Note:* Dashed line indicates sequence of decision process leading to a 101 code.)

for a conversion sequence resulting in the 101 output code. It can be readily shown that only N successive-approximation steps are needed for N-bit resolution. Since each approximation step takes one clock period, the entire conversion process is achieved in N clock cycles. This is a very major speed improvement over integrating or digital-ramp type converters, which take up to 2^N clock cycles to complete a conversion step.

The successive-approximation circuits are among the most widely used A/D converter types. A large majority of monolithic A/D converters also use this technique. The popularity of the successive-approximation converter is due to its high speed and relatively high resolution. The response speed and the resolution of a successive-approximation converter are primarily limited by the speed, resolution, and linearity of the D/A converter section in its feedback path. Using high-speed bipolar logic and current output D/A converter subsections, successive-approximation converters can operate with clock rates of about 1 MHz, and provide conversion times in the range of 8–40 μsec for 8- to 12-bit resolution range.

One of the essential requirements in successive-approximation converter design is the monotonicity of the D/A converter subsection. Any nonmonotonic behavior would result in a missing code (i.e., erroneous reading) in the digital output.

The successive-approximation technique requires the presence of high-speed logic and low-level precision linear circuitry on the same monolithic chip. Thus, it is an excellent example of a complex system function which requires the combination of both analog and digital functions, with the same fabrication technology. At present, a wide choice of commercially monolithic A/D converters, or converter-related products, are available which utilize successive-approximation techniques.[6–8] The high-speed and high-precision circuits, such as those with 10-bit and higher resolution and with 20-μsec or less conversion time, normally use bipolar or combined bipolar/I²L technology. The lower speed and lower resolution circuits are normally implemented using MOS technology. In general, bipolar designs are often used for general-purpose "stand-alone" converters, whereas MOS technology is more often used in designing "dedicated" converters, which function as a subsection of a larger digital system.

Any of the basic D/A converter architectures discussed in Section 14.2 can be used in implementing the D/A subsection of a successive-approximation converter. The main choice depends on the fabrication technology available. Circuits fabricated with bipolar process technology normally use a current-output D/A converter with binary-weighted current sources (see Section 14.4).

Figure 15.17 shows an example of a commercially available monolithic bipolar 10-bit A/D converter (Analog Devices AD-571). The monolithic integrated circuit contains all of the essential functional blocks of a successive-approximation A/D converter system on a single IC chip.[12] The circuit chip, measuring 120 × 151 mils (3 × 3.8 mm) has its analog portions, that is, the D/A converter, the voltage comparator, and the voltage reference sections, designed with conventional linear bipolar devices. Bipolar technology is also used for the output interface devices. The bulk of the digital logic section forming the successive-approximation register portion is implemented with bipolar compatible I²L logic in order to conserve chip area and minimize power dissipation. The 10-bit current output D/A section of the chip is designed using a 10-stage binary-weighted current source array, with a silicon-chromium (SiCr) R-$2R$ resistor ladder, which is laser trimmed at the wafer-sort stage for prescribed linearity and monotonicity. The voltage reference section is implemented as a buried-Zener reference (see Section 4.7), which is then laser trimmed for scale-factor accuracy. The clock signal is generated by a simple nine-stage ring oscillator circuit made up of I²L inverter stages. The oscillator output is then passed through a flip-flop to form a symmetrical square-wave clock signal. Unlike the other A/D conversion techniques, the accuracy of the clock signal is not critical for successive-approximation converters, since the

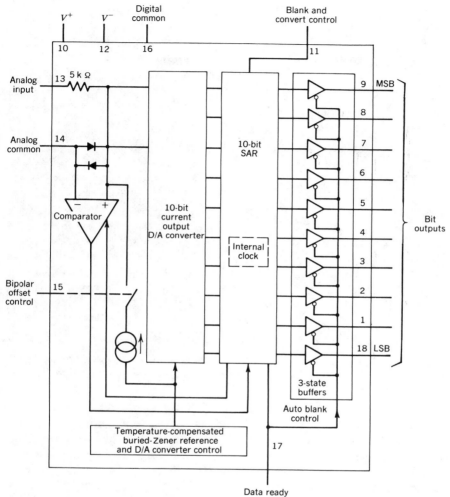

FIGURE 15.17. Functional block diagram of monolithic bipolar 10-bit successive-approximation A/D converter. (Analog Devices AD-571.)

clock frequency does not affect the resolution and the linearity characteristics of the circuit.

The chip photograph of the AD-571 successive-approximation converter is shown in Figure 15.18. Particular features of the chip layout as well as the locations of key sections are also identified. The successive-approximation register section, comprised of I^2L gates, takes up a relatively small portion of the chip area and is located in a single isolation tub. The low-level precision analog circuitry, such as the 10-bit D/A converter and its ladder network and the biasing and the voltage reference circuitry are located in the lower portion of the chip.

FIGURE 15.18.   Photomicrograph of AD-571 10-bit A/D converter. Chip size: 120 mils × 151 mils. Technology used: bipolar and I²L, with laser-trimmed silicon-chromium resistors. (*Photo: Analog Devices, Inc.*)

The laser-trimmed Si-Cr resistors are laid out with large-area trim tabs to facilitate high-resolution trimming during the wafer test stage. The high-level digital circuitry, such as the output buffers, is located at the opposite end of the chip from the analog circuitry in order to avoid possible cross talk or noise problems, and separate ground connections are provided for the analog and digital sections. The monolithic integrated circuit operates with ±15-V power supplies and has a total conversion time of ≤30 μsec.

15.5. SUCCESSIVE-APPROXIMATION CONVERTERS USING MOS TECHNOLOGY

Since A/D conversion techniques require the combination of analog and digital signal processing on the same chip, they are conceptually very well suited for monolithic implementation with MOS technology. This is particularly true if the A/D converter is to serve as a "dedicated" subsystem of a more complex digital signal-processing system such as a coder/decoder (CODEC) or an analog microprocessor. To serve such applications, a number of unique circuit approaches have been developed for designing successive-approximation converters with MOS devices. The most important among these are the *charge-redistribution* and the *potentiometric* A/D conversion techniques.[13]

Charge-Redistribution A/D Converters

The charge-redistribution A/D converter is derived, in principle, from the charge-scaling D/A converter configuration discussed in Section 14.2 (see Fig. 14.8). It uses a charge-scaling D/A converter and a successive-approximation register in feedback around a voltage comparator to perform the A/D conversion.

The charge-redistribution-type A/D converter uses the basic binary-weighted capacitive ladder network shown in Figure 14.8. Note that the ladder contains an additional termination capacitor, equal to the size of the LSB capacitor. The conversion is accomplished in three steps.[14] In the first step, shown as the sample mode in Figure 15.19a, the top plate of the capacitor array is grounded and all the bottom plates are connected to analog input V_A. The charge Q_X, proportional to V_A, is stored on the top plate of the capacitor array,

$$Q_X = V_A C_{tot} = 2CV_A \qquad (15.25)$$

where $C_{tot} = 2C$ is the total capacitance of the array [see Eq. (14.13)].

During the next step, called the hold mode, the top grounding switch S_A is opened, and all the bottom plates are connected to ground individually, as shown in Figure 15.19b. Since the charge Q_X on the capacitor is conserved, the potential V_X of the top plate becomes equal to $-V_A$.

During the third step, called the redistribution mode, the successive-approximation process begins. First, the switch S_1, corresponding to the MSB, is connected to V_{ref}, with the rest of the switches left at ground. Since C is equal to $\frac{1}{2} C_{tot}$, the voltage V_X is increased by an amount equal to $\frac{1}{2} V_{ref}$,

$$V_X = -V_A + \frac{V_{ref}}{2} \qquad (15.26)$$

If this change causes the output comparator to change state, then the MSB is interpreted as 0, and S_1 is brought back to its original position. If the comparator does not change state, then this is analogous to MSB = 1, and S_1 is left connected to V_{ref}. In a similar manner, each of the remaining bits are tried one at a time to determine the digital output code. The switch S_6, corresponding to the termination capacitor, is left permanently connected to ground during this phase. Figure 15.19c shows the example of the switch positions at the end of the redistribution mode, corresponding to a binary code 01100. Notice that all the capacitors corresponding to 0 bits are totally discharged, and the charge Q_X is redistributed between capacitors corresponding to 1 bits.

One important advantage of the charge-redistribution process illustrated in Figure 15.19 is its insensitivity to the bottom-plate parasitics of MOS capacitors. The bottom plate is at all times switched between a voltage source (either V_A or V_{ref}) and ground. The parasitic capacitance associated with the top plate has a

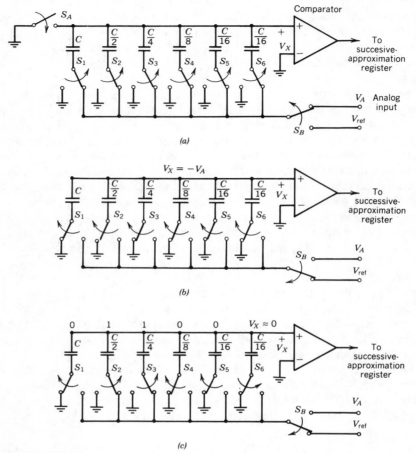

FIGURE 15.19. Sequence of steps in operation of 5-bit charge-redistribution A/D converter: (*a*) Sample mode; (*b*) hold mode; (*c*) end of redistribution mode for 01100 code.

very small but finite error contribution. However, the nature of the conversion process is such that V_X is converged back to its initial value of zero. Therefore, the net voltage across the top-plate parasitic capacitance at the beginning and at the end of the conversion cycle is within 1 LSB of zero. Consequently, the error charge contributed by this parasitic is also nearly zero. For example, the top-plate parasitic capacitance can be 100 times larger than the smallest ladder capacitance, yet cause only an 0.1-bit offset error in a 10-bit converter. This allows the smallest ladder capacitor to be made very small (typically on the order of 1 pF or less), and still remain relatively insensitive to parasitics.[14]

The offset voltage ΔV_{OS} associated with the comparator is a potential source of error during the conversion process. However, this detrimental effect can be largely eliminated by offset cancellation techniques which are described later in this section. The other remaining requirements from the comparator are voltage gain and switching speed. The comparator gain must be high enough to provide sufficient logic swing at its output with 1 LSB of input voltage step. This requires a minimum comparator gain $A_{\min}$,

$$A_{\min} > \frac{V_L}{V_{\text{ref}}} 2^N \qquad (15.27)$$

where V_L is the required logic swing at the comparator output. The comparator response time is an indirect requirement because it affects the conversion rate, rather than the resolution of the converter. Since the comparator response at low input levels is a function of the input drive voltage, it should be measured at an input level of ± 1 LSB to estimate the maximum conversion speed.

The main source of error in charge-redistribution converters is the matching and tracking of the capacitor ratios in the array. As in the case of all successive-approximation converters, monotonicity of the ladder characteristics is essential if all output codes are to be present.

Charge-redistribution techniques are suitable for up to 10 bits of resolution and exhibit excellent temperature-stability characteristics due to very low temperature coefficients of MOS capacitors. The conversion speed of the circuit is somewhat slower than that of comparable bipolar designs. Typical conversion times for 8- or 10-bit operation are in the range of 100–200 μsec. However, due to the small size and high packing density of MOS devices, the charge-redistribution A/D converters can be implemented on a much smaller chip area than comparable bipolar successive-approximation designs.

Potentiometric A/D Converters

The potentiometric A/D converter uses a voltage-scaling D/A converter (see Fig. 14.6) and a successive-approximation register in feedback around a voltage comparator.[15] Figure 15.20 illustrates the conceptual diagram of a 3-bit converter system using this approach. The successive-approximation search is performed in the conventional manner by inserting a trial 1 to the MSB and then examining the state of the comparator output; this is then repeated in descending sequence for each of the lesser bits. When a given bit coefficient is 1, the particular switch or switches associated with it are closed, and their complements are opened.

The potentiometric conversion concept is well suited to MOS designs where analog switches are readily available and the sensing circuitry does not draw any dc bias current. An added benefit of this concept is its inherent monotonicity. The voltage level increases monotonically along the resistor string as one moves from ground toward V_{ref}. Therefore, there can be no missing codes at the output.

The complexity of the simple potentiometric circuit increases rapidly as the bit count N is increased. For example, for an N-bit converter, 2^N resistors and approximately 2^{N+1} switches would be needed. However, this problem can be avoided by using a series connection of smaller ladder sections, and by multiple

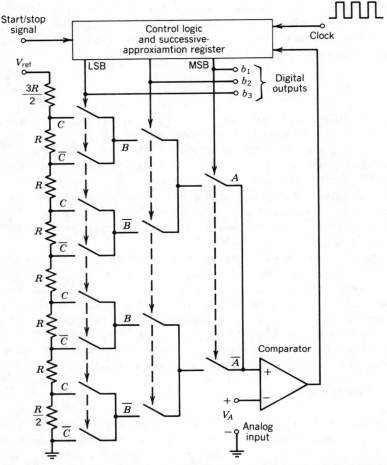

FIGURE 15.20. Conceptual diagram of 3-bit successive-approximation A/D converter using voltage-scaling principle.

decoding of a single ladder during the conversion process. These simplifications become possible if one combines the potentiometric techniques with charge-balancing methods, as will be discussed in the following sections.

In the simplified circuit of Figure 15.20, the comparator offset voltage is a source of error. However, it can be canceled by using self-correcting (auto-zero) sample-and-hold techniques, as described below.

Sampled-Data Comparators

The voltage comparator is one of the essential building blocks of an A/D converter. In MOS technology, poor mismatch between MOS threshold voltages leads to the presence of large offset voltages at the comparator input. These offset voltages can be as high as several hundred millivolts. Unless measures are taken to cancel or compensate for this offset voltage, the usefulness of the comparator is severely limited. Fortunately, the availability of analog switches and the charge-sensing capability of MOS devices readily lend themselves to the use of sampled-data techniques to cancel such offset voltages. This is done by sampling and holding the offset voltage on a sampling capacitor, and subtracting it from the actual input during the measurement cycle. Because of its self-correcting nature, this process is called an *auto-zero* technique.

Figure 15.21*a* shows the principle of operation of the basic sampled-data comparator. The switches S_{S1} and S_{S2} are *closed* during the sample mode and *open* during the hold mode. The voltage levels at the end of the hold mode are

$$V_X = -V_{in} \tag{15.28}$$

$$V_{out} = -A(V_X + \Delta V_{OS}) = A(V_{in} - \Delta V_{OS}) \tag{15.29}$$

where ΔV_{OS} is the offset voltage of the comparator.

Figure 15.21*b* illustrates an alternate circuit configuration which can provide the offset cancellation (i.e., the auto-zero function) by placing the sampling switch S_{S1} from the comparator-inverting input to the output.* Then, during the sample mode, both ΔV_{OS} and V_{in} are sampled simultaneously. During the subsequent hold mode, S_{S1} and S_{S2} open, and the circuit voltages become

$$V_X = -(V_{in} + \Delta V_{OS}) \tag{15.30}$$

$$V_{out} = -A(V_X + \Delta V_{OS}) = AV_{in} \tag{15.31}$$

which is unaffected by the presence of ΔV_{OS}.

Using the offset cancellation technique illustrated in Figure 15.21*b*, the effective value of the comparator offset is typically reduced to less than 1 mV. This residual offset is normally caused by parasitic charge coupling due to imperfect switches.

In the case of the charge-redistribution A/D converter circuit of Figure 15.19, one would normally use a sampled-data comparator to avoid errors due to ΔV_{OS} of the comparator. Since the capacitor array is available to perform the holding function, no additional capacitors would be needed. The comparator can be converted to an offset-canceled sample-data comparator simply by connecting the switch S_A of Figure 15.19*a* to the comparator output, rather than to ground, during the sample mode.

The sampled-data comparator can also be used to sense the difference of two input voltages, as shown in Figure 15.22. In this case, the "phasing" of the switches is important. During the sample mode, sampling switches S_S are closed,

---

*This method assumes that the comparator is stable with unity feedback. If this is not the case, an additional compensation capacitor may have to be switched into the comparator circuit to assure stability during the sample mode.

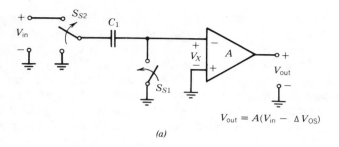

$$V_{out} = A(V_{in} - \Delta V_{OS})$$

(a)

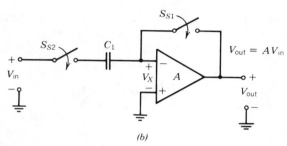

(b)

FIGURE 15.21. Operation of sampled-data comparator: (*a*) Basic circuit; (*b*) operation with offset cancellation.

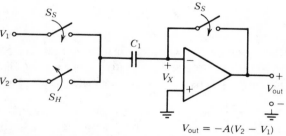

$$V_{out} = -A(V_2 - V_1)$$

FIGURE 15.22.   Differential sampled-data comparator with offset cancellation.

and during the hold mode, holding switch S_H is closed. Thus, during the holding mode, the circuit voltages are

$$V_X = (V_2 - V_1 - \Delta V_{OS}) \tag{15.32}$$

$$V_{out} = -A(V_X + \Delta V_{OS}) = -A(V_2 - V_1) \tag{15.33}$$

A differential sampled-data comparator of the type shown in Figure 15.22 would normally be used to replace the conventional differential comparator circuit shown in Figure 15.20.

The sampled-data comparator can also be used for scaling and summing the differences of several input voltages, as illustrated in Figure 15.23. Again, note

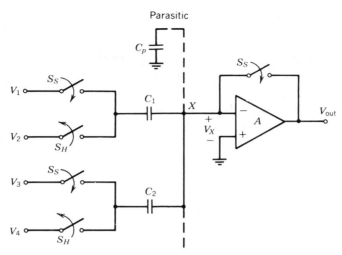

FIGURE 15.23.   Performing differencing and scaling operations with sampled-data comparator.

that switches designated S_S are *closed* during the sample mode, and those designated S_H are closed during the hold mode. The output voltage during the hold mode is

$$V_{out} = \frac{A}{C_{tot}} \left[C_1(V_1 - V_2) + C_2(V_3 - V_4) + \dots \right] \qquad (15.34)$$

where C_{tot} is the total capacitance connected to the input node X. Assuming that a finite amount of parasitic capacitance C_p exists between node X and ground, C_{tot} is

$$C_{tot} = C_1 + C_2 + \dots + C_n + C_p \qquad (15.35)$$

for n separate scaling capacitors connected into the circuit. Equation (15.35) illustrates that, to first order, the circuit is insensitive to the presence of the parasitic capacitor C_p, since it only appears as a part of the comparator gain expression.

The sampled-data comparator circuits all operate on the charge-balancing principle. The net voltage sensed by the comparator is proportional to the total charge delivered to the inverting input node of the comparator. Therefore, this class of circuits are also called charge-balancing circuits. As will be described in the next section, they can greatly simplify the design of potentiometric A/D converters.

Potentiometric Converters Using Charge Balancing

The potentiometric converter circuit of Figure 15.19 gets very complex as the bit count increases. For N-bit conversion, 2^N resistors are required with approximately 2^{N+1} switches. This problem can be greatly simplified by using charge-balancing comparator circuits.[16] Figure 15.24 illustrates the conceptual diagram of a 4-bit successive-approximation A/D converter, combining a potentiometric resistor string with charge-balancing techniques. It is essentially a series connection of two 2-bit resistor strings. The first two significant bits are decoded from the top ladder; the two lesser bits are decoded from the bottom ladder. In the figure, the sample-and-hold switches are identified as S_S and S_H, respectively, and are shown in boxes to separate them from the bit switches. The sample-and-hold switches are driven by the clock signals ϕ and $\bar{\phi}$, whereas the bit-decoding

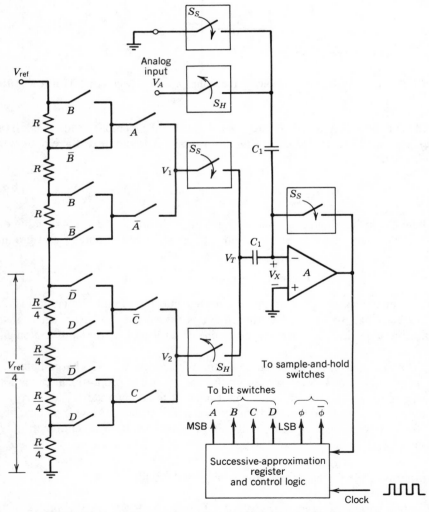

FIGURE 15.24.   4-bit potentiometric A/D converter using charge balancing.

switches are driven by the successive-approximation register output. Switches S_S and S_H are *closed* during the sample and hold modes, respectively.

During the hold period, with switches S_S open and S_H closed, the input voltage V_X at the input of the comparator is*

$$V_X = \tfrac{1}{2}(V_A - V_T) = \tfrac{1}{2}[V_A - (V_1 - V_2)] \tag{15.36}$$

where V_1 and V_2 are the tapped analog voltages associated with the top and bottom halves of the ladder, and correspond to the MSB and LSB pairs, respectively.

During the conversion process, the successive-approximation steps proceed in the conventional manner. During the search for the MSB, the successive-approximation register outputs (A, B, C, and D) would represent the 1000 code. This would result in switch A to be closed, and switches B, C, and D to be open, with their complements in the opposite state. This would result in a ladder output voltage

$$V_T = V_1 - V_2 = \tfrac{3}{4}V_{\text{ref}} - \tfrac{1}{4}V_{\text{ref}} = \frac{V_{\text{ref}}}{2} \tag{15.37}$$

which is then compared with V_A to determine the validity of the MSB. The approximation process continues in this manner until all 4 bits are evaluated.

By combining resistor-string and charge-balancing techniques, one can simultaneously scale voltages by two independent means: either by scaling resistor ratios or by scaling capacitor ratios. Using these two independent degrees of freedom, one can decode two separate sets of voltage levels from the same resistor string at the same time. Figure 15.25 illustrates this method of *double-decoding* a 2-bit ladder to obtain 4-bit resolution. The two most significant bits are decoded from the right-hand side of the ladder, the remaining two less significant bits are decoded from the left-hand side. The voltage level V_X at the input of the comparator is, from Eq. (15.34),

$$V_X = \frac{C_1}{C_{\text{tot}}}\left[V_A - \left(V_1 - \frac{V_2}{4}\right) - \frac{V_{\text{ref}}}{8}\right] \tag{15.38}$$

where V_1 is determined by the first 2-bit coefficients, V_2 is determined by the last 2-bit coefficients, and the $V_{\text{ref}}/8$ term provides the $\tfrac{1}{2}$ LSB offset in the transfer characteristics. Note that the complexity of decoding is greatly simplified in the circuits of Figures 15.24 and 15.25, compared to the simple potentiometric ladder of Figure 15.20.

The double-decoded 2-bit ladder network shown in Figure 15.25 can serve as a building block for higher bit converters. Figure 15.26 shows a convenient symbolic notation for the double-decoded ladder. The symbol corresponds to the 2-bit ladder enclosed within the dashed lines in Figure 15.25. Two such ladders

---

*The comparator offset voltage is neglected since it is self-corrected during sample-and-hold cycles.

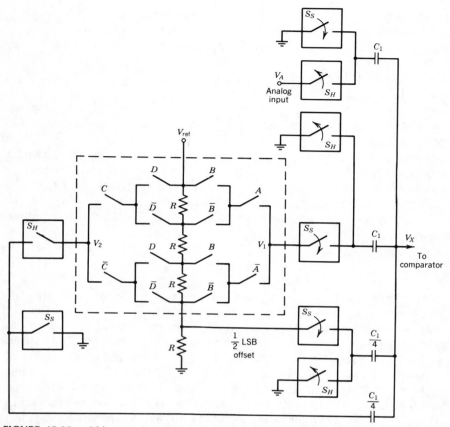

FIGURE 15.25. 4-bit potentiometric converter obtained by double decoding a 2-bit resistor ladder.

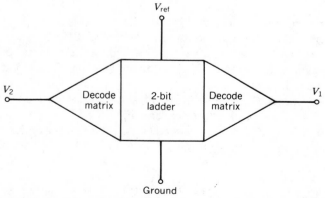

FIGURE 15.26. Symbolic representation of double-decoded 2-bit ladder of Figure 15.25.

can be combined in series to form an 8-bit converter. Figure 15.27 gives a conceptual diagram of such a configuration, using symbolic notation for the ladder and decode matrices. Note that the resistor values in the lower ladder are scaled to provide a total drop of $\frac{1}{4} V_{ref}$ across it. The output voltage appearing at the input of the comparator during the hold mode is

$$V_X = V_A - \left[(V_1 - V_2) + \tfrac{1}{16}(V_3 - V_4) - \frac{V_{ref}}{256} \right] \qquad (15.39)$$

where V_1 is determined by the first 2 most significant bits; V_2 by the next 2, and so on. An additional tap from the bottom ladder, at a voltage level of $V_{ref}/16$, is further scaled down by capacitor ratios to provide $\frac{1}{2}$ LSB offset.

Using conventional decoding techniques and a potentiometric resistor ladder, an 8-bit converter would have required 256 resistors and over 500 analog switches. In the circuit configuration of Figure 15.27, the same function is achieved with 16 resistors, 4 capacitors, and 32 switches, which results in a substantial reduction of circuit complexity and silicon chip area.[16]

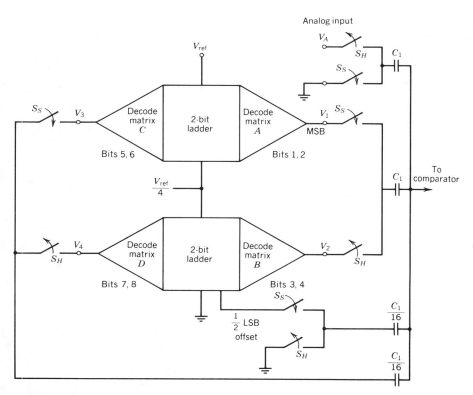

FIGURE 15.27. Conceptual diagram of 8-bit A/D converter using series connection of 2 double-decoded 2-bit ladders.

Improving Resolution of Charge-Redistribution Converters

In conventional charge-redistribution converters (see Fig. 15.19), the component-matching tolerances limit the available resolution to approximately 10 bits. This limitation is primarily due to sheet resistivity and oxide thickness gradients across the chip, as well as to the uncertainty in pattern definition due to limitations of the photomasking process. In a conventional charge-redistribution A/D converter, these errors lead to excessive differential nonlinearity and missing codes (i.e., nonmonotonicity). This problem can be partially solved by combining a charge-redistribution converter with a potentiometric resistor array. Figure 15.28 illustrates the basic circuit diagram of such an A/D converter system, which uses an M-bit resistor array along with a K-bit binary-ratioed capacitor array to provide $(M + K)$-bit resolution.[17]

The advantage of the capacitor array is its simplicity. However, its differential nonlinearity is poor for 10-bit or higher resolution. On the other hand, the potentiometric resistor string provides inherently monotonic behavior (i.e., very small differential nonlinearity) but becomes complex at high bit counts. The

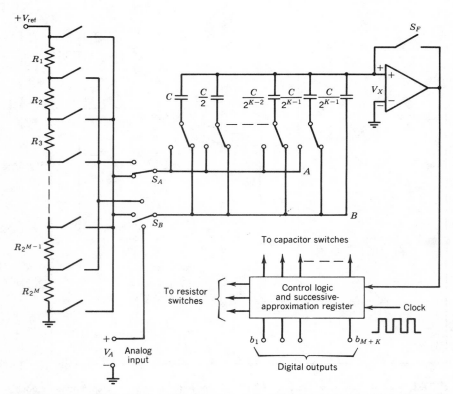

FIGURE 15.28. High-resolution A/D converter configuration combining resistor and capacitor arrays.[17]

circuit concept illustrated in Figure 15.28 is an effective method of combining the benefits of both approaches. Typically, K would be chosen to be in the 8–10-bit range, and M would be in the 3–4-bit range, which keeps both the ladder and the capacitor array configurations relatively simple, but still provides 12 to 14 bits of resolution.

The principle of operation of the circuit in Figure 15.28 can be briefly described as follows. During the initial sample mode, the bit switches of the capacitor array are connected to the B bus, which is then connected to the analog input V_A; the offset-canceling switch S_F is closed. During the subsequent hold mode, S_F is opened and S_B is connected to the resistor string. Then a successive-approximation search is done among the resistor string taps to determine the resistor segment within which the stored sample V_A lies. The result of this search provides the M most significant bits of the conversion process. Next, the switches S_A and S_B connect buses A and B to the top and bottom of the corresponding resistor segments. A second successive-approximation search is performed using the capacitor array switches (see Fig. 15.19c) until the comparator input voltage converges to zero. The result of this second search determines the remaining K bits of the conversion. Note that this concept is essentially equivalent to using a "segmented" D/A converter (see Fig. 14.15) in feedback to form an A/D converter. The first search determines the *segment*, and the second search determines the *step* within the segment.

15.6. PARALLEL A/D CONVERTERS

Parallel or simultaneous A/D conversion is by far the fastest and conceptually simplest conversion. This type of converter uses a separate analog comparator with a fixed reference for every quantization level in the digital word, from zero to full scale. Then the outputs of these comparators are appropriately interconnected by the encoding logic circuitry to produce a parallel digital output.

Figure 15.29 shows a block diagram of a parallel A/D converter system. For N-bit resolution $2^N - 1$ separate comparators and separate reference levels are needed. Thus the system complexity increases very rapidly as the number of parallel bits are increased. In this type of converter, all input bits are processed simultaneously. Therefore, the entire encoding operation can be performed within one clock cycle. Because of this property, such converter systems are also called *flash encoders* or *flash converters*.

The widest range of applications for parallel A/D converters is in video signal processing. They are used in video bandwidth compression, digital video transmission, reception, and recording, radar signal analysis, noise reduction, and digital image enhancement.[18] These applications require conversion rates in the range of 5 MHz (i.e., 5×10^6 conversions per second) to 50 MHz, which can not be accomplished by any other conversion technique.

Figure 15.30 shows the basic architecture of an N-bit parallel A/D converter in somewhat more detail. The $2^N - 1$ decision levels, which correspond to individual quantizing levels are generated by a potentiometric resistor string,

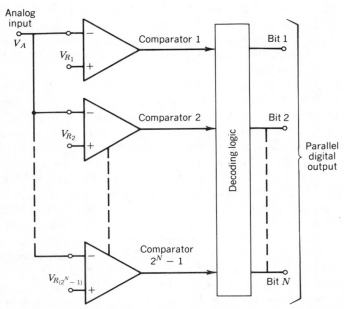

FIGURE 15.29.   Conceptual block diagram of parallel or flash-type A/D converter.

which bias 2^{N-1} latching comparators. All the comparator inputs are connected to a common analog input bus, and are driven simultaneously by the analog input V_A. Thus, at any given analog input level, the outputs of all comparators whose reference voltages are *below* V_A will be at the 1 state, and those above V_A will be at the 0 state. In other words, the output of the comparator array is essentially a bar graph of a total of $2^N - 1$ discrete points whose height is proportional to V_A. The particular resistor segment in which V_A lies is determined by the segment-detection logic made up of a stack of AND gates, as illustrated in Figure 15.30. This level is detected by comparing the logic output of each comparator with the ones immediately above and below it. The resulting information can then be decoded through a combinatorial logic circuitry or a programmable logic array (PLA) to form an N-bit parallel word. Normally, an additional comparator is added to the comparator array to sense the so-called overflow condition if and when the analog level exceeds V_{ref}. Although the entire conversion process can, in theory, be performed in one clock cycle, in practice two clock cycles are used: one to sample the input level and latch the analog comparators, and one to complete the decoding operation.

Although the concept of parallel A/D conversion is straightforward, the overall circuit complexity increases very rapidly. For N-bit resolution, a minimum of $2^N - 1$ comparators and 2^N resistors are required. For example, an 8-bit parallel converter would require over 250 resistors and comparators. Thus, the chip area and the power dissipation requirements increase very rapidly with increasing resolution.

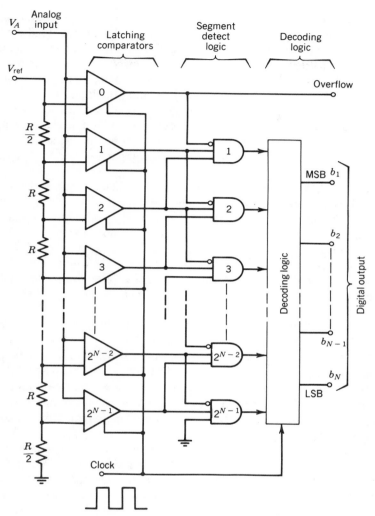

FIGURE 15.30. Basic architecture of N-bit parallel converter.

The parallel A/D converters can be fabricated using either MOS or bipolar technologies. Since conversion speed is the essential attribute of parallel converters, they are normally fabricated using the highest speed versions of the particular technologies. For bipolar designs, emitter-coupled logic (ECL) circuit configurations are normally used for decoding sections.[19] In MOS designs, silicon-gate CMOS technology is used, often in conjunction with silicon-on-saphire (SOS) fabrication techniques, to enhance the conversion speeds.[20]

The design and the layout of parallel converter circuits in bipolar or MOS technologies create problems for circuit complexities of 6 bits or more. The long resistor string and 2^{N-1} comparators must be located with reasonable symmetry on the chip to avoid unequal propagation delays due to unequal signal paths, as well as to minimize nonlinearity errors due to thermal or process gradients across the chip. The comparator arrays are normally laid out in identical *cells* in rows or columns across the chip. To minimize signal delays due to distributed capacitance in the long signal path connected to the comparator inputs, the input signal may be applied in parallel, at several points along the input *bus*.

In CMOS designs, individual offset-cancellation feedback loops are necessary around individual comparators in order to keep the individual comparator offsets well below the $\frac{1}{2}$ LSB level. This requires additional analog switches, sampling capacitors, and clock lines for each comparator, and tends to degrade the conversion speed.

In bipolar designs, the bias currents of individual comparators introduce additional voltage drops, distributed along the resistor string. To compensate for this effect, the dc bias current in the resistor string is normally chosen to be at least three orders of magnitude higher than the comparator bias currents.

Since the analog input node is required to drive a large number of comparator inputs, each having a finite input capacitance associated with it, the input signal is normally buffered by a voltage-follower stage. In order to keep the input drive requirements to a minimum, the full-scale voltage swing V_{FS} is usually limited to 1 or 2 V.

Figure 15.31 shows the chip photograph of an 8-bit parallel A/D converter, fabricated using high-speed bipolar process technology.[19] The monolithic circuit contains 256 latched comparators, a bias string with 255 taps and the associated decoding logic and output buffers on a 261×264-mil chip. The comparators are arranged in eight columns of 32, and take up most of the chip area. The combined input capacitance of the entire comparator array is approximately 300 pF. Thus, an external buffer amplifier is required to provide the necessary analog signal drive.

The bias string with 255 taps is formed by an aluminum strip through which a constant bias current is forced. The circuit operates with 0–2-V input signals, and can convert at speeds of up to 35 megasamples per second, with 8-bit resolution and $\pm\frac{1}{2}$ LSB nonlinearity. Operating with +5-V and −6-V power supplies, it dissipates 2.5 W of power.

Stacking Parallel Converter Sections

Two or more parallel A/D converter sections can be stacked together to improve the converter resolution. Figure 15.32 illustrates a method of stacking two N-bit parallel converters to obtain $(N + 1)$-bit resolution. This can be achieved by connecting the resistor string of both converter sections in series, and by connecting the *overflow* comparator output of the lower unit to the decoding logic section of the upper unit. The N-bit parallel outputs of the individual sections will then need additional decoding to produce $N + 1$ binary outputs.

Analog input
and clock buffers

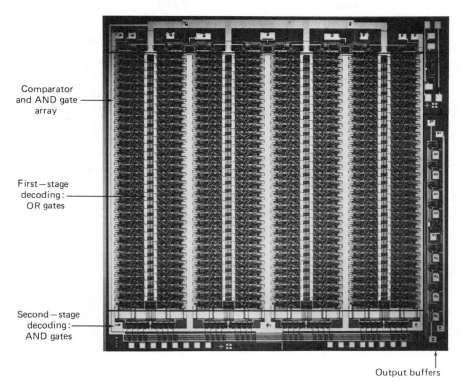

Comparator
and AND gate
array

First–stage
decoding:
OR gates

Second–stage
decoding:
AND gates

Output buffers

FIGURE 15.31. Photomicrograph of 8-bit parallel A/D converter. Chip size: 261 mils × 264 mils. Technology used: high-speed bipolar (ECL). (*Photo:* TRW LSI Products Division.)

The method of combining two sets of converter outputs to obtain $(N + 1)$-bit resolution can be briefly described as follows. The overflow comparator output of the lower converter functions as the MSB output. For input levels less than $V_{FS}/2$, only the lower converter section is active; and its N outputs correspond to the N bits of the $(N + 1)$ bit output word, with the overflow signal providing the MSB state. For input signals greater than $V_{FS}/2$, the N bits of the lower converter are disregarded by the additional decoding logic, and only the N bits of the upper converter are used in conjunction with the overflow output of the lower unit. Note that in this configuration, the size of the voltage step to be sensed by each comparator is reduced by a factor of 2, assuming that V_{ref} remains the same; and the error in threshold voltages due to comparator input bias currents is increased.

The required circuit complexity doubles for each additional bit of resolution required. Thus, if $N + 2$ bits of resolution is required, four converter sections would have to be stacked together.

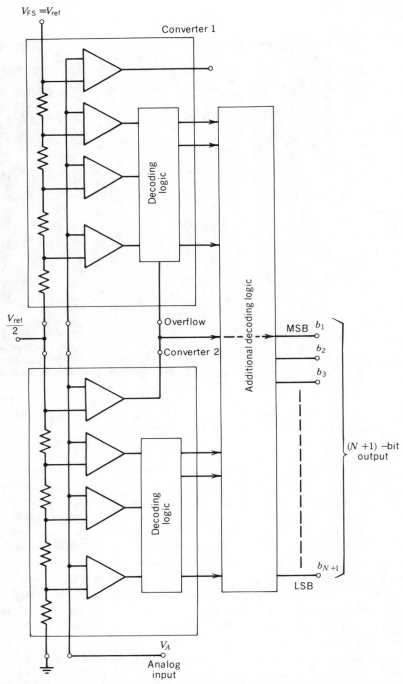

FIGURE 15.32. Stacking two N-bit parallel converters to obtain $(N + 1)$-bit resolution.

870

15.7. OTHER HIGH-SPEED A/D CONVERSION TECHNIQUES

Series–Parallel Converters

Series–parallel A/D converters retain some of the major speed advantages of parallel converters, with somewhat reduced circuit complexity. Figure 15.33 shows a functional block diagram of an $(M + K)$-bit series–parallel converter system. The circuit operates on the input signal in two serial steps. First a coarse conversion is made to determine the M most significant bits; then the results are converted back to an analog signal, V_2, which is subtracted from the actual analog input, amplified by a factor of 2^M, and then converted into K lower bits. In most cases, M and K are chosen to be equal to keep the circuit complexity symmetrical. Thus, for example, for a 6-bit converter, one would use two 3-bit parallel A/D converters, a differencing amplifier with a gain of 8, and a high-speed 3-bit D/A converter. Since parallel A/D converters are used in both steps of conversion, this class of converters is also called *two-step flash* converters.

With reference to Figure 15.33, the principle of series–parallel conversion can be illustrated as follows for the case of a 6-bit converter with $M = K = 3$. The binary output D_1 for the first A/D converter can be expressed as [see Eq. (15.1)]

$$D_1 = \frac{V_{in}}{V_{FS}} = \frac{b_1}{2^1} + \frac{b_2}{2^2} + \frac{b_3}{2^3} \qquad (15.40)$$

The output of the D/A converter is

$$V_2 = D_1 V_{FS} \qquad (15.41)$$

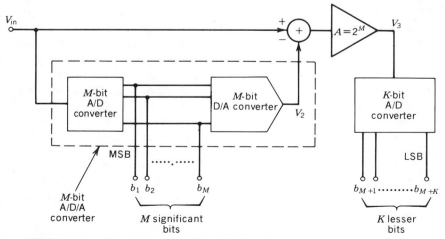

FIGURE 15.33. Block diagram of $(M + K)$-bit series–parallel A/D converter.

which is then subtracted from V_{in} and amplified by gain A to give V_3,

$$V_3 = A(V_{in} - V_2) \tag{15.42}$$

Equation (15.42) is then converted into a 3-bit binary word D_2,

$$D_2 = \frac{V_3}{V_{FS}} = \frac{b_4}{2^1} + \frac{b_5}{2^2} + \frac{b_6}{2^3} \tag{15.43}$$

From the above equations, one can express the input–output relationship of the converter as

$$D_1 + \frac{D_2}{A} = \frac{V_{in}}{V_{FS}} \tag{15.44}$$

which, for $A = 2^M = 2^3$, becomes

$$D = \frac{b_1}{2^1} + \frac{b_2}{2^2} + \cdots + \frac{b_6}{2^6} = \frac{V_{in}}{V_{FS}} \tag{15.45}$$

This is the desired transfer function of a 6-bit A/D converter.

A basic building block of the series–parallel converter is the analog–digital–analog (A/D/A) converter section. A number of specialized circuit techniques have been reported in the literature to implement such a A/D/A function along with the subtractor function.[21] In the implementation of the series–parallel A/D conversion, although each of the converter components has relatively low resolution, their accuracy and linearity characteristics must be sufficient to assure $(M + K)$-bit overall resolution.

Time-Multiplexed Converter Arrays

In this approach, an array of A/D converters with interleaved sampling times are used as if they were effectively equivalent to a single converter operating at a higher conversion rate.[22] The principle of operation of such a converter system is illustrated in Figure 15.34. Assuming that the conversion time associated with each N-bit converter is t_c, then in the case of M parallel paths, the input is sampled every t_c/M time interval, between various converter paths, and while one converter is sampling the input, the rest of the converters proceed with various stages of the conversion process. The output signal is obtained by sampling the converter outputs through a multiplexing switch every t_c/M interval. The net conversion time for the entire converter array is then equivalent to t_c/M, and the system functions as an N-bit converter with M times faster conversion rate.

The principle of operation of a multiplexed converter array relies very strongly on the matching and tracking of converter characteristics on each of the signal paths, and requires very careful control of the clock signals to minimize the *phase-skew* of clocking signals between converters. Any mismatch of converter characteristics, including gain or scale-factor mismatches, appears as additional noise source at the output and degrades the system performance.[22]

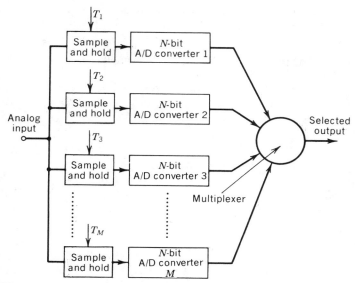

FIGURE 15.34. Time-multiplexing multiple A/D converters to increase conversion speed.

The time-multiplexed converter arrays are particularly attractive for MOS technology where several charge-redistribution successive-approximation converter stages can be employed to form a converter array, which would have conversion rates comparable to a parallel or series-parallel converter system.

15.8. NONLINEAR A/D CONVERTERS FOR TELECOMMUNICATIONS

Nonlinear converter circuits are often used in telecommunication systems to compress or expand the dynamic range of analog signals during the conversion process. This is particularly true for pulse-code-modulated telephone systems, where analog signals must be coded into a digital format by using a nonlinear A/D conversion or *coding* technique. Then, in the receiving end, the digital signal is *decoded* into an analog signal by means of a nonlinear D/A converter having the reciprocal transfer characteristics. A nonlinear coder/decoder system which combines such an A/D and D/A system for pulse-code-modulated telecommunication applications is known as a *codec*.

Particular coding and decoding characteristics of A/D and D/A converters used in codec design must conform to a prescribed form of nonlinearity, called the μ-255 law.[23] The D/A converter structure which closely approximates such a nonlinear transfer characteristic is the *companding* D/A converter, which is discussed in Section 14.4 (see Figs. 14.17 through 14.19).

The encoding portion of a codec requires an A/D converter which has the same transfer function as that illustrated in Figure 14.17, performed in the opposite direction; in other words, producing the digital output shown along the horizontal axis, from an analog input level corresponding to the vertical axis. Such a function can be readily implemented by using a companding D/A converter, such as the one shown in Figure 14.19, in the feedback path of a successive-approximation-type A/D converter.

In most telecommunication applications, the nonlinear converter circuits forming the codec function are required to operate at relatively slow conversion rates, with conversion times on the order of 125 μsec (i.e., 8-kHz sampling rate), and dissipate very low amounts of power. These requirements make such converter circuits nearly ideal for implementation with MOS technology, using charge-redistribution or charge-scaling converters.

Figure 15.35 shows the architecture of a charge-redistribution-type converter which can be used for companded D/A or A/D conversions, in accordance with the basic transfer characteristics shown in Figure 14.17 and the coding format indicated in Figure 14.18.[24]

The principle of operation of the companding converter circuit of Figure 15.35 can be described as follows. The upper capacitor array, designated X array, is comprised of eight binary-weighted capacitors. These capacitors are used in generating the individual segment voltages or pedestals associated with

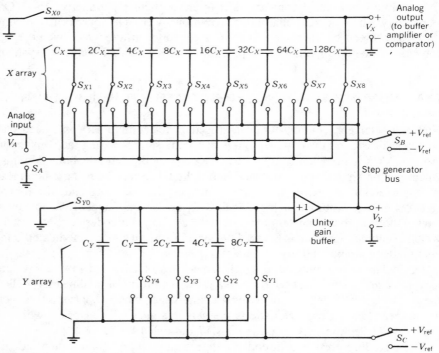

FIGURE 15.35. Nonlinear charge-redistribution converter circuit for companded A/D and D/A conversion.[24]

the companding transfer characteristics given in Figure 14.17. The lower capacitor array, designated Y array, generates submultiples of the reference voltage in units of $V_{ref}/16$, and serves as the step generator to generate the 16 equal-valued steps within a segment. Note that these two capacitor arrays functionally correspond to the segment and step generator subsections of the current-mode companding converter circuit of Figure 14.19. The circuit operates with the 8-bit digital code format shown in Figure 14.18. The first bit, MSB, controls the sign of the output by choosing the proper polarity of V_{ref}. The next 3 bits select the proper combination of eight segment switches S_{X1} through S_{X8} to determine the particular segment. The remaining 4 bits select the particular step within the segment by determining the position of the Y-array switches S_{Y1} through S_{Y4}.

If the circuit is operated as a companding D/A converter, the analog output V_X is obtained from the top plates of the X-array capacitors by means of a voltage-follower buffer amplifier. If the circuit is used as a subsection of an A/D converter, the analog output V_X is connected to a comparator which activates the SAR section to perform the successive-approximation search. The sequence of steps in operating the system in the D/A and A/D modes are briefly explained below.

Operation as D/A Converter. In this mode of operation, there is no analog input. Thus, S_A is permanently connected to ground. The D/A converter operation can be best understood by tracing the transfer characteristics shown in Figure 14.17 from the origin to $+V_{FS}$.

At the reset mode, switches S_{X0} and S_{Y0} are connected to ground, S_B and S_C are connected to $-V_{ref}$. Switches in the X-array and Y-array capacitors are all connected to ground.

In the sample mode, to generate the 16 steps in segment 1, starting from zero, S_{X1} is connected to V_{YS}, and the switches S_{Y1} through S_{Y4} are permutated through their 16 possible states. At the beginning of segment 2, S_{X1} is connected to $-V_{ref}$ through S_B, S_{X2} is connected to V_Y, and the 16 steps in segment 2 are generated in the same manner by Y-array switches. Then S_{X2} is also connected to $-V_{ref}$ and S_{X3} is connected to V_Y for the third segment, and so on. Since the X-array capacitor size is doubled between each segment, the step size and segment height are also automatically doubled, as required by the μ-255 law companding characteristics. In this manner, the output voltage V_X reaches to within 1 LSB of $+V_{FS}$ when all X-array switches and Y-array switches are connected to $-V_{ref}$.

The operation in the negative direction is performed in the same manner, except with S_B and S_C connected to $+V_{ref}$.

Operation as A/D Converter. In this mode of operation, the analog output V_X is connected to successive-approximation register logic through a comparator, and the circuit operates as a charge-redistribution A/D converter (see Fig. 15.19).

During the initial reset mode, all capacitors are discharged by connecting S_{X0}, S_{Y0}, and S_A, to ground. All X-array switches are grounded through S_A, and all Y-array switches are also connected to ground.

During the subsequent sample mode, S_{X0} and S_{Y0} are opened, and S_A is connected to the analog input V_A. This causes the analog output level V_X to equal $-V_A$. The polarity of the comparator output determines the sign bit, b_1, of the output code (see Fig. 14.18 for the code format), which in turn determines whether S_B and S_C are connected to $+V_{ref}$ or $-V_{ref}$ for the rest of the conversion sequence.

In the next step, S_A is connected to ground, and the first phase of the charge-redistribution process is initiated. A successive-approximation search is conducted between eight segment switches in the X-array to determine the segment within which V_A is located. The result determines the bit coefficients b_2, b_3, and b_4 of the output. The next phase of the charge-redistribution process is the determination of the step within the segment. This is performed by a 4-bit successive-approximation search within the Y array, and determines the last 4 bits b_5, b_6, b_7, and b_8 of the digital output code.

It should be noted that due to the successive-approximation nature of the A/D conversion process, the A/D conversion sequence is at least eight times longer than the D/A conversion process. Therefore, in many pulse-code-modulated telecommunication applications where a codec system is used, a single nonlinear converter similar to that illustrated in Figure 15.35 is multiplexed between encoding and decoding applications, while the analog signals are retained in sample-and-hold circuits.

15.9. AN OVERVIEW OF A/D CONVERTER TECHNIQUES

As described in the preceding sections of this chapter, a wide range of A/D conversion techniques are available for monolithic designs, using either bipolar or MOS technologies. The particular conversion technique and the fabrication technology to be used are normally determined by the requirements on two major performance parameters: (1) resolution and accuracy, and (2) conversion speed. These two parameters are in turn dictated by the end product use, or the application of the particular converter. In almost all cases, there is an inherent conflict between the resolution and accuracy requirements on the one hand, and the conversion speed on the other. A significant trade-off is required between these two conflicting parameters within the limitations of circuit complexity and the device fabrication economics (i.e., manufacturing yield).

Almost all of the presently available monolithic A/D converter products fall into the following categories:

1. *Integrating Converters.* These converter circuits, covered in Section 15.2, fall into the high-resolution and accuracy end of the performance spectrum, with conversion times in the range of a few milliseconds to a few hundred milliseconds. They are normally fabricated using MOS (particularly CMOS) technology. The great majority of integrating converters operate on the dual-slope integration principle, or one of its many

derivatives. The main application of integrating converters is in high-resolution dc or in very-low-frequency instrumentation and measurement equipment. Typical end products where they are used are digital multimeters and panel meters, remote telemetering and sensing equipment.

2. *Successive-Approximation Converters.* Successive-approximation converters, covered in Sections 15.4 and 15.5, represent an excellent compromise or trade-off between resolution and converter speed requirements. The great majority of general-purpose A/D converter products used in telecommunication, data processing, and microprocessor interface fall into this category. The conversion rates of successive-approximation converters are in the range of a few microseconds to a few hundred microseconds, and their resolution and accuracy characteristics are primarily determined by the D/A converter subsection of the system.

They are fabricated with either bipolar (and I^2L) or MOS technologies, depending on the speed requirements. In general, most stand-alone or general-purpose successive-approximation converters are fabricated with bipolar technology. On the other hand, those which are dedicated circuits, designed to serve as a subsection of a larger digital system, such as a microprocessor, a microcontroller or a pulse-code-modulated codec, are normally fabricated using MOS process technology and charge-redistribution techniques.

3. *Parallel Converters.* Parallel or flash A/D converters, which are covered in Section 15.6, fall into the high-speed end of the converter performance spectrum. Although the conversion process is conceptually the simplest one of all converter techniques, its resolution capability is limited to 8 bits or less due to rapidly increasing circuit complexity. It is primarily used in converter applications requiring conversion times well under 0.5 μsec. Such applications are encountered in digital processing of video or radar signals, very-high-speed data links, and multiplexed data-processing equipment. Parallel converters can be fabricated using either bipolar or CMOS technology. In specialized cases, more exotic fabrication processes, such as silicon-on-saphire (SOS) technology, may be used. In general, high-speed bipolar technology is preferred because of its speed advantage.

Table 15.1 gives an overview of the resolution and speed characteristics of the three major categories of monolithic A/D converters. In each case, the most applicable IC process technology is also indicated.

The field of data converters, for both A/D and D/A conversion, is a very dynamic and rapidly expanding one. As the capabilities of IC processes and technologies keep increasing, many more converter applications are becoming economically feasible, and new generations of IC products are emerging to serve these needs. The latest and most significant potential growth area in data converters is in the area of digital audio processing, which requires specialized

TABLE 15.1. A Comparison of the Relative Capabilities of Various Monolithic A/D Converter Techniques.

A/D Converter Type	Relative Speed or Performance	Typical Conversion Time					Commonly Used Technology	Common Application Areas
		6 Bit	8 Bit	10 Bit	12 Bit	14 Bit		
Integrating converters (dual-slope and charge balancing types)	Low	—	20 msec	30 msec	100 msec	250 msec	MOS	Digital panel meters
	Medium	—	1 msec	5 msec	20 msec	100 msec	MOS	Low-frequency instrumentation
	High	—	0.3 msec	1 msec	5 msec	30 msec	MOS	Remote-control and telemetry equipment
Successive-approximation converters	Low	60 μsec	100 μsec	120 μsec	150 μsec	—	MOS	General-purpose instrumentation
	Medium	30 μsec	50 μsec	60 μsec	80 μsec	—	MOS or bipolar	Data processing and microprocessor interface
	High	5 μsec	10 μsec	15 μsec	20 μsec	—	Bipolar or MOS	Telecommunications (codecs) Digital Controllers
Parallel or Flash Converters	Medium	100 nsec	200 nsec	—	—	—	MOS or bipolar	Video and radar signal processing
	High	20 nsec	50 nsec	—	—	—	Bipolar	Image enhancement High-speed data links Multiplexed operation

converter circuits with 14–16-bit linearity and with conversion times under 50 μsec. When such converter circuits become available as low-cost mass-produced monolithic integrated circuits, they will revolutionize the high-quality audio broadcasting, recording, or reproduction industry.

Because of the immense potential of the digital audio signal-processing applications, there is a very strong trend toward the development of ultraprecision A/D converter circuits, particularly in the area of high-speed conversion, using successive-approximation techniques. As described in Section 14.12, one significant development in this direction is the evolution of self-correcting or self-calibrating monolithic integrated circuits[25,26], which make ultraprecision circuitry possible without requiring costly trimming procedures.

REFERENCES

1. H. Schmid, *Electronic Analog/Digital Conversions,* Van Nostrand Reinhold, New York, 1970, Chap. 8.

2. D. F. Hoeschele, *Analog-to-Digital and Digital-to-Analog Conversion Techniques,* Wiley, New York, 1968.

3. B. M. Gordon, "Linear Electronic Analog/Digital Conversion Architectures, Their Origins,

Parameters, Limitations and Applications," *IEEE Trans. Circuits and Systems* **CAS-25,** 391–418 (July 1978).

4. E. R. Hnatek, *A User's Handbook to D/A and A/D Converters,* Wiley, New York, 1976.

5. E. L. Zuck, "Where and When to Use Which Data Converters," *IEEE Spectrum,* 39–42 (June 1977).

6. D. J. Dooley, Ed., *Data Conversion Integrated Circuits,* IEEE Press, New York, 1980.

7. E. L. Zuck, Ed., *Data Acquisition and Conversion Handbook,* Datel-Intersil, Mansfield, MA, 1979.

8. D. H. Sheingold, Ed., *Analog-Digital Conversion Notes,* Analog Devices, Norwood, MA, 1977.

9. G. F. Landsburg, "A Charge-Balancing Monolithic A/D Converter," *IEEE J. Solid-State Circuits* **SC-12,** 662–673 (December 1977).

10. Y. Suzuki et al., "A New Single-Chip C^2MOS A/D Converter For Microprocessor Systems—Penta-Phase Integrating C^2MOS A/D Converter," *IEEE J. Solid-State Circuits* **SC-13,** 779–784 (December 1978).

11. A. Barna and D. Porat, *Integrated Circuits in Digital Electronics,* Wiley, New York, 1973.

12. A. P. Brokaw, "A Monolithic 10-Bit A/D Converter Using I^2L and LWT Thin-Film Resistors," *IEEE J. Solid-State Circuits* **SC-13,** 736–745 (December 1978).

13. P. R. Gray, D. A. Hodges, and R. W. Brodersen, Eds., *Analog MOS Integrated Circuits,* IEEE Press, New York, 1980.

14. J. L. McCreary and P. R. Gray, "All-MOS Charge-Redistribution Analog-to-Digital Conversion Techniques—Part I," *IEEE J. Solid-State Circuits* **SC-10,** 371–379 (December 1975).

15. A. R. Hamade, "A Single-Chip All-MOS 8-Bit A/D Converter," *IEEE J. Solid-State Circuits* **SC-13,** 785–791 (December 1978).

16. T. Redfern, J. J. Connoly, S. W. Chin, and T. M. Frederiksen, "A Monolithic Charge-Balancing Successive-Approximation A/D Technique," *IEEE J. Solid-State Circuits* **SC-14,** 912–919 (December 1979).

17. B. Fotouhi and D. A. Hodges, "High-Resolution A/D Conversion in MOS/LSI," *IEEE J. Solid-State Circuits* **SC-14,** 920–926 (December 1979).

18. E. Zuck, "Put Video A/D Converters to Work," *Electronic Design* **16,** 16(1978).

19. J. G. Peterson, "A Monolithic Video A/D Converter," *IEEE J. Solid-State Circuits* **SC-14,** 932–937 (December 1979).

20. A. G. F. Dingwall, "Monolithic Expandable 6-Bit 20 MHz CMOS/SOS A/D Converter," *IEEE J. Solid-State Circuits* **SC-14,** 926–932 (December 1979).

21. R. J. van de Plassche and R. E. J. van der Grift, "A High-Speed 7-Bit A/D Converter," *IEEE J. Solid-State Circuits* **SC-14,** 938–943 (December 1979).

22. W. C. Black and D. A. Hodges, "Time-Interleaved Converter Arrays," *IEEE J. Solid-State Circuits* **SC-15,** 1022–1029 (December 1980).

23. P. R. Gray, D. A. Hodges, and R. W. Brodersen, Eds., *Analog MOS Integrated Circuits,* Part V, IEEE Press, New York, 1980.

24. Y. P. Tsividis, P. R. Gray, D. A. Hodges, and J. Chacko, "A Segmented μ-255 Law PCM Voice Encoder Utilizing NMOS Technology," *IEEE J. Solid-State Circuits* **SC-11,** 740–747 (December 1976).

25. Z. Boyacigiller, B. Weir, and P. Bradshaw, "An Error-Correcting 14b/20 μs CMOS A/D Converter," *Digest of Tech. Papers,* IEEE Int. Solid-State Circuits Conf., pp. 62–63, February 1981.

26. K. Maio et al., "An Untrimmed D/A Converter with 14-Bit Resolution," *IEEE J. Solid-State Circuits* **SC-16,** 616–621 (December 1981).

INDEX